2nd International Conference on Tropical and Coastal Region Eco Development 2016

IOP Conference Series: Earth and Environmental Science
Volume 55

Bali, Indonesia
25 – 27 October 2016

Editor:

Hadiyanto

ISBN: 978-1-5108-3711-9
ISSN: 1755-1307

Printed from e-media with permission by:

Curran Associates, Inc.
57 Morehouse Lane
Red Hook, NY 12571

Some format issues inherent in the e-media version may also appear in this print version.

Printed by Curran Associates, Inc. (2017)

For permission requests, please contact the Institute of Physics
at the address below.

Institute of Physics
Dirac House, Temple Back
Bristol BS1 6BE UK

Phone: 44 1 17 929 7481
Fax: 44 1 17 920 0979

techtracking@iop.org

Additional copies of this publication are available from:

Curran Associates, Inc.
57 Morehouse Lane
Red Hook, NY 12571 USA
Phone: 845-758-0400
Fax: 845-758-2633
Email: curran@proceedings.com
Web: www.proceedings.com

Table of contents

Volume 55

2nd International Conference on Tropical and Coastal Region Eco Development 2016

25–27 October 2016, Bali, Indonesia

Accepted papers received: 30 January 2017
Published online: 1 March 2017

Preface

011001
OPEN ACCESS
2nd International Conference on Tropical and Coastal Region Eco Development 2016

011002
OPEN ACCESS
Peer review statement

Papers

012001
OPEN ACCESS
Past and Future Ecosystem Change in the Coastal Zone

P Gell.....1

012002
OPEN ACCESS
Conserving endangered marine organisms: causes, trends and challenges

Ambariyanto.....9

012003
OPEN ACCESS
Population of *Aedes sp* in Highland of Wonosobo District and Its Competence as A Dengue Vector

Martini Martini, Bagoes Widjanarko, Retno Hestiningsih, Susiana Purwantisari and Sri Yuliawati.....19

012004
OPEN ACCESS
Nutritional value content, biomass production and growth performance of *Daphnia magna* cultured with different animal wastes resulted from probiotic bacteria fermentation

Vivi Endar Herawati, R A Nugroho, Pinandoyo and Johannes Hutabarat.....25

012005
OPEN ACCESS
Symbiotic Fungus of Marine Sponge Axinella sp. Producing Antibacterial Agent

A Trianto, S Widyaningsih, OK Radjasa and R Pribadi.....35

012006
OPEN ACCESS
The Influence of Madden Julian Oscillation on the Formation of the Hot Event in the Western Equatorial Pacific

Anindya Wirasatriya, Denny Nugroho Sugianto and Muhammad Helmi.....42

012007
OPEN ACCESS
Sediment Transport Model In Sayung District, Demak

Aris Ismanto, Muhammad Zainuri, Sahala Hutabarat, Denny Nugroho Sugianto, Sugeng Widada and Anindya Wirasatriya.....51

012008
OPEN ACCESS
Resources Management Strategy For Mud Crabs (*Scylla* spp.) In Pemalang Regency

Aristi Dian Purnama Fitri, Herry Boesono, Agus Sabdono and Nadia Adlina.....57

012009
OPEN ACCESS
Development of Spore Protein of *Myxobolus koi* as an Immunostimulant for Prevent of Myxobolusis on Gold Fish *(Cyprinus carpio* Linn) by Oral Immunisation

Gunanti Mahasri.....63

012010
OPEN ACCESS
The Effectiveness of Extracts Basil Leaves (*Ocimum sanctum* Linn) against *Saprolegnia* sp. by in Vitro

Sudarno, Muhammad Luthfi Hakim and Rahayu Kusdarwati.....73

012011
OPEN ACCESS
Developing groundwater conservation zone of unconfined aquifer in Semarang, Indonesia

T T Putranto, W K Hidajat and N Susanto.....79

012012
OPEN ACCESS
Understanding coastal processes to assist with coastal erosion management in Darwin Harbour, Northern Territory, Australia

S.G. Tonyes, R.J. Wasson, N.C. Munksgaard, K.G. Evans, R. Brinkman and D.K. Williams.....88

012013
OPEN ACCESS
Environmental and Risk Factors of Leptospirosis: A Spatial Analysis in Semarang City

Silviana Nur Fajriyah, Ari Udiyono and Lintang Dian Saraswati.....96

012014
OPEN ACCESS
Study of inundation events along the southern coast of Java and Bali, Indonesia (case studies 4-9 June 2016)

I R Nugraheni, D P Wijayanti, D N Sugianto and A Ramdhani.....106

012015
OPEN ACCESS
The Use of Water Exchange for Feeding Rate and Growth Promotion of Shortfin Eel *Anguilla bicolor bicolor* In Recirculating Water System

N Taufiq-Spj, S Sunaryo, A Wirasatriya and D N Sugianto.....114

012016
OPEN ACCESS
ASSOCIATION BETWEEN MACROMINERALS INTAKE AND CHANGES IN INTERNAL CAROTID ARTERY-INTIMA MEDIA THICKNESS IN POST ISCHEMIC STROKE PATIENTS

Dwi Pudjonarko, Dodik Tugasworo and Rumintang Silaen.....122

012017
OPEN ACCESS
Brain Gym To Increase Academic Performance Of Children Aged 10-12 Years Old (Experimental Study in Tembalang Elementary School and Pedalangan Elementary School Semarang)

M G Marpaung, T P Sareharto, A Purwanti and D Hermawati.....129

012018
OPEN ACCESS
Polymeric Membrane Made of Cellulose Isolated from Tropical Water Hyacinth Blended with Chitosan

Titik Istirokhatun, Richa Rachmawaty, Metty Meriyani, Nur Rokhati and Heru Susanto.....136

012019
OPEN ACCESS
Identification and Antibacterial Activity of Bacteria Isolated from Marine Sponge *Haliclona (Reniera)* sp. against Multi-Drug Resistant Human Pathogen

Meezan Ardhanu Asagabaldan, D Ayuningrum, R Kristiana, A Sabdono, O K Radjasa and A Trianto.....143

012020
OPEN ACCESS
COPE Method Implementation Program to Reduce Communication Apprehension Level in Full Day Yunior High School Students

A R Prasetyo.....154

012021
OPEN ACCESS
Influence of ENSO and IOD to Variability of Sea Surface Height in the North and South of Java Island

Ahmad Fadlan, Denny Nugroho Sugianto, Kunarso and Muhammad Zainuri.....166

012022
OPEN ACCESS
The Roles of Macrobenthic Mollusks as Bioindicator in Response to Environmental Disturbance : Cumulative *k*-dominance curves and bubble plots ordination approaches

Sapto P. Putro, Fuad Muhammad, Amalia Aininnur, Widowati and Suhartana.....176

012023
OPEN ACCESS
The Effect of Early Mosquito Insecticides Exposure on Spraque Dawley Rat Testis: A Histopathological Feature Towards Malignancy?

Tri Indah Winarni, Milzam Auzan Aziman, Anindyo Abshar Andar and Ika Pawitra.....184

012024
OPEN ACCESS
Analysis of heavy metal content of Cd and Zn in ballast water tank of commercial vessels in Port of Tanjung Emas Semarang, Central Java Province

A Tjahjono, A N Bambang and S Anggoro.....189

012025
OPEN ACCESS
Proteomics study of extracellular fibrinolytic proteases from *Bacillus licheniformis* RO3 and *Bacillus pumilus* 2.g isolated from Indonesian fermented food

Diana Nur Afifah, Ninik Rustanti, Gemala Anjani, Dahrul Syah, Yanti and Maggy T. Suhartono.....198

012026
OPEN ACCESS
Screening Antibacterial Agent from Crude Extract of Marine-Derived Fungi Associated with Soft Corals against MDR-*Staphylococcus haemolyticus*

A Sabdaningsih, O Cristianawati, M T Sibero, H Nuryadi, O K Radjasa, A Sabdono and A Trianto.....207

012027
OPEN ACCESS
Exploration of Fungal Association From Hard Coral Against Pathogen MDR *Staphylococcus haemolyticus*

O Cristianawati, O K Radjasa, A Sabdono, A Trianto, A Sabdaningsih, M T Sibero and H Nuryadi.....215

012028
OPEN ACCESS
Isolation, Identification And Screening Antibacterial Activity from Marine Sponge-Associated Fungi Against Multidrug-Resistant (MDR) *Escherichia coli*

Mada Triandala Sibero, Aninditia Sabdaningsih, Olvi Cristianawati, Handung Nuryadi, Ocky Karna Radjasa, Agus Sabdono and Agus Trianto.....223

012029
OPEN ACCESS
Isolation, Characterisation and Antagonistic Activity of Bacteria Symbionts Hardcoral *Pavona* sp. Isolated from Panjang Island, Jepara Against Infectious Multi-drug Resistant (MDR) Bacteria

D. Ayuningrum, R. Kristiana, M.A. Asagabaldan, A. Sabdono, O.K. Radjasa, H. Nuryadi and A. Trianto.....235

012030
OPEN ACCESS
Encapsulation of phycocyanin-alginate for high stability and antioxidant activity

Hadiyanto, Meiny Suzery, Deny Setyawan, Dian Majid and Heri Sutanto.....243

012031
OPEN ACCESS
The Correlation of Upwelling Phenomena and Ocean Sunfish Occurrences in Nusa Penida, Bali

C K Tito and E Susilo.....251

012032
OPEN ACCESS
Initial Study Of Potency Thermal Energy Using OTEC (Ocean Thermal Energy Conversion) As A Renewable Energy For Halmahera Indonesia

Y O Andrawina, D N Sugianto and I Alifdini.....260

012033
OPEN ACCESS
Effect of Reaction Temperature on Biodiesel Production from *Chlorella vulgaris* using CuO/Zeolite as Heterogeneous Catalyst

Dianursanti, M Delaamira, S Bismo and Y Muharam.....269

012034
OPEN ACCESS
Annona muricata modulate brain-CXCL10 expression during cerebral malaria phase

Kis Djamiatun, Sumia M A Matug, Awal Prasetyo, Noor Wijayahadi and Djoko Nugroho.....275

012035
OPEN ACCESS
SOCIO-ECONOMIC SPATIAL FOR THE SUSTAINABILITY OF THE ESTUARY ECOSYSTEM IN PELABUHANRATU COASTAL WEST JAVA

L. Supriatna, J. Supriatna and D. Harmantyo.....286

012036
OPEN ACCESS
The Effect of One-to-one Counseling to Pregnant Women's Knowledge about Anemia in Semarang

N P R Egryani, F Saktini, N Susilaningsih, V D Puspitasari and A R Gumay.....295

012037
OPEN ACCESS
Nutrient Content And Acceptability Of Snakehead-Fish (*Ophiocephalus Striatus*) And Pumpkin (*Cucurbita Moschata*) Based Complementary Foods

Etika Ratna Noer, Aryu Candra and Binar Panunggal.....302

012038
OPEN ACCESS
Isolation and Identification of *Aeromonas hydrophila* and *Saprolegnia* sp. on Catfish (*Clarias gariepinus*) in Floating cages in Bozem Moro Krembangan Surabaya

Rahayu Kusdarwati, Kismiyati, Sudarno, Hendi Kurniawan and Yudha Teguh Prayogi.....309

012039
OPEN ACCESS
The Thermocline Layer and Chlorophyll-a Concentration Variability during Southeast Monsoon in the Banda Sea

Nikita Pusparini, Budi Prasetyo, Ambariyanto and Ita Widowati.....315

012040
OPEN ACCESS
Identification of wave energy potential with floating oscillating water column technology in Pulau Baai Beach, Bengkulu

I Alifdini, D N Sugianto, Y O Andrawina and A B Widodo.....327

012041
OPEN ACCESS
Habitat characteristic of two selected locations for sea cucumber ranching purposes

Retno Hartati, Agus Trianto and Widianingsih.....338

012042
OPEN ACCESS
Social Relation between Businessman and Community in Management of Intensive Shrimp Pond

Indra Gumay Febryano, James Sinurat and Messalina Lovinia Salampessy.....348

012043
OPEN ACCESS
Designing Cognitive Intervention to Improve the Awareness Index of the Residents in the Landslide Area

N. Susanto, T. T. Putranto and E. A. Ulfa.....355

012044
OPEN ACCESS
Genetic diversity of the causative agent of ice-ice disease of the seaweed *Kappaphycus alvarezii* from Karimunjawa island, Indonesia

E Syafitri, S B Prayitno, W F Ma'ruf and O K Radjasa.....364

012045
OPEN ACCESS
Capacity Building Resource Management Of Coastal Areas To Improve The Local Economic Based By Cross-Cutting Partnerships: Case Study on Panjang Beach Bengkulu City

Titi Darmi.....372

012046
OPEN ACCESS
The Dynamics of Sea Surface Height and Geostrophic Current in the Arafura Sea

Umaroh, Sutrisno Anggoro and Muslim.....384

012047
OPEN ACCESS
Vulnerability Assessment: The Role of Coastal Informal Settlement Growth to Social Vulnerability in Genuk Sub-District, Semarang City

Sariffuddin, Khristiana Dwi Astuti, Gustika Farhaeni and Lutfiyatul Wahdah.....392

012048
OPEN ACCESS
Waves Induce Sediment Transport at Coastal Region of Timbulsloko Demak

Purnomo Hawati, Denny Nugroho Sugianto, Sutrisno Anggoro, Anindya Wirasatriya and Sugeng Widada.....404

012049
OPEN ACCESS
Potential impact of climate variability on respiratory diseases in infant and children in Semarang

Budiyono, Rismawati, S P Jati and P Ginandjar.....412

012050
OPEN ACCESS
Early Childhood Environmental Education in Tropical and Coastal Areas: A Meta-Analysis

D R Sawitri.....424

012051
OPEN ACCESS
Diatoms and Water Quality of Telaga Warna Dieng, Java Indonesia

Tri Retnaningsih Soeprobowati, Sri Widodo Agung Suedy and Hadiyanto.....432

012052
OPEN ACCESS
Improvement of Stability and Antioxidant Activities by Using Phycocyanin - Chitosan Encapsulation Technique

Meiny Suzery, Hadiyanto, Dian Majid, Deny Setyawan and Heri Sutanto.....438

012053
OPEN ACCESS
Sea Water Intrusion in Kaligawe Semarang Based on Resistivity Data

Agus Setyawan, Najib, Yoga Aribowo, Agnis Trihadini, Dhana Hastuti, Fitra Ramdhani, Fajar Waskito, Ganap Febrika and Galang Virgiawan.....445

012054
OPEN ACCESS
In Search for Sustainable Coastal Management: A Case Study of Semarang, Indonesia

Sudharto P Hadi.....452

012055
OPEN ACCESS
Effect of economic growth and environmental quality on tourism in Southeast Asian Countries

Firmansyah.....457

012056
OPEN ACCESS
Isolation And Partial Characterization Of Bacteria Activity Associated With Gorgonian *Euplexaura* sp. Against Methicillin-Resistant *Staphylococcus aureus* (MRSA)

R. Kristiana, D. Ayuningrum, M.A. Asagabaldan, H. Nuryadi, A. Sabdono, O.K. Radjasa and A. Trianto.....462

012057
OPEN ACCESS
Antioxidant and antibacterial activity of solid-liquid and enzyme-assisted extraction of phenolic compound from three species of tropical *Sargassum*

M Puspita, M Deniel, I Widowati, O K Radjasa, P Douzenel, G Bedoux and N Bourgougnon.....468

012058
OPEN ACCESS
Low zinc serum levels and high blood lead levels among school-age children in coastal area

Adriyan Pramono, Binar Panunggal, M.Zen Rahfiludin and Fronthea Swastawati.....481

012059
OPEN ACCESS
The need of adequate information to achieve total compliance of mass drug administration in Pekalongan

Praba Ginandjar, Lintang Dian Saraswati, Opik Taufik, Nurjazuli and Bagoes Widjanarko.....490

012060
OPEN ACCESS
Physical characteristics of phycocyanin from spirulina microcapsules using different coating materials with freeze drying method

E N Dewi, L Purnamayati and R A Kurniasih.....500

012061
OPEN ACCESS
MJO (Madden-Julian Oscillation) Analysis of the Chlorophyll-a Distribution in Western Waters Bengkulu

Y D Haryanto, N Fitrianti, A Hartoko, S Anggoro and M Zainuri.....507

012062
OPEN ACCESS
Proximate Content of "Klekap" (Microphytobenthos and Their Associated Meiofauna) from Milk-Fish Pond

Widianingsih, Muhammad Zainuri, Sutrisno Anggoro and Hermin Pancasakti Kusumaningrum.....517

012063
OPEN ACCESS
Effects of Different Heat Processing on Fucoxanthin, Antioxidant Activity and Colour of Indonesian Brown Seaweeds

Eko Susanto, A Suhaeli Fahmi, Tri Winarni Agustini, Septian Rosyadi and Ayunda Dita Wardani.....523

012064
OPEN ACCESS
Determination and Radiocarbon Dating of Marine Mollusc Fossils in Ancient Sea Shelf of Central Java Indonesia

S Aisyah, D Pringgenies, A Hartoko, J T S Sumantyo and H Matsuzaki.....533

012065
OPEN ACCESS
Sustainable Eco Coastal Development Through Corporate Social Responsibility (CSR) Program

Arsi Rakhmanissazly, Yong Mursito Ardy and Abdullah.....540

012066
OPEN ACCESS
The effect of ENSO to the variability of sea surface height in western Pacific Ocean and eastern Indian Ocean and its connectivity to the Indonesia Throughflow (ITF)

H A Rejeki, Munasik and Kunarso.....548

012067
OPEN ACCESS
Antioxidant activity of three microalgae *Dunaliella salina, Tetraselmis chuii* and *Isochrysis galbana* clone Tahiti

Ita Widowati, Muhammad Zainuri, Hermien Pancasakti Kusumaningrum, Ragil Susilowati, Yann Hardivillier, Vincent Leignel, Nathalie Bourgougnon and Jean-Luc Mouget.....555

012068
OPEN ACCESS
Study on the effect of different concentration of *Spirulina platensis* paste added into dried noodle to its quality characteristics

T W Agustini, W F Ma'ruf, Widayat, B A Wibowo and Hadiyanto.....561

012069
OPEN ACCESS
Study on Anticancer Activity of Extracts of Sponges Collected from Biak Water, Indonesia

A. Trianto, A. Ridhlo, D.W. Triningsih and J. Tanaka.....570

012070
OPEN ACCESS
The Application of Ozone and Chitosan as Microbial Inhibitor Prawn Larvae Rearing

Delianis Pringgenies, Muhammad Nur and Rosti Angelia.....575

012071
OPEN ACCESS
Production of Chitosan from *Amusium sp* Scallop Shell Waste

Nur Rokhati, Titik Istirokhatun, Dwi Titik Apriyanti and Heru Susanto.....591

012072
OPEN ACCESS
Achieving Research University: Indonesian Case

Yos Johan Utama and Ambariyanto.....597

Published in IOP Proceeding Series: 2nd International Conference on Tropical and Coastal Region Eco Development (ICTCRED) 2016

PREFACE

The 2nd International Conference on Tropical and Coastal Region Eco Development (ICTCRED) 2016 was held by Research and Community Services Institute, Diponegoro University, Indonesia in Bali during 25-27 October 2016. Bali, the god island amid pleasant surroundings, was a delightful place for this conference. The 165 research papers were presented both by oral and poster presentation, and the conference had delivered many fruitful discussions and exchanges that contributed to the success of the conference. Among these papers, we selected 72 papers through intensive review by scientific committee. Participants from 10 countries made the conference truly international in scope and the presentations were split into two main conference areas, i.e., tropical life sciences and Coastal Region Developments. There were 7 plenary lectures covering the different areas of the conference: Prof. Ocky Karna Radjasa (Ministry of Research, Technology and Higher Education) talked on Policy on Coastal Research in Indonesia, Prof. Ambariyanto (Diponegoro University) on Update in Coastal Development, Prof. Makoto Tsuchiya (University of the Ryukyus – Japan) on Coral Reef Ecosystem, Prof. Peter Gell (Federation University of Australia – Australia) on Pollution and Coastal Environmental Changes in Tropical Region, Prof. Johan C. Winterwerp (TU Delft, Netherlands) on Restoring eroding mangrove-mud coasts through Building with Nature. These plenary speakers have given significant scientific contributions to the conference.

Participants have delivered their talks of valuable research outputs which are varies from extracting valuable compounds from marine environment and convert them to specialty products. During the presentation, marine products and their processing were also elaborated with developments in oceanography technology, coastal environment management and policies. These topics were more fruitful when researches in social science and its developments especially at the coastal area were also presented. These wide range of topics have colorized this conference.

Finally, it is appropriate that we record our thanks to our fellow members of the Organizing Committee and of the Scientific Committee for their work in maintaining high quality of papers and in encouraging participation from those areas. We are also indebted to those who served as chairman of Institute of Research and Community Services, Diponegoro University and Ministry of Research, Technology and Higher Education for their financial supports. Without their support, the conference could not have been the success that it was. We also acknowledge the authors themselves, without whose expert input there would have been no conference. The continuing success of this conference series means that planning can now proceed with confidence for the same event in 2017.

Guest Editors

Hadiyanto,
Chair of Scientific Committee

IOP Conf. Series: Earth and Environmental Science **55** (2017) 011001 doi:10.1088/1755-1315/55/1/011001

Scientific Committee/Editor

Hadiyanto (*Diponegoro University, Indonesia*)
Makoto Tsuchiya (*University of the Ryukyus, Japan*)
Heru Susanto (*Diponegoro University, Indonesia*)
Muhammad Mukhlisin (*Semarang State Polytechnic, Indonesia*)
Jamari (*Diponegoro University, Indonesia*)
Hussein Gasem (*Diponegoro University, Indonesia*)
Ambariyanto (*Diponegoro University, Indonesia*)
Muhammad Zainuri (*Diponegoro University, Indonesia*)
Muhammad Ali (*Mataram University*)
Sultana MH Faraz (*Diponegoro University, Indonesia*)
Gerard Pals (*Vrije Universiteit Amsterdam, Netherland*)
Randi Hagerman (*University of California Davis, United States*)
Flora Tassone (*University of California Davis, United States*)
Herawati Sudoyo (*Eijkman Institute, Indonesia*)
Ocky Karna Rajasa (*MRTHE, Indonesia*)
Craig Starger (*Colorado State University, United States*)
CN Ravishankar (*Central Institute of Fisheries Technology, India*)
Agus Sabdono (*Diponegoro University, Indonesia*)
Maria Barbosa (*Wageningen University, Netherland*)
Yasuhiro Igarashi (*Japan*)
Irwandi Jaswir (*IIUM, Malaysia*)
Michio Hidaka (*University of the Ryukyus, Japan*)

Organizing Committee

Munawar A Riyadi	Chair
Tri Indah Winarni	Co-Chair of TLS
Diah Permata Wijayanti	Co-Chair of CRED
Eko Susanto	Secretary
Desrina	Proceeding
Muh. Arfan	Publication

This conference has been conducted by Institute of Research and Community Services, Diponegoro University, Indonesia 25-27 October 2016

Peer review statement

All papers published in this volume of *IOP Conference Series: Earth and Environmental Science* have been peer reviewed through processes administered by the proceedings Editors. Reviews were conducted by expert referees to the professional and scientific standards expected of a proceedings journal published by IOP Publishing.

Past and Future Ecosystem Change in the Coastal Zone

P Gell[1]

[1] Water Research Network, Federation University Australia, Ballarat, Australia

E-mail: p.gell@federation.edu.au

Abstract. The coastal zone is in a constant state of flux. Long term records of change attest to high amplitude sea level changes. Relative stability though the Late Holocene has allowed for the evolution of barrier dune systems, estuaries and coastal lakes with associated plant and faunal associations. This evolution has been interspersed with changes in the balance between climate driven changes in outflow from catchments. These interactions have been considerably disturbed through the impacts of industrialised people who have diverted and consumed water and invested in infrastructure that has impacted on river flows and the tidal prism in estuaries. This has impacted their provisioning services to humans. It has also impacted their regulating services in that development along the coastline has impacted on the resilience of the littoral zone to absorb natural climate extremes. Looking from the past we can see the pathway to the future and more easily recognise the steps needed to avoid further coastal degradation. This will increasingly need to accommodate the impacts of future climate trends, increased climate extremes and rising seas. Coastal societies would do well to identify their long term pathway to adaptation to the challenges that lie ahead and plan to invest accordingly.

Keywords. estuaries, paleolimnology, climate change, hydroecology, sediments, nutrients, salinization

1. Introduction

The coastal zone is in a constant state of change. This is most evident through the daily cycles of the tides that, particularly in the macrotidal zones of the world, inundate and then strand the marine littoral zones. This alone demands that the biological communities affected by the rise and fall of tides be adaptable to inundation and exposure, and in estuaries, exposed variously by marine waters and those flowing from the hinterland which are usually fresh to oligosaline. This balance varies seasonally with, in temperate zones, wet seasons coinciding with destructive waves regimes leading to the opening of estuary mouths, while the dry season and associated constructive wave regimes lead to mouth closure and the establishment of lagoonal conditions. The consequences are a winter of tidal and river flushing with oxygenated water and a summer with little flow, stratification, hypersalinity and de-oxygenation. These seasonal patterns are taken to extreme under multi-year (e.g. El Nino Southern Oscillation) and even multi-decadal (Inter-decadal Pacific Oscillation) phases of wet and dry conditions exacerbating or subduing the contrasts between seasons. Further the frequency of these cycles extends out to multimillennial scales with the highest amplitude variations coming from the substantial eustatic rises and falls in sea level documented on the Huon Peninsula [1]. These occur on account of the orbital cycles (Milankovitch) of the Earth which, in south-east South Australia [2], have left a series of fossil coastal dunes, each one hundred thousand years older than their more coastal neighbour.

Within the present high sea level stand it is it is tempting to consider that most variations in estuarine condition relate to natural climate and tidal cycles. However, there is also ongoing linear change. Once river valleys were inundated as seas reached maximum level after the last post-glacial transition the river gradients declined and waters in the new coastal zones estuaries became relatively quiescent. This allowed for the transported sediment to settle and so set in place a long extended phase of estuarine evolution that persists into the present[3]. Depending on the balance between the onshore wave environment and river flow an estuary gradually fills with sediment progressing from a wide, open estuary, to a narrow estuary in a wide delta, to an intermittently closed and open lagoon and, ultimately, to a largely closed system with coastal lakes.

A further directional change exists in the form of humanity. Human populations have grown over millennia accelerating after the industrial revolution and more so since the 1950s during the 'Great Acceleration' leading to the identification of a new geological epoch, the Anthropocene [4]. At the coast protective mangrove forests have been removed, coastlines reclaimed, harbours dredged and hard infrastructure commissioned to defend the coast against the tides and waves. Catchments have been cleared leading to increases in the volume of salts and sediments they leak, and nutrients have been mobilised and washed or leached into waterways after deliberate or inadvertent application. River flows have been held behind weirs, diverted for human consumption or regulated with hard infrastructure to allow for navigation and limit flooding. At the coast this results in the reduction of flushing flows, yet increasing loads of nutrients, metals, and sediments. Salinity levels increase owing to increased flux from inland but also due to the reduction in diluting flows[5].

Recent changes, and cyclical variations with short return times, are more familiar to people than those masked by the passing of time. However, because changes are more familiar is no reason why they are greater or have a more substantial impact on coastal ecosystems. The tendency to neglect infrequent cyclical change, or events deeper in the past, leaves society less well informed about the drivers of change and so less well armed to diagnose and treat coastal management issues. Historical evidence of environmental change is reliant on documentary or photographic evidence which tends to be patchy, or on memory which has proven to be unreliable[6]. Deep-in-time and continuous records of change are available, however, in the sediments of inundated coastal systems. Estuaries and lagoons provide a service to historical studies by archiving fossil remains of the biological, chemical and physical nature of waterways as they infill with sediment. The paleolimnological approaches that retrieve this evidence are now widely used globally and provide clear evidence of the influence of climate and people on ecosystems, including in the coastal zone.

2. Paleolimnological approaches

Sedimentary sequences accumulate offshore over long time periods. Onshore they accumulate when depressions are inundated as the still water allows sediment to settle and the anoxia in the littoral zones allow dead remains to be preserved with each deposited layer. As sea levels stabilised ~ 7000 years BP most near-shore records of change extend back to this time. Sophisticated sediment coring techniques are able to extract continuous sediment sequences[7] spanning this period and radiometric dating techniques are able to align a timeline to the extracted core. Being at the end of river catchments the sedimentation rates are often relatively large allowing for high resolution reconstruction of estuary condition through time. The physical nature of the sediment (e.g. particle size) can inform on variations in fluvial energy and so the impact of infilling[8]. Chemical analyses can inform on the source of the sediment (magnetic susceptibility, X-Ray Fluorescence) and the foundations of the food web and trophic status (stable isotopes of nitrogen and carbon) [9]. A wide range of biological remains can be extracted, including gastropods[10], ostracods [11], algal pigments [12], pollen [13], diatom algae [14] and plant fragments [15]. These may provide evidence of changing habitat and substrate availability, the balance between tidal and inland water, salinity and nutrient status, the existence of anoxia and the nature of the light environment of the water body (see [16]). Integrated, multi-proxy investigations of sediment layers provide a means of reconstructing many elements of ever-changing nearshore ecosystems, albeit with the evidence of short time (daily to monthly) variations blended into single samples. Similar proxies may reflect onshore changes to catchments and this evidence, combined with archaeological or ethnohistoric records, can allow for responses to be attributed to particular human or climatic causes [17].

3. South-east South Australia.

The region between Brisbane and Adelaide is the most heavily settled in Australia. There is evidence for indigenous settlement since the Pleistocene and archaeological remains attesting to their activities, living sedentary, rather than nomadic lifestyles and manipulating coastal waterways, at least where resources were plentiful [18]. European settlement commenced in the 18th century and the goldrush of the 1850s brought large numbers of immigrants that saw a rapid expansion in population, much at the expense of the indigenous populations that suffered from disease and conflict. Coastal-flowing catchments were impacted by clearing and grazing and the populations of coastal margins increased dramatically from the mid-20th century and societal wealth increased. Most Australians now live in close proximity to the south-east coastline driving much landscape disturbance and ecosystem change in the near shore zones.

4. Evidence for Ecosystem Change

The first calls for action to mitigate this impact of European activities on coastal ecosystems emerged in the 1960s and 1970s and the first EPA departments was established in 1971 in Victoria [19] but as late as 1991 in NSW. National State of the Environment reporting commenced in 1996[20]. A cap was placed on the extraction of water from the nation's largest basin in 1995 and the Murray River estuary was deemed to be in ecological crisis in 2011[21]. This, however, does not reflect the timing of the impact of the expansion of settlement and associated industries such as mining and agriculture. In fact, long term records of change attest to considerable coastal change from the 1840s [22] and substantial changes in floodplain sedimentation rates and salinity from the 1880s [23].

4.1. Hydrological Change.

Rivers were impounded to retain water supplies for expanding urban populations from the 1860s such as Thorndon Park reservoir in Adelaide. Closer to the coast barrages were established at Geelong in 1840 and at the Murray mouth in 1940 to preserve freshwater resources from the impact of tidal inflows. At Geelong this immediately shifted an estuary to a freshwater lagoon [22] which was gradually colonised by *Phragmites* reedbeds. It also limited the outflow of freshwater from the Barwon River leading to increasing salinity levels in the estuary below the weir [22]. At the Murray mouth regulation constrained the tidal prism leading to the permanent freshening of Lake Alexandrina [24] but lead to rapid sedimentation at the mouth driving the formation of new islands and closing the outlet [25]. This turned the coastal lagoon into a closed system leading to hypersalinity[26], hypoxia and the collapse of critical *Ruppia* seagrass habitats [27],[28]. In the Gippsland lakes a temporary mouth was dredged for navigation leading to a permanent shift in the halo-ecology of the estuary[12]. While this lead to the death of fringing vegetation[29] increased flushing limiting the high cyanobacterial character [12]. The largest engineering feat in Australia, the Snowy River Scheme, diverted ~ 98% of the Snowy River's mean annual flow inland to drive the expansion of the irrigation industry and to generate power. Starved of freshwater supplies, Lake Curlip in the coastal floodplain shifted from a dystrophic lake to a system 50 times more saline than the pre-European baseline[30]. At Tuckean Swamp weirs reduced the tidal prism leading to a freshening of surface waters but deliberate drainage exposed sediments leading to unprecedented acidification [31].

4.2. Pollutants

While the waters at the end of catchments are typically more enriched and saline than their upland counterparts there is clear evidence for the eutrophication of coastal systems through the 20th century (e.g. The Coorong; [26]) and, in the Gippsland Lakes, a return to the cyanobacterial conditions the prevailed before artificial opening, albeit with a novel bloom risk species emerging [12]. Coastal lakes receiving industry waste have seen nutrients rise and the onset of eutrophic conditions [32]. Mining, and industrial development in the coastal zone, are evident in high levels on heavy metals in upper sediments in the lower Barwon River estuary [33] and in Botany Bay in Sydney [34]. The Gippsland Lakes receive drainage from several catchments including the La Trobe River valley which has been long the focus of coal mines and coal-fired power stations, and a native and plantation timber pulp mill, which has lead to fears of high loads of metals including mercury into the estuary sediments and food chain.

4.3. Sedimentation

Catchment development in the Murray River basin has accelerated sediment accumulation rates in floodplain wetlands in the order of 5-30 times and up to 80-fold in the estuary [35]. In parts of the Coorong Lagoon in excess of 50 cm of sediment has accumulated in less than a century and aerial photographs attest to massive deposition of sediments around the river mouth. The barrages have maintained high water levels that have lead to the scouring of exposed littoral margins in Lake Alexandrina and accretion in sheltered zones[25]. The flux of sediments has increased in most situations impacting on the light environment and placing stress on submerged macrophyte communities.

4.4. Salinity

Some lagoons have shown an increase in salinity on account of the reduction in the flushing of freshwater (Reeves et al. 2014) or the construction of openings to the sea ([29],[12]) while others, once saline or tidal, have become relatively fresh on account of barrages restricting inflow of tidal water [33] or the redirection of freshwater into once saline coastal lagoons [26]. In the south lagoon of the Coorong the increase in salinity was from a high base, with the most notable feature being the replacement of salt tolerant thalassic species with euryhaline inland taxa of both diatoms and ostracods ([33]; [24]).

4.5. Ecological Implications

Many estuaries have been impacted by the development of agriculture and intensive industries since the arrival of European settlers. The impact of water quality change is greatest near to large populations e.g. Sydney, or at the end of larger, disturbed catchments e.g. the River Murray. Many of the ecological impacts remain unclear while others had become evident in recent decades and critical with the extended drought period that commenced in 1997. Extreme hypersalinity in the Coorong lead to a shift from a community dominated by *Ruppia megacarpa*, through a *Ruppia tuberosa* ([28]) community phase, to a phytoplankton-brine shrimp state that lead to a decline in ducks and swans and their replacement by Banded Stilt (*Cladorhynchus leucocephalus*) ([24]), as well as declines in fish stocks. Rapid recent accumulation of high acid potential sediments drew an alarming acidification risk in the Murray River's lower lakes while, in the Gippsland Lakes, high river flows in 2007 lead to an unprecedented bloom of the toxic cyanobacterium *Synechococcus* [12] with impacts on seagrasses and bivalve populations. Migratory waterbird populations have declined owing to reduced habitat suitability and, possibly, reduced food stocks, although the impact of modifications to resting sites in their Asian flyway are also implicated.

5. Management Implications

The long term record of change in south-east Australia attests to substantial modifications to coastal waterways from early in European settlement. These shifts are the result of the direct effects of the construction of barrages or the chronic release of sediments, salts and nutrients from various industrial sources and through catchment disturbance from widespread vegetation clearance. While observation and monitoring programs had lead to concerns being raised as to the health of these systems, the onset of record drought conditions rapidly brought forward substantial change in ecosystems and likely ecological regime shifts. Ecosystem managers failed to read the impending crisis and, in several situations, are now left with limited options to halt or reverse the impacts without massive intervention (e.g. [23]).

South-east Australia has been identified as a climate hotspot on account of scenarios of large declines in wet season rainfall and contraction of the winter growing season [36]. This is likely to exacerbate the declining ameliorative effect of freshwater flows into coastal systems leading to further concentration of nutrients and salts. The impact of this in the mobility and accumulation of sediment-bound metals raises issues of concern for the safety of any harvest from these waterways. High nutrients loads, coupled with warmer temperatures and higher concentrations of atmospheric CO_2 are likely to exacerbate cyanobacterial blooms [37] and expand hypoxic zones.

Australia's population is growing at a rapid rate and coastal development is increasing accordingly. There is a seeming reluctance to adequately care for the provisioning, regulating and cultural services these systems provide now, and may do in the future, in the rush to exploit them today. The long term record reveals that the post-1950 Great Acceleration, clearly evident in south-east Australia, rides on the shoulders of a century of ecosystem change wrought by early European settlers. Present practices and population growth is clearly unsustainable. Evidence for non-linear shifts in aquatic ecosystems in highly populated regions elsewhere [38] forewarn of the potential of catastrophic ecological collapses and substantial declines in aquatic resources. This also has the potential to challenge the capacity of society to deal with the risks associated with ongoing warming, a continuing drying trend and the prospects of coastal erosion and inundation with rising sea levels.

These case studies reveal that the provision of evidence of coastal change from long term sedimentary records provides clear insights unavailable from even long term monitoring programs. These provide evidence for long term trajectories of change and so can distinguish cyclical shifts from directional change. This can assist management in assessing the need for action and the level of urgency that measures need to be applied. Societies who are embarking on monitoring programs for coastal system assessment more recently could be well advised to supplement the accumulating survey data with that archived in sediments. In combination these would provide the data bank necessary to both identify slow and low frequency drivers of change and to qualify the importance of monitored variations by setting them in a longer term context.

6. Conclusion

Increasing capacity to observe changes in our environment has enabled humanity to better recognise the opportunity costs of catchment and coastal development to provide services and goods to human economies. In south-east Australia evidence is increasingly being assembled to provide a strong case that many coastal systems have been severely impacted by this development. The assembly of several records of change reaching back further into the past provides an earlier baseline and documents much change before technological advances and concern drew contemporary scientists to monitor long term change. These centennial scale records available in continuous sediment sequences reveal that substantial changes occurred in coastal waterways with years of the building of first settlements. These changes are evident in changed assemblages of fossil biota and the physical and chemical evidence contained in sediment cores. It is clear that coastal systems have been impacted by the increased flux of sediments, nutrients, salts and metals or more than 150 years. Over this time they have been greatly impacted by hydrological change ranging from reduced flows from catchments through inland diversion, increasing or decreasing freshwater inflow or tidal penetration owing to near-mouth impoundments or freshening through the diversion of unwanted fresh floodwaters. All sites studied have changed relative to a pre-settlement baseline in one of these ways, some catastrophically.

Australia is a signatory to the Ramsar Convention on wetlands that obliges it to advise the Secretariat whenever a listed wetland has changed, is changing or is about to change. This convention also requires signatory nations to ensure wise use of all wetlands. Reference to evidence of change (e.g. [29]) from contemporary approaches questions whether Australia is fulfilling those obligations. Independent assessment of condition from long term sedimentary records (e.g. [12]) shows clearly that most wetlands have changed and that regulating, and provisioning services are being compromised. In line with the observations of Finlayson et al. [39] the government should embark on a program that assesses the present condition of its listed wetlands against more realistic, long term baselines to implement measures to avoid them passing irrevocably into degraded states.

7. References

[1]. Chappell J and Polach H 1991 Post-glacial sea-level rise from a coral record at Huon Peninsula, Papua New Gunea *Nature* **349** 147-149

[2]. Schwebel D A 1976 *Quaternary dune systems Natural History of the South East* eds Tyler M J Twidale C R Ling J K and Holmes J W (Adelaide: Royal Society of South Australia) pp 15-24

[3]. Roy P 1984 *New South Wales estuaries: their origin and evolution Coastal Geomorphology in Australia* ed Thom BG (Sydney: Academic Press) p 99-121

[4]. Steffan W Crutzen P J and McNeill J R 2007 The Anthropocene: are humans now overwhelming the great forces of nature *Ambio* **36** 614-621.

[5]. Battarbee R Bennion H Rose N and Gell P 2012 *Human Impact on Freshwater ecosystems The SAGE Handbook of Environmental Change Volume 2* ed JA Matthews (London: SAGE) pp 47-70

[6]. Tibby J Lane M and Gell P A 2008 Local knowledge as a basis for environmental management: a cautionary tale from Lake Ainsworth, northern New South Wales. *Environ. Conserv.* **34** 334-341

[7]. Skilbeck C G Trevathan-Tackett S Apichanangkool P and Macreadie P I in press a *Sediment sampling in estuaries – site slection and sampling techniques Applications of Paleoenvironmental Techniques in Estuarine Studies* (Developments in Paleoenvironmental Research vol 20) eds K Weckstrom P Gell K Saunders and G Skilbeck (London: Springer)

[8]. Skilbeck C G Heap A D and Woodroffe C D in press b *Geology and sedimentary history of modern estuaries Applications of Paleoenvironmental Techniques in Estuarine Studies* (Developments in Paleoenvironmental Research vol 20) eds K Weckstrom P Gell K Saunders and G Skilbeck (London: Springer)

[9]. Leng M J and Lewis J P in press. *C/N ratios and carbon isotope composition of organic matter in estuarine environments Applications of Paleoenvironmental Techniques in Estuarine Studies.* (Developments in Paleoenvironmental Research vol 20) eds K Weckstrom P Gell K Saunders and G Skilbeck (London: Springer)

[10]. Wingard G L and Surge D in press *Application of molluscan analyses to the reconstruction of past environmental conditions in estuaries. Applications of Paleoenvironmental Techniques in Estuarine Studies* (Developments in Paleoenvironmental Research vol 20) eds K Weckstrom P Gell K Saunders and G Skilbeck (London: Springer)

[11]. Reeves J M in press *Ostracods as recorders of palaeoenvironmental change in estuaries Applications of Paleoenvironmental Techniques in Estuarine Studies* (Developments in Paleoenvironmental Research vol 20) eds K Weckstrom P Gell K Saunders and G Skilbeck (London: Springer)

[12]. Cook P L M Jennings M Holland D P Beardall J Briles C Zawadzki A Doan P Mills K and Gell P 2016 Blooms of cyanobacteria in a temperate Australian lagoon system post and prior to European settlement *Biogeosci.* **13** 3677-3686

[13]. Ellison J C in press *Applications of pollen analysis in estuarine systems Applications of Paleoenvironmental Techniques in Estuarine Studies* (Developments in Paleoenvironmental Research vol 20) eds K Weckstrom P Gell K Saunders and G Skilbeck (London: Springer)

[14]. Taffs K H Saunders K M and Logan B in press *Diatoms as indicators of environmental change in estuaries Applications of Paleoenvironmental Techniques in Estuarine Studies* (Developments in Paleoenvironmental Research vol 20) eds K Weckstrom P Gell K Saunders and G Skilbeck (London: Springer)

[15]. Tibby J and Sayer C D in press *Inferring environmental change in estuaries from plant macrofossils Applications of Paleoenvironmental Techniques in Estuarine Studies* (Developments in Paleoenvironmental Research vol 20) eds K Weckstrom P Gell K Saunders and G Skilbeck (London: Springer)

[16]. Weckstrom K Gell P Saunders K and Skilbeck G eds in press *Applications of Paleoenvironmental Techniques in Estuarine Studies* (Developments in Paleoenvironmental Research vol 20). (London: Springer)

[17]. Saunders K M and Gell P A in press *Paleoecological evidence for variability and change in estuaroies: insights for management. Applications of Paleoenvironmental Techniques in Estuarine Studies* (Developments in Paleoenvironmental Research vol 20) eds K Weckstrom P Gell K Saunders and G Skilbeck (London: Springer)

[18]. Builth H 2004 Mt Eccles lave flow and the Gunditjmara connection: a landform for all seasons. Proc. Roy. Soc. Vic. **116(1)** 165-184

[19]. Russ P and Tanner L 1978. *The Politics of Pollution* (Camberwell: Widescope)

[20]. Commonwealth of Australia 1996 *State of the Environment Australia* (Canberra: Commonwealth of Australia)

[21]. Kingsford R T Walker K F Lester R E Young W J Fairweather P G Sammut J and Geddes M C 2011 A Ramsar wetland in crisis – the Coorong, Lower Lakes and Murray Mouth, Australia. *Mar. Freshw Res.* **62** 255-265

[22]. Reeves J M Gell P A Reichman S M Trewarn A J and Zawadzki A 2016 Industrial past, urban future: using palaeo-studies to determine the industrial legacy of the Barwon estuary, Victoria, Australia. *Mar. Freshw. Res.* **67** 837-849

[23]. Gell P in press *The Coorong Applications of Paleoenvironmental Techniques in Estuarine Studies* (Developments in Paleoenvironmental Research vol 20) eds K Weckstrom P Gell K Saunders and G Skilbeck (London: Springer)

[24]. Gell P A 2012 *Palaeoecology as a means of auditing wetland condition Peopled Landscapes: Archaeological and Biogeographic Approaches to Landscapes* (Terra Australis vol 34) eds Haberle SG and David B (Canberra: ANU) pp 445-457

[25]. Bourman R P Murray-Wallace C V Belperio A P and Harvey N 2000 Rapid coastal change in the River Murray estuary of Australia *Mar. Geol.* **170** 141-168

[26]. Fluin J Gell P Haynes D and Tibby J 2007 Paleolimnological evidence for the independent evolution of neighbouring terminal lakes, the Murray Darling Basin, Australia. *Hydrobiol.* **591** 117-134

[27]. Krull E Haynes D Lamontagne S Gell P McKirdy D Hancock G McGowan J and Smernik R 2009 Changes in the chemistry of sedimentary organic matter within the Coorong over space and time. *Biogeochem.* **92** 9-25

[28]. Dick J Haynes D Tibby J Garcia A and Gell P 2011 A history of aquatic plants in the Ramsar-listed Coorong wetland, South Australia *J. Paleolimn.* **46** 623-635

[29]. Boon P Cook P and Woodland R 2016 The Gippsland Lakes: management challenges posed by long-term environmental change *Mar. Freshw. Res.* **67(6)** 721-737

[30]. MacGregor A J Gell P A Wallbrink P J and Hancock G 2005 Natural and post-disturbance variability in water quality of the lower Snowy River floodplain, Eastern Victoria, Australia. *River Res. Appl.* **21** 201-213.

[31]. Taffs K H Farago L J Heijnis H and Jacobsen G 2008 A diatom-based Holocene record of human impact from a coastal environment: Tuckean Swamp, eastern Australia. *J Paleolimnol* **39** 71–82.

[32]. Haynes D Gell P Tibby J Hancock G and Goonan P 2007 Against the tide: the freshening of naturally saline coastal lakes, south east South Australia. *Hydrobiol.* **591** 165-183

[33]. Reeves J M Haynes D Garcia A and Gell P A 2015 Hydrological change in the Coorong Estuary, Australia, Past and Present: Evidence from fossil invertebrate and algal assemblages. *Estuaries and Coasts* **38(6)** 2101-2116

[34]. Kermode S Heijnis H Wong H Zawadzki A Gadd P and Permana A 2016 A Ramsar-wetland in suburbia: wetland management in an urbanised, industrialised area. *Mar. Freshw. Res.* **67(6)** 771-781

[35]. Gell P Bulpin S Wallbrink P Bickford S and Hancock G 2005 Tareena Billabong – a palaeolimnological history of an everchanging wetland, Chowilla Floodplain, lower Murray-Darling Basin *Mar. Freshw. Res.* **56** 441-456

[36]. Giorgi F 2006 Climate change hotspots. Geophys. Res. Lett. **33** L08707. doi:10.1029/2006GL025734

[37]. Visser P M Verspagen J M H Sandrini G Stal L J Matthijs H C P Davis T W Paerl H W and Huisman J 2016 How rising CO_2 and global warming may stimulate harmful cyanobacterial blooms *Harmful Algae* **54** 145-159

[38]. Wang R Dearing J A Langdon P G Zhang E Yang X Dakos V and Scheffer M 2012 Flickering gives early warning signals of a critical transition to a eutrophic lake state. *Nature* doi:10.1038/nature11655.

[39]. Finlayson C M Clarke S J Davidson N C and Gell P 2016 Role of palaeoecology in describing the ecological character of wetlands. *Mar. Freshw. Res.* **67** 687-694

Conserving endangered marine organisms: causes, trends and challenges

Ambariyanto

[1]Department of Marine Sciences, Faculty of Fisheries and Marine Sciences, Diponegoro University, Jl. Prof. H. Soedarto, SH, Tembalang Semarang. 50275
[2]Coastal Disaster Rehabilitation and Mitigation Center, Integrated Laboratory, Diponegoro University, Jl. Prof. H. Soedarto, SH, Tembalang, Semarang. 50275

ambariyanto@undip.ac.id; ambariyanto.undip@gmail.com

Abstract. Increasing demand for marine resources in recent decades for human needs has led to intensified exploitation. This increase accelerates the process of extinction of various marine resources. In order to avoid extinction, it requires conservation measures of marine resources appropriately. This paper provides an overview of causes of extinction, trends and challenges in the conservation of endangered marine organisms. The success of conservation measures is highly dependent on various stakeholders such as governments, communities, the private sector and academics. Differences of the interest of these parties often lead to the failure of conservation programs. In general there is an increasing public awareness of the importance of protecting the diversity of marine resources and avoiding extinction of marine organisms, especially endangered organisms. The existence of comprehensive actions, legislation and improved coordination among government, community, private sector, and academics will significantly improve the success in overcoming all the challenges.

Keywords: conservation, endangered species, extinction

1. Introduction

Two centuries ago, it was believed that marine species would not be extinct simply because being utilized by humans [1]. Human ability to exploit marine resources was considered only a small portion. However, with so many changes taking place in the world, such as science and technology, population growth as well as changes in the sea itself, then the statement is no longer appropriate. Recent report shows that between 20% and 25% of species both marine and non marine taxa are at risk of extinction [2].

Population growth around the world over the past two decades continues to rise significantly [3], it was expected to be close to 9 billion people on the year 2050. Increasing population of the world will affect the carrying capacity of the earth [4]. It also requires increased food availability for consumption [5]. Given that the main food source is a natural resource [6], then certainly there will be increased pressure on the natural resources [7]. Increasing pressure will cause damage to habitats and ecosystems at various places in the world [8].

Destruction on habitats and ecosystems has been widely reported in all forms of the ecosystem and it occurs worldwide [9]. In general, the damage caused by various factors, among others, is the excessive use of natural resources for the purpose of economic growth [10] use of equipment and methods that are not environmentally friendly [11], pollution [12-13] and other nature phenomenon [14]. The negative impact resulting from a variety of habitats and ecosystems destruction was widely reported mainly included in the economic and social aspects [15].One of the effects that result from the above is the increase in the extinction rate of organisms.

The extinction of the organism occurs both marine organisms [16], terrestrial [17] and air [18]. This process occurs primarily caused by human behavior associated with utilizing such an organism as well as habitat destruction. This article looks at the causes, trends and challenges in regard to how we conserve marine species with high possibility to extinct.

2. Extinction of an organism

Many studies have listed causes of the extinction of an organism or population, such as high trophic level, low population density, slow life history and small geographical range size [19], climate change and habitat destruction are the main factors of biodiversity loss [14]. In addition, exploitation, habitat loss, invasive species and other threats (such as climate change, pollution, and disease) are the main causes of extinction [16]. While human population density and proximity to markets have significant influence on vulnerability to extinction of coral reef fishes [20].

The extinction of an organism or population is basically influenced by two factors, namely the level of vulnerability and the existence of the threat. Both of these factors must exist before extinction occurs. The relationship of these two factors is as follows:

$E = V \times T$ (E= Extinction, V= vulnerability, and T= threat)

It means that when the organism or populations in less vulnerable, then despite the threat, extinction is not going to happen, and vise versa.

2.1. Extinction Vulnerability

Vulnerability of extinction is influenced by internal factors of the organisms or populations. These factors are given to the organisms because its part of their biological characteristics.

2.1.1. Size of the animals

Larger body size organisms generally have smaller population density [19]. They also can be easily spotted by predator, including human, and with similar effort predator will get bigger yield. Therefore, larger body size animals have greater vulnerability than smaller ones and possibly also have higher future loss than expected [21, 22]. Although, it was found that intermediate body size organisms attain higher population densities [23, 24], however, these scientists believe that small size organisms have least vulnerable compare with others. This hypothesis has been proved in various kind of organisms, such as birds, mammalian [25], mammalian [26, 27, 28], crustacean [29], as well as other carnivores [30].

2.1.2. Reproduction

Capability to reproduce by an organism will directly influence the vulnerability extinction of the organism [22]. Reproduction rate of an organism will determine its population size. Species with less ability to produce offspring, will have higher vulnerability of extinction then those with higher number of offspring. Although the predation rate is high enough, then it is likely not the entire population will be depleted. Conversely, if the member of the population is low, vulnerability to extinction of organisms would be larger.

2.1.3. Endemic species

Endemic organisms are known as those that exist only in one geographic region. These organisms are found only in certain areas by the distribution of a very limited and easily recognized by predators. With limited range area, then these species are easy prey to predators. Therefore, they are small range species, and tend to have higher risk of extinction. Fish with small ranges have higher vulnerability of extinction [16]. Similarly, extinction of two genera of endemic Mexican fish, Zoogoneticus and Skiffia, as well as endemic plants from Brazil have been reported [31, 32].

2.1.4. Migratory species
Migratory species are species that travel to other locations periodically. This trip could be related to the reproductive process that often occurs seasonally. The trip generally carried long distances and can occur between nations. Migratory species is found in terrestrial, avian and aquatic organisms, and they constitute a significant part in biodiversity [33]. Given the long and frequent migration, then long-distance migrants, are recognized as suffering serious population declines, because its often exposed to a potentially wide array of threats across their range. Approximately 21% of marine migratory species are categorized as Critically Endangered, Endangered or Vulnerable [34]. While 48% of these species are threatened, Near Threatened or Data Deficient. They also found that among all marine migratory species, sea turtles are the most threatened group (85%), followed by seabirds (27%), cartilaginous fish (26%), marine mammals (15%) and bony fish (11%).

2.2. Extinction Threats

Threats of extinction are influenced by external factors or the organisms or populations.

2.2.1. Human population density
Increasing the number of people directly increasing food needs, and the need for it is highly dependent on natural resources [35]. The high demand of natural resources as a result of the increase in population (population pressure) will increase possibility to extinction of species, including fish [20]. Figure 1 shows significant increase in both the number of human population in the world with the increasing number of species extinction [36].

Increase in population and economic growth increases demand for fish products [37]. The increase in population is also considered to affect biodiversity. The extinction rate of organisms increased by 7% in 2020 and 14% in 2050, as a result of population growth alone [38].

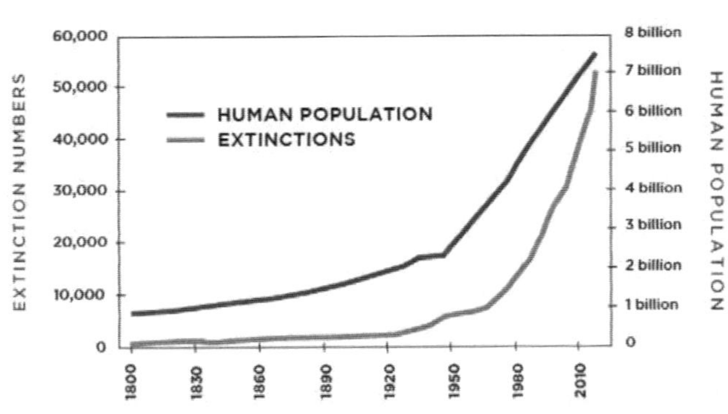

Figure 1. Increasing human population and species extinction [36]

2.2.2. Habitat loss
Increasing demands for human needs especially for shelters as well as other activities have caused habitat change. Changing of habitat will directly reduce the number of organism or even depletion in the worst case. It is believed that habitat destruction is the main cause of organism extinction [39]. For example, it has been reported declining of reptile species in global scale due to several factors including

habitat loss [40]. Changing in habitat due to surface mining activity has threatened plants in Brazil [32]. Similarly, it was found that habitat loss is important cause for organism extinction and recovery in Canada [41]. There is also report that states that among various organisms, fish are organisms that obtain the highest threat of becoming extinct due to habitat loss in Singapore [42].

2.2.3. Over exploitation
Over exploitation of the natural resources will directly increase the rate of extinction of the population. This is mainly caused by the fact that exploitation of natural resources does not take into account the population's ability to recover, it will accelerate the loss of the entire population, and causing the extinction. It has been warned that the damage and loss of terrestrial species can also occur in the oceans [43]. Other scientists found that although fish populations already depleted, however, fishing pressure still continue which will caused fishing induced extinction [44].

2.2.4. Environmental Factors
Living organisms is greatly influenced by environmental conditions that support all biological processes and the life of the organism. Therefore, changes to the surrounding environmental factors, will affect various life processes of the organism. The organism's ability to respond to changes in the environment, will affect susceptibility to extinction. If the changes were drastic and deadly, it is possible there will be a mass mortality of the population of these organisms. Although, environmental factors are included as stochastic threats of extinction, however, some of these factors effects biological processes related to extinction such as reproduction, development and growth. For example, environmental parameters, especially water temperature affects reproduction of European flat oyster (*Ostrea edulis*) [45]. Similarly, it has been reported that salinity, temperature and light conditions affect reproduction and development of a tropical marine copepod *Nitocra affinis f. californica* Lang [46]. In addition, changes in environmental parameters such as water pollution will definitely significantly influence population which will lead to extinction [47].

2.2.5. Invasive species
Invasive species is known as organisms being introduced in an ecosystem that does not occur naturally [48]. These introduced species will survive even reproduce, when they find habitat which similar with their native. They become invasive when they rapidly increase their population and spread and harm the ecosystem. Invasive species is believed as a threat which increases susceptibility to extinction of an organism [49]. This will happen if the new organism affects directly against a variety of existing endogenous organisms. The arrival of a new organism will increase competition for space and food. In addition, if the new organisms are predators on organisms that already exist, it will increase the vulnerability of extinction. This opinion has been questioned by [50] who suggested that this opinion has not been enough proved scientifically. So they suggested to conduct more detail research to get the right answer.

2.2.6. Wildlife Trade
Wildlife trade is "*any sale or exchange by people of wild animal and plant resources*" [51]. Wildlife being traded in many different forms including live animals, meat, bones, or other part of the organisms and they have high economic value. This wildlife being used either as pets or ornamental organisms, source for medicine, food, etc. [51]. Trading on marine organisms have increased with people's interest in ornamental organisms [52], and it is believed that 1.5 and 2 million people in the world keep ornamental fish in their aquarium. They also stated that this live animal trade has high economic value which worth US$200-330 million annually. Ornamental organisms will be more expensive if the organism is unique and rare. The organism is increasingly scarce the price will be higher. The pressure on a population will increase in the event of high interest to the particular organism.

2.2.7. Climate change
One of recently occurring threads of biodiversity is global warming [53] and also induces species extinction. Climate change which is caused by global warming has been known to produce changing in distribution and abundance of organisms. Initially, it was though that physiological tolerance ability of the organism to high ambient temperatures is the cause of extinction related to climate change, however, it was found that it is not the case [54]. Instead, species interaction, in particular decreasing availability of food is the important cause. In other study, predicted that based on mid-range climate-warming scenarios for 2050, approximately 15–37% of terrestrial species will be 'committed to extinction' [55].

3. Conservation Trends
Rapid developments in science and technology around the world have both positive and negative impacts on the environment. Within the last two decades there has been a tendency to increase awareness concerning the environment. Many environmental issues were brought to economic and political matters. Cases related to the destruction of the environment have become the public's attention through the media.

Understanding the community on the importance of conservation issues is increasingly encouraging. This is mainly due to the increased awareness about the importance of protecting the environment itself. In addition, efforts have been made by many parties to improve the environment through conservation. There are several trends noticeable in regard to conservation: Community awareness; Government policy; Development of related science and the interest of scientists; Media cover. These changes were greatly influenced by improvement of education, rapid development of science and technology.

3.1. Community awareness
People are more aware about the importance the environment and conservation. This can be shown among others is the development of environmental groups with all their activities. It has been reported that community awareness on environmental relates issues was significantly influence by their education [56]. They found that perception, knowledge, awareness, and attitude of educated respondents in regard to regional and global environmental problems were much better than those from community group. They also suggested the importance of education either formal or informal to the community on environmental issues. One of the education methods that can be used is by mobile learning [57], or through workplace experiences [58].

3.2. Government policies
Awareness of environmental issues has also increased at the government level. This is shown from the policies that take into account environmental aspects. Environmental understanding by the government is influenced both from within and outside country. The influence of the pioneer country who started their environmental policies have a profound effect on follower countries, so often happens "in common policy" of many countries [59]. Several types of policies that support the conservation of endangered marine organisms include: increasing the number of marine protected areas (MPA); banning all types of destructive fishing methods; as well as providing related and necessary regulations.

3.3. Development of related science and the interest of scientists
Community awareness on environmental issues including endangered species has influence the development of related science. Researchers have a major role, especially for the government in determining environmental policy. A good policy should be based on results of scientific studies. In this regard, it is importance that scientists have to maintain their credibility on negotiating the relationship between scientific objectivity and political advocacy [60]. In addition, the number of scientific publications related to environmental issues has increased significantly. Meaning that related science is developing and more scientists interested in this issue. It has been reported that since the 1990s the number of papers published by four mainstream conservation journals (Conservation Biology, Biological Conservation, Biodiversity and Conservation and Oryx) has increased by 133% [61].

However, the current required more attention to the types of organisms that are less attractive, given that most of the research is still carried out on a charismatic taxa.

3.4. Media coverage

Media (electronic, non-electronic and social media, including television, newspaper, magazine, facebook, etc.) have a major role in spreading environmental issues to the public [62, 63]. Various environmental problems, including issues related to violations of protected organisms, widely reported in the media. Publication in the media covers many issues, among other things on how the protected organism should be treated, including the mechanisms and stages of filing animals to be protected.

4. Conservation challenges

Despite increased awareness related to the environment and conservation of endangered species, it seems there are still some obstacles that must be overcome. Five noticeable challenges that we have to deal: Increasing pressure; Law enforcement; Quality of MPA; Action programs; Genetic resources. In order to overcome these challenges, there are four parties that have to get involve, i.e. government, private sector, community and academics. These parties have to act accordingly related to different challenges.

4.1. Increasing pressure

Although public awareness related to the issue of the environment has improved, as well as the various government policies that takes into account the interests of the environment, but the pressure on the environment and ecosystems remains high. The rapid development of the private sector with various activities, have led to changes or loss of habitat [64]. For example, coastal area reclamation and mangrove deforestation that is usually used for business development, such as tourist area, shopping mall, hotels and resident [65]. Actions that can be done is to increase global awareness and global policy and strategy in order to reduce pressure. This activity will involve several parties namely the government, community and private sector.

4.2. Law enforcement

The government has provided various regulations related to the environment, including with regard to the regulation of organism protected. Various regulations are not only to be disseminated to the public, but they also must be considered and implemented in daily activities of general public. The rapid process of development as well as increased economic demands has caused violations of the laws [66]. Various violations are often committed by many parties concerned including company, whether intentionally or not. This would be detrimental to both the environment and the organism is protected, but also to the general public. Therefore we need a great effort to implement the various rules with all its consequences. Enforcement of environmental laws needs to obtain a special attention by the government. In addition, various other laudable actions such as corruption, collusion, conspiracy, bribery related to environmental issues should also be eliminated. The activities are closely linked to the government, community and private sector

4.3. Quality of MPA

MPA (Marine Protected Areas) have a major role in preserving marine resources [67]. MPA is not only a place of refuge from a variety of marine organisms, but also a place for foraging and reproduction of marine organisms and other biological processes. So the MPA should be a safe area for a variety of marine organism particularly from human disturbance. Another role of the MPA is a source of recruitment for marine organism populations both within and outside the MPA [68]. Therefore, it is necessary to develop a lot of MPA with an area sufficient to function optimally. However MPA in large numbers become less meaningful, if the MPA does not have good quality. For example, the Indonesian government plans to establish MPA for 20 million hectares in 2020 is regarded as a very good policy [69]. The MPA is expected to have good quality, both in terms of the environment, the facilities, as well as management. Therefore, it needs skillful human resources, facilities and funding as well as a suitable regulation. Various parties involved in this case are the government and academic.

4.4. Action programs

Increased awareness of both the public and the government regarding the importance of the environment and marine organisms protected, need to be supported with clear program of action. The program should support the preservation of the environment and reduce the causes of extinction, especially the threat of extinction organism. In addition, this action program should also comprehensively cover various matters related to both technical and non-technical. The program should also develop an activity that can be run at the lowest level in the society to the government and other parties. Some aspects that can be done in this regard is the provision of adequate funding, skillful personnel, activities and good management. The action program will involve all stakeholders including government, community, private sector and academia.

4.5. Genetic resources

Genetic resources have not been given good attention, especially on the regulation and action programs. Policies concerning the determination of organisms protected if only rely on the ability to identify taxonomically based morphology, for now inadequate [70]. This will lead to misguided policy. Therefore it is very important to consider genetic biodiversity on the determination of extinct organisms [71]. Actions that can be done is to the make sure that genetic resources should be included in the new regulation by government prepare. This action requires the involvement of the government and academics (universities).

5. Acknowledgement

Thanks to ICTCRED committee for inviting me as one of the keynote speaker in this important international seminar.

References

[1] Roberts, C.M. and Hawkins, J.P., 1999. Extinction risk in the sea.Trends in Ecology & Evolution, 14(6):241-246.
[2] Webb, T.J. and Mindel, B.L., 2015. Global patterns of extinction risk in marine and non-marine systems. Current Biology, 25(4):506-511.
[3] Cohen, J.E., 2003. Human population: the next half century. science, 302(5648):1172-1175.
[4] Cohen, J.E., 1995. Population growth and earth's human carrying capacity.Science, 269(5222): 341.
[5] Gilland, B., 2002. World population and food supply: can food production keep pace with population growth in the next half-century?.Food policy, 27(1):47-63.
[6] Pimentel, D. and Pimentel, M., 2003.World population, food, natural resources, and survival. World Futures: The Journal of General Evolution, 59(3-4):145-167.
[7] Ehrlich, P.R., Ehrlich, A.H. and Daily, G.C., 1993. Food security, population and environment. Population and development review, pp.1-32.
[8] Hannah, L., Carr, J.L. and Lankerani, A., 1995. Human disturbance and natural habitat: a biome level analysis of a global data set. Biodiversity & Conservation, 4(2):128-155.
[9] Dunlap, R.E. and Jorgenson, A.K., 2012. Environmental problems.The Wiley-Blackwell Encyclopedia of Globalization.
[10] Arrow, K., Bolin, B., Costanza, R. and Dasgupta, P., 1995. Economic growth, carrying capacity, and the environment.Science, 268(5210), p.520.
[11] Turner, S.J., Thrush, S.F., Hewitt, J.E., Cummings, V.J. and Funnell, G., 1999. Fishing impacts and the degradation or loss of habitat structure.Fisheries Management and Ecology, 6(5): 401-420.
[12] Oleksyn, J. and Reich, P.B., 1994. Pollution, habitat destruction, and biodiversity in

Poland.Conservation Biology, pp.943-960.

[13] Cheevaporn, V. and Menasveta, P., 2003.Water pollution and habitat degradation in the Gulf of Thailand.Marine Pollution Bulletin, 47(1):43-51

[14] Travis, J.M.J., 2003. Climate change and habitat destruction: a deadly anthropogenic cocktail. Proceedings of the Royal Society of London B: Biological Sciences, 270(1514):467-473

[15] Cesar, H., Burke, L. and Pet-Soede, L., 2003. The economics of worldwide coral reef degradation. WWF-Netherlands

[16] Dulvy, N.K., Sadovy, Y. and Reynolds, J.D., 2003. Extinction vulnerability in marine populations.Fish and fisheries, 4(1):25-64.

[17] Sheehan, P.M. and Fastovsky, D.E., 1992.Major extinctions of land-dwelling vertebrates at the Cretaceous-Tertiary boundary, eastern Montana.Geology, 20(6):556-560.

[18] Simberloff, D., 1995. Habitat fragmentation and population extinction of birds. Ibis, 137(s1): S105-S111.

[19] Purvis, A., Gittleman, J.L., Cowlishaw, G. and Mace, G.M., 2000. Predicting extinction risk in declining species. Proceedings of the Royal Society of London B: Biological Sciences, 267(1456):1947-1952.

[20] Brewer, T.D., Cinner, J.E., Green, A. and Pressey, R.L., 2013. Effects of human population density and proximity to markets on coral reef fishes vulnerable to extinction by fishing.Conservation Biology, 27(3):443-452.

[21] McKinney, M.L., 1997. Extinction vulnerability and selectivity: combining ecological and paleontological views. Annual Review of Ecology and Systematics, pp.495-516.

[22] Cardillo, M., Mace, G.M., Jones, K.E., Bielby, J., Bininda-Emonds, O.R., Sechrest, W., Orme, C.D.L. and Purvis, A., 2005. Multiple causes of high extinction risk in large mammal species. Science, 309(5738):1239-1241.

[23] Marquet, P.A., Navarrete, S.A. and Castilla, J.C., 1995. Body size, population density, and the energetic equivalence rule. Journal of Animal Ecology, pp.325-332.

[24] Blackburn, T.M., Harvey, P.H. and Pagel, M.D., 1990. Species number, population density and body size relationships in natural communities. The Journal of Animal Ecology, pp.335-345.

[25] Gaston, K.J. and Blackburn, T.M., 1995. Birds, body size and the threat of extinction. Philosophical Transactions of the Royal Society of London B: Biological Sciences, 347(1320):205-212.

[26] Damuth, J., 1981. Population density and body size in mammals. Nature,290(5808):699-700

[27] Cardillo, M. and Bromham, L., 2001.Body size and risk of extinction in Australian mammals.Conservation biology, 15(5):1435-1440.

[28] Fa, J.E., Stewart, J.R., Lloveras, L. and Vargas, J.M., 2013. Rabbits and hominin survival in Iberia. Journal of human evolution, 64(4):233-241.

[29] Arnold, W.S., 1984. The effects of prey size, predator size, and sediment composition on the rate of predation of the blue crab, CallinectesSapidusRathbun, on the hard clam, MercenariaMercenaria (Linné).Journal of Experimental Marine Biology and Ecology, 80(3): 207-219.

[30] Carbone, C., Pettorelli, N. and Stephens, P.A., 2010. The bigger they come, the harder they fall: body size and prey abundance influence predator–prey ratios. Biology Letters, p.rsbl20100996.

[31] De La Vega-Salazar, M.Y., Avila-Luna, E. and Macías-Garcia, C., 2003. Ecological evaluation of local extinction: the case of two genera of endemic Mexican fish, Zoogoneticus and Skiffia. Biodiversity & Conservation, 12(10):2043-2056.

[32] Jacobi, C.M., Do Carmo, F.F. and de Campos, I.C., 2011. Soaring extinction threats to endemic plants in Brazilian metal-rich regions. AMBIO: A Journal of the Human Environment, 40(5): 540-543.

[33] Martin, T.G., Chadès, I., Arcese, P., Marra, P.P., Possingham, H.P. and Norris, D.R., 2007. Optimal conservation of migratory species. PLoS One, 2(8): p.e751

[34] Lascelles, B., Notarbartolo Di Sciara, G., Agardy, T., Cuttelod, A., Eckert, S., Glowka, L., Hoyt, E., Llewellyn, F., Louzao, M., Ridoux, V. and Tetley, M.J., 2014. Migratory marine species: their status, threats and conservation management needs. Aquatic Conservation: Marine and Freshwater Ecosystems, 24(S2):111-127.

[35] Pimentel, D. and Hall, C.W (ed). 1989. Food and natural resources. Academic Press Inc.

[36] Scott, J. Michael, "SLIDES: Threats to Biological Diversity: Global, Continental, Local" (2008). Shifting Baselines and New Meridians: Water, Resources, Landscapes, and the Transformation of the American West (Summer Conference, June 4-6). http://scholar.law.colorado.edu/water-resources-and-transformation-of-American-West/15.

[37] Merino, G., Barange, M., Blanchard, J.L., Harle, J., Holmes, R., Allen, I., Allison, E.H., Badjeck, M.C., Dulvy, N.K., Holt, J. and Jennings, S., 2012. Can marine fisheries and aquaculture meet fish demand from a growing human population in a changing climate?.Global Environmental Change, 22(4):795-806.

[38] McKee, J.K., Sciulli, P.W., Fooce, C.D. and Waite, T.A., 2004. Forecasting global biodiversity threats associated with human population growth. Biological Conservation, 115(1):161-164.

[39] Pimm, S.L. and Raven, P., 2000. Biodiversity: extinction by numbers. Nature, 403(6772):843-845.

[40] Gibbons, J.W., Scott, D.E., Ryan, T.J., Buhlmann, K.A., Tuberville, T.D., Metts, B.S., Greene, J.L., Mills, T., Leiden, Y., Poppy, S. and Winne, C.T., 2000. The Global Decline of Reptiles, Déjà Vu Amphibians Reptile species are declining on a global scale. Six significant threats to reptile populations are habitat loss and degradation, introduced invasive species, environmental pollution, disease, unsustainable use, and global climate change. BioScience, 50(8):653-666.

[41] Kerr, J.T. and Deguise, I., 2004. Habitat loss and the limits to endangered species recovery. Ecology Letters, 7(12):1163-1169.

[42] Sodhi, N.S., Koh, L.P., Brook, B.W. and Ng, P.K., 2004. Southeast Asian biodiversity: an impending disaster. Trends in Ecology & Evolution, 19(12), pp.654-660.

[43] Gross, M., 2015. Can we avert marine mass extinctions?. Current Biology, 25(6):R209-R212.

[44] Castello, L., Arantes, C.C., Mcgrath, D.G., Stewart, D.J. and Sousa, F.S.D., 2015. Understanding fishing-induced extinctions in the Amazon. Aquatic Conservation: Marine and Freshwater Ecosystems, 25(5):587-598.

[45] Cano, J., Rosique, M.J. and Rocamora, J., 1997. Influence of environmental parameters on reproduction of the European flat oyster (Ostrea edulis L.) in a coastal lagoon (Mar Menor, southeastern Spain). Journal of molluscan studies, 63(2):187-196.

[46] Matias-Peralta, H., Yusoff, F.M., Shariff, M. and Arshad, A., 2005. Effects of some environmental parameters on the reproduction and development of a tropical marine harpacticoid copepod Nitocra affinis f. californica Lang. Marine Pollution Bulletin, 51(8):722-728.

[47] Bishop,P.J., A. Angulo, J. P. Lewis, R.D. Moore, G. B. Rabb and J. Garcia Moreno. 2012. The Amphibian Extinction Crisis - what will it take to put the action into the Amphibian Conservation Action Plan?. Sapiens. Vol. 2. https://sapiens.revues.org/1406

[48] Lowe, S., Browne, M., Boudjelas, S. and De Poorter, M., 2000. 100 of the world's worst invasive alien species: a selection from the global invasive species database.

[49] Clavero, M. and Garcı´a-Berthou, E. 2005. Invasive species are a leading cause of animal extinctions. TRENDS in Ecology and Evolution, 20(3)110-110.

[50] Gurevitch, J. and Padilla, D.K., 2004. Are invasive species a major cause of extinctions?.Trends in Ecology & Evolution, 19(9):470-474.

[51] Roe, D. (2008). Trading Nature. A report, with case studies, on the contribution of wildlife trade management to sustainable livelihoods and the Millennium Development Goals. TRAFFIC International and WWF International.

[52] Wabnitz, C., Taylor, M., Green, E., Razak, T. 2003. From Ocean to Aquarium. the global trade

in marine ornamental species (No. 17)UNEP-WCMC, Cambridge, UK.
[53] Malcolm, J.R., Liu, C., Neilson, R.P., Hansen, L. and Hannah, L.E.E., 2006. Global warming and extinctions of endemic species from biodiversity hotspots. Conservation biology, 20(2):538-548.
[54] Cahill, A.E., Aiello-Lammens, M.E., Fisher-Reid, M.C., Hua, X., Karanewsky, C.J., Ryu, H.Y., Sbeglia, G.C., Spagnolo, F., Waldron, J.B., Warsi, O. and Wiens, J.J., 2012, October. How does climate change cause extinction?. In Proc. R. Soc. B (p. rspb20121890).
[55] Thomas, C.D., Cameron, A., Green, R.E., Bakkenes, M., Beaumont, L.J., Collingham, Y.C., Erasmus, B.F., De Siqueira, M.F., Grainger, A., Hannah, L. and Hughes, L., 2004. Extinction risk from climate change. Nature, 427(6970):145-148.
[56] Sudarmadi, S., Suzuki, S., Kawada, T., Netti, H., Soemantri, S. and Tugaswati, A.T., 2001. A survey of perception, knowledge, awareness, and attitude in regard to environmental problems in a sample of two different social groups in Jakarta, Indonesia. Environment, development and sustainability, 3(2):169-183.
[57] Uzunboylu, H., Cavus, N. and Ercag, E., 2009. Using mobile learning to increase environmental awareness. Computers & Education, 52(2):381-389.
[58] Berger, I.E. and Kanetkar, V., 1995. Increasing environmental sensitivity via workplace experiences. Journal of Public Policy & Marketing, pp.205-215.
[59] Jänicke, M., 2005. Trend-setters in environmental policy: the character and role of pioneer countries. European environment, 15(2):129-142.
[60] Horton, C.C., Peterson, T.R., Banerjee, P. and Peterson, M.J., 2016. Credibility and advocacy in conservation science. Conservation Biology,30(1):23-32.
[61] Griffiths, R.A. and Dos Santos, M., 2012. Trends in conservation biology: Progress or procrastination in a new millennium?. Biological Conservation,153:153-158.
[62] Wilkins, L., 1993. Between facts and values: Print media coverage of the greenhouse effect, 1987-1990. Public understanding of science, 2(1):71-84.
[63] Poberezhskaya, M., 2014. Media coverage of climate change in Russia: Governmental bias and climate silence. Public Understanding of Science, p.0963662513517848
[64] Laurance, W.F., 2010. Habitat destruction: death by a thousand cuts. Conservation biology for all, 1(9):73-88
[65] Cabral, R.B. and Aliño, P.M., 2011. Transition from common to private coasts: Consequences of privatization of the coastal commons. Ocean & Coastal Management, 54(1):66-74.
[66] Canfa, W., 2006. Chinese environmental law enforcement: current deficiencies and suggested reforms. Vt. J. Envtl. L., 8:159.
[67] Agardy, M.T., 1994. Advances in marine conservation: the role of marine protected areas. Trends in Ecology & Evolution, 9(7), pp.267-270
[68] Man, A., Law, R. and Polunin, N.V., 1995. Role of marine reserves in recruitment to reef fisheries: a metapopulation model. Biological Conservation, 71(2):197-204.
[69] Wilson J.R. & Green A.L. 2009. Metode Pemantauan Biologi Untuk Menilai Kesehatan Terumbu Karang dan Efektivitas Pengelolaan Kawasan Konservasi Laut di Indonesia (Terjemahan). Versi 1.0. Laporan TNC Indonesia Marine Program No 1/09. 46 hal.
[70] Avise, J.C., 1989. A role for molecular genetics in the recognition and conservation of endangered species. Trends in Ecology & Evolution, 4(9):279-281.
[71] Harte, J., Ostling, A., Green, J.L. and Kinzig, A., 2004. Climate change and extinction risk. Nature, 430(3).

Population of *Aedes sp* in Highland of Wonosobo District and Its Competence as A Dengue Vector

Martini Martini[1,2], Bagoes Widjanarko[1], Retno Hestiningsih[1], Susiana Purwantisari[3], Sri Yuliawati[1]

[1]Public Health Faculty-Diponegoro University
[2]Institute of Research and Community Services, Diponegoro University
[3]Science and Mathematic Faculty-Diponegoro University

E-mail: tinihen65@yahoo.co.id

Abstract. The increased cases of dengue fever have occurred in the highland of Wonosobo District, and the epidemic taken place in 2009 had 59.3 cases per 100,000 populations. This study aimed to describe of vector competence of the mosquitoes as a dengue vector in the highland of Wonosobo District, Central Java Province. The serial laboratory work was done to measure of vector competence complementary with vector bionomic study. The samples were 20 villages, which were located at Wonosobo sub district. Every village was observed about 15-20 houses. The observed variables were vector competition, bionomic and transovarial infection level, and titer of virus on the mosquitoes after injection. Immunohistochemistry or IHC methods were used to identify transovarial infection status. The number of *Ae. aegypti* and *Ae. albopictus* were almost similar and both were found indoors or outdoors. Based on HI and OI index, the larvae density in the highland was enough high than standard of the program. Transovarial infection was found on *Ae. aegypti* and *Ae. albopictus.* Environment parameters such as temperature and relative humidity fulfilled the optimum requirement to support the vectors' life cycle. Transovarial infection has been proven, thus, it indicates that the local transmission has been occurred in this area. Titer of virus was also increasing after day per day. This indicate that the mosquitoes has the ability being vector. As used to do in other area, it is important to conduct breeding places elimination (PSN) indoors as well as outdoors, through active participation of the community in highland area.

Keywords: *Aedes* sp, mosquito, highland, transovarial infection

1. Introduction

Dengue Hemorrhagic Fever (DHF) is a disease caused by dengue virus, transmitted from human to human by *Aedes* mosquitoes (Ae.) sub-genus of Stegomyia. Dengue cases tend to increase and wider spread every year. WHO reports says that more than 40% of the world population are at risk of dengue fever. There was estimated 50 million new infections per year [1]. Intergovernmental Panel on Climate Change (IPCC) predict the incidence of Dengue Hemorrhagic fever (DHF) in Indonesia will increase in the following years [2].

Central Java is the DHF endemic Province since all of its districts (35 districts) reported dengue hemorrhagic case. In the year of 2009 the Incidence Rate (IR) was 5.79/10.000, and the Case Fatality

Rate (CFR) was 1.42%. Wonosobo, a highland district was showed the escalating of DHF incidence (CI was less than 1/10.000 during 2001-2008 and 5.7/10.000 in the year of 2009) and followed with the escalation of CFR (0.43% in the year of 2009 and 2.23 in the year of 2010). In the past, we used to argue that the case in highland area such as Wonosobo happened in others lower area (imported cases) since in highland circumstances *Aedes* spp was not competence to be DHF vector.

Increasing DHF cases in highland indicates the increase of DHF vector's density. There has been an environmental change such as temperature, rainfall, caused by global warming phenomena. Air temperature and humidity affect the viability of the vector [3]. These changes will also affect mosquitoes life cycle, especially the breeding of Aedes spp. The length period from egg to adult mosquito phase is getting shorter. Low temperature can suppress the development of larvae and eggs of Aedes spp [4, 5]. Changes in the environment will affect life of mosquitoes species on ecosystems and the pattern of viral vectors spread and thus can increase the incidence of dengue transmission.

At a temperature of 26^0C virus requires 25 days in mosquito body to get ready to be transmitted to humans (extrinsic period), conversely takes a relatively shorter, 10 days at a temperature of 30^0C. By then, life period could be the critical determinant of *Aedes* spp competency to be DHF vector in higland areas [6]. Increase the capacity of the mosquito to be vector is affected by the mosquito population density, cycle rate gonotropic, extrinsic incubation period as well as the readiness of the virus in the body of the vector for transmission. In [7] proves that the mosquito can live until 1 month in optimum air conditions. By taking into account the DHF virus extrinsic period, the environment changing and mosquitoes lifespan Aedes spp could be the potential vector for Dengue transmission in highland areas [8].

This study try to explore in which extend the *Aedes aegypti* in Wonosobo has competency as DHF vector.

2. Research Method

Entomology survey was done in Wonosobo sub district, the most populated and highest DHF incidence in Wonosobo district. involving 20 houses from 20 vilages. Ovitraps with specific attractant were placed indoor as well as outdoor [9]. Mosquito eggs were incubated to produce mosquito as research material. Mosquitos' age were 7 days used to further measurement. We detect susceptibility *Aedes* sp from dengue virus infection using immunohistochemical (IHC) technique. An laboratory study was applied to examine the mosquito competence as in vivo vectors with injected the mosquito with a virus of Den-2. Mosquito was identified the titer of virus load by using ELISA. Mosquito's age were used this study were 3 days and 10 days after reared from the eggs. Discriptive analysis was applied to explain to elaborate the data.

3. Result and discussion

3.1 Capacity of vector

The DHF incidence in Wonosobo district in the year of 2012 was 0,67/10.000. This incidence is higher than the year of 2006 (0,12/10.000) and in the year of 2008 (0,37/10.000). In 2009, DHF was increasing sharply, 5.79/10.000 population. Until in 2013, DHF still find in this district with incidence around 19,79/10.000 population.

Based on entomology survey, the *Aedes* sp larvae density are, House Index (HI)=8,8%, BI=1,70, Container I=4,65% and OI=5,39% and categorized as low density. Container as mosquito breeding place in household circumstances are batch tube, bucket, and used canned. The average number of egg in inner house ovitrap is 1,16 and 0,73 in outer house. The average number of larvae in village sample are OI 5,4%, and HI 8,8% (higher than standart of the program, 5%).

Mosquito eggs were reared and indentified. Table 1 shows the distribution of Aedes sp. *Aedes aegypti* was founded more inner house (78,4%) than *Aedes albopictus* (54,1%). We also found *Ae. anandalei* during survey using ovitrap.

Table 1. *Aedes* sp from ovitrap survey (rearing) and Mosquito Survey

Height (above sea level))	*Ae. Aegypti*		*Ae. albopictus*		*Ae. anandalei*		Sum	%
	f	%	f	%	f	%		
1. Rearing								
Indoor	58	78,4	15	20,3	0	0,0	74	66,7
Outdoor	9	24,3	20	54,1	8	21,6	37	33,3
2. Survey								
Indoor	34	77,3	10	22,7	0	0,0	0	0,0
Outdoor	1	11,1	8	88,9	0	0,0	0	0,0

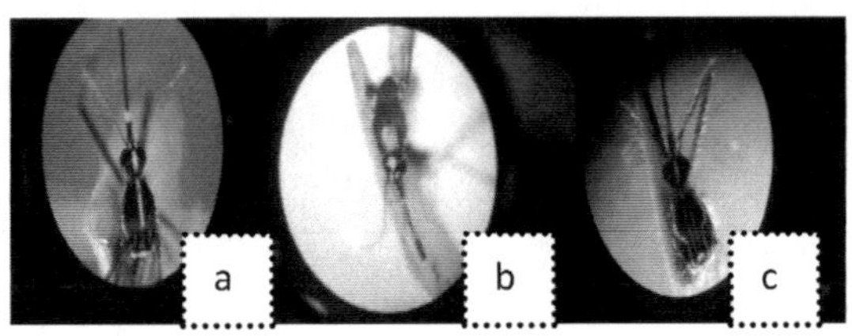

Figure 1. *Aedes* sp was identified in Wonosobo (a. *Ae. Albopictus*, b. *Ae. Anandalei*, c. *Ae. Aegypti*)

The quantity of larvae density and adult *Aedes* sp is linier. This figure indicating the mosquito's dynamic live circle was complete. Environment factors such as temperature, humidity in Wonosobo is adequate to support life circle process [10]. The original habitat of *Ae. albopictus* is forest and have adapted to live in the land use for agriculture surrounding human settlement [11, 12].

The existence of *Ae albopictus* population on the plateau suggests that this species is able to adapt to the environment in the highlands. Surveys both indoor and outdoor proves that there is a shift in the pattern of life from eco-type semi domestic to domestic.

Shifting Ae. albopictus from outside to inside human dwellings is perhap increase the chances of *Ae. albopictus* females in getting the blood [13, 14, 15]. In [15] stated that the shift *Ae. albopictus* into human dwellings can increase the lifespan, and increase its capacity as a vector, both in terms of vector-host contact as well as the density of the vector population. Changes in behavior patterns *Ae. albopictus* from domestic ecotypes into a semi-domestic, also reported in India [16].

Unlike the *Aedes* spp competition in the lowlands, especially in densely populated areas, population *Ae. aegypti* is also found outdoor. This indicates the behavior changing of *Ae. aegypti* to be peri-domestic, bionomics change of the house toward the outdoor [17].

3.2 Aedes sp susceptibility against Dengue virus as *Aedes* marker become vectors of dengue in the highland

Using IHC examination, we found susceptibility of *Aedes* sp in Wonosobo highland area, indicating its capability as DHF vector. Polymerase Chain Reaction (PCR) test failed to identify viral genetic marker. Transovarial infection confirmed the capability of *Aedes* sp as DHF vector in highland area. This phenomena demanding prompt program response to prevent endemic area in highland area. In [18] stated that Dengue virus was persistently ovarial transmitted until F7. We confirmed 7 positive dengue virus among 20 suspect DHF patients (35%) of Wonosobo District Hospital. DEN-1, DEN-2, DEN-3 and DEN-4 were identified moreover 2 patient were mixed infected [19]. Concerning in [20], the existence of four serotype through the year will increase the risk of viral transmission. Many researchers have proved transovarial transmission relating with various level of infection. Umniyati found 13,6-33,8% infection index from transovarial *Ae aegypti* taken wheel in Yogyakarta [21], whilst [7] found 38,5-70,2% from another place in Yogyakarta. Determinant factors of mosquito to become vector are mosquito susceptilbility, geographical variation and mosquito strain [23]. Some researchers stated that a genetic variation so called barrier *infection system* in the midgut, sexes, strain and species as determinant factors of vector susceptibility. Those physiologic barriers in the midgut are *Midgut Infection Barrier* (MIB) and *Midgut Escape Barrier* (MEB) [24], but, the molecular barrier mechanism remain unclear

In our research we found viral DEN-2 titer escalation in experiment group after three days injected compare with control group. In the tenth day, viral titer escalated indicating viral growth after invading physiologic barrier in mosquito body. Based on this finding, dengue hemorrhagic fever cases in Wonosobo plateau could be caused by local transmission as well as imported cases and prevention activities such as mosquito breeding places eradication are needed.

Table 2. Virus titer Den-2 in *Ae. aegypti* in ELISA plate.

	1	2	3	4	5	6	7
A	0,883	0,589	0,305	0,708	1,057	0,227	0,215
	A1	A2	A3	A4	A5	A6	A7
	Dilution 1:4	Dilution 1:10	Dilution 1:20	Incubation for 3 days	Incubation for 10 days	Control (negative)	Control (negative)

Description: A1-A3 is the dilution of the positive samples, A4-A5 mosquitoes injected in accordance with the developmental period of the virus in the mosquito's body, A6-A7: negative control.

4. Conclusion

The dengue incidence in Wonosobo highlands in 2013 amounted to 19.79 / 100,000 people. Increased incidence associated with environmental conditions that have met the requirements to support life optimum vector of dengue.

Competition *Ae. aegypti* and *Ae. albopictus* downloading a balanced approach and both types can be found either in the (indoor) and outside the home (out door). Found another species, namely *Ae. annandale* besides *Ae. aegypti* and *Ae. albopictus*. *Ae. aegypti* strain District of Wonosobo can be a vector of dengue based Den-2 virus multiplication in mosquitoes characterized by increased titers of the virus through the test in vivo,

Transovarial infection occurs in *Ae. aegypti* and *Ae. albopictus*. Transovarial evidence can make repeated epidemics, as happened in 2009.

5. Acknowledgment

We thank to Directorate General of High Education as funding supporter and National Research Vector Centre Laboratory in Banjarnegara for assisting in laboratory work.

References

[1]. Regis L, Souza WV, Furtado AF, Fonseca CD, Silviera Jr PJ, Melo-Santos MA, Carvalho MS, and Monteiro AM. 2009. An Entomological Surveillance System Based on Open Spatial Information for Participative Dengue Control. *An of Acad Braz of Cienc*. **81**(4): 655-662

[2]. Mc Michael, A.J., Champbell-Lendrum, D.H., Corvalan, C.F., Ebi, K.L, Githeko, A.K., Scheraga, J.D., Woodward, A. 2003. Climate Change and Human Health Risk and Response. Geneva : WHO

[3]. Aron JL, and Patz JA. 2001. *Ecosystem Change and Public Health: A Global Perspectiva*. Baltimore : The John Hopkins University Press.

[4]. Patz, J.A., Willem, J.M. Martens, Dana A. Focks,3 and Theo H. Jettend. 1998. Dengue Fever Epidemic Potential as Projected by General Circulation Models of Global Climate Change. *Environmental Health Perspectives*. **6** (30) : 147-153

[5]. Widayani, Prima. 2007. Pemanfaatan Data Pengideraan Jauh dan Sistim Informasi Geografi (GIS) untuk Zonasi Daerah rawan Demam Berdarah Dengue. Workshop Spasial and Temporal Analysis of Malaria Epidemiological Data. Yokyakarta : Fakultas Geografi UGM

[6]. Cook PE, Hugo LE, Iturbe-Ormaetxe, I, Williams CR, Chenoweth, SF, Ritchie SA, Ryan PA, Kay BK, and O'Neill S. 2006. The Use of Transcriptional Profiles to Predict Adult Mosquito Age under Field Conditions. *PNAS* **103** (48): 18060–18065

[7]. Mardihusodo S, Satoto, T.B.T., Mulyaningsih, B., Umniyati, S.R. 2007. Simposium Nasional DBD: Aspek Molekuler, Patogenesis, Manajemen & Pencegahan KLB. Yokyakarta, 16 Mei 2007

[8]. Hidayati R, Hadi UK, Situmeang R, Boer R, Fitriani, danSugiharto. 2009. Kapasitas Vektorial Nyamuk *Ae. aegypti* di Kota Mataram NTB Terkait Iklim. Seminar Nasional & Kongres APNI : "Partisipasi Masyarakat dalam Program Pengendalian Nyamuk Terpadu". Bogor, Senin, 10 Agustus 2009.

[9]. Prihatnolo, A., Martini M, dan Hestiningsih, R. 2011. Efisiensi Ovitrap Modifikasi sebagai Upaya Monitoring Kepadatan Aedes sp sebagai Vektor Demam Berdarah Dengue (DBD). Disampaikan dalam Seminar Nasional Penelitian dan Pengem-bangan Vektor dan Reservoair Penyakit sebagai Lokomotif Pemberantasan Penyakit Bersumber Binatang Dalam Rangka Pembangunan Kesehatan Bangsa, B2P2VRP. Salatiga.

[10]. Wongkoon S, Jaroensutasinee M, Jaroensutasinee K, Preechaporn W, Chumkiew S.2006. Larval Occurrence and Climatic Factors Affecting DHF Incidence in Samui Islands, Thailand. In *J Biol Life Sci* **2** (2): 107-112.

[11]. Santoso Anif B. Hubungan Pengetahuan Sikap dan Perilaku (PSP) Masyarakat terhadap Vektor DBD di Kota Palembang Provinsi Sumatera Selatan. *Jurnal Ekologi Kesehatan*, **7** (2), Agustus 2008: 732-739

[12]. BPS Kabupaten Wonosobo. 2014. Wonosobo Dalam Angka Tahun 2013. Wonosobo: BPS Kabupaten Wonosobo.

[13]. Rozilawati H, Zairi J, and Adanan CR. 2007. Seasonal abundance of *Aedesalbopictus* in selected urban and suburban areas in Penang, Malaysia. *Trop Biomed* **24**: 83–94.

[14]. Nur Aida H, Abu Hassan A, Nurita AT, CheSalmah MR, Norasmah B. 2008. Population analysis of *Aedes albopictus* (Skuse) (Diptera:Culicidae) under uncontrolled laboratory conditions. *Trop Biomed* **25**: 117–125.

[15]. Hamady D, Saifur, R.G.M., Hassan, A.A., Che Salma, M. R., Boots, M., Satho, T., Jaal, Z., and Bakar, S.A.2010. Indoor-Breeding of *Aedes albopictus* in Northern Peninsular Malaysia and Its Potential Epidemiological Implication. *PloS One*. 5 (7): 1-9.

[16]. Rao BB. 2010. Larval Habitats of *Aedes albopictus* (Skuse) in Rural Areas of Calicut, Kerala, India. *J Vector Borne Dis*. **47**:175-7.

[17]. Capinera JL. 2008. *Encyclopedia of Entomology*. 2 nd. Vol 1-4. Springer: Leipzig. 4346pp.

[18]. Joshi V, Morruya DT, and Sharma RC. 2002. Persistence of Dengue 3 Virus through Transovarial Transmission Passage in Successive Generation of *Ae. aegypti* Mosquito. *Am J Trop Med Hyg* **67** (2) : 158-161.

[19]. Martini M, Santoso L, dan Hestiningsih R. 2012. Kajian Karakteristik Virologi Penderita Dengue Dataran Tinggi di Kabupaten Wonosobo. Laporan Penelitian. Tidak Dipublikasikan. Semarang : Fakultas Kesehatan Masyarakat Undip.

[20]. Kusriastuti R. 2005. Epidemiologi Penyakit Demam Berdarah Dengue dan Kebijakan Penanggulangannya di Indonesia. Simposium Dengue Control Up Date. Yogyakarta : Pusat Kedokteran Tropis UGM.

[21]. Umniyati SR. 2004. Prelimenary Investigation on TheTransovarial Transmission of Dengue Virus in The Population of *Ae. aegypti* in The Well. Disampaikan dalam Seminar Hari Nyamuk IV, 21 Agustus 2004. Surabaya.

[22]. Bennett KE, Olson KE, Munoz MDE, Fernandez-Salas I, Farfan-Ale JA, Higgs S, Black WCT, and Beaty BJ. 2002. Variation in Vector Competence for Dengue 2 Virus among 24 Collections of *Aides aegypti* from Mexico and the United States. *Am J Trop Med Hyg* **67**: 85–92.

[23]. Bosio CF, Beaty BJ, and Black WCT. 1998. Quantitative Genetics of Vector Com-petence for Dengue-2 Virus in *Aides aegypti. Am J Trop Med Hyg.* **59**: 965–970.

IOP Conf. Series: Earth and Environmental Science **55** (2017) 012004 doi:10.1088/1755-1315/55/1/012004

Nutritional Value Content, Biomass Production and Growth Performance of *Daphnia magna* Cultured with Different Animal Wastes Resulted from Probiotic Bacteria Fermentation

Vivi Endar Herawati*, R A Nugroho, Pinandoyo and Johannes Hutabarat

Department of Aquaculture, Faculty of Fisheries and Marine Science, Diponegoro University., Semarang 50275, Indonesia

email : anshinvie@yahoo.com

Abstract. Media culture is an important factor for the growth and quality of *Daphnia magna* nutrient value. This study has purpose to find the increasing of nutritional content, biomass production and growth performance of *D. magna* using different animal wastes fermented by probiotic bacteria. This study conducted using completely randomized experimental design with 10 treatments and 3 replicates. Those media used different animal manures such as chicken manure, goat manure and quail manure mixed by rejected bread and tofu waste fermented by probiotic bacteria then cultured for 24 days. The results showed that the media, which used 50% chicken manure, 100% rejected bread and 50% tofu waste created the highest biomass production, population and nutrition content of *D.magna* about 2111788.9 ind/L for population; 342 grams biomass production and 68.85% protein content. The highest fatty acid profile is 6.37% of linoleic and the highest essential amino acid is 22.8% of lysine. Generally, the content of ammonia, DO, temperature, and pH during the study were in the good range of *D. magna's* life. This research has conclusion that media used 50% chicken manure, 100% rejected bread and 50% tofu waste created the highest biomass production, population and nutrition content of *D. magna*.

Keywords: Nutrient value, Fermentation, *D. magna*, Animal wastes, probiotic bacteria

1. Introduction

D. magna is one of the most potential natural feed for fish larvae due to its high nutritional content [1]. Nutritional content of *D. magna's* largely depends on its culture medium as the growth of phytoplankton, which is the feed for *D. magna*, depends on it [2]. The most commonly used culture medium nowadays is the used chicken manure [3]. Another less common one is a combination of chicken dung, bran, and copra waste [4]. The organic materials in bran are highly nutritious for the growth of *D. magna*.

The use of organic fertilizers in culture media including the wastes/feces of chicken, goat and quail mixed with the rejected bread and tofu waste fermented with the probiotic bacteria has not so far been conducted as the use of organic fertilizer could impact the growth performance and content of D. magna. The highest nutrients - particularly for the content of N, P and Ca in organic fertilizer are the food sources of *D. magna*. The chicken waste containing N (4.75%); P (3.57%) and Ca (4.80%); quail waste containing N (4.06%); P (2.96%) and Ca (2.57%); and goat waste/feces containing N (2.36%); P (2.96%) and Ca (3.41%) [1]. Furthermore, the analysis on the dried materials of tofu waste contained crude protein (27.09%), crude fibre (22.85%), fat (7.37%), ash (35.02%), and extract material without nitrogen/BETN (6.87%) [5], [6]. The rejected bread contained the crude protein (12.63%), crude fibre (0.13%); crude fat (4.63%); ash (4.19%) and the extract material without nitrogen (58.42%) [6], [7].

The fermentation of the fertilizer has been proven to be effective in the increase of the nutrient of culture media. The aims of the fermentation are to produce a product (food materials) that contains the nutrients, to have a longer storage time, and to have a better organoleptic characteristics and nutritional components [1]. The mass-cultured *D. magna* in this research used probiotic bacteria as it improves both biomass production and nutritional content of *D. magna*. Probiotic bacteria are supportive for the health of organisms [8]. It also serves to decompose and ferment organic materials [9]. Decomposition is a biological process that makes the most of bacteria's ability to produce growth substances, hormones, vitamins, and other enzymes [3], [10].

The purpose of this study is to find nutritional content, biomass production and growth performance of *D. magna* using different animal wastes fermented by probiotic bacteria to increase its nutrient content, biomass production and the growth. The fertilizer itself is fermented with probiotic bacteria (*Lactobacillus casei* and *Saccharomyces cerevisiae*)

2. Materials and Method

2.1 Fermentation Stage

The fermentation stage is the preparation of the ratio of molasses, water and pro-biotic. The ratio used was 1: 1, i.e. 1mL of molasses, 1 mL of probiotic bacteria and 100 mL of solvent. The organic materials used (chicken manure, goat manure, quail manure, rejected bread and tofu wastes) are dried. The treatments used in this research were: A. rejected bread 100 gr/L + tofu waste 100 gr/L; B. Chicken manure 50 gr/L + rejected bread 50 gr/L + tofu waste 100 gr/L; C. Chicken manure 50 gr/L + rejected bread 100 gr/L + tofu waste 50 gr/L; D. Chicken manure 100 gr/L + rejected bread 50 gr/L + tofu waste 50 gr/L; E. Goat manure 50 gr/L + rejected bread 50 gr/L + tofu waste 100 gr/L; F. Goat manure 50 gr/L + rejected bread 100 gr/L + tofu waste 50 gr/L; G. Goat manure 100 gr/L + rejected bread 50 gr/L + tofu waste 50 gr/L; H. Quail manure 50 gr/L + rejected bread 50 gr/L + tofu waste 100 gr/L; I. Quail manure 50 gr/L + rejected bread 100 gr/L + tofu waste 50 gr/L; I. Quail manure 100 gr/L + rejected bread 50 gr/L + tofu waste 50 gr/L; J. 50 gr/L rejected bread 100 gr/L + tofu waste 100gr/L.

Fertilizers with a combination weight of 200g/L were given pro-biotic bacteria (*L. casei* and *S. cerevisiae*) that were already activated for 3 hours [9], [11]. They were left for fungi to grow and acidic smell to develop. Table 1 and 2 shows the result of nutrient analysis for organic fertilizer before fermented and after fermented for 14 days with pro-biotic bacteria for the mass-cultured *D. magna*.

2.2 Water Quality

The water quality during the research was maintained at 28-29^{0}C temperature, 0.3 ppm DO and 8,1-8,2 pH, which is ideal. This is in line with the statement of previous research [1], [12], [13] that the proper temperature for *D. magna* culture is 25-30°C, DO at 0.3-0.6 ppm, and pH at 6.5-9. Excellent water quality helps grow phytoplankton and algae for *D. magna* to propel its growth.

2.3 D. magna Culture

The 100 ind/l *D. magna* was spread for each pool containing 200 g/L fermented organic fertilizer. Observation for the abundance of D. *magna* was conducted every two days to monitor the population of *D. magna*. The water (20-25%) of this culture was replaced, and its pH level was monitored every morning at around 7 to maintain the quality. The pH was maintained at its maximum with the addition of 1 L of dolomite /1.000 L of water.

2.4 Statistical Analysis
This research employed the complete random design with 10 treatments and 3 repetitions. The weight of biomass was analyzed using variant analysis to figure out differences among treatments. The parameters analyzed were growth, biomass production, and nutritional content of *D. magna*.
Proximate analysis. The proximate chemical composition of the samples was determined using a standard procedure [6], [14]. The crude protein content was calculated by multiplying the total nitrogen factor. The carbohydrate content was estimated by the difference.
Essential amino acid profile. The amino acid composition of the sample was determined using HPLC (Shimadzu LC-6A) [6], [14].
Fatty acid profile. The fatty acid composition of the sample was determined using a gas chromatograph (Shimadzu) [6], [14].

3. Results and Discussion

The results of the research found the culture media before and after being fermented experiencing an enhancement in which N before the fermentation was in the range of N: 1.19 - 2.75%; P: 0.19 - 0.54%; and K (Potassium): 0.15 – 0.59% and after the fermentation was in the range of 2.12-3.84%; P: 0.27-1.74%; and K (Potassium): 0.69-1.21%. Table 1 and Table 2 show the results of the analysis on N, P and K (Potassium) culture medium before and after fermentation.

Table 1. The results of the analysis on N, P and K (Potassium) culture medium before fermentation.

Parameters	Analysis Result										Method
	Poultry manure (%)	Chicken manure (%)			Goat manure (%)			Quail manure (%)			
	A	B	C	D	E	F	G	H	I	J	
Nitrate (N)	1,54± 0.06	2,24± 0.09	2,55± 0.07	1.23± 0.02	1,25± 0.08	2,23± 0.01	1,78± 0.08	1,19± 0.05	2,19± 0.03	2,26± 0.09	Kjeldahl
Phosphor (P)	0,19± 0.03	0,18± 0.07	0,25± 0.01	0,23± 0.03	0.17 ± 0.05	1.03± 0.09	0.45± 0.06	0,41± 0.02	0,23± 0.08	0.54± 0.07	AOAC 958.01. 2000
Potassium (K)	0,39± 0.02	0,27± 0.01	0,59± 0.05	0.26± 0.06	0.33± 0.09	0.45± 0.03	0,15± 0.03	0,21± 0.09	0,13± 0.03	0,54± 0.02	AOAC

Table 2. The results of the analysis on N, P and K (Potassium) culture medium after fermentation.

Parameters	Analysis Result										Method
	A	B	C	D	E	F	G	H	I	J	
Nitrate (N)	2,74± 0.05	3,40 ± 0.05	3,84 ± 0.03	2.43 ± 0.09	2,12 ± 0.08	3,29 ± 0.02	2,98 ± 0.06	2,89 ± 0.01	3,29 ± 0.03	2,37 ± 0.09	Kjeldahl
Phosphor (P)	0,27± 0.02	0,48 ± 0.04	1,43 ± 0.06	0,59 ± 0.03	1,14 ± 0.02	1,30 ± 0.01	1,16 ± 0.01	1,21 ± 0.07	1,33 ± 0.02	1,24± 0.09	AOAC 958.01.200 0
Potassium (K)	0,69± 0.09	1,21 ± 0.02	0,89 ± 0.08	0,86 ± 0.03	1,56 ± 0.03	1,11 ± 0.09	1,05 ± 0.05	1,41 ± 0.01	0,34 ± 0.03	0,96 ± 0.06	AOAC 958.01.200 0

This research is the development of the research on the massive culture of *D. magna* using the chicken manure, rice bran and residue of copra fermented using the probiotic bacteria in 2015. The research found that the organic material in the form of a variety of the fermented animal manures mixed with rejected bread and waste tofu experienced an increase for its nutrient quality. The increases of the content included N at 1.2%; P and K at 1% (Table 1 and Table 2), the improvement of the nutrient quality on the media fermented is the process of anaerobic dissimilation of organic compounds by the microorganisms activity or the extract of microorganisms cells. The factors affecting biomass and the nutrient content of *D. magna* included the nutrient quality of the media, availability of food in the form of phytoplankton, bacteria detritus and the environment [1], [2]. The organic materials in the fermented media can increase the number of bacteria and organic particles from decomposition result to increase the availability of nutrients in the media. This then influenced the population growth and biomass of *D. magna*. This is supported by the results of previous research [8] that fermentation is aimed to increase the number of microorganisms and activate the metabolism in food, resulting in new food products using microorganisms.

The increase of nutrient in the media, especially nitrate, functions to determine the level of phytoplankton present in the culture medium – in addition to bacteria and detritus - as the food source of *D. magna*. The results of this study stated that the promotion and the growth of phytoplankton population in waters are associated with the nutrient availability, especially nitrates and the sunlight confirmed by previous research [6], [15]. This argument has been proven through the results of this research that the high level of nutrient of media brought an influence on the growth and absolute biomass of *D. magna* the highest growth was found in media of the fermentation of 50 g/L of chicken manure + 100 g/L of rejected bread + 50gr/l tofu waste with a content of N and P at 3.84% ± 0.03 and 1.43% ± 12:06 respectively providing the growth of 211788.9 ind /L (Chart 1 and Table 3) and the highest biomass was at 345 g ± 2.64 (Graph 2) .The quality of the nutrient culture media is very influential for the supply of plankton and bacteria to boost the population growth and biomass of *D. magna*. The high level of organic materials can bring an effect on the density and biomass of *D. magna* [2]. Growth performance *D. magna* cultured using the different manures of chicken, goat, quail, rejected bread and tofu waste fermented using probiotic bacteria as illustrated in Figure 1:

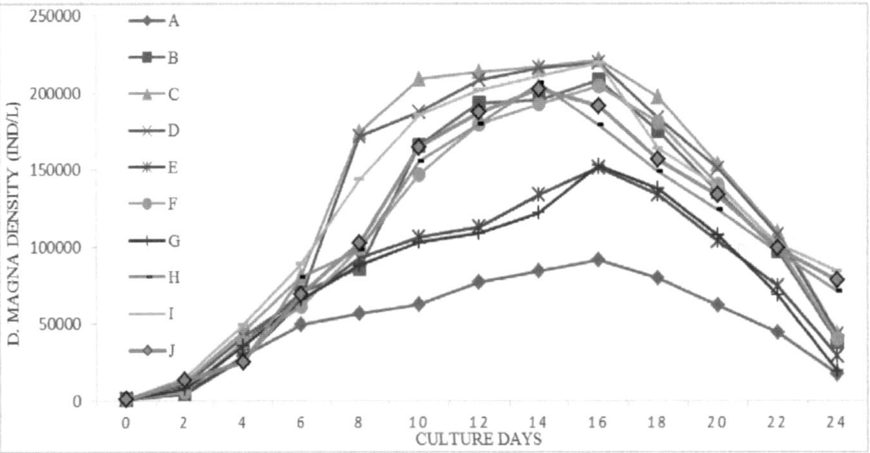

Figure 1. Growth performance of *D. magna* cultured using the different manures of chicken, goat, quail, rejected bread and tofu waste fermented using prebiotic bacteria.

The research found that the growth of *D. magna* massively cultured by means of a mix of animal different animal wastes with the rejected bread and tofu waste at the lag phase and death phase did not give any significant influence ($p<0.01$) among treatments in contrast to the exponential and stationary phases showing a significant influence ($p>0.01$) between treatments. The results of the highest growth on *D. magna* mass massively cultured using the fermentation of 50 g/L of chicken manure + 100 g/L of rejected bread+ 50 g/L waste tofu (C) with a dense population of the lag phase at 46251.85 ind/L on day 4; exponential phase at 198955 ind/L on day 10; stationer phase at 211788.9 ind /L on day 16 and the death phase at 102177.8 ind/L on day 26. Table 3 presents the results of the growth of *D. magna* massively cultured using the fermentation of a mix of different animal wastes with rejected bread and tofu waste.

Table 3. Growth of *D. magna* massively cultured using the fermentation of a mix of different animal wastes with rejected bread and tofu waste.

Treatment	A (%)	B (%)	C (%)	D (%)	E (%)	F (%)	G (%)	H (%)	I (%)	J (%)
Lag	2809.63 ± 0.05^{a}	37622.22± 0.01 ^{a}	46251.85± 0.01 ^{a}	40048.15 ± 0.05 ^{a}	32437.04 ± 0.05^{a}	35085.19 ± 0.02^{a}	25646.3 ± 0.01^{a}	39578.9 ±0.04 ^{a}	42312.2 ±0.04^{a}	42256.7 ±0.02^{a}
Exponential	65418.52 ± 0.07^{bdc}	153233.3± 0.03 ^{a}	198955.6± 0.02 ^{a}	189140.7 ± 0.04 ^{a}	100863 ± 0.02^{a}	141863 ± 0.01^{dba}	90362.96 ± 0.06^{a}	167423.3 ±0.03 ^{a}	209656.7 ±0.02^{c}	174723.3 ±0.01^{a}
Stationer	85270.37 ± 0.03^{a}	208000 ± 0.02 ^{ac}	211788.9± 0.05 ^{ab}	209603.7 ± 0.02^{ba}	139548.1 ± 0.09^{da}	192714.8 ± 0.05^{bda}	90770.37 ± 0.05^{a}	178101.1 ± 0.06^{a}	195778.9 ± 0.07^{c}	189478.9 ±0,04^{a}
Death	41400 ± 0.02^{a}	90640.74± 0.09 ^{a}	102177.8± 0.04 ^{a}	100.418.5 ±0.08^{a}	69270.37 ±0.01^{a}	93788.89 ± 0.04^{a}	45323.93 ± 0.03^{a}	974233.3 ± 0.02^{a}	101356.7 ±0.02^{a}	100678.9 ±0.01^{a}

The growth of *D. magna* massively cultured using the fermentation of various manures, rejected bread and tofu waste did not show a significant effect on the lag phase (P<0.01) between the treatments. The lag phase occurred on the fourth day and the highest one occurred in the media of fermentation of 50g/L of chicken manures + 100 g/L of rejected bread + 50gr /L of tofu waste at 46251.85 ± 0.01 ind/L. This was because at the lag phase *D. magna* began to adapt to the new environment in which if the concentration of media was equal to the nature, it would provide the fast growth and if there would be any differences in the culture medium and in the growth at a concentration of media in nature, *D. magna* would require more time for growing. The difference between the concentration of culture media and liquid cell in plankton would bring an effect on the restitution of enzyme and substrate concentrations to the further level for growth and the presence of nutrients in the cells was through the diffusion process as a result of differences in the concentration between the culture medium and body fluid [1].

The exponential phase occurred on the 10th day followed by a stationary phase on day 16. At the exponential phase and stationary phase, there was a significant influence (P <0.01) between the treatments. The length of the stationary phase was correlated with the duration of the *D. magna* to adapt to a new culture media. This was because the length of the stationary phase had an effect on the nutrient absorption into the culture medium by *D. magna*. The results of this research were consistent with previous research [1], [15], showing the discontinuation of the exponential phase due to the lack of nutrients in the increase of cell density. *D. magna* massively cultured using a mixture of rejected bread , tofu waste and various animal manures fermented using the probiotic bacteria provided a significant influence (P >0.01) at absolute biomass. *D. magna* massively cultured using the media of fermentation of 50 g/L of chicken manure + 100 g/L of rejected bread + 50 g/L of tofu waste (C) provided the highest absolute biomass at 345 g ± 2.64. Figure 2 presents the biomass of *D. magna* massively cultured using fermentation of a mixture of different manures, rejected bread and tofu waste.

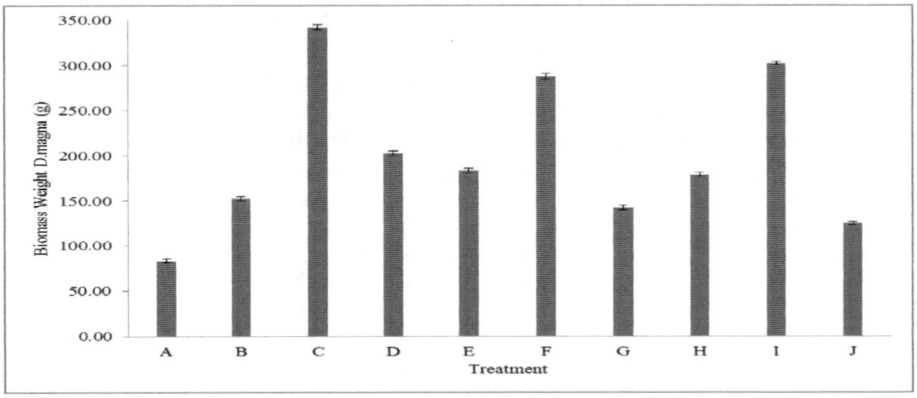

Figure 2. Biomass of *D. magna* massively cultured using fermentation of a mixture of different manures, rejected bread and tofu waste.

The media of fermentation of 50 g/L of chicken manure + 100 g/L of rejected bread + 50 g/L of tofu waste also resulted in a highest content of nutrients of *D. magna* that are 68.85 % ± 0.23 for protein and 8.84% ± 0.01 for fat. While, for the lowest level of nutrient content of *D. magna* massively cultured using the media of fermentation A (100 g/L of rejected bread + 100 g/L of tofu waste) with 60.32 % ± 0:04 and 7.16 %± 0:08 for protein and fat respectively. Table 4 presents the results of the analysis on the proximate of *D. magna* massively cultured using the fermentation of the mix of various animal manures, rejected bread and waste tofu.

Table 4. The result of the proximate composition of *D. magna* in mass culture using the different animal wastes resulted from probiotic bacteria fermentation.

Proximate	Poultry manure (%)	Chicken manure (%)			Goat manure (%)			Quail manure (%)		
	A	B	C	D	E	F	G	H	I	J
Crude ash	8.43 ± 0.07	8.67 ± 0.02	6.11 ± 0.08	6.52 ± 0.06	6.57 ± 0.09	6.60 ± 0.09	7.75 ± 0.23	7.75 ± 0.09	6.11 ± 0.09	7.99 ± 0.23
Crude lipid	7.16 ± 0.08	8.71 ± 0.03	8.84 ± 0.01	8.16 ± 0.09	8.34 ± 0.02	8.65 ± 0.03	6.95 ± 0.06	7.25 ± 0.02	8.14 ± 0.03	7.78 ± 0.06
Crude fiber	2.15 ± 0.03	3.05 ± 0.06	2.91 ± 0.02	2.59 ± 0.07	2.73 ± 0.01	2.81 ± 0.05	2.65 ± 0.19	3.37 ± 0.01	2.54 ± 0.05	2.89 ± 0.19
Crude protein	60.32 ± 0.04	65.18 ± 0.08	68.85 ± 0.23	65.55 ± 0.01	61.47 ± 0.03	63.56 ± 0.06	62.44 ± 0.03	62.75 ± 0.03	67.13 ± 0.06	63.04 ± 0.03
Carbohy Drate	21.94 ± 0.10	14.39 ± 0.01	13.29 ± 0.19	17.18 ± 0.07	20.89 ± 0.09	18.38 ± 0.10	20.21 ± 0.05	18.88 ± 0.09	16.08 ± 0.10	18.30 ± 0.05

From the nutrient content based upon the analysis on proximate as shown in Table 4 it was found that the highest level of the protein and fat on *D. magna* massively cultured using the media of fermentation of 50 g/L of chicken manure + 100 g/L of rejected bread + 50gr / L of tofu waste were at 68.85% ± 0:23. The result of this research was lower than previous research [1] in which on protein it was 73.90% ± 0.04 higher than the results obtained from the previous research reaching at 68.12% [16] and the other research reaching 4% [13] of wet weight. In this research, the highest level of fat was 8.84%± 0, 01 higher than previous research [1] at 7.89% ± 0.02. The high level of protein content and the low level of fat from the study were due to the high level of nutrient in the culture media of *D. magna* where the nitrate and phosphate levels were getting higher. The higher the N and P content, the higher the protein in the cultivar [17]. The fat content was inversely related to the protein content. The results of this study were supported by the results of previous research that a higher protein content was always conversely comparable to fat because the fat work doubled in the body compared to the protein [18].

The profile of fatty acid of *D. magna* massively cultured using the mix of d rejected bread, tofu waste and different animal manures fermented using the highest probiotic bacteria on the media of 50 g/L of chicken manure + 100 g/L rejected bread + 50 g/L of tofu waste; those are linoleic fatty acid at 6.37% ± 0.02; while the lowest one was found in DHA at 0.06% ± 0.02 in treatment *D. magna* massively cultured using the media of fermentation of 100 g/L of rejected bread + 100 g/L of tofu waste. The profile of the fatty acid of *D. magna* massively cultured using the fermentation of a mixture of different animal manures, rejected bread and tofu waste is presented in Table 5.

Table 5. Profile of fatty acid of *D. magna* in mass culture using the different animal wastes resulted from probiotic bacteria fermentation.

Fatty acids profile (%)	A	B	C	D	E	F	G	H	I	J
Miristic	0,28±0.04	0,49±0,04	0,41±0,09	0,48±0,02	0,01±0,05	0,01±0,05	0,02±0,03	0.65±0.03	0.25±0.05	0.25±0.02
Pentadecanoic	0,10 ± 0,08	0,18±0,06	0,15±0,08	0,17±0,04	0,08±0,02	0,09±0,06	0,04±0,02	0.12±0.02	0.12±0.07	0.02±0.03
Palmitic	2,01 ± 0,06	2,29±0,08	3,59±0,04	1,97±0,08	0,12±0,01	1,14±0,09	0,14±0,06	1.52±0.01	3.52±0.03	2.23±0.01
Stearic	0,41 ± 0,07	1,65±0,02	2.91±0,09	0,52±0,03	0,08±0,05	1,11±0,07	0,10±0,08	0.50±0.037	2.50±0.05	1.75±0.05
Oleic/ω9	1,62 ± 0,05	0,95±0,03	2,61±0,01	0,89±0,08	0,05±0,03	1,07±0,02	0,06±0,07	1.04±0.08	2.04±0.06	0.23±0.08
Linoleic/ω6	0,54 ± 0,02	5,46±0,07	6,37±0,02	4,49±0,07	3,97±0,02	4,73±0,09	3,52±0,01	5.13±0.09	6.13±0.01	4.48±0.04
Linolenic/ω3	0,19 ± 0,06	2,38±0,09	4,32±0,01	3,39±0,03	156±0,07	2,54±0,05	1,90±0,04	2.38±0.02	3.38±0.08	3.45±0.03
Arachidic	0,02 ± 0,07	2,83±0,02	3,05±0,03	1,02±0,04	1.25±0,05	2,30±0,08	1,27±0,05	2.78±0.01	3.26±0.04	2.26±0.05
Arachidonic	0,07 ± 0,01	0,15±0,05	0,13±0,08	0,15±0,02	0,06±0,08	0,07±0,02	0,03±0,09	2.19±0.03	0.17±0.08	0.23±0.01
Eiksapentaenoic	0,27 ± 0,05	2,53±0,09	3,52±0,06	0,50±0,04	0,05±0,02	0,06±0,07	0,01±0,02	1.19±0.05	3.19±0.05	1.19±0.07
Omega 3	0,03 ± 0,02	4,01±0,04	5,91±0,04	3,99±0,08	2,02±0,04	4,06±0,08	3,04±0,08	4.25±0.05	5.25±0.08	3.17±0.02
Omega 6	0,63 ± 0,08	5.64±0,07	7,53±0,08	5,68±0,02	5,03±0,05	6,05±0,02	5,03±0,05	4.08±0.05	6.26±0.01	5.26±0.08
Omega 9	1,62 ± 0,05	0,95±0,01	0,61±0,01	0,89±0,08	0,60±0,03	0,48±0,02	0,63±0,03	0.59±0.06	0.89±0.04	0.29±0.09
Unsattu rated fatty Acid	0,88 ± 0,09	4,46±0,06	3,56±0,07	4,24±0,02	2,68±0,04	3,65±0,08	1,05±0,04	0.68±0.08	2.18±0.09	1.18±0.03
Satturated fatty acid	0,97 ± 0,01	1.09±0,09	2,80±0,08	1,49±0,07	0,97±0,02	1,24±0,02	1,52±0,05	0.14±0.02	1.17±0.01	0.17±0.09
Mono unsaturated fatty acid (MUFA)	1,63 ± 0,05	2,642±0,02	4,97±0,06	2,40±0,04	3,57±0,05	3,93±0,03	4,67±0,07	2.74±0.08	3.05±0.03	2.07±0.01
Polyunsaturated fatty acid (PUFA)	1,24 ± 0,08	3,82±0,08	4,58±0,04	2,83±0,02	3,17±0,01	4,31±0,09	4,50±0,03	3.65±0.03	4.25±0.05	3.25±0.02
AA	0,08 ± 0,09	0,15±0,04	0,13±0,07	0,15±0,09	2,18±0,03	2,71±0,03	2,89±0,01	2.12±0.02	6.12±0.07	2.02±0.03
DHA	0,06 ± 0,02	0,08±0,04	1,07±0,03	0,07±0,01	0,39±0,08	0,83±0,05	0,77±0,04	0.52±0.01	0.52±0.03	0.23±0.01
EPA	0,27 ± 0,03	1.53±0,02	2,52±0,06	1,50±0,07	0,09±0,02	0,08±0,08	0,11±0,06	0.50±0.037	0.50±0.05	0.75±0.05

The total profile of fatty acid (Table 5) found the most acidic saturated fats (SAFA) on *D. magna* massively cultured using the fermentation media of 50 g/L of chicken manure + 100 g/L of rejected bread+ 50 gr/L of tofu waste included the palmitic fatty acid and unsaturated fatty acids (PUFA) at the linoleic fatty acids. The palmitic fatty acid content was found at highest at 3.59% ± 0.04. The palmitic fatty acids serve as the energy storage for the phytoplankton and zooplankton. This statement is supported by the results of previous research that the saturated palmitic fatty acid has the function for energy storage later used for the biosynthesis process of saturated fatty acids (SAFA) [1], [18]. The palmitic fatty acid is the substrate of biosynthesis process of SAFA fatty acid [1], [19]. The highest linoleic fatty acid was at 6.37% ±0.02 and the results of this study were higher than that of previous research [1] at 0.2%. The linoleic fatty acid serves as a substrate in the formation of long chain of PUFA. The results of this study were confirmed through a statement of previous research that the linoleic fatty acid acts as the base substrate to form a long chain of Omega 6 and Omega 3 [1], [19], [20].

The results of research on the profile of amino acid showed the highest level of D. *magna* massively cultured using the fermentation of 50 g/L of chicken manure + 100 g/L of rejected bread + 50 g/L of tofu waste at the amino acid of non-essential glutamic acid of 31.2% ± 0.02 and the essential lysine amino acids at 22.8% ± 0.03. While, the lowest levels of *D. magna* massively cultured with the media of the fermentation A (100 g/L of rejected bread + 100 g/L of tofu waste) was the non-essential amino acid 2.7% ± 0.06 for arginine and essential amino acids of cysteine at 0.3% ± 0.01. The profile of amino acid of *D. magna* massively cultured using the fermentation of a mixture of different animal manures, rejected bread and tofu waste is presented in Table 6.

Table 6. Profile of amino acid of *D. magna* in mass culture using the different animal wastes resulted from probiotic bacteria fermentation.

Amino Acid Profile	A (%)	B (%)	C (%)	D (%)	E (%)	F (%)	G (%)	H (%)	I (%)	J (%)
Histidine	1.9±0.09	4.6± 0.08	8.0± 0.05	4.6± 0.06	3.8 ± 0,07	6.6± 0.06	2.2 ±0,05	1.18±0.07	1.38±0.08	4.31±0.06
Serine	3.8±0.04	7.9± 0.09	12.1 ± 0.08	6.9± 0.01	4.9 ± 0,06	5.9± 0.01	4.6± 0,09	0.93±0.08	2.97±0.01	2.71±0.09
Arginine	2.7±0.06	7.3± 0.06	10.6 ± 0.06	8.4 ±0.02	3.1 ± 0,05	4.4± 0.03	3.4 ±0,07	6.96±0.03	8.33±0.02	7.99±0.07
Glycine	7.6±0.03	12.2± 0.04	17.60± 0.09	12.0±0.08	11.3 ± 0,04	13.0± 0.08	8.5 ± 0,04	1.66±0.04	2.09±0.23	1.66±0.01
Aspartic Acid	6.1±0.08	14.5± 0.09	16.5± 0.04	16.9±0.04	9.5 ± 0,03	15.9± 0,04	7.8 ± 0,07	1.44±0.10	1.41±0.19	1.08±0.07
Glutamic Acid	15.4±0.09	24.4 ± 0.05	31.2± 0.02	27.9±0.01	19.5 ± 0,01	20.9± 0.01	17.5± 0,01	21.15±0.23	23.44±0.02	21.29±0.10
Threonine	5.4±0.03	8.9± 0.07	15.2± 0.08	8.7± 0.02	6.9 ± 0.04	9.7± 0.02	6.8 ± 0,05	0.16±0.26	1.16±0.03	0.11±0.13
Alanine	14.6±0.06	19.4± 0.06	23.1± 0.03	17.1±0.06	15.5 ± 0.01	18.1± 0.06	17.2±0,08	1.53±0.19	2.62±0.17	1.42±0.09
Proline	5.7±0.04	9.3 ± 0.09	6.8±0.09	8.7 ±0.04	7.1 ± 0.09	9.7± 0.04	6.8 ± 0,09	0.64±0.10	3.63±0.10	0.48±0.08
Cystine	0.3±0.01	0.4 ± 0.05	0.8±0.07	0.3 ±0.05	0.3 ± 0.07	0.4± 0.05	0.4 ± 0,07	0.36±0.03	0.45±0.26	0.29±0.03
Lysine	10.1± 0.09	16.8 ± 0.07	22.8± 0.03	15.± 0.03	14.5± 0.04	17.2± 0.07	12.3±0,09	19.43±0.45	21.46±0.09	17.94±0.4
Tyrosine	3.5±0.08	7.1 ± 0.04	10.9± 0.05	7.0 ±0.08	6.7 ± 0.06	5.7± 0.07	3.5 ± 0,07	6.88±0.19	8.83±0.15	7.48±0.01
Metheonine	1.8±0.05	5.0± 0.08	7.4 ±0.06	5.6 ±0.01	4.2± 0.08	4.7± 0.06	2.5 ± 0,06	6.06±0.03	8.08±0.02	7.79±0.05
Valine	7.7±0.08	13.1 ± 0.05	16.5± 0.04	11.5±0.05	11.5±0.01	13.1± 0.03	9.7 ± 0,03	8.05±0.12	12.53±0.11	11.69±0.0
Isoleucine	6.1±0.05	9.9 ± 0.02	13.33± 0.02	8.9 ±0.03	8.8± 0.02	10.3± 0.08	6.9 ± 0,01	10.62±0.09	12.64±0.17	11.53±0.1
Leucine	6.9±0.09	15.8 ± 0.08	19.2± 0.08	13.8 ±0.06	8.2 ± 0.04	10.8± 0.02	7.9 ± 0,04	8.94±0.01	9.37±0.45	7.88±0.09
Phenylalanine	6.6±0.05	12.8 ± 0.09	17.2± 0.09	13.6 ±0.06	12.1 ± 0.06	11.3 ± 0.03	7.1 ± 0,09	10.49±0.12	13.72±0.07	11.53±0.30
Tryptophan	2.17±0.09	2.5 ± 0.08	2.9±0.05	1.5± 0.01	2.2± 0.02	2.9± 0.02	2.2 ± 0,07	1.23±0.23	2.23±0.15	1.19±0.12

D. magna massively cultured using the media of fermentation of 50 g /L of chicken manure + 100 g/L of rejected bread + 50 g/L of tofu waste also provided the highest profile of amino acids in which for the non-essential amino acids of glutamic amino acid reached 31.2% ± 0:05 which was higher than previous research [1] at 0.76%. Glutamic amino acid acts as the natural component in the mixing in the metabolism and as the building block for proteins. This statement is in line with previous research stating that the glutamic amino acid is a natural component present in every living thing and plays an important role in metabolic processes [1], [21]. *D. magna* massively cultured using the fermentation medium of 50 g/L of chicken manure + 100 g/L of rejected bread + 50gr / L tofu waste provided the highest lysine essential amino acids at 22.8% ± 0.03. This result was higher than that of the previous research [1], reaching at 0:44%. The function of the lysine amino acid was as a building framework of vitamin B1 and anti-virus, helped in calcium absorption, stimulated the appetite, and helped in the production of carnitine to convert fatty acids into energy [1], [21], [22].

D. magna massively cultured using the media fermentation of 50 g/L of chicken manures + 100 g/L of rejected bread + 50gr / L of tofu waste based upon the research results provided an increase to the growth, and the highest biomass production. The nutritional quality was based on the analysis on proximate, profile of amino acid and amino fatty acid on *D. magna* massively cultured using the same media as well.

Acknowledgements
This research was funded by the Ministry of Education and Culture, PNBP (Non-Tax State Revenues) of Diponegoro University in the budget year of 2016 through the DIPA (Budget Implementation Checklist) of Diponegoro University Number DIPA – 042.01.2.400898/2016 dated 7 Desember 2015.

References

[1] Herawati VE, Hutabarat J, Pinandoyo, Radjasa OK. 2015. Growth and Survival Rate of Tilapia (Oreochromis niloticus) Larvae Fed by Daphnia magna Cultured with Organic Fertilizer Resulted from Probiotic Bacteria Fermentation. HAYATI J Biosci; 22:169–73.

[2] Damle DK, Chari MS. 2011. Performance Evaluation of Different Animal Wastes on Culture of Daphnia sp. J Fish Aquat Sci; 6:57.

[3] Zahidah, Gunawan W, Subhan V. 2012. Analysis of Population and Growth of Daphnia sp. In Floating Cages Culture at Cirata Reservoirs with Waste Fertilizers Fermented EM4. J Aquat Sci; 3:84–94.

[4] Herawati VE, Agus M. 2014. Analysis growth and survival of catfish larvae feed Daphnia sp. in mass culture using fermented organic fertilizer. J Sci Tech; 26:1–11.

[5] Liswahyuningsih E, Khotimah AU, Febriana DT. 2011. Utilization of tofu waste (dregs and liquid) as a basic materials substitute organic fertilizer production as fertilizer greener chemical. J Ind; 2:57–66.

[6] Herawati VE, Hutabarat J, Radjasa OK. 2016. Growth and Survival Rate of Tilapia (Oreochromis niloticus) Larvae Fed by Daphnia magna Cultured With Organic Fertilizer Resulted From Probiotic Bacteria Fermentation. HAYATI J Biosci:4–8. doi: 10.1016/j.hjb.2015.08.001.

[7] Purbowati E, Sutrisno CI, Baliarti E, Budhi SPS, Lestariana W. 2007. The effect of complete feed with different protein and energy levels on feed conversion of male local sheep fattened on feedlot system. Pros Semin Nas Teknol Peternak Dan Vet:394–401.

[8] Nwachi. 2013. An Overview of the Importance of Pro-biotic in Aquaculture. J Fish Aquat Sci; 8:30-2.

[9] Yuniwati, Iskarima M, Padulemba. 2012. Optimization of Compost Production with Organic Waste from Fermentation Method. J Tech; 5:172–81.

[10] Asadi Rad M, Zakeri M, Yavari V, Mousavi SM. 2012. Effect of Different Levels of Dietary Supplementation of Saccharomyces cerevisiae on Growth Performance, Feed Utilization and Body Biochemical Composition of Nile Tilapia (Oreochromis niloticus) Fingerlings. J Pers Gulf;3.

[11] Abu-Elala N, Marzouk M, Moustafa M. 2013. Use of different Saccharomyces cerevisiae biotic forms as immune-modulator and growth promoter for Oreochromis niloticus challenged with some fish pathogens. Int J Vet Sci Med; 1:21–9. doi: 10.1016/j.ijvsm.2013.05.001.

[12] Jusadi D, Sulasingkin, Mokoginta I. 2005. Effect of Different Concentrations Against Yeast Growth and Population Daphnia Sp. J Aquat Ind; 12:17–21.

[13] Nina S, Givskov J, Martin D. 2012. The Potential of Dietary Polyunsaturated Fatty Acid to Modulate Ercosanoid Synthesis and Reproduction in Daphnia magna. J Physicol; 162:449-54.

[14] [AOAC] Association of official analytical chemists. 2005. Official Methods of Analytical Chemists. Washington, DC, USA. n.d.

[15] Fogg GE. 1965. Algae Culture and Phytoplankton Ecology. The University of Winconsin Press. Madisson Milk. n.d.

[16] Mokoginta, Jusadi D, Pelawi T. 2003. Effect of Daphnia sp. Enriched with Different Fat Sources On The Survival and Growth of Tilapia Larvae (Oreochromis niloticus). J Aquat Ind; 2:7–11.

[17] Widianingsih, Hartati R, Endrawati, Yudiati E, Valentina. 2011. Nutrient Reduction Effect of Phosphate and Nitratee Concentrations Against Total Lipid Content Nannocloropsis oculata. J Mar Sci; 16:24-9.

[18] Lim C, Aksoy Y, Klesius P. 2011. Lipid and Fatty Acid Requirements of Tilapia, North Amerika. J Aquat Int; 7:188–19.

[19] Zengin H, Vural N, Çelik VK. 2013. Comparison of Changes in Fatty Acid Composition of Starved and Fed Rainbow Trout, (Oncorhynchus mykiss) Larvae. Turkish J Fish Aquat Sci;13.

[20] Pratiwi AR, Syah D, Hardjito L, Panggabean LMG, Suhartono MT. 2009. Fatty Acid Synthesis by Indonesian Marine Diatom, Chaetoceros gracilis. HAYATI J Biosci; 16:151–6. doi:10.4308/hjb.16.4.151.

[21] Valverde S, Liorens M, Vidal AT, C R, Estevanel, Gairin JL. 2013. Amino Acids Composition and Protein Quality Evaluation of Microalgae and Meals for Feed Formulations In Cephalopods. J Aquat Int; 21:413–33.

[22] Ovie SO, Eze SS. 2011. Fisheries and Aquatic Science. Fish Aquat Sci; 6:186–93. doi: 10.1146/annurev.ecolsys.110308.120220.

Symbiotic Fungus of Marine Sponge Axinella sp. Producing Antibacterial Agent

A Trianto[1,2], S Widyaningsih[3], OK Radjasa[1,2], and R Pribadi[1]

[1]Natural Product Laboratory, Centre of Research and Services (Laboratorium Terpadu) Diponegoro University, Indonesia
[2]Department Marine Science, Faculty of Fisheries and Marine Science, Diponegoro University, Semarang, Indonesia
[3]Master of Marine Sciences, Faculty of Fisheries and Marine Science, Diponegoro University, Semarang, Indonesia

E-mail: Agustrianto.undip@gmail.com

Abstract. The emerging of multidrug resistance pathogenic bacteria cause the treatment of the diseases have become ineffective. There for, invention of a new drug with novel mode of action is an essential for curing the disease caused by an MDR pathogen. Marine fungi is prolific source of bioactive compound that has not been well explored. This study aim to obtain the marine sponges-associated fungus that producing anti-MDR bacteria substaces. We collected the sponge from Riung water, NTT, Indonesia. The fungus was isolated with affixed method, followed with purification with streak method. The overlay and disk diffusion agar methods were applied for bioactivity test for the isolate and the extract, respectively. Molecular analysis was employed for identification of the isolate. The sponge was identified based on morphological and spicular analysis. The ovelay test showed that the isolate KN15-3 active against the MDR *Staphylococcus aureus* and *Escherícia coli*. The extract of the cultured KN15-3 was also inhibited the *S. aureus* and *E. coli* with inhibition zone 2.95 mm and 4.13 mm, respectively. Based on the molecular analysis, the fungus was identified as *Aspergillus sydowii*. While the sponge was identified as *Axinella* sp.

Keywords: Fungi, sponge, *Aspergillus s*ydowii, multidrug resistace, *Axinella* sp.

1. Introduction

Marine sponge is a source of bioactive compounds the pharmaceutical properties such as antioxidants [1], antibacterial [2], antifungal [3], anti-inflammatory [4,5], anti-malarial [6], and anticancer [7].

However, only few of them passed to the preclinical and clinical stages [8]. Compounds supply is a major problem for drug development from marine source [9]. One of the most serious bottlenecks in developing natural products from coral reefs has been the availability of biomass to gain sufficient amounts of substances for preclinical and clinical studies. Exploitation is further complicated by the fact that most of these metabolites possess highly complex structures, making them difficult to be produced economically via chemical synthesis. Sea culture is one alternative that may be used to solve the need of bioactive compounds from sponge though the sponge grows slowly [10,11]. Burgess *et al.* [12], mentioned that bacteria symbiotically associated with soft corals can synthesize secondary metabolites similiar to host.

Sponge is a filter feeder animals so that microorganisms accumulate in large quantities as mesophyll. Some researchers believe that microorganisms play an important role, such as providing food, bioactive compounds, or precursors to their host [13]. Association number of bacteria in the sponge more than fungal symbionts, but fungal symbionts produce clinically active compounds is more important than bacteria [14].

This paper describes the isolation and identification of the associate fungus of marine sponge collected from Riung Water, East Nusa Tenggara, Indonesia.

2. Materials and methods

2.1. Sample collection

Marine sponges were collected from the Tujuh Belas Pulau, Riung, East Nusa Tenggara, Indonesia by SCUBA diving at 3-15 m depth. The specimens were kept in a cool box until inoculation process [15].

2.2. Isolation of fungi

Isolation of fungi on a sponge conducted with modification methods of Subramani *et al.* [16]. The sponges were washed with sterilized sea water prior to inoculation process to removed loosed associated microorganisms from its surface. Then, the specimen was cut into small piece (1 x 1 x 1 cm). The specimen laid on the surface of malt agar medium. After 7 days incubation, the fungus colonies were separated based on their morphological characteristics, and each colony was inoculated on a new agar plate contain MEA.

2.3. Antibacterial screening

Isolates of MDR bacteria, *Staphylococcus aureus* and *Escherichia coli*, was obtained from the Laboratory of Microbiology, Kariadi Hospital, Semarang. Culture of each MDR bacterium in the logarithmic phase was mixed with Zobell soft agar medium (1% v/v), which were then poured on to the respective agar surface previously inoculated with fungi symbionts sponge and incubated for 24 h [17]. The inhition zone indicated that isolates were active against the antagonistic bacteria (the data were not shown in this paper).

2.4. Sponge Identification

Identification sponge symbiont host of fungal isolates that produce antibacterial MDR *E. coli* and *S. aureus* based on morphological characteristics and the characteristics of spicules based methods of Hooper [18], with slight modifications. Morphological identification is done by color, surface shape, and the shape of growth.

Identification of the spicules is done by cutting a small fragment of the sponge and put in eppendorf microtub, then sodium hypochlorite was added. After the organic components were dissolved and leaving only the mineral skeleton, the bleach was diluted with ethanol and replaced carefully. The washing process was repeated several time. Spicules were taken and placed on a glass slide, and then observed under a microscope. the observations were documented by camera.

2.5. Extraction

The pure isolated shown active against *S. aureus* and *E. coli* were cultured on 1 L MEB media for 7 days or until the maximum growth at ambien temperature. The media was filtered to obtain the micellium that was extracted with methanol at room temperature. The solvent was filtered using filter paper and then dried with rotary evaporator [19].

2.6. Bioassay methanolic extracts

Bioassay methanolic extracts based on the method proposed by Safaeian *et al.* [20] with slight modification The extracts were tested againts the *S. aureus* and *E. coli* using difussion agar method at the concentrations 400, 200, 100, 50 and 25 μg/disk with two times repeated. The pathogenic bacteria were inoculated by spread method on MEA media, after 30 minutes incubation, the paper disks contained the methanolic extract were placed on the agar media surface.

2.7. Fungi KN-15-3 identification

2.7.1. Morphology identification

Fungi symbionts aged 7 days taken by ose needle and placed on a glass slide that has been etched with a solution of lactic acid then observed under a microscope with a magnification 40x. The results are documented using camera [21].

2.7.2. Molecular identification

The DNA isolate was extracted by *Wizard Genomic DNA purification Kit* following the instruction provided by the company (PROMEGA).

DNA engine thermal cycler (MyCycler, Biorad), PCR run comprised 34 cycle with cycle: initial denaturation at 95 ^{0}C for 5 minutes, annealing (57,1 ^{0}C for 1 min), extension (72 ^{0}C for 7 min), denaturation (94 ^{0}C for 40sec), and 42 ^{0}C for 1 min, 70 ^{0}C for 5 min, and the last 4 ^{0}C.

The DNA in supernatant was extracted with isopropanol for further analyses with electrophoresis on agrose gel. Electrophoresis was run at 100 volt with TAE (Tris-Acetate-EDTA) as running buffer. Ethadium bromida was utilized as pigment prior to visualization under UV light. The primers used for PCR amplification of 18S rRNA is universal primers ITS 1 (5'TCCGTAGGTGAACCTGCGG-3') and primer ITS 4 (5'TCCTCCGCTTATTGATATGC-3') [22]. Sequencing process was conducted in Macrogen, Korea.

3. Result and Discussion

From Riung Water of Nusa Tenggara Timur Province-Indonesia, a total of 18 sponges were collected that provided 33 fungus isolates. After antagonistic bioassay, the most active isolate, KN-15-3, was chosen for further study (the screening process was not performed in this paper).

The host of the isolate was also be identified base on morphological characteristics and spiculas analyses. Sponges host KN-15-3 has typical characteristicssuch such as in natural environment has a bright orange red color, while in alcohol brown or light brown, corrugated, with vertical or horizontal and branches in bush-shaped (Figure 1).

Figure 1. The photograph of the sponge K-N-15 (a). In situ (b). in the laboratory.

The identification of host sponge spicules KN-15-3 have type: style, subtylostyle, and strongyle (Figure 2). Based on the characteristics identification and spicules, sponges included in Genus Axinella [18, 23]. Genus Axinella known to be rich bioactive compounds such as anti-virus, cytotoxic, neurological activity, antioxidants and antifungals [24-28].

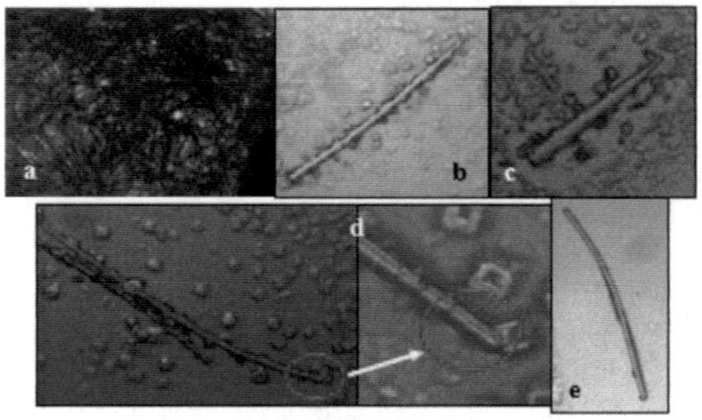

Figure 2. Spicules of the sponge K-N-15 (a = 10x, b-e = 40x) a= skeleton; b.,c. = Style; d = subtylostyle; e.= strongyle

On Malt extract agar media, fungal isolates have white mycelium and spores darker in color. On the observation by a microscope isolates KN-15-3, spores are round and surrounded by conidia thus forming like a flower (Figure 3).

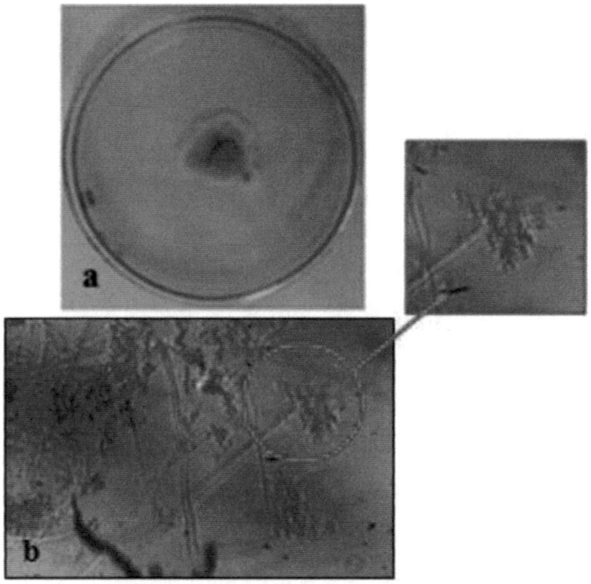

Figure 3. a. Fungi symbionts K-N-15-1. b. Mycelium of fungi symbiont sponge K-N-15-1

Based on the sequencing of fungal symbionts KN-15-3 has 99% similarity with the fungus Aspergillus sydowii (GenBank accession no. JN851052.1) (Figure 4). The isolate has been registered to GenBank with accession number LC094427.

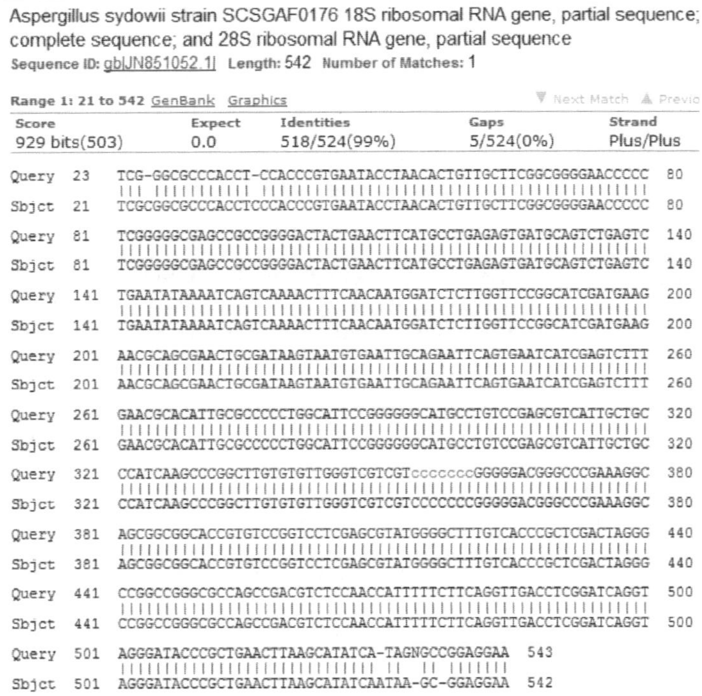

Figure 4. Results of sequence homology analysis isolates KN-15-3using the BLAST database.

The genus Aspergillus is the macroscopic filamentous fungi and one of the causes contaminants in food. This is caused by Aspergillus grows colonize and reproduce mycotoxins [29]. Aflatoxins are secondary metabolites produced by *A. flavus*, A. *parasiticus*, *A. nomius,* dan *A. tamarii* [29-32]. Aflatoxins can be found in many types of food and feed ingredients such as wheat, wheat flour, maize, cereals, rice, beans, spices, and beer [33-34].

The genus of Aspergillus is also known and recognized as a source of bioactive compounds in pharmacological field. Some bioactive compounds produced by Aspergillus genus such as antibacterial varixanthone, anticancer asperazine, antiparasitic gliotoxin 6-methoxyspirotryprostatin B, and antifungal Amphotericin B [35-38].

A.sydowii known to have some activity such as tyrosinase inhibitors, anthraquinone, as antimicrobial, sydowiols A and sydowiols C as anti *M. tuberculosis*, and as acetylcholinesterase inhibitor [39, 40]. Acetylcholinesterase inhibition resulted in increased concentrations of acetylcholine. This further result in increased communication between nerve cells that use acetylcholine as a chemical messenger producing a therapeutic effect in patients with Alzheimer's disease

In antagonistic test the isolate KN-15-3 showed active against two MDR bacteria *Staphylococus aures* and *Escherichia coli.* The methanolic extract of the isolate also exhibited strong activity against both pathogens with inhibition zone and 15.10 mm (*E. coli*) and 18.75 mm (*S. aureus*) at the concentration 400 μg/disk, so it can be developed as a source of new drug.

4. Conclusion

Isolates KN-15-3 produces antibiotic that inhibited the MDR *S. aureus* and *E. coli.* Based on the morphological analysis and molecular analysis the fungus isolate KN-15-3 identified as *Aspergillus sydowii.* The morphological and spicular characteristics indicated that the sponge host, KN-15 was Axinella sp.

5. Acknowledgement

We thank Marine Diving Club members for their assistance during sponge collection. This Research was supported by The Ministry of Research, Technology, and Higher Education through Post Graduate Scheme Research Grant with contract number :139-01/UN7.5.1/PG/2015.

References

[1] Trianto A, Hermawan I, de Voogd N J, and Tanaka J 2011 Halioxepine, a new meroditerpene from an Indonesian sponge *Haliclona* sp. *Chemical & Pharmaceutical Bulletin*, 59(10), 1311–3.

[2] Li Z 2009 Advances in Marine Microbial Symbionts in the China Sea and Related Pharmaceutical Metabolites. *Mar. Drugs*, 7: 113-129.

[3] Sik W, Ki H, Young K, Am S, Soo Y, and Hee I 2006 Antifungal activity of synthetic peptide derived from halocidin, antimicrobial peptide from the tunicate, Halocynthia aurantium. *FEBS Letters*, 580: 1490–1496.

[4] Qaralleh H, Idid S, Saad S, Susanti D, Taher M, and Khleifat K 2010 Antifungal and Antibacterial Activities of Four Malaysian Sponge Species (Petrosiidae). *J. Med Mycology*, 20(4), 315–320.

[5] Yang L and Andersen R J 2002 Absolute Configuration of the Antiinflamatory Sponge Natural Product Contignasterol, 1924–1926.

[6] Fattorusso C, Persico M, Basilico N, Taramelli D, Fattorusso E, Scala F, andTaglialatela-Scafati O, 2011, Antimalarials based on the dioxane scaffold of plakortin. A concise synthesis and SAR studies. *Bioorganic & Medicinal Chemistry*, 19(1), 312–20.

[7] Trianto A, N J de Voogd, and J Tanaka 2014 Two new compounds from an Indonesian spons Dysidea sp., *J. Asian Nat. Prod. Res.,* 16(2):163-168.

[8] Mayer, A M S, Glaser K B, Cuevas C, Jacobs, R S, Kem W, Little R D, and Shuster D E, 2010 The odyssey of marine pharmaceuticals : a current pipeline perspective. *Trends Pharmacol Sci.*, 31(6): 255-65

[9] Hadas E, Shpigel M, and Ilan M 2005 Sea ranching of the marine sponge Negombata magnifica (Demospongiae, Latrunculiidae) as a first step for latrunculin B mass production. *Aquaculture*, 244(1-4), 159–169.

[10] de Caralt S G, Agell M J and Uriz 2003 Long-term culture of sponge explants: conditions enhancing survival and growth, and assessment of bioactivity, Biomolecular Engineering 20. 339-347.

[11] Trianto A, Radisya N N, Wijayanti D P, Rifai A, Ismunarti D H, and Destio, 2013 Laju Pertumbuhan dan Kelulushidupan Transplan Spons Amphimedon sp., IJMS. Vol. 18(4):225-230

[12] Burgess J G, Boyd K G, Amstrong E, Jiang Z, Yan L, Berggren M, May U, Pisacane T, Granmo A, and Adams D R 2003 Development of a marine natural product-based antifouling paint. *Biofouling.*, 19:197- 205.

[13] Hochmuth T and Piel J 2009 Polyketide synthases of bacterial symbionts in sponges--evolution-based applications in natural products research. *Phytochemistry*, 70(15-16), 1841–1849.

[14] Vasanthabharathi V and Jayalakshmi S 2011 Bioactive potential of symbiotic bacteria and fungi from marine sponges. *Afric J. Biotech.*, 11(29): 7500-7511.

[15] Li Q and Wang G 2009 Diversity of fungal isolates from three Hawaiian marine sponges. *Microbiological Research*, 164(2): 233–41.

[16] Subramani R, Kumar R, Prasad P, Aalbersberg W, and Retheesh S T 2013 Cytotoxic and antibacterial substances against multi-drug resistant pathogens from marine sponge symbiont: Citrinin, a secondary metabolite of Penicillium sp. *Asian Pacific J Trop. Biomedicine*, 3(4), 291–296.

[17] Radjasa O K, Martens T, Grossart H P, Brinkhoff T, Sabdono A, and Simon M, 2007. Antagonistic Activity of a Marine Bacterium Pesueodaltromonas luteoviolacea TAB 4.2 Associated with Coral Acropora sp. *J. Bio. Sci.*, 7(2): 239-246.

[18] Sponguide. Guide To Sponge Collection and Identification. *Order A Journal On The Theory Of Ordered Sets And Its Applications*, (Version).

[19] Almeida A P, Dethoup T, Singburaudom N, Lima R, Vasconcelos M H, Pinto M and Kijjoa A 2010 The in vitro anticancer activity of the crude extract of the sponge-associated fungus Eurotium cristatum and its secondary metabolites. J. Nat. Pharm. 1, 25–29.

[20] Safaeian S, Hosseini H, Abbas P A A, and Farmohamadi S 2009 Antimicrobial activity of marine sponge extracts of offshore zone from Nay Band Bay, Iran. *Journal de Mycologie Médicale / J. Med. Mycology*, 19(1), 11–16.

[21] Watanabe T 2002 Pictorial Atlas of Soil and Seed Fungi, Morphologies of Cultured Fungi and Key to Species. 2nd Ed. CRC Press. https://books.google.co.id/books/about/Pictorial_Atlas_of_Soil_and_Seed_Fungi.html?id 5ElPPhLN8JYC&redir_esc=y.

[22] Nursid M, Fajarningsih N D, and Chasanah E 2011 Screening of Antitumor Bioactivity of Fungi Associated with Macro Algae and Sponge from Indrayanti Beach , Jogjakarta, *Squalen Bulletin of Marine & Fisheries Postharvest & Biotech.*, *8*(2), 47–56.

[23] Alvarez B, Van Soest R W M, and Rutzler K 1998 A Revision of Axinellidae (Porifera: Demospongiae) of the Central West Atlantic Region. *Smithsonian Contributions to Zoo.*, 598(598): 1–47.

[24] Hamill P, Hudson D, Kao R Y, Chow P, Raj M, Xu H, Richer M J, and Jean F 2006 Development of a red-shifted fluorescence-based assay for SARS-coronavirus 3CL protease: identification of a novel class of anti-SARS agents from the tropical marine sponge Axinella corrugata. *Biol Chem.*387(8): 1063-74.

[25] Dai J, Fishback J A, Zhou Y D, and Nagle D G 2006 Sodwanone and yardenone triterpenes from a South African species of the marine sponge Axinella inhibit hypoxia-inducible factor-1 (HIF-1) activation in both breast and prostate tumor cells. *J. Nat Prod*, 69(12): 1715-20.

[26] Aiello A, Fattorusso E, Giordano A, Menna M, Müller W E, Perovic-Ottstadt S, and Schroder H C 2007 Damipipecolin and damituricin, novel bioactive bromopyrrole alkaloids from the Mediterranean sponge Axinella damicornis. *Bioorg. Med. Chem*, 15: 5877–5887.

[27] Sakai R, Matsubara H, Shimamoto K, Jimbo M, Kamiya H, and Namikoshi M 2003 Isolations of N-methyl-d-aspartic acid-type glutamate receptor ligands from Micronesian sponges. *J. Nat. Prod.*, 66:784–787.

[28] Yalçın F N, Ohno T, Ersöz T, Saracoğlu İ, Çalış İ, Inoue M, and Ogihara Y 2002 Cytotoxic and Antioxidant Activities of the Eastern Mediterranean Sponges Ircinia and Axinella. Hacettepe University, *J Faculty of Pharmacy*, 22: 19.

[29] Moťková P and Vytřasová J 2011 Comparison of Methods for Isolating Fungal DNA. *Czech J. Food Sci*, 29: S76–S85.

[30] Bennett J W and Papa K E 1988. The aflatoxigenic Aspergillus spp. In: Sidhu G.S. (ed.): Advances in Plant Pathology, No. 6. Academic Press, London: 263–280.

[31] Mayer Z , Bagnara A, Färber P, and Geisen R , 2003 Quantification of the copy number of nor-1, a gene of the aflatoxin biosynthetic pathway by real-time PCR, and its correlation to the cfu of Aspergillus flavus in fous. *Int. J. Food Microb*, 82: 143– 151.

[32] Goto T, Wicklow D T, and Ito Y 1996 Aflatoxin and cyclopiazonic acid production by a sclerotium-producing Aspergillus tamarii strain. *Applied and Environmental Microbiol*, 62: 4036–4038.

[33] Villa P and Markaki P 2009 Aflatoxin B1 and ochratoxin A in breakfast cereals from Athens market: Occurrence and risk assessment. *Food Control*, 20: 455–461.

[34] Yang Z, Shim W B, Kim J H, Park S-J, Kang S, Nam B, and Chung D 2004 Detection of aflatoxin-producing molds in Korean fermented foods and grains by multiplex PCR. *J. Food Protection*, 67: 2622–2626.

[35] Malmstrøm J, Polson S C, Polson S W, Smith G W, and Frisvad J C 2001 Study of secondary metabolites associated with virulent and non-virulent strains of Aspergillus sydowii: Sea fan pathogen. Pages 48-52 in Proceedings of the Symposium on the Natural History of the Bahamas. Gerace Research Center, San Salvador, Bahamas.

[36] Bugni T S and Ireland C M 2004 Marine-derived fungi: a chemically and biologically diverse group of microorganisms. *Nat. Prod.*, 21: 143–163.

[37] Swathi J Narendra K Sowjanya K M Satya A K 2013 Marine fungal metabolites as a rich source of bioactive compounds. *Nat. Prod*, 21: 143–163.

[38] Choudhary M I Musharraf S G Mukhmoor T Shaheen F Ali S Rahman A 2004 Isolation of Bioactive Compounds from Aspergillus terreus. *Z. Naturforsch*. 59b: 324 – 328.

[39] Blunt J W Copp B R Keyzers R A Munro M H G Prinsep M R 2015 Marine natural products. *Nat. Prod. Rep.*, 32 : 116-211.

[40] El-hady F K A Abdel-aziz M S Shaker KH. El-shahid ZA 2014 Tyrosinase, Acetylcholinesterase Inhibitory Potential, Antioxidant and Antimicrobial Activities of Sponge Derived Fungi with Correlation to Their GC/MS Analysis. *Int. J. Pharm. Sci. Rev. Res.* 26(58): 338–45.

The Influence of Madden Julian Oscillation on the Formation of the Hot Event in the Western Equatorial Pacific

Anindya Wirasatriya[1,2], Denny Nugroho Sugianto[1,2], Muhammad Helmi[1,2]

[1]Department of Oceanography, Faculty of Fisheries and Marine Science, Diponegoro University, Semarang, Indonesia
[2]Center for Coastal Disaster Mitigation and Rehabilitation Studies, Diponegoro University, Semarang Indonesia

E-mail: aninosi@yahoo.co.id

Abstract. Hot event (HE) is the high SST phenomena higher than about 30°C, occur in an area of more than 2×10^6 km^2 and last for a period more than 6 days. HE develops only under the condition of high solar radiation and low wind speed. The indication of the relation between HE and MJO has been described in the previous study for one HE case. In the present study, the more case of MJO-HE relation is collected for the period of 2003-2011 and the possible mechanisms is examined. New Generation Sea Surface Temperature for Open Ocean (NGSST-O-Global-V2.0a) was used to identify HEs. Precipitation from TRMM were bandpass filtered with cut off period of 30-60 days for MJO identification. Observation data from TAO/TRITON buoy were used for investigating the possible mechanism of MJO-HE relation. Off 48 HE cases located along the equatorial band, the development of 29 HE cases was related to the suppressed phase of MJO whereas the high solar radiation occurred. High precipitation during the active phase of MJO may contribute to stabilize the upper mixed layer. The stable upper water column fasten the heating process during the suppressed phase of MJO, generating HE.

Keyword: Hot Event, MJO, freshening effect

1. Introduction

Sea surface temperature (SST) is a fundamental parameter for the air-sea interaction processes. In the tropical region, SSTs distribution is negatively skewed which shows the different atmospheric processes before and after 29.5°C Before 29.5°C, the deep convection represented by the highly reflective cloud (HRC) increase following SST. Conversely, after 29.5°C the convection decrease as the SST increase [1]. For decades, study on SSTs were focused on the high SSTs less than 29.5°C since it has significant influence on the climate system. The high possibility of the rain formation occurs under the condition of SSTs between 26 and 29.5°C. For example, typhoon is generated under the condition of SSTs> 26°C [e.g., 2]. Moreover, [3] emphasized that the deep convections frequently occur when their bottom SST is higher than 28°C. Investigations on SSTs higher than 30°C were rarely conducted due to the requirement of high spatial and temporal resolution of SST dataset i.e. daily and less than 0.25° grid resolution. The signal of SSTs higher than 30°C may be dissembled in the lower resolution of SST dataset. Thus, many aspects of high SSTs phenomena in the tropics are still left to be observed.

The investigation of short timescale and high SST phenomena (i.e. less than a month and higher than 30°C) called Hot Event (HE) in the tropical region have been conducted by [4, 5, 6, 7, 8, 9, 10, 11]. They used daily satellite based SST data with grid interval of less than 0.25°×0.25°. They found that HE occur under the condition of low wind speed and high solar radiation associated with high diurnal SST amplitude. Through observation of 10 years SST data, [4] found 31 HEs distributed in whole equatorial region. Satellite SSTs has systematic positive bias against buoy SSTs for the SSTs higher than 30°C due to the large in-situ diurnal SST [5]. Therefore, [10, 11] used diurnally corrected SST dataset for the investigation of HE in the western equatorial Pacific. Through observation of 9 years SST data, [10] found 71 HEs distributed in the western equatorial Pacific. In the case study of HE began at 28 May 2003 called HE030528, [11] found that the development of HE030528 was related to the subsidence air over HE area induced by the deep convection at the northern and southern side of HE area which created tropospheric flywheel effect.

In the larger scale, the subsidence air over HE area may be influenced by the suppressed phase of the Madden Julian Oscillation (MJO). MJO is the overturning global circulation connecting the region of strong deep convection with high precipitation known as the active phase and the region of weak deep convection with low precipitation known as the suppressed phase. This circulation moves eastward with the speed of 5 m/s and period of 30-90 days [12]. The development stage of HE030528 was related to the suppressed phase of MJO [11]. However, one case study cannot represent the general relation between HE and MJO. Thus, this paper is aimed to identify the more cases of HE-MJO relation and investigating the possible mechanisms.

2. Material and Method

2.1. Material

We used the daily New Generation Sea Surface Temperature for Open Ocean (NGSST-O-Global-V2.0a) to identify HEs in the western equatorial Pacific [13]. This SST data has applied diurnal correction to reduce positive bias against buoy SST as described in [5]. This dataset also has been examined for HE identification [10]. The grid interval of the data is 0.25°×0.25°. The bias (root mean square error) against 1 m depth temperature measured by TAO/TRITON buoys in the western equatorial Pacific is -0.107°C (0.428°C).

MJO was observed by using precipitation data. We used The Tropical Rainfall Measuring Mission (TRMM), the daily data of satellite based global gridded precipitation with the grid interval of 0.25°×0.25°. This rain rate products are derived using the algorithm described in [14] and its validation is described by [15].

For investigating the mechanisms underlying the MJO-HE relation, we used observational measurements from the TAO/TRITON buoys [16] located at the western equatorial Pacific. The parameters were SST, solar radiation, and surface salinity. Since the data consist of 6 quality grades, only the highest and default quality were used in the analysis.

2.2. Method

HE is defined as a connected region with a SST higher than the space-time dependent threshold (~30°C), an areal size larger than 2×10^6 km, and lasting for a period longer than 6 days. The detail detection method for HEs is explained in [10]. Following the track of MJO, the present study used 48 HE cases located along the equatorial band identified by [10]. For capturing MJO signal, precipitation data from TRMM was filtered with cut-off period of 30-60 days and presented in Hovmöller diagram. The relation between HE and MJO was indicated by the coincident event between the suppressed phase of MJO and the development stage of HE.

3. Result and Discussion

3.1. Identification of MJO-HE relation

For detecting MJO signal, [11] used vertical pressure velocity data from JRA-55 without any filtering process which made the MJO signal was noisy. In this study we used satellite based precipitation data and then filtered with cut-off period of 30-60 days. The result is shown in Hovmöller diagram (Fig. 1). The active (suppressed) phase of MJO signal is clearly shown by eastward movement of positive (negative) precipitation. Every year, about 8 MJO events were identified passing the western equatorial Pacific during 2003-2011.

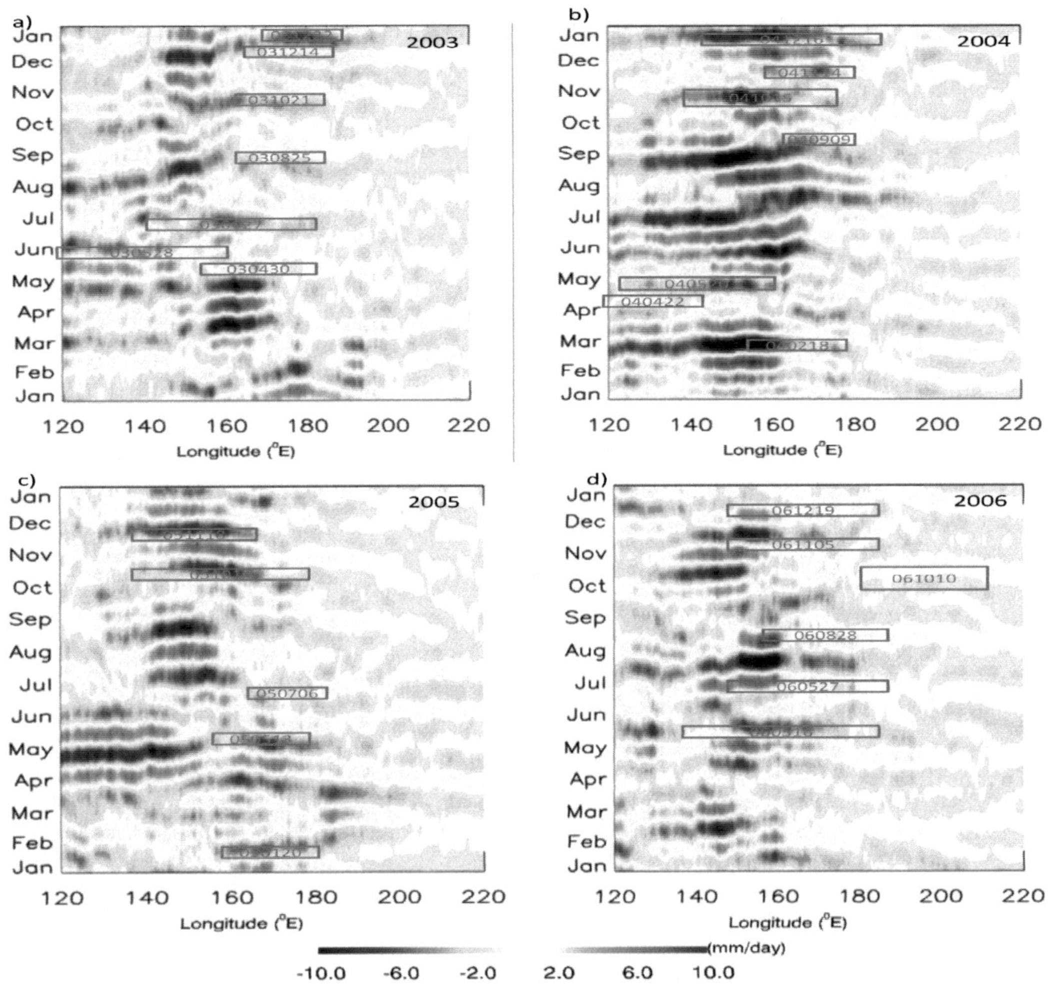

Fig. 1. Hovmöller diagram (5°S-5°N) of precipitation after filtering with cutting period from 30 to 60 days. Red and blue boxes denote the coincidence of HE to suppressed phase and active phase of MJO respectively. HE cases are taken from [10].

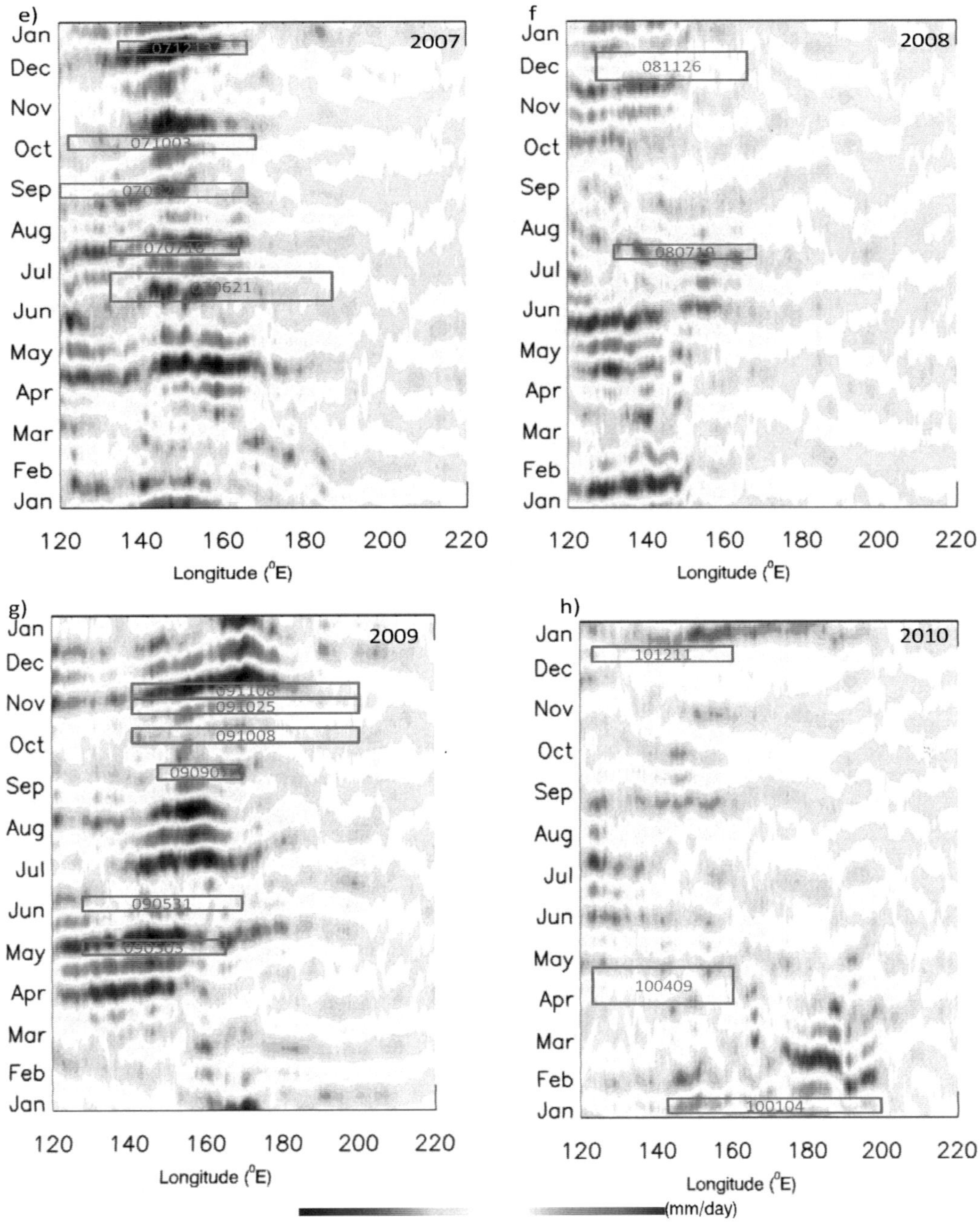

Fig. 1. Continues

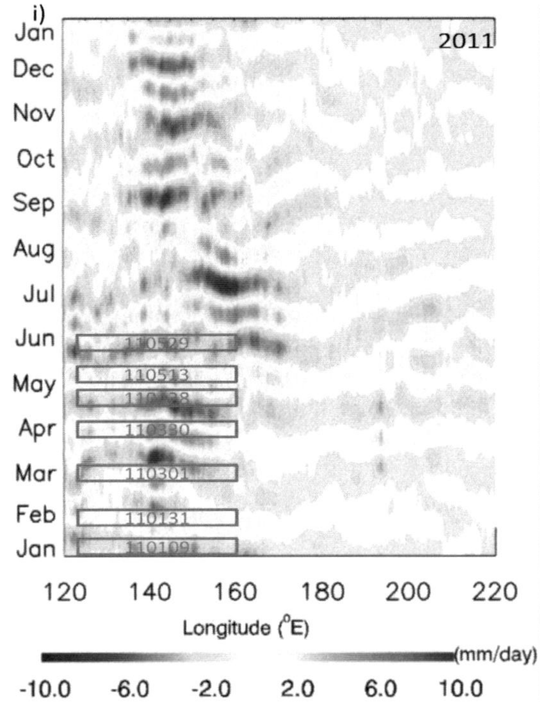

Fig. 1. Continues.

The oceanic response on MJO event has been observed by many researchers. The fluctuation of SST related to MJO is vary e.g. 0.25°C [17], 0.5°C [18, 19] and even higher i.e, 2°C [20, 21]. Figure 2 shows the evidence of increasing SSTs during the suppressed phase of MJOs. Through case study of HE030528, [11] found the increase of SST about 0.5°C during the suppressed phase of MJO.

The relation between MJO and HE was indicated by the coincide event between the period of suppressed phase of MJO and the period of the development stage of HE denoted by the red box (Fig.1). The blue boxes denote the development stages of HEs that do not coincide with the suppressed phases of MJOs. Off 48 HE cases during 2003-2011, we identified 29 cases of the development stage of HE coincide with the suppressed phase of MJO. This number is much more than the number of the HE generated by the positive anomaly of subsurface temperature during El Niño as identified by [6] i.e., 3 cases during 1993-2003. Thus, the influence of MJO to the development of HE is more frequent than ENSO.

3.2. Possible mechanism of MJO-HE relation.

The mechanisms on how HE developed and decayed was exposed by [11] in terms of climatological analysis and case study. Solar radiation variability due to the remote convection mechanisms and surface wind speed variability influenced by the topography determine the development and decay stage of HE. In the present study, the higher solar radiation due to the subsidence air during the suppressed phase of MJO were the most prominent signal on how MJO influence the HE development. These phenomena occurred for all HEs related to MJO. Moreover, we also introduce precipitation as a new parameter affecting HE development.

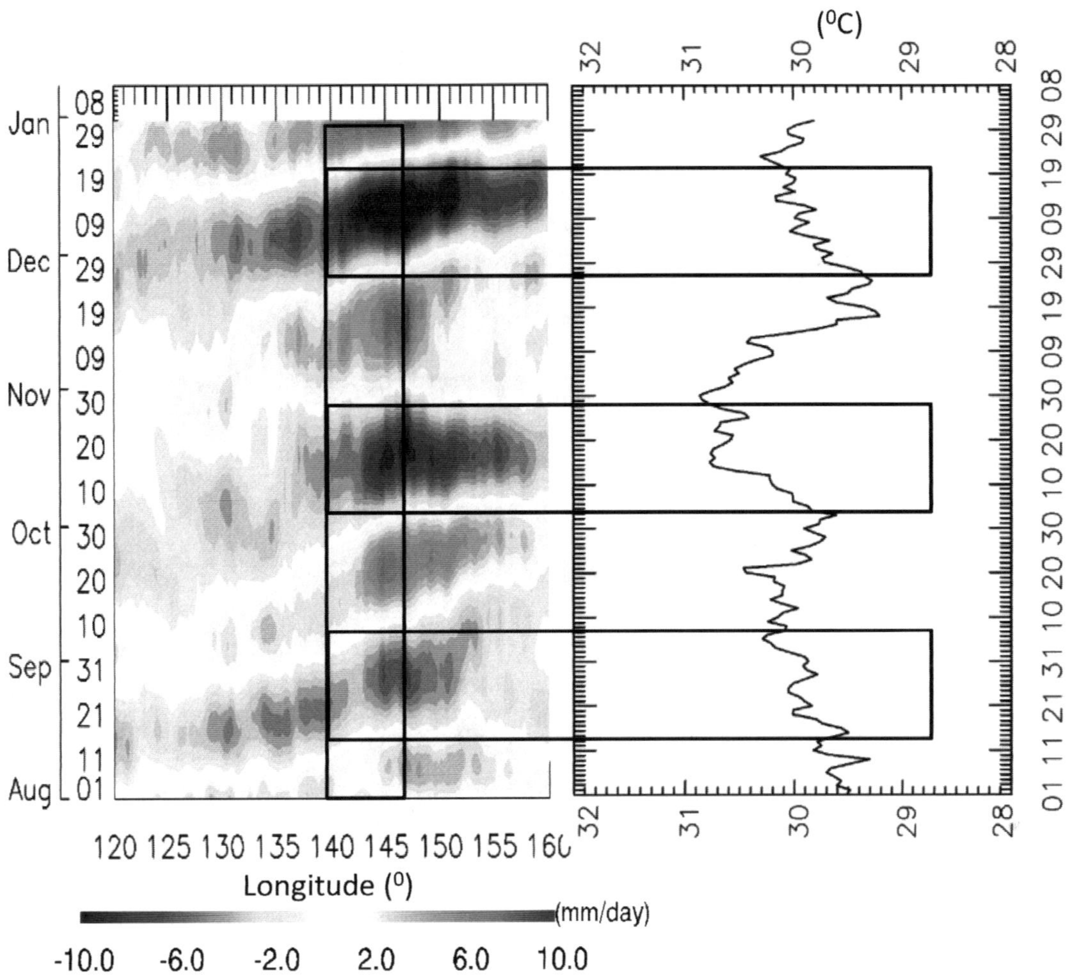

Fig. 2. Hovmöller diagram (5°S-5°N) of precipitation after filtering with cutting period from 30 to 60 days (left figure) and the corresponding mean SST value inside the black box (right figure).

MJO influence the SST through the buoyancy flux into the ocean. It increases (decreases) with surface warming by solar radiation (cooling by latent heat flux and nighttime infrared radiation) and freshening (salinizing) by precipitation (evaporation) [12]. In the present study, the influence of freshwater flux was examined for the development of HE090503. HE090503 is one of HE cases related to MJO event. The amplitude map is shown in Fig. 3a. Figure 3b shows the Hovmöller diagram of filtered precipitation and time series of SST, solar radiation and salinity observed by TAO/TRITON buoy at 147°E, 5°N denoted by the green dot in Fig. 3a. The active phase of MJO was indicated by the positive filtered precipitation in the Hovmöller diagram, lower and fluctuated solar radiation and the decrease of salinity in the time series data. The decrease of salinity denoted the freshening process. The suppressed phase of MJO was indicated by the negative filtered precipitation in the Howmoller diagram and the increasing solar radiation in the time series data. Salinity remained low in this period. This period coincided to the increasing SST which indicated the development stage of HE. The freshening process during the active phase of MJO may stabilize the water column, prevented the mechanical mixing,

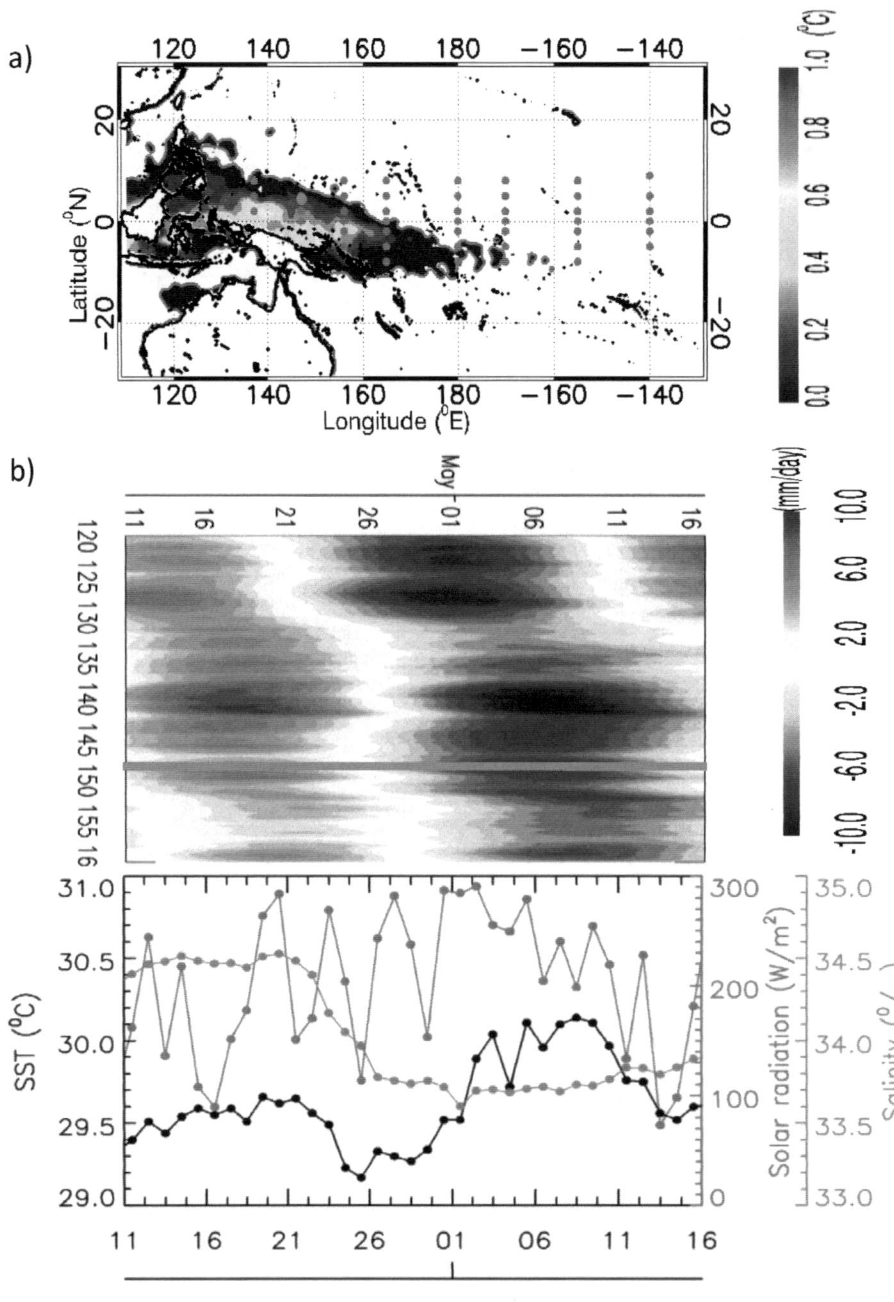

Fig. 3. (a) The amplitude map of HE090503. Amplitude map is the composite of daily amplitude throughout the period of HE090503. Daily amplitude is calculated by subtracting daily SSTs from space-time dependent threshold as shown in [10]. The green dot is the position of TAO/TROTON buoy used in the time series analysis of Fig. 3b. (b) The Hovmöller diagram of 2°N-7°S of TRMM filtered precipitation and the time series of SST, solar radiation and surface salinity of TAO/TROTON buoy at 5°N, 147°E. The green line in Hovmöller diagram points 147°E.

fastened the warming in the sea surface, and generating HE. Thus, beside high solar radiation during the suppressed phase of MJO, this study firstly shows the contribution of high precipitation during the active phase of MJO to the development of HE by stabilizing the upper water column. However, this phenomena occurred only in some cases. Strong oceanic processes, such as horizontal advection by ocean currents, cold water entrainment at the bottom of the mixed layer or the effect of the diurnal cycle may blurred the mixed layer responses to MJO forcing [12]. Thus, the stabilizing effect of precipitation were dissembled in some other HE cases.

4. Conclusion

The study of the influence of Madden Julian Oscillation on the formation of the Hot Event in the western equatorial Pacific can be concluded as follows :

a. Off 48 HE cases located along the equatorial band during 2003-2011, the development of 29 HE cases was related to the suppressed phase of MJO.
b. The high solar radiation due to the subsidence air during the suppressed phase of MJO were the most prominent factor on how MJO influence the HE development.
c. The high precipitation during the active phase of MJO may contribute to HE development by stabilizing the upper mixed layer. The stable upper water column fastened the heating process during the suppressed phase of MJO, generating HE.

Acknowledgement
NGSST-O-Global-V2.0a is produced by Center for Atmospheric and Oceanic Studies, Tohoku University, Japan. Contact for data inquiry and request is Kohtaro Hosoda (hosoda@ocean.caos.tohoku.ac.jp). TRMM data were produced by Remote Sensing Systems and sponsored by the NASA Earth Sciences Program. Data are available at www.remss.com/missions/tmi. TAO/TRITON buoy data is provided by TAO Project Office of NOAA/PMEL. First author thank to Dr. Kohtaro Hosada and Dr. Teruhisa Shimada for the valuable discussion. This research is funded by PNBP DIPA Diponegoro University No: SP DIPA-042.01.2.400898/2016.

References

[1] Waliser DE, and Graham NE 1993 *J. Geophys. Res.* **98 (D7)** 12881
[2] Gray WM 1968 *Mon. Weather Rev.* **96** 669
[3] Fasullo J and Webster PJ 1999 *J. Clim.* **12** 1292
[4] Qin H, Kawamura H, and Kawai Y 2007 *J. Geophys. Res.* **112** C07015 doi:10.1029/2006JC003969
[5] Kawamura H, Qin H, and Ando K 2008 *J. Oceanogr.* **64** 847
[6] Qin H, Kawamura H, Sakaida F, and Ando K 2008 *J. Geophys. Res.* **113** C08045, doi:10.1029/2007JC004640
[7] Qin H and Kawamura H 2009 *J. Oceanogr.* **65** 605
[8] Qin H and Kawamura H 2009 *Adv. Space. Res.* **44** 395 doi:10.1016/j.asr.2009.03.003
[9] Qin, H., and H. Kawamura (2010), Air-sea interaction throughout the troposphere over a very high sea surface temperature, Geophys. Res. Let., 37, 1-4. doi:10.1029/2009GL041685.
[10] Wirasatriya A, Kawamura H, Shimada T, and Hosoda K 2015 *J. Oceanogr.* **71** 77 doi: 10.1007/s10872-014-0263-3
[11] Wirasatriya A, Kawamura H, Shimada T, and Hosoda K 2016 *J. Geophys. Res. Atmos.* **121** doi:10.1002/2016JD025268
[12] Zhang C 2005 *Rev. of Geophys.* **43** 1
[13] Hosoda K and Sakaida F 2015 *Proc. 30th Int. Symp. on Space Technology and Science* (2015-n-37, Kobe, Japan) Online available from http://archive.ists.or.jp/upload_pdf/2015-n-37.pdf.
[14] Hilburn KA and Wentz FJ 2008 *J. of Appl. Meteor. And Clim.* **47** 778 DOI: 10.1175/2007JAMC1635.1
[15] Hilburn K and Smith DK 2012 *AGU Fall Meeting*, online available from http://www.remss.com/papers/hilburn/Hilburn_RSS_V7_Rain_Validation_FallAGU2012.pdf
[16] McPhaden MJ, Ando K, Bourles B, Freitag HP, Lumpkin R, Masumoto Y, Murty VSN, Nobre P, Ravichandran M, Vialard J, Vousden D, Yu W 2010 *Proceedings of the "OceanObs '09: Sustained Ocean*

Observations and Information for Society" conference, vol 2 (Venice, Italy, 21-25 Sep 2009: ESA publication) WPP-306
[17] Shinoda T, Hendon H and Glick J 1998 *J. Clim.* **11** 1685
[18] Zhang C 1996 *J. Atmos. Sci.* **53**,739
[19] Jones C, Waliser DE, and Gautier C1998 *J. Clim.* **11** 1057
[20] Krishnamurti TN, Oosterhof DK and Mehta AV 1988 *J. Atmos. Sci.* **45** 1304
[21] McPhaden MJ 2002 *J. Clim.* **15** 2632

Sediment Transport Model In Sayung District, Demak

Aris Ismanto[1], Muhammad Zainuri[1], Sahala Hutabarat[1], Denny Nugroho Sugianto[2,3], Sugeng Widada[2,3] , Anindya Wirasatriya [2,3]

[1]Doctorate Programme of Coastal Resources Management/Aquatic, Diponegoro University, Semarang, Indonesia
[2]Department of Oceanography, Faculty of Fisheries and Marine Science, Diponegoro University, Semarang, Indonesia
[3]Center for Coastal Disaster Mitigation and Rehabilitation Studies, Diponegoro University, Semarang Indonesia

E-mail: aris.ismanto@gmail.com

Abstract. Demak has 34,1 km coastline and located in 6⁰43'26" - 7⁰09'43" South Latitude and 110⁰27'58" - 110⁰48'47" East Longitude. In the last few years rapid shoreline and erosion has threatened Demak coastal area. No less than 3000 villages on Java suffer similar problems. Hard structures such as dykes and breakwaters is one of the method that is commonly used to solve this problem. However, this method may fail to provide adequate protection to the environment and become counterproductive. One of the alternative to solve the problem is using hybrid engineering concept.This study aims is to assess the distribution model of the sediment on the application of technology as a hybrid structure for the mitigationand rehabilitation of coastal areas in Demak. This research using quantitative method, including field surveys and mathematical modeling methods. The model show that the sedimention is quite big in highest flood condition and must have the right structure for the hybrid engineering. This study is expected to answer the question of the erosion problem in the District Sayung, Demak.

Keywords : Sediment Transport, Experimental Method, Demak, Erosion

1. Introduction

Demak, Central Java is one of the area that has many problems related with erosion and shoreline degradation. The lost of the flood protection that the mangrove forests once offered has caused coastal retreat from a few hundred meters up to kilometer (at some places) in recent decades [1] No less than 3000 villages on Java suffer similar problems [2]. These vulnerabilities will further exacerbate if the growing degradation and erosion problems are not addressed. Establishment of aquaculture ponds along low-lying sedimentary reaches resulted in the near total destruction of mangrove forests since the 1980's [3]. Mangrove Forest were converted, mostly in western Indonesia [4]. The removal of the mangroves and confinement of the intertidal range due to construction of earth bunds around the shrimp ponds caused changes in sediment dynamics. This triggered massive erosion and the related land loss, inundation and saltwater intrusion problems[2]. In order to solve this problem hard structures such as dykes and breakwaters is one of the method that is commonly used. Locally, these hard structures were even found yo intensify erosion by blocking sediment transport towards the coast and enhancing wave effests due to wave reflection. [5] However, this method may fail to provide adequate protection to the environment and become counterproductive. For that reason hybrid enggineering is one of the alternative to solve the problem. Hybrid engineering is a friendly environment method that accommodates economic and livelihood development needs, and combines technical and ecosystem-based solutions [6]. Semipermeable dams made of bamboo piles and brushwood have succesfully reduced erosion and stimulated sedimentation in comparable muddy environments.[7]

The first step is identify the mechanisms controlling mixing and stratification as a product of physical input such as bathymetry, tidal, wind driven circulation, and sedimentation. The next step is identify the sediment source domination that is happening in this coastal area, predicting the suspended sediment concentration and fine sediment fluxes that is happening in the area would be the study that has to be done. Than characterize the dominant currents and quantify the associated residual transport on sediment.

After all the physical parameters have been analysed hopefully it will give the best elements for the Hybrid Engineering. The study aims is to know the ocean curents pattern and to predict the sediment transport dispersion as if the dredging process.

2. Material and Method

2.1 Material

2.1.1 Hydrodynamics Model

The hydrodinamic model is used to simulate the currents in the research area. In general, a two-dimensional equation of hydrodynamic consists of the main components which include:
conservation equations of mass (continuity) and momentum conservation equation. Circulation caused by tidal currents and wind in coastal waters and assume as a homogen water mass by depth. In this model the sea water assume as an incompresible fluid.

The continuity equation is [8]:

$$\frac{\partial \xi}{\partial t}+\frac{\partial U}{\partial x}+\frac{\partial V}{\partial y} = Q_s$$

The momentum equation is:

$$\frac{\partial U}{\partial t}+\frac{U}{H}\frac{\partial U}{\partial x}+\frac{V}{H}\frac{\partial U}{\partial y}+gH\frac{\partial \xi}{\partial x}+rU\frac{\sqrt{U^2+V^2}}{H^2}$$
$$+A_h\left(\frac{\partial^2 U}{\partial x^2}+\frac{\partial^2 U}{\partial y^2}\right) = \lambda W_x\sqrt{W^2{}_x+W^2{}_y}\,2$$

$$\frac{\partial U}{\partial t}+\frac{U}{H}\frac{\partial V}{\partial x}+\frac{V}{H}\frac{\partial V}{\partial y}+gH\frac{\partial \xi}{\partial y}+rV\frac{\sqrt{U^2+V^2}}{H^2}$$
$$+A_h\left(\frac{\partial^2 V}{\partial x^2}+\frac{\partial^2 V}{\partial y^2}\right) = \lambda W_y\sqrt{W^2{}_x+W^2{}_y}$$

where :

x,y : space coordinat grew to the east and north (m).

u,v : current velocitu in x and y (m/s)

U : transport in x direction (m^2/s)

$$U = \frac{1}{h+\xi}\int_{-h}^{\xi} udz$$

V : transport in y direction (m^2/s)

$$V = \frac{1}{h+\xi}\int_{-h}^{\xi} vdz$$

t : time (s)

ξ : elevation from sea water level from mean sea level (m)

g : gravity (m/s^2)

H : Real depth = h + z (m)

h : fixed depth (m)

r : friction coefficient

A_h : eddy frictionn coefficient (m^2/s)

λ : wind friction coefficient

Wx,Wy : Wind Speed in x and y (m/s)

Q_s : debit (m^2/s)

The hydrodinamics model is solved using the semi implisit where the variabel is calculate from the space cell in each time step. This method is chosen because it doesn't depend in time step option in the Courant-Friedric-Lewy stability criteria like in the explisit methods, so the computer memory can be use effective and efficient. This model using bathimetri data, tides data, wind and the physics condition in the research area (inlet outlet, beach building, etc).

2.1.2 Sediment Transport Model

The sediment trasport model in a canal and sea basically is determine as a suspension material dispersion in water, such as :

a) With the velocity of fluid flow result in a sediment transport dispersed directly by the flow. This movement is called advection process.
b) Turbulent diffusion process relating to variations in the speed of the flow direction of the cross section by the motion of turbulence

Both the process and the exchange between the air-sea sediments formulated in the following equation[8]:

$$\frac{\partial C}{\partial t} + u\frac{\partial C}{\partial x} + v\frac{\partial C}{\partial y} = K_x \frac{\partial^2 C}{\partial x^2} + K_y \frac{\partial^2 C}{\partial y^2} + \frac{A}{\rho A_p h}(C - C_E) + Q_C$$

where :

C = Actual sediment Concentration (kg/m^3)
C_E = Natural sediment concentration (kg/m^3)
Q_C = rate of change in sediment concentration (mg/l s)
K_x, K_y = koefisien difusi turbulen arah x, dan y (m^2/detik)
$A = (4{,}48 + 0{,}049T) + f(1{,}12 + 0{,}0180T + 0{,}00158T^2)$ = sea air interaction coeffiecient(kkal/m^{2}s °C)
$f = 3{,}6 + 2{,}5W_3$; wind factor for A coefficient
W_3 = wind speed that is measured in 3 m above water level (m/s)
ρ = Sea water density (kg/m^3)
A_p = specific sediment in constant pressure (kkal/kg °C)
h = depth (m)

For practical purposes it can be assumed the water area is small enough to be reviewed, the value C_E is taken from the value of the natural sediment concentration of sea water in places far away from the point source pollutant concentrations.

Vertical distribution of sediment concentration reasonably assumed homogeneous or variation is not great because there has been a turbulent mixing is almost perfect from the surface to the seafloor, so that two-dimensional sediment transport models can be used horizontally and approached the dispersion process at the pool / canal / sea. 2D model equations are solved by numerical methods upstream and sediment exchange coefficient between the sea and air using empirically derived formulation. The model simulated sediment distribution in accordance with the scenario that is designed the same as the current model simulation scenarios as described in advance, due to hydrodynamic models and models of sediment run concurrently or so-called model coupled from the two models.

2.2 Method

Bathymetry Sayung coastal waters, Demak ranging from 1 meter to 15 meters to the north (offshore). Broadly speaking, the location of the model can be seen in Figure 1 below.

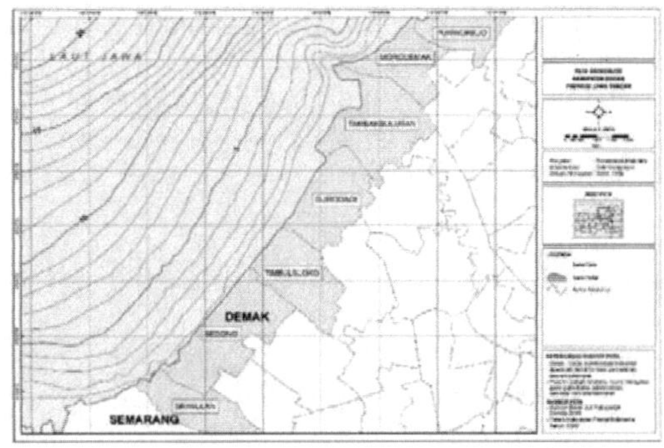

Fig 1. Research area in Sayung District, Demak (Source : Data Processing,2015)

The area of this model is divided into 455 x 422 grids (455 grid east-west direction and 422 grid north-south direction) with the size of each grid dx = dy = 50 m. Calculation time step (dt) is 1 second. By simulating 15 days old.

Hydrodynamics and sediment dispersion models simulated by inserting force generating tidal, and wind simulation was performed under various scenarios with tidal conditions

3. Result and Discusion

3.1 The validation of result

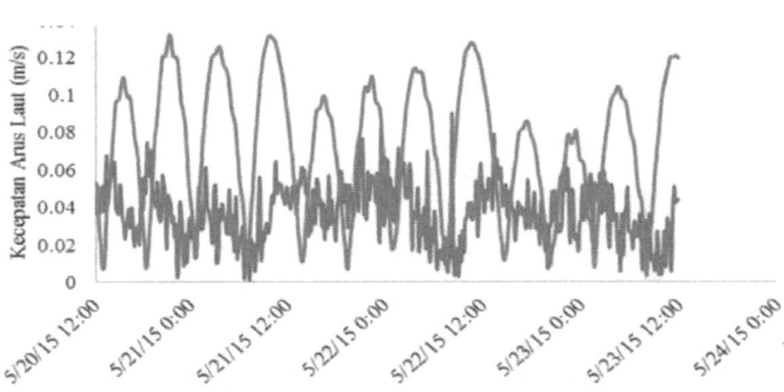

Fig 2. Validation result graphic between model and field current (Source : Data Processing, 2015)

CF = 1.1258. CF (Cost Function) is a non-dimensional value to indicate a match two kinds of data. The criteria used are CF <1 is very good, good 1-2, 2-3 enough,> 3 less. The results validate the model results with field results indicate that the ocean current modeling results Demak has a good criterion.

3.2 Hydrodynamics Model Result

The pattern of ocean currents in the period of flood toward ebb tide has current velocity ranges from 0.01 to 0.5 m / s with the movement direction leading toward south. In the period of lowest ebb flow velocity ranges from 0.01 to 0.4 m / s with the direction of movement leads to the south. On the condition of ebb tide toward flood current velocity ranges from 0.02 to 0.4 m / s with the direction of movement in the north. In the period of the highest flood the current dominance of speed ranges from 0.1 to 0.5 m/s, during this period the flow rate appear higher than the speed of currents in the other tidal period. Distribution of the current direction in this period also just leads to the north of Demak waters.(see figure

3). Bappenas & KOICA (2012)[9] reported a wave-induced current of about 0,5 m/s when west-northwest in Demak Coastal area., but expert think the current velocity is an order of magnitude lower.

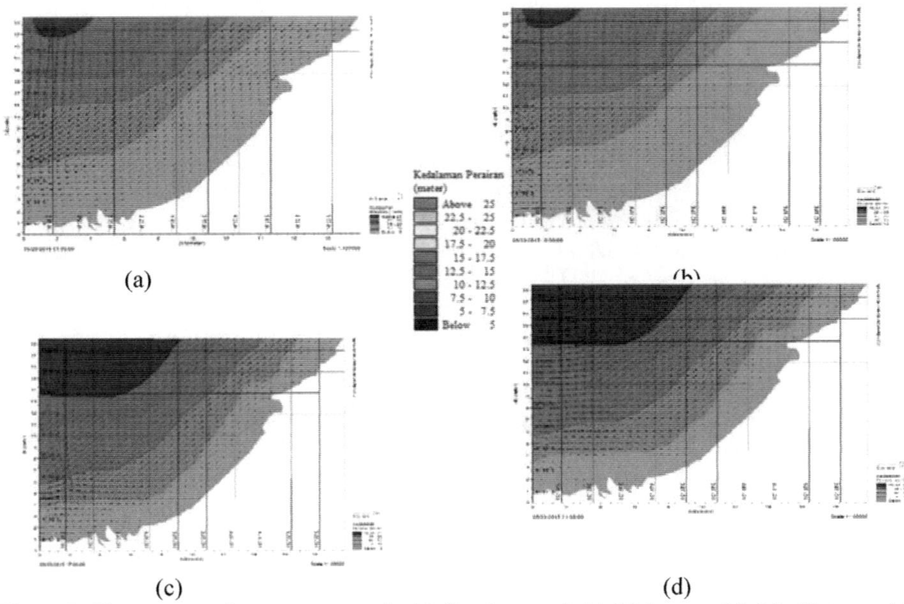

Figure 3. The pattern of ocean currents in (a) flood toward ebb(b) lowest ebb (c) ebb toward flood (d) highest flood

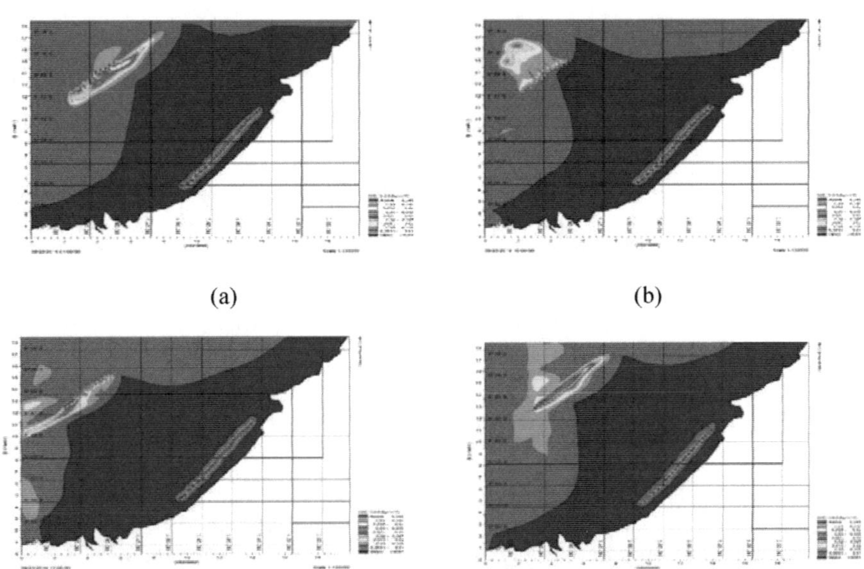

Figure 4. The pattern of sediment transport in (a) flood toward ebb(b) lowest ebb (c) ebb toward flood (d) highest flood

3.3 Sediment Transport Model Result

Sediment concentration distribution model results in Figure 3. Seen from tides scenario that the highest flood tide (Figure 3a) change the concentration of sediment dominant with a distance of 6 km from the source of pollutants where sediment concentration values ranging between 0,015 - 0,045 kg / m3. While the condition of the lowest tide ebb sediment concentrations scattered small enough dominance in which the value of sediment concentration ranges beetween 0,01 -0,03 kg / m3 with a distance of 3 km distribution (Figure 3b). This sediment transport patterns affected by the tidal patterns, where the pattern of sediment transport and the direction of ocean currents towards to near the shore in accordance with the conditions of simulated at high tide. It is adapted to the statement of Gyr and Hoyer [10] that sediment transport patterns that are influenced by the mass of water flow patterns also depend on monsoon different wind conditions. The simulation shows that the distribution of sediment is quite small because of the small ocean currents velocity (See figure 4).

4. Conclusion

The sediment transport modeling has been done in Sayung District Demak for the existing condition as well as after the hybrid engineering concept, with the following

1. In the period of the highest flood has the biggest ocean current speed appear than the speed of currents in the other tidal period where the speed ranges from 0.1 to 0.5 m/ s
2. The simulation shows that the distribution of sediment is quite small because of the small ocean currents velocity that happen in Demak Waters

Acknowledgements

Authors would like to thank to the Lembaga Penelitian dan Pengabdian kepada masyarakat (LPPM) Diponegoro University and Center for Coastal Disaster Mitigation and Rehabilitation Studies (PKMBRP), Diponegoro University. Thanks also to all the respondents, the village chiefs, and all those who have helped so done research.

References

[1] Ecoshape. 2015. Building with nature Indonesia (securing eroding delta coastlines): Design and engineering plan. Tech rep. Ecoshapes.

[2] Winterwerp J., Van Weesenbeeck B., Van Dalfsen J., Tonneijck, F., Astra, A., Verschure, S., & Van Eijk, P. 2014. A Sustainable solution for massive coastal erosion in Central Java. Tech rep, Deltares and Wetland International.

[3] Van Wesenbeeck, B., Balke, T., Van Eijk, P., Tonneijck, F., Siry, H., Rudianto, M., & Winterwerp, J. 2015. Aquaculture induced erosion of tropical coastline throws coastal communities back into poverty. Ocean & Coastal Management, 116, 466-469.

[4] Departmen Kehutanan. 2005. Pemetaan dan Kajian untuk Penyusunan Rencana Rehabilitasi Hutan dan Lahan dengan Mangrove dan Tanaman Pantai di Daerah Bekas Bencana di Propinsi Nanggroe Aceh Darussalam dan Sumatera Utara. Direktorat Pengelolaan DAS dan Rehabilitasi Lahan. Jakarta.

[5] Winterwerp, J., Erftemeijer, P., Suryadinaputra, N., Van Eijk, P. & Zhang, L. 2013. Defining Eco-morphodynamic requirements for rehabilitating eroding mangrove mud-coasts. Wetlands. 33, 515-526.

[6] Sugianto,D.2015.Introduksi Konsep Building with Nature, Kasus Eksisting, dan Pentingnya Kaitan antara Sains, Kebijakan dan Praktek. Workshop dan Launching Building With Nature. March 4, 2015. Semarang.

[7] Saengsupavanich, C. 2013. Erosion Protection option of a muddy coastline in Thailand: Stakeholders shared responsibilities. Ocean &Coastal Management, 83, 81-90.

[8] Sugianto, D. 2009. Simulasi Model Transpor Sedimen Tersuspensi untuk Mendukung Perencanaan Pelabuhan Teluk Bayur Sumatera Barat, Jurnal Teknik Lingkungan, Vol. 5, No. 2, Des. 2009, pp. 46-544.

[9] Bappenas&KOICA. 2012. Coastal protection and management policy adressing climate change impact in Indonesia. Tech. Rep.

[10] Gyr, Albert and Hoyer, Klaus., 2006. Sediment Transport. Fluid mechanics and its application Volume 82. Springer-Verlag Berlin Heidelberg.

IOP Conf. Series: Earth and Environmental Science **55** (2017) 012008 doi:10.1088/1755-1315/55/1/012008

Resources Management Strategy For Mud Crabs (*Scylla* spp.) In Pemalang Regency

Aristi Dian Purnama Fitri[1], Herry Boesono[1], Agus Sabdono[1], Nadia Adlina[1]

[1]Faculty of Fisheries and Marine Science, Diponegoro University, Semarang, Indonesia
Email: aristidian.undip@gmail.com; aristidian@fisika.undip.ac.id

Abstract. The aim of this research is to develop resources management strategies of mud crab (*Scylla* spp.) in Pemalang Regency. The method used is descriptive survey in a case study. This research used primary data and secondary data. Primary data were collected through field observations and in-depth interviews with key stakeholders. Secondary data were collected from related publications and documents issued by the competent institutions. SWOT Analysis was used to inventory the strengths, weaknesses, opportunities and threats. TOWS matrix was used to develop an alternative of resources management strategies. SWOT analysis was obtained by 6 alternative strategies that can be applied for optimization of fisheries development in Pemalang Regency. The strategies is the control of mud crab fishing gear, restricted size allowable in mud crab, control of mud crab fishing season, catch monitoring of mud crab, needs a management institutions which ensure the implementation of the regulation, and implementation for mud crab aquaculture. Each alternative strategy can be synergized to optimize the resources development in Pemalang Regency.

1. Introduction

Mangrove forest in Pemalang Regency is producing mud crab and has an area measuring 120.39 hectares of mangrove ecosystem to be used. In addition, the location is in coastal areas or boundaries to the Java Sea and can be utilized in the development of the fisheries sector especially in captured fisheries. The potential for sustainable (Maximum Sustainable Yield / MSY) fisheries resources in Indonesia is estimated at 6.4 million tons per year, whereas potential that can be exploited (allowable catch) of 80% of the MSY is 5.12 million tons per year [1]. The production of mud crab catches in Pemalang Regency in 2006 reached 11.89 tons [2].

There are 4 types of fishing gears to catch mud crab in Pemalang Regency and mud crabs have become a target catch or by catch. This research used two types of fishing gears; gill net and Badong (trap) because they are the dominant fishing gears. Rod and Cone were not analyzed because the limitations of the data sources and difficulty in identification, and because Ulujami sub district also provide a mangrove tour in which visitors can freely capture mud crab. The catches from this activity is difficult to identify as the fish are not directly captured by local fishermen.

In addition, mud crabs caught by fishermen generally measure less than 15 cm. Minister of Marine and Fishery (PERMEN-KP) Regulation No. 1 year 2015 states the prohibition of catching mud crab in spawning condition and carapace width of less than 15 cm. Therefore there needs to be a strategy to manage resources of mud crab in Pemalang Regency in a sustainable way so that resources and can provide high economic benefits for fishermen.

2. Method

The method used is descriptive survey in a case study. The aim of a descriptive study is to create descriptive, factual and accurate information of facts, and the relationship among the phenomena investigated. In other words, the survey method is to obtain facts and phenomena that exist and seek explanations of fact, either from social, economic, or political institutions in an area [3]. The case study is to select in depth cases only at a specific time and place, while the results do not necessarily apply to different times and places although the target object is the same [4].

2.1. Data Collection

This research used both primary and secondary data. Primary data were collected from field observations and in depth interviews with stakeholders. Respondents in this study include civil servants from the marine and fisheries department (two persons), employees of Asem Doyong fishing port (two persons), fisherman (two persons), fish traders (one person), and lecturer (one person). On the other hand, secondary data were collected from publications and documents issued by the competent authority.

2.2. SWOT Analysis

To gather information about mud crab management resources, we conducted an analysis of the strengths, weaknesses, opportunities, and threats (SWOT). A SWOT analysis is a simple and flexible tool, consisting of gathering opinions from a knowledgeable body of people familiar with a particular business or industry to help evaluate internal strengths and weaknesses, as well as external opportunities and threats.

Applications of SWOT to marine and fresh water fisheries have appeared in recent years in peer-reviewed journals [5-7] and government reports [8]. Recently, SWOT analysis was used to compare the strengths and weaknesses of alternative discard mitigation approaches to achieve mandates under the 2012 reform of the European Union's Common Fisheries Policy that bans future fishery discards [9].

Furthermore, researches using SWOT analysis to develop these strategies have been carried out by several researchers, both in terms of business strategies [10-16] and public policy strategies [17-21]. Some other research developed fisheries strategies using the TOWS matrix. In the TOWS matrix, there are four strategy types, i.e. SO strategy (integrating the strengths and opportunities), ST strategy (integrating the strengths and threats), WO (integrating the weaknesses and opportunities), and WT (integrating the weaknesses and threats).

3. Results and Discussion

3.1. Strengths

Ulujami Sub district is a coastal area or boundary to the Java Sea and can be utilized in the development of the fisheries sector. Its geographical condition supports fishing activities, thereby allowing the development of the fisheries sector, especially captured fisheries. Ulujami Sub district also has a mangrove ecosystem area measuring 120.39 hectares. The production of mud crab catches in Pemalang area in 2006 reached 11.89 tons [2]. This research made use of gill net and Badong (trap) because they are considered legal and not listed in the regulation the Minister of Marine and Fishery No. 2 of 2015 on the restriction of the use trawls and seine nets fishing gear at all fishery management areas in Indonesia.

Based on the aforementioned facts, several strengths that can be optimized for the resource management strategy of mud crab in Pemalang Regency are formulated. Some of these strengths are as follow:

1. The potential geography in Pemalang regency (Code: S1)
2. The availability of fisheries potential resource (Code: S2)
3. The fishing gears that are considered legal (Code: S3)

3.2. Weaknesses

There are many regulations regarding catching mud crab with the purpose of regulating these fishery activities. On the other hand, many fishermen do not understand those regulations because their implementation by the authorities in charge is not firm. Supervision and control of the stock resources especially mud crab needs to be done by the authorities so that the waters are not over-exploited. In addition, mangrove tour activities are a particular concern because mangrove tours allow visitors to freely capture mud crab. Yet another problem is the difficulty faced by fisherman to obtain capital from either banks or financial institution.

Based on those aforementioned facts, several weaknesses that can be optimized for the resource management strategy of mud crab in Pemalang regency are formulated. Some of these strengths are as follow:

1. Less attention to regulation implementation (Code: W1)
2. Less attention to supervision (Code: W2)
3. High cost of fishing operational (Code: W3)

3.3. Opportunities

Mud crab is one of the marine commodities with important economic values. Based on interviews with local fishermen, mud crab sold for high prices. Mud crab is of higher value than blue swimming crab when they are fresh. Despite having been harvested for a longer period than blue swimming crab, mud crab has better freshness level. Consumer demands for mud crab continue to increase. One of the policies in fishery in 2015 is to improve the prosperity of fishermen. This is in addition to the Regulation of the Minister of Marine and Fishery (PERMEN-KP) No. 1 year 2015 on prohibition of catching mud crab in spawning condition and carapace width of less than 15 cm, which is a good policy to preserve the habitat and maintain the stock of particular mud crab for sustainable fisheries.

Based on the aforementioned facts, several opportunities that can be optimized for the resource management strategy of mud crab in Pemalang regency are formulated. Some of these opportunities are as follow:

1. Increased income for mud crab fisherman (Code: O1)
2. Consumer demand vs increasing yield of mud crab (Code: O2)
3. The national fishery policy (Code: O3)

3.4. Threats

Based on interviews with local fishermen, mud crab production is already decreasing. CPUE value calculation that decreased in 2010 and 2015 were analyzed in this research. The market price is uncertain depending on the extent of mud crab supplies. This situation is detrimental to fishermen. At peak season prices are low and there is no guaranteed market price for mud crab.

Based on the aforementioned facts, several threats that can be optimized for the resource management strategy of mud crab in Pemalang regency are formulated. Some of these threats are as follow:

1. Reduced production of mud crab (Code : T1)
2. Less capital support (Code : T2)
3. No guaranteed market price for mud crab (Code : T3)

3.5. Alternative Strategies

Results of SWOT analysis can be used to develop an alternative strategy using the TOWS matrix. Alternative strategies can be divided into four groups, i.e. SO, ST, WO and WT in Table 1 and Table 2.

IOP Conf. Series: Earth and Environmental Science **55** (2017) 012008 doi:10.1088/1755-1315/55/1/012008

Table 1. Alternative Strategies for Mud Crab in Ulujami sub-district,Pemalang using the TOWS Matrix

IFAS / EFAS	**Strength (S)** 1. Potential geography of Pemalang regency (Code: S1) 2. Available fisheries potential resource (Code: S2) 3. Legal fishing gears (Code: S3)	**Weakness (W)** 1. Less attention to regulation implementation (Code: W1) 2. Less attention to supervision (Code: W2) 3. High cost of fishing operational (Code: W3)
Opportunities (O) 1. Increased income for mud crab fishermen (Code: O1) 2. Consumer demand vs increased mud crab yields (Code: O2) 3. National policy on fishery (Code: O3)	Strategy SO 1. Control of mud crab fishing gears (S1, S2 , S3, O1 , O2, O3)	Strategy WO 1. Restricted allowable size of mud crab (W1, W2, O1, O2, O3)
Threats (T) 1. Reduced mud crab yields (Code: T1) 2. Less capital support (Code: T2) 3. No guaranteed market price for mud crab (Code : T3)	Strategy ST 1. Control of mud crab fishing season (S2, S3, T1, T3) 2. Catch monitoring of mud crab (S1, S2, S3, T1, T3)	Strategy WT 1. A management institution to ensure implementation of regulation (W1, W2, T1) 2. Implementation of mud crab aquaculture (W3, T2, T3)

Table 2. Review of Alternative Strategies for Mud Crab in Ulujami Sub district, Pemalang

No.	SO Alternative Strategy	Review
1.	Control of mud crab fishing gears	The purpose of control is to reduce incidence of unintentional catch of small-sized mud crab. The fishing gear should come with escape gaps, especially for badong (trap), so that small-sized mud crabs can run away and grow to their mature size for future consumption.
	Alternative Strategy	
2.	Restricted size of allowable mud crab	The size of mud crabs before mature or gonad is 10cm (Wijaya et al., 2010). Regulation of the Minister of Marine and Fishery (PERMEN-KP) No. 1 year 2015 prohibits the catching mud of crab in spawning condition and carapace width of less than 15 cm to ensure the sustainability of mud crab resource. The Solutions to Badong fishing gear is by designing the escape gap so that mud crab carapace with width of less than 6cm can get away from Badong. On the other hand, the gill net fishing gear needs to be used properly via proper communication with the local fishermen.
	ST Alternative Strategy	
3.	Control of mud crab fishing season	Based on the results of interviews with local fishermen, it is known that mud crab's peak season is from January to April, while its regular season is in May, July, Nov, and December, and the lean season is from August to October. In addition, April and June show mud crabs laying mature gonads [22].

		So, the alternative is that this is the close season for mud crab harvesting.
4.	Catch monitoring of mud crab	Research shows that mud crab catch in the peak season is abundant that some are discarded prior to selling them to traders as the handling proses is not proper. Hence, the alternative is to conduct socialization and counseling concerning the proper handling of mud crab catch. This will improve its selling value, which may include not only for consumption but also for soft-shelled crab farming (aquaculture).
	WT Alternative Strategy	
5.	A management institution that ensures implementation of regulation is required.	Management institution here is a group consisting of several elements of fishermen, fish traders, businessmen, and government officials for the supervision and control of regulation and the local wisdom that aims to manage mangrove crabs farming to be sustainable. One good example of this is the POKMASWAS (Community Supervising Group). Meetings among the entities here should be done regularly on a yearly basis for necessary reviews to be made.
6.	Implementation of mud crab aquaculture	Mud crabs size of less than 6 cm sis till worth selling for consumption. The Solution for this is to use this carapace for soft-shelled mud crab farming activities (aquaculture), so it will have a higher reselling value.

4.Conclusion

It can be inferred from the analyses above that there are 6 alternative strategies that can be applied for optimization of the resource management strategy in Pemalang regency. These strategies include controlling mud crab fishing gears, restricting mud crab size for consumption, controlling mud crab fishing season, monitoring mud crab catch, founding a management institution that ensures the implementation of the regulation, and implementing mud crab aquaculture.

Acknowledgments

This research was funded by Diponegoro University, World Wide Fund for Nature (WWF), and TAKA organization.

References

[1] Adam, L. 2012. *J.of Fisheries and Marine Resources.* **2** 115
[2] Agus M. 2008. Analysis of Carrying Capacity on Sentra Aquaculture Mud Crab Scylla sp. In Pemalang, Central Java. Thesis. Diponegoro University. Semarang
[3] Nazir, Moh. 2011. Research Methods: Edition 6. Ghalia Publisher. Bogor pp 103
[4] Dantes, N. 2012. Research Methods. Andi Offset Publisher. Yogyakarta pp 84
[5] Çelik, A., Metin, _I., Çelik, M., 2012. Taking a Photo of Turkish Fishery Sector: a SWOT Analysis. *Procedia Social Behaviour Science.* 58 pp 1515-1524
[6] Panigrahi, J.K., Mohanty, P.K., 2012. Effectiveness of the Indian Coastal Regulation Zones Provisions For Coastal Zone Management and Its Evaluation Using SWOT Analysis. *Procedia Ocean Coastal Management* 65 pp 34-50
[7] Stead, S.M., 2005. Changes In Scottish Coastal Fishing Communities Understanding Socio-Economic Dynamics to Aid Management, Planning and Policy. *Procedia Ocean Coastal Management.* 48 pp 670-692

[8] Loefflad, M.R., Wallace, F.R., Mondragon, J., Watson, J., Harrington, G.A., 2014. Strategic Plan for Electronic Monitoring and Electronic Reporting in the North Pacific. *US Dept Commer, NOAA Tech Memo.* NMFS-AFSC-276

[9] Sigurðard_ottir, S., Stef_ansd_ottir, E.K., Condie, H., Margerisson, S., Catchpole, T.L., Bellido, J.M., Eliason, S.Q., Go~ni, R., Madsen, N., Palialexis, A., Uhlmann, S.S., Vassilopoulou, V., Feekings, J., Rochet, M.J., 2015. *J. of Marine Policy* **51** 366

[10] Abdi, A., Ashouri, M., Jamalpour, G., Sandoosi, S. M., 2013. *Singaporean Journal of Business Economics and Management Studies* **1** 69

[11] Agarwal, R., Grassl, W., Pahl, J., 2012. *J. of Business Strategy* **33** 1

[12] Chan, X., 2011. *International Journal of Business and Social Science* **2** 147

[13] Dyson, R.G., 2004. *European Journal of Operational Research* **152** 631

[14] Ommani, A. R., 2011. *African Journal of Business Management* **5** 9448

[15] Shojaei, M.R., Taheri, N.S., Mighani, M.A., 2010. *Asian Journal of Management Research* 759

[16] Wang, X.P., Zhang, J., Yang, T., 2014. *Journal of Applied Research and Technology* **12** 230

[17] Adepoju, T. L., Famade, O. A., 2010. *Educational Research and Reviews* **5** 354-361

[18] Chen, M., 2014. *World Transactions on Engineering and Technology Education* **12** 671

[19] Jamshidi, A., Sajjadi, S. N., Honari, H., 2012. *International Journal of Academic Research in Business and Social Sciences* **2** 106

[20] Kantawateera, K., Naipinit, A., Sakolnakorn, T. P. N., Churngchow, C., Kroeksakul, P. 2013. *Asian Journal of Social Science* **9** 226

[21] Sayyed, M. R. G., Mansoori, M. S., Jaybhaye, R. G., 2013. SWOT Analysis of Tandooreh National Park (NE Iran) for Sustainable Ecotourism. *Proceedings of the International Academy of Ecology and Environmental Sciences* 3 pp 296-305

[22] Wijaya, N.I., Yulianda, M., and Juwana S. B. 2010. *J. of Oceanology and Limnology* **36** 439

Development of Spore Protein of *Myxobolus koi* as an Immunostimulant for Prevent of Myxobolusis on Gold Fish (*Cyprinus carpio* Linn) by Oral Immunisation

Gunanti Mahasri[1]

[1,] Department of Fish Health Management and Aquaculture Faculty of Fisheries and Marine, Airlangga University, 031-5914151,

mahasritot@gmail.com

Abstract. Production of Gold fish (*Cyprinus carpio* Linn) in Indonesia has always increased from 2013 to 2015 year by year with increasing average 2% per year. The amount of production was respectively 571.892 tonnes, 1129.273 tonnes, and 1186.674 tonnes. There were almost no problems to sale of gold fish because it had a good enough prospect. The aims of this research were Isolation of spore protein of *Myxobolus koi* by using SDS-PAGE to analyze immun respons and survival rate gold fish that immunized with spore protein of *Myxobolus koi*. The method of this research used experimental method, and belonged to 4 treatments that are: Control, the group of gold fish not immunized with protein spore of *Myxobolus koi* neither infected by *Myxobolus koi* (T1). The group immunized and infested by spore of *Myxobolus koi* (T2). The group which immunized and not infested by *Myxobolus koi* (T3), and The group only infested by *Myxobolus koi* (T4). The dose of immunostimulant was 5 ml in 1 kg of food. The result showed that there were two bands of whole spore protein with molecule weight (MW) 150 kDa and 72 kDa and one band of crude protein *Myxobolus koi* with molecule weight 73 kD and the optical density point was 0.132 on the first day and increased to 0.769 on the 56 th day. The result also showed that the immun respons and survival rate increased from 27% to 86% in chellence test. The protein spore of *Myxobolus koi* can used to develop material for immunostimulant and to prevent the myxobolusis.

1. Introduction

One of the diseases that plagues often become parasiter is a Protozoan disease caused by Myxobollus koi called myxobolusis. This disease started from 2009 was included into quarantine disease pest (HIPK) class I, because the disease causes illness and death of fish. The disease is called Myxobolusis which cause serious problems on the goldfish, koi and lead to the death up to 60-90% with prevalence reaching 100%. Furthermore it is said that in 1974 and 1978 Myxobolus attacked has occurred in Indonesia that led to the deaths of up to 100% of the koi fish mainly on stadia seeds.

The fish that attacked by the myxobolusis disease showed in difficulties to breathe because it found the existence of a nodule or cyst or nodule on the Gill filaments. In Blitar reported by farmers that in 2010 has occurred an outbreak myxobolusis on CARP koi in size 3-5 cm with mortality reached up to 90%.

While in 2002 goldfish mass death has occurred in Kulon Progo and Sleman, caused by a parasite *Myxobolus* sp. and *Henneguya* so that the losses experienced by the fish farmers were quite big. *Myxobolus* sp. was also found in Ngrajek, Magelang Regency in 2006 with prevalence reaching up to 91%. Followed the gold fish ponds koi in Blitar prevalence reaches up to 86% in 2010 [1]. The prevention efforts and counter measures against myxobolusis have been done using a disinfectant as well as other chemicals, but not yet able to meet the target, even cause resistance and residues in the body of the fish. It is necessary to look for the alternative prevention efforts which not cause a negative impact. One of the efforts that have already begun to be developed at this time is vaccination that can be done by immersion or injection.

The characterization of protein can be done by using SDS-Page. By using this method we know its molecular weight protein wether suitable for vaccine or not. Itabashi et al[2] and Mahasri[3] have managed to detect a protein core and anti-Zoothamnium arbuscula with cytoplasmic protein spasmin-1 on spasmonema. Immunoblotting analysis of results showed that the protein has a molecular weight of antigennya 68 kDa, 55 kDa and a 71 kDa. Chavda *et al* [4] suggested the results of the analysis of SDS-PAGE electrophoresis of Myxobolus cerebralis spores which infect fish Catla catla retrieved six expressions Ribbon protein with molecular weight 130 kDa and 60 kDa as well as four other 7 kDa protein Ribbon up to 45 kDa. Asri et al [5] showed the result of Myxobolus spores protein isolation koi which infects fish Cyprinus carpio with a method electrophoresis SDS-PAGE was obtained by two Ribbon proteins with molecular weight 70.22 kDa and 22.83 kDa. More Insariani *et al* [6] put forward the results of the insulating surface glycoproteins M. koi using the method of polymerase chain reaction (PCR) obtained a protein with molecular weight of 12 kDa, 25 kDa and 27 kDa.

The function of the receptor cell responses of fish is portrayed by some parameters of immune cells to produce a response form. MHC (Major Hstocompatibility Complex) is one of the molecules that are instrumental in the immune system of the fish. If antigen (vaccine) enters the host cell or body fish then the antigens presented by MHC, antigen will be captured by receptors on T helper cells (2), and helper T cells (2) going to secrete cytokines, namely IL-2, IL-4, IL-6 and aimed at the differentiation and proliferation of B cells, differentiation of B cells and plasma cells will produce memory cells. Next the cell plasma will synthesize specific antibody binds antigens so as to prevent the movement of antigen and eases the process of phagocytosis.

Anderson [7] argued that the immunostimulant is injected into the body of the fish will get into the kidneys and can increase the Adaptive immune response against the vaccine. In the front part of the kidney, macrophages and neutrophil antigens or vaccines will feed that goes into the body of the fish through the process of phagocytosis and then carried towards the thymus for T cell activation triggers. T cells expressing these antigens to special receptor known as the major histocompatibility complex or Major histocompatibility complex (MHC) antigen, then these are carried towards spleen, splenic release will occur in sitokinin form the B cells that cause an increase of the immune cells as a whole. B cells most do poliferasi or reproduce themselves and partly differentiated body cells into plasma B cells and memory as the humoral immune system. Based on that background the purpose of this research is to develop spore protein of *Myxobolus koi* as immunostimulant material to prevent the death of carp (*Cyprinus carpio Linn*).

2. Research Methodology

2.1. Time and Location

This research was carried out from March to November 2013, with the location of the research in the laboratory of Dry of the Faculty of Fisheries and Marine, Airlangga University and the laboratory of molecular biology of the Faculty of Sciences the university of Brawijaya, Malang.

2.2. Research Methods

This research is descriptive research experiments of laboratory, with the purpose to analyze immune response (an overview of fish blood)) of koi fish that vaccinated with protein *Myxobolus koi* as a sub unit material of vaccine to suppress the death of koi fish due to myxobolusis.

2.3. Research Materials

Carps were tested by using 4 aquariums with a capacity of 5 lt of water. Fish stocking density was 5 fish/aquarium with fish size 7-10 cm. Fish used for treatment were healthy fish marinated first in Methylene blue 3-5 g/m3 of water for 5 minutes to clear the organism that clings to the body of the fish. The treatment given is: T1 = 20 goldfish injected with a solution of PBS and without infection spores m. Koi (control), T2 = 20 carps applied with protein whole spores m. koi and infected with 80 spores/fish m. ko, T3 = 10 goldfish vaccinated with whole protein spores m. koi without spores infected m. Koi and T4 = as many as 20 carps without the vaccinated whole protein spores m. koi. Granting of immunostimulant protein from whole spores of Myxobolus koi done in person with a dose of 1 ml protein/kg feed, which refers to research conducted by [8].

The main materials for the isolation and characterization of a protein used are physiological NaCL, ethanol, solvents of percoll gradient, pepsin, HCL, EDTA, KCL, Na2HPO4, KH2PO4, trypsin, sodium citrat, NaHPO4, H2), NaHCO3, glucosa, phenols red 0.5%, NaOH, filter 0.22 um, bovineserum albumin, dextrosa, ETOH, proteinase, ForwardERIB1 primer pairs 5 '-ACCTGGTTGATCCTGCCAG-3'(2-20) and Reverse ERIB10 5'-CCTCCGCAGGTTCACCTACGG-3 ' (2069-2070), 400 UM DNTP, 3 um Mgso4 , yellow and blue dye, agarose, TAE buffer, sybrsafe, 100 DNAladder bp and bp 1, loading dey, tris-HCL, 2-mercaptoethanol, sodium dodecyl sulfate (SDS), bromophenol blue, glycerol, SDS loading buffer, polyacrylamide, stacking gels, ammonium peroxidaisulphate (APS), glycine and TEMED. Sedangkan peralatan utama yang digunakan adalah Haemocytometer, rubber pollicemen, microscope micrometer, autoclaf, tabung centrifuse, swinging rotor, water bath sonicator dan 1 set peralatan SDS-PAGE electrophoresis [9].

While the main equipments used are the Haemocytometer, rubber pollicemen, microscope tube micrometer, autoclaf, centrifuse, swinging rotors, water bath sonicator and 1 set of tools the SDS-PAGE electrophoresis [9].

2.4. Work Procedures

2.4.1. *Samples preparation of spores of Myxobolus koi*

The fish stricken by myxobolusis are washed with aquadest so that the dirt that clings to the body of the fish is gone, and then nodule myxobolus stuck in gills is taken by using a scalpel and tweezers slowly so that the nodules containing spores are not destroyed. The nodules already taken are then placed in Petri disc and given PBS to taste. Nodules are cut several pieces to remove spores by using a scalpel then added aquadest, included in the test tubes and centrifuged at the speed of 2500 rpm for 15 minutes to separate the cells of Myxobolus. Supernatan formed is then discarded and sediment plus aquadest, centrifuged again up to 4000 rpm with speed solidified for 10 minutes. Moreover pellet is calculated by haemocytometer, then added as much as 2 ml PBS and stored at 4oC in the freezer.

2.4.2. *Isolation of whole proteins of spores of Myxobolus koi*

Spores that have been calculated are given PBS to taste then centrifuged at speed 5000 rpm for 10 minutes. Pellets are added buffer lisis 500 μl of Lysis then sonicated in ice (1 minute sonication ½ minute break), carried out repeatedly 10 times. The results of it and then is vortexed (½ minute vortex I minute break) in ice, carried out repeatedly 15 times. The results of a vortex with a speed of 12000 rpm centrifuged for 5 minutes, the supernatan formed were collected and then carried out the analysis of SDS-PAGE. Determination of the concentration of the whole protein Myxobolus spores used Bio-Rad Protein Assay and read using UV-Visible Spectrophotometer with a wavelength of 600 nm.

2.4.3. *Analysis of whole proteins of spores of Myxobolus koi with SDS-PAGE*

The purpose of this activity is to find out the pattern of the molecular weight of each fraction protein. Analysis of the protein is done with a method electrophoresis SDS-PAGE gel separating composition of 12.5% and 5% stacking gel. This is done by electrophoresis method: Running a gel is created and inserted into the plate glass. After hardened on the top put stacking gel prepared.

A total of 10 μ g samples that added Laemly buffers with a comparison of 2:1 is done boiling at 100oC for 5 minutes, put in well located in the stacking gel. As a marker used protein with molecular weight in the range 10-180 kDa (New England Bio-Labs). Then do a running on the chamber has filled Electrode Buffers 1 x with 100 Volts, 40 mA. The running process is stopped after a blue marker reaches the bottom plate of the gel. Next gel inserted into the washer solution consisting of 25 ml methanol, acetic acid, 3.7 ml Aquades 100 ml. and Rocked above the shaker for 30 minutes. Repeated washing is done with a solution that is similar to the reduction of the composition of ethanol and acetic acid addition half of previously for 30 minutes. Subsequent washing with a solution of 10% glutaraldehyde and aquadest for 30 minutes. After the washed gel stained with silver nitrate (AgNO3) for 15 minutes, then do the washing with aquades twice each for 2 minutes. Given the color development solution consisting of formaldehyde 3.7%, zitronsauce 5% and aquades. After the tape is visible then the reaction is stopped by adding acetic acid 10%. The results of the gel that has a Band-Band protein looks ready documented.

2.4.4. *Observation on the blood picture (immune response)*

Observation on blood picture of the fish is done as complementary parameters for analyzing immune response koi fish being exposed with crude protein spores Myxobolus koi. The parameters of the image of blood is observed counts (differensial) leukocytes by way of making the preparations of blood review [8]. The making of preparations is carried out by placing the blood review a drop of blood on a glass object first, object second pair of glasses with glass objects against angle 27o first, then pulled up to touch the blood, the blood is allowed to spread along the edge of the glass, and then the second object glasses of the second object is pushed along a surface of glass objects first so that it forms a thin layer of blood evenly.

Furthermore preparations were drained at room temperature then fixed with absolute methanol for 3 minutes and dried at room temperature back before tinged with Giemsa dye 10% for 15 minutes, then wash it back with aquadest preparations to reduce excess color and dried at room temperature. Once it is observed under a microscope with 1000x zoom. Leukocyte percentage calculated by way of observing 10 viewpoint and each type of Leukocyte that counted was grouped and presented according to type, in percent (%).

2.4.5. *Determination of Survival Rate (SR)*

Determination of level of survival rate is performed as the supporting parameter for analyzing immune response of koi fish in each treatment. The level of survival rate is expressed in the form of a percentage of the number of koi fish that lives up to the 14th day post treatment trial against the total number of fish kept. Fish survival rate is calculated by using the formula:

$$\mathrm{SR} = \frac{\mathrm{Nt}}{\mathrm{No}} x100\%$$

Description:
SR = survival Rate
NT = the number of fish that live at the end of the observation
No = the number of fish that lived in the early test challenge

Data obtained consists of qualitative and quantitative data. Qualitative data is character data of protein in the form of protein molecules that Spore is shown as an image (a protein Band) results from elektroferesis of Gel SDS. While quantitative data is data analysis results description of blood and Survival Rate were analyzed using the statistical test ANOVA (analysis of Variance) and if there is a difference, it is continued by Duncan's Multiple Range Test) with a confidence level of 5%.

IOP Conf. Series: Earth and Environmental Science **55** (2017) 012009 doi:10.1088/1755-1315/55/1/012009

3. Results and Discussion

3.1 The Characterization of Protein Spores with SDS-PAGE

Identification of the results showed that the morphology of spores observed in accordance with the identification of Alvin and Matt [10] is spore M.koi. Spores are elliptical-shaped or Cone, had two twin polar capsules located in the anterior portion, pinkish white nodules, irregular round nodules form and located in the gills. While the results of the counting of the number of spores collected from carp (Cyprinus carpio) is 8 x 10^8 spores. The morphology of the spores found in the goldfish can be seen in Figure 1.

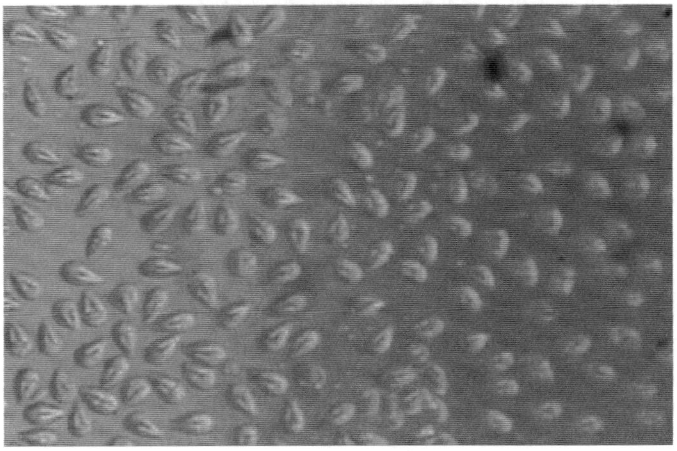

Figure 1. Myxobolus spores on Goldfish (400X Zoom)

3.2 Myxobolus Spores Protein by SDS-PAGE

Analysis results of Myxobolus koi spores protein with SDS-PAGE can be seen in Figure 2. that shows that 2 bands of soluble protein are found with molecular weight 150 kDa and 72 kDa and a Band of protein crude of Myxobolus koi with molecular weight 73 kDa.

3.3 Examination an Overview of White Blood Cells

Examination results of blood overview of Goldfish that is applied by Myxobolus koi spores protein are presented in table 1, which shows that the highest number of erythrosit occurred at the treatment K1 i.e. 88.21% and at lowest of at K2 i.e. 30,43%. While the highest number of leukocytes occurs at the K2 treatment i.e. 71.57% K2 and clearly different from treatment K3that is 63,08%.

Table 1. The results of the determination of the Blood of Goldfish exposed by Spore Proteins After 21 days of maintenance.

Treatments	Description of gold fish blood after 21 days of maintenance (%)	
	Erythrosit	Leukosit
control, not infected and not exposed by protein (K1)	88.21	11.79
Infected by Myxobolus and exposed with Spore Protein dose 600 µl/fish (K2)	28.43	71.57
Infected by Myxobolus and not exposed with Spore Protein dose 600 µl/fish (K3)	34.92	63.08
Not infected by Myxobolus and exposed with Spore Protein dose 600 µl/fish (K4)	59.63	41.23

Description: different superscript on the same column indicates the existence of the real difference ($p < 0.05$)

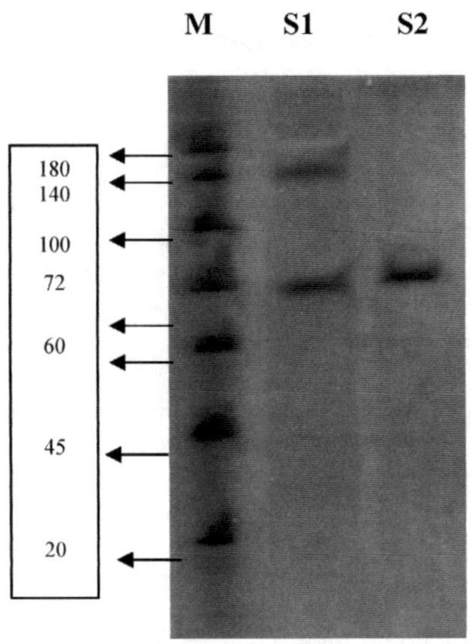

Figure 2. Profile of protein with SDS-PAGE method. M = marker 20-180 kDa, S1 = samples of whole protein, S2 = sample of solubel protein.

3.4 Determination of Differensial Leukocytes of Carps exposed by Protein Spores.

The results of the examination of white blood cells (leukocytes) can be seen in table 2, which indicates that the highest percentage of lymphocytes occurring on protein without the exposure of infection treatment Myxobolus koi (K2) 59.40%, then followed the treatment K3, K1 and K4 40, consecutive 58.20; 56.30 and 53,42%. The highest percentage of monocytes in the K2 's treatment of 21.20%, then followed in the treatment of 20,40% K3, K4 40 treatment of 19.23% and lowest occurred on treatment K1 of 13.20%. The highest percentage of neutrophils in the treatment of 20.20% K1 and K2 treatment occurs at the lowest (8, 14%) i.e. precisely on the treatments being with pprotein. While the highest percentage basophils occurs at the treatment i.e. K1 1.50% and lowest on K3 's treatment of 0.40%.

Table 2. The image of Blood white (Leukocytes) in Carp (*Cyprinus carpio*)

TREATMENT	KINDS OF LEUKOCYTES (%)				
	LIMPHOCYITE	MONOCYTES	NEUTROPHILS	EOSINOPHILS	BASOPHILS
K1	52.40	16.40	20.20	8.60	1.50
K2	59.40	21.20	8.14	10.55	0.71
K3	58.20	20.40	9.60	11.40	0.40
K4	54.00	19.23	14.30	11.26	1.21
NORMAL	**52.00**	**18.00**	**20.00**	**8.00**	**2.00**

3.5 Determination of Survival rate of Carp Exposed by Spore Protein.

Survival rateof koi fish is expressed in percentage and calculated on day fourteen presented in table 3. In the table 3. the test shows that in M. Koi spore protein protection test, there is a noticeable difference ($p < 0.05$) between the treatment of average survival rate between being and not being with spores of Myxobolus koi after 21 days of maintenance.

Table 3. The Results of The Determination of Survival Rate of Koi Fish Exposed by Spore Protein After 21 days of Maintenance

Treatment Group	Survival rate of Carps after imminized
Control, not infected and not exposed with Protein (K1)	$96.00^{a} \pm 2.24$
Infected by Myxobolus and exposed with Dipapar spore Protein doses 600 μl/fish (K2)	$86.00^{b} \pm 6.52$
Infected by Myxobolus and not exposed by spore Protein dose 600 μl/fish (K3)	$27.00^{c} \pm 3.54$
Not Infected by Myxobolus and exposed by spore Protein dose 600 μl/fish (K4)	$98.00^{a} \pm 2.94$

Description: different Superscript on the same column indicates the existence of the real difference ($p < 0.05$)

3.6 Examination of Water quality on maintenance media

The results of examination of water quality of maintenance of water quality parameters: temperature, pH, dissolved oxygen, nitrite, nitrate and ammonia can be seen in Table 4.

Table 4. The results of examination of water quality of maintenance during studies

No.	Parameter	Treatment			
		K1	K2	K3	K4
1.	Temperature (Celsius degree)	30	30	30	30
2.	Dissolved Oxygen (ppm)	5	5	4.0	4.0
3.	pH	8	8	8	7.5
4.	NH3	0.2	0.2	0.3	0.3
5.	NO2	0.3	0.3	0.4	0.4

Table 4 indicates that the water quality parameters during maintenance in optimal conditions for the maintenance of the fish, although in media exposure with the maintenance of the protein and the presence of infection with myxobolus koi showed a decline, but still in the normal range.

The results of the analysis of the whole protein spores M. koi used SDS-PAGE 2 band which seem obvious, namely protein with molecular weight (BM) 150 kDa and 72 kDa and one band of crude protein Myxobolus koi 73 kDa molecular weight .The results showed that proteins found have a high molecular weight. Other studies conducted by Chavda et al. (2010) shows the results of the SDS-PAGE electrophoresis analysis of Myxobolus cerebralis spores obtained expressions band protein with molecular weight 130 kDa and 60 kDa as well as four bands other protein with BM 7 kDa to 45 kDa. Protein molecular weight differences on both the research allegedly because of differences of origin, preparation techniques and sampling technique of insulation used. immunogenic proteins are proteins that have an molecular weight between 20,000-100,000 Dalton. A molecule has munogen properties when its weight is more than 10 kDa. Based on these statements, whole spore protein molecular weight of Myxobolus koi of the analysis result by SDS-PAGE BM has more than 10 kDa, but to ascertain the nature of immunogenic, it needs to do test again with imunoblotting so that potential of immunogenic properties of whole spore proteins of Myxobolus koi can be utilized as a candidate of vaccine in tackling myxobolusis.

Fish have a defense system of the body to fight various diseases. Fish immune defences consists of non-specific and specific defences. Non-specific defenses consist of skin, scales and mucus which is the body leading the defense in the face of the onslaught of various microorganisms that enter. Specific defences on a fish that comprise the macrophage cells, Natural Killer cells and leukocytes. Specific defence system takes time to identify antigens before able to give a response [11].

According to Guyton [12], leukocytes or white blood cells that make up the immune system is the most active units because it acts against various infectious diseases and foreign materials. Leukocytes are found in the bone marrow (mieloid tissue) and partly on the lymph tissue, then remained stored in the bone marrow and will increase in number when required. Leukocytes of fish composes of granulocytes (neutrophils, eosinophils, and basophils) and agranulosit (lymphocytes and monocytes).

Changes in the total number of and types of leukocytes can be used as an indicator of the presence of certain infectious diseases that occur in fish. Leukocytes in blood components is one that serves as a specific body defenses that will neutralize and destroy pathogens through phagocytosis. Lymphocytes is one form of leukocytes. The observations of lymphocytes in the table 1 showed the highest percentage in the treatment of K2 (vaccinated by whole spore protein M. koi and not infected by 80 spores/fish of M. koi) of different real 59.40% with treatment K3 (exposed by whole spore protein of M. koi and infected by M. koi) and K1 (control). but the treatment did not differ markedly with K4 treatment (without exposed with whole spore protein of M. koi and infection with 80 spores/fish of M. koi). Normal lymphocyte percentage of the fish ranges between 60-70% [13].

An increased number of lymphocytes in vaccinated and infected fish with spores of M. koi is the body's defense system response of fish over the entry of pathogens. This is in accordance with the opinion of Bastiawan et. al. [14] lymphocytes functions as antibody producer for immune to interference from disease. Lymphocyte cells essentially consists of two populations i.e. B cells and T cells. B cells have the ability to transform into plasma cells i.e. cells which produce antibodies and memory cells. T cell has important role for immune (cytotoxic T cells) and control the immune response (suppressor T cells). After the binding of an antigen with antigen receptor cell lymphocytes, lymphocyte cells will then divide and differentiate into effector cells and memory cells.

Monocytes of fish are round oval, located in the middle of the nucleus and the cytoplasm has no bergranula. There are monocytes in blood circulation and in the number of little limfonodus, embedded in bone marrow and lymph. Monocytes migrate from blood circulation towards to the tissue when it receives the appropriate stimulation with its reseptor. Immature monocytes leave the blood circulation, to and settled in the tissue, and then develop into macrophages (Feldman et al., 2000). The observations of each treatment are shown on monocytes Table 2. The highest percentage of monocytes in treatment (K2) of 21.20% followed by treatment of K4 and K3. An increase in the number of monocytes in treatment of K2, K4 and K3 when compared with controls.

This increase is due to the monocytes fagosits spores M. koi in the body of the fish. This is in accordance with the opinion of Ardelli and Woo [15] stating that the monocyte with macrophage will fagosit along with disease-causing agents that enter the body. Bastiawan et. al. [14] explains, the monocytes function as phagocytes against foreign objects which act as the agent of the disease. Further Kresno (2001) argued the role of monocytes is very important, as the cells phagocytes to destroy a variety of main pathogens and acts as an antigen presenting cells (APC) that function to present antigen to the lymphocytes cells.

Neutrophils in fish is formed in kidney organ in the front. The highest percentage of neutrophils in the treatment of the K1 (control) of different real 21.20% with treatment of the K2, K3 and K4 as shown in Table 2. A decrease in the percentage of neutrophils in the treatment of the K2, K3 and K4 when compared to control due to the increase in lymphocytes and monocytes neutrophils so decreased. Affandi and Tang [13] suggests that the percentage of neutrophils in the blood of fish ranged between 6-8% of total leukocytes. In addition neutrophils do not too plays a role in responding to an infection caused by a parasite. More neutrophils play a role in the infections caused by bacteria.

The observations of the number of eosinophils showed the highest percentage in treatment of K3 11.40% that tends to be decreased in treatment of K2, K1 and K3. Spore infection of M. koi increases the number of eosinophils in the blood of fish. This is in accordance with the opinion of the Tizard [16] stating that eosinophils is one of the body's defense cells that are dominant in the blood and will increase sharply in number in case of an infection of parasitic diseases. Eosinophils formed on haematopoiesis process that occurs in the bone marrow before migrating into blood circulation. They contain a number of chemical substances include histamine, eosinophils peroxidase, Ribonuclease, deoksiribonuklease, lipase, and some amino acids are produced through a process of degranulation after they are activated. These substances are toxins against parasites and body tissues.

Observation of basophils on treatment indicated a small percentage compared to the other types of leukocytes. The highest percentage basophils in treatment of 1.5% K1 also tend to decline treatment K3, K2 and K3. Affandi and Tang [13] stated that the percentage basophils in the blood of fish ranged from 0.17-0.19% and measuring 8-12 μm. The existence of basophils in blood circulation has been observed only in a small number of species of fish. Even more rarely basophils are found on examination of blood compared to eosinophils.

4 Conclusion

Catfish (Clarias gariepinus) cultured in Moro Krembangan Surabaya, East Java, are mostly infected with Aeromonas hydrophila and *Saprolegnia* sp., It is caused by the environmental conditions that are unfavorable. The percentage of catfish (Clarias gariepinus) infected with *Aeromonas hydrophila* is 95% and infected with *Saprolegnia* sp. is 90%. The conclusions that can be presented from this research is that there are 5 profiles of spore protein bands of Myxobolus sp that are obtained spores with molecular weight. Whole spore proteins of Myxobolus koi is found with molecular weight 150 kDa and 72 kDa. This spore protein of Myxobolus koi can enhance the immune response and survival rate of koi fish from 10% up to 86% and can increase the body's defenses against infection of Myxobolus koi , so that it could be developed as immunostimulant.

Acknowledgement

Financial support for this research was provided by Ministry of Research Technology and Hightly Academic National. The authors wish to thank for Prof.H. Hery Purnobasuki, Drs.,M.Si.,Ph.D. Head of Research and Inovation Institute, Airlangga University.

References

[1]. Anugrahi, T. 2010. Pemanfaatan Daun Ketepeng Cina (Cassia alata) Untuk Pengobatan Myxobolus pada Ikan Mas (*Cyprinus capio*). Fakultas Perikanan dan Kelautan, Universitas Airlangga, Surabaya.
[2]. Itabashi T, Terasaki T and Asai H, 2004. Novel Nuclear and Cytoplasmic Proteins Detected by Anti-*Zoothamnium arbuscula* (Protozoa) Spasmin 1 Antibody In Mammalian Cells Are Dependent on the Cell Cycle. *Dalam* : G. Mahasri. Respon Imun Udang Windu (*Penaeus monodon* Fabricus) yang Diimunisasi Dengan Protein Membran Imunogenik MP 38 dari *Zoothamnium penaei*. Surabaya. Hal 23.
[3]. Mahasri, G. 2008. Ilmu Penyakit Ikan Protozoa Pada Ikan dan Udang.Surabaya. p. 25-29.
[4]. Chavda, D., 2010. Pathogenicity of Myxobolus Infection and Its Effect on Protein Expression in Catla Catla in Central Gujarat Region. P. 1-11. *http://faqs.org*
[5]. Asri, A. A. P. 2012. Isolasi Protein Spora *Myxobolus* pada Nodul Insang Ikan Koi (*Cyprinus carpio*) dengan Metode Elektroforesis SDS-PAGE. Skripsi. Universitas Airlangga. Surabaya. Hal 28.
[6]. Insariani, K. Puspasari, Haririyah, E. Widiyastuti, C. Chotimah dan H. Wuryastuti. 2012. Isolasi, Identifikasi dan Karakterisasi Glikoprotein Permukaan *Myxobolus koi*. Jurnal Sain Veteriner. Fakultas Kedokteran Hewan. Universitas Gadjah Mada. Hal 7.
[7]. Anderson, D. P. 1997. Adjuvant and Immunostimulants For Enhancing Vaccine Potency In Fish. 257-256 p.
[8]. Abbas, Maswan, N. A. Pengujian Efektivitas Dosis Vaksin Dna dan Korelasinya Terhadap Parameter Hematologi Secara Kuantitatif. 2009. Skripsi. Teknologi dan Manajemen Akuakultur. Fakultas Perikanan dan Ilmu kelautan. Institut Pertanian. Bogor. Hal 33.

[9]. Rantam. 2003. Metode Imunologi. Dalam : A. Fajriah. Karakterisasi Protein Lernea sp. Dengan menggunakan metode SDS-PAGE. Surabaya. 2009.

[10]. Alvin, C. C. And J. G. Matt. 2010. Molecular Characterization an dan Histopathology of My*xobolus koi* Infecting the Gills of Koi, *Cyprinus carpio*, With an Amended Morphological Description of the Agent. Journal of Parasitology, 96(1):116-124

[11]. Purwanto A. 2006. Gambaran Darah Ikan Mas Cyprinus carpio Linn yang Terinfeksi Koi Herpes Virus. Skripsi. Fakultas Perikanan dan Ilmu Kelautan, Institut Pertanian Bogor. Hal 42-42.

[12]. Guyton Arthur C. and Hall John E. 1999. *Buku Ajar Fisiologi Kedokteran.* Setiawan I, Tengadi KA, Santoso A, penerjemah: Setiawan I, editor. Jakarta : EGC. Terjemahan dari : Textbook of Medical Physiology. Hal 25-27.

[13]. Affandi R dan Tang UM. 2002. *Fisiologi Hewan Air.* Unri Press. Pekanbaru. 38 - 40.

[14]. Bastiawan D, A. Wahid, M. Alifudin dan I. Agustiawan. 2001. Gambaran Darah Lele dumbo *(Clarias spp*) yang Diinfeksi Cendawan *Aphanomyces* sp pada pH yang Berbeda. *Jurnal Penelitian Indonesia* 7(3) : 44-47.

[15]. Ardelli BF and Woo PTK. 2006. Immunocompetent Cells and Their Mediators in Fin Fish. Vol 3. Ed ke-2. UK : CABI Pub. 702-724p.

[16]. Tizard I. 1988. *Pengantar Imunologi Veteriner.* Ed ke-2. Partodirejo M, Hardjosworo S, penerjemah; Surabaya: Airlangga University Press. Terjemahan dari: *An Introduction to Veterinary Immunology.*

The Effectiveness of Extracts Basil Leaves (*Ocimum sanctum* Linn) against *Saprolegnia* sp. by in Vitro

Sudarno[1], Muhammad Luthfi Hakim[2] and Rahayu Kusdarwati[3]

[1]Department of Fish Health Management and Aquaculture Faculty of Fisheries and Marine, Airlangga University- 031-5914151,
[2] Study Program Aquaculture Faculty of Fisheries and Marine, Airlangga University
[3]Department of Fish Health Management and Aquaculture, Faculty of Fisheries and Marine, Airlangga University

darnosudarno1@gmail.com

Abstract. *Saprolegnia* SP. is a fungi which is opportunistic and generally as a secondary pathogen on fish. *Saprolegnia* sp. infects epidermis tissue that begins at the head or fins and can spread over the entire surface of the body. The result of the using of chemicals to control infections of *Saprolegnia* spp. can cause pollution of the environment and harm the consumer. The purpose of this research was to determine the potential and the minimum concentration of extracts basil leaves (*Ocimum sanctum* Linn) as antifungi against the growth of *Saprolegnia* sp. by vitro. The research was held in Fish Quarantine Kelas I Juanda Suarabaya in January 2015. A positive result was obtained in the test of the effectiveness of basil leaves in inhibiting the growth of the fungus *Saprolegnia* sp. Concentration of the extract given to treatment 90% and 100% was able to inhibit the growth of *Saprolegnia* sp., indicated by the formation of the inhibitory zones at a concentration of treatment, and had the best results on the concentration of 100%.

1. Introduction

Fish disease is one of the serious problems that must be faced in the aquaculture business. One of the diseases that attack the fish cultivation is caused by fungus, *Saprolegnia* sp. [1]. This fungal infection can be triggered by several factors, one of which is the high fish density. Saprolegniasis is the disease in fish and fish eggs caused by *Saprolegnia* SP. or watermolds. The infected fish show the presence of white patches on the body and there is a bloody wound observed on the gills and skin. These fungi attack may cause death in fish eggs or fish itself which is significantly harmful to the survival of the fish farming business. Many farmers and businessmen fish use a variety of chemicals such as formaldehyde or antibiotics in fish disease control. However, the use of chemicals and antibiotics is continuous with the dose or concentration that is not appropriate. It will cause a problem recently in the form of increased resistance to microorganisms against these materials. In addition, another issue is the danger posed to the environment surrounding it, the fish is concerned, and human consumption [2]. Therefore, it is an alternative way to control the disease. The use of medicinal plants is a safe way to inhibit microbial growth and to kill as well as to be environmentally friendly. One of them is by using plant basil leaves (*Ocimum sanctum* Linn). The part that can be used is part of the leaf. Basil leaves can be used as antifungi because it contains essential oils with a population amounting to 0.8 mg/100 g. The content of essential oil compounds in basil leaves as methyl chavicol antifungi and linalool which reacts with the cell membrane and a significant reduction in the amount of ergosterol. Ergosterol is a vital component of fungal cell membranes that keep [3]. Other compounds in the basil leaves are alkaloids, tannins, flavonoids and saponins [4]. On the research that has been done from extracts of essential oils essential oil, kemangi leaves have antifungi activity against *Fusarium solani*, *Penicillium funicolusum*, *Trichorderma reesi,* and *Rhizomucor auricus* [5], and *Aspergillus fumigatus* and *Aspergillus niger* [6]. Based on the background, this research aims to discover the influence of extracts of basil leaves (Ocimum sanctum Linn) against the growth of the fungus *Saprolegnia* sp. in vitro.

2. Materials and Methods

The research was conducted in the laboratory of Fish Quarantine Kelas I Juanda Surabaya in January 2015.

2.1. Equipments and materials

2.1.1. Equipments

Equipment needed in this study are reaction tubes, racks of reaction tubes, petri dish, bunsen burners, ose, microscopes, scalpel, glass objects, cover glass, autoclave, hot plate, paper discs, a micropipette, pipette drops, mixer, erlenmeyer flask, measuring cup, haemocytometer, digital scales and Rotary vacum evaporator.

2.1.2. Materials

Materials required are pure *Saprolegnia* sp. isolates obtained from the laboratory of Fish Quarantine Kelas I Juanda Surabaya, basil leaves (*Ocimum sanctum* L.), aquades, ethanol 96%, Formalin, Dimethyl sulfoxide (DMSO) 10% , Sabouraud Dextrose Agar (SDA), and antibiotics.

2.2. Media Preparation

Use of the media is very important for the isolation, identification or differentiation to get a suitable environment for the growth of fungi. Erlenmeyer flask is placed in an autoclave along with petri dish at a temperature of 121 ° C 1 atm pressure for 15 minutes. The SDA materials in liquid state included as many as 20 ml in each Petri dish.

2.3. Solvent extracts of basil leaves

Basil leaves extracts is diluted with the solvent Dimethyl sulfoxide (DMSO). The solvent DMSO used is concentrations of 10%, to make the solution 10 ml of DMSO is required 100% added to 90 ml of aquades [7]. Basil leaves is dried by Sun light for 3 days and mashed into powder of basil. Basil powder as much as 500 grams is macerated using ethanol 96% over the 3 x 24 hours in room temperature. Maceration is a way of extracting the most simple because the materials would be extracted simply dissolved in the solvent. Besides solvent used in the study is 96% ethanol. especially those that have antifungal properties. The solution obtained is then filtered using filter paper and then evaporated using a Rotary evaporator vacum. The resulting extract is then diluted in accordance with the concentration 100% i.e. 5 ml ingredient extracts of basil, 90% i.e. 4.5 ml ingredients plus 0.5 ml of DMSO 10%, 80% i.e. 4 ml ingredients plus 1 ml of DMSO, 70% that is 3.5 ml ingredients coupled with 1.5 ml of DMSO, 60% that is 3 ml ingredients coupled with 2 ml of DMSO, 50% i.e. 2.5 ml ingredients coupled with 2.5 ml of DMSO, 40% IE 2 ml ingredients coupled with 3 ml of DMSO, 30% i.e. 1.5 ml 3.5 ml plus material of DMSO, 20% i.e. 1 ml ingredients coupled with 4 ml of DMSO, 10% i.e. 0.5 ml ingredients coupled with 4.5 ml of DMSO, control the positive i.e. 0.2 ml formalin and control the negative i.e. 5 ml DMSO.

2.4. Culture and identification of *Saprolegnia* sp.

Saprolegnia sp. obtained from laboratory stocks of Fish Quarantine Kelas I Juanda Surabaya reproduced by culturing mildew on SDA media. The process begins with the way of inoculating mushroom on one SDA media made by using a scalpel in aseptis.

The purification process begins by taking one type of colony using a scalpel on the old SDA media which has a similar color and texture then inoculated on SDA new media and incubated at temperatures of 25° C for 2-7 days to obtain pure isolates. After done, culture media of water for easy observation. Start with cutting edge of jelly that has a colony of mold approximately 0.5-1.0 cm using a scalpel blade set on fire and then move jelly piece to aquades in a petri dish and incubate for several days at temperatures of 25° C [8].

2.5. Yeast Suspension

Isolate pure fungi *Saprolegnia* SP. which has been cultured on the media of jelly to be moved in a tube.Tto get the suspense of mushrooms, spores calculation is done using haemocytometer. The density of spores at a minimum to be used is around 10^5 sel/ml [8].

2.6. The Effectiveness of Antifungal

Testing of the effectiveness of the disc diffusion method used antifungal. Testing is conducted to find out the minimum concentration of a solution of an antifungal that may inhibit the growth of mold. Paper disc diffusion method begins by soaking paper discs that will be used during CA. 10 minutes into the basil leaf extract with a concentration of 10%, 20%, 30%, 40%, 50%, 60%, 70%, 80%, 90% and 100% with a size paper discs of 6 mm, media control (-) is given a 10% DMSO. Then the study used the concentration of 0.4% in controls (+). Incubation is done at a temperature of 25 ° C for 48 hours. The magnitude of the clear zones produced antifungal potential comparable to that produced by the active substances contained in materials and measured using a ruler.

2.7. Data Analysis

Data research results are in the form of drag zone diameter growth of mold *Saprolegnia* sp. by basil leaves extracts antifungi (*Ocimum sanctum* Linn) and formalin 0.4%. The diameter of the zones of drag is the diameter of the area which is not covered by fungus around the paper discs with a diameter reduced paper discs.

3. Result and Discussion

3.1. Identification of *Saprolegnia* sp.

Identification process is done to ensure the truth against fungal isolates obtained from Fish Quarantine Hall class I Surabaya I that the fungus is a fungus *Saprolegnia* SP. Identification can be done by looking at the macroscopic and microscopically [9].

Identification of the fungus microscopically can be done by looking at the morphology of *Saprolegnia* sp. i.e. aseptat and Hypha, has tube-shaped, with a system of branching in sporangial (internal proliferation), and has two forms of zoospore (*dimorphic*), namely primary and secondary zoospore. Primay zoospore is pyriform-shaped with no flagellate. Secondary zoospore is cysts-shaped (reniform) like nuts with flagella and swim more strongly than primary zoospore. The characterisitics of the suit stated by [10].

Macroscopicly *Saprolegnia sp.* can be seen with the characteristics of a colony of white-yellowish with forms such as cotton. The growth of a colony is very fast (optimum at 15-30 ° C) [11]. Observation of results of test extract potential basil leaves against the growth of *Saprolegnia* SP. is done by looking at the onset of drag that results from the zone mechanism of antifungi.

3.2. The Minimum Inhibitory Concentration (MIC)

Observation of results of test extract potential kemangi leaves against the growth of *Saprolegnia* SP. is done by looking at the onset of drag that results from the zone mechanism of antifungi in inhibiting the growth of fungus. The results obtained in this study demonstrate positive results because it occurs at concentrations inhibitory zones given. Basil leaf extract that is produced is able to inhibit the growth of *Saprolegnia* sp. It was indicated by the formation of the inhibitory zones at a concentration of 90% and 100% control in conjunction with the positive (+) in the form of a solution of formalin 0.4% which is able to inhibit the *Saprolegnia* sp. by 1 cm.

Table 1. The results of observations of the formation of the inhibitory zones basil leaves extracts against *Saprolegnia* sp.

No.	Concentration basil leaves extracts (%)	Inhibitory zone diameter (cm)
1.	10	0,0
2.	20	0,0
3.	30	0,0
4.	40	0,0
5.	50	0,0
6.	60	0,0
7.	70	0,0
8.	80	0,0
9.	90	0,8
10.	100	1,7
11.	K (-)	0,0
12.	K (+)	1,0

The results of the assay activity antifungi extracts of basil leaves (Ocimum sanctum Linn) against the fungus *Saprolegnia* sp. has been carried out in vitro. Antifungi activity test includes testing the Minimum Inhibitory Concentration (MIC) concentration of 10% up to 100% concentration. The method used is the method of diffusion paper discs that show Basil leaf extract has the ability to inhibit the growth of mold *Saprolegnia* sp. This can be seen with the formation of the zone around the mushroom paper drag discs, these results correspond to the positive (+) control as a comparison which is also visible for the presence of inhibitory zones around the paper discs.

The difference between the concentration of inhibitory power extracts are inoculated with the fungus and results positive control then, approaching the concentration of extract, basil leaves are considered capable of inhibiting the growth of *Saprolegnia* sp. [7]. The greater the concentration of the extract of the basil leaves, the greater the drag the extract power anyway against the growth of *Saprolegnia* sp. [12] state that the higher the concentration of extract, the bigger the death or inhibition of the growth of a fungus.

From the pictures the minimum inhibitory concentration test results show that the growth of *Saprolegnia* sp. colonies. Is white like cotton and dominions of the round. This is in accordance with [10] that the colony of *Saprolegnia* sp. is white with yellowish surface such as cotton, round and prominent. The growth of colonies of different sizes is due to the difference in concentration of the basil leaves extract that can inhibit the growth of *Saprolegnia* sp. the greater the concentration of extract the basil leaves, the smaller the size of the colony of *Saprolegnia* sp. or not even growing colonies.

Not growing colonies of fungus on the positive control containing 0, 2 ml formalin 10% proves that formaldehyde has the ability in inhibitting and killing the fungus. Formaldehyde (formalin) is highly reactive chemicals that interact with proteins, DNA, and RNA in vitro. Formaldehyde was chosen because according to [13] that formaldehyde is a versatile antimicrobial compounds, one of which can kill fungus. Formaldehyde is considered sporicidal based on its ability to penetrate into the inside of the mold spores. Formaldehyde also reacts extensively with nucleic acids [14].

While the growth of a colony of mold on negative control containing 5 ml of DMSO 10% indicates that the solvent DMSO does not affect the growth of mold. This is in accordance with [15], States that DMSO is not toxic and carcinogenic at concentrations of 5%-10% so that DMSO does not enter reacts with microbial testing.

Basil leaves were chosen because it has the active ingredient as antifungi. Content of active ingredients in basil leaves already examined to be able to fight fungi is essential oil, flavonoids, and saponins. On the research that has been done from extracts of essential oil of basil leaves, exposing the antifungi activity against *Fusarium solani*, *Penicillium funicolusum*, *Trichorderma reesi* and *Rhizomucor auricus* [5]. The active compounds that can inhibit and kill *Saprolegnia* sp. obtained from results of extraction using ethanol. Essential oils in basil leaves most of the lot contains ethyl p-metoksisinamat (EPMS) are generally insoluble in solvents of ethanol, ethyl acetate, methanol and heksan [16].

Flavonoid is compound of polar so that flavonoids are soluble in polar solvents such as ethanol, methanol, acetone, dimethyl sulfoksida (DMSO) and dimethyl fonfamida (DMF) [17]. According to [19] it is a secondary metabolite compound found in plants. Flavonoids found in plants are flavones and a flavonol. Besides flavonoids have compound genestein function to inhibit the cleavage of cell proliferation or mushrooms. These compounds bind to proteins on microtubules in the cell and disrupts the function of mitosis giving rise to inhibition of fungal growth. Influence of phenol compounds as an antifungal protein bonding denature is with the membranes of the cell so that the cell membrane Lysis and allowed the phenol to penetrate into the cell nucleus that cause fungi does not develop [18]. The Saponins are a work mechanism with antifungi that can form a complex with sterol that is lowering the surface tension of the membrane permeability of the cell membrane so that the sterol molds and yeasts [19]. This result in the cell walls becomes permeable and fungal cell structure is destroyed due to the antifungal compounds inhibiting ergosterol biosynthesis of sterols which is to maintain the integrity of the cell membrane of yeast [20].

4. Conclusion

Based on the results of this research, it can be concluded that solution of extracts of basil leaves (Ocimum sanctum Linn) has the potential of antifungi against *Saprolegnia* sp. Concentration of 90% (0.9 g/ml) with zone drag 0.8 cm and 100% (10 g/ml) with zone drag 1.7 cm from extracts of basil leaves is able to inhibit the growth of *Saprolegnia* sp. while formalin 0.4% as control (+) is able to inhibit the inhibitory zones with a 1 cm.

References

[1] Subandi. 2010. Mikrobiologi. PT Remaja Rosdakarya. Bandung.
[2] Sugianti, B. 2005. Pemanfaatan Tumbuhan Obat Tradisional dalam Pengendalian Penyakit Ikan. PPS-702 Institut Pertanian Bogor.
[3] Kadian, R. and M. Parle. 2012. Therapeutic potential and phytopharmacology of tulsi. International Journal of Pharmacy & Life Sciences, 3(7): 1860.
[4] Shafqatullah, M. Khurram, Asadullah, Khaliqurrehman and F. A. Khan. 2013. Comparative Analyses of Ocimum santum Stem and Leaves for Phytochemicals and Inorganic Constituents. Middle-East Journal of Scientific Research 13 (2): 236-240.
[5] Dharmagadda, V. S. S., M. Tandonb, and P. Vasudevan. 2005. Biocidal activity of the essential oils of Lantana

camara, Ocimum sanctum and Tagetes patula. Journal of Scientific & Industrial Research. 64 : 53-56.

[6] Bansod, S. and M. Rai. 2008. Antifungal Activity of Essential Oils from Indian Medicinal Plants Against Human Pathogenic Aspergillus fumigatus and A. Niger. World Journal of Medical Sciences 3 (2) : 81 – 88.

[7] Assidqi, K. 2012. Potensi Ekstrak Daun Patikan Kebo (*Euphorbia hirta*) Sebagai Antibakteri Terhadap Aeromonas hydrophila Secara In Vitro. Skripsi. Universitas Airlangga. Surabaya.

[8] Yuasa K., Novita P., Meliya P., dan Edy B.K, 2003. Teknik Diagnosis Penyakit Ikan Budidaya Air Tawar di Indonesia. Japan International Cooperation Agency dan Panduan Diagnosis Penyakit Ikan. Balai Budidaya Air Tawar Jambi. Jambi. Hal. 52-53.

[9] Ilyas, M. 2007. Isolasi dan Identifikasi Mikoflora Kapang pada Sampel Serasah Daun Tumbuhan di Kawasan Gunung Lawu, Surakarta, Jawa Tengah. Biodiversitas, 8 (2): 105-110

[10] Irianto, A. 2005. Patologi Ikan Teleostei. UGM Press. Yogyakarta.

[11] Neish, G.A., and Hughes, G.C. 1980. Diseases of fishes, Book 6, Fungal Diseases of Fishes. T.W.F. Publications, Neptune, New Jersey. 159 p.

[12] Kosenda, L. 2009. Penampisan Senyawa Bioaktif dari Daun Jambu Biji (*Psidium guajava L.*) sebagai Anticandidiasis dan Profil Kromatografi Lapis Tipisnya. Skripsi. Fakultas Kedokteran dan Ilmu-Ilmu Kesehatan. Purwokerto.

[13] Hastuti, S. 2010. Analisis Kualitatif dan Kuantitatif Formaldehid pada Ikan Asin di Madura. Bangkalan. AGROINTEK, 4 (2) : 132-137

[14] McDonnell, G *and* A. Denver Russell. 1999. Antiseptics and Disinfectants: Activity, Action, and Resistance. Clinical Microbiology Reviews. pp. 153.

[15] Akoes, H., William, R.W., Bela, U., Kent, M.W., Jan, C.D and Terrence, R.T. 2005. The Relationship of The Cryoprotectant Methanol and Dimethylsulfoxid and Hyperosmotic Extenders on Sperm Cryopservation of Two North- American Sturgeon Species. Aquaqulture, pp. 245-251

[16] Barus, R. 2009. Amidasi Etil P-Metoksisinamat yang diisolasi Dari Kencur (*Kaempferia galangal Linn*). Tesis. Universitas Sumatera Utara. Medan.

[17] Markham K.R. 1988. Cara Mengindentifikasi Flavonoid. Terjemahan Kosasih Padmawinata. ITB: Bandung

[18] Sulistyawati, D. dan Mulyati, S. 2009. Uji Aktivitas Antijamur Infusa Daun Jambu Mete (*Ana cardium occidentale*, L.) terhadap *Candida albicans*. Biomedika. 2(1): 47-51.

[19] Wulandari, A.R. 2012. Uji Daya Efektifitas Antifungi Ekstrak Biji Tanjung (*Mimusops elengi Linn.*) Terhadap Pertumbuhan *Candida albicans* Secara In Vitro Dengan Metode Difusi. Skripsi. Fakultas Kedokteran. Universitas Pembangunan Nasional "VETERAN". Jakarta.

[20] Brennan B. and Leyden JJ. 1997. Overview of Topical Theraphy for Common Supperficial Fungal Infection and The Role of New Topical Agents. Journal of American Academy of Dermatology Part 1. Volume 36 : 2.

Developing groundwater conservation zone of unconfined aquifer in Semarang, Indonesia

T T Putranto[1], W K Hidajat[1], N Susanto[2]

[1]Geological Engineering Department, Faculty of Engineering, Diponegoro University, Jl Prof. Soedharto SH, Tembalang Semarang 50275 Indonesia
[2]Industrial Engineering Department, Faculty of Engineering, Diponegoro University, Jl Prof. Soedharto SH, Tembalang Semarang 50275 Indonesia

E-mail: putranto@ft.undip.ac.id

Abstract. The increasing exploitation of groundwater affects some environmental problems such as declining groundwater levels and its quality, land subsidence, seawater intrusion, and flooding in Semarang. Nowadays, groundwater protection and its management programs are crucial in Semarang due to raising several environmental problems. Semarang has two different morphologies i.e. lowlands, and coastal plains are in the north while hilly areas spread in the south. The main objective of this research is delineating groundwater conservation zone based on its current status of groundwater both in quantity and quality. The hydrogeology mapping was performed by measuring water table as well as physical parameters of groundwater samples from both dug wells and springs. Results conduct that the groundwater depth is approximately 0.03 m to 34 m below ground surface and flowing from the south to the north. The pH varies between 4.63 and 8.95. The Dissolved Oxygen is around 0.06-3.37 mg/L while salinity has yielded up to 3,600 mg/L in the north of Semarang. The EC value is in a range from 39-6,370 μS/cm. Based on lowering the water table, deterioration of groundwater quality, and land subsidence there are four groundwater conservation zones, i.e. secure, vulnerable, critical, and damaged zones.

1. Introduction

The issue of urban groundwater management is growing significance for groundwater-dependent cities due to its urbanisation has a major impact on aquifer beneath cities [1]. One of the consideration transformations in Indonesia is about urbanization. The proportion of people living in urban areas exceeds 50% by 2010 and will reach up to 60% by 2025. Badan Pusat Statistik (BPS-Statistics Indonesia) [2] stated that the increasing population is unevenly distributed in Indonesia. Java Island is the most populated with an annual population growth around 0.37-1.9% in the years 2000-2010. The highest percentage of urban growth in Java Island occurs in the coastal cities.

Semarang, as an urban coastal city and a capital city of the Central Java Province in Indonesia, is facing some enviromental problems (e.g. declining groundwater level and its quality, land subsidence, seawater intrusion, and flooding) related to the groundwater used which is in response to population growth [3] – [5]. Statistic Center Bureau of Semarang City [6] reported the rate of population growth in the last five years (2010-2015) increased around 0.85%. Recently, it is up to 1.59 million people live in Semarang. According to Ditjen Cipta Karya [7], the standard of fresh water in Semarang as a metropolitan city is around 150 L/person/day. Hence, it needs around 239x10^6 L/person/ day or up to 87.3 million cubic meters per year (MCM/yr). However, the local water board (PDAM) can only provide 44.5 MCM/yr or around 50% of water needs [6]. Thus, people employ groundwater to fullfil their fresh water need. Groundwater plays a major role in the total of the water supply of much of in Semarang. One of the main aquifer system in Semarang is the unconfined aquifer [8] – [10]. Groundwater is exploited by numerous dug wells, mainly for domestic water supply in the unconfined aquifer. However, overexploitation and poor groundwater management contribute to disreputable groundwater depletion problems and deterioration in groundwater quality.

Given the steadily expanding groundwater dynamics in the urban coastal city, it is thoughtful to take an adjusting concept to groundwater resource management guided by developing groundwater conservation zone. The conservation zone is advanced by monitoring of groundwater levels and its quality trends. This should authorize the assessment of future prognoses and precede to more robust and sustainable solutions to municipal water supply [11]. Moreover, there may be a need to examine the potential loss of water supply access of dug wells in areas of very shallow water table. Thus, this study aims at delineating groundwater conservation zone of the unconfined aquifer by considering such criteria as lowering the water table, deterioration of groundwater quality, and land subsidence.

2. Hydrogeologic Setting

Semarang has two different morphologies, namely coastal plains and lowlands in the north, and hilly area in the south. The northern part, where the economic and trade centre such as an airport, bus and train stations, and harbour are located, is relatively flat to gentle with slopes ranging between 0 and 5° and altitude up to 30 m sea level datum (masl). Meanwhile, in the southern part, slopes reach up to 45°, and altitudes reach up to 440 masl [12]. The land use in the north is mainly residence with higher density than in the south, industrial zones, ponds, and transportation areas. Open space areas (such as plantation, forest, moor) and residence are the main land use in the south of Semarang.

According to the regional hydrogeological map of Semarang [13], which the study area is located in Semarang (a capital of Central Java Province, Indonesia), the occurrence and productivity of aquifer can be divide into three as shown in Figure 1. In the northern part, the aquifer is mainly flowing in the intergranular system. It consists of the extensive aquifer with higly to the productive aquifer. The main litology in this aquifer system is alluvial deposits. Meanwhile, in the southern part, aquifer mainly flows both through fissures, and interstices which extensive spreading and moderately productive, but locally also available productive aquifer. It is located in the hilly area where volcanics and sedimentary rocks presence. There are areas with poor productivity as well as a region without exploitable groundwater due to the presence of sedimentary rocks such as claystone and marl appear.

3. Methods

The integrated approach of the hydrogeological mapping, physical parameters assessment, and spatial analysis were the basis methods to achieve the main objective of this research. Measuring water table of the dug wells and its geographical position, and investigating surface lithology were performed in the hydrogeological mapping. There were three hundred and eighteen (318) measurements of groundwater table and physical parameters of water sample (pH, Dissolved Oxygen/DO, salinity, and Electrical Conductivity/EC). The values were then interpolated to visualize the groundwater level and its quality maps. Lowering water table, deterioration of groundwater quality, and land subsidence rates were the parameters to delineate groundwater conservation zone [14]. All parameters were depicted into raster images and overlaid into a groundwater conservation zone map by using spatial analysis in ArcGIS. There were four (4) zones namely secure, vulnerable, critical, and damaged zone (Table 1). The secure zone was represented by low of EC value (<1,000 µS/cm) and groundwater level decline up to 40%. Meanwhile, the vulnerable zone was depicted by a scale of EC value between 1,000-1,500 µS/cm and lowering groundwater level in the range 40-60%. Areas where the EC value up to 5,000 µS/cm and above with lowering groundwater level more than 60% including the subsidence rates >8 cm/yr were in the critical zone and damaged zone. In the damaged zone, heavy metals were a presence to enhance unfavourable areas.

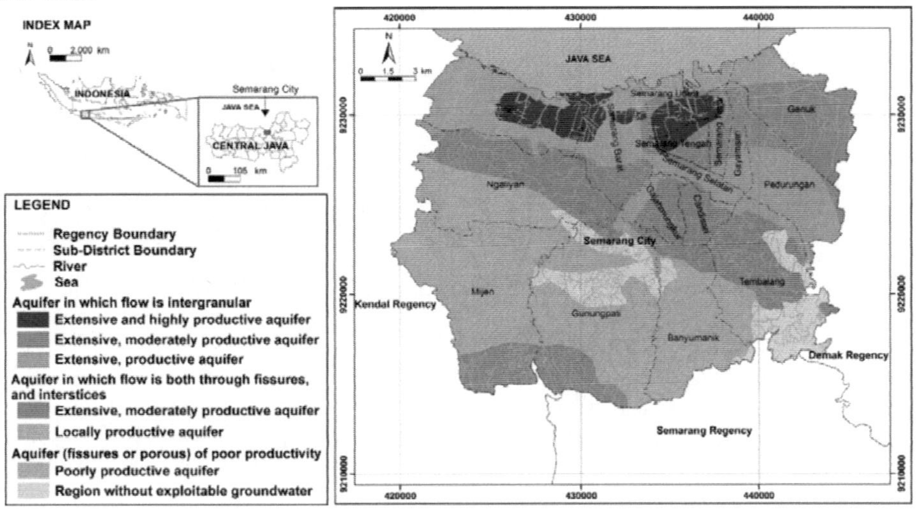

Figure 1. Hydrogeological map of Semarang.

Table 1. Matrix for delineating groundwater conservation zone

Groundwater Quality	Lowering groundwater level < 40%	40%-60%	60%-80%	> 80%
EC< 1,000 μS/cm	secure			
EC 1,000-1,500 μS/cm		vulnerable		
EC 1,500-5,000 μS/cm and subsidence			critical	
EC>5,000 μS/cm and the toxic heavy metal presence				damaged

4. Results and discussion

4.1 Groundwater level

Groundwater from the unconfined aquifer is abstracted by numeourus dug wells in Semarang. According to the hydrogelogical mapping, the groundwater depth is approximately 0.03 m to 34 m below ground surface and flows from the south to the north. The minimum depth is found in the north to the east of Semarang i.e. Genuk Sub-District, Semarang Utara Sub-District, and Semarang Timur Sub-District. Water table depth is below 2 m. Meanwhile, the water table depth in the centre to the south (Semarang Tengah Sub-District to Gunungpati Sub-District) varies between 2 and 34 m. The maximum depth is found in Kembangarum (Semarang Barat Sub-District). Groundwater levels in Semarang are around below 1 meter to 376 masl. To assess declining the groundwater level, the groundwater level in 2016 compare to the report of Direktorat Tata Lingkungan Geologi dan Kawasan Pertambangan [15] that provides water table measurement in 2003 (Figure 3). The declining groundwater level in the north and the east of Semarang is up to 15 m (80-85% of the lowering water table). Meanwhile, In the centre and the west, it is around 5-10 m drop of the water table. There is insignificance water table diffence in the south of Semarang. It means the groundwater used of the unconfined aquifer in these areas less than in the north and the east. Furthermore, it shows in the spreading of dug wells location which is very dense in the north and the east of Semarang (Figure 2).

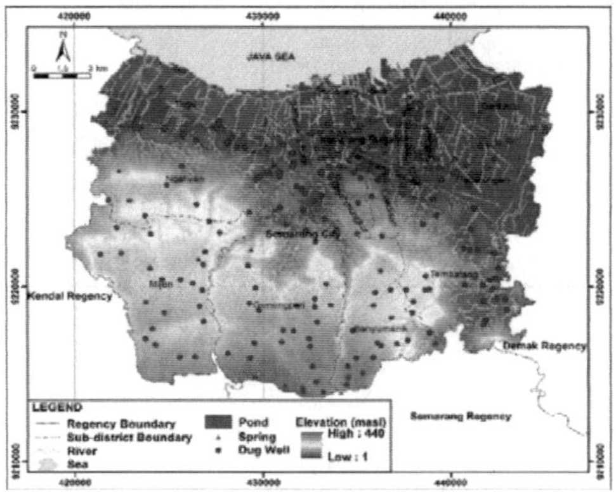

Figure 2. Location of groundwater samples. The elevation is retrieved from Digital Elevation Model [18].

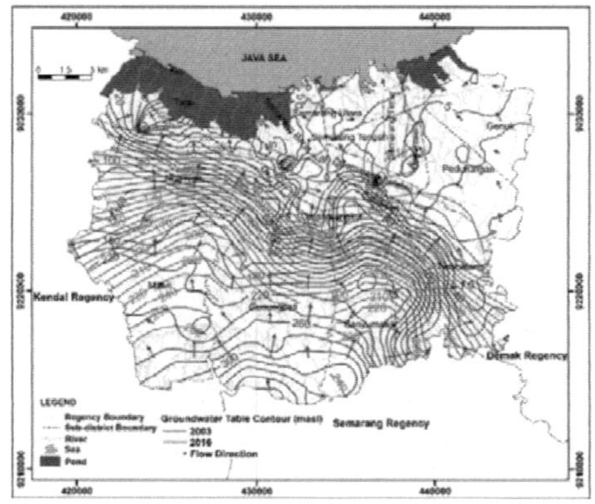

Figure 3. Groundwater level and its direction. Equipotensial lines in 2003 and 2016 are represented by green and purple respectively.

4.2 Groundwater quality

pH on groundwater samples provides a range from 4.88 to 8.9 (Figure 4). The minimum and maximum pH values are found in Mijen Sub-District and Gunungpati Sub-District respectively. Both areas are located in the highlands. Meanwhile, the pH values of the water in coastal plains and lowlands of Semarang are neutral around 6.5-8.2. As stated by WHO [16], the pH values are in desirable range for pH in water-drinking quality. Although pH usually has no direct impact on water consumers, it is one of the most fundamental operational water-quality parameters.

EC is a measure of water capacity to conduct electrical current. The EC value is varying from 39 to 6,370 μS/cm with an average of 855 μS/cm (Figure 5). The acceptable limit of EC in drinking water is recommended as 1,500 μS/cm [16]. The minimum and maximum of EC values are 39 and 1,313 μS/cm respectively. The EC value in the highlands of Semarang shows in an average of 406 μS/cm. Thus, groundwater in this region is permissible for drinking water. Meanwhile, the EC has a range 274-6,370 μS/cm with the average of 1,154 μS/cm in the coastal plains and lowlands. The highest of EC is found in the northeast of Semarang. High EC value in that area indicates the enrichment in groundwater. The EC value also represents the index of the total content of dissolved solids in water. It can be classified as type I, if the EC<1,500 μS/cm indicates the enrichments of salts are low; type II if EC 1,500-3,000 μS/cm represent the enrichments of salts are medium, and type III if EC above 3,000 μS/cm expresses the enrichments of salts are high [17]. According to with EC classification above, 84% of samples in the coastal plains and lowlands are in type I (low enrichments of salts), 13% of type II, and around 3% samples are type III. The latest is mainly located in the north (Genuk Sub-District), and the east (Semarang Timur Sub-District). The effect of seawater intrusion leads the medium to high level of the enrichments of salts in those regions.

Accordance with EC, groundwater samples in coastal plains and lowlands indicates high salinity from 1,500 mg/L up to 3,600 mg/L in the north and the east of Semarang (Figure 6). It is obviously that sea water coming in those areas. Meanwhile, in the center and the west, salinity ranges between 400 and 800 mg/L. Low salinity appears in the south of Semarang which is varying 0-200 mg/L. It illustrates fresh water predominant in the highlands.

DO is a major indicator of groundwater quality which often controls the fate of dissolved organic contaminants by constraining the types and numbers of microorganisms present within an aquifer [19]. Dissolved oxygen is available into the water by diffusion from the atmosphere, aeration of the water, and as a waste product of photosynthesis. The DO level ranges from 0.06-3.37 mg/L in the study area

(Figure 7). The low DO is found in the lowlands and coastal plains in the north and the east of Semarang which is below 1 mg/L. Several factors affect the low DO level such as the water is too warm, many bacteria and an excess amount of Biological Oxygen Demand (BOD). It means in these regions no supporting for aquatic life organisms due to the DO level below 3 mg/L.

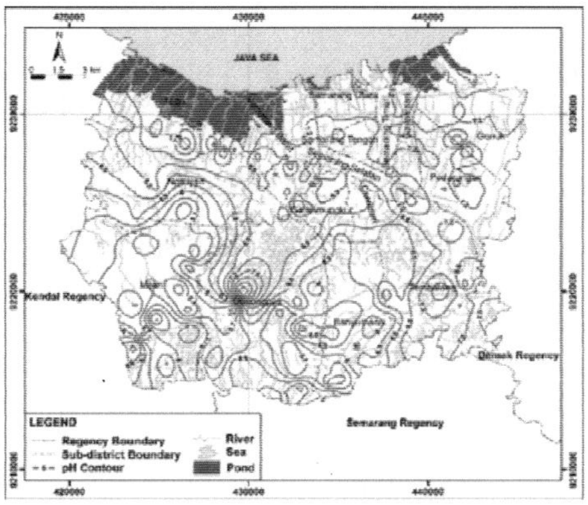

Figure 4. pH map of the study area.

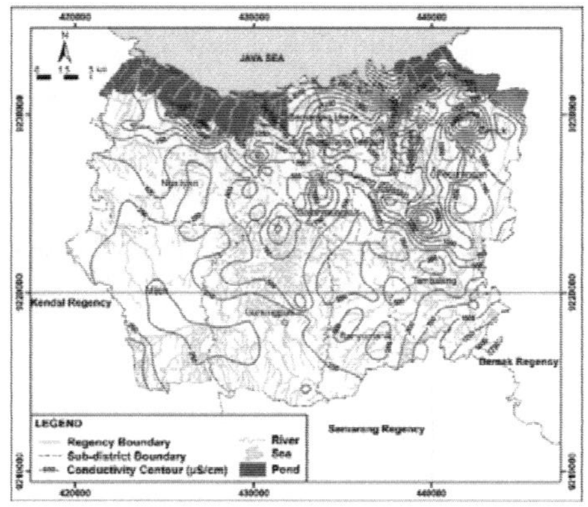

Figure 5. EC map of the study area.

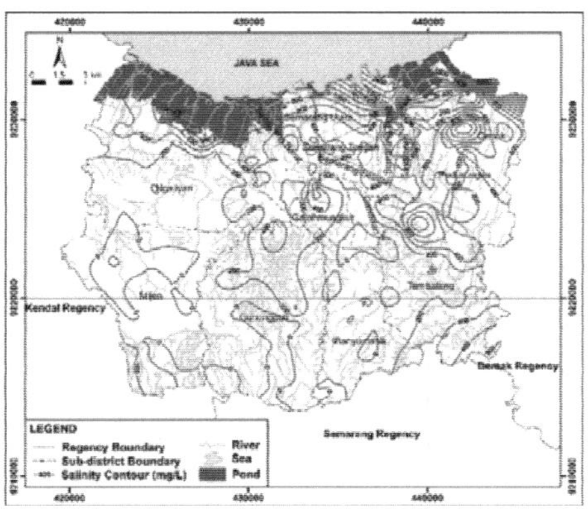

Figure 6. Salinity map of the study area.

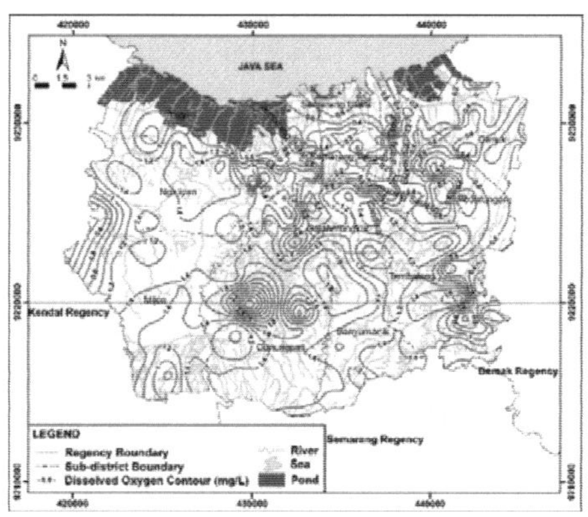

Figure 7. DO map of the study area.

4.3 Land subsidence

Several researchers [3],[12],[20] investigated land subsidence in Semarang using several geodetic monitoring methods such as levelling surveys, Persistent Scatterer Interferometry (PSI) technique, GPS surveys, and microgravity surveys.

The land subsidence in Semarang is supposed to be resulting mainly from natural compaction in alluvial sediments, excessive of groundwater abstraction and load of construction (i.e. settlement on high compressibility soil). The lowlands and coastal plains of Semarang consist of unconsolidated alluvium with alternating layers of medium grained sands with gravel and clayey sands and clayey materials. They are the aquifer composition in these regions which groundwater is excessive abstracted. The consolidation occurs as a consequence the water pressure in the aquifer declines and compression occurs within the sand layers. The number of registered deep wells in Semarang-Demak groundwater basin increase sharply in the 1980s. In 1900, the total number of deep wells was 16, become 260 in the 1990s and the first decade of the new millennium it goes to 1,127 wells with total abstraction more than 30 million cubic meters per year [21].

Based on the results of Kuehn et al. [20] investigation of land use using Persistent Scatterer Interferometry (PSI) as shown in Figure 8, the annual land subsidence rates detected vary from a millimeter to above 8 cm per year. Distribution and visualization of the ground motion data evidently shows the boundary between preferably stable ground in the south and the increasingly land-subsidence affected land towards the harbor area (Semarang Utara Sub-district) in the north.

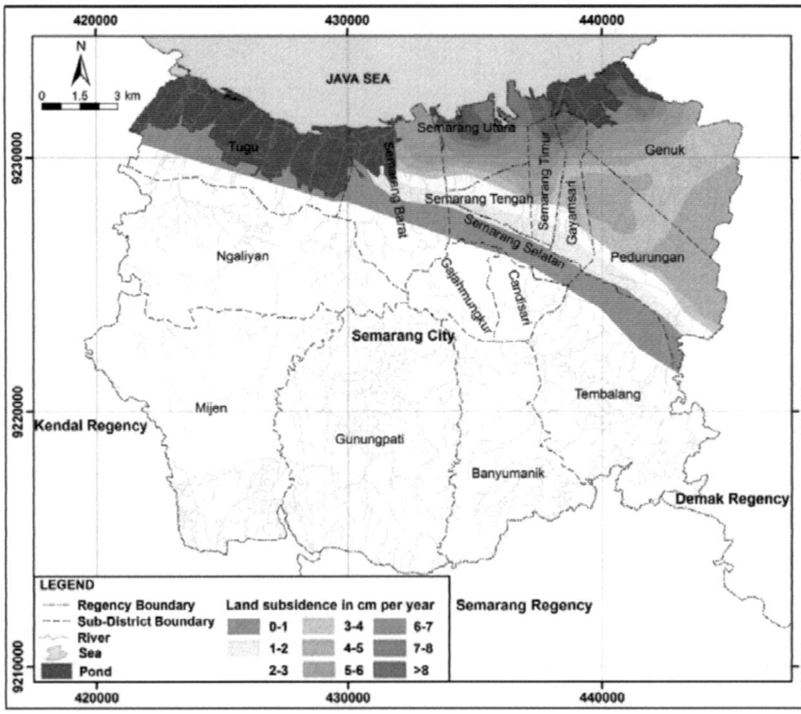

Figure 8. Land subsidence in Semarang (adapted from Kuehn et al. [20]).

4.4 Groundwater conservation zone

Three parameters which are groundwater quality, lowering groundwater level, and land subsidence rates must be concerned to develop groundwater conservation zone of an unconfined aquifer. Groundwater quality is represented by the classification of water quality based on EC value while lowering groundwater level depicts groundwater used in some periods. Furthermore, land subsidence interprets a geological hazard as an impact of natural and human activities is a major constraint to the city development.

All parameters to develop groundwater conservation zone are explained clearly above, a spatial analysis in ArcGIS is then applied to overlay those parameters become a groundwater conservation zone map (Figure 9). There are four zones which are secure, vulnerable, critical, and damaged. The characteristics of each zone is shown in Table 2.

The secure zone is mainly located in the center to the south of Semarang (gentle slope to highlands), groundwater is favorable for water drinking with low of EC value (<1,000 μS/cm), insignificant of lowering water table (<10%), and land subsidence rate maximum 2 cm/year in the center of Semarang. The vulnerable zone widespread in the city center to the east and the west, and locally in the middle slope due to the EC value reaches up to 1,500 μS/cm, but it is acceptable for water drinking. The water table in this zone is around 30-40% decline in the land subsidence rate up to 4 cm/yr. Lowlands in the north to the east of Semarang are classified into the critical zone due to EC value up to 5,000 μS/cm. It means groundwater quality in these regions permissible to unsuitable for water drinking. Moreover, in the area which is the EC value above 2,000 μS/cm, it is classified as medium to high salinity water. Thus, groundwater in this area has no recommendation for water drinking. Furthermore, this critical zone has the impact of land subsidence with the rate up to 7 cm/yr, and water table drops up to 60%. The damaged zone spreads in the coastal plains which is located in the north of Semarang. Groundwater in this region is unsuitable for water drinking due to sea water intrusion occurs in this region. It is reflected by the EC value above 5,000 mg/L. Water table declines more than 80%, and land subsidence rate more than 7 cm/yr. Indeed, Suryono et al. [22] reported that toxic heavy metals detected in shallow groundwater of Semarang coastal region. Toxic heavy metals levels (mg/L) in the water were Cr^{2+} 5.083±1.59, and Pb^{2+} 5.52±1.34. Those levels were exceeded the maximum permissible limit for drinking water by WHO [16].

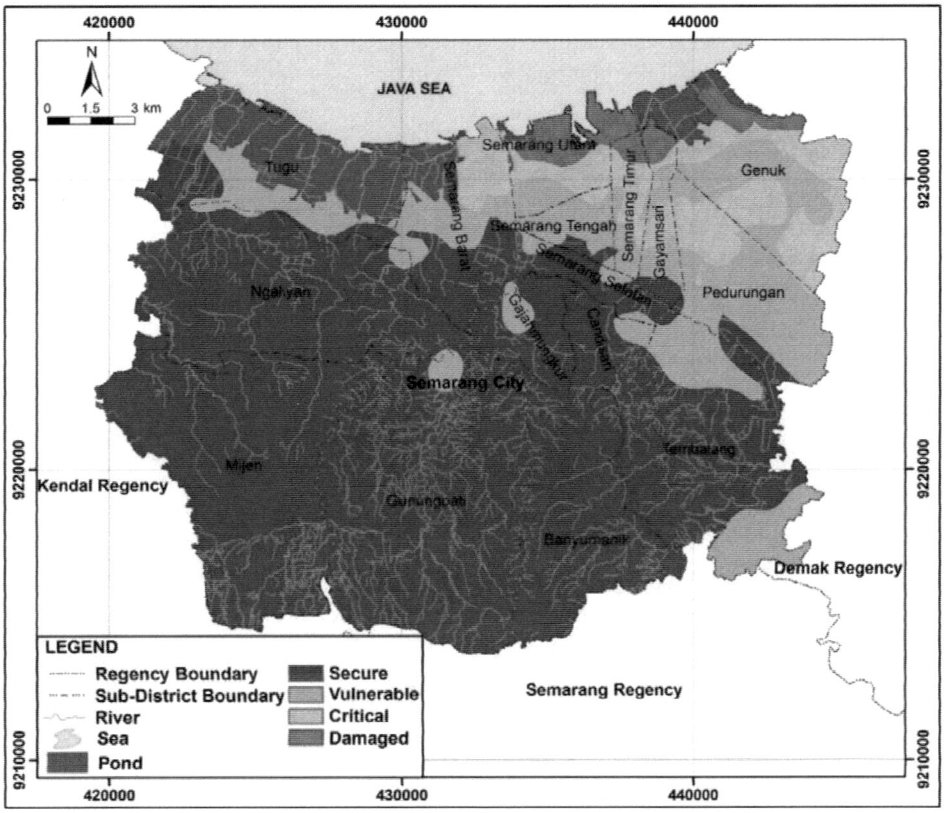

Figure 9. Map of groundwater conservation zone of unconfined aquifer in Semarang

Table 1. Parameter for defining groundwater conservation zone of unconfined aquifer in Semarang

Parameter	Groundwater conservation zone			
	Secure	Vulnerable	Critical	Damaged
Groundwater quality (EC μS/cm)	<1,000	1,000-1,500	1,500-5,000	>5,000
Lowering groundwater level (%)	5-10	30-40	50-60	80-85
Land subsidence rate (cm/yr)	0-2	2-4	4-7	>7

5 Conclusion

Groundwater quality, lowering of water table, and land subsidence rates are the main parameters to develop groundwater conservation zone of unconfined aquifer in Semarang, Indonesia. There are four conservation zones which are secure, vulnerable, critical, and damaged. In the secure zone, groundwater is pleasant for drinking water in the south of Semarang. Human activity needs to be a concern in the vulnerable zone (the center of Semarang) to preserve groundwater quality. Groundwater in this zone is permissible for drinking water. Meanwhile in the critical zone which widespread in the north to the east of Semarang, groundwater is moderate to high salinity water. Groundwater is classified as doubtful to unsuitable for drinking water related to its quality, and sea water occurs in this region. Furthermore, the damaged zone strictly has no recommendation for water drinking. The presence of toxic heavy metals and sea water intrusion predominant affect that groundwater is unsuitable for water drinking in the north of Semarang.

Acknowledgments

Authors are whishing thankful to the Faculty of Engineering at the Diponegoro University for the funding with the scheme of "Penelitian Strategis"with the Contract No. 2933/UN7.3.3/PG/2016.

References

[1] B Morris 1999 Earthwise Sustainability *Urban Groundwater: Monitoring the contamination of aquifers below cities* ed J. Thomas (Nottingham: British Geological Survey) p 23
[2] Badan Pusat Statistik (BPS-Statistics Indonesia) *Population Growth by Province*, Retrieved from http://www.bps.go.id/tab_sub/view.php?kat=1&tabel=1&daftar=1&id_subyek=12&notab=2 (Citation date: January 12th, 2014)
[3] M A Marfai and L King 2007 Monitoring Land Subsidence in Semarang, Indonesia *Environmental Geology*. **53(3)** 651 – 659
[4] N Rahmawati and M A Marfai 2013 Salinity pattern in Semarang Semarang Coastal City *Indonesian Journal on Geoscience*. **8(2)** 111-120
[5] E Suhartono, P Purwanto, and S Suripin 2015 Seawater Intrusion Modeling on Groundwater Confined Aquifer in Semarang *Procedia Environmental Sciences*. **23(2015)** 110 – 15
[6] Statistic Center Bureau of Semarang City 2015 *Semarang City in Figures* (Semarang: Badan Pusat Statistik Kota Semarang)
[7] Ditjen Cipta Karya 1996 *Standar Kebutuhan Air Bersih Untuk Keperluan Rumah Tangga* (Jakarta: Departemen Pekerjaan Umum)
[8] Spitz K and Moreno J 1996 *A practical guide to groundwater and solute transport modeling* (Canada: John willey & sons Inc.)
[9] Arifin M B and Wahyudin 2000. *Penyelidikan potensi cekungan air tanah Semarang, Jawa Tengah; data dan analisis* (Bandung: Direktorat Geologi Tata Lingkungan)
[10] T T Putranto and T R Rüde Hydrogeological model of an urban city in a coastal area, case study: Semarang, Indonesia. *Indonesian Journal on Geoscience*. **3(1)** 17-27
[11] S D Foster, R Hirata and K W F Howard 2010 Groundwater use in developing cities: policy issues arising from current trends. *Hydrogeology Journal*. **19** 271-274
[12] H Z Abidin, H Andreas, I Gumilar, T P Sidiq, M Gamal, D Murdohardono, Supriyadi, and Y Fukuda 2010 Studying Land Subsidence in Semarang (Indonesia) Using Geodetic Methods *FIG Congress Facing the Challenges – Building the Capacity (Sydney, Australia, 11-16 April 2010)*
[13] H D Said and Sukrisno 1988 *Hydrogeological Map of Indonesia Scale 1:250.000 Sheet VII Semarang (Jawa)* (Bandung: Directorate of Environmental Geology)
[14] Danaryanto, Djaendi, Harnadi D, Mudiana W and Budiyanto 2007 *Kumpulan panduan teknis pengelolaan airtanah* (Bandung: Pusat Lingkungan Geologi)
[15] Direktorat Tata Lingkungan Geologi dan Kawasan Pertambangan 2003 *Laporan akhir kajianzonasi konfigurasi dan tata guna air bawah tanah pada Cekungan Semarang-Demak, Subah, dan Karanganyar-Boyolali* (Bandung: Direktorat Tata Lingkungan Geologi dan Kawasan Pertambangan)
[16] WHO/World Health Organization 2011 *Guidelines for drinking-water quality 4th Ed.* (Geneva, Switzerland)
[17] S V S Prasanth, N S Magesh, K V Jitheshlal, N Chandrasekar, and K Gangadhar 2012 Evaluation of groundwater quality and its suitability for drinking and agricultural use in the coastal stretch of Alappuzha

District, Kerala India. *Appl Water Sci.* **2**(165-175)

[18] Jarvis A, Reuter H I, Nelson A, Guevara E 2008 *Hole-filled SRTM for the globe Version 4*, available from the CGIAR-CSI SRTM 90m Database. Retrieved from http://srtm.csi.cgiar.org. (citation date: June 26th, 2010)

[19] S Rose and A Long 1988 Monitoring dissolved oxygen in groundwater: some basic consideration *Ground Water Monitoring & Remediation.* **8(1)** 93-97

[20] F Kuehn, D Albiol, G Cooksley, J Duro, J Granda, S Haas, A Hoffman-Rothe, and D Murdohardono 2010 Detection of land subsidence in Semarang, Indonesia, using stable points network (SPN) technique. *Environmental Earth Scince.* **60(5)** 909-921

[21] Putranto T T 2013 *Hydrogeological and numerical groundwater flow model in Semarang/Indonesia* (Aachen: Dissertation RWTH Aachen University Germany)

[22] C A Suryono, B Rochaddi, A Sabdono, and B T Susanti, 2007 Physico-chemical characteristics and heavy metal contents in shallow groundwater of Semarang coastal region, *Journal Ilmu Kelautan.* **12(4)** 227-232

Understanding coastal processes to assist with coastal erosion management in Darwin Harbour, Northern Territory, Australia

S.G. Tonyes[1,2], R.J. Wasson[1,3], N.C. Munksgaard[1], K.G. Evans[1], R. Brinkman[4], D.K. Williams[5]

[1]Charles Darwin University, Darwin, Northern Territory 0909, Australia
[2]Udayana University, Bukit Jimbaran, Bali 80361, Indonesia
[3]Lee Kuan Yew School of Public Policy, National University of Singapore, Singapore 259772
[4]Australian Institute of Marine Science, Townsville, Queensland 4810, Australia [5]Australian Institute of Marine Science, Arafura Timor Research Facility, Casuarina NT 0811, Australia

Email: Silvia.Tonyes@cdu.edu.au

Abstract. Sand transport pathways in Darwin Harbour, Northern Territory, Australia, are being investigated to assist with coastal management. Coastal erosion, which threatens public and private infrastructure, is one of the major problems along the harbour beaches. A study of sediment transport is essential to identify the challenges encountered by the stakeholders in coastal management. Darwin Harbour, located in the tropical, cyclone prone area of Australia, was, until recently, considered a near pristine estuary. A semi-diurnal macro-tidal embayment, the tidal variation in the harbour reaches up to 8 m with a mean tidal range of 3.7 m. The beach morphology consists of sandy pocket beaches between coastal cliffs, sandbars, rocky shore platforms, tidal flats and mangrove fringes. A two-dimensional depth averaged finite-element hydrodynamic model (RMA-2), coupled with a sediment transport model (RMA-11) from Resource Modelling Associates, has been used to infer the sources and the depositional areas of sand in the harbour. Grain size distributions and geochemical analysis are also used to characterize the sand and its source(s). Initial results show that the beach sand is mostly of offshore origin with small sand input from the rivers. Potential supplementary sand sources are the eroded materials from the shore platforms and the rocky cliffs. Due to the rapid development in Darwin Harbour, this study is fundamental in understanding coastal processes to support decision making in coastal management, particularly in a macro-tidal, tropical estuary.

Keywords: Darwin Harbour, macro-tidal, sand transport, coastal erosion, RMA

1. Introduction

Coastal erosion is a natural phenomenon. In fact, coast lines change continually, controlled by the interaction of the local hydrodynamics and their morphology. Coastal change is a longstanding problem that mankind has had to deal with to provide safety from flooding and to protect transportation infrastructure. Conventionally, coastal erosion is managed locally using hard engineering approaches, such as sea walls or breakwaters, which do not guarantee good outcomes and often create erosion in other areas [1, 2]. These consequences often stem from engineering decisions that only consider the immediately affected area, underestimating the processes that are occurring in the wider coastal zone.

IOP Conf. Series: Earth and Environmental Science **55** (2017) 012012 doi:10.1088/1755-1315/55/1/012012

The coastal zone is composed of diverse complex environments, shaped by coastal processes, coastal geology, variations in coastline characteristics and coastal sediment budgets [3]. A more technically and environmentally satisfactory coastal engineering design should include each element and its interaction, incorporating how they affect the whole system [1]. Coastal processes relate to physical processes such as tides, waves, currents and winds that act upon and shape the coastline, while coastal geology determines the origin, structure and characteristics of the sediments that make up the coastal region. Considering that coastal erosion essentially indicates an imbalance in the sediment supply and removal in the area, an ideal way to deal with the problem is to study the complete cycle of sedimentation in a theoretically confined coastal area called a sediment/coastal cell [3]. The boundaries of a sediment cell can be marked by features such as headlands, submarine canyons or river mouths, where the sources, transport paths and sinks of sediment occur.

The sources of coastal sediment can be examined by means of provenance indicators using mineralogical or geochemical properties. Among a number of suitable indicators of provenance, carbonate minerals and rare earth elements (REEs) are often used. In coastal sediments, carbonate mineral constituents, such as calcite and aragonite, can be used to distinguish whether a sediment is of marine or of terrestrial origin [4]. REEs are excellent in determining sediment sources because the ratios of individual REE are not easily fractionated during transport and show consistent behaviour during weathering [5].

This study examines the coastal processes in Darwin Harbour, a macrotidal estuary situated in cyclone prone tropical northern Australia. The semi-diurnal tides record the highest astronomical tide at 8 m and the smallest low tide at 0.3 m with a mean range of 3.7 m [6, 7]. A large embayment, Darwin Harbour covers the area from Charles Point in the west to Gunn Point in the east with Blackmore, Elizabeth and Howard Rivers, the major streams that flow into the harbour. Presently coastal cells have not been defined in Darwin Harbour, therefore this study was carried out within two prominent headlands of the harbour i.e. Charles Point in the west and Lee Point in the east (Figure 1).

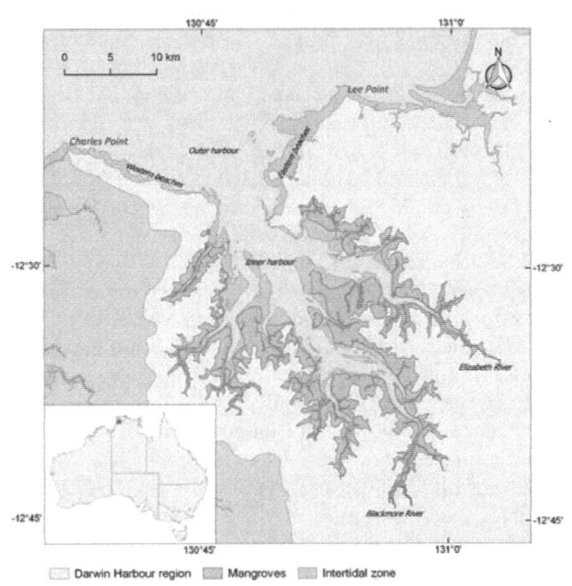

Figure 1. Darwin Harbour, the study area.

The western part of Darwin Harbour consists of mainly rocky shore platforms interspersed by sandy pocket beaches. The eastern part of Darwin Harbour comprises longer stretches of sandy beach and less extensive rock flats. The beaches are backed by either coastal cliffs or sand dunes. Mangrove forests border the intertidal areas of the inner harbour and to a lesser extent on the west and east beach areas.

Similar to other areas in the world, coastal problems encountered in Darwin Harbour arise from mixed uses of the area and conflicting concerns among the stakeholders. Previous coastal erosion studies mainly focused on the eastern part of the harbour. The studies documented coastal cliff erosion rates averaging 30 cm y^{-1} [8], while the sandy beaches experience seasonal changes in both climatic and oceanographic events [9]. Visual observation at several sites and anecdotal information indicates that the eastern beach dunes and the western beaches have also experienced substantial erosion in recent decades. Sand dynamic studies were carried out on relatively small areas in the harbour for specific purposes, e.g. shipping channels development work. Despite long term beach erosion, no study on sand dynamics, incorporating coastal processes, has been carried out for Darwin Harbour. This study is an attempt to make a start at filling this gap, and aims to contribute to understanding the role of coastal processes occurring in the area, i.e. to determine the sand pathways in the harbour, to infer the sources of beach sand, and thereby assist with coastal management in Darwin Harbour. Two potential sand sources are investigated: the inflow from offshore and the contribution of Elizabeth and Blackmore Rivers, while the potential sand sinks are the western and eastern beaches of the harbour.

2. Methods

A 2-D depth-averaged hydrodynamic (RMA-2) and sand transport modelling (RMA-11) software package from Resource Modelling Associates [10, 11] was used to simulate the hydrodynamics of the study area and to infer the sources and sinks of sand in Darwin Harbour. The simulations were run on the calibrated and validated Darwin Harbour modelling-mesh constructed by the Australian Institute of Marine Science (AIMS) based on 2012 bathymetry. Numerous surveys of tidal currents profiling by AIMS have confirmed that a 2D modelling method is valid for Darwin Harbour hydrodynamic simulations, i.e. the vertical profiles of currents are of similar magnitude and direction during the tidal cycle. In multiple AIMS projects it was also proven that the computation of bed shear gives similar values compared to a 3D model.

The model mesh consists of 10,227 elements and 21,219 nodes. The cell sizes range from 20 m^2 at the wharf area to 3,000 m^2 at the offshore boundary. The mesh was divided into three element types, each assigned with different bed roughness in Manning's 'n' values, i.e.: 1) Submerged/water area, 'n' = 0.030; 2) Mangrove area, 'n' = 0.100; and 3) Intertidal area, 'n' = 0.025. The model was run for a 12-month period, from May 2012 to April 2013, covering both the dry and the wet seasons. Tide forces and river inflow were used to run the model with a 15-minute time step.

The results from the RMA-2 hydrodynamic simulations were used as the input to simulate sand transport pathways in the harbour with RMA-11. The fine, medium and coarse sand transport were simulated using the sand transport potential method based on Van Rijn's 1984 computation, which is most suitable for sand with diameter > 0.100 mm (fine sand size and greater). The size distribution of sand used in the simulations was determined from terrestrial and marine samples from the study area. Sub samples were also taken for geochemical analysis.

In order to simulate the transport pathways in the harbour, 5 mg L^{-1} of sand was introduced from the network boundaries with no initial bed thickness in the harbour. The sand transport pathways were inferred by the bed level changes in the modelling area. Positive bed change indicates sand deposition, which shows the sand sink areas, therefore, any positive bed change in the model domain can be inferred as the sand direction from the source to the sink area. In order to distinguish between offshore and terrestrial sources area, offshore and river sand simulations were run separately. River sand simulations were carried out from Elizabeth and Blackmore Rivers (Figure. 1).

The sediment transport pathways were also inferred using the Calcium Carbonate concentration and the REE composition of the sand samples. The $CaCO_3$ concentration was determined using cold acid digestion, while REE compositions were determined using a semi-quantitative ICP-MS method [5].

3. Results and Discussion

The sand transport pathways analysis is discussed based on the beach areas depicted in Figure 2.

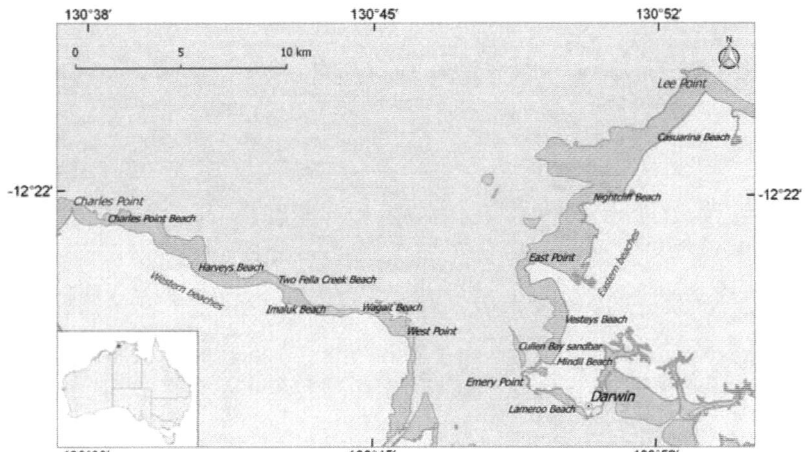

Figure 2. Beach area under study in Darwin Harbour.

3.1. Sand pathways from offshore

The simulation results showed that offshore sand deposited predominantly on the northern parts of the western and eastern beaches and decreased further into the harbour (Figure 3). The extent of deposition was governed by beach morphology, more deposition occurred in embayment areas, particularly on the lee side of the headlands due to the sheltering effect of the headlands [12]. This trend was very apparent in the western beach areas, on the lee side of Charles Point (node 99). The deposition level at this point was significantly higher compared to other beach areas. Apart from the deposition at this node, the long term deposition of offshore sand at the western and eastern beaches showed an analogous trend (Figure 4), inferring similar pathway patterns of offshore sand to the beach area.

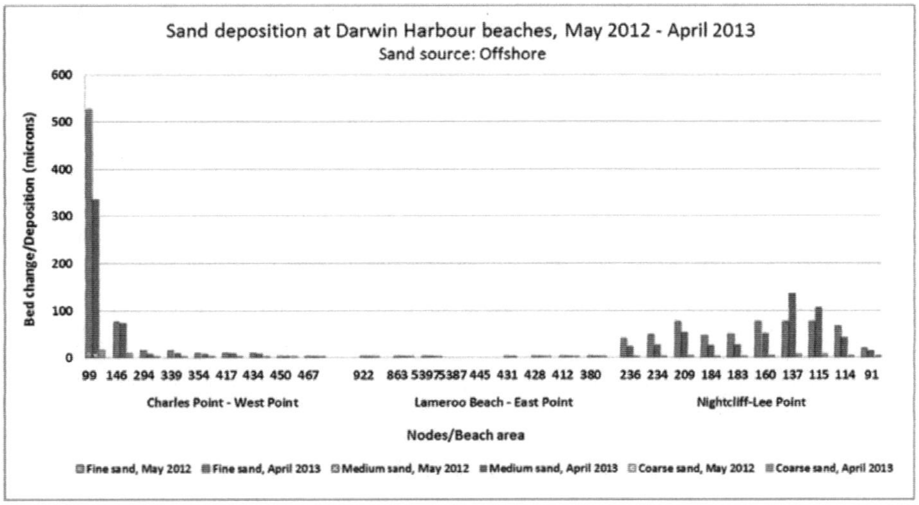

Figure 3. Offshore sand deposition on Darwin Harbour beaches.

As indicated in Figure 4, offshore sand deposition at the beaches in the eastern part of the harbour, between Lameroo Beach and East Point, was substantially lower than at the northern beaches, suggesting that East Point headland, Cullen Bay sandbar and Emery Point prevent the offshore sand from being transported to the area.

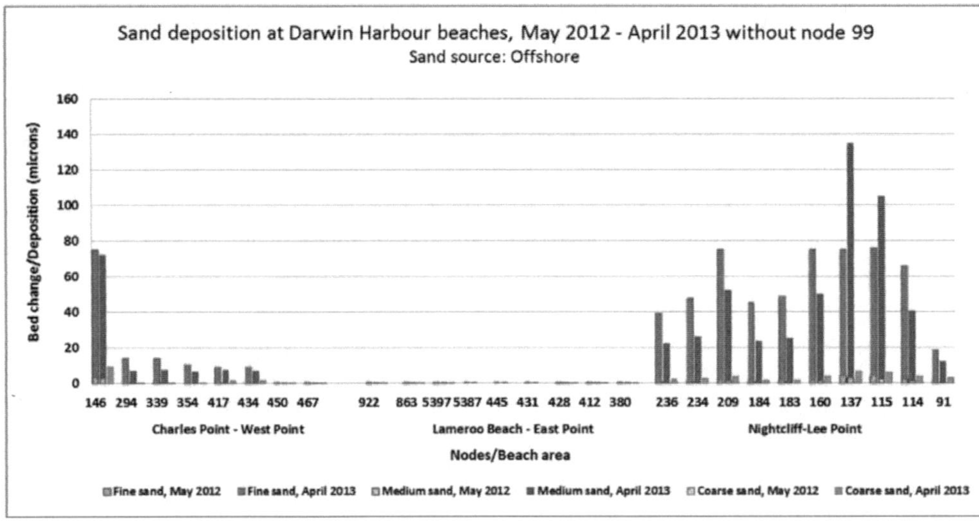

Figure 4. Offshore sand deposition on Darwin Harbour beaches excluding node 99.

3.2. Sand pathways from Rivers

The sand transport simulations showed that the contribution of river sand to the beaches in Darwin Harbour was significantly lower than offshore sand. Albeit of very small quantity, river sand deposited mostly in the embayment facing the inner harbour (Lameroo Beach, Figure 5).

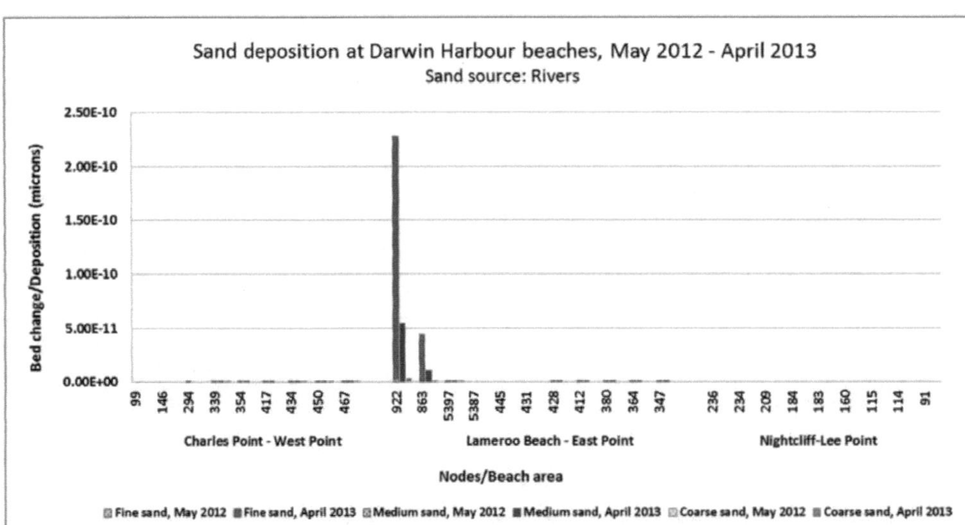

Figure 5. River sand deposition on Darwin Harbour beaches.

After passing Emery Point, the river sand was transported more to the western beaches rather than eastward, confirming that Cullen Bay sandbar and East Point constrain the sand transport pathways eastward (Figure 6).

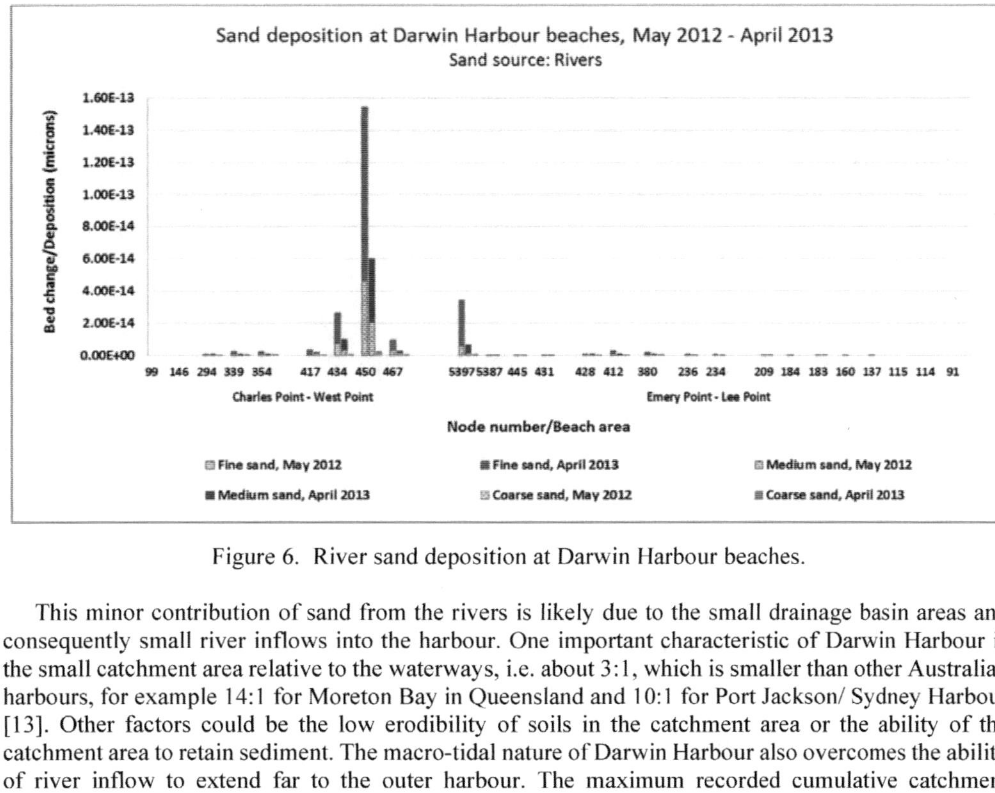

Figure 6. River sand deposition at Darwin Harbour beaches.

This minor contribution of sand from the rivers is likely due to the small drainage basin areas and consequently small river inflows into the harbour. One important characteristic of Darwin Harbour is the small catchment area relative to the waterways, i.e. about 3:1, which is smaller than other Australian harbours, for example 14:1 for Moreton Bay in Queensland and 10:1 for Port Jackson/ Sydney Harbour [13]. Other factors could be the low erodibility of soils in the catchment area or the ability of the catchment area to retain sediment. The macro-tidal nature of Darwin Harbour also overcomes the ability of river inflow to extend far to the outer harbour. The maximum recorded cumulative catchment discharge into the harbour during floods was estimated to be about 1% of the peak spring tide discharge [14].

Notwithstanding the low contribution, the simulation results showed that the sand transport pathways from rivers in Darwin Harbour were initially inclined eastward in the inner harbour and turned westward after passing the 'neck' of the harbour. These pathways are obviously influenced by the current strength and directions, while the location of the rivers, which are on the east side of the harbour, is another important factor.

3.3. The provenance of beach sand

Based on sedimentary composition, sand on Darwin Harbour beaches appears to be a mix of marine and terrigenous sources, with $CaCO_3$ concentrations in the sand samples from eastern beaches substantially higher compared to the sand samples from the western beaches (Figure 7).The carbonate content in sand from the eastern beaches is likely derived from offshore sources, in-situ sources and reworked sediment within the harbour. There are scattered hard substrates in the harbour providing carbonate sand sources from marine organisms, e.g. corals, molluscs, echinoderms and foraminifera [15] that may contribute carbonate sand to the area. The majority of foraminifers' biotopes located in the eastern part of Darwin Harbour contains foraminifera species typically found on the shallow continental shelf [6]. The fact that the local coral reefs are mostly located in the eastern part of the harbour (Lee Point and East Point), demonstrates that a more detailed study confirming the source of carbonate sand is necessary.

The low $CaCO_3$ concentration in the western beach samples shows a more terrestrial origin compared to offshore elements in the samples. Considering that the river contribution to beach sediment is relatively low, the signature may also be contributed by attrition of the local shore rock platforms, which occur extensively in the western beach area.

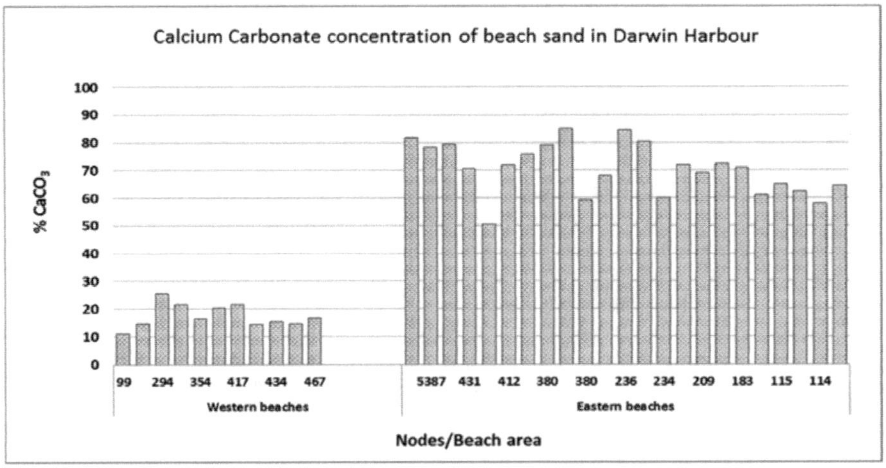

Figure 7. Calcium Carbonate concentration of beach sand samples in Darwin Harbour.

The REE composition of the beach sand samples, as depicted by a light- and heavy-REE bi-plot, shows that the eastern beach sand clusters closer to the outer harbour sand, while the western beach sand shows closer similarities with river sand (Figure 8).

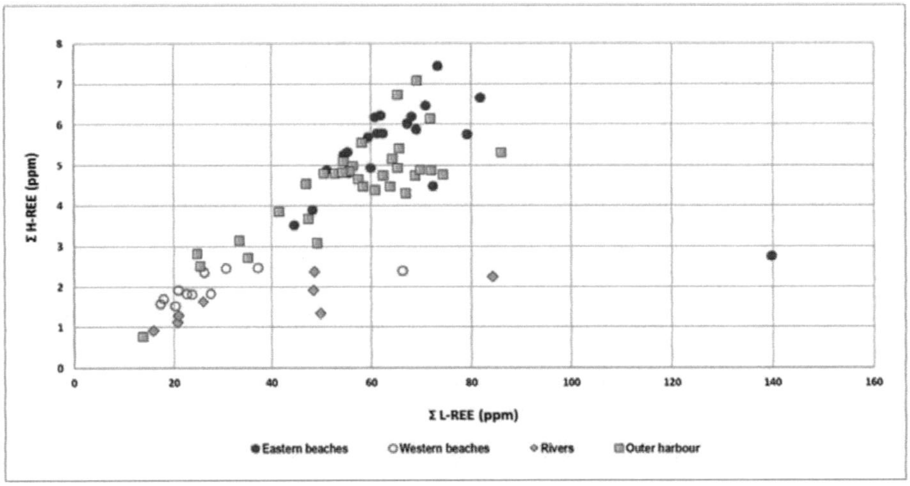

Figure 8. Light- versus Heavy-REE concentration of sand samples in Darwin Harbour.

With the assumption that outer harbour sand is largely of offshore origin, it is evident from Figure 8 that the sand in the east beach area indicates an offshore source, while the west beach sand shows a more river/terrestrial origin. Once again, as the sand transport simulations show that river input to beach sand is low, the characteristics of the west sand may also be due to the contribution from the local bedrock formations.

Notwithstanding the high signature of an offshore sand source on the eastern beaches, considering the presence of eroding cliffs in the eastern part of Darwin Harbour [7], we cannot overlook the possibility that the rocky cliffs at the back of the beach are also a source of sand to the beach. Future study of the possibility of the rock flats and cliffs as sources of beach sand is necessary.

4. Conclusion

Two-dimensional hydrodynamic and sand transport simulations showed that beach sand in Darwin Harbour is mainly of offshore origin. On the other hand, the parallel geochemical analysis showed a slightly different pattern. The $CaCO_3$ concentration in beach sand suggests that the eastern beaches received substantially more offshore sand compared to the western beaches. Similarly, parallel Rare Earth Element analysis indicates a closer relationship between sand from eastern beaches and the outer harbour area. While the Rare Earth Element analysis shows that the western beach sand has more similarities with Elizabeth and Blackmore Rivers, a contribution from the local shore platform bedrock is possible.

Acknowledgement

The first author would like to acknowledge Charles Darwin University, the Indonesian Directorate General of Research, Technology and Higher Education (Kemenristekdikti) and the Australian Institute of Marine Science (AIMS) for supporting this research.

References

[1] Kamphuis J W 2000 *Introduction to Coastal Engineering and Management* (Singapore: World Scientific) p 437
[2] Nordstorm K F 2014 Living with shore protection structures: A review *Estuar Coast Shelf Sci* **150** Part A pp 11-23
[3] Woodroffe C D 2003 *Coasts: Form, process and evolution* (Cambridge: University Press) p 623
[4] Siever R 1988 *Sand* (New York: Scientific American Library) p 237
[5] Munksgaard N C, Lim K and Parry D L 2003 Rare earth elements as provenance indicators in North Australalain estuarine and coastal marine sediments *Estuar Coast Shelf Sci* **57** pp 399-409
[6] Michie M G 1987 Distribution of Foraminifera in a macrotidal tropical estuary: Port Darwin, Northern Territory of Australia, *Austr. J. Mar. Freshw. Res.,* **38** pp 249-59
[7] Williams D, Wolanski E and Spagnol S 2006 Hydrodynamics of Darwin Harbour in: *The Environment in Asia Pacific Harbours* ed E Wolanski (Dordrecht: Springer) p 497
[8] Jones G, Baban S and Pathirana S 2008 *Coastal Erosion Issues in the East Point and Nightcliff areas of Darwin – a report prepared for Darwin City Council* (Lismore: School of Environmental Science and Management, Southern Cross University) p 5
[9] Comley B W T 1996 *Darwin Beach Erosion Survey, Technical Report R96/1, Land Resource Conservation Branch* (Darwin: Department of Lands Planning and Environment) p 108
[10] King I P 2013 *A Two-dimensional Finite Element Model for Flow in Estuaries and Streams; RMA-2, Users Guide, Version 8.4T* (Sydney: Resource Modelling Associates) p 135
[11] King I P 2015 *A Three-dimensional Finite Element Model for Water Quality in Estuaries and Streams; RMA-11, Users Guide, Version 9.0G* (Sydney: Resource Modelling Associates) p 171
[12] Van Rijn L C 2011 Coastal erosion and control, *Ocean Coast Manage* **54** pp 867-87
[13] Australian Government 2006 *Australia's National Programme of Action for the Protection of the Marine Environment from Land-Based Activities, Case Study 23: Darwin Harbour regional plan of management* (Canberra: Dept. of the Environment and Water Resources) p 5
[14] NTEPA 2014 *A stormwater strategy for the Darwin Harbour Region version 1.0,* (Darwin: Northern Territory Environment Protection Authority) p 32
[15] Smit N 2003 Marine invertebrate life in Darwin harbour region and management implications, in *Proc. Darwin Harbour Region: Current knowledge and future needs* ed Working Group for the Darwin Harbour Advisory Committee pp 37-64

Environmental and Risk Factors of Leptospirosis: A Spatial Analysis in Semarang City

Silviana Nur Fajriyah, Ari Udiyono, and Lintang Dian Saraswati,

Department of Epidemiology and Tropical Diseases, Faculty of Public Health, Diponegoro University, Jl. Prof Sudharto SH Tembalang Semarang, Indonesia. Zip code: 50275

lintang.saraswati@live.undip.ac.id

Abstract. Leptospirosis is zoonotic potentially epidemic with clinical manifestations from mild to severe and can cause death. The incidence of leptospirosis in Indonesia tends to increase by the year. The case fatality rate in Semarang was greater than the national's (9.38%). The purpose of this study was to describe the environmental risk factors of leptospirosis in Semarang spatially. The study design was descriptive observational with cross sectional approach. The population and samples in this study were confirmed leptospirosis in Semarang from January 2014 until May 2015, 88 respondents in 61 villages of 15 sub-districts in Semarang. The variables were environmental conditions, the presence of rats, wastewater disposal, waste disposal facilities, the presence of pets, the presence of rivers, flood's profile, tidal inundation profile, vegetation, contact with rats, and Protected Personal Equipment/PPE utilization. Based on the spatial analysis, variables that found in the big half area of Semarang are environmental conditions, the presence of rats, wastewater disposal, waste disposal facilities, contact with rats, and PPE utilization. The presence of pets at risk, the presence of rivers, flood's profile, inundation profile, and vegetation were found only in small half of Semarang area. People are expected to maintain their personal and environmental hygiene to prevent the transmission.

Keywords: Leptospirosis; risk factor; spatial; mapping.

1. Introduction

The risk factors contributing to its perceived re-emergence or fluctuations in prevalence are not completely understood. Some factors that are commonly reported to play a role were living within urban areas [1][2] and peri-urban areas [3], flooding events [3,4], contact with infected wild and peridomestic vectors [3], less than ideal socio-economic conditions in areas where dogs reside [5–7], environmental conditions, the presence of rodents, wastewater disposal, garbage disposal facilities, presence of potential pets as reservoir, the presence of rivers, flood history, a history of tidal inundation, vegetation, history of contact with rats, and use of personal protective equipment [8–11].

Dimentional space picture Information related risk factors for leptospirosis cases will provide benefits for the leader in the Health Service in the planning decisions and the implementation of control programs and the prevention of leptospirosis in the city of Semarang. One way to map risk factors for leptospirosis cases is to use spatial analysis. Spatial analysis is a data analysis technique that refers to the position, the object and the relationship between them in the space of the earth. Spatial analysis of the health sector needs to be done to know the perception of health-related issues or based space and can help analyze the appropriate control measures and proper treatment [12]. Modern spatial analysis capability of a Geographic Information Systems (GIS) enables us to study the associations of environmental factors with diseases; this, combined with the abundant availability of spatial data in the public domain allows us to assess such associations with more than one perspective. The purpose of this study was to describe the environmental and behavioral risk factors of leptospirosis in Semarang spatially.

2. Method

2.1. Study Design and Minimum Sample Size:
We conducted an explanatory study with cross sectional design. The sample size was calculated by the formula for cross-sectional design. Using $Z_{1-\alpha}$=0,196, $Z_{1-\beta}$=1,282, test value of the population proportion =0.5, and anticipated value of the population proportion=0.3, the minimum sample size was 74 patients.

2.2. Subject of Study:
The subject was confirmed a case of leptospirosis which is suspected persons with clinical symptoms of leptospirosis and laboratory-confirmed diagnosis. For this study, only confirmed cases were included. DHO Semarang uses the same definition recommended by WHO that a case of leptospirosis requires laboratory confirmation which taken consecutively during January 2014-May 2015 (101 cases). The samples were all members of the population study who live in Semarang and willing to become respondents. Exclusion criteria were patients who died. Total samples were 88 subjects.

2.3. Variables:
Variables in this study consisted of the environment, the presence of kinds of rodent, waste water disposal, garbage disposal facilities, the presence of potential pets as reservoir, presence of river, history of flood, history of tidal inundation, vegetation, history of contact with kinds of rodents and habitat of rodents, personal protective equipment.

2.4. Ethics:
Ethical clearance was obtained from the Commission of Ethics of Medical and Public Health Research, Faculty of Public Health, Diponegoro University

2.5. Statistics:
Distribution of data was determined by Kolmogorov-Smirnov test. The risk factors were analyzed by frequencies and percentages. Mapping used GIS software to produce graphic displays of geographical information of cases leptospirosis and the risk factors. Map of Semarang was obtained from *Dinas Tata Ruang Kota Semarang*. Coordinate of cases was obtained from GPS

3. Result and Discussion
Cases of leptospirosis in Semarang City mostly suffered by men with age at most in the range of 51-55 years (19.1%) where the education level was high school graduates as many as 33.0%. A study explored several socio-economic and demographic characteristics of Sao Paulo, Brazil with historical human leptospirosis cases and found significant associations with average monthly income, literacy rate, and some people living in a household, among other factors [13]. Likewise, education, income, housing type, and some people living per household were risk factors for human leptospirosis in a different study from urban Recife in Brazil [14].

Table 1. Risk Factors of Leptospirosis in Semarang City (n=88)

	Variable	f	%
1.	Environment		
	1. Potentially at risk	88	100,0
	2. Not at risk	0	0,0
2.	Presence of rodents		
	1. Yes	88	100,0
	2. No	0	0,0
3.	Waste water disposal		
	1. Potentially at risk	84	95,5
	2. Not at risk	4	4,5
4.	Garbage Disposal facilities		
	1. Potentially at risk	84	95,5
	2. Not at risk	4	4,5
5.	Presence of potential pet as reservoir		
	1. Yes	12	13,6
	2. No	76	86,4
6.	Presence of river		
	1. Yes	32	36,4
	2. No	56	63,6
7.	History of floods		
	1. Yes	20	22,7
	2. No	68	77,3
8.	History of tidal inundation		
	1. Yes	7	8,0
	2. No	81	92,0
9.	Vegetation		
	1. $\geq$ 3 kinds	31	35,2
	2. < 3 kinds	57	64,8
10.	History contacts with rodents		
	1. Yes	67	76,1
	2. No	21	23,9
11.	The use of personal protective equipment		
	1. No	85	96,6
	2. Yes	3	3,4

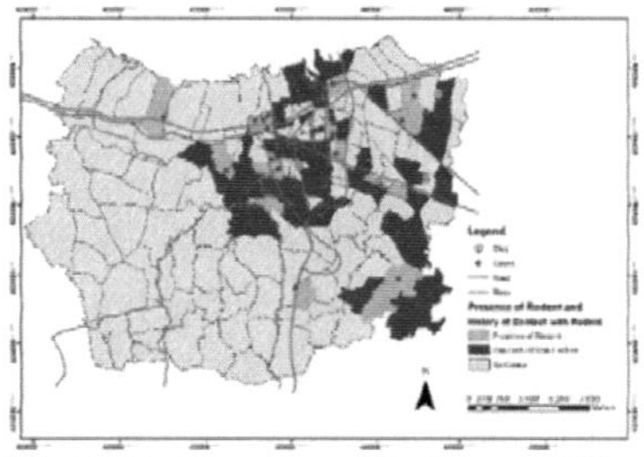

Fig 1. The Presence of Rodent and History of Contact

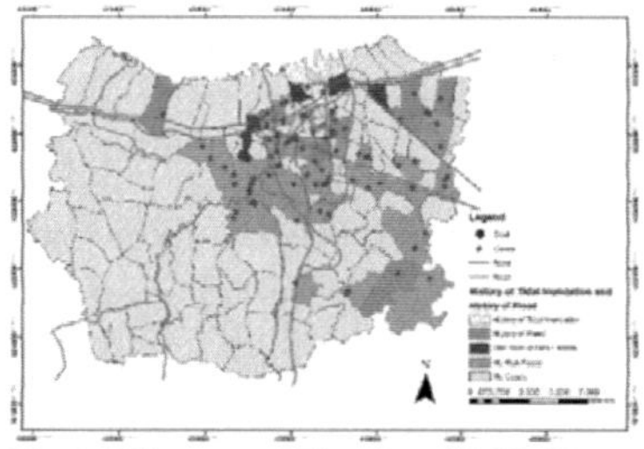

Fig 2. History of Flood and Tidal Inundation

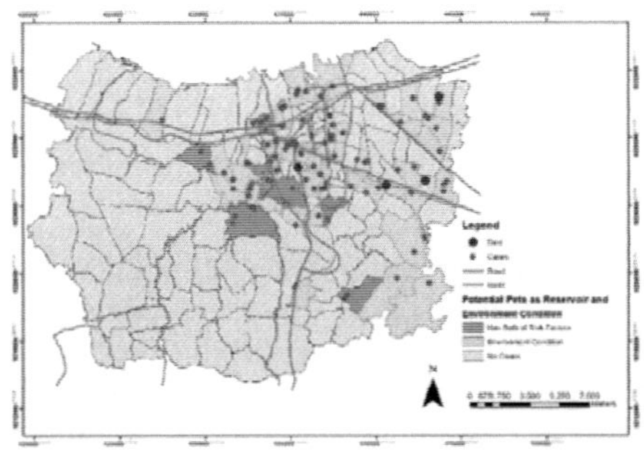

Fig 3. Environment and Presence of Potential Reservoir

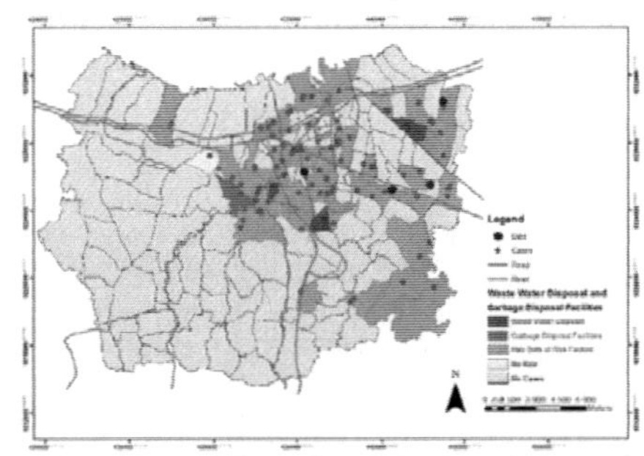

Fig 4. Waste Water l and Garbage Disposal

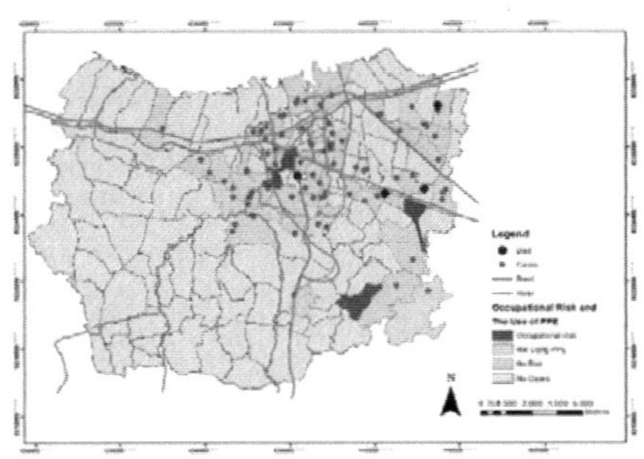

Fig 5. The High Risk Occupation and The Use of PPE

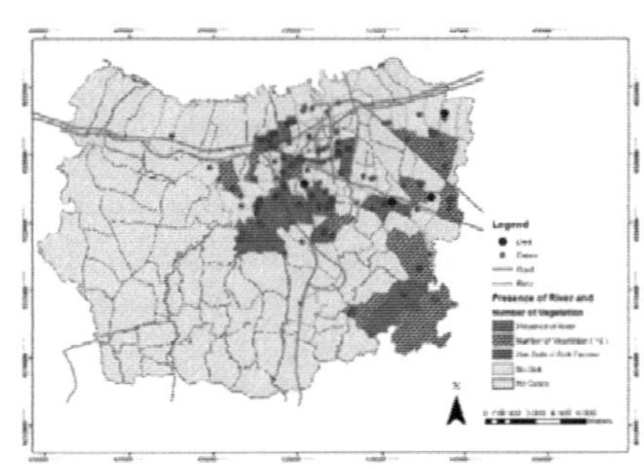

Fig 6. The Presence of River and Vegetation

3.1. The presence of rodent and history of contact

The risk factors existence of rats found in all regions of leptospirosis cases, such as in urban area, rural area as well as in the port area with a history of tidal inundation (Fig. 1). The existence of rodents known from the presence of rat droppings in the home of respondents and the emergence of the rat. Also found a rat hole on the inside and outside of a few respondents. Rats were all over the respondent's house because the environment that supports the life of rats and mice given the reproductive rate was very high.

Indeed, rats are usually considered as the major rodent reservoirs for leptospirosis world wide[15]. Rats reproduce as much as 5-7 times a year with 5-8 tails in the breed, and almost all respondents houses were in densely populated areas causing rats can easily move from one house to another. Rats can get into the house through a small gap and waterways, as well as being able to climb the wall to get into the house through a gap in the roof. Rats move in an area with a radius of 60-80 meters from the nest to the spread follows the pattern of spread of the human. Bacteria of Leptospira lives in the kidneys of mice and can be transmitted through urine and rat tissue, and therefore the presence of a rat is a risk factor of incident leptospirosis [15–17].

Research in Demak and Semarang stated that the presence of the rat is a risk factor leptospirosis [18]. Research in Sleman states of the total 61 respondents there were 54 respondents had the presence of rats inside the house[19]. Similarly, research conducted in France in 2011 which concluded the incidence of leptospirosis is associated with the presence of rodents [20].

History contacts with rodents found in most of the areas with leptospirosis cases (Fig.1). Contact with rodents is one of direct transmission of leptospirosis, from urine or tissues [21–23]. The similar study in Brazil that states contact with rodents was a risk factor for leptospirosis [24]. However, a systematic review of leptospirosis in Indonesia stated that eight out of 13 Indonesian studies and about a quarter of studies from other Asia-Pacific countries had reported a significant association between the existence of rats near human habitation and the occurrence of human leptospirosis infection or disease. Surprisingly, none of the 11 Indonesian studies showed a significant association between the existence of domestic animals/livestock close to residential premises and leptospirosis infection [11].

3.2. History of flood and Tidal Inundation

Only a relatively small area with leptospirosis who have a history of flooding (Fig.2). Flooding can cause the rats come out of hiding and wandering in a residential neighborhood. The flood waters can be the medium transmission of Leptospira Sp, and can be a place to live Leptospira Sp. Flooding can expand Leptospira transmission so that it can increase the risk of exposure to human [19]. From this study, only a small part of the leptospirosis cases have a history of flooding risk factors, but by previous studies that concluded that flooding is a risk factor for leptospirosis. A study in Jakarta said that the outbreak of leptospirosis occurred after the flood [25].

Reports have suggested an association between canine and human leptospirosis with flooded areas and flood events [4]. These associations are likely due to several factors: leptospiral organisms in the soil (e.g., river banks) deposited there by reservoir wildlife enter the floodwaters; spread of floodwaters and the organism; displacement of infected peridomestic wildlife to dry land; and subsequent increased direct and indirect contact with the organism for dogs [4]. Frequently flooded areas within urban boundaries may include impervious surfaces and stream banks and are especially important in the context of leptospirosis transmission due to the higher density of dogs and peridomestic wildlife that live there. The association of leptospirosis with rain is mentioned in several publications and measured in a few others even large outbreaks have been reported during seasonal periods of heavy rainfall and flooding [24,26–30]

History of tidal inundation occurs only in a small area with leptospirosis (Fig. 2). Tidal inundation occurs along the north side of the north coastlines of Semarang. Semarang had a small area with a history of tidal inundation. *Leptospira Sp* cannot survive in salt water with pH out of the range 6.2 to 8.0. While the tidal water pH range between 8.3-8.5. It is also found at Semarang in 2008, history of tidal inundation is not a risk factor of leptospirosis [8].

3.3. Environment condition and The Presence of Pet as Potential Reservoir

The risk factors environmental conditions were evenly spread throughout the region with leptospirosis cases, among others in urban areas, the center of government and commerce, and in the suburb which was a port area with dense housing (Fig. 3). Environmental conditions in the region at risk, because it had an open sewer and was less than two meters away from the house with stagnant sewer water. Likewise with cases scattered on the outskirts of the city, such as in Sub-districts in Gunung Pati, Tembalang, Pedurungan, Genuk, and Tugu. The area was a rural areas that were not as close to housing in urban areas, but the environmental conditions in the region were also considered at risk because of the distance between home and the sewer less than two meters, with an open sewer, sewer water is stagnant and there were piles of garbage around the house.

Environmental conditions as above were by research in Semarang in 2014 which stated environmental conditions with open sewers, and garbage was a risk factor of leptospirosis in Sleman [31]. Research in 2011 also stated that 47 respondents out of 61 respondents leptospirosis had a ditch around the house[19]. Research in Marseille France in 2011 also concluded that the cause of leptospirosis was a pile of garbage inviting the presence of rodents [20]. Indonesian studies frequently (13; 92.9%) point to flooding, stagnant water surrounding the house, poor sewer condition, and poor sanitation of the house and surroundings as potential environmental risk or modifiable determinants for leptospirosis infection [11]. In developing countries, high infection rates were also found in cities, essentially within disadvantaged urban areas that usually show poor sanitation and where rodents are numerous [6,26,32–34].

The result of spatial analysis known that the presence of a potential pet as reservoir was a risk factor only in a small area, only 13.6% of respondents had potential pets as a reservoir (Fig. 3). Potential pets owned by respondents include cats and dogs. Cats and dogs can be a reservoir of the bacterium *Leptospira Sp.* A study in Jakarta found several types of the Leptospira serovar from the urine of dogs and cats in Java Island [25]. The similar result was found in Banyumas, out of 13 respondents only three respondents (23.1%) who have pets in their home [9]. However, a systematic review found that none of 14 Indonesian studies showed a significant association between contact with the domestic animal and human leptospirosis infection [11].

3.4. Waste Water Disposal and Garbage Disposal facilities

The results of this study suggest that almost the entire city has a wastewater disposal (Fig. 4) which is at risk (95.5%), as well as with the results of research in Jakarta concluded sewerage is a risk factor leptospirosis [31]. Almost all respondents had Wastewater Disposal Facility is open and cannot be absorbed which causes puddles of water that can be a place to live *Leptospira* bacteria. Urban slums are lacking access to adequate sewage disposal and water treatment infrastructure [6,34–37].

Proximity to open sewer and public waste disposal sites has been associated with human leptospirosis from other countries [14,24,38]. Wastewater disposal is open can be a pathway to enter into houses. A study about leptospirosis outbreaks in dogs primarily occur due to exposure to water contaminated with urine of an infected animal, and in both dogs and humans, exposure to an open body of water is commonly described as a potential risk for leptospirosis [6].

The result of spatial analysis found that almost all regions with leptospirosis cases in the Semarang city had garbage disposal facilities at risk (95.5%) (Fig.4). Landfills/lay still/dumpster in most of the respondent's houses did not have a lid and is ≤500 meters from the residence; so as to trigger the appearance of mice/rat/rodents because they have broad range distance which is affected by some food resources and their population. Rats will migrate with the longest distance of 1-2 km. The lack of adequate sewage systems, trash deposits, and poor housing favor high rodent densities which in turn lead to environmental contamination with pathogenic. Leptospira and high-level transmission of leptospirosis in these communities [6,24,26,27,39,40]. The same thing was found in studies in Jamaica in 2014 which stated that the incidence of leptospirosis occurs in respondents with an open landfills/lay still/dumpster (50%) [41]. Eight out of 13 Indonesian studies and about a quarter of studies from other Asia-Pacific countries have reported a significant association between the existence of rats near human habitation and the occurrence of human leptospirosis infection or disease. Surprisingly, none of the 11 Indonesian studies showed a significant association between the existence of domestic animals/livestock close to residential premises and leptospirosis infection [11].

3.5. The High Risk Occupation and The use of Personal Protective Equipment (PPE)

The personal protective equipment in this study were long shirts, long pants, shoes, and gloves. Almost all respondents did not use personal protective equipment when doing risky activities such as during the flood, tidal inundation, sewage, garbage/disposal treatment, or even contact with rodents or habitat of rodents. This kind of behavior may extend the possibility of the body surface to exposed by *Leptospira Sp.* The similar study found that in Banyumas from 12 out of 13 respondents did not use personal protective equipment [9]. A systematic review of leptospirosis in Indonesia stated that not wearing personal protective equipment was significantly associated with human leptospirosis infection [11]. These findings of systematic review were from studies that investigated specific study sites and activities, such as an outbreaks of leptospirosis after flooding [4], among town service workers [42], in a slum and flood-prone area [37], in an endemic area, occur in disaster situations such as hurricanes and monsoons [43], is increasingly recognized as an emerging infectious disease with cyclic climatic events [44], and in contact with animal excreta [45].The protective benefit of wearing long trousers or long skirts, instead of shorts, in watery places, has been shown in pond-cleaning activities in Thailand [46].

3.6. The presence of river and Vegetation

The presence of the river as a risk factor of leptospirosis was only found in some areas with leptospirosis (Fig. 6).The river water can be medium of transmission of leptospirosis. Distance to the river with the respondent's house closest was one meter. The distance was so close to the river may increase the risk of flooding, the vegetation such as shrubs and bamboo on the outskirts of the river can be a habitat for rodents [9]. The incidence of leptospirosis in this study as well as the incidence of leptospirosis in the city of Banjarnegara in 2013 that occurred in the settlement/residence with the distance radius 50-600 meters from the river [9].

In Semarang, only some areas in Semarang which had three types of vegetations (Fig. 6). However, according to study conducted in Brazil that concluded the positive relationship between the numbers of leptospirosis cases with the existence of vegetation [24]. Any types of vegetation can be habitat for kinds of rodent. Semarang consists of lowland, highland, coastal and hills. Vegetation in the low-land areas in Semarang were shrubs, ornamental plants, and shade trees.

The type of vegetation in the highlands and hills was higher than that in the lowlands. Such as paddy-fields, gardens, bushes, shrubs and shade trees. Shrubs and paddy-fields were the habitats of rodents; knaggy trees make possible of kind of rodents comes into the house by climbing through branches or twigs which adjacent to the house.

3.7. Distribution of leptospirosis cases and risk factors at Semarang

Leptospirosis cases were spread in 15 sub-districts at Semarang (Fig. 1-6). Most cases in downtown which is an urban area, the center of government, commerce, and industrial areas. Sub-district with most cases was South of Semarang, and the fewest cases were in Mangkang and Banyumanik. Sub-district of South Semarang had more cases and the risk factors because of the areas a nice habitat for rodents due to the environment which is a commerce area and a public cemetery which has a variety of vegetation (Fig. 1,3,6). The human behavior–related risk factors included in Indonesian leptospirosis studies were contact with stagnant water (4; 28.6%); contact with river or flood water, muddy areas (4; 28.6%); swimming in a river (2; 14.3%); taking a bath in a river (6; 42.9%); washing in a river (6; 42.9%); contact with animal urine, bodies, or tissues (4; 28.6%); not wearing personal protection equipment (4; 28.6%); walking barefoot (2; 14.3%); and using streams as a source of drinking water (2; 14.3%) [11]. Urban leptospirosis is a consequence of disorganized urbanization and lack of investment in adequate housing, sewage systems and refuses collection services. The most effective interventions will, therefore, be those that directly address the underlying conditions of poverty, such as lack of access to proper sanitation, which is responsible for the emergence of this urban health problem [26,47].

4. Conclusion

The risk factors were found in more than half of the incidence of leptospirosis include environmental conditions, the presence of rodents, wastewater disposal, garbage disposal facilities, history of contact with rats, and use of PPE. The risk factors are found only in less than half the area with leptospirosis include the presence of pets at risk, the presence of the river, history of flood, history of tidal inundation, and vegetation. Districts with the most cases and the risk factor is Southern District of Semarang. Districts with the fewest cases and the risk factor is Sub-district Banyumanik. District Health Office is expected to conduct cross sectorial cooperation to provide personal protective equipment such as gloves and boots for communities. Optimize the role of health workers always to urge people to maintain a healthy environment. The Community is expected to keep the environment clean. To be carried out further studies with variable density of rats.

References

[1] Raghavan R K, Brenner K M, Higgins J J, Hutchinson J M S and Harkin K R 2012 Evaluations of hydrologic risk factors for canine leptospirosis: 94 cases (2002–2009) *Prev. Vet. Med.* **107** 105–9

[2] Raghavan R, Brenner K, Higgins J, Van der Merwe D and Harkin K R 2011 Evaluations of land cover risk factors for canine leptospirosis: 94 cases (2002-2009). *Prev. Vet. Med.* **101** 241–9

[3] Ward M P, Guptill L F and Wu C C 2004 Evaluation of environmental risk factors for leptospirosis in dogs: 36 cases (1997–2002) *J. Am. Vet. Med. Assoc.* **225** 72–7

[4] Ward M P 2002 Seasonality of canine leptospirosis in the United States and Canada and its association with rainfall. *Prev. Vet. Med.* **56** 203–13

[5] Raghavan R K, Brenner K M, Higgins J J, Shawn Hutchinson J M and Harkin K R 2012 Neighborhood-level socioeconomic and urban land use risk factors of canine leptospirosis: 94 cases (2002-2009). *Prev. Vet. Med.* **106** 324–31

[6] Reis R B, Ribeiro G S, Felzemburgh R D M, Santana F S, Mohr S, Melendez A X T O, Queiroz A, Santos A C, Ravines R R, Tassinari W S, Carvalho M S, Reis M G and Ko A I 2008 Impact of environment and social gradient on Leptospira infection in urban slums. *PLoS Negl. Trop. Dis.* **2** e228

[7] Tassinari W S, Pellegrini D C P, Sá C B P, Reis R B, Ko A I and Carvalho M S 2008 Detection and modelling of case clusters for urban leptospirosis. *Trop. Med. Int. Health* **13** 503–12

[8] World Health Organization 2003 *Human leptospirosis : guidance for diagnosis, surveillance and control* (Geneva: World Health Organization)

[9] Rejeki D S S, Nurlaela S and Octaviana D 2013 Pemetaan dan Analisis Faktor Risiko Leptospirosis *Kesmas J. Kesehat. Masy. Nas.* **0** 179

[10] Lau C L, Smythe L D, Craig S B and Weinstein P 2010 Climate change, flooding, urbanisation and leptospirosis: fuelling the fire? *Trans. R. Soc. Trop. Med. Hyg.* **104** 631–8

[11] Sakundarno M, Bertolatti D, Maycock B, Spickett J and Dhaliwal S 2014 Risk factors for leptospirosis infection in humans and implications for public health intervention in Indonesia and the Asia-Pacific region. *Asia. Pac. J. Public Health* **26** 15–32

[12] Pfeiffer D U, Robinson T P, Stevenson M, Stevens K B, Rogers D J and Clements A C A 2008 *Spatial Analysis in Epidemiology* (Oxford University Press)

[13] Soares T S M, Latorre M do R D de O, Laporta G Z and Buzzar M R 2010 Análise espacial e sazonal da leptospirose no município de São Paulo, SP, 1998 a 2006 *Rev. Saude Publica* **44** 283–91

[14] Oliveira D S C, Guimarães M J B, Portugal J L and Medeiros Z 2009 The socio-demographic, environmental and reservoir factors associated with leptospirosis in an urban area of north-eastern Brazil. *Ann. Trop. Med. Parasitol.* **103** 149–57

[15] Himsworth C G, Parsons K L, Jardine C and Patrick D M 2013 Rats, Cities, People, and Pathogens: A Systematic Review and Narrative Synthesis of Literature Regarding the Ecology of Rat-Associated Zoonoses in Urban Centers *Vector-Borne Zoonotic Dis.* **13** 349–59

[16] Collares-Pereira M, Mathias M L, Santos-Reis M, Ramalhinho M G and Duarte-Rodrigues P 2000 Rodents and Leptospira transmission risk in Terceira island (Azores). *Eur. J. Epidemiol.* **16** 1151–7

[17] Doungchawee G, Phulsuksombat D, Naigowit P, Khoaprasert Y, Sangjun N, Kongtim S and Smythe L 2005 Survey of leptospirosis of small mammals in Thailand. *Southeast Asian J. Trop. Med. Public Health* **36** 1516–22

[18] DH F and Ristiyanto R 2008 DISTRIBUSI DAN FAKTOR RESIKO LINGKUNGAN PENULARAN LEPTOSPIROSIS DI KABUPATEN DEMAK, JAWA TENGAH *Media Penelit. dan Pengemb. Kesehat.* **18**

[19] Febrian F and Solikhah . 2013 ANALISIS SPASIAL KEJADIAN PENYAKIT LEPTOSPIROSIS DI KABUPATEN SLEMAN PROPINSI DAERAH ISTIMEWA YOGYAKARTA TAHUN 2011 *J. Kesehat. Masy. (Journal Public Heal.* **7**

[20] Dupouey J, Faucher B, Edouard S, Richet H, Kodjo A, Drancourt M and Davoust B 2014 Human leptospirosis: An emerging risk in Europe? *Comp. Immunol. Microbiol. Infect. Dis.* **37** 77–83

[21] Levett P N 2001 Leptospirosis. *Clin. Microbiol. Rev.* **14** 296–326

[22] Dobigny G, Garba M, Tatard C, Loiseau A, Galan M, Kadaouré I, Rossi J-P, Picardeau M and Bertherat E 2015 Urban Market Gardening and Rodent-Borne Pathogenic Leptospira in Arid Zones: A Case Study in Niamey, Niger. *PLoS Negl. Trop. Dis.* **9** e0004097

[23] Benacer D, Thong K L, Verasahib K Bin, Galloway R L, Hartskeerl R A, Lewis J W and Mohd Zain S N 2016 Human Leptospirosis in Malaysia: Reviewing the Challenges After 8 Decades (1925-2012). *Asia. Pac. J. Public Health* **28** 290–302

[24] Sarkar U, Nascimento S F, Barbosa R, Martins R, Nuevo H, Kalofonos I, Kalafanos I, Grunstein I, Flannery B, Dias J, Riley L W, Reis M G and Ko A I 2002 Population-based case-control investigation of risk factors for leptospirosis during an urban epidemic. *Am. J. Trop. Med. Hyg.* **66** 605–10

[25] Minisitry of Health of Indonesia 2015 *Health Profile of Indonesia in 2014* (Jakarta)

[26] Ko A I, Galvão Reis M, Ribeiro Dourado C M, Johnson W D and Riley L W 1999 Urban epidemic of severe leptospirosis in Brazil. Salvador Leptospirosis Study Group. *Lancet (London, England)* **354** 820–5

[27] Barcellos C and Sabroza P C 2001 The place behind the case: leptospirosis risks and associated environmental conditions in a flood-related outbreak in Rio de Janeiro. *Cad. saúde pública* **17** **Suppl** 59–67

[28] Amilasan A-S T, Ujiie M, Suzuki M, Salva E, Belo M C P, Koizumi N, Yoshimatsu K, Schmidt W-P, Marte S, Dimaano E M, Villarama J B and Ariyoshi K 2012 Outbreak of leptospirosis after flood, the Philippines, 2009. *Emerg. Infect. Dis.* **18** 91–4

[29] Dechet A M, Parsons M, Rambaran M, Mohamed-Rambaran P, Florendo-Cumbermack A, Persaud S, Baboolal S, Ari M D, Shadomy S V., Zaki S R, Paddock C D, Clark T A, Harris L, Lyon D, Mintz E D, Chawla V, Trivedi T, Yeolekar M, Levett P, Levett P, White F, Hospedales C, Gaynor K, Katz A, Park S, Nakata M, Clark T, Kawaguchi L, Sengkeopraseuth B, Tsuyuoka R, Koizumi N, Akashi H, Ko A, Reis M G, Dourado C R, Jr W J, Riley L, Motie A, Myers D, Silverman M, Aronson L, Eccles M, Eisenstat J, Gottesman M, Myers D, Ruiz A, Applewhaite L, Gonsalez C, Casseb J, Monteiro F, Paula-Neto J, Fernandez R, Sehgal S, Sugunan A, Murhekar M, Sharma S, Vijayachari P, Takafuji E, Kirkpatrick J, Miller R, Karwacki J, Kelley P, Brett-Major D, Lipnick R, Liverpool J, Francis S, Liverpool C, Dean G, Mendez D, Bajani M, Ashford D, Bragg S, Woods C, Aye T, Jr J C, Sulzer C, Pursell A, Zaki S, Shieh W, Alexander A, Genrich G, Guarner J, Paddock C, Shieh W, Greer P, Galloway R, Levett P, Tumeh J, Flowers C, Sanders E, Rigau-Perez J, Smits H, Deseda C, Vorndam V, Karande S, Gandhi D, Kulkarni M, Bharadwaj R, et al 2012 Leptospirosis Outbreak following Severe Flooding: A Rapid Assessment and Mass Prophylaxis Campaign; Guyana, January–February 2005 ed U G Munderloh *PLoS One* **7** e39672

[30] Sehgal S C, Sugunan A P and Vijayachari P Outbreak of leptospirosis after the cyclone in Orissa. *Natl. Med. J. India* **15** 22–3

[31] Auliya R 2014 Hubungan Antara Strata PHBS Tatanan Rumah Tangga dan Sanitasi Rumah Tangga dengan Kejadian Leptospirosis *Unnes J. Public Heal.* **3** 1–10

[32] Halliday J E B, Knobel D L, Allan K J, de C Bronsvoort B M, Handel I, Agwanda B, Cutler S J, Olack B, Ahmed A, Hartskeerl R A, Njenga M K, Cleaveland S and Breiman R F 2013 Urban leptospirosis in Africa: a cross-sectional survey of Leptospira infection in rodents in the Kibera urban settlement, Nairobi, Kenya. *Am. J. Trop. Med. Hyg.* **89** 1095–102

[33] Costa F, Ribeiro G S, Felzemburgh R D M, Santos N, Reis R B, Santos A C, Fraga D B M, Araujo W N, Santana C, Childs J E, Reis M G, Ko A I, Ko A, Reis M G, Dourado C R, Jr W J, Riley L, Ko A, Goarant C, Picardeau M, Costa F, Martinez-Silveira M, Hagan J, Hartskeerl R, Reis M Dos, Riley L, Ko A, Unger A, Reis M, Sarkar U, Nascimento S, Barbosa R, Martins R, Nuevo H, LaRocque R, Breiman R, Ari M, Morey R, Janan F, Karande S, Kulkarni H, Kulkarni M, De A, Varaiya A, Amilasan A, Ujiie M, Suzuki M, Salva E, Belo M, Ganoza C, Matthias M, Collins-Richards D, Brouwer K, Cunningham C, Reis R, Ribeiro G, Felzemburgh R, Santana F, Mohr S, Felzemburgh R, Ribeiro G, Costa F, Reis R, Hagan J, McBride A, Athanazio D, Reis M, Ko A, Gouveia E, Metcalfe J, Carvalho A de, Aires T, Villasboas-Bisneto J, Faria M de, Calderwood M, Athanazio D, McBride A, Hartskeerl R, Costa F, Porter F, Rodrigues G, Farias H, Faria M de, Carter M, Cordes D, Hathaway S, Blackmore D, Kajdacsi B, Costa F, Hyseni C, Porter F, Brown J, Barocchi M, Ko A, Ferrer S R, Faria M T, Reis M G, Lee S, et al 2014 Influence of Household Rat Infestation on Leptospira Transmission in the Urban Slum Environment ed P L C Small *PLoS Negl. Trop. Dis.* **8** e3338

[34] Ganoza C A, Matthias M A, Collins-Richards D, Brouwer K C, Cunningham C B, Segura E R, Gilman R H, Gotuzzo E and Vinetz J M 2006 Determining risk for severe leptospirosis by molecular analysis of environmental surface waters for pathogenic Leptospira. *PLoS Med.* **3** e308

[35] Karande S, Gandhi D, Kulkarni M, Bharadwaj R, Pol S, Thakare J and De A 2005 Concurrent outbreak of leptospirosis and dengue in Mumbai, India, 2002. *J. Trop. Pediatr.* **51** 174–81

[36] Kaur I R, Sachdeva R, Arora V and Talwar V 2003 Preliminary survey of leptospirosis amongst febrile patients from urban slums of East Delhi. *J. Assoc. Physicians India* **51** 249–51

[37] Johnson M A S, Smith H, Joeph P, Gilman R H, Bautista C T, Campos K J, Cespedes M, Klatsky P, Vidal C, Terry H, Calderon M M, Coral C, Cabrera L, Parmar P S and Vinetz J M 2004 Environmental exposure and leptospirosis, Peru. *Emerg. Infect. Dis.* **10** 1016–22

[38] Krøjgaard L H, Villumsen S, Markussen M D K, Jensen J S, Leirs H and Heiberg A-C 2009 High prevalence of Leptospira spp. in sewer rats (Rattus norvegicus). *Epidemiol. Infect.* **137** 1586–92

[39] Yersin C, Bovet P, Mérien F, Wong T, Panowsky J and Perolat P 1998 Human leptospirosis in the Seychelles (Indian Ocean): a population-based study. *Am. J. Trop. Med. Hyg.* **59** 933–40

[40] Romero E C, Bernardo C C da M and Yasuda P H 2003 Human leptospirosis: a twenty-nine-year serological study in São Paulo, Brazil Leptospirose humana: estudo sorológico de 29 anos em São Paulo, Brasil *Rev. Inst. Med. Trop. Sao Paulo* **45** 245–8

[41] Mwachui M A, Crump L, Hartskeerl R, Zinsstag J and Hattendorf J 2015 Environmental and Behavioural Determinants of Leptospirosis Transmission: A Systematic Review. *PLoS Negl. Trop. Dis.* **9** e0003843

[42] Sulong M R, Shafei M N, Yaacob N A, Hassan H, Daud A, Mohamad W M Z W, Ismail Z and Abdullah M R 2011 Risk Factors Associated with Leptospirosis among Town Service Workers *Int. Med. J.* **18** 83–8

[43] Watson J T, Gayer M and Connolly M A 2007 Epidemics after natural disasters. *Emerg. Infect. Dis.* **13** 1–5

[44] Hunter P R 2003 Climate change and waterborne and vector-borne disease. *J. Appl. Microbiol.* **94 Suppl** 37S–46S

[45] Leal-Castellanos C B, Garcia-Suarez R, Gonzales-Figueroa E, Fuentes-Allen J L and Escobedo-De La Pena J 2003 Risk factors and the prevalence of leptospirosis infection in a rural community of Chiapas, Mexico *Epidemiol. Infect.* **131** S0950268803001201

[46] Phraisuwan P, Whitney E A S, Tharmaphornpilas P, Guharat S, Thongkamsamut S, Aresagig S, Liangphongphanthu J, Junthima K, Sokampang A and Ashford D A 2002 Leptospirosis: skin wounds and control strategies, Thailand, 1999. *Emerg. Infect. Dis.* **8** 1455–9

[47] Mcbride A J, Athanazio D, Reis M and Ko A I 2005 Leptospirosis *Leptospirosis Curr. Opin. Infect. Dis.* **18** 376–86

Study of inundation events along the southern coast of Java and Bali, Indonesia (case studies 4-9 June 2016)

I R Nugraheni[1,4], D P Wijayanti[1], D N Sugianto[2,3], A Ramdhani[4]

[1] Faculty of Fisheries and Marine Science, Diponegoro University, Jl. Prof. Soedarto SH. UNDIP, Tembalang, Semarang 50275, Indonesia
[2]Departement of Oceanography, Fisheries and Marine Science Faculty, Diponegoro University, Jl. Prof. Soedarto SH. UNDIP, Tembalang, Semarang 50275, Indonesia
[3]Center for Coastal Disaster Mitigation and Rehabilitation Studies, Diponegoro University, Jl. Prof. Soedarto SH. UNDIP, Tembalang, Semarang 50275, Indonesia
[4]Indonesian Agency for Meteorology Climatology and Geophysics, Jl. Angkasa I No.2, Kemayoran, Jakarta 10610, Indonesia

E-mail: imma.redha@yahoo.com

Abstract. It is important to understand the causes of costal inundation in order to supporting coastal hazard mitigation strategies, especially for the island nation like Indonesia. We investigated the interaction of waves, swell, tides and sea level variability at the coastal inundation events that occurred along southern coast of Java and Bali on June 4 to 9, 2016. The increase of high water level and low water level clearly seen compared to the same astronomic phase tides data before and after the inundation events. The high value of sea level anomaly (SLA) also indicates that there is sea level variability. The wave and swell analysis suggests that extreme waves and swell which occurs during the early of June 2016 has a contribution to the rising of the sea level during the inundation time span. Swell has a major contribution in the wave action that triggering this inundation event given that the significant waves that occur in the time span of the inundation were not affected by wind, but by the swell. This research has shown that the interaction of astronomical tides, SLA height, and high waves and swell along the southern coast of Java and Bali was causing the inundation events at 4-9 June 2016.

Keywords: Coastal Inundation, Tides, SLA, Wave, Swell

1. Introduction

Fifth Assessment Report (AR5) Intergovernmental Panel on Climate Change (IPCC) states that climate change had an impact on the changes of surface winds and sea waves, rise of sea levels and the increase of strong-intensity cyclones/hurricanes frequency in the ocean that potentially increases the incident of inundation / flooding in coastal areas [1]. Coastal inundation events may altering the coastline, damaging the infrastructure, causing contamination of freshwater reserves, and in the worst cases can leading to the loss of human life [2]. Coastal inundation generally caused by astronomical tides, tropical cyclone storm surges [3,4] and the regionally sea level variability such as due to El Niño/Southern Oscillation (ENSO) [5,6]. Understanding the processes that lead to coastal inundation/ flooding is very important to coastal hazard mitigation strategies, especially for the island nation.

In early June 2016, Indonesian Agency for Meteorology Climatology and Geophysics (hereafter referred to as BMKG) reported on the incident of high waves (>2 m) along the southern coast of Java and Bali. Furthermore, according to data from the Indonesian National Disaster Management Agency (hereafter referred to as BNPB), at least 5 coastal inundation incidents and 6 wave tides and abrasion incidents occur along the south coast of Java and Bali during June 4 to 9, 2016. This incident leads to the damage of infrastructure along the coast and severe abrasion [7].

The events reported here are by no means unprecedented. A similar case occurred in the Pacific Island in which the widespread coastal inundation occurred over several consecutive days in December 2008. These events were not associated with generally identified cause (such as: tropical cyclone, or unusually large astronomical tides), but instead the swell wave as the main cause [8].

Research on the inundation and extreme waves at the south coast of Indonesia has been done by previous researchers. The study on modeling of storm tide flooding along the southern coast of Java was conducted which the results showed that although the Cyclones Jacob and George in March 2007 did not generate high storm surges along the coast of Java (surge heights <20 cm), the southern coast of Java was still severely flooded [3]. This suggests that there are factors beyond the hurricane / cyclone that may contribute to the occurrence of the inundation. Furthermore, the simulation of extreme wave events in May 2007 was conducted using the wave model WaveWatch-III (hereafter referred to as WW3) [9] which occurred simultaneously with inundation along western coast of Sumatra and southern coast of Java to Nusa Tenggara. The results showed that the dominant factor causing these extreme events are swell that propagate from the Cape of Good Hope, Africa which caused by the hurricane /cyclone [10]. However, these researches have not shown how the interaction of wave, tides and swell contributed to the inundation incident.

Here, we present the documentation and analysis of the coastal inundation that occurred along southern coast of Java and Bali on June 4 to 9, 2016. We draw on data from a variety of sources, including the disaster report from BNPB, waves and swell modeling data from BMKG, oceanographic observation data by satellite altimetry from AVISO and tides data from Indonesian Geospatial Information Agency (hereafter referred to as BIG). The objective of this research is to investigate the interaction of waves, swell, tides and sea level variability at the inundation incident. The results of this present study would be provide the insight into the causes of coastal inundation events and could be significantly valuable for adaptation and mitigation strategies in coastal inundation hazard.

2. Methodology

2.1. Study area

Study areas of this research are the southern coastal areas of Java and Bali (figure 1). The southern coastal areas of Java and Bali are facing directly to the Indian Ocean. Bathymetric profile of southern Java and Bali are parallel to the coastline which relatively shallow near the coast, gradually deeper toward the open sea and at some point varies drastically drop due to the sea trench [11].

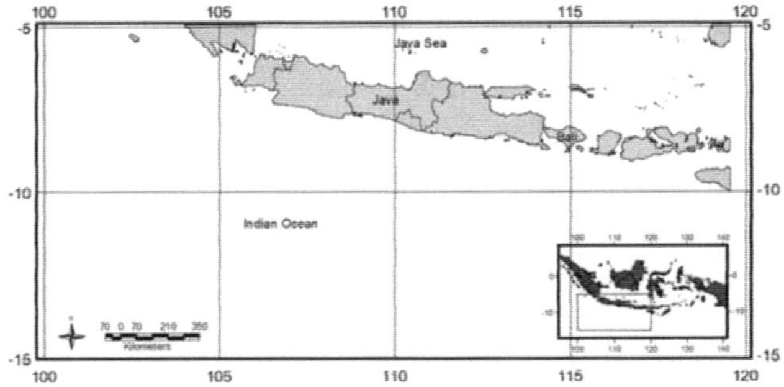

Figure 1. Map of study area (the southern coast of Java and Bali), the red dots show the location of inundation during 4-9 June 2016. Respectively: (1)Pandeglang; (2)Pangandaran;(3)Cilacap; (4)Lumajang; (5)Jembrana.

Oceanographic dynamics in the coastal waters area in the southern coast of Java and Bali are strongly influenced by oceanographic and atmospheric dynamics that occurred in the Indian Ocean. Observed strong swell transmitted from the Southern Indian Ocean (due to hurricane/cyclone) moves in from the south-west to the south coasts of Sumatra and Java [12]. In some cases, the swell propagation leads to the incident of high waves and inundation on the west coast of Sumatra and the southern coast of Java to Nusa Tenggara [10]. Furthermore, the wave actions in the southern coast of Java and Bali are strongly influenced by monsoon activities. Between April and November (east monsoon) south-easterly winds are dominant over sea areas of the southern coast of Java and Bali and it leads to the relatively high wave occurrence [12]. In general, sea level variations of ± 16 cm are seen in the waters area of southern coast of Java and Bali [13,14]. Sea surface height anomaly follows a sinusoidal pattern, which reached the peak in May and the lowest in September [14].

2.2. Inundation reports

Reports of inundation (table 1) were collated from a variety sources including the Disaster Data and Information of Indonesia (hereafter referred to as DIBI) by BNPB (http://dibi.bnpb.go.id/) [7], as well as reports from BMKG. Information contained within the reports varied widely and significant interpretation was required, e.g. precise location of incident was sometimes difficult to ascertain and actual inundation may have occurred somewhat earlier than reported.

Table 1. Inundation reports in early June 2016.

No	Province	District	Date the incident	Source of information
1	Bali	Jembrana	4/6/2016	BMKG report
2	Central Java	Cilacap	6/6/2016 and 8/6/2016	DIBI-BNPB
3	East Java	Lumajang	7/6/2016	DIBI-BNPB
4	West Java	Pangandaran	9/6/2016	DIBI-BNPB
5	Banten	Pandeglang	9/6/2016	DIBI-BNPB

The inundation events occurred almost simultaneously in most areas of the southern coast of Java and Bali. In some areas, the inundation incident was also reported in conjunction with tidal waves and abrasion (in Cilacap and Lumajang) [7]. The report indicates that the incidents are not triggered by local causes, but regional. In addition, reports from BMKG shows that the southern coast of Bali was also experienced an extreme waves (South Karangasem, Gianyar, Sanur and Kuta) and inundation (Perancak - Jembrana District) on June 4 to 9, 2016.

2.3. Tides

The 6 minutely sea level which recorded by encoder sensor from BIG tide gauge data is used as the main data in the analysis of the tides. Tide gauge stations on the south coast of Java and Bali are found in four locations: Pangandaran, Cilacap, Prigi and Benoa. However, we only analyses the location of Cilacap (7.75 ° S - 109 ° E) and Benoa (8.77 ° S - 115.22 ° E) to represent the tides in southern coast of Java and Bali due to the lack of data availability, despite the location of inundation are in Pandeglang, Pangandaran, Cilacap, Lumajang and Jembrana.

Tide gauges data in April, May, July, and August in the same year was also used in the analysis. Tide data on the inundation incidents was compared with tides data in the months before and after the inundation incident which has a same astronomical tides phases to see whether there is any anomaly in the period of inundation events.

2.4. Sea level anomaly (SLA)

The satellite altimetry AVISO delayed-time gridded (1/4° x 1/4°) sea level anomaly daily product (DT-MSLA-H, hereafter referred to as SLA) [15] was used to examine regional sea level variation during the

IOP Conf. Series: Earth and Environmental Science **55** (2017) 012014 doi:10.1088/1755-1315/55/1/012014

inundation events. This satellite product is gridded sea level anomalies data which computed with respect to a twenty-year mean. Hovmöller diagram was used to show the spatial and time distribution of SLA in period of 1-15 June 2016. Hovmöller diagram shows time on one axis of a picture and either latitude or longitude on the other.

2.5. Wave and swell

In situ wave and swell observations do not available in area and time of the study. Numerical hindcast provide the only source of wave and swell information at the study areas. Three-hourly significant wave height (Hs) and primary swell from BMKG [16] on 1-15 June 2016 was averaged to daily data. In addition we also used six-hourly wind speed and direction data to compare to the wave direction. The numerical hindcast of Hs and primary swell was modelled using wave model WW3 with a resolution of 0.125° x 0.125° [17] which has been operationalized by BMKG as a source of wave and swell information in Indonesia.

Parameterization used in the modelling and its validation can be referred to the research of Ramdhani (2015). Validation of the WW3 model show a good agreement with satellite altimetry and buoy data for the high wave condition (bias of 0.12 - 0.21 m to buoy data and 0.02 – 0.06 m to satellite altimetry data) [17].

3. Result and discussion

3.1. Tidal analysis

Spring tides (at the new moon phase) occur in between the inundation events, which was on June 5, 2016. In this condition, the rise of tide is a natural thing. However, on the two tides observation point, the increase of high water level (HWL) and low water level (LWL) during the inundation events (June 4-9) are clearly seen compared to the same astronomic phases tides data before and after the events (figure 2). This indicates the occurrence of sea level height variability during the inundation events.

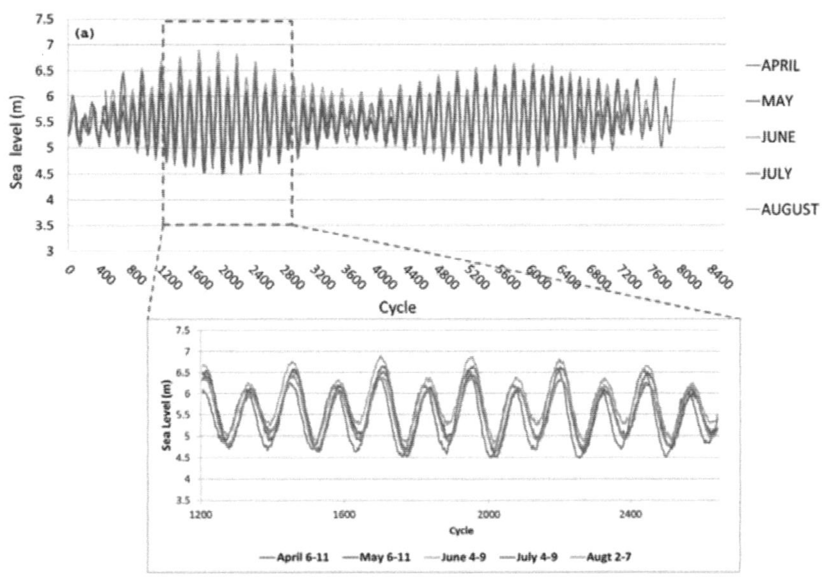

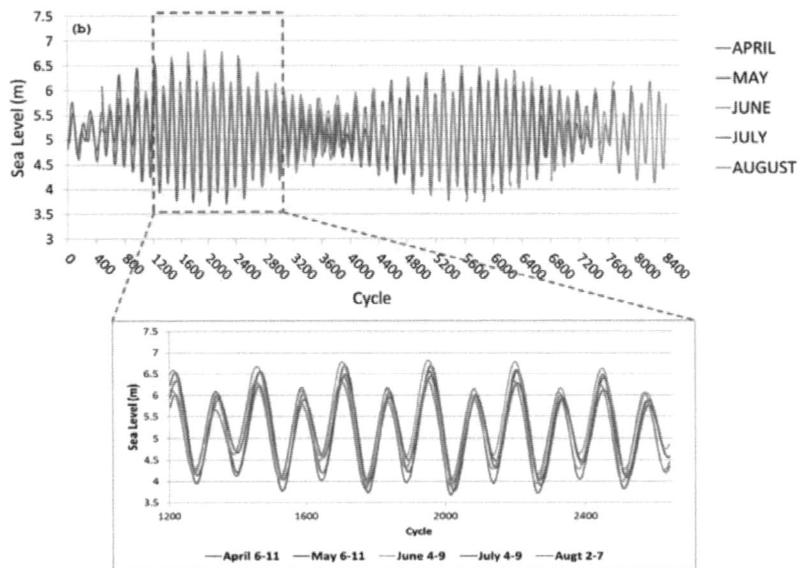

Figure 2. Sea level in (a) Cilacap and (b) Benoa. The main graphic shows the sea level at same astronomic tides phases in April, May, June, July and August and the insert shows the emphasis on June 4-9 as reference time (April 6-11, May 6-11, July 4-9 and August 2-7).

3.2. SLA analysis

The general condition of SLA in the southern coast of Java and Bali can be shown in Hovmöller diagram (figure 3). It shows the highest SLA occurred in the southern coast of Banten, West Java, Central Java and East Java (partially in eastern part). In general, the increase of SLA on the south coast of Banten and West Java (105°E - 108° E) began to occur on June 1 to 8 with the peak occurred on June 3 to 7. On the southern coast of Central Java (108°E - 110° E) to East Java (partially in eastern part, 110°E – 111.5° E), the increase of SLA began to occur on June 4 to 12 and the peak occurred on June 5 to 11. Next on the rest of southern coast of East Java, the SLA is relatively lower compared to the other area that has been mention before with the peak occurred on June 5-8. Furthermore, the highest SLA on the southern coast of Bali occurred in southern part of Bali Strait.

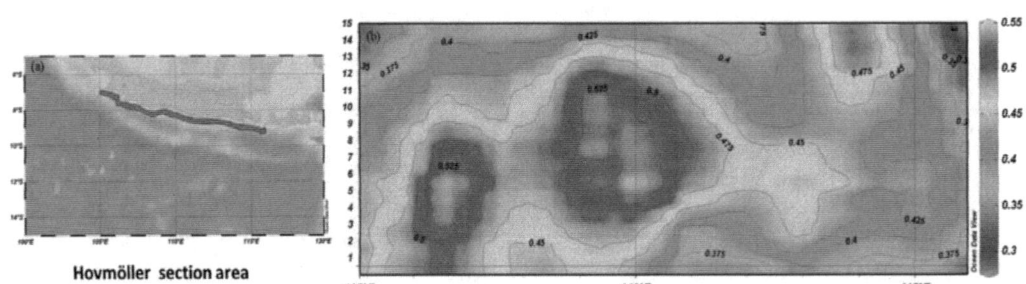

Figure 3. (a) Domain of Hovmöller diagram, red area shows the section area (30 km width) (b) Hovmöller diagram of SLA 1-15 June 2016 (x axis is longitude, y axis is days and the color contour is SLA height).

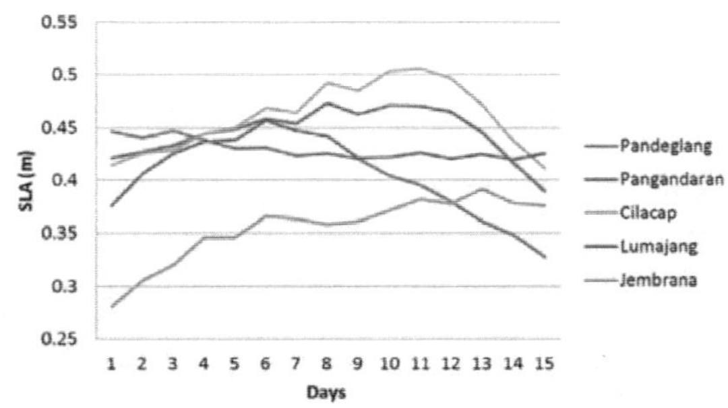

Figure 4. SLA height at the inundation locations during 1-15 June 2016.

SLA height at the location of inundation events (figure 4) respectively ranged between 33-45 cm in Pandeglang; 39-47 cm in Pangandaran; 41-51 cm in Cilacap; 38-46 cm di Lumajang; and 28-29 cm in Jembrana. Overall, the high of SLA in each of location was relatively high compared to the average of SLA variation in southern coast of Java that is ± 16 cm [13, 14].

3.3. Wave and swell analysis

In general, the condition of Hs and swell in 1-15 June 2016 have the same pattern in with the highest peak occurred on June 8, 2016 at all locations (figure 5). High Hs (wave height over 2 m) occur throughout 1-11 June in Pangandaran, Cilacap and Lumajang and at 1-4 and 8-10 June in Pandeglang. On the other hand, Hs in Jembrana is not included on the criteria of high waves.

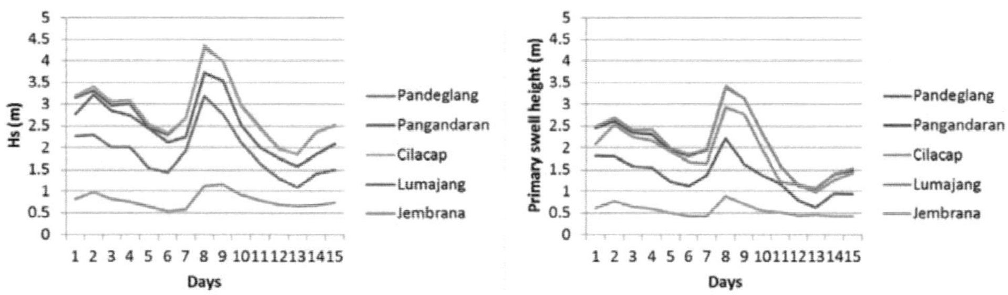

Figure 5. Daily average (a) significant wave height and (b) primary swell height on 1-15 June 2016.

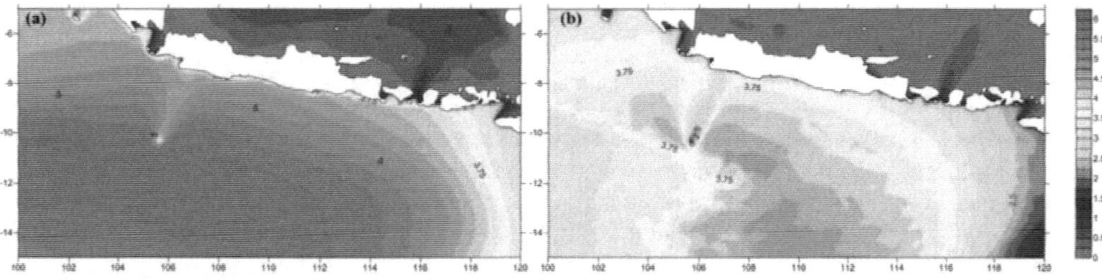

Figure 6. Spatial conditions of daily average (a) significant wave height (b) primary swell height on June 8, 2016.

The highest of Hs and swell height occurred on June 8, 2016, respectively ranging between 1.1 - 4.4 m for Hs and 0.9 - 3.4 m for swell. Spatial condition in the peak of the Hs and swell (8 June) is shown in figure 6. The propagation direction of Hs and swell is relatively same, that is from south to southwest.

Figure 7 is clearly shows that the propagation direction of Hs at 00:00 UTC and 12:00 UTC on 8 June are from south and southwest. In other hand, the wind direction is generally from east and southeast (figure 8). The difference of Hs propagation direction with the wind direction shows that the Hs condition on 8 June 2016 is dominantly influenced by the swell and not by the wind.

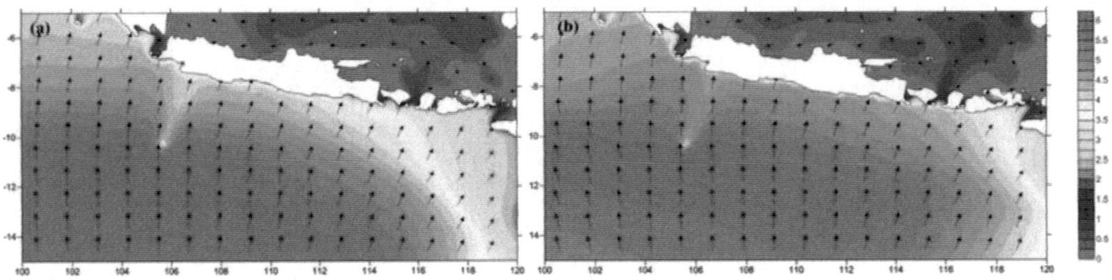

Figure 7. Overlay of Hs and peak wave direction on June 8, 2016 at (a) 00:00 UTC (b) 12:00 UTC.

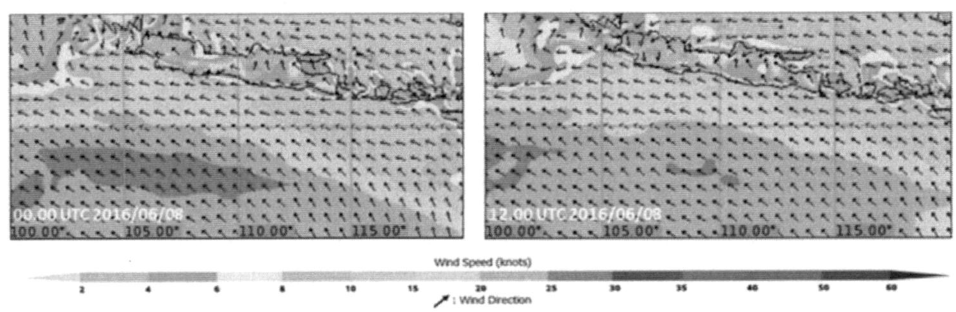

Figure 8. Wind conditions on June 8, 2016 at 00:00 UTC and 12:00 UTC [16].

3.4. Interaction of tides, SLA, wave and swell

The increase of high water level (HWL) and low water level (LWL) that observed in tides data show that there is sea level variability during the inundation events. The values of SLA also strengthen that information. Nonetheless, the rise in sea level that correlated with astronomical tides during this inundation event show a higher value than the sea level on the same astronomical tides phases in the months before and after the events. This suggests that there is another phenomenon that contributes to the rise of sea level. Furthermore, on the time span of the inundation events, the incidence of wave and swell that are relatively high was also observed in almost of all locations. Wave set-up, or the local elevation of the mean still water surface due to the breaking (dissipation) of waves has the potential to be a significant driver of extreme sea levels along the coastlines [8]. In addition, swell dissipation typically generates infra-gravity waves that causes the wave run-up (maximum level of the waves that reach on the beach relative to the still water level) [8]. This suggests that extreme waves and swell which occurs during the early of June 2016 may have a big contribution to the rising of the sea level during the inundation time span. Regardless, swell has a major contribution in wave actions that triggering these inundation events given that the waves that occur in the time span of the inundation were not affected by wind, but by the swell.

4. Conclusion

Our analysis has shown that the inundation events at 4-9 June 2016 was cause by the interaction of astronomical tides (spring tides), anomalously sea level height, high waves and swell along the southern coast of Java and Bali. Superposition of spring tide and SLA in the early June 2016 was causing the rising of sea level. Furthermore, the interaction of high waves that causing the higher wave set-up and

IOP Conf. Series: Earth and Environmental Science **55** (2017) 012014 doi:10.1088/1755-1315/55/1/012014

swell that causing the higher wave run-up with rising of sea level due to spring tides and SLA, lead to the overflow/surging of sea water towards the mainland and causing coastal inundation. Regardless that the exact contribution by each phenomenon / parameter is beyond the scope of this work, this study provides insight that the superposition of these phenomena need to be considered in making a mitigation strategies in coastal inundation hazard.

References

[1] IPCC, 2014: Climate Change 2014: Synthesis Report. Contribution of Working Groups I, II and III to the Fifth Assessment Report of the Intergovernmental Panel on Climate Change [Core Writing Team, R.K. Pachauri and L.A. Meyer (eds.)]. IPCC, Geneva, Switzerland, 151

[2] Barnett J 2011 Dangerous Climate change in the Pacific island: food production and food security. *Journal of Regional Environmental Change* **11** (1), 229-237

[3] Ningsih, N.S., Hadi, S., B. Harto, A., Utami, M.D., P. Rudiawan, A. 2010. Modelling of storm tide flooding along the southern coast of Java, Indonesia. Advances in Geosciences (24), 87-103

[4] Walsh, K.J.E., McInnes, K.L., McBride, J.L., 2012. Climate change impacts on tropical cyclones and extreme sea levels in the South Pacific—a regional assessment. Global and Planetary Change (0), 149–164

[5] Church, J.A., White, N.J., Hunter, J.R., 2006. Sea-level rise at tropical Pacific and Indian Ocean islands. Global and Planetary Change 53 (3), 155–168

[6] Lowe, J.A., Woodworth, P.L., Knutson, T., McDonald, R.E., McInnes, K.L., Woth, K., von Storch, H., Wolf, J., Swail, V., Bernier, N.B., Gulev, S., Horsburgh, K.J., Unnikrishnan, A.S., Hunter, J.R., Weisse, R., 2010. Past and Future Changes in Extreme Sea Levels and Waves, Understanding Sea-Level Rise and Variability. Wiley-Blackwell 326–375

[7] BNPB. 2016. The Indonesian Disaster Data and Information-DIBI. Available at http://dibi.bnpb.go.id/

[8] Hoeke, Ron K., McInnes, Kathleen L., Kruger, Jens C., McNaught, Rebbeca J., Hunter, John R., Smithers, Scoot G. 2013. Journal of Global and Planetary Change 108 (2013), 128-138

[9] Tolman, H.L., 2009.User Manual and System Documentation of WaveWatch III Version 3.14. Tech. Note 276, NOAA/NWS/NCEP/MMAB

[10] Habibie, M.N., Permana, D.S., Suratno. 2013. The Simulation of Extreme Wave due to Swell in Indonesia Using WAVEWATCH-III. Journal Meteorology and Geophysics 14, 99-108

[11] Soeprapto, T.A.. 2004. The bathymetri of southern Java. Bandung: Centre for Research and Development of Marine Geology

[12] Bird, E.C.F. and Ongkosongo, O.S.R. 1980. Environmental changes on the coasts of Indonesia. Japan : United Nations University Press, 510 p

[13] Syamsudin, F. and Kaneko, A. 2013. Ocean variability along the southern coast of Java and Lesser Sunda Islands. Journal Oceanography. Published online: 16 July 2013, The Oceanographic Society of Japan and Springer Japan 2013

[14] Marpaung, S., Harsanugraha, W. K. 2014. Distribution characteristics of sea surface anomalies height in the waters of the north and south part of Java Island. National Seminar on Remote Sensing. Proceedings Sinasindraja 2014. LAPAN, 569-575

[15] AVISO. 2016. Aviso global sea surface height products. Available at http://www.aviso.altimetry.fr/en/data/products/sea-surface-height-products/global.html

[16] BMKG. 2016. Ina-Waves. Available at http://peta-maritim.bmkg.go.id/2.0/

[17] Ramdhani, A. 2015. Effects of tropical cyclone and the Madden-Julian Oscillation (MJO) to the high wave events in Indonesian waters. Doctoral Dissertation Study of Earth Sciences, Bandung Institute of Technology

The Use of Water Exchange for Feeding Rate and Growth Promotion of Shortfin Eel *Anguilla bicolor bicolor* In Recirculating Water System

N Taufiq-Spj[1,2,3], S Sunaryo[1], A Wirasatriya[1,2], and D N Sugianto[1,2].

[1] Fac. of Fisheries & Mar. Scie., Diponegoro Univ. Semarang, Cent. Java, Indonesia.
[2] Member of Center for Coastal Disaster Mitigation and Rehabilitation Studies, Center of an Exellence Research University, Diponegoro Univ. Semarang, Cent. Java, Indonesia.
[3] E-mail: taufiqspj_1999@yahoo.com

Abstract. The shortage of eel's seed *Anguilla* spp. and it's consumption size will induce the hunting of this species. Thus, stocks are critically endangered. Central Java have two indigenious species i.e. *A. marmorata* and *A. bicolor bicolor.* These two species will be endanger if management of stock do not conduct properly. The use of water exchange by replacing water volume for feeding rate and growth promotion of Indonesian shortfin eel *A. bicolor bicolor* in RWS will at least give a basic information in appropriate aquaculture systems and as a part of disaster mitigation of indegenious species extinction. The study used 3 replications and 3 treatments of water exchange, i.e. T_1: 50, T_2: 100, and T_3: 200 % of water removed d^{-1}. The fish were fed with approx. 2.2% dry basis of formulated feed. Tanks with volume of 1 m^3 contain a density of approx. 20 kg seed stock of shortfin eel fingerlink size per m^3 water in RSW. The result shows that feeding rate among treatments are significantly different ($\alpha < 0.001$), where T_1= 1.62±0.23% d^{-1}, T_2 = 1.89±0.06% d^{-1} and T_3 = 1.99±0.06% d^{-1}. Individual and Biomass weight gain increase (IWG: T_1= 0.039±0.003, T_2= 0.050±0.001, T_3= 0.076±0.002 kg and BWG: T_1= 7.087±0.443, T_2 = 11.051±0.894, T_3= 19.722±0.831 kg respectively) by increasing water exchange ($\alpha < 0.001$). Meanwhile the absolute growth was found higher in individual (IAG) compare to biomass (BAG) and both of them increase by increasing water exchange ($\alpha < 0.001$). Where for IAG: T_1= 55.42±3.69, T_2= 71.02±2.40, T_3= 106.12±1.98% and BAG: T_1= 35.11±1.13, T_2= 52.94±6.17, T_3= 90.77±3.14% respectively.

Key words: Eel, *Anguilla bicolor bicolor*, water exchange, feeding rate, growth promotion, mitigation.

1. Introduction

Eels, *Anguilla* sp. is a catadromous species and a popular fish for the worldwide dishes known as *unagi, kabayaki* and *unadon* in Japan, jellied eel in England. Some dishes of *sidat* (*panggang* and *bothok*) in Indonesia, dishes of elver or smoked eel also found in some countries of Europe and Australia. European eel *Anguilla anguilla* as well as Canadian/American eel *A. rostrata,* and Japanish eel *A. japonica* stocks are critically endangered [1,2]. This condition will push the eel traders to hunt the seed or consumption size in any places of countries. Whilst the Indonesian eels, *A. marmorata, A. celebensis, A. ancentralis, A. borneensis, A. bicolor bicolor,* and *A. bicolor pacifica* [3,4], are in a condition of disaster mitigation especially in stock and their population [5,6].

The shortage of eel's seed in Japan caused tracking and hunting seed to Canada, US, Indonesia, Phillippines, and Malay. The price of eel's seed (glass eel) increases very steeply up to US$ 2,600 per pound of eel and known as "gold rush" [7]. Moreover, the price of Indonesian glass eel in fisherman firsthand just only US$ 150 to 220 per kilogram (in 2016). However, this condition will also induce big hunting for this species. Hence, the population of these species (Central Java have only two sepecies of *A. marmorata* and *A. bicolor* bicolor) now is decreasing due to rush hunting and improper environment especially at *Bendung* (dam) Gerak Banyumas and *Bendung* (dam) Menganti, Cilacap – Central Java (Muslim, pers. com., 2016). This condition requires mitigation and rehabilitation of the species especially in lower and mid estuaries before extinction disaster occur. The statistical fisheries of Indonesian long and shortfin eels unfortunately were not officially recorded.

From 2009 to 2011 the activity of fisherman in catching glass eel was going down after the application of government regulation (No PER 18/MEN/2009) about the banning of eel seeds export decreased the activity of glass eel hunting during that time. In 2012 there are a few culture in this species and become bigger in the following years. The culture techniques in the society nowadays slightly develop and become a strong desire during 2015. This is due to the increase demand of eels in Indonesian market. Nevertheless, culture systems which is used by society commonly still use traditional earthen imponding water system. For this reasons, the study on manipulating water exchange will hopefully improve the feeding rate of organism as well as weight gain.

Eel farming, which is responsible for over 90% of all Anguilla production worldwide, is reliant on growing out wild-caught juvenile eels (glass eels or eel fry). Historically, farming and trade in East Asia involved the Japanese eel *A. japonica*, native to the region; from the 1990s large quantities of European eel *A. anguilla* glass eels were also imported, due to decreasing availability of *A. japonica* (Shiraishi and Crook, 2015). International trade caused *A. Anguilla* to be listed in the Convention on International Trade in Endangered Species (CITES) of Wild Fauna and Flora in 2007. In addition, the European Union banned all trade in *A. anguilla* from the EU in 2010. As a consequence, the Americas and South-East Asia have become increasingly important sources of juvenile eels of other *Anguilla* spp. for farms in East Asia [8]. The aims of this study were to evaluate water exchange for feeding rate and growth performance (i.e. absolute growth and weight gain) as well as to compare effectiveness of water coloumn interm of biomass density.

2. Methodology

Big pencil sizes (elver) of eel *A. bicolor bicolor* were obtained from seed collector farm in Purwokerto (Central Java). Biomass of fish seeds (approx. 20 kg) had been distributed to every tank with 1 m^3 water volume, hence various masses will slightly differ among the tanks. Completely random design and analysis co-variance were used with 3 treatments and 3 replications. In this basic recirculating system [9] nine tanks of high density poly ethylene (hdpe), 1 water reservoir, single biofilter and single patch to collect water were used in every treatment. In order to recirculate water, submerged pump was also used in every group of treatment. In term of water exchange from the tank, additional fresh water of 50, 100 and 200 % d^{-1} of water replacement was applied in recirculating water system. The fresh water was flown through perforated hose with valve for adjusting water volume. A mixture of commercial feeds (Table 1) was given by 2.2 % d^{-1} dry basis and given twice a day in the morning (06.00 a.m) and before dark (05.00 pm), uneaten food was collected and weighted after two hours feeding. This feeding percentage is mainly due to initial study that maximum feeding rate never attain 2.2 % and as a concern to save the environment.

Table 1. Feeds proximate analysis given to fish

Feeds	Water (%)	Ash (%)	Protein (%N)	Lipid (%)
1 (50%)	10,51	13,18	39,67	12,17
2 (50%)	9,25	10,98	27,71	12,04

Estimated marginal mean was used in Feeding Rate, while individual and biomass growth were analyzed by using generalized linear model repeated measurement ANOVA and post hoc test using multiple comparison based on observed means. Feeding Rate (%) were calculated by using this formula, i.e. amount of food given per day minus uneaten food divided by fish biomass in every tank. Absolute growth of individual and biomass of fish per tank in group of treatments (T_1, T_2 and T_3) was calculated by the difference of weight at "t" time and initial weight divided by initial weight in percent. Water temperature, pH and dissolve oxygen were recorded every second day, while salinity, NH_3, NO_2^- and alkalinity measured in every 20 days. As an addition, in order to evaluate water coloum efficiency, biomass density l^{-1} was calculted by amount of biomass in the end of measurement divided by water volume in the captive tank.

3. Result and Discussion

Anguilla sp. is a taft organism, they can survive even captive in a limited water condition. The results of this study show that fish culture was commonly found healthy in every tank of treatment, fish behaviour look active in T_1 (50%) and T_2 (100%), and look very active in T_3 (200% water exchange). Daily feeding, absolute growth both of individual and biomass of *A. bicolor bicolor* generally increase due to increasing water exchange. Water quality seems to show in normal range of temperatures, pH, salinity, and alkalinity, but increased consistenly in NH_3 and NO_2^- concentration during 100 days culture.

3.1. Feeding Rate

The given paste feed assemblied with fish biomass (approx 2.2% of body weight) in the tanks, uneaten food was decreased by increasing water exchange and feeding rate is increased by increasing water exchange. The increase of feed rate shows significant value on the time series of observation ($\alpha = 0 < 0.01$) in 100 days culture (Table 2 and Fig. 1). The influencing factors for feeding rate are food and feeding habit. Concerning food habit, it will be difficult to distinguish the treatments due to the feed given used simply from similar sources, form, and type of food.

Nevertheless, from the factor of feeding habit, differenciation among treatments can be seen from the environmental exchange.These conditions show that increasing the water exchange can reduce uneaten food. Hence water qualities in the rearing tank were better in higher water exchange (Table 3). Eels living in a better environment took more food for their consumption (Table 2). Moreover in a high food consumption, it will also leave uneaten food and faeces, and will induce deteroiration of organic matter, especially in NH_3 as a results of protein degradation (Table 3).

Table 2. Mean of Feeding Rate (% d^{-1}) of eel during 100 days observation

Treat-	Days of observations						
ments	01-20	21-40	41-60	61-80	81-100	Mean	±SD
T_1	1.30	1.75	1.54	1.92	1.57	1.62	0.23%
T_2	1.79	1.90	1.89	1.91	1.96	1.89	0.06%
T_3	1.93	2.07	1.93	1.98	2.04	1.99	0.06%

Table 2. show that T_3 with 200% of water exchange gave best value of FR (1.99±0.06% d^{-1}) and followed by T_2 (1.89±0.06% d^{-1}) and T_1 (1.62±0.23% d^{-1}). These conditions were also indicated by post hoc test base on observed means, where the mean different of FR are significantly different at 0.05 level or 95% confidence interval. Mean of feeding rate during 100 days observation increased by increasing the water exchange, but as seen from the rhythm for all treatments, the feed rate seems to be fluctuates (Table 2).

This fluctuation of feeding rate may correspond with water quality measured, where at treatment with 50 % water exchange (T_1), temperature had highest in fluctuation (±SD = 1.06°C), lowest dissolve oxygen, highest in NH_3 and NO_2^- concentrations compare to T_2 and T_3 (Table 3). In term of water temperature, [10] used different temperatures for racing *A. marmorata* and *A. bicolor pacifica* in recirculating water system. A better FR for *A. bicolor pacifica* found was 2.41±0.04% day^{-1} and for *A. marmorata* (2.23±0.10% day^{-1}) occured at 33°C, while temperature at 28°C stated have lower FR.

Table 3. Water quality (Mean ±SD) measured in every 20 days during 100 days culture system.

Treatments		T°C	S‰	pH	O_2	NH_3	NO_2^-	Alkalinity
T_1	Mean	28.93	1.27	7.43	3.73	0.0089	0.0078	66.17
	±SD	1.06	0.06	0.06	0.15	0.0028	0.0025	6.79
T_2	Mean	29.53	1.27	7.47	4.40	0.0088	0.0078	70.33
	±SD	0.36	0.06	0.06	0.10	0.0016	0.0014	6.43
T_3	Mean	29.63	1.13	7.64	4.70	0.0071	0.0062	82.00
	±SD	0.26	0.06	0.05	0.10	0.0007	0.0006	1.00

3.2. Growth performance

3.2.1. Absolute growth

Absolute growth is normally used by farmers in order to get figure wether the culture had been efficiently conducted or not. During 100 days observation the cultured organisms have grown significantly different either individually or biomass by the time. ANOVA analysis of those growth measured during 6 times observation the grown show significantly different by the time ($\alpha = 0 < 0.01$). Figure 2 and 3 show that the treatment 3 (T_3) gave the best growth either in individual and biomass of fish followed by T_2 and T_1. The growth pattern of both individual and biomass seems to be similar. When continued to conduct post hoc test, the different of individual and biomass mean weight indicated significantly difference ($\alpha = 0 < 0.01$). These growth performance seems related to the feeding rate above, the highest feeding rate gave the best growth, whilst, the lowest feeding rate followed by low growth of *A. bicolor bicolor*.

Relationship between feeding rate, growth and FCR was described by [11]. They reported that, there is a rapid increase in growth rate when the feeding rate is over maintenance requirement since the energy consumed equals the energy required by fish to maintain itself without mobilizing endogenous energy reserves. Then, growth increases at a decreasing rate to a maximum point of feeding rate. Moreover, FCR decreases as the feeding rate increases towards to the maximum rate. However, this level varies according to the fish species, water temperature and fish size [11].

In holding *A. bicolor bicolor* in this study, the rearing tank with lower water exchange caused rapid exchange of water qualities, hence organism that live in this habitat required a higher adaptation to its environment. Lower water exchange almost provided high NH_3 and NO_2^- and became consistently bigger during 100 days observation. Alkalinity will be related to hydrogen ion concentration (pH), and the pH will affect to NH_3 and NO_2^- concentration. The pH concentrations was between 7.43 to 7.64, these correspond with [12] stating that *Nitrosomonas europaea* can grow better in pH between 7 to 8. Nevertheless, increasing NH_3 and NO_2^- along the culture, can be related to dissolve oxygen which is concerned to have relatively low concentration (3.73 to 4.70 ppm) (Table 3).

IOP Conf. Series: Earth and Environmental Science **55** (2017) 012015 doi:10.1088/1755-1315/55/1/012015

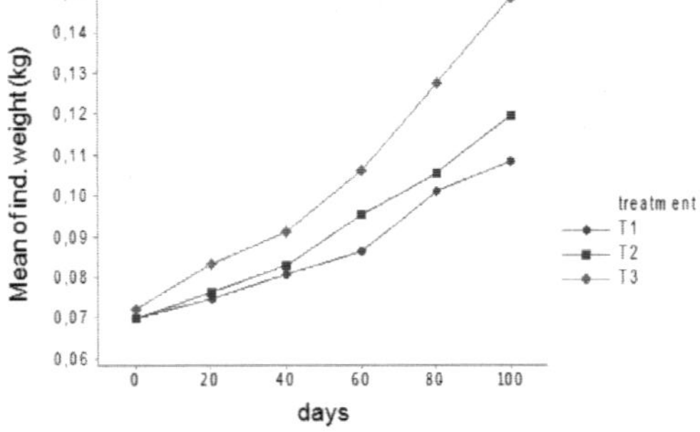

Figure 1. Growth performance in weight (kg) (individual absolute growth) of fish (*A. bicolor bicolor)* measured in every 20 days during 100 days culture in different volumes of water exchange.

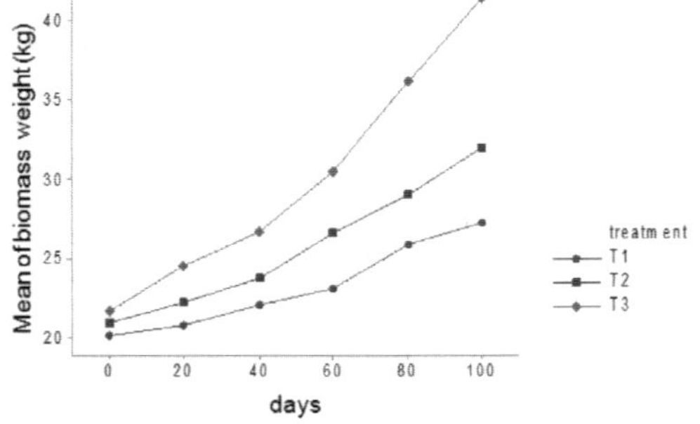

Figure 2. Growth performance in weight (kg) (Biomass absolute growth) of fish (*A. bicolor bicolor)* measured in every 20 days during 100 days culture in different volumes of water exchange.

This adaptation processes require high energy of captive fish. Protein is sparing effect of lipid, if energy in the feed for organism is inadequate to be used for environmental adaptation. Hence protein energy will be used for adaptation, not for growth. Hereafter the graph of feeding rate looks fluctuate inconsistently especially for T_1 of 50% water exchange d^{-1}(Table 2) compared to growth performance (on Figure 1 & 2).

Low growth in T_1 occured during the first 40 days culture and start to grow exponentially after 60 days culture (Figure 1 & 2). These correspond to stress during intial culture where the death rate was bigger than T_2 and T_3 and in fact that initial weight was smaller than T_2 and T_3 (Table 3.). The eels seem to face in a new environment and have to adapt with different water exchange, where the previous environment of eels were captive in earthen pond with quite enough running water of water exchange. The initial decrease in weight of *A. anguilla* has previously been observed by [13], and can be regarded as a common phenomenon caused by the time needed for the eels to complete their metamorphosis and to adapt to their new environment. [14] describes an initial reduction in length and weight of European glass eels in aquaria, fed or unfed, while pigmentation advances. These explanation may correspond with decreasing weight in one of replication tank of T_1 and T_2 in the early culture of 20 days. On the other hand, T_1 has water temperature mean of 28.93°C with slightly higher in ±SD (1.06°C) than others (Table 3).

This temperature fluctuation will lead to unstable metabolic rate of aquaculture organisms. The organism will take more energy and oxygen consumption as the water temperature increases [15]. For this reason, mean values of dissolve oxygen in T_1 almost lower compare to T_2 and T_3. Since *A. marmorata* and *A. bicolor pacifica* are tropical and subtropical species, they can tolerate higher water temperatures [10], while some of Indonesian shortfin eel can tolerate to 41°C temperature [16]. The optimal growth temperatures of the juveniles of *A. bicolour pacifica* and *A. marmorata* are estimated to be 30.1°C and 27.6°C respectively [10]. On the other hand, *A. Anguilla* entered a state of torpor at water temperatures of 1°C and 3°C when fish acclimated at 23°C and 29°C respectively while temperature decreases 1°C every 10 min., *A. rostrata* could still survive when acclimated at ≤ 5°C for more than five weeks though the fish ceased feeding and displayed a dramatic decreased in oxygen consumption rates [17, 18]. In this temperature fluctuation perspective, *A. bicolor bicolor* still has a normal feeding habit while some time protruding their head onto water surface may due to limited dissolve oxygen.

In calculation during 100 days culture, Table 4 shows that individual absolute growth (IAG) consistently higher in every treatment than biomass (BAG), while both of them increase by increasing water exchange. Comparison among treatments have sighnificantly different either individual or biomass ($\alpha = 0 < 0.01$). The best percent IAG attain in T_3 = 106.12±1.98% and BAG also in T_3 = 90.77±3.14% during 100 days culture (Table 5). [10] used 45 days culture of two species, with initial and final mean of body weights of 5.62 ± 0.30 g and 11.59 ± 0.03 g for *A. bicolor pasifica* and 4.97 ± 0.36 g and 7.18 ± 0.80 g for *A. marmorata* at 28°C, hence the BAG are approx. 106,2% and approx. 44,5 % respectively. Moreover, Angelidis *et al.* (2005) using *A. anguilla* have mean of daily absolute growth 5.5 % observed between 76th and 105th day of culture, and decreased to 2.9 % observed during 52 days latter. For these facts, the individual absolute growth of *A. bicolor* bicolor is quite similar to *A. bicolor pasifica*, much higher than *A. Marmorata* [10] and much lower than *A. anguilla* [19].

3.2.2. Weight gain

The weight gain of *A. bicolor bicolor* increased by increasing water exchange ($\alpha = 0 < 0.01$) either individually or in biomass after cultured during 100 days. Mean of individual weight gain in T_1 indicate only has 0.039±0.003, T_2 = 0.050±0.001, and T_3 = 0.076±0.001 kg. Mean while the biomass gain shows 7.087±0.443, 11.051±0.894, and 19.722±0.831 kg for T_1, T_2, and T_3 respectively (Table 4). It means that IWG of of this study equal to 0.39, 0.50, and 0.76 g d^{-1} in T_1, T_2, and T_3 respectively (calculated from Table 4).

Table 4. Individual and biomass weight gain (IWG & BWG in kg), individual and biomass absolute growth (IAG & BAG in %) of *A. bicolor bicolor* race during 100 days in recirculating water system (where: $I_{0..100}$: Individual weight of fish at day 0 to 100, $B_{0..100}$: biomass weight of fish in tank at day 0 to 100)

Treat-ments	I_0 (kg)	I_{100} (kg)	IWG (kg)	IAG (%)	B_0 (kg)	B_{100} (kg)	BWG (kg)	BAG (%)
$T_{1.1}$	0.070	0.112	0.041	58.77	20.877	28.475	7.598	36.40
$T_{1.2}$	0.070	0.107	0.036	51.46	19.858	26.667	6.809	34.29
$T_{1.3}$	0.069	0.107	0.038	56.04	19.789	26.643	6.854	34.64
Mean±SD			0.039±0.003	55.42±3.69			7.087±0.443	35.11±1.13
$T_{2.1}$	0.069	0.119	0.050	73.09	20.479	32.130	11.651	56.89
$T_{2.2}$	0.071	0.119	0.048	68.40	21.869	31.892	10.023	45.83
$T_{2.3}$	0.071	0.121	0.050	71.59	20.465	31.944	11.479	56.09
Mean±SD			0.050±0.001	71.02±2.40			11.051±0.894	52.94±6.17
$T_{3.1}$	0.072	0.148	0.076	104.67	21.672	41.452	19.779	91.27
$T_{3.2}$	0.073	0.152	0.079	108.37	21.917	42.439	20.522	93.63
$T_{3.3}$	0.071	0.145	0.075	105.32	21.580	40.444	18.864	87.41
Mean±SD			0.076±0.002	106.12±1.98			19.722±0.831	90.77±3.14

A. bicolor pasifica can attain DWG of 2.63±0.12 g d^{-1}, while 1.19±0.38 g d^{-1} for *A. marmorata* which both achieved at 28°C of captive water [10]. Nevertheless, [20] reported that European eel *Anguilla anguilla* fed with different types of food i.e. trash fish (TF), slaughterhouse waste (SW), pelleted trout feed (PF), and combination of them (TF+PF and SW+PF) only have daily growth of 0.21±0.01, 0.32±0.03, 0.29±0.01, 0.30±0.17, and 0.0.36±0.23 g respectively. [21] use different sources of diet, diet from fish (CB) source seems to be effective to give better weight gain instead of combination between fish and beef tallow (BT) or fish with soybean (SB). The results show that the CB give 42% absolute growth, while BT and SB only 37 and 30 % respectively. [22] conducted an eel farm in recirculating water system, the 20,000 elvers of *A, anguilla* with mean initial weight of 5.89g (total weight 117.8 kg) produced a total weight of 478.8 kg after 335 days of farming with mortality 27.1%. It means that the biomass weight gain only attain 361 kg for almost one year with 14,580 elver survive. Hence individual weight gain only 24.76 g or additional growth 0.074 g d^{-1}. In this study daily weight gain seems to have lower result than [10], but much better results than [20] and much better than [22].

Nevertheless, further comparison of the results conducted by [10], they used *A marmorata* with individual initial weight range 4.97±0.36 g, *A. bicolor pasifica* 5.62±0.30 g and stock with 20 eels in each tank of 105 l water volume. This means that [10] used initial biomass density 0,95 g l^{-1} for *A. marmorata*, and 1,07 g l^{-1} for *A. bicolor pasifica,* whilst for this study used *A. bicolor bicolor* with initial biomass density approx. 21.72 g l^{-1}. For this fact, the initial density of this study is 22 times more than that applied to *A. marmorata* and almost 21 times more than that applied to *A. bicolor pasifica*. When using equal comparison of culture duration, during 100 days culture, *A. bicolor pasifica* will attain total biomass of approx. 377.73 g, while *A. marmorata* 197.62 g in 105 l water volume. Hence in term of water column eficiency, during 100 days culture of eels species, this study can prove that T3 can maintain the biomass density of *A. bicolor bicolor* as much as 41.45±0.99 g l^{-1}, whilst for *A. bicolor pasifica* and *A. marmorata* only can hold approx. 3.60 and 1.88 g l^{-1} (calculated from [10].

4. Conclusion

Feeding rate of *A. bicolor bicolor* apparently follows the volume of water exchange in maximum of 200% water exchange. Higher water exchange will lead higher feeding rate and growth performance. These water exchange will correspond the water quality as well as growth performance either absolute growth and weight gain. Mass culture using at least 1000 l of water column can perform stable aquaculture in term of water column efficiency. Efficiency value of this study represents by biomass density per litre of water volume, which having a value of 41.45±0.99 g l^{-1} during 100 days culture.

Acknowledgements

Authors thank to every single person who help and involve for this study. This work was financially supported by Diponegoro University, Funding source of PNBP DIPA UNDIP fiscal year of 2016. Special thanks to Mr. Tri Ananto who help in setting up the recirculating system.

References

[1] Able K W, Smith J M and Caridad J F 2015 American Eel Supply to an Estuary and Its Tributaries; Spatial Variation in Barnegat Bay, New Jersey *J. of Northeastern Naturalist* **22** (1): 53-68.

[2] Bevacqua D, Meli_a P, Gatto M, De-Leo G A 2015 A global viability assessment of the European eel. *Glob. Change Biol.*, doi **10** 1111/geb 12972.

[3] Aoyama J 2009 Life History and Evolution of Migration in Catadromous Eels *(Genus Anguilla) Aqua-BioSci. Monogr.* **1**(2): 1-3.

[4] Teng H-Y, Lin Y-S and Tzeng C-S 2009 A new Anguilla species and a reanalysis of the phylogeny of freshwater eels *Zoological Studies* **48** 808-822.

[5] Jacoby, D. and Gollock M 2014 *Anguilla bicolor, Anguilla celebesensis, Anguilla japonica, Anguilla luzonensis* and *Anguilla marmorata. The IUCN Red List of Threatened Species.* Version 2014.2: www.iucnredlist.org.

[6] Watanabe S, Aoyama J and Tsukamoto K 2006 Confirmation of morphological differences between *Anguilla australis australis* and *A. australis schmidtii New Zealand Journal of Marine and Freshwater Research* **40** (2) 325-331.

[7] Berita KKP 2016 RPP Sidat. http://djpt.kkp.go.id/ditsdi/arsip/c/899/Penyusunan-Dokumen-Rencana-Pengelolaan-Perikanan-RPP-Sidat/

[8] Shiraishi H and Crook V 2015 Eel market dynamics: An analysis of Anguilla production, trade and consumption in East Asia *TRAFFIC Report*. Japan.

[9] Lee J, Jo J Y, Lee SM, Hutabarat J and Taufiq-Spj N 2009 Biological Estimation of Waste Products from Olive Flounder (*Paralichthys olivacens*) Fed on Three Different Feed Types *Fish Aqua Sci.* (*KOSFAS*) **12** (4) 317-323.

[10] Luo M, Guan R, Li Z and Jin H 2013 The effect of water temperature, on the survival, feeding, and growth of the juveniles of *Anguilla marmorata* and *Anguilla bicolor pasifica*. *Aquaculture* 400-401: 61-64.

[11] De-Silva S S and Anderson T A 1995 Fish Nutrition (1st Ed.Chapman and Hall) pp 32-35.

[12] Hill D R and Webster T 2008 High pH Inhibit Nitrifying Bacteria *Des Moines Water Works* (www.dmww.com) *American Water Works Association*. Opflow July 2008.

[13] Holmgren K, Wickstrô M H and Fredg A K 1991 Difficulties when estimating growth of eel in a mesoscale experiment *European Inland Fisher y Advisor y Commission (EIFAC) Working Party on Eel*, Dublin, Ireland, 20-25 May 1991, 1-19.

[14] Tesch E W 1977 The Eel: Biology and Management of Anguillid Eels. (*Chapman and Hall*), London.

[15] Clark A and Johnston N.M 1999 Scaling of metabolic rate of body mass and temperatures in teleost fish *J. of Animal Ecology* **68** 893-905.

[16] Fahmi M R 2010 Phenotypic Plastisity Kunci Sukses Adaptasi Ikan Migrasi: Studi Kasus Ikan Sidat *(Anguilla sp.) In*: Fahmi MR (ed) *Prosiding Forum Inovasi Teknologi Akuakultur*; Balai Riset Budidaya Ikan Hias, Desember 2010. Depok.

[17] Sadler K 1979 Effect of temperature on the growth and survival of the European eel, *Anguilla anguilla* L. *J of Fish Biology* **15** 499-507.

[18] Walsh P J, Foster G D and Moon T W 1983 The effects of temperature on the metabolism of the American eel *Anguilla rostrata* (LeSueur): compensation in the summer and torpor in the winter *Physiological Zoology* **56** 532-540.

[19] Angelidis P, Pournara I, and Photis G 2005 Glass eels (*Anguilla angilla*) growth in recirculating system, *Mediterranean Marine Science* **6** (1) 99-106.

[20] Altun T, Takelioğlu N, Nevşat E and Sağat Y 2005 Some growth parameters on European eel (*Anguilla anguilla* L., 1758) fed with different feeds *E.U. Su Urunleri Dergisi. EU J. of Fisheries and Aquatic Sciences* **22** (1-2) 215–219.

[21] Luzzana U, Scolari M, Dall'orto D C, Caprino F, Turchini G, Orna E, Sinesio F and Valfre F 2003 Growth product quality of European eel (*Anguilla anguilla*) as affected by dietary protein and lipid sources *J Appl. Ichtyology* **19** 74-78.

[22] Fotis G, Florou-Paneri P and Gavriildou E 2000 Effects of closed recirculating systems on growth of eel *Anguilla Anguilla* (L.). *Animal Science Review* (in Greek) **27** 55-66.

Association of Macro-minerals Intake and Changes in Internal Carotid Artery-Intima Media Thickness in Post Ischemic Stroke Patients

Dwi Pudjonarko[1,2], Dodik Tugasworo[1], Rumintang Silaen[1]

1. Department of Neurology, Faculty of Medicine Diponegoro University / Kariadi Central Hospital. Jl. Dr. Soetomo 14 Semarang, Central Java, Indonesia.
2. Corresponding Author, E-mail: dr.onang@yahoo.co.id

Abstract. Carotid Intima Media Thickness (C-IMT) has been widely used as marker for atherosclerosis. Previous studies on minerals intake and its association with C-IMT revealed various. Most of the studies showed inconsistent results. The aim of this study is to determine wether macro minerals intake is related to internal carotid-intima media thickness (IC-IMT). This is a longitudinal study, pre test post test design conducted in Neurology clinic, dr. Kariadi hospital, Semarang from June to December 2014. Subjects were 22 post ischemic stroke patients. Minerals intake and IC-IMT was measured using Food Frequency Questionnaire and Duplex Carotid Ultrasonography. Statistical analysis was performed using Chi-Square, Fisher Exact and Logistic Regression test. Subjects included in this study were 17 male subjects (77.3%) and 5 female subjects (22.7%). Mean of IC-IMT in female subjects was found to be higher than in male. Mean of total IC-IMT was increased after a period of six months (0.96±0.80 to 0.97±0.21 mm). There were significant association between calcium as well as sodium intakes and IC-IMT. In contrast, there were no association between magnesium as well as potassium intake and IC-IMT. Multivariate analysis suggest that sodium intake (OR=26.828) was the most influencing factor for IC-IMT, followed by calcium intake (OR=0.042). Calcium as well as potassium intake were independently associated with IC-IMT. Magnecium as well as sodium intake were not independently associated with IC-IMT changes. Sodium intake was the most influencing variable to IC-IMT changes, followed by calcium intake.

Keywords: calcium, potassium, sodium, magnesium, IC-IMT

1. Introduction

Atherosclerosis is the main pathway underlying stroke occurrences. Previous long term studies had identified risk factors for atherosclerosis including family history, age, sex, obesity, dyslipidemia, hypertension, diabetes, physical inactivity, and smoking [1][2][3][4][5]. Carotid intima media thickness (C-IMT), measured noninvasively using high resolution ultrasound imaging, had been widely used as a marker for atherosclerosis. A previous large scale study has revealed that increased C-IMT was a predictor for stroke occurrence [2][6][7][8][9].Early detection as well as modification of risk factors for increased C-IMT can bring significant influence to prevention of atherosclerotic diseases such as stroke. Hypertension, diabetes mellitus (DM), dyslipidemia, and smoking have been associated with increased C-IMT, though it was incompletely explain how these risk factors cause high proportion in C-IMT difference [10].

Some studies concerning mineral intake and C-IMT still revealed various and showed inconsistent results. A study involving 5181 healthy subjects revealed that there was no association between magnesium intake and internal carotid-intima media thickness (IC-IMT). In contrast, there was association between magnesium intake and common carotid-intima media thickness (CC-IMT) [11]. Another study involving 351 healthy subjects revealed that sodium intake as well as ratio of sodium/potassium have negatively correlated with C-IMT[12][13]. A study involving 592 healthy subjects showed that magnesium intake was associated with C-IMT, however calcium, sodium and potassium intake were not associated with C-IMT[14]. Studies on mineral intake and the association with C-IMT changes were relatively scarce. Moreover, C-IMT studies in post ischemic stroke patients focusing measurements in internal carotid artery segment (as a site of advanced phase of atherosclerosis) were also scarce [15]. This study was aimed to determine the association between macro minerals intake and internal carotid intima media thickness (IC-IMT) changes in post ischemic stroke patients with IC-IMT above normal value on the first measurement.

2. Methods

This study was a research in the field of Neurology, Nutrition and Radiology. This study was performed using one group pretest-posttest design. Targeted population in this study was post ischemic stroke patients. Study population was post ischemic stroke patients who visited outward Neurology clinic, dr. Kariadi hospital from June to December 2014. Subjects were the same participants included in the study of Influence of Risk Factors to Atherosclerosis Progression in post ischemic stroke patients, performed in Dr.Kariadi hospital at the same period of time. Inclusion criterias were post ischemic stroke with the onset of more than 1 month and having IC-IMT above normal value on the first ultrasound examination. Exclusion criterias were renal dysfunction, diuretic consumption, history of steatorhhea, inflammatory bowel disease (IBD) and incomplete data. Study was performed after approval from Ethical commission of dr.Kariadi hospital. Interview was performed to all patients followed by physical examination, filling the food frequency questionnaire (FFQ), then Carotid ultrasonography, two times in the period of 6 months.

Mineral intake measurements

Measurement of minerals intake (in milligram) was performed using a semiquantitative food frequency questionnaire (FFQ) and Nutrisurvey Software Programme. FFQ was performed based on the study of 20 subjects who underwent food recall for 3x24 hours. Data showed the most common foods containing the macro minerals consumed based on the food recall.

The tunica intima media thickness measurement

Left and right carotid arteries of each subject were measured using ultrasonography Siemens soniline Omnia FBE 0322®. The tunica intima media carotid artery was examined from the transverse projection and carotid artery along the cervical area starting from the supraclavicular region toward the cranial area of mandible angle. Tunica intima-media thickness of the internal carotid artery is the maximum thickness of 1 cm distal of bifurcation. The maximum left or right intima-media thickness of internal carotid artery (IC-IMT) were included to statistical analysis[9][16][17].

Statistical Analysis

Primary data obtained from patients after ischemic stroke include gender, age, history of hypertension, diabetes mellitus, dyslipidemia, smoking, overweight, intake of minerals (magnesium, calcium, sodium, potassium) and intima media thickness of the internal carotid artery. Statistical analysis to examine association of each factor of atherosclerosis were performed using Chi Square and Fisher Exact test, followed by logistic regression multivariate analysis. Significant result if p value <0.05.

3. Results and Discussion

In this study, 22 subjects whose IC-IMT above normal level had the interview to fill FFQ form. Most of the subjects were male (17 subjects/77.3%). Only 5 subjects were female (22.7%). Mean of total IC-IMT on the first measurement was 0.96 mm. Mean of IC-IMT was 0.96 mm for male and 1 mm for female. On the second measurement, mean of total IC-IMT was 0.97 mm, 0.98 for male and 1.08 mm for female.

Mean of age of subjects included in this study was 62.8 years old. The youngest subject was 51 years old and the oldest one was 81 years old. Seven subjects (31.8%) were > 65 years old. In regards to risk factors for stroke, 15 subjects (68.2%) in this study had hypertension, 14 subjects (63.6%) had dyslipidemia, 9 subjects (40.9%) had diabetes mellitus, 11 subjects (50%) had history of smoking and 2 subjects (9.1%) were overweight. From all subjects included in this study, there were 10 subjects (45.5%) whose IC-IMT increased from first to second measurement (Table 1).

Table 1. Frequency Distribution of Subject's Characteristic

Age		Frequency	%
	>= 65	7	31.8
	< 65	15	68.2
	Total	22	100.0
Sex			
	male	17	77.3
	female	5	22.7
	Total	22	100.0
Hypertension			
	Yes	15	68.2
	No	7	31.8
	Total	22	100.0
Dyslipidemia			
	Yes	14	63.6
	No	8	36.4
	Total	22	100.0
DM type II			
	Yes	9	40.9
	No	13	59.1
	Total	22	100.0
Smoking			
	Yes	11	50.0
	No	11	50.0
	Total	22	100.0
Obesity			
	Yes	2	9.1
	No	20	90.9
	Total	22	100.0

The description of minerals intake as well as IC-IMT of each subjects in this study were showed in Table 2.

Chi-Square bivariate test performed in this study suggested there was significant association between calcium intake and IC-IMT changes ($p=0.043$). Additionally, there was also significant association between potassium intake and IC-IMT changes ($p=0.031$). However, there was no association between magnesium intake and IC-IMT changes ($p=0.087$). There was also no association between sodium intake and IC-IMT changes ($p=0.074$) (Table 3).

IOP Conf. Series: Earth and Environmental Science **55** (2017) 012016 doi:10.1088/1755-1315/55/1/012016

Table 2. Frequency Distribution of minerals intake and IC-IMT

Macrominerals		Frequency	%
Kalium			
	< 2516.7	8	36.4
	>= 2516.7	14	63.6
	Total	22	100.0
Natrium			
	>= 1431.7	8	36.4
	< 1431.7	14	63.6
	Total	22	100.0
Calcium			
	< 493.7	10	45.5
	>= 493.7	12	54.5
	Total	22	100.0
Magnesium			
	< 276.5	11	50.0
	>= 276.5	11	50.0
	Total	22	100.0
IC-IMT 1			
	≥ 0.8	22	100.0
	< 0.8	0	0.0
	Total	22	100.0
IC-IMT 2			
	≥ 0.8	16	72.7
	< 0.8	6	27.3
	Total	22	100.0
Δ IMT			
	Increase	10	45.5
	Constant/decrease	12	54.5
	Total	22	100.0

Table 3. Bivariate Analysis of Magnesium, Calcium, Sodium and Potassium With IC-IMT Changes.

Intake	Bivariate						
	Alteration IC-IMT		Statistic	*P*	RO	IK 95%	
	Increase (%)	Constant/ Decrease (%)				Min	Max
Magnesium							
< 276.5	7 (70)	4 (33.3)	X^2 =0.087	0.087	4.667	0.765	28.466
≥ 276.5	3 (30)	8 (66.7)					
Calcium							
< 493.7	2 (20)	8 (66.7)	X^2 = 0.043	0.043	0.125	0.018	0.887
≥ 493.7	8 (80)	4 (33.3)					
Sodium							
≥ 1431.7	6 (60)	2 (16.7)	X^2 =0.074	0.074	7.5	1.039	54.116
< 1431.7	4 (40)	10 (73.3)					
Potassium							
< 2516.7	1 (10)	7 (58.3)	X^2 =0.031	0.031	0.079	0.007	0.843
≥ 2516.7	9 (90)	5 (41.7)					

Statistical analysis using Fisher Exact and Chi Square test showed that there was no association between some traditional risk factors (as such hypertension, dyslipidemia, diabetes mellitus, smoking, obesity, age, and sex) and IC-IMT changes (see Table 4).

Table 4. Bivariate analysis association of traditional risk factors and IC-IMT changes.

Variables	Bivariate						
	IC-IMT		Statistic	*p*	OR	95% CI	
	Increase (%)	Constant/ Decrease (%)				Min	Max
Sex							
Male	7 (70)	10 (83.3)	Fisher= 0.624	0.624	0.467	0.061	3.565
Female	3 (30)	2 (16.7)					
Age							
≥ 65	2 (20)	5 (41.7)	Fisher = 0.381	0.381	0.350	0.051	2.407
< 65	8 (80)	7 (58.3)					
Hypertension							
Yes	6 (60)	9 (75)	Fisher= 0.652	0.652	0.5	0.081	3.082
No	4 (40)	3 (25)					
Dyslipidemia							
Yes	7 (70)	7 (58.3)	Fisher= 0.675	0.675	1.667	0.283	9.822
No	3 (30)	5 (41.7)					
DM type II							
Yes	7 (28)	19 (40)	Fisher= 0.296	0.296	0.573	0.201	1.637
No	18 (72)	28 (59.6)					
Smoking history							
Yes	3 (30)	8 (66.7)	$X^2 = 0.087$	0.087	0.214	0.035	1.307
No	7 (10)	4 (33.3)					
Overweight							
Yes	0 (0)	2 (16.7)	Fisher= 0.481	0.481	–	–	–
No	10 (100)	10 (83.3)					

Multivariate analysis was performed using logistic regression test. There were 5 variables included, history of smoking, magnesium intake, calcium intake, sodium intake, potassium intake. 2 variables influencing IC-IMT changes, including calcium intake (*p=0.037*) and sodium intake (*p=0.049*). The most influencing variable was sodium intake (OR=26.828) followed by calcium intake (OR=0.042). Subjects with sodium intake ≥ 1431.7 mg/day were 26.828 times more likely to have an increase in IC-IMT than subject with sodium intake < 1431.7 mg/day. Meanwhile, subjects with calcium intake < 493.7 mg/day were 0.042 times more likely to have an increase in IC-IMT than subjects with calcium intake ≥ 493.7 mg/day (table 5).

Table 5. Logistic regression analysis of association between risk factors and IC-IMT changes.

Variables	Bivariate				Multivariate			
	p	OR	95% CI		*Sig*	OR	95% CI	
			Min	Max			Min	Max
Sex	0.624	0.467	0.061	3.565				
Age ≥ 65	0.381	0.350	0.051	2.407				
Hypertension	0.652	0.5	0.081	3.082				
Dyslipidemia	0.675	1.667	0.283	9.822				
Diabetes mellitus	0.296	0.573	0.201	1.637				
Smoking history	0.087	0.214	0.035	1.307	0.162	0.106	0.005	2.464
Overweight	0.481	–	–	–				
Magnesium intake < 276. 5 mg/day	0.087	4.667	0.765	28.466	0.055	19.784	0.936	418.26
Calcium intake< 493.7 mg/day	0.043	0.125	0.018	0.887	0.037	0.042	0.002	0.826
Sodium intake≥ 1431.7 mg/day	0.074	7.5	1.039	54.116	0.049	26.828	1.015	0.826
Potassium intake< 2516.7 mg/day	0.031	0.079	0.007	0.843	0.507	0.333	0.013	8.584

In this study, though there was no association between sex and IC-IMT changes theorytically, data showed that mean of IC-IMT in female subjects was greater than male. On the first measurement, mean of IC-IMT in male subjects was 0.96 mm and female 1 mm. On the second measurement, performed 6 months later, mean of IC-IMT in male subjects was 0.96 mm and 1.08 in female subjects. This study differed from previous ones in term of subjects. In the previous study, the subjects were healthy subjects while in this study, the subjects were post ischemic stroke patients whose IC-IMT already above normal value [3][4][14].

Bivariate analysis performed in this study showed that there were no associations between some traditional risk factors, analyzed as confounding factors (such as hypertension, diabetes mellitus, dyslipidemia, smoking, obesity, age, and sex) and IC-IMT changes (table 2). This finding was different to other previous studies which revealed the association of the traditional risk factors with IC-IMT [5][14][18]. Explanation to this findings were that previous studies were conducted in huge number of healthy subjects, who were free from cerebrovascular disease. Variables were measured once (cross section) [4][5]. While this study was performed to 22 post ischemic stroke subjects whose IC-IMT already above normal value at first measurement. Measurement of IC-IMT was performed twice. The second measurement was performed six months later.

In this study, there was significant association between calcium intake and IC-IMT changes in post ischemic stroke patients ($p=0.043$). This finding was different to previous study by Masley et al, which was performed cross section in 592 healthy subjects and revealed that calcium intake was not associated to IC-IMT [8].Some possibilities concerning this finding included that subjects in this study were already patients with atherosclerotic disease (post ischemic stroke patients and had abnormal IC-IMT in the first measurement).

Significance association were found between potassium intake and IC-IMT changes in post ischemic stroke patients in this study ($p=0.031$). This finding was similar to previous study by Ortiz et al in 351 subjects without cerebrovascular disease which revealed that association between potassium intake and C-IMT resembled J-shaped curve [12]. However, this finding was different to study by Masley et al which performed cross section in 592 healthy subjects. Masley et al concluded that potassium intake was not associated with C-IMT. Same consideration was revealed by writer, it was due to the situation that subjects in this study were already in atherosclerotic disease (post ischemic stroke patients whose IC-IMT already above normal value in first measurement) while in Masley's study the subjects were still "free" form atherosclerotic disease [8].

Similar to the finding of Otto et al, there was no association between magnesium intake and IC-IMT changes. However it was different to the result of Masley's study that revealed magnesium intake was associated with C-IMT [8][11].

There was also no association between sodium intake and IC-IMT changes in post ischemic stroke patients in this study ($p=0.074$). This result was different to study of Ortiz et al that concluded sodium intake was negatively correlated with C-IMT[12]. The Difference between this study and previous one was that previous study was done in 351 subjects without cardiovascular disease and C-IMT measurement was performed in common carotid artery, while in this study was performed to 22 post stroke patients and measurement was performed at the internal carotid artery segmented as the site for advanced phase of atherosclerosis.

Multivariate analysis using logistic regression test showed that the most influencing variable to IC-IMT changes was sodium intake, followed by calcium intake. This finding was similar to previous study that already confirmed sodium intake was negatively correlated with C-IMT.[12] Person with sodium intake $\geq$ 1431.7 mg/day tend to have increased IC-IMT changes 26.828 times compare to those with sodium intake < 1431.7 mg/day. This could be a warning to reduce sodium consumption in daily diet.

4. Conclusion

Calcium as well as potassium intake in post ischemic stroke patients were independently associated with IC-IMT changes. In contrast, magnecium as well as sodium intake are not independently associated with IC-IMT changes. Sodium intake is the most influencing variable to IC-IMT changes, followed by calcium intake.

References

[1] Nelms R, Sucher K, Lacey K and Roth S 2010 *Nutrition Theraphy and Pathophysiology* (Belmont: Wadworth Cengage Learning)

[2] Naqvi T Z and Lee M-S 2014 Carotid Intima-Media Thickness and Plaque in Cardiovascular Risk Assessment *JACC Cardiovasc. Imaging* **7** 1025–38

[3] Stensland-Bugge E, Bønaa K H, Joakimsen O and Njølstad I 2000 Sex differences in the relationship of risk factors to subclinical carotid atherosclerosis measured 15 years later : the Tromsø study. *Stroke.* **31** 574–81

[4] Iglseder B, Cip P, Malaimare L, Ladurner G and Paulweber B 2005 The metabolic syndrome is a stronger risk factor for early carotid atherosclerosis in women than in men *Stroke* **36** 1212–7

[5] Stein J H, Douglas P S, Srinivasan S R, Bond M G, Tang R, Li S, Chen W and Berenson G S 2004 Distribution and cross-sectional age-related increases of carotid artery intima-media thickness in young adults: The Bogalusa Heart Study *Stroke* **35** 2782–7

[6] Lorenz M W, Von Kegler S, Steinmetz H, Markus H S and Sitzer M 2006 Carotid intima-media thickening indicates a higher vascular risk across a wide age range: Prospective data from the Carotid Atherosclerosis Progression Study (CAPS) *Stroke* **37** 87–92

[7] Wang S 2013 Correlation analysis between the carotid artery atherosclerosis severity of patients with lacunar infarct and the level of serum uric acid, high-sensitivity C-reactive protein *Chinese J. Contemp. Neurol. Neurosurg.* **13** 66–70

[8] Masley S, Roetzheim R, Masley L, McNamara T and Schocken D 2015 Emerging risk factors as markers for carotid intima-media thickness scores *J Am Coll Nutr* **34** 100–7

[9] Groot E, Hovingh G, Wiegman A, Duriez P, Smit A and Fruchart J 2014 Measurement of Arterial Wall Thickness as a Surrogate Marker for Atherosclerosis *Circulation* **109** 33–8

[10] Rundek T, Blanton S H, Bartels S, Dong C, Raval A, Demmer R T, Cabral D, Elkind M S V, Sacco R L and Desvarieux M 2013 Traditional risk factors are not major contributors to the variance in carotid intima-media thickness *Stroke* **44** 2101–8

[11] de Oliveira Otto M C C, Alonso A, Lee D-H, Delclos G L, Jenny N S, Jiang R, Lima J a, Symanski E, Jacobs D R and Nettleton J a 2011 Dietary micronutrient intakes are associated with markers of inflammation but not with markers of subclinical atherosclerosis. *J. Nutr.* **141** 1508–15

[12] García-Ortiz L, Recio-Rodríguez J I, Rodríguez-Sánchez E, Patino-Alonso M C, Agudo-Conde C, Rodríguez-Martín C, Castaño-Sánchez C, Runkle I and Gómez-Marcos M A 2012 Sodium and potassium intake present a J-shaped relationship with arterial stiffness and carotid intima-media thickness *Atherosclerosis* **225** 497–503

[13] Ferreira-sae M C S, Cipolli A A, Corne E, Fernandes M N, Schreiber R, Costa F O, Franchini K G, Rodrigues R C, Gallani M C and Jr W N 2011 Sodium Intake Is Associated with Carotid Artery Structure Alterations and Plasma Matrix Metalloproteinase-9 Upregulation in Hypertensive Adults 1 – 3 *J Nutr* **141** 877–82

[14] Stamatelopoulos K S, Armeni E, Georgiopoulos G, Kazani M, Kyrkou K, Stellos K, Koliviras A, Alexandrou A, Creatsa M, Papamichael C and Lambrinoudaki I 2012 Recently postmenopausal women have the same prevalence of subclinical carotid atherosclerosis as age and traditional risk factor matched men *Atherosclerosis* **221** 508–13

[15] Deverts D, COHEN S, Matthews K, Jacobs D and Adler N 2011 Occupational Mobility and Carotid Artery Intima-Media Thickness: Findings from the Coronary Artery Risk Development in Young Adults (CARDIA) Study *Psycosom med* **73** 795–802

[16] Dogan S, Yvonne P, Crouse J, Evans G, Raichien J and O'Leary D 2011 Measurement of CIMT in trials. Algorithms to measure carotid intima–media thickness in trials: a comparison of reproducibility, rate of progression and treatment effect *Hypertension* **29** 2181–93

[17] Vila A, Mamani E, Gilabert R, Nunez I, Torre R D La and Gutierrez V R 2013 Changes in Ultrasound-Assessed Carotid Intima-Media Thickness and Plaque With a Mediterranean Diet. A Substudy of the PREDIMED Trial *Arterioscler. Thromb. Vasc. Biol.* **34** 439–45

[18] Qu B and Qu T 2015 Causes of changes in carotid intima-media thickness: a literature review *Cardiovasc. Ultrasound* **13** 1–10

Brain Gym To Increase Academic Performance Of Children Aged 10-12 Years Old (Experimental Study in Tembalang Elementary School and Pedalangan Elementary School Semarang)

M G Marpaung[1], T P Sareharto[1], A Purwanti[1], D Hermawati[2]

[1]Pediatric Departement of Medical Faculty Diponegoro University 2012. Prof.Soedarto Street,Semarang,Indonesia.
[2]Biology Departement of Medical Faculty Diponegoro University. Prof.Soedarto Street,Semarang,Indonesia.

Email: galatiamona@gmail.com

Abstract. Academic performance becomes an important determinant of individual quality. it is determined by the function of affective, cognitive, psychomotor, and intelligence. Brain gym can improve learning processes and integrate all areas that related to the learning process. To prove the effect of brain gym towards academic performance of children aged 10-12 years. This study was a quasy experiment study with one group pre and post test design. Samples (n=18 male=7 and female=11) were taken from five and six grader and conducted in Tembalang and Pedalangan Elementary School, Semarang. Pretest were administered, followed by brain gym, and post test administered in the end of study. The measurement of Intelligence Quotient pre and post test using Culture Fair Intelligence Test Scale 2. Among the 18 subjects (male=7 and female=11) the average of academic performance and IQ score after brain gym showed improvement. The Improvement of IQ score with Culture Fair Test Scale 2 was analyzed by Dependent T test showed significant results (p=0,000). The improvement of Bahasa score was analyzed by Wilcoxon test showed significant results (p=0,001), an unsignificant result were shown in Mathematics p=0,079 and natural sciences p=0,306. Brain gym can increase academic performance of children aged 10-12 years old.

Keywords: Brain gym, affective, cognitive, psychomotor, intelligence.

1. Introduction

The middle period of children's grwoth is at the aged of 10-12 years old. In this period there is physical development and nervous system development. The period of 10-12 years old can be mentioned as operational stages to manage and to control emotional expression such as an ability to knowing their own emotion, managing mood, motivating their own self, building and maintaining relationships with others. Physical development and motorical at the age of 10-12 years old is already well coordinated. In this period their development marked by sprightful motorical activity. Therefore, this periode is an ideal time to learn skill that related to motorical activity such as writing, drawing, painting, computing, swiming, and playing athletically. Besides that emotional aspect and physical development, children in this period can react to intellectual stimulus, conduct tasks that insist intellectual ability or cognitive ability[1]. Therefore, intellectual ability in this period is enough as a basic for giving stimulus to develop their mindset or intuition.

The quality of human brain not only depend on genetical aspect but also depend on the quality of stimulus that receives by children[2]. Stimulation is an activity to stimulate the ability of children so that children can grow and develop optimally[2,3].Stimulation is the pioneer of learning process that can support educational process[4]. The enhancement of brain quality needs spesific stimulus because the nature of stimulus is a part of learning process that related to individual's intellectual ability[2,3,4].

Brain gym is a collection of simple movements and aims to connect or to unite the body and mind[5]. Brain gym is a part of kinesiology education[5,6] . Kinesiology is a study about body's movements and the relation between brain and body's posture to brain function. All the movements of brain gym can increase learning process and integrate all areas that related to learning process[6,7].

The research about brain gym such as, Watson [8] did research about brain gym everyday to fourth grade of children in elementary school during seven to ten minutes for about three months. At the end of the research, with STAR reading test, evaluation of the comprehensive understanding about reading before brain gym and after brain gym showed a significant result[8] . While Drabben[9] did brain gym to Alzheimer's patients during eight weeks everyday. All subjects showed a significant increase in conversion numbers, the repetition of words, and the ability to remind previous objects.

Relaxation technique if perform with reguler time can help children to control stress so that they can achieve a peaceful situation. Brain gym can be expected to help children control their stress because all technique in brain gym such as massages, meditation, stretching, and abdominal respiration that can help children to achieve peaceful situation and can help the way of intelligence[10]. Because of so many factors that related to the development of academic performance on children aged 10-12 years old then it is important to do the research about the relationship of brain gym to increase academic performance of children aged 10-12 years old. By knowing the relationship of brain gym to academic performance, it can be expected all children can develop their academic performance maximally.

2. Research Method

The design of this research is quasi experimental pre and post test design without control group to know the difference academic performance before and after given brain gym. The data that be used in this research is primary data. it is contain of intelligence result and secondary data is reports of academic score. Subjects must meet the inclusion criteria, children in aged 10-12 years old at that time research and be ready to be responden in this research. The method of this research is purposive sampling. To count the sample of this research by using average test of two populations and it also estimate the drop out criteria for about 10%. The independent variable of this research is brain gym and dependent variable are academic performance and intelligence quotient (IQ). The instrument of this research are gym audio and Culture fair intelligence test. The first step of this research is giving intelligence test, measure the report score of academic performance, intervention of brain gym, and it ends with post test intelligence test and measure final score of academic performance. The treatment group was given brain gym stimulation for three times a week during two months. The duration time for every stimulation is thirty minutes.

The analysis of hypothesis to differentiate academic performance before and after intervention of brain gym using Wilcoxon test because the distribution of the data is normal (Test of Normality Shaphiro-Wilk, $p<0,05$) and after that data will be transform, after transformation of the data still the distribution of data is not normal. While analysis of hypothesis to differentiate intelligence before and after intervention of brain gym using Paired T-test because all the requirements of the data meet the normal distribution (Test of Normality Shapiro-Wilk, $p<0,05$)

3. Result and Discussion

3.1. Characteristics of Research Subject

This research involves eighteen students that come from two different elementary schools and meet the inclusion criteria of the research. The location of two elementary school are in Semarang disctricts that have the same characteristics. The students that meet all the criteria are given intelligence quotient test (IQ) and then intervention of brain gym. From all twenty students that related in this research just eighteen students that meet all criteria in this intervention research. There are two subjects that meet exclusion criteria because two of them aged thirteen years old. The average of their age is 11,67±0,485 with the youngest is eleven years old and the oldest is twelve years old. Gender of research subject is male (n=7) and female (n=11). The characteristics of both group are shown in the table 1.

Table 1. Description of subject's characteristics (n=18)

Characteristic	n(%)	Mean±SD(min-max)
Age		11,67±0,485(11-12)
Gender		-
- Male	7 (38,9%)	
- Female	11(61,1%)	
Brain defect		-
- Yes	0(0%)	
- No	18(100%)	
Brain gym		-
- Ever	0(0%)	
- Never	11(100%)	
Severe illness		
- Yes	0 (0%)	
- No	18 (100%)	

3.2. The Result of Academic Performance

3.2.1. Measurement of Intelligent Quotient (IQ)

Tabel 2. Score Pre dan Post test Intelligent Quotient

	IQ Test		P
	Pre Treatment	**Post Treatment**	
IQ Score mean±SD;median(min-max)	94,94±12,859 ;98,50 (64-113)	104,44±12,463 ;100 (81-128)	0,000

The result of research show there is increase in the mean of pretest score and post test score IQ, from 94,94 to 104,44. It means there is increase in intelligence of the subjects. This data using Saphiro Wilk

Test. The distribution is normal that show the increase in the ability of IQ significantly, p=0,000. Academic performance can be defined as academic achievement, the involvement that aims education, getting knowledge, skill and competency, and also perseverance in learning process[11]. Academic performance can be known with measurement and evaluation to the education that given by the teacher and can be measured through intelligence quotient (IQ) [12.13.14] .The analysis of data pretest and posttest intelligence quotient with Paired T-Test show that there is increase in posttest score of intelligence quotient, Z=-3,535 and significance level 0,000>0,05. it increases significantly. Intervention of brain gym was given three times a week during nine weeks. The increase in posttest indicate that intervention that given in the form of brain gymnastic movements to children aged 10-12 years old give the effect to the increase in children intelligence. The research by Chandra et.al to six hundred fourteen students aged 13-17 years old mentioned that intelligence affect academic performance significantly. In that research mentioned that children with high intelligence have better academic performance than children with average intelligence [15]. It is appropriate with this research that mentioned children that have good categorical of IQ have a good academic performance too. It is because the role of stimulation in the process of neurotransmitter that related to signalling dopamin receptor that produce brain derived neurotrophic factor (BDNF) through Ca^{2+} signalling process. In the brain, BDNF is found actively in hippocampus and cortex that is vital area in learning process, reminding, and thoughtful process. BDNF can increase synaps fosforilation through receptor activation that cause releasing of neurotransmitter. Not only neurotransmitter process, but also the intteligence ability to cortex function such as abstraction, logical, language, and every area that have spesific function [16].

3.1.2 Measurement of Bahasa Score

Table 3. Pretest and Posttest Bahasa Score

	Score		**P**
	Pre Treatment	**Post Treatment**	
Score Mean±SD;median(min -max)	43,44 ± 34,904; 60 (5-83)	46,26 ± 36,817; 60 (7-92)	0,001

The result of research show there is increase in the mean of pretest score and post test Bahasa score, from 43,44 to 46,26. It means there is increase in Bahasa score of the subjects. This data using Wilcoxon Test because the distribution is not normal. It shows the increase in the ability to solve problem significantly, p<0,001. The analysis of data pretest and posttest Bahasa score with paired T-test showed there was increase in posttest score of intelligence quotient, Z=-3,410 and significance level 0,001. It means that score increase significantly. The increase of Bahasa score is in accordance with the previous researchs about brain gym although those researchs were used other tests that differ with this research. Andrea (2014) with percent academic engagement found the significant increase in the ability to understand the text.[17] Thompson (2007) with STAR reading test found there was increase in the understanding comprhensively about text compare before practice brain gym.[18] This case relates to the effect of brain gym to cerebri cortex function to assosiation of parieto-occipitotemporal that include the understanding of language, called Wernicke area and area that include reading process, called angular gyrus that interpretate words that receive visually and then it will continue to wernicke area.

3.1.3. Measurement of Mathematic Score

The result of research show there is increase in the mean of pretest score and post test Mathematic score, from 41,85 to 44,82. It means there is increase in Mathematic score of the subjects. This data using Wilcoxon Test because the distribution is not normal. However, It does not increase logical ability significantly, p>0,05 (p=0,079)

Table 4. Pretest and Posttest Mathematic Score

	Score		P
	Pre Treatment	**Post Treatment**	
Score Mean±SD;median(min -max)	41,85 ± 32,843; 60 (6-90)	44,82 ± 36,324; 60 (5-100)	0,079

Sutoro [19] did research about brain gym to evaluate mathematical ability during four weeks and there was an increase in arithmetic and counting ability. However, in this research there is no significant increase to mathematic (p=0,079) and Natural Science (p=0,306) although still show the increase in the mean of pretest and posttest Mathemathic and Natural Science. This could be any factors that affect daily score of children such as the ability of teacher, how to teach, even the interest of children in that subjects. Other causes could related to research method, with different method the increase of mean Mathemathic and Natural Science could be significant.

3.1.4. Measurement of Natural Science

Table 5. Pretest and Posttest Natural Science Score

	Score		P
	Pre Treatment	**Post Treatment**	
Score Mean±SD;median(min -maks)	44,69 ± 35,86; 60 (7-92)	51,39 ± 41,802; 68 (6-100)	0,306

The result of research show there is increase in the mean of pretest score and post test Natural Science score, from 44,69 to 51,39. It means there is increase in cognitive ability of Natural Science of the subjects. But statistically using Wilcoxon Test The distribution data is not normal show the increase in logical ability of Natural Science is unsignificant, $p>0,05$ ($p=0,306$)

The unsignificant result in Mathemathic and Natural Science could be caused by other factors that has not been evaluate strictly in this research. Other factors could be stress factor because most of elementary students do not like mathematic because too complicated and often makes children become bored.

4. Conclusion and Recommendation

Conclussion

There is a significant increase in intelligence quotient (IQ) after intervention of brain gym three times a week during nine weeks with duration time thirty minutes every intervention in children aged 10-12 years old. There is a significant increase in Bahasa score after Intervention of brain gym three times a week during nine weeks with duration time thirty minutes every intervention in children aged 10-12 years old. There is no significant increase in Mathematic and natural science score after Intervention of brain gym three times a week during nine weeks with duration time thirty minutes every intervention in children aged 10-12 years' old

Recommendation

In this research, intervention of brain gym during nine weeks, three times a week with thirty minutes duration time. How long brain gym can withstand in order to remain effective without repetition is still unknown. Therefore, researcher propose the next research to analyze how long brain gym can withstand without the repetition of brain gym itself. Require further research to further anaylze the discipline of research's subject when they are doing brain gym in terms of time, frequency, concentration, and environmental factors that can related to confounding variables of the research. Require further research about brain gym on various pupulations not only children aged 10-12 years old but also all school-age children so that it can be expected all achool age children will be able to develop ther academic performance maximally. Require further research about braun gym in elementary school to increase academic performance.

Acknowledgement

The researcher deepest gratitude goes to Tembalang and Pedalangan elementary school, psychologist's team of Soegijapranata Catholic University, and to all subjects that have been participated and provide them self to be a responden in this research, and also to the others that cannot be mentioned one by one.

References

1. Reber M. Development during Middle Childhood 1986 *The Years from Six to Twelve.* (Washington: National Academy Press) p137-8.
2. Jääskeläinen IP, Sams M, Andermann ML 2012 *Brain stat-triggered stimulis delivery: An efficient tool for probing ongoing brain activity* Open J Neurscience [Internet] [cited 2015 Nov 11];**18**(2):1–19 Available from: Ross Science
3. Raine A, Reynolds C, Venables PH, Mednick S a 2002 *Stimulation seeking and intelligence: A prospective longitudinal study.* J Pers Soc Psychol [Internet] [cited 2015 Nov 11];**82**(4):663–74 Available from: http://doi.apa.org/getdoi.cfm?doi=10.1037/0022-3514.82.4.663
4. Jeng-Fung Hung, Jen-Chin Lin 2002 *The Development of the Simulation Modeling System and Modeling Ability Evaluation* Int J U- E-Service, Sci Technol [Internet]. 2009 [cited 2015 Nov 11];**2**(4):1–16 Available from: www.sersc.org/journals/IJUNESST/vol2_no4/1.pdf
5. Official Brain gym ® Website. 2002 [cited 2015 Nov 11] Available from: http://www.braingym.org
6. Spaulding LS, Mostert MP, Beam AP 2010 *Is Brain Gym® an Effective Educational Intervention? Exceptionality* [Internet] [cited 2015 Nov 11];**18**(1):18–30 Available from : http://digitalcommons.liberty.edu/cgi/viewcontent.cgi?article=1167&context=educ_fac_pubs
7. Hoffman SJ, Harris JC, Hoffman SJ. Introduction to Kinesiology 2009 *Studying Physical Activity* (United States of America:*Human Kinetic Press*) p570
8. Watson A, Ginger L Kelso 2014 *The Effect of Brain Gym ® on Academic Engagement For Children with Developmental Disabilities.* International Journal of Special Education [Internet] [cited 2015 Nov 11];**29**(2): 7. Available from : Proquest
9. Drabben-Thiemann, D. Hedwig, M. Kenklies, A. Von Blomberg, G. Marahrens, A. Marahrens, K. Hager 2002 *The effect of Brain Gym® on the cognitive performance of Alzheimer's patients* Brain Gym Journal [Internet] [cited 2015 Nov 11] Available from : www.braingym.org/brochures/BG_Research.pdf
10. *Brain Gym; Learning Empowered by Movement; Brain Gym Exercises Wake Up Brain and Body* 2009 Pediatrics Week.
11. Sukardi 1985 *Neuroanatomia Medica* (Jakarta: Universitas Indonesia Press) p232-255.
12. Crede M, Kuncel NR 2008 *Study habits skills and attitudes* Psychological Journal [Internet] [cited 2015 Nov 16];**3**(6):425–54. Available from : Psychological Science
13. Wolf PJ 2007 *Academic Improvement Through Regular Assessment* Peabody J Educ [Internet] [cited 2015 Nov 16];**82**(4):690–702. Available from : https://www.researchgate.net/publication/254310132_Academic_Improvement_Through_Regular_Assessment

14. Official Wisconsin Department of Public Instruction Website 2004 [cited 2015 Nov 16]. Available from : http://dpi.wi.gov/dpi/oea/kce.html
15. Nuraeni 2012 *Inteligence test* (Yogyakarta: Purwokerto Press) p23-85.
16. Berg DA, Belnoue L, Song H, Simon A 2013 *Neurotransmitter-mediated control of neurogenesis in the adult vertebrate brain* Development (Cambridge, England) [cited 2016 Jan 19];**140**(12):2548-61. Available from: http://www.ncbi.nlm.nih.gov/pmc/articles/PMC3666382/
17. Andrea Watson, Ginger L Kelso 2014 *The Effect of Brain Gym ® on Academic Engagement For Children with Developmental Disabilities* International Journal of Special Education [Internet] [cited 2015 Nov 11];**29**(2): 7. Available from : Proquest
18. Thompson, H.L 2007 *Impacts of Educational Kinesiology Activities on Fourth Graders Reading Comprehension Achievement* [dissertation] (Washington : United States Walden University)
19. Sutoro 2014 *Influence of brain gym on mathematical achievement of children* Asian Soc Sci;**10**(5):114–7.

Polymeric Membrane Made of Cellulose Isolated from Tropical Water Hyacinth Blended with Chitosan

Titik Istirokhatun *[1,3], Richa Rachmawaty[2], Metty Meriyani[2], Nur Rokhati[2,3], Heru Susanto[2,3]

[1]Environmental Engineering Department, Engineering Faculty, Diponegoro University
[2]Chemical Engineering Department, Engineering Faculty, Diponegoro University
[3]Membrane Research Centre (Mer-C), Centre Of Research and Services-Diponegoro University
Jl. Prof Soedharto, Tembalang, Semarang, Indonesia 50275
*Presenting author, telephone: +62 24 76480678, fax +62 24 76918157,

titik.istirokhatun@live.undip.ac.id

Abstract. In Indonesia, membrane application is limited by the use of high cost imported membranes. Therefore, efforts to produce membrane using local materials are very important to be performed. Cellulose and chitosan are two materials, which their availability in Indonesia is abundant. In this work, cellulose acetate (CA) has been successfully produced from tropical water hyacinth via isolation and acetylation processes. By using FTIR, a comparative analysis of functional groups between the resulted CA and commercial CA was also investigated. The result shows that the IR spectra of the resulted CA and commercial CA are similar. The produced CA can further be processed into membrane. The resulting membranes were then characterized using SEM. In addition, the membranes were examined for filtering humic acid solution as surface water model. To improve the performance, CA membranes prepared from water hyacinth were modified by addition of chitosan (Chi) via blending in phase separation method. The resulting CA-Chi membranes were characterized using SEM and FTIR.

Keywords: polymeric membrane, local materials, tropical water hyacinth, chitosan, phase separation

1. Introduction

Membranes have increasingly been accepted as promising technology in a wide range of applications including high-quality water process production, removal or recovery of toxic or valuable components from various industrial effluents, dairy, biotechnological, pharmaceutical industries, food and beverage processing, and medical applications [1-2]. The emergence of membrane technology has made unit operations such as separation, and purification become industrially viable because of the high efficiency of separation, low energy of operation, simplicity of operation with modern compact modules [3]. As consequence of this increasing demand, efforts to use natural resources for membrane material are gaining more and more importance. Several previous membrane studies more focused on the process of membrane preparation, while the analysis of alternative materials that have economic value has been largely ignored even the presence of these materials are abundant in nature. In addition, environmental concerns such as pollution and decreases in natural resources, have led in recent years to an increased demand for renewable materials.

Cellulose, as natural renewable polymer, the main structural component of plant, is considered as an attractive raw material for the production of membrane, because of its availability in large quantities. Cellulose Acetate (CA), cellulose which has undergone acetylation process, is suitable as membrane material, since its advantages such as moderate flux, high salt rejection properties, cost effectiveness, relatively easy manufacture, renewable resource of raw material, and non-toxic [4] besides strongly hydrophilic and thus demonstrate lower fouling characteristic. CA membrane was the first high-performance asymmetric membrane [5]. It has been widely used for reverse osmosis (RO), microfiltration (MF) [6] and gas separation [7]. On the other hand, CA membranes yield a low flux and are susceptible to chemical and bacteriological agents in their first generation. In addition, the disadvantages of CA membrane are its stability over a narrow pH and their poor resistance to microbial attack [8] low oxidation, thermal and chemical resistances, [9, 10]. For this reason, the modification of CA membrane by blending CA with an appropriate polymer may be improves the performance of CA membrane.

The abundant aquatic weed species, Water hyacinth (*Eichhornia crassipes*) in tropical weather conditions such as Indonesia, is a major problem in open water bodies like power generation, irrigation and boating. It is considered as worst aquatic weed, an incredibly fast growing plant. Under the climatic conditions, it is reported a daily average water hyacinth biomass productivity of 0.26 ton of dry biomass per hectare [11]. Water hyacinth constitutes a potential biomass resource for cellulose isolation, while it is rich in fibre content and has no major differences in the percentage of cellulose content in the shoot and root [12]. Therefore, water hyacinth has great potential to be used as cellulose source, which then further be converted into cellulose acetate (CA) via acetylation process.

Chitosan as biopolymer from renewable resources -which its existence is also very plentiful in Indonesia- has been used extensively to form membranes [13]. Owing to its high biodegradability, biocompatibility, nontoxicity and antimicrobial properties, chitosan is widely used as antimicrobial agent either alone or blended with other natural polymers [14]. The antimicrobial properties of chitosan can cover CA membrane infirmity. Our previous research has succeeded to synthesize CA membranes prepared from water hyacinth [15]. The produced CA was in accordance with commercial CA.

This research was intends to investigate CA membrane performance and to study the effect of chitosan addition on the CA membrane. The use of CA and chitosan for membrane preparation are expected to be a solution for the problem of membrane application in Indonesia which is still limited by the high cost of imported membrane.

2. Materials and Methods

2.1. Materials

Water hyacinth stem were collected from Rawa Pening lake, Semarang, Indonesia. $NaClO_2$, NaOH, H_2SO_4, CH_3COOH, HCl, Toluene 99.9%, ethanol 99.8%, acetone 99.8 %, $(CH_3CO)_2O$ (acetic anhydride) 98,5 % and PEG (polyethylene glycol) were purchased from Merck, (Hohenbrunn, Germany). While cellulose acetate as a reference was used for membrane preparation, bought from Aldrich Chemistry, USA. Chitosan $(C_6H_{11}NO_4)n$ with deacetylation degree of 80.4% and a viscosity of 118.4 cp, produced by Biotech Surendo, Cirebon, Indonesia. Distilled water for all experiments produced from home-made pure water unit.

2.2. Methods

2.2.1. Cellulose Acetate Preparation Method. Cellulose isolation was performed by extraction followed by acetylation resulting CA. The method was adapted from previous studies [15-16].

2.2.2. CA and CA-Chi Membrane Preparations. Preparation of CA membrane was conducted by wet phase inversion. CA solutions with concentration of 13-15 % were prepared using acetone as solvent then stirred for 3 hours until all the cellulose acetate dissolved. Polyethylene glycol (5%) was added drop by drop. The homogenous polymer solution was left without stirring until no bubbles were observed. The polymer solution was cast using a steel casting knife on a glass substrate, then left in room temperature with time varies i.e. 0, 5, 10 and 15 seconds. Thereafter, the proto-membrane was solidified by inserting into a coagulation bath containing water for 1 day at room temperature. The resulting membranes were washed to remove excess solvent. The method used to prepare CA-Chi membranes was similar with CA membrane preparation.

3. Results and discussions

3.1 Investigation of separation performance of cellulose acetate membrane
The applicability of the obtained membrane in the separation of representative surface water (humic acid of 25 ppm) was evaluated.

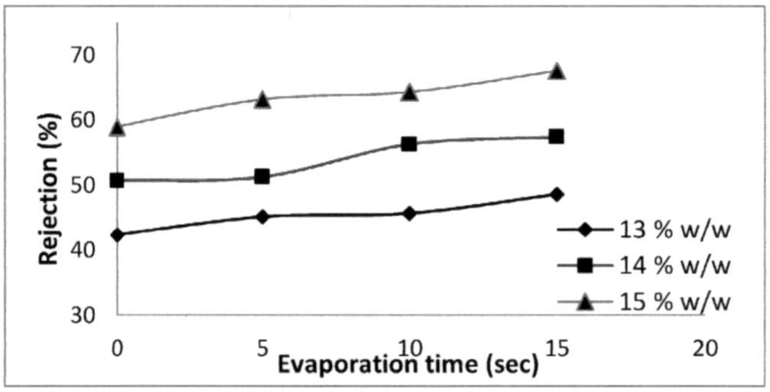

Figure 1. Effect of CA concentration and evaporation time on membrane separation efficiency.

The increase in concentration of cellulose acetate increase rejection. This phenomenon can be explained by the higher concentration of cellulose acetate, the higher number of polymer molecules per unit volume; consequently, the number of solvent molecules which will be replaced by non-solvent into the pore decreases [17]. In addition, the increase in CA concentration also strengthens the thermodynamic stability of the solution to be formed into films. Thereby, the membrane pore sizes resulted from a higher polymer concentration should be narrower. Figure 1 presents measurement data with the variation of evaporation time and CA concentration. On the other hand, rejection also increase along with the increasing evaporation time.

3.2 SEM imaging of CA membranes
Visualization of membrane surface by using SEM supports the preceding explanations that the increased evaporation time produced denser membrane. Fig 2 revealed that the membrane pore diameter has a size of 1-10 μm wherein the pore size is categorized as microfiltration membranes. With the same CA concentration (13% w/w) but with different time evaporation (0, 10 and 15 seconds), revealed that increasing time of evaporation reduced membrane pore diameter and quantity of membrane, so that the surface of membrane tends to denser.

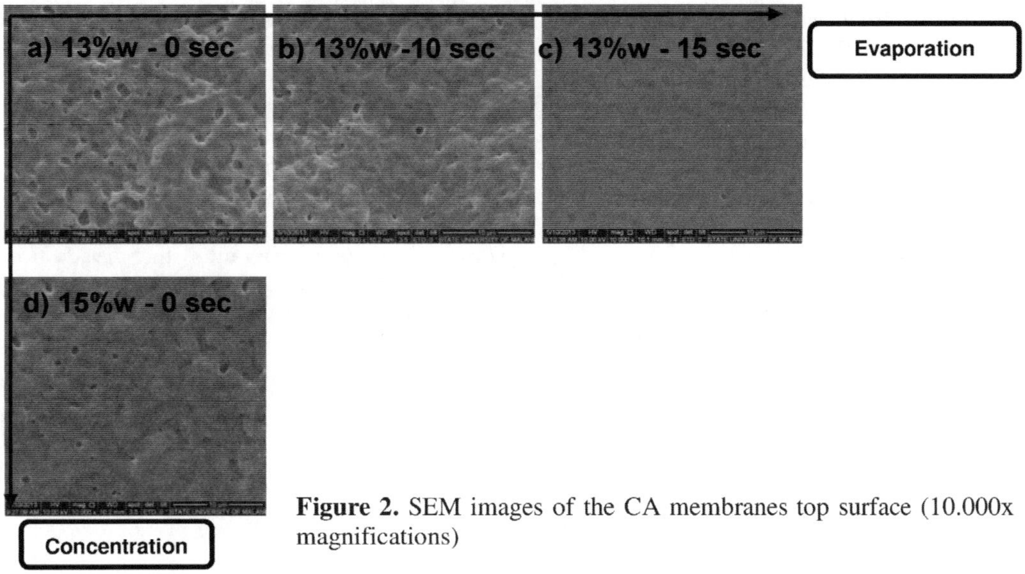

Figure 2. SEM images of the CA membranes top surface (10.000x magnifications)

Longer evaporation time causes an increase in the polymer concentration in the upper layer of the membrane and caused the membrane pores become smaller or narrower, due to interaction switchover solvent with the ambient air which is shorter compared to switchover solvent and non-solvent [18]. Likewise, comparison of membranes with different polymer concentrations, i.e. (concentration of 13% w and 15% w), it can be seen that the greater the concentration of the polymer, the number of membrane pore decreases and its diameter narrows. It proves that the more the polymer concentration causes increasing in rejection.

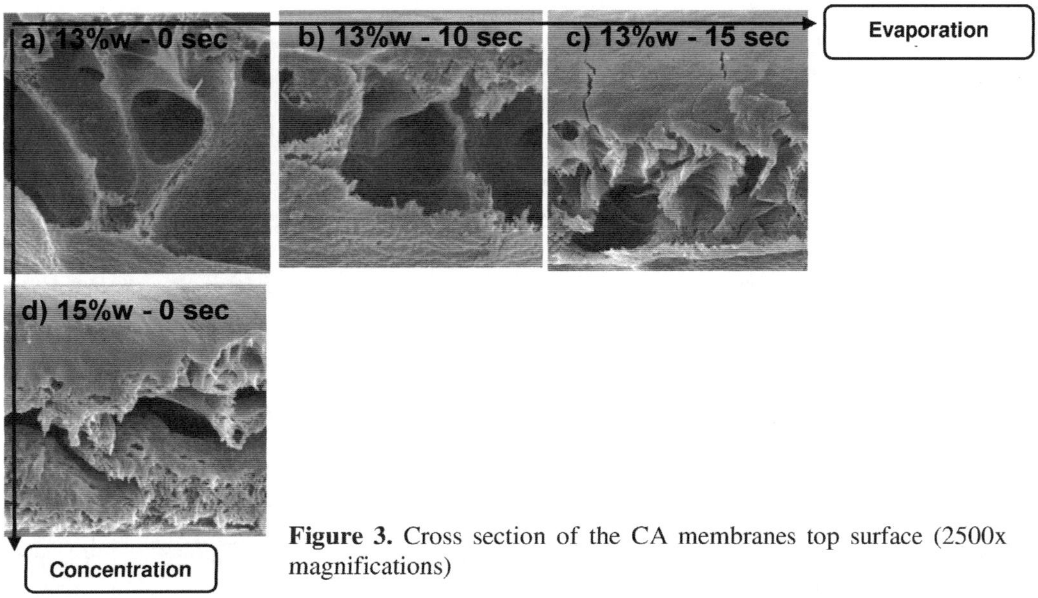

Figure 3. Cross section of the CA membranes top surface (2500x magnifications)

Figure 3 reinforces previous explanation that evaporation time affect the changes morphology of membrane structure. The larger the area of absorption (absorbance value) indicates that the polymer concentration in the membrane increases, resulting in more dense membrane. On the other hand, the concentration of polymer also affects the morphology of the membrane structure. Higher polymer concentration resulted better morphology of membrane structure (catchment area of membrane with 15 % w polymer concentration greater than 13% w).

3.3. Cellulose acetate-chitosan (CA-Chi) membrane

CA-Chi membranes were prepared by polymer blending between cellulose acetate and chitosan. It is clearly observed from Figure 4 that pure water flux decreased with the addition of chitosan composition. The highest value, 68.79 L/m^2. hour was in flux membranes without chitosan (Chi 0%) was and lowest flux value was 5.70 L/m^2, produced by 1% Ca-Chi membrane. In other word, we can say that the addition of chitosan decreased water permeability of CA-Chi membranes. As the concentration of chitosan was increased, the water flux decreased. This phenomenon can be explained that addition of chitosan increases polymer concentration and solution viscosity.

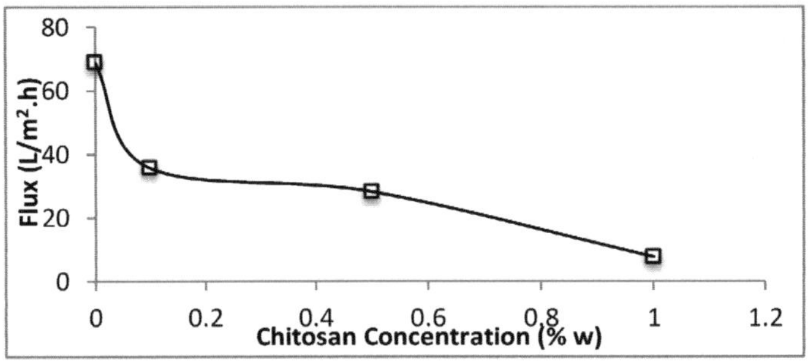

Figure 4. Water permeability of CA-Chi membranes as a function of chitosan concentration

The increase in polymer concentration increases polymer fraction in polymer solution, which will decrease the rate of liquid-liquid demixing resulting a membrane with lower porosity. The increase in solution viscosity causes reduction of mutual diffusivities between non-solvent and solvent in the system during solidification of the casting solution. Similar results have been reported in previous literatures [3, 19, 20]. In addition, the flux decline may occur due to the effect of chitosan addition in membrane preparation causing membrane properties being changed. Characteristic of cellulose acetate is more hydrophilic than chitosan, so this reason caused membrane flux reduction [21].

3.4. FTIR Analysis

Figure 5 shows infrared spectrophotometer images CA membrane without chitosan as control membrane and Ca-Chi membranes with various concentrations of chitosan. Interaction between cellulose acetate and chitosan in CA-Chi membranes can be seen from new or shifted peaks which are different from existing condition.

Special characteristics of cellulose acetate could be observed from the peaks at 1730 cm^{-1} representing the presence of carbonyl group (C=O), the highest peak at 1025 cm^{-1} representing C–O–C bond, and the peaks at 1360 and 1220 cm^{-1}, which indicate C–H bond from (CH_3). In addition, CA could also be observed by minor peaks at ~3500 and ~2840 cm^{-1} indicating O–H stretching and C–H bond stretching, respectively. This result agrees well with previous publications [21, 22, 23].

The addition of chitosan could clearly be observed by the peaks at ~1514 cm^{-1} (marked by black oval shaped). These peaks represent absorption band of amine bond (NH_2). The formed peaks in the range of 3687-3793 cm^{-1} indicate the presence of hydrogen bonds (OH) (also marked by black oval shaped). Moreover, slight movement of peak (from 1550 to 1514 cm^{-1}) is believed because both CA and chitosan present on the membrane surface. Similar results were reported by previous authors [24-27].

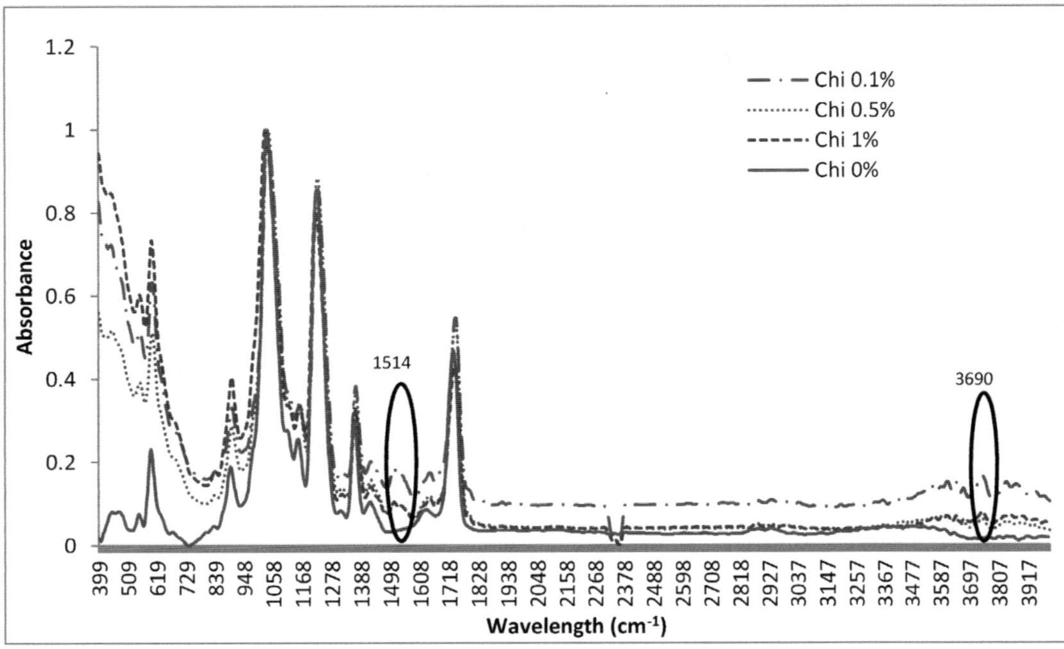

Figure 5. FTIR spectra of CA-Chi membranes.

3.5. SEM imaging of CA-Chi membranes

Visualization of membrane surface by using SEM (Figure 6) supports the preceding explanations (Section 3.4). It was observed qualitatively that addition of chitosan decreased membrane pore size as well as pore density. In addition, CA-Chitosan membrane (Fig 6.b.) revealed a rougher surface than CA membrane (without chitosan) (Fig 6.a.).

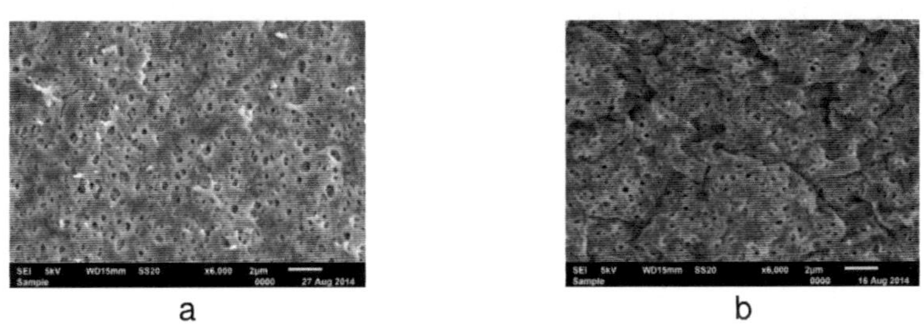

Figure 6. SEM images of membrane surface morphology (a) CA membrane, (b) CA-Chi membrane

4. Conclusion

This work provides a simple approach for obtaining modified nature polymer materials. Cellulose has been succesfully isolated from water hyacinth. The yield obtained can be processed into cellulose acetate (CA) which has similar chemical structure with commercial CA. Therefore, it possible to use it as material to prepare membranes. CA membrane with various concentration and time evaporation were also prepared and investigated. Addition of chitosan in the preparation of cellulose acetate membranes decreased membrane pore size as well as pore density.

Acknowledgement

The authors thank to Diponegoro University for the financial support of this research through Non Tax National Revenue Budget execution list Nr. 023.04.2189185/2014 year 2014.

References

[1] Susanto H, Stahra N and Ulbricht M, 2009 *J. Membr. Sci.* **342** (1), pp.153-164
[2] Abedini R, Mousavi S M and Aminzadeh R, 2011 *Desalination* **277** (1), pp.40-45
[3] Susanto H and Ulbricht M, 2009 *J. Membr. Sci.* **327**(1), pp.125-135
[4] Mahdavi H and Shahalizade T, 2015 *J. Membr. Sci.* **473**, pp.256-266
[5] Idris A, Yee HK, Kee CM, 2009 *J. Technol.* **51** F pp.67-76
[6] Perera DHN, Nataraj SK, Thomson NM, Sepe A, Hüttner S, Steiner U, Qiblawey H and Sivaniah E, 2014 *J. Membr. Sci.* **453** pp 212-220
[7] Moghadassi AR, Rajabi Z, Hosseini SM, Mohammadi M. 2014 *J. Ind. Eng. Chem.* **20** (3):1050-1060
[8] Wang LK, Chen JP, Hung YT, Shammas NK, editors. 2010 *Membrane and desalination technologies.* Humana press
[9] Sivakumar M, Malaisamy R, Sajitha CJ, Mohan D, Mohan V, Rangarajan R, 2000 *J. Membr. Sci.* **169** (2) pp. 215-228
[10] Sivakumar M, Mohan DR, Rangarajan R. 2006 *J. Membr. Sci.* **268** (2) pp.208-219
[11] Kamal H, Abd-Elrahim FM, Lotfy S, 2014 *J. Radiat. Res. Appl. Sci.***7** (2) pp. 146-53
[12] Zhou W, Zhu D, Langdon A, Li L, Liao S, Tan L. 2009 *Bioresour. Technol.* 2009 **100** (21), pp. 5366-5369
[13] Krajewska B. 2005. *Sep. Purif. Technol.* **41** (3) pp. 305-312.
[14] Kong M, Chen XG, Xing K, Park HJ. 2010 *Int. J. Food Microbiol.* **144** (1) pp.51-63
[15] Istirokhatun T, Rokhati N, Rachmawaty R, Meriyani M, Priyanto S, Susanto H. 2015 *Procedia Environ. Sci.* **23** pp. 274-281
[16] Sundari MT, and Ramesh A, 2012 *Carbohydr. Polym.* **87** (2), pp.1701-1705
[17] Mohan D, Arthanareeswaran G, Thanikaivelan P, Srinivasn K, Mohan D, Rajendran M. 2004 *Eur. Polym. J.* **40** (9) pp. 2153-2159
[18] Strathmann H, Scheible P and Baker, RW, 1971 *J. Appl. Sci.* **415** (4), pp.811-828
[19] Saljoughi E and Mohammadi T, 2009 *Desalination* **249** (2), pp.850-854
[20] Rahimpour A, Jahanshahi M, Rajaeian B and Rahimnejad M, 2011 *Desalination* **278** (1), pp.343-353
[21] Liu C, Bai R 2005 *J. Membr. Sci* **267** (1) pp. 68-77
[22] Waheed S, Ahmad A, Khan SM, Jamil T, Islam A and Hussain T, 2014 *Desalination* **351**, pp. 59-69
[23] Ali M, Zafar M, Jamil T and Butt MTZ, 2011 *Desalination* **270** (1), pp.98-104
[24] Li XF, Feng XQ, Yang S, Fu GQ, Wang TP and Su ZX, 2010 *Carbohydr. Polym.* **79** (3), pp. 493-499
[25] Liu CX, Zhang DR, He Y, Zhao XS and Bai R, 2010 *J. Membr. Sci.* **346** (1), pp.121-130
[26] Sun Y, Liu Y, Li Y, Lv M, Li P, Xu H and Wang L, 2011 *Carbohydr. Polym.* **84** (3), pp. 952-959
[27] Ni L, Meng J, Li X and Zhang Y, 2014 *J. Membr. Sci.* **451**, pp.205-115

IOP Conf. Series: Earth and Environmental Science **55** (2017) 012019 doi:10.1088/1755-1315/55/1/012019

Identification and Antibacterial Activity of Bacteria Isolated from Marine Sponge *Haliclona (Reniera)* sp. against Multi-Drug Resistant Human Pathogen

Meezan Ardhanu Asagabaldan[1] , D Ayuningrum[1], R Kristiana[1], A Sabdono[2], O K Radjasa[2,3], and A Trianto[2]

[1] Department of Coastal Resources Management, Diponegoro University, St. Imam Bardjo, SH, No. 15, Semarang 50241, Indonesia

[2] Department of Marine Science, Faculty of Fisheries and Marine Science, Diponegoro University, St. Prof. H. Soedharto,SH, Tembalang, Semarang 50275, Indonesia

[3] Director of Research and Public Services at Ministry of Research, Technology and Higher Education, Indonesia

meezan.asagabaldan@gmail.com

Abstract. The marine sponge *Haliclona (Reniera)* sp. was a potential source of natural bioactive compounds. This sponge widely distributed along the coast of Panjang Island, Jepara, Indonesia. The aims of this research were to isolate the associated bacteria with *Haliclona (Reniera)* sp. and to screen the antibacterial activity against Multi-Drug Resistant (MDR) bacteria. Amount five bacteria were isolated using media selective for bacteria. The antibacterial activities of bacteria were performed by overlay methods. The bacteria strain PSP. 39-04 had the best activity against *Pseudomonas aeruginosa, Staphylococcus aureus, Acinetobacter baumannii,* and *Enterobacter cloaceae.* Based on colony morphology and phylogenetic characterization using 16S rRNA gene sequencing, PSP 39-04 was closely related with *Chromohalobacter salixigens strain DSM3043.*

Keywords: *Haliclona (Reniera)* sp., bacteria, screening, Multi-Drug Resistant, antibacterial

1. Introduction

Infectious disease agents such as *Pseudomonas aeruginosa, Klebsiella pneumonia, Escherichia coli, Staphylococcus aureus, Acinetobacter baumannii,* and *Enterobacter cloaceae,* became a serious health problem in many developing countries, including in Indonesia [1],[2],[3], because they led into morbidity and high mortility in humans [4]. In recent years, those pathogen bacterial became into organisms resistant to multiple antibiotics, or also called the Multi Drug Resistant (MDR). The increasing number of victims have been reported [5],[6],[7]. Moreover, these problems led to higher maintenance costs and only a few hospitals that could solved this problems [4]. The emergence of discovery the newer antibiotics were needed with a promise for combating MDR strains [8],[9],[10].

Many studies have been reported that the natural resources from the sea have abundant potential bioactive compounds against human pathogens [11],[12],[13],[14],[15]. Sponge had into the richest sources bioactive compounds and could be used in the field of pharmacy [16],[17],[18],[19],[20]. Since more than 50 years, sponge organism had been attracted the attention of researchers because of its ability to produce the natural bioactive compounds that derived from the secondary metabolisms [13]. Several studies have been reported that the sponge *Haliclona (Reniera)* sp. produced Haliclonin-A, as the antibacterial compounds caused by human disease bacterial [21]. In addition, the compounds was from sponge *Haliclona crassiloba* has been reported to inhibit the activity of pathogenic bacteria in humans both negative and positive grams [14]. Other studies have shown that extracts of Sponge *Haliclona* sp. had some activity against gram-positive bacteria *Staphylococus aureus*, and *Bacillus subtilis* [15].

However, the main problems was the limitations of the sponge organisms. Exploration of bioactive compounds from sponge required a lot of biomaterial that might happened *over-exploitation* and low levels of conservation [22],[23]. Bacteria associated with sponge was a new breakthrough for searching the bioactive compounds, because it was easily produced with rapid and large scale, without having to take excessive sponge [24],[25]. Sponge became one of the residences of marine microorganisms where located in mesohil layer of sponge or commonly called *High Microbial Abundance Sponge* (HMAS) including bacteria. Role of bacteria in the sponge was as a defense from predators [26] and also helped in the metabolism of a sponge [27],[28]. Another important role of the bacteria associated with the sponge has produced bioactive compounds which acted as antimicrobial, antiviral, antioxidant and antitumor activity [19,29,30]. Several studies have shown that extracts of some bacteria associated with sponge *Haliclona vansoesti* have antibacterial activity against human diseases such as *S. aureus, P. aeruginosa, E. coli* and *B. subtilis* [31]. The sponge *Cliona* sp. associated bacteria *Comamonas testosteroni* and *Citrobacter freundii* hade a compound that can be used as a candidate against Ebola virus [32].

In the present study, we identify the associated bacteria isolated from sponge *Haliclona (Reniera)* sp. from Panjang Island, Indonesia and determine the antibacterial activity against MDR human pathogen bacteria.

2. Matherial and Methods

2.1. Sample collection of sponge

The sponge *Haliclona (Reniera)* sp. (figure 1) were collected from Teluk Awur (06°36'57.4"S, 110°38'19.4"E) and Panjang Island (06°34'35.2"S, 110°37'52.7"E), Jepara, North Java, Indonesia by snorkling from a depth of approximately 3 m. The collection sponge were put into sterile plastic bags underwater to avoid contact with air and brought to Laboratory of Tropical Marine Biotechnology, Diponegoro University. The Specimens were processed under aseptic conditions as follows: sponge were rinsed with sterile seawater and 1 cm^3 of sponge tissue was excised from the middle of the whole sponge with sterile knife. The sponge were macerated at room temperature (RT, 24 ± 2°C) for 5 min.

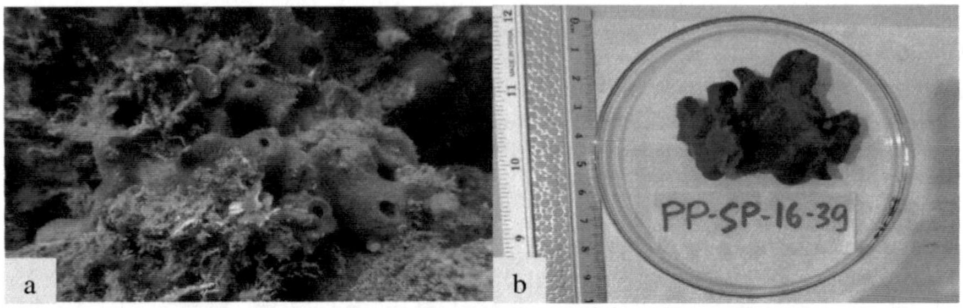

Figure 1. (a) selected Sponge *Haliclona (Reniera)* sp. from Panjang Island, Jepara; (b) sample collection (PP-SP-16-39) was under aseptic conditions

2.2. Isolation and purification of bacteria

The macerated sponge were serially diluted, spread on ½ strenght Zobell 2216E marine agar medium [33],[34],[35] and incubated at room temperate (RT, 24 ± 2°C) for 48 h. On the basis of morphological features, colonies were randomly picked and purified by making streak plates [36]. The strains of bacteria associated were named with the initials of sponge from which they were isolated. To purify the bacteria isolates, the colonies with different morphological characteristics were picked up and transferred onto freshly prepared media until pure cultures were obtained.

2.3. Screening for antibacterial activities against MDR pathogen bacteria

Screening biological activity of bacteria associated with sponge were conducted by testing against isolate pathogenic bacteria *P. aeruginosa, K. pneumonia, E. coli, S. aureus, A. baumannii,* and *E. cloaceae,* that were obtained from Kariadi Hospitals, Semarang. A total of five bacteria isolates from sponge were tested by using overlay methods [36]. Those isolates of bacteria were inoculated in agar medium and were incubated at optimum temperate growth for bacteria (37 ± 2°C) for 18 – 20 h. On the other sides, the isolates of MDR pathogenic bacteria were inoculated in soft agar medium (70% agar compotition) with the concentration were 0,5 McFarland, then it were poured into the surface of bacteria isolates (after incubation). The plates were incubated at optimum temperate growth for bacteria (37 ± 2°C) for 24 h. The antibacterial activity of MDR pathogen was defined by the formation of Diameter of Inhibitation (DOI) around the bacteria isolates.

2.4. Morphological and biochemical characterization of associated bacteria

Morphological characterization of potential bacteria isolates was performed by gram staining according to method previously described [37]. Biochemical tests (indole, oxidase, urease, Citrate Simmons, nitrate reduction, lysine decarboxylase, maltose, and sucrose) were carried out at Laboratory of Tropical Marine Biotechnology and Microbiology Laboratory, Medical Faculty, Diponegoro University.

2.5. Amplification of partial 16S ribosomal RNA of associated bacteria

The molecular-based works were carried out at Laboratory of Tropical Marine Biotechnology, Diponegoro University. Genomic DNA of potential bacteria isolates was used as a template for polymerase chain reaction (PCR) amplification. The amplification of 16S ribosomal RNA (rRNA) was performed using primers 27f (5′-AGAGTTT- GATCMTGGCTCAG-3′) and 1492r (5′-TACGGY-TACCTTGTTACGACTT-3′) [38]. The PCR was conducted in thermal cycler (Perkin – Elmer, USA) under the following cycling conditions according to methods that previously described [36][37], it consisted of an initial denaturation at 95 °C for 3 min and then 30 cycles with denaturation at 94 °C for 1 min, annealing at 55 C for 1 min, and extension at 72 °C for 1 min. A final extension was performed at 72 °C for 7 min. The PCR products were analyzed by agarose (1%) gel electrophoresis and the digital images were obtained using UVI Doc HD5 (UVITEC Cambridge).

2.6. DNA sequencing and Phylogenetic analysis

The DNA sequencings were carried out by Molecular Biology Laboratory, Agency for the Assesment and Application Technology (BPPT), Jakarta, Indonesia. To identify the closest neighbors of potential bacteria isolates, the 16S rRNA gene sequences were analyzed using BLAST [40]. The 16S rRNA gene sequences were aligned and a neighbor-joining phylogenetic tree was constructed using MEGA 6 according to methods that previously described by Tamura *et al.* [41]. Bootstrap tests were performed 1,000 times using MEGA 6.

3. Results and discussion

3.1. Isolation of Reniera sp. associated bacteria

The associated bacteria were obtained from three dilution series of macerated sponge samples. The results are shown in table 1. Only five cultured bacteria were associated with *Reniera* sp. with different morphological characteristics.

Table 1. Morphological characteristics of *Reniera* sp. associated bacteria

No.	Isolates	Colour	Size	Shape	Margin	Elevation
1	PSP.39-01	White	Small	Round	Entire	Convex
2	PSP.39-02	White	Pin point	Round	Entire	Convex
3	PSP.39-03	White	Small	Round	Entire	Convex
4	PSP.39-04	Red	Small	Round	Entire	Convex
5	PSP.39-05	Opaque yellow	Small	Round	Entire	Convex

3.2. Screening of anti-MDR bacterial activities

Associated bacteria of *Reniera* sp. were tested antibacterial activities against MDR Pathogens. Only the isolates bacteria with code PSP.39-04 shown the positive antibacterial activities against *P. aeruginosa* (21.20 ± 0.26 mm), *E. cloaceae* (17.13 ± 0.31 mm), *A. baumannii* (15.90 ± 0.26 mm), and *S. aureus* (13.10 ± 0.26 mm) (figure 2) and shown negative against *K. pneumonia* and *E. coli.* The results are shown in table 2.

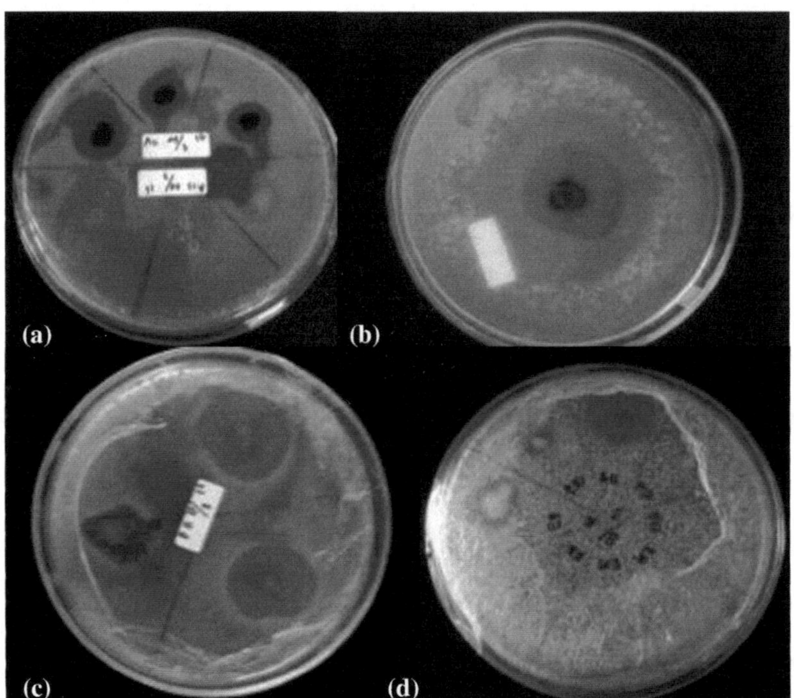

Figure 2. The antibacterial activity of PSP.39-04 isolated bacteria against MDR pathogen; (a) against MRSA *S. aureus,* (b) against MDR *E. cloaceae,* (c) against MDR *P. aeruginosa,* (d) against MDR *A. baumannii*

Table 2. Screening of *Reniera* sp. associated bacteria against MDR pathogen bacteria

Isolates	Zone of Inhibition (mm)					
	P.aeruginosa	*K. pneumonia*	*E. cloaceae*	*A. baumannii*	*E. coli*	*S. aureus MRSA*
PSP-39-01	-	-	-	-	-	-
PSP-39-02	-	-	-	-	-	-
PSP-39-03	-	-	-	-	-	-
PSP-39-04	21.20 ± 0.26	-	17.13 ± 0.31	15.90 ± 0.26	-	13.10 ± 0.26
PSP-39-05	-	-	-	-	-	-

3.3. Morphological and biochemical characterization of PSP.39-04 bacteria isolates
The morphology of isolated bacteria PSP.39-04 was belonged to gram negative (red colour) bacteria and rod shapes, that shown in figure 3.

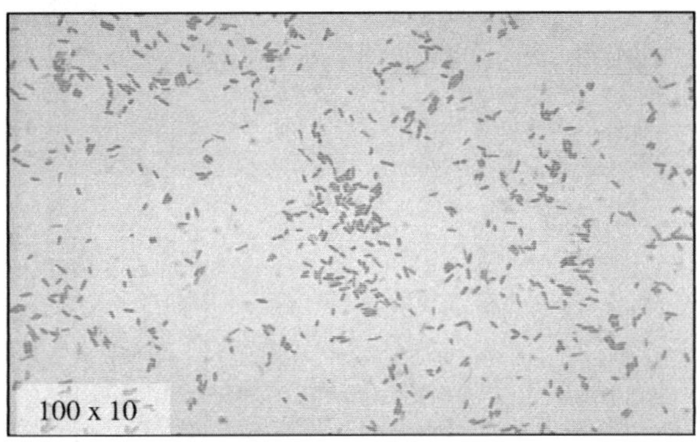

Figure 3. The morphological caracterization of PSP.39-04 bacteria isolates with 100 x10 magnification

The result of biochemical characterization shown that PSP.39-04 isolates in table 3.

Table 3. The biochemical characterization of PSP.39-04 bacteria isolates from *Reniera* sp.

No.	Biochemical characteristics	Tests
1	Oxidase	-
2	Nitrate reduction	+
3	Anaerobic growth with nitrate	+
4	Urease	+
5	Indole	-
6	Citrate (Simmons)	+
7	Hydrolysis of casein	+
8	Ornithine decarboxylase	-
9	Lysine decarboxylase	-
10	Arabinose	+
11	Maltose	+
12	Sucrose	-
13	d –Trehalose	+
14	H5S production	+
15	d-Cellobiose	-
16	Sucrose	+

IOP Conf. Series: Earth and Environmental Science **55** (2017) 012019 doi:10.1088/1755-1315/55/1/012019

3.4. Amplification of partial 16S ribosomal RNA of PSP.39-04 bacteria isolates

Amplification and gel electrophoresis of 16S rRNA showed that PSP.39-04 bacteria isolates had approximately 1500bp and it was belonged to bacteria groups. PSP.39-04 DNA amplification were be able to do the sequencing of DNA. The results of 16S rRNA analysis shown in figure 4.

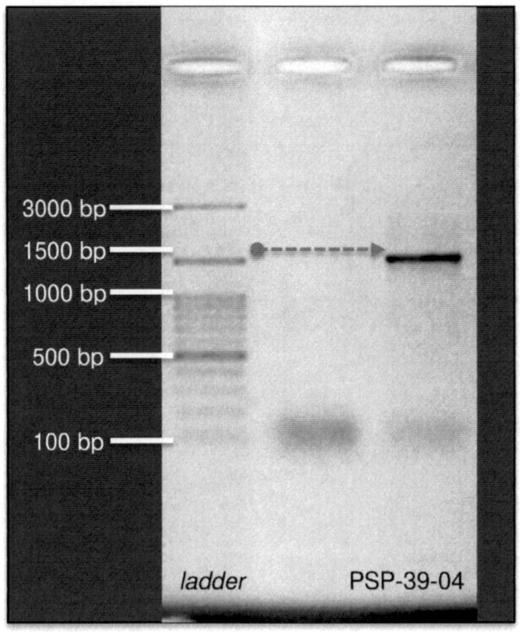

Figure 4. The 16S rRNA analysis of PSP.39-04 bacteria isolates

3.5. DNA sequencing and Phylogenetic contructions of PSP.39-04 bacteria isolates

Based on DNA sequencing, the DNA amplification of PSP.39-04 Isolates represented around 1376bp and it shown in figure 5.

CAGTCGAGCGGTAACAGGTCCAGCTTGCTGGACGCTGACGAGCGGCGGACGGGTGAGTAATGCATAGGAATCTACCCAGTCGTGGGGGATAAC
CTGGGGAAACCCAGGCTAATACCGCATACGTCCTACGGGAGAAAGCGGGGGCTCTTCGGACCTCGCGCGATTGGATGAGCCTATGTCGGATTA
GCTGGTTGGTGGGGTAACGGCTCACCAAGGCGACGATCCGTAGCTGGTCTGAGAGGATGATCAGCCACACTGGGACTGAGACACGGCCCAGAC
TCCTACGGGAGGCAGCAGTGGGGAATATTGGACAATGGGCGAAAGCCTGATCCAGCCATGCCGCGTGTGTGAAGAAGGCTTTCGGGTTGTAAA
GCACTTTCAGTGGGGAAGAAGGCTTGTCGGCCAATACCCGGCAAGAGCGACATCACCCACAGAAGAAGCACCGGCTAACTCCGTGCCAGCAGC
CGCGGTAATACGGAGGGTGCAAGCGTTAATCGGAATTACTGGGCGTAAAGCGCGCGTAGGCGGCTTGTCACGCCGGGTGTGAAAGCCCCGGGC
TCAACCTGGGAACGGCATCCGGAACGGGCAGGCTAGAGTGCAGGAGAGGAAGGTAGAATTCCCGGTGTAGCGGTGAAATGCGTAGAGATCGG
GAGGAATACCAGTGGCGAAGGCGGCCTTCTGGACTGACACTGACGCTGAGGTGCGAAAGCGTGGGTAGCAAACAGGATTAGATACCCTGGTAG
TCCACGCCGTAAACGATGTCGACTAGCCGTTGGGTCCCTTGAGGACTTAGTGGCGCAGTTAACGCGATAAGTCGACCGCCTGGGGAGTACGGCC
GCAAGGTTAAAACTCAAATGAATTGACGGGGGCCCGCACAAGCGGTGGAGCATGTGGTTTAATTCGATGCAACGCGAAGAACCTTACCTACCCT
TGACATCCTGCGAACCCGGAAGAGATTCCGGGGTGCCTTCGGGAGCGCAAAGACAGGTGCTGCATGGCTGTCCTCAGCTCGTGTTGTGAAATGT
TGGGTTAAGTCCCGTAACGAGCCTAACCCTTGTTCCTATTTGCCAGCGATTCGGTCGGGAACTCTAGGGAAACTGCCGGTGACAAACCGGAGGA
AGGTGGGGACAACTTCAAGTCATCATGGCCCTTACGGGTAGGGCTACACACGTGCTACAATGGCCGGTACAAAGGGTTGCGAAGCCGCGAGGT
GAAGCCAATCCCAGAAAGCCGGCCTCAGTCCGGATTGGAGTCTGCAACTCGACTCCATGAAGTCGGAATCGCTAGTAATCGTGCATCAGAATGG
CACGGTGAATACGTTCCCGGGCCTTGTACACACCGCCCGTCACACCATGGGAGTGGACTGCACCAGAAG

*A=Adenin; G=Guanine; T=Timin; C=Cytosin

Figure 5. The DNA sequencing results of PSP.39-04 isolates

The construction of phylogentic trees was analysed with BLAST homology and the PSP.39-04 isolates was closely related with *Chromohalobacter salexigens* strain DSM 3043 as much as 99%, Halomonadaceae. The phylogenetic trees of PSP.39-04 isolates shown in figure 6.

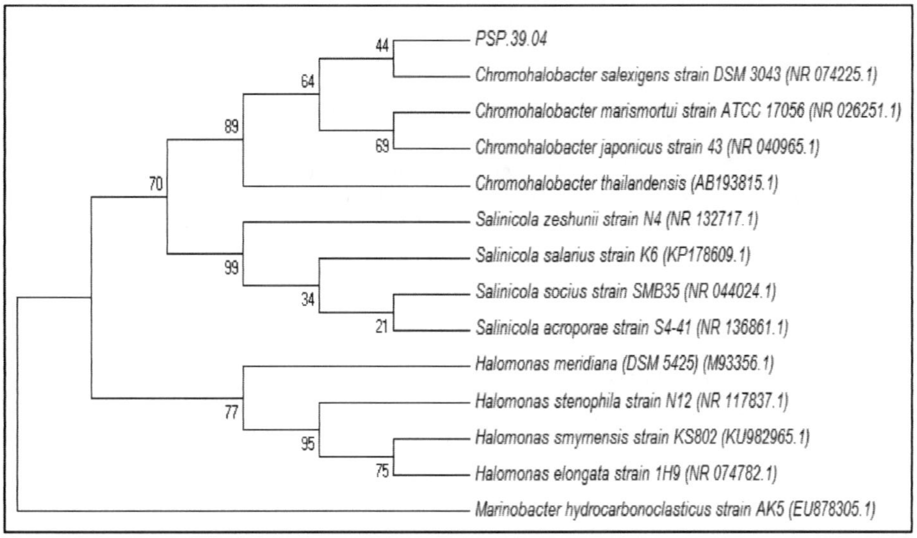

Figure 6. The phylogenetic trees of PSP.39-04 isolates

The antimicrobial activity of sponge *Reniera* sp. had been done in 1979 with Renierone [42] and in 2007, Renierosides [43] as antimicrobial metabolites. We assume that secondary metabolites were derived to associated bacteria, according to study previously reported [44],[45],[46],[47],[48]. The presence of sponge plays as host to high density of microorganisms (virus, bacteria, fungi) or as known as High Microbial Abundance Sponge (HMAS) with 40-60% of biomass of sponge consists of microorganisms. We reported only five bacteria can be isolated in sponge samples (*Reniera* sp.). The results was high contrast with the nature of sponge which high density of microbial. This is due to of the characteristic of associated bacteria, that isolation of bacteria just used one culture media (Zobell) where it is a non-specific or non-selective media. The sponge associated bacteria have been investigated that less than 1% of taxa are be able to culture under laboratory conditions [49]. Hence, we can use a wide range of culture condition approaches in next study.

Screening of Antibacterial activity is the one of useful and effective ways for the preliminary assessment to know which associated bacteria have antibacterial activity. Thus, we are able to keep focus on it. The present study reported that the associated bacteria of sponge *Haliclona (Reniera)* sp. from Pulau Panjang, Indonesia showed high antibacterial activity against MDR pathogen. Only the isolates of PSP.39-04 showed strong growth inhibition of MDR-pathogen *P. aeruginosa, A. baumannii, E. cloaceae,* and *E. coli.* The growth inhabitation was performed by the role of secondary metabolites from PSP.39-04, which will be active when the presence of MDR-pathogens.

The pathogenic bacteria *P. aeruginosa, E. cloaceae, A. baumannii,* and *E. coli* are gram negatives bacteria, while *S. aureus* is gram positive which resistant with methicillin groups (MRSA). The secondary metabolites of PSP.39-04 isolates can inhibit the activity of gram negatives bacteria which have lipopolysaccharides (LPS) cell walls and we assume it have inhibitor for beta lactamase enzymes were produced by *S. aureus.* The previous research by Radjasa *et al.* [34] investigated that the associated bacteria of *Haliclona* sp. had an antimicrobial activities against MDR-pathogen. Another research showed the sponge *Haliclona* sp. associated bacteria have antibacterial activities against pathogenic bacteria [19],[48],[50],[51]. This result offers using [52] sponge *Haliclona* sp. associated bacteria as the source of antibacterial compounds for controlling the MDR-pathogenic bacteria.

The phylogenetic tree analysis showed that PSP.39-04 bacteria isolates is closely related to *Chromohalobacter salexigens strain DSM3043* which isolated from solar salt facilities [53]. It is a halophilic or salt needed for growth and displays a high halotolerant, as the ability survive in wide range of salt concentration [54]. Thus, from this feature, *C. salexigens* has special metabolites that perform for to survive. *C. salexigens* is included in Gammaproteobacteria that produces many bioactive compounds against MDR-pathogens [55],[56],[57],[58] and. Genus Chromohalobacter also produces halo beta lactamase enzymes and efflux pump that can excrete toxic compounds from cells [59]. The recent study showed that *Chromohalobacter* sp. produced bioactive compounds (cyclic peptides) against MDR pathogens [60] and another research was done by making *Chromohalobacter* sp. as biogenic to synthesize nanoparticles for combating against MDR pathogens [61]. We assume that PSP.39-04 bacteria isolates from sponge *Haliclona (Reniera)* sp. is the potential bacteria for developing new bioactive compounds against MDR-pathogens.

4. Conclusion

In this present study, we revealed that the marine sponge *Haliclona (Reniera)* sp. associated bacteria with PSP.39-04 codes performed the antibacterial activity against MDR pathogenic bacteria, which are *P. aeruginosa, E. cloaceae, A. baumannii,* and *S. aureus (MRSA)*. The PSP.39-04 was closely related to *Chromohalobacter salexigens* with 99% homology. It could be a vital source for promoting as antibacterial. Hence, the sponge *Haliclona (Reniera)* sp. associated bacteria is considered as a good source for the isolation of bioactive compounds and development of new drugs against MDR pathogens.

Acknowledgment

Dr. Karyadi Hospital, Semarang, contributed MDR-pathogenic bacteria. Thanks also go to Handung Nuryadi, who assisted with the sponge sample collections.

This work was supported by PMDSU (Master Program of Education Leading to Doctoral Degree for Excellent Graduates) scholarship from Ministry of Research and Higher Education, Indonesia.

References

[1] Buford V R, Kumar V and Kennedy B R 2016 Relationship of various infection control interventions to the prevalence of multidrug-resistant Pseudomonas aeruginosa among U.S. hospitals *Am. J. Infect. Control* 13–7

[2] Devasahayam G, Scheld W M and Hoffman P S 2010 Newer antibacterial drugs for a new century. *Expert Opin. Investig. Drugs* **19** 215–34

[3] Radji M, Agustama R A, Elya B and Tjampakasari C R 2013 Antimicrobial activity of green tea extract against isolates of methicillin-resistant Staphylococcus aureus and multi-drug resistant Pseudomonas aeruginosa *Asian Pac. J. Trop. Biomed.* **3** 663–7

[4] O'Neill J 2014 Antimicrobial Resistance : Tackling a crisis for the health and wealth of nations *Rev. Antimicrob. Resist.* 1–16

[5] Lambert P A 2002 Mechanism of antibiotic resistance in Pseudomonas aeruginosa *J R Soc Med* **95** S22-26

[6] Meletis G and Bagkeri M 2013 Pseudomonas aeruginosa : Multi-Drug-Resistance Development and Treatment Options *Infect. Control* 33–56

[7] Djordjevic Z, Folic M M, Zivic Z, Markovic V and Jankovic S M 2013 Nosocomial urinary tract infections caused by Pseudomonas aeruginosa and Acinetobacter species: Sensitivity to antibiotics and risk factors *Am. J. Infect. Control* **41** 1182–7

[8] Mayer A M S and Hamann M T 2005 Marine pharmacology in 2001--2002: marine compounds with anthelmintic, antibacterial, anticoagulant, antidiabetic, antifungal, anti-inflammatory, antimalarial, antiplatelet, antiprotozoal, antituberculosis, and antiviral activities; affecting the cardiova *Comp. Biochem. Physiol. Toxicol. Pharmacol. CBP* **140** 265–86

[9] Mayer A M S, Rodríguez A D, Taglialatela-Scafati O and Fusetani N 2013 Marine pharmacology in 2009-2011: Marine compounds with antibacterial, antidiabetic, antifungal, anti-inflammatory,

antiprotozoal, antituberculosis, and antiviral activities; affecting the immune and nervous systems, and other miscellaneous mechanisms of *Mar. Drugs* **11** 2510–73

[10] Qi S H, Xu Y, Gao J, Qian P Y and Zhang S 2009 Antibacterial and antilarval compounds from marine bacterium Pseudomonas rhizosphaerae *Ann. Microbiol.* **59** 229–33

[11] Huang R M, Chen Y N, Zeng Z, Gao C H, Su X and Peng Y 2014 Marine nucleosides: Structure, bioactivity, synthesis and biosynthesis *Mar. Drugs* **12** 5817–38

[12] Lee J, Han C, Lee T G, Chin J, Choi H, Lee W, Paik M J, Won D H, Jeong G, Ko J, Yoon Y J, Nam S-J, Fenical W and Kang H 2016 Marinopyrones A–D, α-pyrones from marine-derived actinomycetes of the family Nocardiopsaceae *Tetrahedron Lett.* **57** 1997–2000

[13] Genta-Jouve G and Thomas O P 2012 *Sponge Chemical Diversity. From Biosynthetic Pathways to Ecological Roles* vol 62

[14] Cheng Z Bin, Xiao H, Fan C Q, Lu Y N, Zhang G and Yin S 2013 Bioactive polyhydroxylated sterols from the marine sponge Haliclona crassiloba *Steroids* **78** 1353–8

[15] Nazemi M, Alidoust Salimi M, Alidoust Salimi P, Motallebi A, Tamadoni Jahromi S and Ahmadzadeh O 2014 Antifungal and antibacterial activity of Haliclona sp. from the Persian Gulf, Iran *J. Mycol. Med.* **24** 220–4

[16] Indraningrat A, Smidt H and Sipkema D 2016 Bioprospecting Sponge-Associated Microbes for Antimicrobial Compounds *Mar. Drugs* **14** 87

[17] Steinert G, Whitfield S, Taylor M W, Thoms C and Schupp P J 2014 Application of Diffusion Growth Chambers for the Cultivation of Marine Sponge-Associated Bacteria *Mar. Biotechnol.* **16** 594–603

[18] Taylor M W, Radax R, Steger D and Wagner M 2007 Sponge-Associated Microorganisms: Evolution, Ecology, and Biotechnological Potential *Microbiol. Mol. Biol. Rev.* **71** 295–347

[19] Santos O C S, Pontes P V M L, Santos J F M, Muricy G, Giambiagi-deMarval M and Laport M S 2010 Isolation, characterization and phylogeny of sponge-associated bacteria with antimicrobial activities from Brazil *Res. Microbiol.* **161** 604–12

[20] Santos-Gandelman J, Giambiagi-deMarval M, Oelemann W and Laport M 2014 Biotechnological Potential of Sponge-Associated Bacteria *Curr. Pharm. Biotechnol.* **15** 143–55

[21] Jang K H, Kang G W, Jeon J E, Lim C, Lee H S, Sim C J, Oh K B and Shin J 2009 Haliclonin A, a new macrocyclic diamide from the sponge Haliclona sp *Org. Lett.* **11** 1713–6

[22] Jana S, Gandhi A, Chakraborty S, K S K and . S K B 2013 Marine Microorganisms and Their Versatile Applications in Bioactive Compounds *Mar. Microbiol. Bioact. Compd. Biotechnol. Appl.* 379–87

[23] Leal M C, Sheridan C, Osinga R, Dionísio G, Rocha R J M, Silva B, Rosa R and Calado R 2014 Marine microorganism-invertebrate assemblages: Perspectives to solve the "supply problem" in the initial steps of drug discovery *Mar. Drugs* **12** 3929–52

[24] Thakur A N, Thakur N L, Indap M M, Pandit R A, Datar V V. and M??ller W E G 2005 Antiangiogenic, antimicrobial, and cytotoxic potential of sponge-associated bacteria *Mar. Biotechnol.* **7** 245–52

[25] Muscholl-Silberhorn A, Thiel V and Imhoff J F 2008 Abundance and bioactivity of cultured sponge-associated bacteria from the Mediterranean Sea *Microb. Ecol.* **55** 94–106

[26] Unson M D and Faulkner D J 1993 Cyoanobacterial symbiont biosynthesis of chlorinated metabolites from *Dysidea herbacea* (Porifera) *Experientia* **49** 349–53

[27] Wilkinson C R 1980 Nutrient translocation from green algal symbionts to the freshwater sponge Ephydatia fluviatilis *Hydrobiologia* **75** 241–50

[28] Schmidt E W and Donia M S 2010 Life in cellulose houses: Symbiotic bacterial biosynthesis of ascidian drugs and drug leads *Curr. Opin. Biotechnol.* **21** 827–33

[29] Usher K M, Sutton D C, Toze S, Kuo J and Fromont J 2005 Inter-generational transmission of microbial symbionts in the marine sponge Chondrilla australiensis (Demospongiae) *Mar. Freshw. Res.* **56** 125–31

[30] Khan S T, Komaki H, Motohashi K, Kozone I, Mukai A, Takagi M and Shin-Ya K 2011 Streptomyces associated with a marine sponge Haliclona sp.; biosynthetic genes for secondary metabolites and products *Environ. Microbiol.* **13** 391–403

[31] van Soest R W M, Boury-Esnault N, Vacelet J, Dohrmann M, Erpenbeck D, de Voogd N J, Santodomingo N, Vanhoorne B, Kelly M and Hooper J N A 2012 Global diversity of sponges (Porifera) *PLoS One* **7**

[32] Skariyachan S, Rao A G, Patil M R, Saikia B, Bharadwaj Kn V and Rao Gs J 2014 Antimicrobial potential of metabolites extracted from bacterial symbionts associated with marine sponges in coastal area of Gulf of Mannar Biosphere, India *Lett. Appl. Microbiol.* **58** 231–41

[33] Radjasa O K, Marterns T, Grossart H P, Brinkhoff T, Sabdono A and Simon M 2007 Antagonistic activity of a marine bacterium Pseudoalteromonas luteoviolacea TAB4.2 associated with coral Acropora sp *J. Biol. Sci.* **7** 239–46

[34] Radjasa O K, Sabdono A, Zocchi J and Zocchi E 2007 Richness of secondary metabolite-producing marine bacteria associated with sponge Haliclona sp *Int. J. Pharmacol.* **3** 275–9

[35] Radjasa O K, Khoeri M M, Darusallam C C, Trimasanto H and Sudoyo H 2013 Bacterial symbionts of reef invertebrates: screening for anti-pathogenic bacteria activity *Biodiversity* **14** 80–6

[36] Madigan, Michael T.; Martinko, John M.; Bender, Kelly S.; Buckley, Daniel H.; Stahl D A 2013 *Brock Biology of Microorganisms* vol 53

[37] Dobbins J J 2010 Prescott's Microbiology, Eighth Edition *J. Microbiol. Biol. Educ.* **11**

[38] Lane D J 1991 16S/23S rRNA sequencing. *Nucleic Acid Techniques in Bacterial Systematics, xxix (ed)* p 329

[39] Lee Y K, Kim H W, Liu C L and Lee H K 2003 A simple method for DNA extraction from marine bacteria that produce extracellular materials *J. Microbiol. Methods* **52** 245–50

[40] Altschul S F, Madden T L, Schäffer A A, Zhang J, Zhang Z, Miller W and Lipman D J 1997 Gapped BLAST and PS I-BLAST: a new generation of protein database search programs *Nucleic Acids Res* **25** 3389–402

[41] Tamura K, Dudley J, Nei M and Kumar S 2007 MEGA4: Molecular Evolutionary Genetics Analysis (MEGA) software version 4.0 *Mol Biol Evol* **24** 1596–9

[42] McIntyre D E, Faulkner D J, Van Engen D and Clardy J 1979 Renierone, an antimicrobial metabolite from a marine sponge *Tetrahedron Lett.* **20** 4163–6

[43] Mansoor T A, Park T, Luo X, Hong J, Lee C O and Jung J H 2007 A new sphingosine from a marine sponge Haliclona (Reniera) sp. *Nat. Prod. Sci.* **13** 247–50

[44] Anand T P, Bhat A W, Shouche Y S, Roy U, Siddharth J and Sarma S P 2006 Antimicrobial activity of marine bacteria associated with sponges from the waters off the coast of South East India *Microbiol. Res.* **161** 252–62

[45] Phelan R W, O'Halloran J A, Kennedy J, Morrissey J P, Dobson A D W, O'Gara F and Barbosa T M 2012 Diversity and bioactive potential of endospore-forming bacteria cultured from the marine sponge Haliclona simulans *J. Appl. Microbiol.* **112** 65–78

[46] Thiel V and Imhoff J F 2003 Phylogenetic identification of bacteria with antimicrobial activities isolated from Mediterranean sponges *Biomol. Eng.* **20** 421–3

[47] Margassery L M, Kennedy J, O'Gara F, Dobson A D and Morrissey J P 2012 Diversity and antibacterial activity of bacteria isolated from the coastal marine sponges Amphilectus fucorum and Eurypon major *Lett. Appl. Microbiol.* **55** 2–8

[48] Skariyachan S, Acharya A B, Subramaniyan S, Kulkarni S and Narayanappa R 2015 Secondary metabolites extracted from marine sponge associated Comamonas testosteroni and Citrobacter freundii as potential antimicrobials against MDR pathogens and hypothetical leads for VP40 matrix protein of Ebola virus : an in vitro and in silico inves *J. Biomol. Struct. Dyn.* **18** 1–19

[49] Matsunaga T 2015 *Springer Handbook of Marine Biotechnology*

[50] Santos O C S, Soares A R, Machado F L S, Romanos M T V, Muricy G, Giambiagi-deMarval M and Laport M S 2015 Investigation of biotechnological potential of sponge-associated bacteria collected in Brazilian coast *Lett. Appl. Microbiol.* **60** 140–7

[51] Hoppers A, Stoudenmire J, Wu S and Lopanik N B 2015 Antibiotic activity and microbial community of the temperate sponge, Haliclona sp. *J. Appl. Microbiol.* **118** 419–30

[52] Tokunaga H, Mitsuo K, Ichinose S, Omori A, Ventosa A, Nakae T and Tokunaga M 2004 Salt-inducible multidrug efflux pump protein in the moderately halophilic bacterium Chromohalobacter sp. *Appl. Environ. Microbiol.* **70** 4424–31

[53] Arahal D R, Garc??a M T, Vargas C, C??novas D, Nieto J J and Ventosa A 2001 Chromohalobacter salexigens sp. nov., a moderately halophilic species that includes Halomonas elongata DSM 3043 and ATCC 33174 *Int. J. Syst. Evol. Microbiol.* **51** 1457–62

[54] Copeland A, O'Connor K, Lucas S, Lapidus A, Berry K W, Detter J C, Del Rio T G, Hammon N, Dalin E, Tice H, Pitluck S, Bruce D, Goodwin L, Han C, Tapia R, Saunders E, Schmutz J, Brettin T, Larimer F, Land M, Hauser L, Vargas C, Nieto J J, Kyrpides N C, Ivanova N, Göker M, Klenk H-P, Csonka L N and Woyke T 2011 Complete genome sequence of the halophilic and highly halotolerant Chromohalobacter salexigens type strain (1H11T) *Stand. Genomic Sci.* **5** 379–88

[55] Radjasa O K, Sabdono A, Zocchi J and Zocchi E 2007 Richness of secondary metabolite-producing

marine bacteria associated with sponge Haliclona sp *Int. J. Pharmacol.* **3** 275–9

[56] Schneemann I, Wiese J, Kunz A L and Imhoff J F 2011 Genetic approach for the fast discovery of phenazine producing bacteria *Mar. Drugs* **9** 772–89

[57] Graça A P, Bondoso J, Gaspar H, Xavier J R, Monteiro M C, De La Cruz M, Oves-Costales D, Vicente F and Lage O M 2013 Antimicrobial activity of heterotrophic bacterial communities from the marine sponge Erylus discophorus (Astrophorida, Geodiidae) *PLoS One* **8**

[58] Elahwany A M D, Ghozlan H a., Elsharif H a. and Sabry S a. 2013 Phylogenetic diversity and antimicrobial activity of marine bacteria associated with the soft coral Sarcophyton glaucum *J. Basic Microbiol.* 1–9

[59] Tokunaga H, Ishibashi M, Arakawa T and Tokunaga M 2004 Highly efficient renaturation of beta-lactamase isolated from moderately halophilic bacteria. *FEBS Lett.* **558** 7–12

[60] Furtado I J and Furtado I J 2014 " Large Scale Production and Characterization of Antibacterial Compound Produced by Chromohalobacter SP . of Biont of Sponge as 3-Isobutyl- **2** 745–57

[61] P.Tharanya, Kayeen Vadakkan, J.Hemapriya V R K and S V 2015 Biogenic Approach for the Synthesis of Titanium Dioxide Nanoparticles Using a Halophilic Bacterial Isolate - Chromohalobacter salexigens Strain PMT-1 *Int. J. Curr. Res. Acad. Rev.* **3** 334–42

COPE Method Implementation Program to Reduce Communication Apprehension Level in Full Day Yunior High School Students

A R Prasetyo

Faculty of Psychology, Diponegoro University
Jl. Prof Sudharto. SH, Kampus Tembalang, Semarang, 50275

anggun.resdasari@gmail.com

Abstract. This study was aimed to explore the effect of COPE method to reduce communication apprehension level of students in Early Adolescence who become Full Day Junior High School students. Full Day Junior High School students, especially in Surabaya coastal area, have more demands to develop the communication aspects such as group discussions and presentations and extracurricular activities. Higher demands to develop such aspects of communication may cause them to experience communication apprehension. The subject was Full Day School students totaling 31 students. The design of the research was experimental design. The experimental method used was a non-randomized pretest posttest control group design and purposive sampling was also used. COPE method is a process that consists of four main stages where people are trying to deal with and control of stressful situations as a result of the problem being faced by conducting cognitive and behavioral changes. Four main stages COPE method is Calming the nervous system, Originating an imaginative plan, Persisting in the face of obstacles and failure, and Evaluating and adjusting the plan. Results of quantitative analysis based on U-Mann Whitney Test shows significant effect on the COPE Method to decrease anxiety levels of communication (0.000 <0.005).

Keywords: Early Adolescence, Full Day Junior High School Students, COPE Method, Communication apprehension

1. Introduction

Almost 70% of Indonesian territory consists of ocean which results in various maritime potentials. Indonesian maritime territory consists of 5,8 million square of coastal areas, territorial sea, and exclusive economic zone. The coastal line stretches along 81.000 kilometers [1]. Coastal and marine areas are part of our environment that has great potential in providing living space and living resources. These marine potentials of course, need to be addressed in comprehensive human resources as well. The structure of people living in coastal areas has high potential of social vulnerability which is caused by two things; First, those people have the issue of low income level or poverty and limited job opportunity and second, people living in coastal area are outspoken and temperamental [2]. Based on those backgrounds, one of human resources that need to be handled carefully is adolescents living in coastal area. Adolescents as social beings have always interacted to parents and all of their life is spend to adapt to their environment due to changing life situations. Early adolescents (age 12-14 years old) have several social needs for their existence of self-development which is related to social skills like sosiogenic need, need for novel experience, need for physical and social adequacy, need for status, and security need [3].

Sosiogenic need includes group acceptance, social and personal identification, and participation. Group and peer acceptance is important during early adolescence. Early adolescents usually put their priority to their group rather than to their family. Early adolescents who prioritize group acceptance are considered normal as long as the group conduct positive activities and do not harm others. To develop and fulfill s*osiogenic need,* early adolescents should be introduced to their environment as early as possible. Early introduction to environment is beneficial for them so they will be able to acknowledge bigger social environment besides their parents, relatives, and grandparents.

Need for novel experience is related to early adolescents' desire to gain experience and enjoy fun activities. They are interested to join various groups to exchange information about new and interesting things. They also have the desire to be accepted by their reference group (need for physical and social adequacy). Early adolescents who lack of social skills and are not accepted by their group will experience anxiety, inferiority feelings, low self esteem and gradually suffer shyness of withdrawn. *Need for status* in early adolescents is related to the desire to be recognized by family, peers, school and community.

Security need is the strongest motivation in early adolescents which is related to social status, power and worth, self-confident, and moral integrity. Those aspects are important for early adolescents since they often experience confusion or ambivalences like the desire to be independent or dependent to others, selfishness or altruism, individuality or conformity. They will be in conflict if they cannot solve their confusion or ambivalences so they will feel less confidence to their competence and at the end, will have negative self-concept.

Social needs of early adolescents are in line with one of their tasks namely developing interpersonal communication skills and practice to get acquaintance with their peers and others, in terms of individual interactions or group interactions [3]. Early adolescents who are less confident in performing interpersonal communication will experience communication apprehension [4]. Jalaluddin Rakhmad [4] states that the reasons why individuals experience communications anxiety are: first, they do not know what to do; second, they know that they will be assessed; third, they are faced to unfamiliar situation and they feel unprepared.

Burgoon and Ruffner [5] mention the characteristics of communication apprehension which consist of *Unwillingness* to communicate, *Avoiding* or rejection to participate and *Control* or lack of control to communication situation. Individuals who experience communication apprehension will withdraw themselves from association, afraid that others will be mocking him and blame him, speak haltingly during speech, only say something if forced to do so, often choose irrelevant topics because relevant topics usually attract others' responses and lead to longer speech. [5]. Communication skills are essential for early adolescents because they should start to get acquainted with peers, teachers, people surrounding their house, and others. They also have to develop communication skills related to school assignments and activities like discussion, presentation, and art exhibition [4]. Early adolescents who suffer from impaired communication apprehension will lead to shyness or social anxiety [6] which defines as the fear of being unaccepted by their group. Early adolescents who suffer from shyness or social anxiety tend to avoid others, easily get scared, skeptical, reserved and reluctant to talk to others, even early adolescents do not have the initiative in social situations, tend to talk quietly, avoid eye contact and less able to communicate. If the communication apprehension remains untreated, those adolescents will face difficulties in completing next developmental tasks when they continue their study and have a job. In a study conducted by Rahayu et al [7], it was observed that 45,56% of college students suffer from high communication apprehension. In another study conducted by Suwandi [8] in Sanata Darma University, it was observed that 48,3% of the students experience high communication apprehension.

Communication apprehension experienced by early adolescents should be treated so that it will not affect next developmental tasks. Therefore, by conducting this study, researcher was interested in applying COPE method to lower communication apprehension in early adolescents studying in Full Day Junior High School which is located in West Surabaya which is also a coastal area. Early adolescents studying in Full Day Junior High School are required to develop their communication aspects by having longer study-time at school which is based on educational program having the purpose to develop communication aspects like group discussion and class presentation as well as extracurricular activity. Their communication skills should be ready to be applied not only in schools but also in community to create better public relations since education is an effort to make students realize their role as part of the environment. Education also has a role to establish the social value of adolescents to conform to the value of adults as well as to develop social skills and social competence.

COPE method was introduced by Dacey & Ficore [9] which explains that 90% of children and adolescent suffering from anxiety can be treated using that method (2000). COPE method is a process which consists of four main phase where individuals attempt to handle stressful situations as a result of their problems by applying cognitive and behavioral changes to feel comfortable [4]. Those four phases in COPE method are "Can COPE method be applied to lower communication apprehension of Full Day School Students?". The objective of the study was assessing hypothesis that COPE method can be applied to lower communication apprehension of Full Day School Students.

2. Method

The independent variable in the study was COPE Method and communication apprehension acted as dependent variable. Purposive sampling was used as sample collection technique to take sample based on particular considerations. In the study, 31 out of 42 research subject from 7th grade students in Full Day Nur Hidayah Junior High School ware selected as sample. Those subjects were selected based on low, medium, high, and very high anxiety score. Two groups were then formed from a total of 31 students which consist of 15 students act as experimental group and 16 students in control group.

The study used quasi experimental design which adapted nonrandomized pretest-posttest control group design consisting experimental and control group. Experimental group was treated with COPE method while no intervention applied to control group as the comparative group.

The intervention given to experimental group is the implementation of COPE Method given for 4 days using the duration of 3 hours per day. COPE Method is a process which consists of four main phases where individuals attempting to handle and manage stressful situations by making cognitive or behavioral changes to overcome communication apprehension and increase self-control by cognitive orientation [9]. Those four main phases are:

- C = *Calming the nervous system.*

In this phase, students experiencing communication apprehension are guided to acknowledge the reaction of the anxiety by calming them down and feeling comfortable in facing their problems. Students try to think clearly about what has happened, say what they feel and feel the difference when they feel anxious and when they feel relaxed.

- *O = Originating an imaginative plan.*

In this phase, when individuals have calmed down, students are directed to think creatively and find an insight about the reason of their anxiety and identifying their strengths and weaknesses.

- *P = Persisting in the face of obstacle and failure.*

During this phase, students are motivated to strive and accept their failure and obstacles so that they will be stronger in facing any situations. Students are also taught and supported to have positive thinking and willing to solve their problems.

- *E = Evaluating and adjusting the plan.*

In this phase, therapist and students are making plans for their problem solutions and making evaluation about their plan. Therapist should provide support, persuasion and feedback for clients to make students believe in their ways in solving problems.

COPE Method were designed based on method stated by Dacey & Ficore [9] which includes role play of special technique application, lecture, group discussion on problems experienced by participants, homework and feedback and reinforcement from facilitator. Those methods can be described as follows:

a. Role Play
Role play is important to provide particular situations for participants. In this training, role play is applied at the beginning of the training by creating a fun situation so that participants are able to having fun by laughing together and at the end of training, a relaxation session is applied. The practice was done so participants are able to study the techniques and apply them when having stressful situations.

b. Lecture
Lecture is a face to face activity involving communication between trainer and participants. The purpose is conveying information about knowledge, opinion or new approach to the participants. During this activity, participants are expected to digest the information conveyed by the trainer and respond to that information.

c. Feedback and Reinforcement
Feedback and reinforcement are applied to reflect story shared by participants in group discussion during the training. Participants receive inputs, comments, and reinforcements for their feeling, attitude and behavior conducted when experiencing stress at work. Based on the feedback, participants should be able to establish proper behavior.

d. Homework
Homework is given to monitor personal behavior. The purpose is to strengthen process applied in the previous session.

e. Group discussion
In group discussion, participants review all their problems, analyze, and making simple conclusion from their own problem. The purpose of this training is to give opportunities to participants to apply various skills needed in particular situations.

Measurement tools used in this study was communication apprehension scale which has reliability value (rxx) as much as 0.9012. The scale was created based on the theory stated by Burgoon [5]:

1. *Unwillingness* or individual unwillingness to participate in communication, which is characterized by:
   a. Showing high anxiety reaction which results in unwillingness to communicate, like anxiety in starting communication, group discussion or presentation in front of the class.
   b. The tendency to stay away from others which is characterized by low participation in the communication situation by remaining silent and sowing little talk or participation in communication
   c. Avoidance of participation in communication by avoiding to talk, to participate in group discussions and presentation to the class, and delegating presentation task to others..
   d. The perception of not accepted by others by feeling that his/her opinion is not accepted, cornered, thinking that others are not enthusiastic, lost others' trust, and unwanted.
   e. Feeling unconfident to talk directly with others, afraid to talk with people who are smarter or have a certain authority, afraid not being able to present yourself well, afraid making mistake
2. Lack of control to communication situations which is characterized by:
   a. Inability to adapt to the environment, whether its supply, room or the communication atmosphere which is rowdy or crowded, formal, non-formal and unfamiliar or unobserved.

b. Inability to adapt to others which consists of inability to adapt in public, inability to adapt to wishes and requirements from others, inability to start communication with strangers, friends from other classes and seniors.
c. Inability to adapt to unpleasant reactions from conversation partner like receiving many challenging questions, unpleasant comments and unsupported responses.

3. Result and Discussion
The study used Mann Whitney U-Test data analysis. The result of *Mann Whitney U-Test* data analysis to observe COPE method implementation to reduce communication apprehension can be seen in the following Table 1 and 2:

Table 1. Sum of rank Pretest and Postest

			Ranks	
	Group	*N*	*Mean Rank*	*Sum.of.Ranks*
Postes	Experiment	15	8.00	120.00
	Control	16	23.50	376.00
	Total	31		
Pretest	Experiment	15	16.67	250.00
	Control	16	15.38	246.00
	Total	31		

Table 2. Mann Whitney U Test

	Post-test	*Pre-test*
Mann-Whitney U	.000	110 000
Wilcoxon W	120 000	246 000
Z	-.4.745	-.396
Asymp. Sig. (2-tailed)	.000	.692

a Grouping Variable: group

U-Test calculation on posttest data of experimental and control group shows that calculated p value based on z statistics is 0,000. The calculated p value based on z statistics is lower than significance level (p) which is measured at 0,05 which means that posttest data of experimental and control group is significantly different. It explains that the hypothesis of the study (Ha) is accepted, or there is a significant result from COPE method implementation to reduce communication apprehension in students in early adolescence studying in Full Day Junior High School.

The difference of sum of rank (in U-Test data) of communication apprehension posttest data score of experimental and control group also shows significant difference compared to sum of rank of communication apprehension pretest data score of experimental and control group.

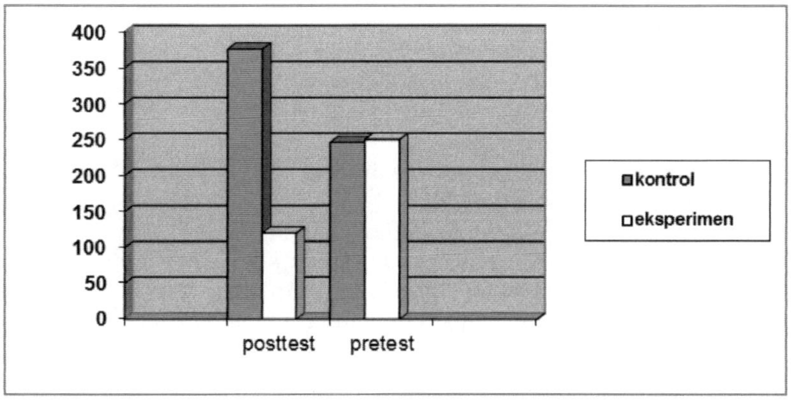

Figure 1. The histogram of *sum of rank* of control and experimental Group

Besides quantitative analysis, the study presents another supporting data which were obtained from each subject per stage during the process of COPE method, which is summarized in the table below:

Table 3. Qualitative Data of Each Subject in COPE method

Subject	The Summary of Data per Subject in *COPE Method*
1	**1. *Calming the nervous system*** **Pleasant Situation**: feels happy, relief and confident **Unpleasant Situation**: feels afraid, confused, and worried when the subject is required to do presentation **2. *Originating an imaginative plan*** Subject feels anxious: - Subject does not get clear information on the situation at hand. - Subject is less prepared to complete the task given to him **3. *Persisting in the face of obstacle and failure*** Subject experiences communication apprehension because of less preparation **4. *Evaluating and adjusting the plan*** **Solution from subject**: attempts to get others well, tried to have small talks, practiced to speak and memorizing presentation material

Subject	The Summary of Data per Subject in *COPE Method*
2	1. ***Calming the nervous system*** **Pleasant Situation**: feels happy, satisfied and having fun activities with friends. **Unpleasant Situation**: feels sad, worried, cried, and feeling unsatisfied. 2. ***Originating an imaginative plan*** Subject feels anxious when the information is not clear. 3. ***Persisting in the face of obstacle and failure*** Subject experiences communication apprehension: - Getting unfamiliar communication task - Getting little information to complete the task 4. ***Evaluating and adjusting the plan*** Practicing to speak fluently and feel relaxed when starting conversation with others and receiving a call from someone else, trying to be relax when seeing the Headmaster and studying how to convey opinion, practicing hard when appointed as the officer of the ceremony, conducting presentation in front of the class and joining competition.
3	1. ***Calming the nervous system*** **Pleasant Situation**: feels happy and comfortable **Unpleasant Situation**: feels afraid and stressed 2. ***Originating an imaginative plan*** Subject feels anxious: - Faced to first time experience - When undergoes frightening situation, subject has negative thoughts that may occur when facing the situation. 3. ***Persisting in the face of obstacle and failure*** Subject feels anxious: - Faced to first time experience - When undergoes frightening situation, subject has negative thoughts that may occur when facing the situation. 4. ***Evaluating and adjusting the plan*** **Solution from subject:** making preparation and constructing positive thoughts that subject is able to present in front of the class, perform well when appointed as the officer of the ceremony, seeing nice people in school and being able to join competition well.
4	1. ***Calming the nervous system*** **Pleasant Situation**: feels happy **Unpleasant Situation**: feels afraid, nervous, worried and trembling 2. ***Originating an imaginative plan*** Subject feels anxious when faced to first time experience, frightening situation and subject has negative thoughts about the situation. 3. ***Persisting in the face of obstacle and failure*** Subject feels anxious: - Faced to first time experience - When undergoes frightening situation, subject has negative thoughts that may occur when facing the situation. 4. ***Evaluating and adjusting the plan*** **Solutions from Subject:** trying to listen to people, practicing to communicate better
5	1. ***Calming the nervous system*** **Pleasant Situation**: feels happy, excited **Unpleasant Situation**: feels afraid, nervous, worried, and trembling

IOP Conf. Series: Earth and Environmental Science **55** (2017) 012020 doi:10.1088/1755-1315/55/1/012020

Subject	The Summary of Data per Subject in *COPE Method*
	Originating an imaginative plan The subject experiences anxiety when less well-prepared in during task completion or facing certain situations. **2.** ***Persisting in the face of obstacle and failure*** The subject experiences anxiety when less well-prepared in during task completion or facing certain situations. **3.** ***Evaluating and adjusting the plan*** **Solutions from Subject:** making preparation to answer question well, feeling confident when entering the teachers' room, making preparation before doing presentation and practicing regularly, accepting the opportunity to be the group coordinator.
6	**1.** ***Calming the nervous system*** **Pleasant Situation**: feeling happy and able to do the assignments **Unpleasant Situation**: feeling very scared, very nervous and tensed. **2.** ***Originating an imaginative plan*** Subject feels anxious when facing first time assignment and according to subject's opinion, the assignment is challenging. **3.** ***Persisting in the face of obstacle and failure*** Subject experiences communication apprehension when performing challenging tasks, first-time experience, or first-time assignment. **4.** ***Evaluating and adjusting the plan*** **Solutions from Subject:** learning to speak fluently and respecting teachers, minimizing feeling nervous and struggling to be confident during classroom presentation, actively asking questions to teachers, and starting to make friends
7	**1.** ***Calming the nervous system*** **Pleasant Situation**: feeling happy and contented. **Unpleasant Situation**: feeling terrified and unable to sleep well **2.** ***Originating an imaginative plan*** Subject feels anxious when having the obligation to completing tasks given by people who have authority 3. ***Persisting in the face of obstacle and failure*** Subject experience communication apprehension when accepting challenging assignment and given by people who have authorities. 4. ***Evaluating and adjusting the plan*** **Solutions from Subject:** obeying parents, studying for presentation, and considering that making mistake is a part of the process
8	1. ***Calming the nervous system*** **Pleasant Situation**: feels excited. **Unpleasant Situation**: feels afraid and unconfident **2.** ***Originating an imaginative plan*** Subject feels anxious when subject is afraid to make mistake and thinking about things that is possible to occur because subject feels that she/he is a reckless person. **3.** ***Persisting in the face of obstacle and failure*** Subject experiences communication apprehension when obtaining unclear information about particular situation and thinking about the negative effects of the situation **4.** ***Evaluating and adjusting the plan*** **Solutions from Subject:** starting to make friends, boosting self-confidence, being impartial by not focusing on bad things

Subject	The Summary of Data per Subject in *COPE Method*
9	1. ***Calming the nervous system*** **Pleasant Situation:** feeling happy, laughing easily, and sharing freely **Unpleasant Situation:** feeling afraid, uncomfortable and nervous **2. *Originating an imaginative plan*** Subject experiences anxiety when subject feels less well-prepared in completing tasks **3. *Persisting in the face of obstacle and failure*** Subject experiences communication apprehension when subject has less knowledge to make friends and how to speak English fluently **4. *Evaluating and adjusting the plan*** **Solutions from Subject:** starting conversation, making small talk, studying English
10	1. ***Calming the nervous system*** **Pleasant Situation:** feeling happy, laughing easily, and sharing freely **Unpleasant Situation:** feels afraid, nervous, trembled, and anxious 2. ***Originating an imaginative plan*** Subject feels anxious when being irresponsible 3. ***Persisting in the face of obstacle and failure*** Subject experience communication apprehension when feeling afraid of making mistakes and seeing people with authority. 4. ***Evaluating and adjusting the plan*** **Solutions from Subject:** obeying and asking permission from parents, preparing well for presentation, preparing well for Leadership Basic Training to boost self-confidence.
11	**1. *Calming the nervous system*** **Pleasant Situation:** feeling happy, laughing easily, and sharing freely **Unpleasant Situation:** feels afraid, embarrassed, sad, and cries **2. *Originating an imaginative plan*** Subject feels anxious when accepting a task or facing particular situation which being subject's first time experience and subject has negative thoughts about it. **3. *Persisting in the face of obstacle and failure*** Subject experiences communication apprehension because subject is afraid of making mistakes when performing the task, first-time experience or first-time assignment so subject has negative thoughts about that situation or tasks. 4. ***Evaluating and adjusting the plan*** **Solutions from subject:** seeing others positively, practicing dance more regularly and feeling more confident, being well-prepared for presentation, making friends, and practicing to convey opinions clearly.
12	**1. *Calming the nervous system*** **Pleasant Situation:** feels excited and relieved. **Unpleasant Situation:** feels afraid and anxious. **2. *Originating an imaginative plan*** Subject feels anxious when completing challenging tasks and having negative thoughts about the situation. **3. *Persisting in the face of obstacle and failure*** Subject experiences communication apprehension when having minimum information about a situation and having negative thoughts about that. **4. *Evaluating and adjusting the plan*** **Solutions from Subject:** seeing others positively and starting to make friends.
13	**1. *Calming the nervous system***

Subject	The Summary of Data per Subject in *COPE Method*
	Pleasant Situation: feels happy and excited. **Unpleasant Situation:** feels afraid and anxious. **2. *Originating an imaginative plan*** Subject feels anxious when seeing people with certain authority like parents, teachers, and headmaster. **3. *Persisting in the face of obstacle and failure*** Subject experience communication apprehension when making friends and talking to people with particular authority. **4. *Evaluating and adjusting the plan*** **Solutions from Subject:** practicing to speak fluently, practicing on how to make friends, being calm during phone calls, practicing on how to convey opinion to teachers
14	1. ***Calming the nervous system*** **Pleasant Situation:** feels excited **Unpleasant Situation:** feels nervous and trembling **2. *Originating an imaginative plan*** Subject will experience anxiety when subject should complete challenging task and facing crowd **3. *Persisting in the face of obstacle and failure*** Subject will experience communication apprehension because subject has negative thoughts about things possible to happen when subject is completing challenging tasks and should be in front of public, having conversation with people having certain authorities and having new experience **4. *Evaluating and adjusting the plan*** **Solutions from Subject:** practicing more regularly, thinking positively, studying well, and respecting teachers and headmaster
15	**1. *Calming the nervous system*** **Pleasant Situation:** subject feels happy. **Unpleasant Situation:** subject feels anxious and afraid.. **2. *Originating an imaginative plan*** Subject experiences anxiety when subject is being irresponsible in completing task from his/her parents **3. *Persisting in the face of obstacle and failure*** Subjects will experience communication apprehension because subject has negative thoughts which are possible to happen when completing challenging tasks and when subject is irresponsible in completing the task. **4. *Evaluating and adjusting the plan*** **Solutions from Subject:** trying to reduce nervous feeling gradually, and asking permission to go home late.

From the Table 3, the effect of COPE Method can be more explained based on students exploration during COPE Method process. Communication apprehension experienced by students, based on COPE method process, it can be concluded that communication apprehension in yunior high school students involved subjective interpretation which was caused by inappropriate cognitive processing. It was confirmed by data collected from subject while joining COPE method process which can be related to the source of communication apprehension which was explained by DeVito [10]:

a. *Lack of communication skills and experience.* Students do not prepare well, receive communication tasks which were rarely done, and cannot get proper information to complete the assignment or solve problem so they have no clue to complete the assignments.

b. *Degree of evaluation* (the feeling of being over-evaluated by others). Students feel that others get negative emotion at them because they are irresponsible and have a thought that others will evaluate the result of their learning process.
c. *Prior successes and failures* (previous communication experience which create pleasant and unpleasant feeling). Students have bad experience because of being irresponsible and get unwanted results.
d. *Subordinates status.* Students should complete assignments from those who have authority like teachers, parents, and principals.
e. *Degree of conspicuousness* (the feeling of being the center of the attention and the obligation to demonstrate good communication skill) Students are challenged for difficult assignments, need to perform well and enjoy being center of the attention.
f. *Degree of unpredictability* (being in unexpected situations). Students face distressing assignments or situations and have negative thoughts which are possible to occur when facing the situations.
g. *Degree of dissimilarity* (unusual and unfamiliar situation). Students are faced to unfamiliar situation like visiting other school and meeting new people.

The process of each subject when undergo COPE method, which is changing the cognitive thought of each subject in the experimental group to be more positive and gaining skills to solve problems needs further treatment. The process of COPE method implementation can be explained as follows:
1. Students are calmed down from their anxiety through the implementation of the first phase, *Calming the nervous system*.
2. Students are convinced to think creatively and find the insight of their source of anxiety, find their strengths and weaknesses by applying the second phase, *Originating an imaginative plan*.
3. Students are given a positive stimulus and direction that the feelings of anxiety can be addressed through the implementation of the third phase, *Persisting in the face of obstacle and failure*.
4. Students are invited to construct their point of view of the world (environment) and can implement coping strategy through the implementation of the fourth phase, *Evaluating and adjusting the plan*.

From COPE, students can explore and manage positive affection. Positive affection is important because it promoted coping responses which led to finding meaning in various stressful experiences Positive affection may have made them more resilient by protecting them from maladaptive responses, which occur during communication apprehension. COPE Method involves strategies for reframing an event to see it in a more positive light. How an individual copes with a communication experience depends upon how it is interpreted with respect to the significance of the stressful situation for students. So, this has important implications for regulating emotions successfully. Positive reappraisal in COPE method help to generate positive emotion in students, which enables one to effectively, regulate negative emotional experiences.

COPE Method implementation can be beneficial and is important to apply because human relations are built through communication. Individuals who communicate with themselves will be able to use their own behavior and others' behavior to choose proper behavior [11]. Barriers in education cannot be separated from communication. This is based on the statement, "In the education process we often encounter failures which is usually caused by weak communication system" [12].

4. Conclusion

Based on the study, it can be concluded that quantitative analysis used in this study has provided evidence that there has been a significant result from the implementation of COPE method to the decrease in communication apprehension of students in early adolescence in Full Day Junior High School. The process of each subject when undergo COPE method (based on the table) and the result of the qualitative analysis according to the main objective of COPE method which is changing the cognitive thought of each subject in the experimental group to be more positive and gaining skills to solve problems needs further treatment.

For further research, it is necessary to pay attention to the process of COPE method implementation. Future studies are necessary to monitor and do in depth exploration about the reaction of both physiological and psychological anxiety experienced by the subject. Researchers also need to conduct a review or repetition of the previous material to students. The process is intended to provide a review to the students about the previous process on an ongoing basis so that students can really benefit from COPE method.

References

[1] Koya I 2009 *Indonesian Fisheries Book* (Jakarta: Ministry of Marine Affairs and Fisheries & Japan International Cooperation Agency (JICA)

[2] Pakpahan H T, Lumintang R W and Santosa D 2006 *Institut Pertanian Bogor Jurnal Penyuluhan Hubungan Motivasi Kerja dengan Perilaku Nelayan Pada Usaha Perikanan Tangkap* **2** 1858-2664

[3] Hurlock B E 2007 *Psikologi Perkembangan Suatu Pendekatan Sepanjang Rentang Kehidupan.* (Jakarta: Penerbit Erlangga)

[4] Rakhmat J 2008 *Psikologi Komunikasi Edisi Rev*isi (Bandung:Rosdakarya)

[5] Burgoon M and Ruffner M 1978 *Human Communication* (New York: Rinehart and Winston)

[6] Tubbs L S 2003 *Human Communication* 9rd ed (New York:MCGraw Hill)

[7] Rahayu I. T, Ardani T A and Sulistyaningsih 2004 Hubungan Pola Pikir Positif dengan Kecemasan Berbicara di Depan Umum. *Jurnal Psikologi UNDIP* 131-143.

[8] Suwandi 2004 Hubungan efikasi diri dan konsep diri dengan kecemasan berbicara di depan umum mahasiswa teologi terapan universitas sanata dharma. *Tesis (Tidak Diterbitkan).* Yogyakarta : Fakultas Psikologi Universitas Gadjah Mada

[9] Dacey, J S., and Fiore L B 2000 *Your anxious child (*San Fransisco: Jossey Bass Publisher)

[10] Devito, J A 2011 *Komunikasi Antar Manusia* (Pamulang-Tangerang Selatan: Karisma Publishing Group)

[11] Hardjana A 2007 *Komunikasi Intarpersonal dan Interpersonal* (Yogyakarta: Kanisius)

[12] Gunawan 2009 Tiga Pola Komunikasi dalam Proses Belajar Mengajar. *Jurnal Pendidikan*, **2** 86

Influence of ENSO and IOD to Variability of Sea Surface Height in the North and South of Java Island

Ahmad Fadlan[1,4], Denny Nugroho Sugianto[2,3] Kunarso[2,3],and Muhammad Zainuri[1]
[1] Faculty of Fisheries and Marine Science, Diponegoro University. Jl. Prof. Soedarto, SH. UNDIP, Tembalang, Semarang, 50275, Indonesia
[2]Departement of Oceanography, Fisheries and Marine Science Faculty, Diponegoro University. Jl. Prof. Soedarto, SH. UNDIP, Tembalang, Semarang, 50275, Indonesia.
[3]Center for Coastal Disaster Mitigation and Rehabilitation Studies, Diponegoro University. Jl. Prof. Soedarto, SH. UNDIP, Tembalang, Semarang, 50275, Indonesia.
[4] Indonesian Agency for Meteorology Climatology and Geophysics. Jl. Angkasa I No. 2, Kemayoran, Jakarta 10610, Indonesia

Email: afadlan0@gmail.com

Abstract. Indonesia is one of the largest archipelagic countries in the world. Besides being in the tropics, these waters are also located between two continents and two oceans that making this area are heavily influenced by the global atmospheric phenomena such as El Niño Southern Oscillation (ENSO) and Indian Ocean Dipole (IOD). Sea level rise is one of the main threats faced by Indonesia and other island countries in the world. Not only global warming, sea level rise in the tropics also caused by inter annual variability such as ENSO and IOD. This research has been aimed to determine the influence of ENSO and IOD to variability of sea surface height in the north and south of Java. This research used Satellite altimetry data were constructed by multi-mission satellite. These results showed that the sea level anomaly generally inversely related to ENSO and Dipole Mode Index. The phenomenon that most influence a decrease sea level anomaly in the study area is a phenomenon IOD + and the phenomenon that most influenced the increase sea level anomaly is La Niña. South of Java is area that has the large effect of changes in sea levels than north of Java. During La Niña, sea level anomaly can be increase up to 0.3 m and sea level anomaly can be decrease up to -0.28 m during IOD+.

Keyword: Sea Level Anomaly, El Niño, La Niña, IOD

1. Introduction

Indonesia is one of the largest archipelagic countries in the world. Besides being in the tropics, it's also located between two ocean and two continents that making it as a center of atmospheric and global ocean circulation [1]. One of the effects of the circulation is change in sea level of Indonesian sea especially in the waters area of north and south of Java. Sea level conditions are currently linking to the phenomenon of global warming [2] which causes melting of the volume of ice in the Polar so increasing sea level in some area of the world [3]. Not only global warming, sea level rise in the tropics also caused by inter annual variability [4] such as ENSO [5][6] and IOD [7] especially in bordering region of the Pacific Ocean and the Indian Ocean. The waters area of north and south of Java has dynamics of the atmosphere and oceans which affected by ENSO and IOD [8]. Sea level rise is one of the main threats faced by Indonesia and other island countries in the world [9]. To observe of sea level changes, required the availability of adequate spatially and temporally data. The presence of satellite altimetry is a solution to comply of oceanographic data especially sea levels data, such as global and regional data [10]. The validation result of the altimetry data and tide observations were produced sea level measurement with very good accuracy and concluded that the altimetry data can represent the field observations data [11]. This research has been aimed to determine the influence of ENSO and IOD to variability of sea surface height in the north and south of Java. This research is important because changes in sea level can be effect for susceptibility of coast condition and coastal development especially in major cities in Java which has a quite high coastal development.

2. Methodology

2.1. Area of Study

The study was conducted in the waters of north and south of Java island with coordinate ± 3° – 11° S and 105° – 115° E (Figure 1) which includes six station points such as Jakarta (6.17° S, 106.67° E), Semarang (6.94° S, 110.42° E), Surabaya (7.21° S, 112.73° E), Prigi (8.28° S, 111.73° E), Cilacap (7.75° S, 109.02° E) and Pelabuhan Ratu (7.00° S, 106.50° E). Selection of a point by the big cities in Java and can represent the north and south coast of Java Island. The waters south of Java generally influenced by several factors phenomenon of ocean and atmospheric systems such as Monsoon, Indian Ocean Dipole (IOD), the El Niño Southern Oscillation (ENSO), a Kelvin wave and the Madden Julian Oscillation (MJO), Indonesian Trough Flow (ITF), South Equatorial Flow (SEF) and the flow of the west coast of Sumatra Island [12][13][14]. In the waters of Java Sea, the dominant factor affecting oceanography parameter is the activity of monsoon [15].

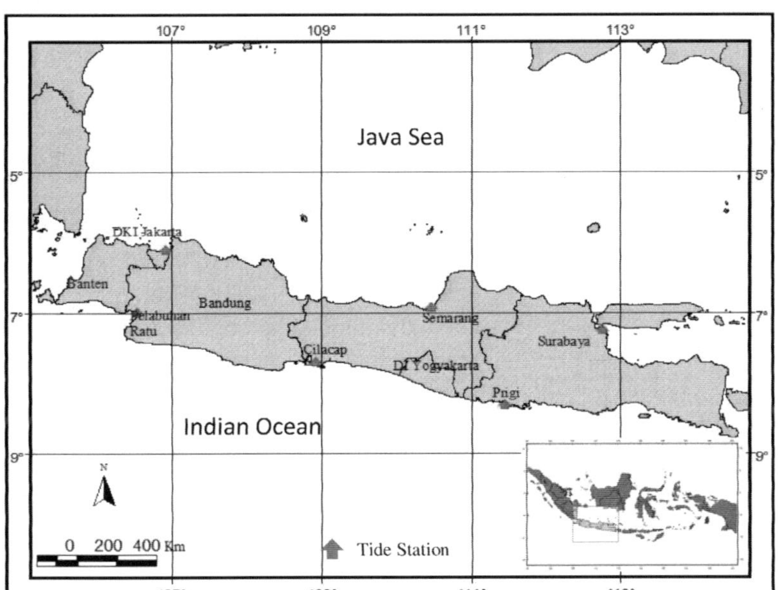

Figure 1. Area of Study

2.2. Satellite Altimetry Data

Satellite altimetry data used in this research were constructed by multi-mission satellite. The data downloaded from the AVISO web site (*ftp://ftp.aviso.altimetry.fr/global/delayed-time/grids/ climatology/monthly mean/*) the reference version of gridded sea level anomaly (SLA) data. The data are mapped on a 1/4° Mercator grid and extracted for the area of study from the global SLA field for the period 1993-2015 (23 years). Sea level anomaly data are obtained from the following calculation:

$$SLA = SSH - MSSH \tag{1}$$

SLA is sea level anomaly, SSH is sea surface height and MSSH is mean of sea surface height.

2.3. ENSO Index

ENSO indices used in the determination of the value equation sea surface temperature (SST) anomalies in the Niño 3.4 region (120 ° W-170 ° W, 5 ° N-5 ° S) as follows:

$$\Delta SST_{3.4} = SSTa - SSTm \tag{5}$$

$\Delta SST_{3.4}$ is sea surface temperature anomalies in Niño 3.4 region, $SSTa$ is actual sea surface temperature in Niño 3.4 and $SSTm$ is mean of sea surface temperature from long time series in Niño 3.4 region. The result annual cycle for the Niña 3.4 region is shown in figure 2a. When SST anomalies are positive, then the phenomenon that occurs is El Niño and conversely if SST anomalies are negative then the La Niña phenomenon that will occur. Sea surface temperature anomaly data in 3.4 regions sourced from (*cpc.ncep.noaa.gov/data/indices/*).

2.4. Dipole Mode Index (DMI)

In determining the DM index, calculated the SST IOD [16] based on the following equation:

$$DMI = SSTmW - SSTmE \tag{2}$$

DMI is Dipole Mode Index, $SSTmW$ is mean of SST anomaly in west box area (10°N – 10°S, 50°E – 70°E) and $SSTmE$ is mean of SST anomaly in east box area (0° – 10°S, 90°E – 110°E). SST data sourced from *extreme.kishou.go.jp/itacs5/* (ITACS 5) in 1993-2015. The result of difference between anomaly west box and east box area is shown in figure 2b.

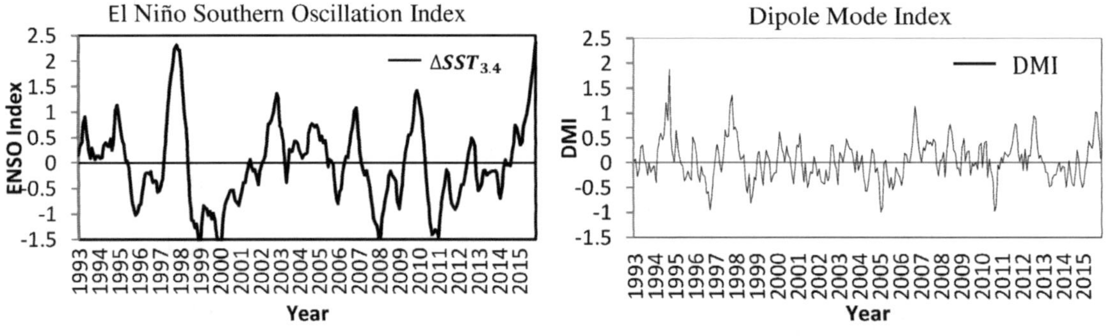

(a) (b)

Figure 2. ENSO Index from SST anomaly in 3.4 region (a) and Dipole Mode Index (DMI) from difference of SST in Indian Ocean (b).

Based on the calculation above anomaly, furthermore do analysis the data to determine the relationship between sea surface height variability and the atmosphere phenomenon. Stages of analysis to be performed is describing the criteria for El Niño/La Niña in the Pacific Ocean and phenomenon of positive IOD (DM+) and negative IOD (DM-) in the Indian Ocean region. Analysis of the relationship between sea surface height variability and atmospheric phenomena by comparing the sea surface height anomaly with normal condition.

2.5. *Surface Wind*

Wind data show the vector represented by the direction and wind speed. To calculate the average wind at each point, the wind data must be parsed into an east-west wind component or commonly called zonal wind component (U) and the north-south wind component or commonly called the meridional wind component (V). Wind direction data indicated by the angle formed by the wind vectors with North [17]. The surface component wind data downloaded from the European Centre for Medium-Range Weather Forecasts (ECMWF) web site (*http://apps.ecmwf.int/datasets/data/*). The data are mapped on a 0.75^{0} Mercator grid and extracted for the area of study from 1993-2015 (23 years).

3. Result and Discussion

Based on the index chart of ENSO and DMI (Figure 2) and then have been selected four cases of occurrence of ENSO and IOD such as ENSO + in 2002, ENSO- in 2007, IOD + in 2011, IOD- in 2005.

3.1. *The Influence of ENSO on Sea Surface Height Anomaly in the Water of North and South of Java.*

ENSO phenomenon includes the El Niño and La Niña is phenomenon that not only affects climate variability and weathers in Indonesia but also its affect the variability of marine phenomena and ocean parameters in Indonesian waters. In 2002, El Niño (ENSO+) that occurs indicated a moderate level in September to December (Figure 3).

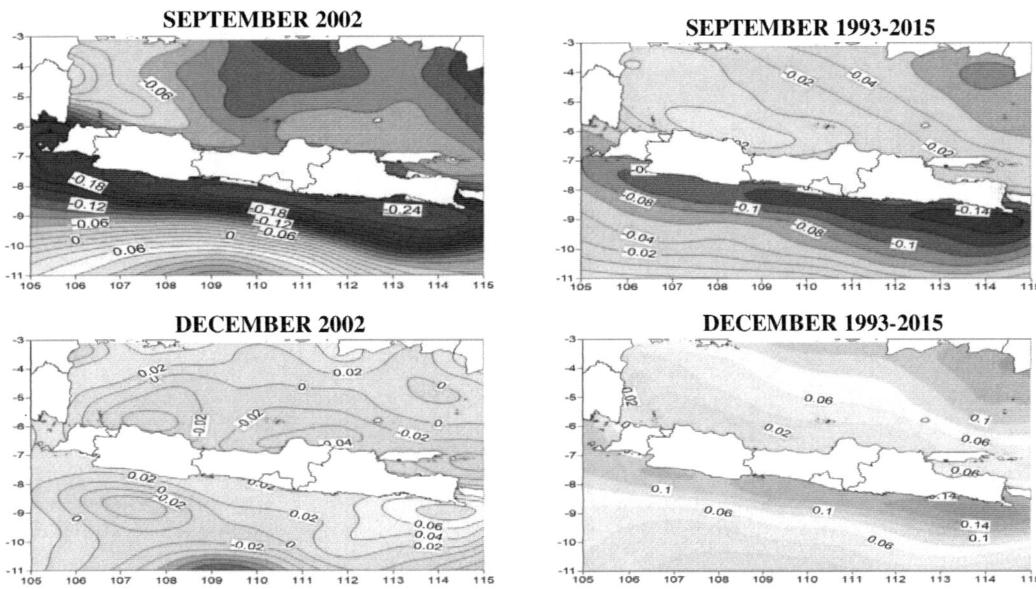

Figure 3. SLA in the water of north and south of Java at El Niño and average from 23 years

In 2002, it has shown quite obvious influence of El Niño on sea level anomalies in the north and south of Java. If compared with average conditions, when the El Niño phenomenon in 2002, the value of SLA tends to be lower than the average for 23 years. Based on these conditions, sea level anomaly in south of java decreased up to -0.24 m in September. In December that should anomalous sea surface height is positive in the southern region of Java however due to El Niño is active so it's causing SLA in southern of Java tends to be negative. Besides, almost all spatial regions decreased anomalously sea surface height.

When El Niño occurs, the entire system of trade winds weaken especially in west Tropical Pacific Ocean boundary. Consequently, the water that has accumulated in the west (warm pool) will flow turned to the east. Along with the warm pool shifting to the east and the weakening of the trade winds cause more shallow thermocline in Indonesia which accompanied by the increased upwelling. During El Niño, upwelling along the southern East Nusa Tenggara waters to the west of Sumatra has longer time and a wider distribution of spatial [18]. Based on mechanisms of upwelling, sea surface in upwelling region will be lower than other regions [19].

In September, October, November (SON) generally there is a change of wind easterlies become westerlies. When El Niño in 2002, wind condition showed an increase of easterlies wind in Java areas than normal conditions in September to December (Figure 5). Increased easterly wind to reach 1 to 1.5 m/s is followed by the weakening of the westerlies wind that moves from the equator [20]. Based on the Ekman transport theory where Ekman transport moves water to the left of the wind direction in the Southern Hemisphere. Combination of wind and the Coriollis Effect on Ekman transport creates areas of upwelling and downwelling in coastal locations of the Southern hemisphere. Therefore, when the easterlies wind dominant in the southern of Java, Ekman transport will creates area of upwelling in southern coastal.

In contrast to El Niño conditions, when La Niña (ENSO-) that occurred in late 2007 and early 2008, sea surface anomaly generally increased compared to the average conditions (Figure 4). In 2007/2008 La Niña has known occurred in November 2007 to March 2008.

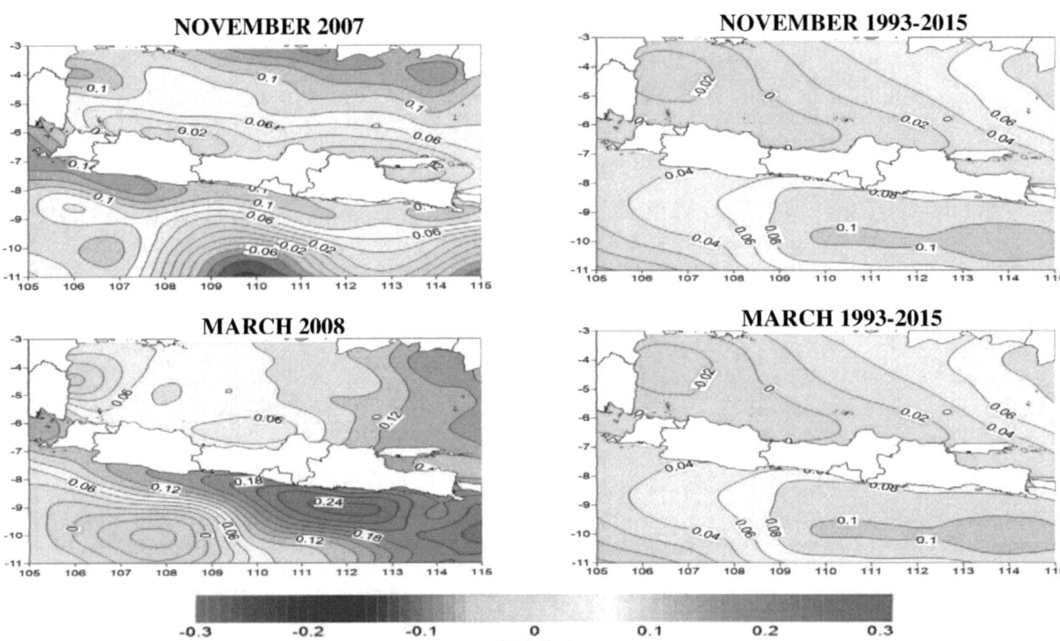

Figure 4. SLA in the water of north and south of Java at La Niña and average from 23 years

During La Niña phenomenon in 2007/2008, sea level anomaly in the waters north and south of the island of Java in general has increased. A significant increase occurred in January to March where along the southern coast of Java sea level anomalous increase reach 0.3m. This condition is higher than the average condition for 23 years is only 0.17 m. When normal conditions, in the northern waters of Java

generally tend anomaly has particularly negative in the westerly months, whereas during La Niña (2007/2008), nearly all waters including in the northern waters of Java has positive anomaly.
Changes in sea level will be increased during the period of transition from El Niño to La Niña, as well as during La Niña [6]. This is because the trade winds in the Pacific Ocean strengthen and bring the masses of water of the eastern Pacific around Peru to the Indonesian Waters. This condition is characterized by the movement of warm pool of the Central Pacific to Indonesia. As a result of heaping sea water mass in Indonesian waters which resulting in the majority of the Indonesian waters regions experienced downwelling and followed by an increase sea level anomaly. Based on, mechanisms of downwelling, sea surface in downwelling region will be higher than sea surface in other regions [19]. The impact of these conditions make the sea level anomaly in Indonesian waters generally increased from the average conditions. Besides that, different from El Niño conditions, there was a strengthening of the westerly wind in Java area during La Niña 2007/2008 (Figure 5). This condition makes downwelling area in southern coastal of Java and causes sea level anomaly increase.

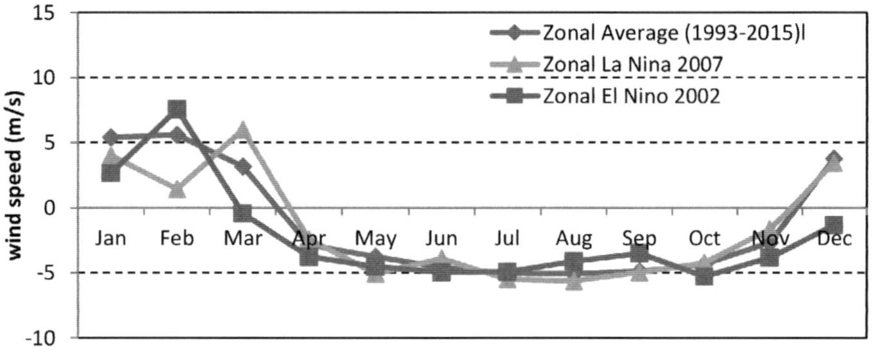

Figure 5. Zonal wind average in study of area during ENSO

Based on the observation station point (Table 1), in study area generally increase of the average conditions especially in January and March during La Niña. The highest increase occurred at stations in the southern of Java. At Prigi station in East Java observed increase in sea level can reach 0.23 m. As for the Cilacap and Pelabuhan Ratu station, sea level anomaly can reach 0.21 m. While on station in the northern of Java, the increase of sea surface height anomaly ranges from 0.07m until 0.12 m.

Table 1. Sea Level Anomaly in Station Point during ENSO

Station	ENSO + (2002)				ENSO - (2007/2008)				
	Sep	Oct	Nov	Des	Nov	Dec	Jan	Feb	Mar
Jakarta	-0.09	-0.03	-0.01	-0.02	0.04	0.01	0.04	0.05	0.07
Semarang	-0.09	-0.05	-0.07	-0.04	0.04	0.04	0.08	0.05	0.07
Surabaya	-0.10	-0.08	-0.05	-0.02	0.00	0.02	0.11	0.08	0.12
P. Ratu	-0.20	-0.11	-0.06	0.03	0.14	0.14	0.13	0.20	0.12
Cilacap	-0.22	-0.11	-0.08	0.03	0.11	0.11	0.14	0.21	0.16
Prigi	-0.21	-0.16	-0.07	0.02	0.08	0.10	0.19	0.23	0.20

3.2. The Influence of IOD on Sea Surface Height Anomaly in the Water of North and South of Java
IOD (Indian Ocean Dipole) phenomenon which covers the DMI+ and DMI- basically has the same principle with ENSO related relationship with the condition of the atmosphere and ocean in Indonesia. The difference between both of them is location of this phenomenon formed. IOD is formed in the Indian

Ocean while ENSO is formed in the Pacific Ocean. Based on SST anomaly difference charts between east box and west box of Indian Ocean (Figure 2) DM+ known have occurred in 2011 and DM- occurred in 2005. In 2011, DM+ indicated occurs in August to October and DM- indicated from February to March and July to September in 2005.

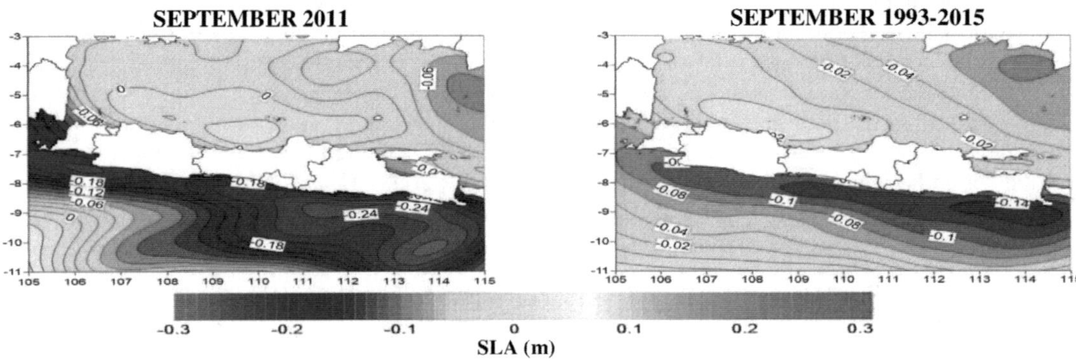

Figure 6. SLA in the water of north and south of Java at IOD+ in September 2011 and average from 22 years

IOD+ makes sea level anomaly in southern Java to be lower than the average conditions (Figure 6). In September 2011, IOD+ influence on sea level anomaly in the southern of Java can be seen. Under normal condition, sea level anomaly in September is concentrated in the southern area of East Java. While at IOD+ almost any area of the study had negative anomalies, especially in Southern of Java where the sea surface height anomaly is lower than the monthly average condition for 23 years. Sea level anomaly has detected down to -0.28 m in September, -0.22 m in August and -0.18 m in October 2011. If compared with El Niño in 2002, IOD+ 2011 more impact on the declining sea level in the southern of Java because the location of IOD + is in the Indian Ocean and this phenomenon is greatly affecting the dynamics of the waters in southern area of Indonesia.

Based on sea surface temperatures and chlorophylls data from another research showed the temporal and spatial variation of upwelling associated with variations in ENSO and IOD. However, based on variations in the case of ENSO and IOD to the variability that occurs in the area of upwelling waters south of Java to Timor indicated that IOD+ phenomenon more influence than ENSO to variability upwelling area occurred [21]. Significant decline in sea surface height that occurred in September caused by IOD+ was higher activity in this month. The activity caused wind speed anomaly peaked and upwelling along southern of Java and West Sumatra more intensified [16]. IOD+ caused upwelling intensity in easterly periods more increased and continued until transition period [22]. This condition make the sea level anomaly will decrease until October and November.

Different from IOD+, sea level anomaly generally increased than the average condition around Java area in July 2005 (Figure 7). This increase occurred because of Indonesia in general has downwelling due to the mass of water flowed to East Indian Ocean. Viewed from the increase in anomalies, IOD- phenomenon not sees significant changes over the average condition in 2005. Sea level anomaly only increased 0.09 m in February and March. This condition is even smaller than average sea level anomalies which reached 0.15 m. However in July and September sea level anomaly has increased about 0.13 m while the normal condition that only 0.08 m.

Although there is accumulation of water masses in the waters area of East Indian Ocean on IOD + but because of the weakening of the zonal winds in February and March (Figure8) made in the downwelling formation not quite strong. This condition caused the increased of sea level anomaly is below average condition. Meanwhile in July and September, due to the zonal wind are generally the same as the average conditions and couple with the mechanism of downwelling in the East Indian Ocean that result an increase in sea surface height anomaly becomes higher than the average condition.

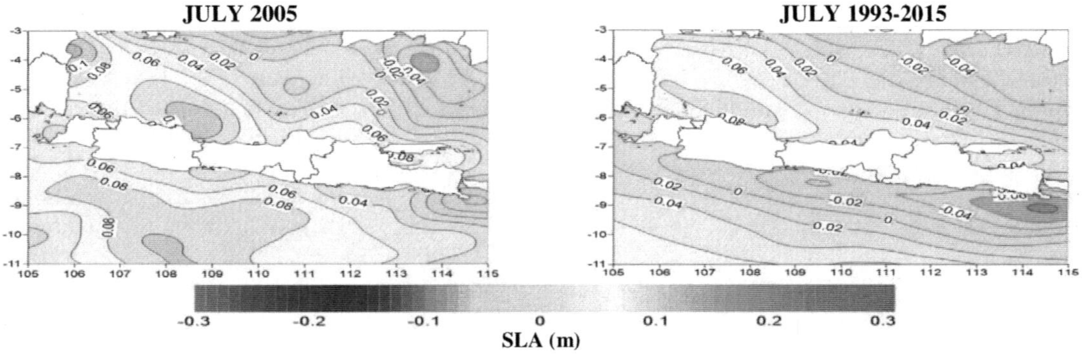

Figure 7. SLA in the water of north and south of Java at IOD- in July 2005 and average from 22 years

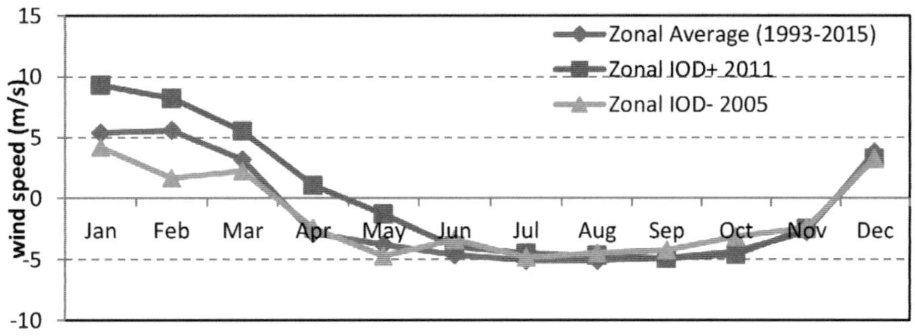

Figure 8. Zonal wind average in study of area during IOD

Based on the observation station point (Table 2), in study area generally decrease of the average conditions especially in the last year in IOD+ or IOD- period. The highest decrease occurred at stations in the southern of Java. At Pelabuhan Ratu station in Westt Java observed decrease in sea level can reach -0.19 m in October during IOD+. As for the Cilacap and Prigi station, sea level anomaly can reach -0.17 m and -0.19. While on station in the northern of Java, sea surface height anomaly has decrease ranges from -0.04m until -0.06 m.

Table 2. Sea Level Anomaly in Station Points during IOD

Station	IOD + (2011)			IOD- (2005)			
	Sep	Oct	Nov	Feb	Mar	Jul	Sept
Jakarta	0.0349	-0.0403	0.0496	-0.02007	-0.00922	0.068526	0.001248
Semarang	0.0172	-0.0075	0.0245	0.005635	0.010226	0.040417	-0.01002
Surabaya	0.0239	-0.0631	0.0046	0.025309	0.026965	0.054978	-0.02169
P. Ratu	-0.1872	-0.2148	-0.0659	0.04013	0.033713	-0.00623	-0.10523
Cilacap	-0.1705	-0.1787	-0.0534	0.072065	0.066265	-0.0291	-0.11301
Prigi	-0.1527	-0.1944	-0.0839	0.088296	0.075952	-0.02767	-0.11308

4. Conclusion

Based on the discussion that has been described, it can be concluded that the sea level anomaly generally inversely related to ENSO and Dipole Mode Index. The phenomenon that most influence a decrease sea level anomaly in the study area is a phenomenon IOD + and the phenomenon that most influenced the increase sea level anomaly is La Niña. South of Java is area that has the large effect of changes in sea levels than north of Java. During La Niña, sea level anomaly can be increase up to 0.3 m and sea level anomaly can be decrease up to -0.28 m during IOD+. The parameters that directly impact is the zonal wind conditions that moves above water areas north and south of Java. Southern of Java is area that has the effect of changes in sea levels very large. By knowing the influence of La Niña that caused sea level anomaly rise in the north and south of Java, we are expected to be alert of the impact that would be caused. La Niña not only causes sea level rise, but also cause high waves and rainfall increase significantly.

Acknowledgement

We thank to Indonesian Agency for Meteorological, Climatological and Geophysical for supporting us with Master Degree Scholarship in Diponegoro University and for funding our research.

References

[1] Aldrian, E. 2008. *Meteorologi Laut Indonesia.* BMKG: Jakarta
[2] Gregory, J. 2013. *Projection of Sea Level Rise.* IPCC AR5.
[3] Unnikrishnan, A.S. and Shankar D. (2007) Are sea level rise trends along the coasts of north Indian Ocean coasts consistent with global estimates? *Global and Planetary Change*, 57:301-307
[4] Church, J. A., White, N. J., Hunter, J. R. 2006. Sea-level Rise at Tropical Pacific and Indian Ocean Islands. *Global and Planetary Change.* 53:155-168
[5] Sofian, I., dan Antonius, B. W. 2008. Proyeksi Kenaikn Tinggi Muka Laut di Jakarta Berdasarkan Skenario IPCC AR4. *Jurnal Ilmiah Geomatika.* 14(2): 71-80
[6] Sofian, I. 2007. Simulation of The java Sea using an Oceanic Feneral Circulation Model. *Jurnal Ilmiah Geomatika.* 13(2): 1- 14.
[7] Antomy, P., Jumarang, M. I., Ihwan, A. 2014. Kajian Elevasi Muka Air Laut di Selat Karimata Pada Tahun Kejadian El Niño dan Dipole Mode Positif. *Prima Fisika.* 2(1): 01-05
[8] Kunarso, Hadi, S., Ningsih, N. S., Baskoro, M. S. 2012. Perubahan Kedalaman dan Ketebalan Termoklin Pada Variasi Kejadian ENSO, IOD dan Monsun di Perairan Selatan Jawa Hingga Pulau Timor. *Ilmu Kelautan.* 17(2): 87-98
[9] Syah, A. F. 2010. Indikasi Kenaikan Muka Air Laut di Pesisir Kabupaten Bangkalan Madura. [Thesis] Marine Enginering, Trunojoyo University, Madura.
[10] Handoko, E.Y. 2004. Satelit Altimetri dan Aplikasinya Dalam Bidang Kelautan. Scientific. *Journal Pertemuan Ilmiah Tahunan I.* Teknik Geodesi – ITS: Surabaya, 137-144.
[11] Fu, L.L., Stammer, D., Leben, R.R., Chelton, D.B. 2003. Improved Spatial Resolution of Ocean Surface Topography From The T/P Jason-1 Altimter Mission. *EOS, Transaction American Geophysical Union.* 84(26): 241-248
[12] Purba, M. 2007. Dinamika Perairan Selatan P. Jawa – P. Sumbawa Saat Muson Tenggara. *Torani,* 17(2):140-150
[13] Tubalawony, S. 2008. Dinamika Massa Air Lapisan Ekman Perairan Selatan Jawa – Sumbawa Selama Muson Tenggara. *Torani.* 17(2):140-150.
[14] Yoga, R. B., Setyono, H., Harsono, G. 2014. Dinamika Upwelling dan Downwelling Berdasarkan Variabilitas Suhu Permukaan Laut dan Klorofil-A di Perairan Selatan Jawa. *Jurnal Oseanografi.* 3(1):57-66
[15] Oki, D. dan Putri, M. R. 2014. Identifikasi Awal Eddies di Perairan Laut Jawa. *Jurnal Oseanografi.* 1(1): 12-20
[16] Saji, N. H., B. N. Goswani, P. N. Vinayachandran, and T. Yamagata. 1999. A Dipole in The Tropical Indian Ocean. *Nature.* 401:360-363.
[17] Hartoko, A., Sulistya, W. 2010. Meteorologi & Sifat Lautan Indonesia. Pusat Penelitian dan Pengembangan Badan Meteorologi Klimatologi dan Geofisika. Jakarta.

[18] Kunarso, Ningsih, N. S., Supangat, A. 2005. Karakteristik Upwelling di Sepanjang Perairan Selatan NTT Hingga Barat Sumatera. *Ilmu Kelautan.* 10(1): 17-23

[19] Lentz J.A. 2010. *Ekman Transpor and Coastal Sea Level*. Oceanography & Coastal Sciences.LSU.

[20] Mulyana, E. 2002. Analisa Angin Zonal di Indonesia Selama Periode ENSO. *Jurnal Sains dan Teknologi Modifikasi Cuaca.* 3(2): 115-120

[21] Kunarso, Hadi, S., Ningsih, N. S., Baskoro, M. S. 2011. Variabilitas Luasan Upwelling Intensitas Kuat pada Variasi Kejadian ENSO dan IOD di Perairan Selatan Jawa Hingga Timor. *Prosiding Seminar Nasional Kelautan POSEIDON.* 2012. ITB, 102-113

[22] Gaol, J. L., Nurjana, I. W., Amri, K. 2014. Dampak Perubahan Iklim Terhadap Kondisi Oseanografi dan Laju Tangkap Tuna Mata Besar (*Thunnus obesus*) di Samudera Hindia Bagian Timur. Simposium Nasional Pengelolaan Perikanan Tuna Berkelanjutan. Bali, 10-11 Desember 2014. *WWF Prosding*. 2015. Kementrian Kelautan dan Perikanan. Bali, VI-109.

The Roles of Macrobenthic Mollusks as Bioindicator in Response to Environmental Disturbance : Cumulative *k*-dominance curves and bubble plots ordination approaches

Sapto P. Putro[1], Fuad Muhammad[1], Amalia Aininnur[2], Widowati[3], and Suhartana[4]

[1] Center of Marine Ecology and Biomonitoring for Sustainable Aquaculture (Ce-MEBSA), Integrated Laboratory, 2nd floor, Diponegoro University, Jl. Prof. H. Soedarto, SH, Semarang, 50275, Indonesia
[2] Biology Department, Science and Mathematics Faculty, Diponegoro University, Prof. H. Soedarto, SH street, Semarang, 50275, Indonesia
[3] Mathematics Department, Science and Mathematics Faculty, Diponegoro University, Prof. H. Soedarto, SH street, Semarang, 50275, Indonesia
[4] Chemistry Department, Science and Mathematics Faculty, Diponegoro University, Prof. H. Soedarto, SH street, Semarang, 50275, Indonesia

E-mail : saptoputro@gmail.com

Abstract. Floating net cage is one of the aquaculture practice operated in Indonesian coastal areas that has been growing rapidly over the last two decades. This study is aimed to assess the roles of macrobenthic mollusks as bioindicator in response to environmental disturbance caused by fish farming activities, and compare the samples within the locations using graphical methods. The research was done at the floating net cage fish farming area in the Awerange Gulf, South Sulawesi, Indonesia at the coordinates between 79°0500'- 79°1500' LS and953°1500'- 953°2000' BT, at the polyculture and reference areas, which was located 1 km away from farming area. Sampling period was conducted between October 2014 to June 2015.The sediment samples were taken from the two locations with two sampling time and three replicates using Van Veen Grab for biotic and abiotic assessment. Mollusks as biotic parameter were fixed using 4% formalin solution and were preserved using 70% ethanol solution after 1mm mesh size. The macrobenthic mollusks were found as many as 15 species consisting of 14 families and 2 classes (gastropods and bivalves). Based on cumulative *k*-dominance analysis projected on each station, the line of station K_3T_1 (reference area; first sampling time) and KJAB P_3T_2 (polyculture area; second sampling time) are located below others curves, indicating the highest evennness and diversity compared to the other stations, whereas station K_2T_1 (reference area; first sampling time) and K_3T_2 (polyculture area, second sampling time) are located on the top, indicate the lowest value of evenness and diversity. Based on the bubble plots NMDS ordination, the four dominant taxa/species did not clearly show involvement in driving/shifting the ordinate position of station on the graph, except *T. agilis*. However, the two species showed involvement in driving/shifting the ordinate position of two stations of the reference areas from the first sampling time by *Rynoclavis sordidula,* and three stations of the polyculture areas from the first sampling time by *Gemulla* sp. away from other stations, thus are considered as bioindicators of environmental disturbance. These results imply that cumulative *k*-dominance curves and bubble plots ordination may be effective method to assess the indication of disturbance of the area.

Keywords : Macrobenthic molluks, polyculture, bubble plots, cumulative *k*-dominance

1. Introduction

Over the last two decades, aquaculture in form of floating net cage has been developed well around the globe. The cage is a pen/impound for fish farming that usually placed in a large flow of water such as lakes, rivers, and sea. The use of floating net cage has proven to facilitate the people for its efficiency of land and food, easier to harvest either selective or total, business units can be determined according to the funding ability by utilizing materials available on farms and easily monitored and do not require water management specifically as in ponds [1]. Floating net cage is one of the aquaculture practice operated in Indonesian coastal areas that has been growing rapidly. The operation of the cage may be done using monoculture and polyculture techniques. Monoculture is growing up a species of farmed biota in one cage over one period of harvest time, while the polyculture is growing up more than one biota in one cage over one period of harvest time. The various types of fish require different types of food so that each type of fish will not compete for food [2]. The use of monocultures system for a long time can cause environmental degradation and can lead to eutrophication due to a buildup of nutrient excretion and inefficient feeding system. Polyculture fish farming system also has the potential to impact on the farm's environment. Intensive mariculture of fish, shrimp, shellfish or the other economic aquatic animals may not only result in an increase of nutrient concentrations in coastal waters, but also lead to changes of the dissolved inorganic nutrient structures and sedim entary environments [3].

Benthic animals are animals that spend some or all of their life in the bottom of waters, either sessile, crawling or digging holes. Benthic animals have an important role in the process of decomposition and mineralization of organic material in the water, as well as occupying several trophic levels in the food chains [4], [5]. Macrobenthos generally cannot move fast, has large body size that make it easy to be identified and spend their lives in the bottom of waters. Their abundance and diversity are highly depended on changes in water quality and substrate tolerance, and also their activity and sensitivity to environmental changes. Tolerance range of macrobenthos in the environment is different one to each other [6]. Macrobenthos provide advantages as biological indicators because of its characteristics, as they are ubiquitous animals, able to provide response spectrum to environmental stress, live a sedentary life in the habitat that may explain the spatial changes and also has longer life cycle that can explain temporal changes. Bioindicators include biological processes, species, or communities and are used to assess the quality of the environment and how it changes over time. Changes in the environment are often attributed to anthropogenic disturbances (e.g., pollution, land use changes) or natural stressors (e.g., drought, late spring freeze), although anthropogenic stressors form the primary focus of bioindicator research [7], [8].

One of macrobenthos assessed as a good bioindicator is mollusks group. Mollusks are invertebrates that have the ability as bio-indicators of water pollution, because it has sensitivity to the compound of pollutants in ecosystems. Awerange Gulf is one of the gulfs located in Barru District, South Sulawesi. The gulf has been used as floating net cage over the last 15 years. The water conditions can be changed as a result of farming activities in the bay, and subsequently affect the lives of organisms that live in these waters, one of which is benthos. This study is aimed to assess the roles of macrobenthic mollusks as bioindicator in response to environmental disturbance caused by fish farming activities, and compare the samples within the locations using graphical methods.

2. Research Method

The study was conducted on October 2014 until April 2015 in the Awerange Gulf, South Sulawesi. The research was done at the floating net cage polyculture system in the Awerange Gulf, South Sulawesi, Indonesia at the coordinates between 79°0500'- 79°1500' LS and 953°1500'- 953°2000' BT. The sediment samples were taken from the two locations with two sampling time and three replicates using Van Veen Grab for biotic and abiotic assessment. Samples of mollusks were used as biotic parameters, whilst sediment and water samples were used as abiotic parameters. Mollusks as biotic parameter were fixed using 4% formalin solution and were preserved using 70% ethanol solution after 1mm mesh size. Sediment samples were taken and then sieved using a sieve with a net benthos measuring 1 mm. Siftings then washed under water for subsequent sorting. Results sorting then inserted into a bottle containing 70% ethanol solution and labeled. The next stage is the identification and analysis of data.

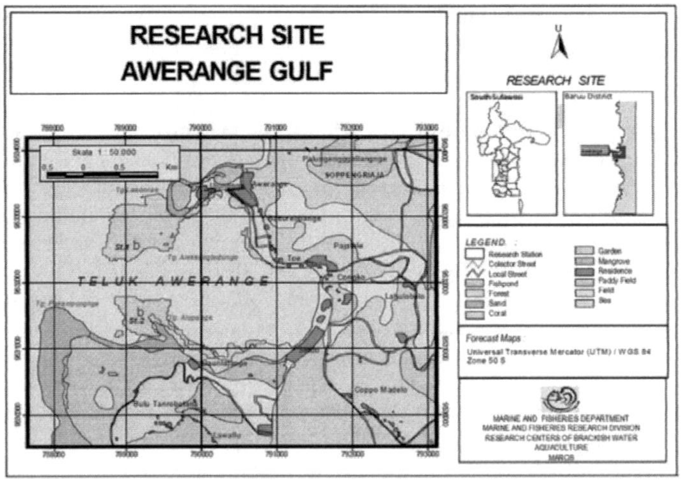

Figure 1. Map of Awerange Gulf, Barru, South Sulawesi as a study site (Source: BPAP Maros, 2014).

The changes in the dominance pattern of macrobenthic mollusk assemblages were assessed using cumulative *k*-dominance curves for each station plotted on the same graph [9]. The extent of disturbance was evaluated by a comparison of the samples between sites and times based on the curves. The dominance (and intrinsically diversity) of the assemblages was generated by plotting cumulatively ranked abundances against log species/taxa rank [10]. Non-metric Multi Dimensional Scaling (NMDS) of Bray-Curtis similarities on log (X + 1) transformed data was used to provide a visual representation of differences between sites over time on the ordination. The analysis incorporates all sites at all times to observe if there was any tendency of separation between polyculture and reference sites. The four dominant taxa were then superimposed on the MDS plots using bubble plots to assess their roles in configuring the ordinations and degree of disturbance. The correlations between physical-chemical factors, substrate granular composition and the carbon and nitrogen content with the abundance of mollusks were analyzed using BIO-ENV. All multivariate and graphical method were analysed using the PRIMER 6.1.5 software packages.

3. Results and Discussion

The macrobenthic mollusks were found as many as 15 species consisting of 14 families and 2 classes (gastropods and bivalves). *Turritella* sp. (Fam. Turritellidae) was the most dominant taxa both at the reference area and the polyculture cage area, followed by *Tellina agilis* (Fam. Tellinidae), *Nassarius castus* (Fam. Nassariidae), and *Anadara* sp. (Fam. Arcidae), as seen in Table 1. Based on sampling location, mollusks abundance at the polyculture area ($\overline{X}$ = 140.67 , SD = 8.71) exhibited significantly different (ρ = 0.001) compared to those at reference area ($\overline{X}$ = 49.2 , SD = 13.9).

Table 1. The average of abundance of mollusks in the reference and polyculture areas on the first and second sampling time at Awerange Gulf, Barru District, South Sulawesi (per grab).

NO.	CLASS	FAMILY	SPECIES	Polyculture Area		Reference Area	
				Sampling I	Sampling II	Sampling I	Sampling II
1	Gastropod	Turritellidae	*Turritella*sp.	25.67	21.33	126.67	93.00
		Chilodontidae	*Perrinia* sp.	0.00	0.00	0.67	0.00
		Mitridae	*Mitra* sp.	0.33	1.67	0.67	0.67
		Nassariidae	*Nassari uscastus*	5.33	13.33	7.67	16.00
		Cylichnidae	*Cylichnella* sp.	2.33	3.33	5.00	6.67
		Turbinidae	*Guildfordi ayoka*	0.00	0.00	0.33	0.00
		Calliostomatidae	*Photinula caerulescens*	2.00	1.00	0.67	3.67
		Costellaridae	*Vexillum*sp.	0.00	1.67	0.00	0.33
		Epitoniidae	*Epitonium scalare*	0.00	0.33	0.00	0.33
		Cerithiidae	*Rhinoclavis sordidula*	0.00	0.00	0.67	0.00
		Turridae	*Gemmula* sp.	1.00	0.00	0.00	0.00
2	Bivalve	Cardiidae	*Fulvia scalata*	0.67	3..00	3.00	3.67
		Arcidae	*Anadara* sp.	1.00	8.00	1.00	3.33
		Tellinidae	*Tellina agilis*	0.67	5.67	0.67	6.67
Abundance				**39**	**60**	**149**	**134.33**
Total Species				**8**	**10**	**10**	**8**

At the polyculture area, the composition of species abundance in the first and second sampling was dominated by *Turritella* sp. (Gastropoda: Fam. Turritellidae), 70.22 % and 49.19 % respectively, while the lowest abundance was belonged to *Mitra* sp. (Gastropoda: Fam. Mitridae) in the first sampling time, and *Epitonium scalare* (Gastropoda: Fam. Epitoniidae) in the second sampling time, 0.82 % and 0.69 %, respectively. At the reference sites, the composition of gastropods abundance during the study period was dominated by *Turritella* sp. (Fam. Turritellidae), i.e. 88.91% in the first sampling and 61.47% in the second sampling.

Meanwhile, the composition of species abundance of Class Bivalvia was dominated by 63.83% of *Fulvia scalata* Vidal, 1994 (Fam. Acrididae) in the first sampling time and 51.33% of *Tellina agilis* Stimpson, 1857 (Fam. Tellinidae) in the second sampling time, indicating that the species composition of Class Gastropoda is higher than members of Class Bivalvia. *T. agilisis* is a suspension feeder that lives in shallow water with sandy or muddy buttoms [11]. The data of macrobenthic mollusk abundance was highlighted only on the dominant taxa for multivariate analysis using bubble plots of NMDS. They display how the abundance of the four dominant taxa relates to overall assemblage differences between sites. Based on the bubble plots NMDS ordination, the four dominant taxa/species did not clearly show involvement in driving/shifting the ordinate position of station on the graph, except *Tellina agilis* (Fam. Tellinidae), as shown in Figure 2. The differences in relative proportions of *T. agilis* are responsible for the horizontal spread of stations in that the higher proportions of this taxa are plotted to the left on the ordination. However, based on Figure 3, the two species showed involvement in driving/shifting the ordinate position of two stations of the reference areas from the first sampling time by *Rynoclavis sordidula* (Fam. Cerithidae), and three stations of the polyculture areas from the first sampling time by *Gemulla* sp. (Fam. Turridae) away from other stations, thus are considered as bioindicators of environmental disturbance. The differences in relative proportions of *Gemulla* sp. and *R. sordidula* are responsible for the horizontal spread of stations in that the higher proportions of these taxa were plotted to the bottom and the upper right of the configurations, respectively.

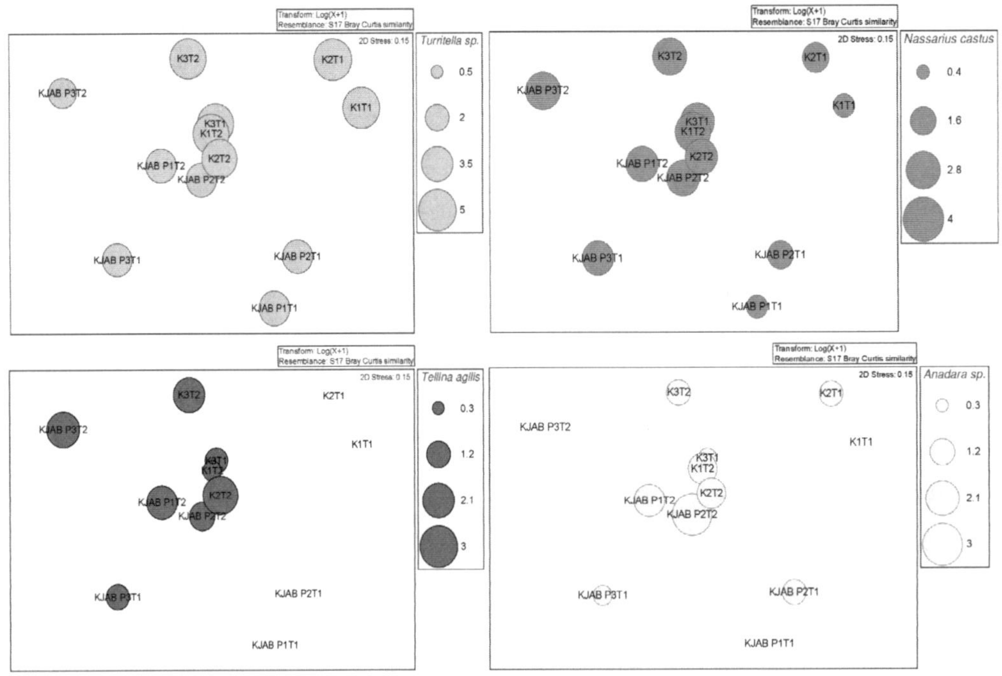

Figure 2. The role of dominant taxa in driving ordinate position on the 2d NMDS superimposed using bubble plots.

Members of Family Cerithidae is a worldwide marine mollusks, most commonly inhabiting shallow waters in tropical and subtropical regions. They are commonly found in a range of habitats such as sand flats and are abundant on trees and roots on the seaward side of mangroves which serve as microalgal or detritus feeders [12], [13], [14]. *R. sordidula* is mostly found in mud at 40 m water depth, few shells in muddy sand and sand between coral patches. In sampled mud station it is frequently colonized by solitary corals. Conservative position of corals on the backside of the shells may suggest that the gastropods were colonized by the coral during lifetime. They can be found in clearer shallow water in coral reefs, preferring sandy to muddy bottoms, feed on deposits in the sediment. *Gemmula* thrive mostly in the deeper waters of the tropics. They can survive in muddy environment. They are carnivores that live by eating polychaetes and seaworms.

The table 2 shows the correlations between physical-chemical factors, substrate granular composition and the carbon and nitrogen content with the abundance of mollusks were analyzed using BIO-ENV on PRIMER V.6.1.5 software. Based on the table, the correlation between the abundance of mollusks with DO, C content and N content have the greatest degree of correlation, which is 0.457. This indicates that the value of DO, the carbon and nitrogen content are the abiotic components that has the most influence in the abundance of mollusks in the Awerange Gulf. The shape of the rank/abundance plot can also provide an indication of dominance or evenness, for example, steep plots signify assemblages with high dominance and shallower slopes indicate higher evenness [15], by then the tendency of environmental disturbance may be assessed [3], [16]. Figure 4 shows cumulative dominance curves between stations at polyculture and reference areas projected based on sampling time.

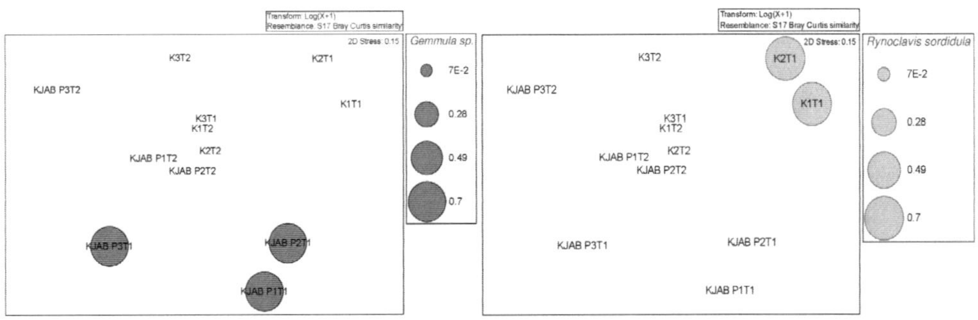

Figure 3. The Tendency of *Gemulla* sp. and *Rynoclavis sordidula* which Present at Certain Location and Absent at Other Location in Driving Ordinate Position on 2D NMDS Bubble Plot.

Table 2. Analysis of the correlation between biotic and abiotic component in reference area and polyculture cage area

No.	Correlation (r)	Variable	Variable amount
1	0.457	4,7,8	3
2	0.447	1,7,8	3
3	0.437	1,4,7,8	4
4	0.416	2,4,7,8	4
5	0.41	1,3,4,7,8	5
6	0.401	1,2,4,7,8	5
7	0.394	1,2,7,8	4
8	0.391	3,4,7,8	4
9	0.388	2-4,7,8	5
10	0.388	1-4,7,8	6

Note: $^{1)}$ pH, $^{2)}$ Temperature, $^{3)}$ Salinity, $^{4)}$ DO, $^{5)}$ Sand, $^{6)}$ Silt, $^{7)}$ C Content, $^{8)}$ N Content

Station KJAB P1T1 (polyculture area, first sampling time), KJAB P2T1 (polyculture area, first sampling time), and KJAB P1T2 (polyculture area, second sampling time) represent curves with steep plots, indicate the dominance value of the stations are high. On the other hand, curve with shallower slope as shown for K1T1 (reference area, first sampling time) indicates higher evenness.

In a stressful condition, macrobenthos community are dominated by organisms with *k*-selection strategy in their lives, called opportunistic species, with characteristics such as relatively small body, short lifetime, dominance in species quantity with low biomass, has high potential reproductive level and early maturity [17]. Based on cumulative *k*-dominance analysis projected on each station by sampling time, station K3T1 (reference area; first sampling time) and KJAB P3T2 (polyculture area; second sampling time) located below others curves, indicate the highest evenness and diversity rather than the other stations and station K2T1 (reference area; first sampling time) andK3T2 (polyculture area, second sampling time) located on the top, indicate the lowest value of evenness and diversity [18].

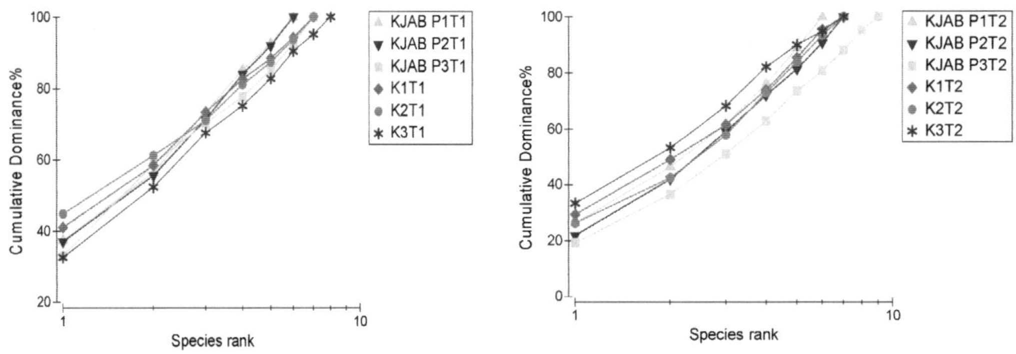

Figure 4. Cumulative dominance curves between stations at polyculture and reference areas projected based on sampling time: (A) First sampling time (October 2014); and (B) Second sampling time (April 2015).

4. Conclusions

Based on the bubble plots NMDS ordination, the four dominant taxa/species did not clearly show involvement in driving/shifting the ordinate position of station on the graph, except *T. agilis*. Furthermore, the two species of *Rynoclavis sordidula* (Fam. Cerithidae) and *Gemulla* sp. (Fam. Turridae) showed involvement in driving/shifting the ordinate position of the stations. The higher proportion of *Gemulla* sp. was plotted to the bottom, whilst the higher proportions was plotted to the upper right of the configurations. Based on abiotic-biotic correlation analysed using BIO-ENV, the value of DO, the carbon and nitrogen content are the abiotic components that has the most influence in the macrobenthic mollusk abundance in the Awerange Gulf. Based on cumulative *k*-dominance analysis projected on each station by sampling time, station K3T1 (reference area; first sampling time) and KJAB P3T2 (polyculture area; second sampling time) located below others curves, indicate the highest evenness and diversity rather than the other stations and station K2T1 (reference area; first sampling time) and K3T2 (polyculture area, second sampling time) located on the top, indicate the lowest value of evenness and diversity. These results imply that the graphical methods using cumulative *k*-dominance curves and bubble plots ordination on the macrobenthic mollusk abundance are effective and sensitive method to assess the indication of disturbance of the area.

Acknowledgement

We would like to thank Ministry for Research, Technology, and Directorate of Higher Degree (Simlitabmas - RISTEKDIKTI) for supporting research on the development of sustainable productive aquaculture through funding the research grant under the scheme 'National Strategic of Excellent Research (PUSNAS)', year II/2016.

References

[1] Espineira R M, Chopin T, Robinson S, Noce A, Knowler D and Yip W 2014 Estimating The Biomitigation Benefits of Integrated Multi-Trophic Aquaculture: A Contingent Behavior Analysis. *Aquaculture,* **437**, 182-194.

[2] Afrianto E and Liviawaty E 200 .[*Pakan Ikan*] Fish Food. Kanisius Press. Yogyakarta. Pp. 11-15. [Bahasa Indonesia].

[3] Putro S P 2007 Spatial and Temporal Patterns of The Macrobenthic Assemblages in Relation to Environmental Variables. *Journal of Coastal Development,* **10**(3),15-22.

[4] Odum E P 1993 [*Dasar-dasar Ekologi*]. Fundamental of Ecology. Gadjah Mada University Press.Yogyakarta. Pp. 46-79. [Bahasa Indonesia]

[5] Putro S P 2014 [*Metode Sampling Penelitian Makrobenthos dan Aplikasinya*] Research Sampling Method of Macrobenthos and Its Application. Graha Ilmu. Yogyakarta. Pp. 97-117. [Bahasa Indonesia]

[6] Marsaulina L 1994 [*Keberadaan dan Keanekaragaman Makrozoobenthos di Sungai Semayang Kecamatan Sunggal*]. Macrozoobenthic Presence and Diversity at Semayang River, Sunggal District. Lembaga Penelitian USU. Medan. [Bahasa Indonesia]

[7] Rosenberg R 2001 Marine benthic faunal successional stages and related sedimentary activity. *Scientia Marina*, **65** (2), 107-119.

[8] Mann K H 1982 *Ecology of Coastal Waters*. University of California Press. USA.

[9] Warwick R M 1986 A new method for detecting pollution effects on marine macrobenthic communities. *Marine Biology,* **2**, 557-562.

[10] Lambshead P J D, Platt H M and Shaw K M 1983 The detection of differences among assemblages of marine benthic species based on an assessment of dominance and diversity. *Journal of Natural History*, **17**: 859-874.

[11] Abbott AT and Morris P A 1995. Shells of the Atlantic and Gulf Coasts and the West Indies. Houghton Mifflin Company. Ney York. Pp. 36-55.

[12] Houbrick R S 1992 Monograph of the genus *Cerithium* Bruguiere in the Indo-Pacific (Cerithiidae: Prosobranchia). Smithsonian Institution Press, Washington D. C.. Pp 209.

[13] Poutiers J M 1998 Gastropods. In: FAO species identification guide for fishery purposes. The living marine resources of the western central pacific volume 1. Carpenter K. E., Niem V. H., (eds), Food and Agriculture Organization of the United Nations, Rome, Italy. Pp. 364-648.

[14] Beechey D 2005 Family Cerithiidae. Seashells of New South Wales. Available at: http://seashells-ofnsw.org.au/Cerithiidae/Pages/Cerithiidae_intro.htm. Accessed on June, 2015.

[15] Clarke and Gorley 2006 Primer v6: User Manual/Tutorial. PRIMER-E Ltd. Plymouth, pp. 150-155.

[16] Putro S P 2010. *Environmental Quality Assessment of Fish Farming: Solutions Toward Sustainable Aquaculture*. Lambert Academic Publishing, Saarbrucken. Germany. Pp. 110-122.

[17] Diaz R J and Rosenberg R 1995 Marine benthic hypoxia: a review of its ecological effects and the behavioural responses of benthic macrofauna. *Oceanography and Marine Biology Annual Review*, **33**: 245-303.

[18] Clarke K R 1990 Comparisons of Dominance Curves. *Journal of Experimental Marine Biology and Ecology*, **138**: 143-157.

The Effect of Early Mosquito Insecticides Exposure on Spraque Dawley Rat Testis: A Histopathological Feature Towards Malignancy?

Tri Indah Winarni[1], Milzam Auzan Aziman[1], Anindyo Abshar Andar[1], Ika Pawitra[2]

[1]Department of Anatomy, Faculty of Medicine, Diponegoro University, Semarang
[2]Department of Patology Anatomy, Faculty of Medicine Diponegoro University, Semarang

Coresponding Author: Tri Indah Winarni.Email: tri.winarni@fk.undip.ac.id OR triindahw@gmail.com

Abstract. The incidence of health problems associated with endocrine-disruption have increased. Many studies suggesting that endocrine disruptor chemicals (EDC) do contribute to cancer through estrogen-related receptors. Many chemicals have EDCs properties including insecticides. Early life exposure to EDCs can increased the risk of testicular cancer have been reported in the last decade. This study was aimed to determine the effect of insecticides exposure on histopathological tumor cell development of germ and Leydig cell. True experiment research design with posttest only control group design was applied. Sprague Dawley (SD) rat (n = 25) were randomly divided into 5 groups (control group, 25 mg β estradiol 3-benzoate, spiral mosquito coil repellent, 3 ml of liquid mosquito repellent, and 4 ml of liquid mosquito repellent). The exposure were administered for 20 days started at aged 3 days. At the age of 100 days (older adult), testis was stained using Hematoxyllin Eosin (HE) and histological features predicting malignancy were observed. The number of tumor cell development in both testicular germ cells and Leydig cells significantly increased in all treated group compared to those of control and the changes towards malignancy were also observed in all treated group. Exposure to mosquito insecticides causes significant changes in testicular germ and Leydig cell histological features that leads to malignancy.

Keywords: endocrine disrupting chemicals, insecticides, testicular germ cell, Leydig cell, malignancy.

1. Introduction

Since the late 80s, there has been increasing concern about the impact of environmental pollutants, called endocrine disruptor chemicals (EDCs), substance that may interfere with the biosynthesis, metabolism, and/or an action of endogenous hormones. Alterations of male reproductive development and health that have been reported associated with EDCs, which includes demasculinization or feminization, cryptorchidism, hypospadia, in situ germ cell testicular carcinoma, reduced semen quality, and progression of prostate cancer[1-5].
Skakkebaek proposed the existence evidence called testicular dysgenesis syndrome (TDS) which includes poor semen quality, hypospadia, undescensus testis, and testicular cancer[6], which may increased due to adverse environmental exposure[3, 6]. The most common form of TDS is a mild form which may present with slight impairment of spermatogenesis counted as much as 20% [7], in contrast, the most severe form of TDS is testicular cancer appear to be uncommon (~5%)[6]. Although it appear to be uncommon form, testicular cancer (TC) is the most common malignancy in 20 to 30-year-old male. With prevalence rate is diverse among countries (Finland 2.5/100,000 cases versus Denmark 9.2/100,000)[8]. Numerous publications have shown an

increasing incidence of testis cancer in the last 50 years. The increase in the TC incidence was associated with environmental factor with a key role of endocrine disrupters chemicals (EDCs) has been point it out by several group of study. Recent increase in the TC rate in most industrialized countries should lead more attention to the association of EDC and testicular cancer[9, 10].

In tropical country like Indonesia, mosquitoes are the most common transmitters of vector-borne human diseases in tropical countries due to environmental conditions (temperature and humidity) conducive for reproduction. In addition, human cultural behavior also plays a role in diseases-transmission control programming. Children under the age of 5 have a higher risk to mosquito-borne diseases. Accordingly, to protect children from mosquito bites, mosquito repellent usage is prevalent, often without consideration of potential side effects[11, 12]. Based of epidemiological studies, suggest that TDS is a result of adverse environmental exposure during fetal and early life. Every organ has a specific window of development, which refers to the critical period when the developing tissue is susceptible to the effects of EDC exposure[13]. The critical window of reproductive tract development is between 7 and 40 weeks of gestation, in which cells undergo rapid mitotic division and differentiation, eventually resulting maturation. The exposure of EDCs during the critical window period will affect cell reprogramming[14]. This study was done to elucidate the adverse effect of mosquito insecticides exposure in early life on testis spraque dawley rat.

2. Materials and methods

A posttest-only control group was designed to assure the objectivity of the study. Male one-day-old Sparque Dawley (SD) rats were acclimatized for 2 days in standard cages under standard conditions (room temperature at 26 ± 3°, 12-hour light/ dark cycle) and kept in standard cages at the Animal Care Unit in Gadjah Mada University, Yogyakarta. The rats were randomly selected and allocated into 5 groups: control (C), treated groups including 25 μg β estradiol 3-benzoat (P1), and insecticide groups [spiral mosquito coil repellent (P2) contains transfluthrin 0.03%/coil; 3-ml (P3), and 4-ml mosquito liquid repellent (P4) contains transfluthrin 0.162 mg and propoxur 4.05 mg/ml]. After weaning, the rats were housed up to 5 per cage. Drinking water was available in plastic bottles and the rats were fed pellet diet *ad libitum* from weaning to postnatal day (PND) 100. Pesticide exposure was designed to resemble the natural setting of mosquito repellent usage.

Testis were removed and fixed in 10% buffered neutral formaldehyde solution for 24 hours at room temperature before being embedded into paraffin blocks. The haematoxylin and eosin (HE) staining method was used to explore tumor cell development in both testicular germ cells and Leydig cells (large nucleus, having an irregular size and shape, the nucleoli are prominent and irregular, the cytoplasm is scarce and intensely colored or, on the contrary, is pale)[15]. HE slides were aseessed by two experienced pathologist using microscope *Olympus CX21* in a blinded manner. *Kruskal-Wallis* test was used to determine the difference between groups, while *Mann Whitney* test was used to compare the difference between groups.

3. Results

Tumor cell development features were observed significantly in both germ and Leydig cells with p=0.024 and p=0.001, respectively. The difference number of germ cell development between groups can be seen in Table 1. Germ tumor cell development in estradiol group was significantly increased compare to those of control group (p=0.008), the same result was seen also in mosquito repellent (p=0.019) and 4-ml mosquito liquid repellent (p=0.025)

Table 1. The difference number of germ cell tumor development between groups.

Group	P1	P2	P3	P4
C (Control)	p = 0,008*	p = 0,019*	p = 0,589	p = 0,025*
P1(Estradiol)	-	p = 0,207	p = 0,059	p = 0,142
P2 (Spiral mosquito coil repellent)	-	-	p = 0,246	p = 0,674
P3 (3-ml mosquito liquid repellent)	-	-	-	p = 0,295

$p < 0.05$= significancy (*)

The difference number of Leydig cell between groups can be seen in Table 2. Leydig tumor cell development in estradiol group was significantly increased compare to those of control group (p=0.007), the same result was seen also in mosquito repellent (p=0.008) and 4-ml mosquito liquid repellent (p=0.008).

Table 2. The difference number of Leydig cell tumor development between groups.

Group	P1	P2	P3	P4
C (Control)	p = 0,007*	p = 0,008*	p = 0,827	p = 0,008*
P1(Estradiol)	-	p = 0,119	p = 0,008*	p = 0,013*
P2 (Spiral mosquito coil repellent)	-	-	p = 0,011*	p = 0,197
P3 (3-ml mosquito liquid repellent)	-	-	-	p = 0,026*

p < 0.05= significancy (*)

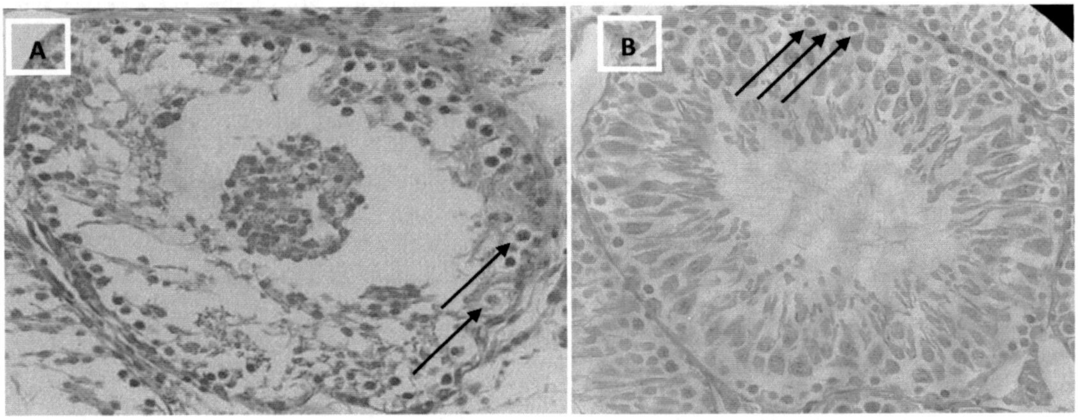

Figure 1. Histotogocal feature of tumor germ cell development (A) and (B) normal germ cell. A: Germ cell with large, irregular size and shape nucleus, the nucleoli are prominent and irregular with rough chromatin, the cytoplasm is scarce and pale (arrow). B: Germ cell with regular size and shape nucleus with smooth chromatin (arrow)

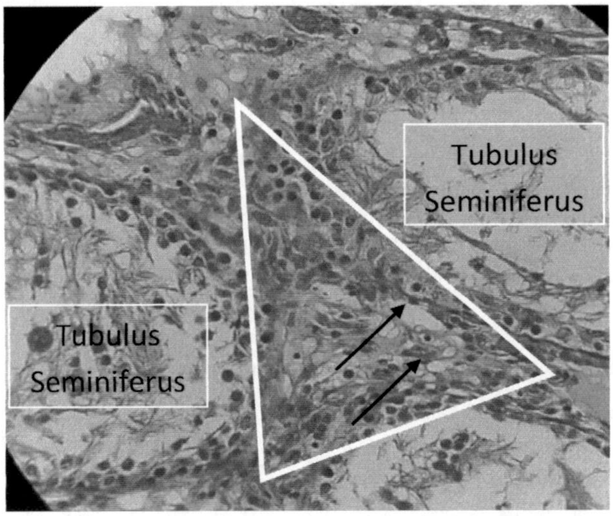

Figure 2. Histotogocal feature of tumor cell development of Leydig cells. Leydig cell with large, irregular size and shape nucleus, the nucleoli are prominent and irregular with rough chromatin, the cytoplasm is scarce and pale (indicate by arrows).

4. Discussion

These study indicated that there was significant number of both germ and Leydig cells lead towards malignancy in experimental groups i.e. estradiol, spiral mosquito coil repellent, and 4-ml mosquito liquid repellent. The change of Sprague Dawley rat germ cells toward malignancy may due to neonatal exposure of the mosquito insecticide repellent that act as a EDC, where EDCs bind to estrogen receptor (ER)-β in spermatogonia cells. This binding can also cause oxidative damage to deoxyribonucleic acid (DNA), it may results germ cells changing towards malignancy. This result is consistent with previous research conducted by Anja Wellejus et al in 2002 [16], and Vernon Pais et al in 2003[17]. ER-β has an important role in the process of change into a germ cell malignancy, it's due to the exposure to EDCs[17]. The exposure of estrogen to the ER- β can mediates the carcinogenic effects through the oxidative damage process to DNA with a marked by increasing significant levels of 8-OxodG. This damage also more commonly found in spermatogonia cells, because these cells express many sort of estrogen receptor and are located outside of the blood testis barrier[16, 17]. The time of exposure during early critical window period of development, when the cells undergone differentiation and maturation, has also been found will affect cell reprogramming[14]. Beside of those, the exposure during early postnatal period also give a greater impact other area such as environmental induced disease etiology, epigenetic transgenerational inheritance, and the general systems biology of organisms and evolution[18].

The number of germ cell that lead to malignancy was found higher in the group of spiral mosquito coil repellent compared to those of liquid repellent. This may also cause due to the additional exposure of smoke of mosquito coil that can cause increase level of free radicals, it can create damage lipid membranes of cells and integrity of the nucleus. This damage can modulate germ cells exposure to the chemical compounds. Thus, all together stimulate the changes in histopathological features of germ cells towards malignancy[19].

The exposure of EDCs can cause disturbing negative feedback mechanism of hypothalamic-pituitary-adrenal (HPA) axis, resulting an adverse effect of hormonal balance including the excessive production of luteinizing hormone (LH) that results Leydig cell hyperplasia. If the exposure is persistence and within the critical period, the change of Leydig cells can cause of severe form of hyperplasia ranging from dysplasia to malignancy[20]. Mostly, EDCs was designed have longer half-life for industrial benefit, it will unfortunately give long-term effect on disturbing negative feedback mechanism[21]. In previous studies, chronic progressive of excessive LH production can cause the Leydig cell more easily enters into the mitosis phase. The role of LH in mitogenic process also observed, wherein an administration in rats for chronic period will cause Leydig cell adenoma[22]. Limitation of study. Active ingredients of mosquito insecticides both air transfluthrin concentration inside coil repellent group cage; and air transfluthrin and propoxure concentration inside 3-ml and 4-ml liquid repellent group cages are not measured.

5. Conclusions

Mosquito insecticides exposure in early life will leads histopathological features of tumor cell development in both germ and Leydig cells in adult life. With regards to this study, mosquito insecticides usage in a daily basis should accompany with precautions especially when it exposed to the infant.

Acknowledgement:

This works was funded by DIKTI Postgraduate Scholarship, Ministry of Education Republic Indonesia. Thanks to the Department of Pathology Anatomy Faculty of Medicine Diponegoro University/dr. Kariadi Hospital for processing the materials. The author declares there is no conflict of interest.

References

[1] Carlsen E, Giwercman A, Skakkebaek NE. Declining sperm counts and increasing incidence of testicular cancer and other gonadal disorders: is there a connection? 1993 *Ir Med J* 86 85.

[2] Jacobsen R, Bostofte E, Skakkebaek NE, Hansen J, Moller H. Offspring sex ratio of subfertile men and men with abnormal sperm characteristics. 2000 *Hum Reprod* 15 2369.

[3] Schiffer C, Muller A, Egeberg DL, Alvarez L, Brenker C, Rehfeld A, et al. Direct action of endocrine disrupting chemicals on human sperm. 2014 *EMBO Rep.*

[4] Sharpe RM, Skakkebaek NE. Are oestrogens involved in falling sperm counts and disorders of the male reproductive tract? 1993 *Lancet* 341 1392.

[5] Prins GS, Tang W-Y, Belmonte J, Ho S-M. Perinatal Exposure to Oestradiol and Bisphenol A Alters the Prostate Epigenome and Increases Susceptibility to Carcinogenesis. 2008 *Basic & Clinical Pharmacology & Toxicology* 102 134.

[6] Skakkebaek NE, Rajpert-De Meyts E, Main KM. Testicular dysgenesis syndrome: an increasingly common developmental disorder with environmental aspects. 2001 *Hum Reprod* 16 972.

[7] Andersen AG, Jensen TK, Carlsen E, Jorgensen N, Andersson AM, Krarup T, et al. High frequency of sub-optimal semen quality in an unselected population of young men. 2000 *Hum Reprod* 15 366.

[8] Huyghe E, Matsuda T, Thonneau P. Increasing Incidence of Testicular Cancer Worldwide: A Review. 2003 *The Journal of Urology* 170 5.

[9] Soto AM, Sonnenschein C. Environmental causes of cancer: endocrine disruptors as carcinogens. 2010 *Nat Rev Endocrinol* 6 363.

[10] McGlynn KA, Quraishi SM, Graubard BI, Weber JP, Rubertone MV, Erickson RL. Persistent organochlorine pesticides and risk of testicular germ cell tumors. 2008 *J Natl Cancer Inst* 100 663.

[11] Ohta K, Ogawa T, Suzuki T, Ohta S, Endo Y. Novel estrogen receptor (ER) modulators: carbamate and thiocarbamate derivatives with m-carborane bisphenol structure. 2009 *Bioorg Med Chem* 17 7958.

[12] Go V, Garey J, Wolff MS, Pogo BG. Estrogenic potential of certain pyrethroid compounds in the MCF-7 human breast carcinoma cell line. 1999 *Environ Health Perspect* 107 173.

[13] Calafat A, Needham L. Human Exposures and Body Burdens of Endocrine-Disrupting Chemicals. In: Gore A, editor. *Endocrine-Disrupting Chemicals*. Contemporary Endocrinology: Humana Press; 2007. p. 253.

[14] UNEP, WHO. State of the science of endocrine disrupting chemicals 2012 Geneva, Switzerland: WHO Press; 2013.

[15] Baba AI, Catoi C. Tumor Cell Morphology. *Comparative Oncology*. Bucharest: The Publishing House of the Romanian Academy; 2007.

[16] Wellejus A, Loft S. Receptor-mediated ethinylestradiol-induced oxidative DNA damage in rat testicular cells. 2002 *FASEB J* 16 195.

[17] Pais V, Leav I, Lau KM, Jiang Z, Ho SM. Estrogen receptor-beta expression in human testicular germ cell tumors. 2003 *Clin Cancer Res* 9 4475.

[18] Skinner MK. Role of epigenetics in developmental biology and transgenerational inheritance. 2011 *Birth Defects Res C Embryo Today* 93 51.

[19] Madhubabu G, Yenugu S. Effect of continuous inhalation of allethrin-based mosquito coil smoke in the male reproductive tract of rats. 2012 *Inhal Toxicol* 24 143.

[20] Naughton CK, Nadler RB, Basler JW, Humphrey PA. Leydig cell hyperplasia. 1998 *Br J Urol* 81 282.

[21] Patisaul HB, Adewale HB. Long-term effects of environmental endocrine disruptors on reproductive physiology and behavior. 2009 *Front Behav Neurosci* 3 10.

[22] Clegg ED, Cook JC, Chapin RE, Foster PM, Daston GP. Leydig cell hyperplasia and adenoma formation: mechanisms and relevance to humans. 1997 *Reprod Toxicol* 11 107.

Analysis of heavy metal content of Cd and Zn in ballast water tank of commercial vessels in Port of Tanjung Emas Semarang, Central Java Province

A Tjahjono[1,3], A N Bambang[2] and S Anggoro[2]

[1] Graduate Program of Environmental Science, School of Post Graduate Studies, Diponegoro University, Semarang, Central Java, 50241, Indonesia.

[2] Program of Environmental Science, School of Post Graduate Studies, Diponegoro University, Semarang, Central Java, 50241, Indonesia.

[3] e-mail: a_agus_tjahjono70@yahoo.co.id

Abstract. Commercial vessels that do not conduct ballast water exchange, in accordance with International Convention Ballast Water Management, will endager the environment of ports. This research is aimed to know the metal content in ballast water tank of commercial vessels that have not performed ballast water exchange, in accordance with regulations of International Maritime Organization (IMO). The present research is focused on the heavy metal content of ballast water of commercial vessels, both passenger or cargo vessells, berthing in Port of Tanjung Emas Semarang (PTES). Water sample in ballast tank is collected by method of AAS (Atomic Absorption Spectrophotometer). Results of the research show that the content of Cd is about 0.001-0.46 mg l^{-1}, and Zn is about 0.001-2.464 mg l^{-1}. Based on the Decree of Minister of Environment Number 51/2004, the heavy metal content of Cd and Zn has exceeded quality standards of sea water for port water, which is 0.1 mg l^{-1} both.

Key words : commercial vessels, ballast water tank, heavy metal content

1.Introduction

International Maritime Organization (IMO) is a United Nations organization that attempts to increase the safety of shipping and protection of marine environment. One of of comittees in the organization is the Commitee of Marine Environment Protection (MEPC) in 2004 that has legalized regulations of controlling management of vessel ballast water. The regulation of ballast water set by IMO which aims to minimalize the entrance of indigenous species and sediments to other water areas.

To minimalize the risk, based on D1 standards, each commercial vessel entering to port water areas is requested to do ballast water exchange in ballast water tank that is located 200 miles from the port, in depth of more than 200 meters, and by level of efficiency of at least 95%. There are three methods used in ballast water exchange. First, sequential method, a process of emptying and refilling ballast water to get at least 95% of volumetric exchange. Second, flow-through method, a process of pumped ballast water exchange so that the water comes through overflow pipes or other structures. Third, dilution method, a process of ballast water exchange that is filled through the upper parts of ballast water tank by simultaneous release from bottom of the tank [1].

The stability of the vessel depends on the cargo and stabilizer of the vessel. Recently, it uses sea water that is pumped in ballast water tank. When the cargo is empty,the vessel needs to pump sea water to ballast water tank to get positive GM and to avoid the instability of the vessel. When the vessel load cargo, sea water is pumped out of ballast water tank to stabilize the vessel and get positive GM. When the vessel does not pump out sea water, it can transfer sea water into the empty ballast water tank so that the vessel can be in *even keel* condition with the same mean draft and balanced listing. If ballasting is not conducted, the unstable equiribillium condition happens so that is easy to get influenced by other factor [2].

The research about the effectiveness of mid ocean ballast water exchange has been conducted on 34 container vessels (20 vessels are from Oakland, California) in Hong Kong Port. The research shows that in the beginning of September and in the middle of August, *Skeletonema costatum* dominates the ballast water tank (14,000 cells l^{-1}). There are dangerous *diatoms*, *Chaetoceros caoncavicornis* and *Alexandrium catenella*. Other researches are conducted to four vessels from Manzanillo, Mexico. The result of the research shows that effectiveness of mid ocean ballast exchange is about 48% in reducing the abundance of *diatom* and *dinoflagellata*. Old container vessels ($\geq$ 15 years) are not effective to release *diatom* and *dinoflagellata* compared to the new vessels due to sediments on the bottom of ballast water tank [3].

Semarang estuary water is not deep. It is about 14 meters, dirty . It has high solid suspense. The bottom of the sea is muddy. The pollution comes from household waste originated from two main canals [4]. The heavy metal content of Pb on the sediments of Semarang Port pond is higher than the estuary water in the canals of Banjir Kanal Timur and Banjir Kanal Barat. The concentration of Zn in the sediments is about 84.14-131.74 mg kg^{-1}, with the average of 97.11 mg kg^{-1} [5].

Port of Tanjung Emas Semarang (PTES) is located in the north coast of Central Java Province, Indonesia ($6^0$53' S, $110^0$24' E). The port is situated in the narrow side of the coast, less than 10 km in width. It gets wider in the eastern side, maximum speed of general streams is 0,31 knot with the direction of 324^0 in the afternoons, when the water is going to lowest tidal , and in the late afternoon, for about 0,28 knot with the direction of 164^0. The stream of both the lowest and highest tidal are weak. There is hindered and weak stream in front of port entrance [6].

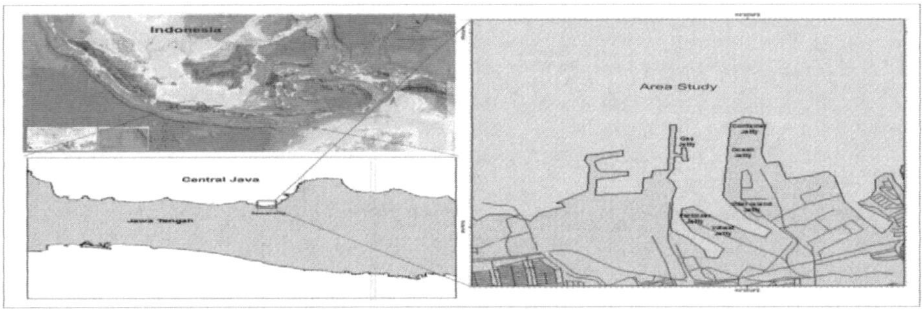

Figure 1. Research location

The study aims to determine implemention of Convention of Ballast Water Management 2004 through the measurement heavy metal content of Pb, physical parameters in ballast water, the origin of the commercial vessels, DWT, and vessels ballast tank capacity.

2.Material and Methods

The sample was gathered from vessels berthed in PTES, dated on December 18th 2014 to October 21st 2015. In the commercial vessels, sample collected twice on one ballast water tank with DWT more than 400 tons. Research in ballast water tank of commercial vessels is conducted by using purposive sampling that covers 6 wharfs, like Pusri fertiliser and oil, passenger, domestic, wheat, container and LPG wharfs.

The capacity of ballast water for each vessel berthed in port is calculated by using the formula of 36,5% of general cargo, 35% of solid cargo, 35% of liquid cargo, 30% of container, 33% of mixed cargo , and 33% mixed cargo, and 33% of Ro-Ro vessels from DWT [7].

The water sample in ballast tank is collected by using the method conducted by putting sounding meter in ballast water tank through sounding pipes. Portable pump and suction hoses are used to collect the water sample [8]. The portable pump is from Sanyo brand, model P-WH137C, 220 V~50 Hz of voltage source, 125 W of output power, 30 l min^{-1} of maximum water capacity. The suction hose has 0.019 m width of diameter and 10 m in length. The end of suction hose is completed by foot valve. The water sample of the ballast water in the surface of the tank can be collected by rising the end of suction hose, while the water sample from bottom of the tank can be collected by lowering the end of the suction hose.

Sample of the ballast water, that is on the surface of the tank, can also be collected by opening manhole in the ballast tank of commercial vessels. Sampel of water , that is at the bottom of the tank, is collected by using 10 liter bucket.

The water sample is filtered by using 40 (0.42 µm) whatman filter paper and washed by using HNO_3 concentrated to PH < 2. Then, it is placed in water sample of 5 l volume. The water sample is then brought to the laboratory. The water (100 ml) is mixed thoroughly and put in the beaker. Then, it is added by 5 ml of citric acid and heater, 50 ml of distilled water is added, and the mixture is put in 100 ml of graduated flask [9] [10]. The level of heavy metal Cd and Zn in the sample of ballast water of commercial vessel is determined by AAS (Atomic Absorption Spectrophotometer) type Shimadzu AA-6300 by using flame-mixed of air acetylene.

3.Results and Discussion

Deadweight of vessels are about 537 to 46,769 MT, while the capacity of ballast tank is about 196 to 16,269.2 tons. The sample of ballast water is mostly gathered from domestic ports in Palembang (JM, SB, MP, OK PI, IZ, Ce 8 and S vessels), and Jakarta (A, MB2, DF2 and GNA vessels). Sample of ballast water gathered from foreign ports is PA, SB and HL vessels (Figure 2)

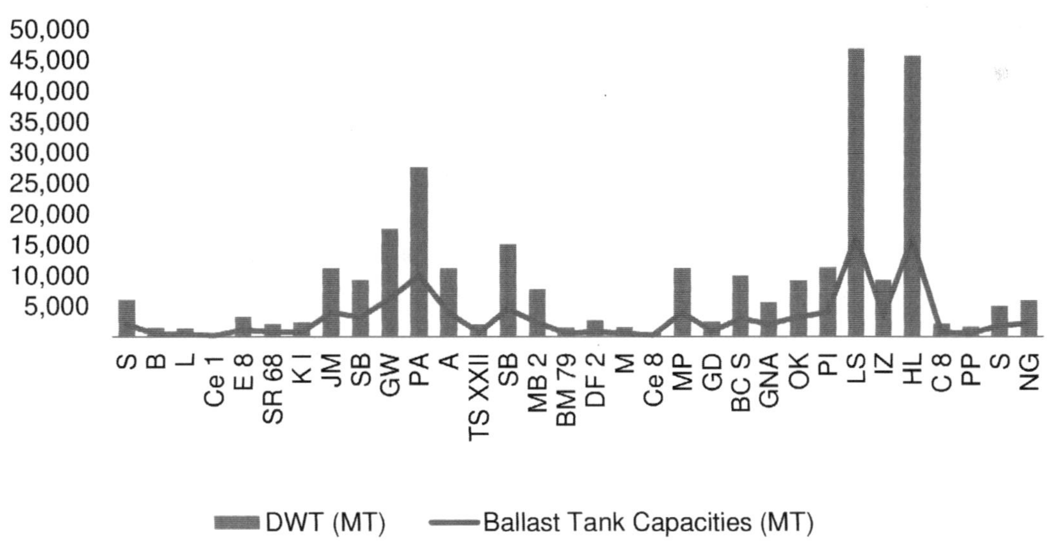

Figure 2. DWT and ballast tank capacities of vessels

The temperature of ballast water researched is about 28.4 to 32.4 ^{0}C. Ph is between 6.2 to 8.16. TSS and salinity ranges respectively are 1 to 686 mg l^{-1} and 1.6 to 36.3‰. DO is about 3.51 to 7.24 mg l^{-1}. The odor of all ballast water is nature, except for ballast water in S vessel, caused by oil content from the origin port, Jakarta (Table 1). Different salinity shows the different origin waters. The salinity of

4.8-12‰ is brackish waters, while the salinity of 30-40‰ is sea water. TSS above 400 mg l^{-1} shows that waters are dangerous for fishery [11].

Table 1. Results of ballast water of commercial vessels from in situ and ex situ

No.	Vessels/Parameter	Odors	TSS (mg l^{-1})	Temp. (^{0}C)	Ph	Salinity (ppm)	DO (mg l^{-1})
1.	S	Oily	686	30.6	6.92	36.3	NA
2.	B	Nature	28	31.2	7.71	34.7	NA
3.	L	Nature	30	29.6	7.39	34.1	6.24
4.	Ce 1	Nature	8	29.8	7.13	4.8	7.24
5.	E 8	Nature	7	29.1	6.3	28.7	6.72
6.	SR 68	Nature	7	29.1	6.72	20.4	6.52
7.	K1	Nature	6	28.4	8.16	1.6	6.52
8.	JM	Nature	7	30.8	6.71	31.1	5.27
9.	SB	Nature	6	30.6	6.92	31.2	5.07
10.	GW	Nature	6	30.3	7.01	30.8	5.37
11.	PA	Nature	3	32.4	7.97	27.5	4.83
12.	A	Nature	3	28.6	7.80	28.3	4.03
13.	TS XXII	Nature	7	30.3	7.24	12	4.23
14.	SB	Nature	3	28.4	7.59	27.1	5.24
15.	MB 2	Nature	1	28.9	6.63	11.3	6.14
16.	BM 79	Nature	2	29.6	6.68	31.3	5.85
17.	DF 2	Nature	2	29.4	7.65	30.8	6.22
18.	M	Nature	2	29.5	7.34	30.9	6.09
19.	Ce 8	Nature	2	29.5	7.44	22.2	5.49
20.	MP	Nature	2	29.8	7.26	11.6	5.09
21.	GD	Nature	1	31.5	7	31.5	5.01
22.	BC S	Nature	1	31.4	7.61	32	3.96
23.	GNA	Nature	1	31.4	7.18	29.5	5.77
24.	OK	Nature	1	31.4	7.05	28.8	5.40
25	PI	Nature	3.5	30.2	7.6	29.6	6.24
26	LS	Nature	1.5	30.8	7.4	28.7	6.12
27	IZ	Nature	2	30.3	8.0	28.7	4.03
28	HL	Nature	1	30.3	8.0	33.5	6.16
29	C 8	Nature	2.5	30.6	7.7	33.7	3.51
30	PP	Nature	37	30.5	7.4	32.4	4.29
31	S	Nature	6.5	30.6	6.2	7.7	6.52
32	NG	Nature	2	30.8	7.9	32.1	5.8

Note : NA (Not Available)

The passenger vessel of S, that has dead weight of 1,400 tons and berthed on December 19th 2014, has 11 of ballast water tank. They have total ballast water capacity of 818 tons. Sample was collected from ballast water tank number 61, starboard and port side tank. The tanks are interconnected. Her ballast water capacity was 76.3 tons. The ballast water origin is from Jakarta dockyard waters, oil polluted waters. The high content of TSS shows that ballast water contains high level of mud.

The passenger vessel of B, DWT of 1450 tons, was berthed on December 18th 2015. She has 11 ballast water tank with the total capacity of 819.3 tons. Sample is collected from ballast water tank number 32 (port side and starboard side tank). It has capacity of 218 tons. Ballast water origin is from Pontianak.

The passenger vessel of L, with DWT of 1,400 tons, was berthed on January 7th 2015. She also has 11 ballast water tank with the total capacity of 819.3tons. Sample is colleted from Double Bottom Deep tank number 32 (port side and starboard tank). She has ballast water capacity of 218.6 tons, its ballast water from Pontianak. Ce 1 vessel that loads rice is also from the same waters.

The fertilizer carriers of J and SB are from the same port of origin, Palembang. The vessels have dead weight of 11,615 and 9,237.4 tons respectively. Sample of ballast water from each of the vessel is collected from ballast water tank number 4 (port side) and number 3 (starboard side). While another fertilizer carrier, A, is from Jakarta.

Heavy metal is metal with the density of five gr cm^{-3}or more. It has atom number from 22 to 96 [6]. The characterictics of heavy metal are, it has bigger gravitational specification (more than 4), it has atom number responds of specific biochemistry on living organism [12]. The source of heavy metal pollutants are from natural and artificial sources. The natural source is from the coastal area (from rivers and coastal abrasion due to wave activity), metal released from volcanoes activity, and chemistry process. The artificial source is from industrial process of mining activity.

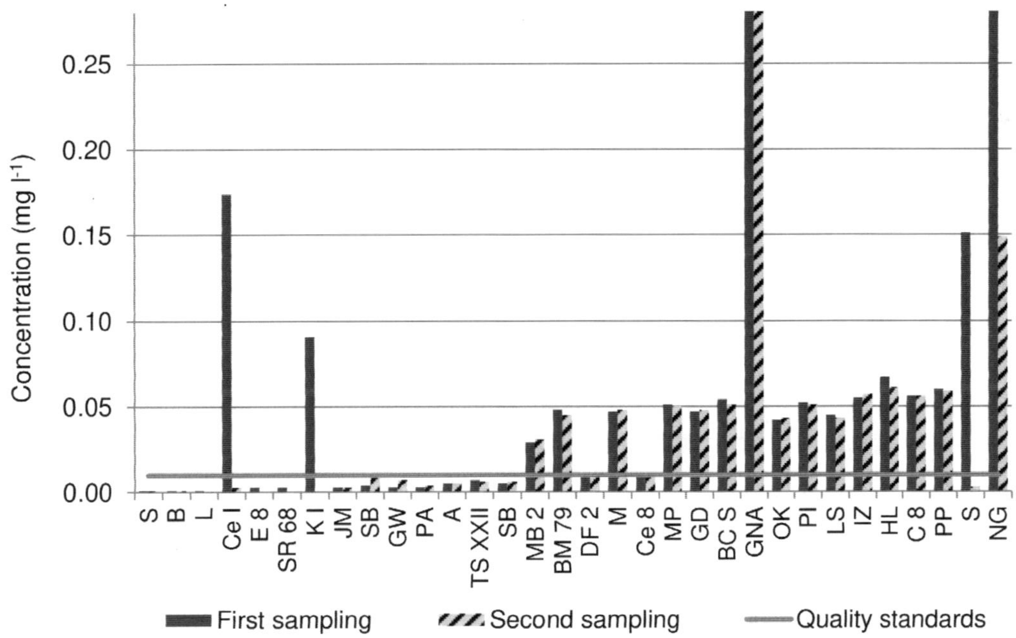

Figure 3. Cd concentration in ballast water of commercial vessels

Kadmium is widely spread out in the nature. The type of metal is mostly used in indutries as a stabilizer and alloy for battery industries, airplanes, weapons, and materials for welding. In waters, Cd solubility in certain concentration can be kill waters biota. In the concentration of 0.005-0.15 ppm, crustaceans will be dead in 24-504 hours. In the concentration of 0.003-18 ppm, insects will be dead in 24-672 hours. In the concentration of 1.092-1.104 ppm, carps will be dead in 96 hours [13].

Kadmium concentration in ballast water of the vessel researched is from 0,001 to 0,46 mg l^{-1}. Passengger or Ro-Ro vessels studied are S, B, L, K1, and DF2. All of them still follow the quality standard, except K1 vessel. She exceeds the quality standard because there is only little water volume in the ballast water tank of the vessel. The vessel with Cd concentration of 0,003 to 0,057 mg l^{-1} are JM, SB, A, MP, OK, PI and IZ. Among other vessels that are from the same port, C8 and S also have the highest Cd content. TS XXII and GW tankers have Cd content of 0,003-0,007 mg l^{-1}. In addition, PA and SB vessels have Cd content of 0,003-0,006 mg l^{-1}. The Cd content in those vessels still fulfil the a standard required, except for C1 that is from Pontianak. The highest Cd content level in this research is in GNA on shipping dockyard because its ballast water is in minimum condition. NG vessel has the highest Cd content, it from Merak-polluted waters (Figure 3).

Zinc is plenty in nature. The level of Zinc in earch crust is about 70 mg kg^{-1}. The level of Zinc in waters is about 0.01 mg l^{-1}.Zinc is an essential source for living things. It can help enzime to work properly, it becomes an agent in transferring hydrogen in plant's photosynthesis, it has a part in protein fromation. Zinc is widely used in industries of iron and steel, paint, rubber, textile, paper and pulp [13].

The concentration of Zinc dissolved in ballast water of commercial vessels is 0.001 to 2.464 mg l^{-1}. DF2, it is a passenger vessel with the highest concentration of Zinc as well as it has content that of 21 times higher than quality standard, because there is only little volume that can be examined. OK has the highest content of Zn amomg all fertilizer carriers from same port. For container vessels and tankers, Zn content in SB and TS XXII exceeds the quality standard, about 0.56 to 0.65 mg l^{-1}. The content of high Zinc in ballast water of the vessels can be caused by age of the vessels. The passenger vessels, B and L are 21 years old. The fertilizer carriers, A and S, are about 29 to 30 years old, while the other vessels are about 10 years old. The second possibility is the origin port, has the highest concentration i.e. NG is from Merak waters (Figure 4).

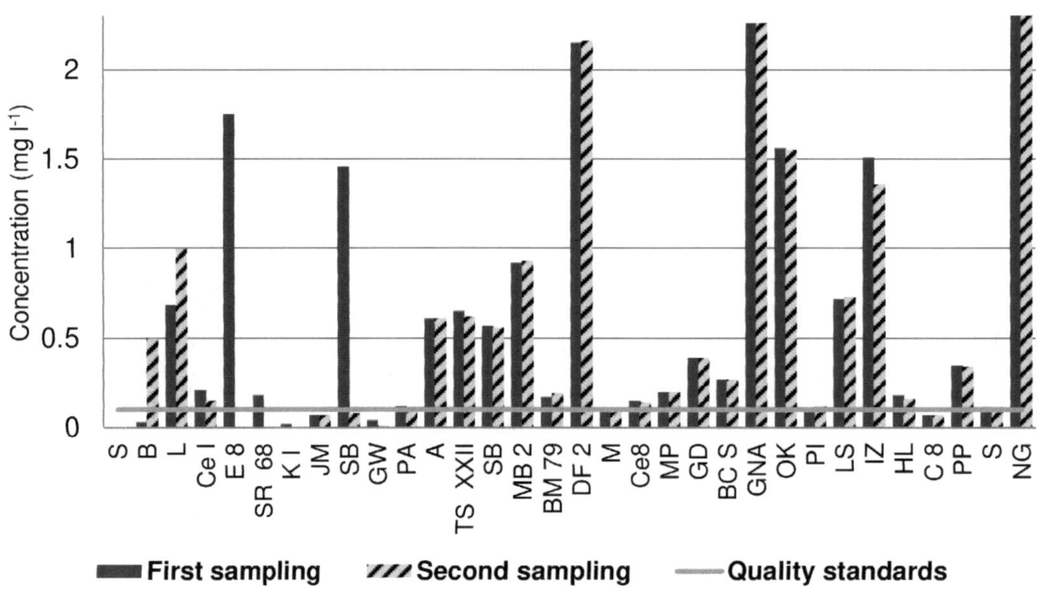

Figure 4. Zn concentration in ballast water of commercial vessels

The study shows the mean concentration of heavy metals of Cd and Zn respectively in a ballast tanks of commercial vessels are 0.04894 ± 0.07874 mg l^{-1} and 0.59122 ± 0.70373 mg l^{-1}. It shows that mean content of Cadmium from the ballast water of commercial ships is 49 times higher than the Bay of Jakarta and Cisadane estuaries, and 16 times higher than the Bangka waters. The mean content of Zinc in the ballast water of commercial ships showed 84 times higher than the Bay of Jakarta, 197 times higher than Cisadane estuaries and 147 times higher than the Bangka waters (Table 2).

Heavy metal pollution in the middle of Jakarta Bay in June and September 2003 was dominated by Zinc metal in order of Zn>Cu>Cd. Tanjung Priok port area becomes a place for vessels to dump their ballast water, industry and domestic wastes [14] (Table 3). The sample of ballast water from Tanjung Priok port is collected from A, GNA, M and MB2 vessels. The ballast water was dominated by Zinc with the average level of 0.9725 mg l^{-1}. It has exceeded the quality standard of the port 0.1 mg l^{-1} [17].

Table 2. Comparison of Research on Commercial Vessels and Other Research

Vessels/Waters	Metal concentration (mgl^{-1})		References
	Cd	Zn	
Vessels	0.001-0.46	0.001-2.464	Tjahjono, 2016*
Jakarta bay	< 0.001	< 0.001-0.007	Lestari & Edward, 2004
Cisadane estuaries	< 0.001-0.001	< 0.001-0.003	Rochyatun *et. al.*, 2006
Bangka waters	< 0.0001-0.003	0.001-0.004	Arifin, 2011

Table 3. Comparison of heavy metals (Zn and Cd) (mg l^{-1}) in sediments in the Bay of Jakarta and commercial vessels from Jakarta

Parameter	Research location (mg l^{-1})	
	Middle of Jakarta bay	Vessels from Jakarta
Zn	71.13-230.54	NA
Cd	0.001-0.28	0.005-0.282

Note : NA (Not Available)

4.Conclusion

The result of the study shows that the Cd content of ballast water vessels is a about 0.001-0.46 mg l^{-1}, and Zn is 0.001-2.464 mg l^{-1}. Heavy metal content of ballast water vessels are in order Zn>Cd. It happens because the vessels are quite old. It is also influenced by sea water of origin port. The content of heavy metal in the ballast water vessels shows that the vessels have not conducted ballast water exchange before entering to Tanjung PTES. The more frequency vessels entry and more dead weight, more dangerous the ballast water discharged into the port waters. It will put the marine environment around the port in danger.

Acknowledgement. The authors want to show his deepest gratitude to those who have helped in this research. They are the Chief of the Harbor Master and Port Authority of PTES, the ship's officers who give permission to the technicians on board, and the staffs of Regional Health Laboratories Semarang in analyzing the samples.

References

[1] International Maritime Organization 2009 *Ballast Water Management Convention and the Guideliness for Its Implementation 2009 Edition* (London: International Maritime Organization)

[2] van Dokkum K 2005 *Ship Knowledge Covering Ship Design, Construction and Operation : 2nd edition* (Dokmar : Enkhuizen)

[3] Zhang F and Dickman M 1999 Mid-Ocean Exchange of Container Vessel Ballast Water 1 : Seasonal Factors Affecting The Transport Of Harmful Diatoms And Dinoflagellates *J Marine Ecology Progress Series* **176** (243-251)

[4] Hartoko A 2013 *Oceanographic Characters and Plankton Resources of Indonesia* (Yogyakarta : Graha Ilmu)

[5] Rositasari R L 2013 Evaluasi Lingkungan Perairan Pesisir Semarang *Jurnal Ilmu dan Teknologi Kelautan Tropis* **5** (1), pp. 112-121.

[6] Jawatan Hidro-Oseanografi 1983 *Informasi Pelabuhan Indonesia* (Jakarta : Jawatan Hidro-Oseanografi)

[7] Butron A, Orive E and Madariaga I 2011 Potential Risk of Harmful Algae Transport by Ballast Waters : The Case of Bilbao Harbour *Journal Marine Pollution Bulletin* **62** (747-75)

[8] Garret M J, Wolny J L, Williams B J, Dirks M D, Brame J A, Richardson R W 2011 Methods for Sampling and Analysis of Marine Microalage In Ship Ballast Tanks : A Case Study from Tampa Bay, Florida, USA *Journal Algae* **26** (2),181-192

[9] Badan Standardisasi Nasional 2005 *Cara Uji Kadmium dengan Spektrofometri Serapan Atom (SSA)-Nyala* SNI 06-6989.51

[10] Badan Standardisasi Nasional 2005 *Cara uji Seng dengan Spektrofometri Serapan Atom (SSA)-Nyala* SNI 06-6989.51

[11] Effendi H 2006 *Telaah Kualitas Air Bagi Pengelolaan Sumber Daya dan Lingkungan Perairan* (Yogyakarta : Penerbit Kanisius)

[12] Rhidowati S 2013 *Mengenal Pencemaran Logam* (Yogyakarta : Graha Ilmu)

[13] Palar H 2008 *Pencemaran dan Toksilogi Logam Berat* (Jakarta : Rineka Cipta)

[14] Rochyatun E dan Rozak A 2007 Pemantauan Kadar Logam Berat dalam Sedimen di Perairan Teluk Jakarta *Jurnal Makara Sains* **11**(1), 28-36

[15] Lestari dan Edward 2004 Dampak Pencemaran Logam Berat Terhadap Kualitas Air Laut dan Sumber Daya Perikanan (Studi Kasus Kematian Massal Ikan-Ikan di Teluk Jakarta) *Jurnal Makara Sains* **8** (2), 52-58

[16] Arifin Z 2011 Konsentrasi Logam Berat di Air, Sedimen dan Biota di Teluk Kelabat, Pulau Bangka *Jurnal Ilmu dan Teknologi Kelautan Tropis* **3**(1), 104-114

[17] Republik Indonesia 2014 *Keputusan Menteri Negara Lingkungan Hidup Nomor 51 Tahun 2004 tentang Baku Mutu Air Laut*

Glossary and Abbreviation

1. Portside : the left-hand side of or direction from a vessel, facing forward
2. Starboard : right-hand side of or direction from a vessel, facing forward
3. GM : measurement of the initial static stability of a floating body. It is calculated as the distance between the centre of gravity of a ship (G) and its meta center (M). A larger meta centric height implies greater initial stability against overturning.
4. Listing : the degree to which a vessel heels (leans or tilts) to either port or starboard. A listing vessel is stable and at equilibrium, but the distribution of weight aboard (often caused by uneven loading or flooding) causes it to heel to one side.
5. LPG : Liquid Petroleum Gas
6. Ro-Ro vessel : Roll-on/roll-off vessel, vessels designed to carry wheeled cargo, such as cars, trucks, semi-trailer trucks, trailers, and railroad cars, that are driven on and off the ship on their own wheels or using a platform vehicle, such as a self-propelled modular transporter.
7. Even-keel : balance condition of vessel with the same between forward (bow) and after (stern) draft.

8.	Draft	: the vertical distance between the waterline and the bottom of the hull (keel), with the thickness of the hull included; in the case of not being included the draft outline would be obtained. Draft determines the minimum depth of water a ship can safely navigate.
9.	Dead weight	: the difference between the displacement and the lightweight at any givin draft, measured in ton. Deadweight is the weight of cargo, fuel, stores, etc. That a vessel can carry.
10.	DWT	: Dead Weight Ton (ton)
11.	TSS	: Total Suspended Solid (mg l^{-1})
12.	DO	: Dissolved Oxigen (mg l^{-1})
13.	MT	: Metric Ton

Sources of Glossary and Abbreviation

1. Taylor D A 1980 *Merchant Ship Construction Fourth Edition* (London: The Institute of Marine Engineering, pp. 262-268)
2. van Dokkum K 2005 Ship *Knowledge Covering Ship Design, Construction and Operation : 2nd Edition* (Enkhuizen : Dokmar, , pp. 356-366)
3. Istopo 1999 Kamus Istilah Pelayaran & Ensiklopedi Maritim (Jakarta:Yayasan C.A.A.I.P/P.L.A.P, pp. 1-294)

Proteomics study of extracellular fibrinolytic proteases from *Bacillus licheniformis* RO3 and *Bacillus pumilus* 2.g isolated from Indonesian fermented food

Diana Nur Afifah[1], Ninik Rustanti[1], Gemala Anjani[1], Dahrul Syah[2], Yanti[3], Maggy T. Suhartono[2]

[1]Department of Nutrition Science, Faculty of Medicine, Diponegoro University, Indonesia.
[2]Department of Food Science and Technology, Faculty of Agricultural Technology Bogor Agricultural University, Indonesia.
[3]Faculty of Biotechnology, Atma Jaya Catholic University, Indonesia.

Email: d.nurafifah.dna@fk.undip.ac.id

Abstract. This paper presents the proteomics study which includes separation, identification and characterization of proteins. The experiment on Indonesian fermented food such as extracellular fibrinolytic protease from *Bacillus licheniformis* RO3 and *Bacillus pumilus* 2.g isolated from *red oncom* and *tempeh gembus* was conducted. The experimental works comprise the following steps: (1) a combination of one- and two-dimensional electrophoresis analysis, (2) mass spectrometry analysis using MALDI-TOF-MS and (3) investigation using protein database. The result suggested that there were new two protein fractions of *B. licheniformis* RO3 and three protein fractions of *B. pumilus* 2.g. These result has not been previously reported.

1. Introduction

Proteomics is defined as a comprehensive protein analysis to a particular cell or organism. This study includes separation, identification and characterization of proteins. One of the most effective methods to separate the proteins contained in the complex mixture is by using two-dimensional (2D) gel electrophoresis [1]. This method is employed to separate proteins with two following steps (1) isoelectric point (pI) in the first dimension (IEF, *isoelectric focusing*) and (2) molecular weight (MW) in the second dimension (SDS-PAGE, *sodium dedocyl sulfate polyacrilamide gel*)[2].

2D electrophoresis have ability to separate a large number of proteins, including post-translational modifications. The 2D electrophoresis caused the changes in the molecular charge or weight and unique forms of proteolysis-resulted proteins [3,4]. This technique employed *Immobilized pH Gradient* (IPG) strips made from copolymerization of acids and bases derivatives with different concentrations of acrylamide in polyacrylamide matrix [5].

2D electrophoresis is used to separate the complex protein into single polypeptide component which is the major analysis in proteomics study. To date, 2D electrophoresis technique is the only method for separating thousands of proteins in a single separation stage. Few samples for the analysis are used such as biofluid, tissues, cells, and organelles. A comparison analysis of sample pairs , for example comparing normal cells with cells experiencing transformation or comparing multiple cells at different growth stages is presented by [2]. This technique can be used for proteomic study of cancer and a variety specific gene expression analysis from mixed community of prokaryotic microorganisms in the environment [6]. The combination of the method and mass spectrometry analysis of *Matrix Assisted Laser Desorption Ionization Time-of-Flight* (MALDI-TOF), two extracellular serine protease from *Bacillus subtilis* 168, namely WprA (52 kDa) and Vpr (68 kDa) can be detected. These proteins havefibrinolytic activity [7].

Bacillus subtilis produces several extracellular p0roteases in the late exponential growth phase [8]. The main extracellular proteolytic enzymes are subtilysin and neutral protease which are coded as apr and npr gene [9]. Other extracellular proteases are *bacillopeptidase* F, EPR, and Mpr. *Bacillopeptidase* F [10] and protein Epr, which are coded as bpr and epr genes [11], both are serine proteases, while Mpr, coded as mpr gene, include metalloprotease [12].

In this study, proteomics study on extracellular enzymes of *Bacillus licheniformis* RO3 and *Bacillus pumilus* 2.g isolated from Indonesian fermented food, red oncom and tempeh gembus was conducted.

2. Materials and Methods

2.1 Microorganisms

Bacillus licheniformis RO3 (AB968524) was isolated from fermented food of red oncom while *B. pumilus* 2.g (AB968523) was isolated from tempeh gembus [13].

2.2 Preparation of samples

2.2.1 B. licheniformis RO3

B. licheniformis RO3 was grown in ½ *Luria-bertani broth* (LB) + 1% oncom flour (w/v), hereinafter called LBO, for 0, 12, 24, 36, 48, 60, and 72 hours on 120 rpm shake incubator at 37°C. Extracellular enzyme was obtained by 6000 g centrifugation for 15 min at 4°C. The enzymes was then freeze-dried to increase the protein concentration. The protein concentration in the sample was calculated by the method of Bradford [14]. Protease activity was calculated with the method of Bergmeyer et al. [15] with casein substrate. Before conducting mass spectrometry analysis using MALDI-TOF-MS, the protein fractions of extracellular enzyme of this *B. licheniformis* RO3 were analyzed using one-dimensional electrophoresis (SDS-PAGE) and two-dimensional (2D) electrophoresis.

2.2.2 *B. pumilus* 2.g

B. pumilus 2.g was grown in a medium (nutrient broth) (NB) for 72 hours at 120rpm shake incubator at 37°C. Extracellular enzyme was obtained by 12,000 g centrifugation for 10 min at 4°C. Enzymes were then purified with 80% ammonium sulfate precipitation, ion exchanger chromatography with *CM-Sephadex* C-50 (Amersham Pharmacia Biotech, Uppsala, Sweden), and hydrophobic interaction chromatography with *Phenyl Sepharose 6 Fast Flow* (Amersham Pharmacia Biotech, Uppsala, Sweden). At every stage, dialysis was conducted. Pure enzyme was then freeze-dired and ready for further analysis. The protein concentration in the sample was calculated by the method of Bradford [14]. Fibrinolytic activity was calculated by the method of Jeong et al.[16] with fibrin substrate. Before conducting mass spectrometry analysis using MALDI-TOF MS, the protein fractions of pure enzyme were only analyzed with one-dimensional electrophoresis (SDS-PAGE).

2.3 One-dimensional electrophoresis (SDS-PAGE) and zimogram

Sodium dodecyl sulfate-polyacrylamide gel electrophoresis (SDS-PAGE) was conducted with separator gel with concentration of 12% and barrier gel of 4% [17]. Zimogram was conducted with separator gel with concentration of 12 containing 0.02% fibrinogen or fibrin.

2.4 Two-dimensional (2D) electrophoresis

Electrophoresis method conducted in this study was performed according to *Protean* IEF *Cell Instruction Manual* from BioRad.

Rehydration of Immobilized pH gradient (IPG) strips. A total of 50 µl samples with protein concentrations up to 1 mg mixed with 150 mL of rehydration buffer (8 M urea, 2% CHAPS, 0.2% Bio-Lytes, and 50 mM DTT). The sample mixture was inserted into afocusing tray . Protected surface of IPG strip sheet was opened with tweezers and then inserted into the sample mixture solution, location of positive and negative poles in IPG strip and focusing tray were adjusted, and the surface of IPG strip was facing down. The position of the IPG strip on the *focusing tray* must be balanced. IPG strip wasthen coated with mineral oil as much as 2.5 ml. Furthermore, wicks paper soaked with ddH_2O was pasted on the electrode surface of focusing tray. After ascertaining that there was no bubbles, focusing tray was sealed and IPG strips were incubated for 16 hours in the *Protean* IEF *cell.*

Isoelectric Focusing (IEF). After 16 hours, *running* IEF was performed with the following settings: Voltage I: 250 V, 30 minutes; Voltage II: 4000 V, 1 hour; Voltage III: 4000 V, 3 hours at 10°C. After *running* IEF for 4.5 hours, IPG strips was removed and preparation before the two-dimensional electrophoresis was done.

Prior to one-dimensional electrophoresis (SDS-PAGE) IPG strips were incubated for 15 minutes in 2.5 ml equilibration buffer I with composition as follow: 6 M urea, 0.375 M Tris-HCl, 2% SDS, 20% glycerol, 2% DTT).The incubation process was followed by second incubation process for 15 minutes in 2.5 ml of equilibration buffer equilibrasi II with composition as follow: 6 M urea, 0.375 M Tris-HCl, 2% SDS, 20% glycerol, 2.5% iodoacetamida. During the incubation process in the equilibration buffer, polyacrylamide separator gel was prepared initially for SDS-PAGE without using barrier gel. Furthermore, equilibration buffer was discarded and IPG strip was ready to be running on a second dimension with SDS-PAGE electrophoresis at a voltage of 100 volts for 2 hours in electrophoresis buffer. After electrophoresis, staining with *Coomassie Brilliant Blue* R-250 dye solution was conducted.

2.5 Spectrometric analysis of MALDI-TOF-MS

Digestion from proteins point resulted from 2D Electrophoresis or protein ribbons resulted from SDS-PAGE in gel was performed using the method of Lee et.al. [18]. All mass spectrometry analyzes were performed using MALDI-TOF-MS Voyager Biospectrometry Workstation (PE Biosystems, Foster City, CA, USA). Spectra was analyzed with data Explorer software (PE Biosystems) and compared to databases fromNational Center of Biotechnology Information (NCBI) using *MASCOT peptide mass fingerprinting* program (http://www.matrixscience.com). Protein score more than 74 was stated as identical to particular protein ($p < 0.05$).

2.6 Estimation of molecular weight (MW) and isoelectric point (pI)

The molecular weight of protein fraction was calculated based on the standard curve equation of protein molecule weight, the protein fraction isoelectric was calculated based on the standard curve equation of isoelectric point. The calculation results of MW and pI were used to verifythe results of MALDI-TOF-MS, i.e. by entering the calculation result of MW and pI into ExPASy-TagIdent program (http://web.expasy.org/tagident/).

3. Results and Discussion

The result of one-dimensional electrophoresis study (SDS-PAGE) of extracellular protease from *B. licheniformis* RO3 with LBO production media is depicted in Figure 1. The result shows that the protein fraction ribbon was not visible in 0 hour fermentation.Some ribbons began visible in the 12th hour fermentation. The significant appearance of the protein ribbons was observed during the 24th hour fermentation. However, a number of protein fraction ribbons was stable during the 36th to 72nd hour fermentation..

IOP Conf. Series: Earth and Environmental Science 55 (2017) 012025 doi:10.1088/1755-1315/55/1/012025

Critical parameter that determines the success of proteomics study is the ability to obtain a single protein in a complex mixture.

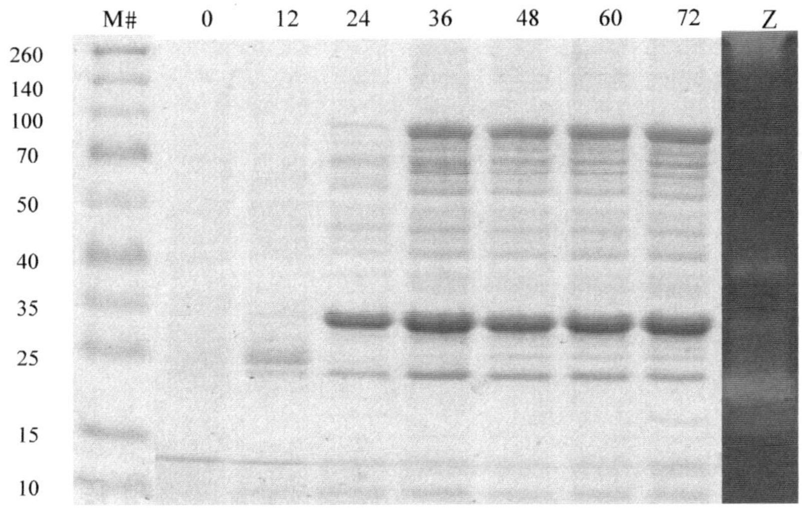

Figure 1. The results of one-dimensional electrophoresis (SDS-PAGE) of extracellular protease from *B. licheniformis* RO3 with LBO media. M: Fermentation marker, Z: zimogram results with fibrinogen substrate

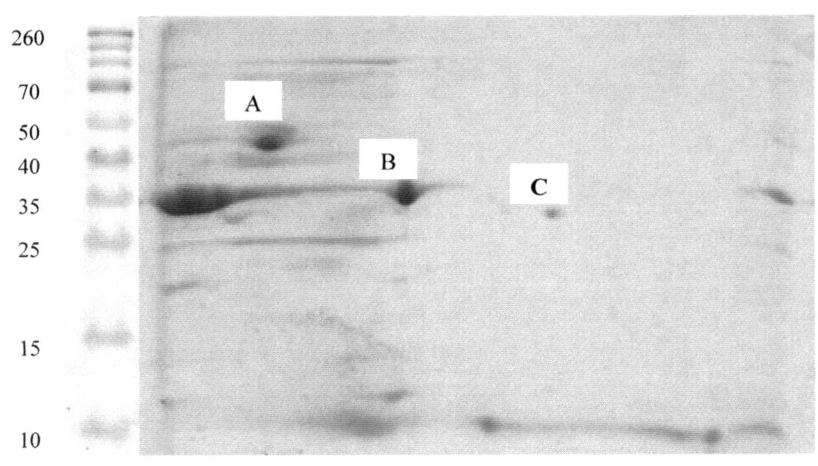

Figure 2. The results of two-dimensional electrophoresis of *B. licheniformis* RO3 with LBO media

One of the most effective ways to separate proteins in a complex mixtures is by using two-dimensional electrophoresis. In this study, two-dimensional electrophoresis was performed on extracellular protease which was fermented in LBO media for 48 hours, which was when they achieved the highest enzyme activity, 0.283 U/mg and several fibrinogenolitik extracellular protease enzymes were formed (Figure 1). Identification using MALDI-TOF resulted from two-dimensional electrophoresis was applied to three protein points: A, B, and C (Figure 2).

The results of MALDI-TOF-MS, followed by the identification of proteins showed fromthe three points that were tested, only one point which had protein score of more than 74, i.e point B with protein score of 120 and it was identical to *flagellin* of *Bacillus licheniformis* DSM 13. *Flagellin* is globular protein that regulates itself in the hollow cylinder to form filaments in bacterial flagellum. *Flagellin* has molecular weight of 30,000 to 60,000 dalton. Flagellin is a major substituent of bacterial flagella, and presents in large numbers in almost all flagelled-bacteria [19].

Table 1. Comparison of m/z protein of protease *B. licheniformis* RO3 with database

Code	MW(kDa)	pI	Similarity				
			Accsess Code	Protei n Score	Name	MW (kDa)	pI
A	48	4.80	A4IQG0	31	*Superoxide dismutase* [*Geobacillus thermodenitrificans* NG80-2] YP_001126309.1 [a]	52	4.96
			G4NXJ1	29	*Response regulator aspartate phosphatase* F [*Bacillus subtilis* subsp. *spizizenii* TU-B-10] YP_004879327.1 [a]	46	5.34
			Q65JX1	27	*Bacillopeptidase* F [*Bacillus licheniformis* DSM 13 = ATCC 14580] YP_006713127.1 [a]	155	5.18
B	37	5.38	Q65EB8	120	*Flagellin* [*Bacillus licheniformis* DSM 13 = ATCC 14580] YP_006715058.1 [a]	33	5.41
C	33	5.99	C8WV29	43	*Short-chain dehydrogenase/reductase* SDR [*Alicyclobacillus acidocaldarius* subsp. *acidocaldarius* DSM 446] YP_003184407.1 [a]	29	6.02
			L0BRU6	29	*Hypothetical protein* B938_15200 [*Bacillus amyloliquefaciens* subsp. *plantarum* AS43.3] YP_007187718.1 [a]	13	5.08

[a] NCBI *Reference Sequence*

Points A and C which are resulted from 2D electrophoresis have certain similarities with particular protein, with protein score of less than 74. Point A, hereinafter referred to as fraction A is possibly *bacillopeptidase* DSM F because it has similarities with bacillopeptidase F of *B. licheniformis* DSM 13 with protein score of 27. Based on the result of molecular weight (MW) and pI with standard curves, fraction A has MW of 48 kDa and pI of 4.80. While bacillopeptidase F from *B. licheniformis* DSM 13 has MW of 155 kDa and pI of 5.18.

Bacillopeptidase F was first discovered by Roitsch & Hageman [10] which was serine endopeptidase excreted by *Bacillus subtilis* 168 after the end of exponential growth. The enzyme is active at pH 10 and inhibited by PMSF inhibitor. There were two bacillopeptidase F enzymes which were successfully isolated, one of them hasmolecular weight of 33 kDa, and the other had molecular weight of 50 kDa. Both forms of enzyme is glycoprotein, which is smaller and having pI of 4.4, while the larger has pI of 5.4. Bacillopeptidase F successfully isolated by Roitsch & Hageman [10] evidentlydifferent from all the other proteases from *B. subtilis. Bacillopeptidase* F of *B. licheniformis* KJ-31, because it was fibrinolytic as identified by Hwang et al. [20] and it hasMW of 37 kDa. *Bacillopeptidase* F of B. *subtilis natto* that was fibrinolytic also has been identified by Hitosugi et al. [21] and it hasMW of 34.1 kDa.

Point C, hereinafter referred to as fraction C is probably uncharacterized proteins because it is similar to *hypothetical protein* from *B. amyloliquefaciens* subs. *plantarum* AS43.3 with protein score of 29. Fraction C has MW of 33 kDa and pI of 5.99.

The result of identification using MALDI-TOF-MS also verifiedby comparing the result of the molecular weight (MW) and pI calculations using ExPASy-TagIdent program (http://web.expasy.org/tagident/).

It turns out that fractions A and C do not have similarities to proteins that exist in UniProtKB/Swiss-Prot database. These results indicate that both protein fractions in extracellular protease of *B. licheniformis* RO3 is new or different from those already reported.

In one-dimensional electrophoresis study (SDS-PAGE) of extracellular protease from *B. pumilus* RO3 with NB production medio, some protein fractions characterized by the presence of several ribbons were visible. However, only three factions that were continued to be identified. The measured fibrinolytic protease activity of this enzyme is 1,442.3 U/mg.

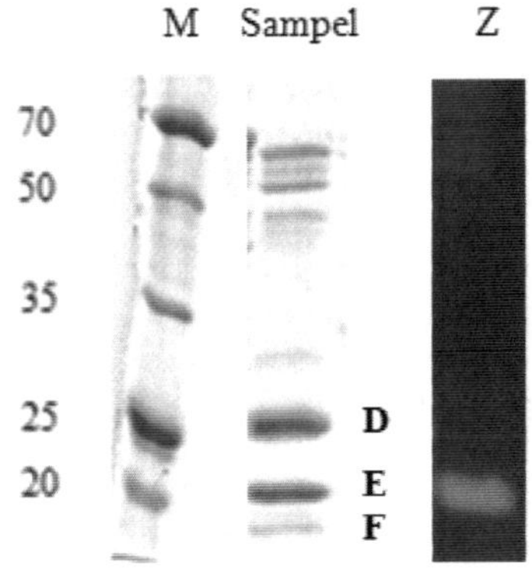

Figure 3. Extracellular fibrinolytic protease of *B. pumilus* 2.g
M: marker, Z: zimogram with fibrin substrate

The result of identification showed that of the three ribbons that were tested, none of theprotein score more than 74. Thus, none of the ribbon has similarities to the existing proteins database. This is inline with the characterization results of fibrinolytic enzyme produced by *B. pumilus* 2.g [22]. Fibrinolytic enzyme is a serine protease but has MW of 20 kDa. Generally, protease serine having fibrinolytic nature has MW of 27 kDa [23,24]

A number of researchers has presenteda genomic study to *Bacillus subtilis*. Moreover, the researchers has identified 11 different extracellular proteases [25]. Study conducted by Park et.al. [7] was able to detect two extracellular serine protease having fibrinolytic activity of *Bacillus subtilis* 168, i.e. WprA (52 kDa) and VPR (68 kDa).

In this study, one protein fractions namely flagellin was identified. However flagellin is not extracellular protease. Thus, the other potential protein to be tested are new protein fractions which

IOP Conf. Series: Earth and Environmental Science **55** (2017) 012025 doi:10.1088/1755-1315/55/1/012025

have not been reported and have fibrinolytic protease activity. This new protein fraction may be *bacillopeptidase F* and other extracellular proteases.

Table 2. Comparison of m/z protein of protease from *B. pumilus* 2.g with database

Code	MW (kDa)	Similarities				
		Protein score	Access Code	Name	MW(kDa)	pI
D	23	52	C8WXL1	*Septum site-determining protein* MinD [*Alicyclobacillus acidocaldarius* subsp. *acidocaldarius* DSM 446] YP_003185221.1 [a]	29	6.02
		42	M4HMH2	*Hypothetical protein* BCK_26788 [*Bacillus cereus* FRI-35] YP_006599369.1 [a]	21	8.80
		41	F8IDH7	*Septum site-determining protein* MinD [*Alicyclobacillus acidocaldarius* subsp. *acidocaldarius* Tc-4-1] YP_005518349.1 [a]	29	6.85
		40	G4NRL1	*Hypothetical protein* GYO_1184 [*Bacillus subtilis* subsp. *spizizenii* TU-B-10] YP_004876473.1 [a]	25	7.58
E	20	49	Q65117	*Zn-finger protein* [*Bacillus licheniformis* DSM 13 = ATCC 14580] YP_079408.1 [a]	19	4.83
		43	K0G197	CTP *synthase* [*Bacillus thuringiensis* MC28] YP_006831249.1 [a]	47	6.24
F	15	25	F2F9E3	*Permease* [*Solibacillus silvestris* StLB046] YP_006462816.1 [a]	94	9.48

[a] NCBI *Reference Sequence*

It has been mentioned earlierthat the most important thing of proteomics is in the separation process of complex proteins into a single protein fraction. Referring to the further study, it is needed to ensure the existence of some new protein fraction, especially on the stage of protein separation.

4. Conclusion

An experimental study of extracellular fibrinoltic proteases proteomics from *Bacillus licheniformis* RO3 and *Bacillus pumilus* 2.g isolated from Indonesian fermented food has been presented. The *Bacillus licheniformis* RO3 and *B. pumilus* 2.g can produce several extracellular proteases. The result of identification using MALDI-TOF-MS shows that there are five new protein fractions contained in extracellular fibrinolytic protease produced by *Bacillus licheniformis* RO3 (2 fractions) and *B. pumilus* 2.g (3 fractions).

5. References

[1] Philips CI, Bogyo M. 2005. Micro review proteomic meets microbiology technical advance in the global mapping of protein expression and function. *Cell microb* **8**:1061-1076.

[2] Gravel P. 2002. Two-Dimensional Polyacrylamide Gel Electrophoresis of Proteins Using Carrier Ampholyte pH Gradients in the First Dimension. Di dalam: Walker JM, editor. *The Protein Protocols Handbook.* Ed ke-2. Totowa, NJ: Humana Press. p 163-168.

[3] Anderson NL, Anderson NG. 1998. Proteome and proteomics: new technologies, new concepts, and new words. *Electrophoresis* **19**:1853–1861.

[4] Cordwell SJ, Nouwens AS, Walsh BJ. 2001. Comparative proteomics of bacterial pathogens. *Proteomics* **1**:461–472.

[5] Gianazza E. 2002. Casting immobilized pH gradients (IPGs). In: Walker JM, editor. *The Protein Protocols Handbook.* Ed ke-2. Totowa, NJ: Humana Press. p 169-180.

[6] Wilmes P, Bond PL. 2004. The application of two-dimensional polyacrylamide gel electrophoresis and downstream analyses to a mixed community of prokaryotic microorganisms. *Env Mikrobiol* **6**:911-920.

[7] Park SG, Kho CW, Cho S, Lee DH, Kim SH, Park BC. 2002. A functional proteomic analysis of secreted fibrinolytic enzymes from *Bacillus subtilis* 168 using a combined method of two-dimensional gel electrophoresis and zymography. *Proteomics* **2**:206-211.

[8] Priest FG. 1997. Extracellular enzyme synthesis in the genus *Bacillus. Bacteriol. Rev.* **41**: 711–753.

[9] Kawamura F, Doi RH. 1984. Construction of a *Bacillus subtilis* double mutant deficient in extracellular alkaline and neutral proteases. *J. Bacteriol.* **160**: 442–444.

[10] Roitsch CA, Hageman JH. 1983. Bacillopeptidase F: two forms of a glycoprotein serine protease from *Bacillus subtilis 168. J. Bacteriol.* **155**: 145–152.

[11] Sloma A, Ally A, Ally D, Pero J. 1988. Gene encoding a minor extracellular protease in *Bacillus subtilis. J. Bacteriol.* **170**: 5556–5557.

[12] Rufo GA, Sullivan BJ, Sloma A, Pero J. 1990. Isolation and characterization of a novel extracellular metalloprotease from *Bacillus subtilis. J. Bacteriol.* **172**: 1019–1023.

[13] Afifah DN, Sulchan M, Syah D, Yanti, Suhartono MT. 2014. Isolation and identification of fibrinolytic protease-producing microorganisms from *Red Oncom* and *Gembus*, Indonesian fermented soybean cakes. *Malaysian Journal of Microbiology* **10**(4): 273-279.

[14] Bradford MM. 1976. A rapid and sensitive method for quantification of microgram quantities of protein utilizing the principles of protein dye-binding. *Anal Biochem* **72**:234-254.

[15] Bergmeyer HU, Grassl M. 1983. *Methods of enzymatic analysis.* Vol 2. Weinheim: Verlag Chemie. p 1007-1009

[16] Jeong SJ, Kwon GH, Chun J, Kim JS, Park CS, Kwon DY, Kim JH. 2007. Cloning of fibrinolytic enzyme gene from *Bacillus subtilis* isolated from *Cheonggukjang* and its expression in protease-deficient *Bacillus subtilis* Strains. *J. Microbiol. Biotechnol.* **17**: 1018–1023.

[17] Laemmli UK. 1970. Cleavage of structural proteins during the assembly of the head of bacteriophage T4. *Nature* **227**: 680-685.

[18] Lee SG, Lee KW, Park TH, Park JY, Han NS, Kim JH. 2012. Proteomic analysis of proteins increased or reduced by ethanol of *Lactobacillus plantarum* ST4 isolated from *Makgeolli*, traditional Korean rice wine. *J. Microbiol. Biotechnol.* **22**: 516-525.

[19] Veith B, Herzberg C, Steckel S, Feesche J, Maurer KH, Ehrenreich P, Bäumer S, Henne A, Liesegang H, Merkl R, Ehrenreich A, Gottschalk G. 2004. The complete genome sequence of *Bacillus licheniformis* DSM13, an organism with great industrial potential. *J. Mol. Microbiol. Biotechnol.* **7**(4): 204-11.

[20] Hwang KJ, Choi KH, Kim MJ, Park CS, Cha J. 2007. Purification and characterization of a new fibrinolytic enzyme of Bacillus licheniformis KJ-31, Isolated from Korean traditional Jeot-gal. *Journal of Microbiology and Biotechnology* **9**:1469–1476.

[21] Hitosugi M, Ikeda M, Zhu X, Kato H, Omura K, Nagai T, Tokudome S. 2007. Anticoagulant and fibrinolytic effects of functional food materials produced by *Bacillus subtilis natto. Journal of Japanese Society of Biorheology* **21**(1): 35-40.

[22] Afifah DN, Sulchan M, Syah D, Yanti, Suhartono MT. Kim JH. 2014. Purification and characterization of a fibrinolytic enzyme from *Bacillus pumilus* 2.g isolated from *Tempeh Gembus*, an Indonesian fermented food. *Preventive Nutrition and Food Science* **19**(3):213-219.

[23] Jo HD, Lee HA, Jeong SJ, Kim JH. 2011. Purification and characterization of a major fibrinolytic enzyme from *Bacillus amyloliquefaciens* MJ5-41 isolated from *Meju. J Microbiol Biotechnol.* **21**: 1166–1173.

[24] Jo HD, Kwon GH, Park JY, Cha J, Song YS, Kim JH. 2011. Cloning and overexpression of aprE3-17 encoding the major fibrinolytic protease of *Bacillus licheniformis* CH3-17. *Biotechnol Bioprocess Eng.* **16**: 352-359.

[25] Kunst F, Ogasawara N, Moszer I, *et al.* 1997. The complete genome sequence of the gram-positive bacterium *Bacillus subtilis. Nature* **390**: 249–256

Screening Antibacterial Agent from Crude Extract of Marine-Derived Fungi Associated with Soft Corals against MDR-*Staphylococcus haemolyticus*

A Sabdaningsih[1], O Cristianawati[1], M T Sibero[1], H Nuryadi[2], O K Radjasa[3, 4], A Sabdono[4], and A Trianto[4]

[1]Coastal Resources Management, Faculty of Fisheries and Marine Sciences, Diponegoro University, 50275 Tembalang, Semarang, Indonesia
[2] Tropical Marine Biotechnology Laboratory, Central Laboratory of Research and Services, Diponegoro University, 50275 Tembalang, Semarang, Indonesia
[3]Ministry of Research, Technology and Higher Education of the Republic of Indonesia, 10270
[4]Department of Marine Sciences, Faculty of Fisheries and Marine Sciences, Diponegoro University, 50275 Tembalang, Semarang, Indonesia

E-mail: aninditia@student.undip.ac.id

Abstract. Multidrug resistant *Staphylococcus haemolyticus* is a Gram-positive bacteria and member of coagulase negative staphylococci (CoNS) which has the highest level of antimicrobial resistance. This nosocomial pathogen due to skin or soft tissue infections, bacteremia, septicemia, peritonitis, otitis media, meningitis and urinary tract infections. The ability to produce enterotoxins, hemolysins, biofilm, and cytotoxins could be an important characteristic for the successful of infection. Marine-derived fungi have potency as a continuous supply of bioactive compound. The aim of this research was screening antibacterial agent from crude extracts of marine-derived fungi associated with soft corals against MDR-*S. haemolyticus*. Among 23 isolates of marine-derived fungi isolated from 7 soft corals, there were 4 isolates active against MDR-*S. haemolyticus*. The screening was conducted by using *agar plug diffusion* method. Isolate PPSC-27-A4 had the highest antibacterial activity with diameter 23±9,6 mm. The crude extract from this isolate had been confirmed to antibacterial susceptibility test and it had the highest antibacterial activity in 12.2 mm with concentration of 300μg/ml from mycelia extract. PPSC-27-A4 had been characterized in molecular, based on the sequence analysis of 18S rRNA, PPSC-27-A4 isolate was identified as *Trichoderma longibrachiatum*.

Keywords: antibacterial, associated marine-derived fungi, Multidrug resistant, nosocomial pathogen, soft coral, *Staphylococcus haemolyticus*

1. Introduction
Staphylococcus haemolyticus infection is commonly occurred in the hospital as a nosocomial infection and related to implanted medical devices [1][2]. This bacteria include in a gram positive, non-spore forming, non-motile, facultative anaerobic bacteria. The other biochemistry reactions are negative for coagulase (CoNS), DNase, ornithinine decarboxylase, phosphatase, urease and oxidase [3]. This

nosocomial pathogen due to skin or soft tissue infections, bacteremia, septicemia, peritonitis, otitis media, meningitis and urinary tract infections [4][5]. The ability to produce enterotoxins, hemolysins, biofilm, and cytotoxins could be an important characteristic for the successful infection [6] [7]. According to [8] the biofilm formation of *S. haemolyticus* has different profile from *S. epidermidis.* 72 clinical isolates of *S. haemolyticus* have been analysed their antibiotic resistant, 89% isolates were detected resistant to all beta-lactam antibiotics encoded by *mecA* gene and 85% isolates were detected resistant to aminoglycoside encoded by [*aac(6')-Ie-aph(2")*] [8]. Not only in the hospital this pathogen was also found in the veterinary clinic. [9] mentioned that all isolates resistant to tetracycline, macrolides, and chloramphenicol. Thus, *S. haemolyticus* has the highest level of antimicrobial resistance among all CoNS species [10]. Based on this evidence, the exploration of new anti-MDR *S. haemolyticus* agent should be investigated in order to develop new therapy.

In this recent decades, [11] explained that investigation of natural product such us antibacterial agent were expanded to marine environment. One of the prospective sources for searching antibacterial agent from marine environment is soft coral [12][13]. Cladidiol is a new compound which was successfully isolated from methanolic extract of *Cladiella* sp. Its ability against several pathogens at a concentration of 50 μg/ml. Unfortunately they needed 4.5 kg dry weight [14]. A new approach to obtain a sustainable source and answer the scarcity of source is marine-derived fungi isolated from soft corals. This was supported by a statement from [15] and [16] that the microorganisms associated with macroorganisms have a symbiotic relationship with its specific host, so the fungus associated with soft corals can be used as a sustainable source of bioactive compounds because they grow faster than macroorganisms. Therefore, exploitation of soft corals could be reduced.

This research was designed to screening and characterized fungi associated with soft corals which had the crude extract potential against MDR-*S. haemolyticus.* The objective of this research may serve a pioneer of antibacterial agent with sustainable source and may provide the information to a pharmaceutical industry about marine natural product that is environmental friendly.

2. Materials and Methods

2.1. Collection of Sample

Soft corals were collected from the vicinity of Panjang Island (06°34'S and 110°37'E) using snorkeling technique. Samples were taken only 3 – 5 cm. Preservation of samples were created in a zip lock plastic bag to be stored temporarily in a cool box.

2.2. Isolation and Purification

Isolation of associated fungi was conducted using method of [17]. Fresh samples were initially sprayed with sterile sea water and cut approximately 1 cm^2. They were slashed, so that fungi on the inside of the tissue could be grown. In this research, samples were transferred as a duplo in Malt Extract Agar (MEA) added by sea water. Samples were incubated at room temperature for 3 days. Purification was conducted by separation of colony morphology (colours, shapes, and sizes).

2.3. Antibacterial Preliminary Test

MDR *S. haemolyticus* was obtained from Clinical Microbiology Laboratory of Kariadi Hospital Semarang. This test was conducted using agar plug diffusion method [18]. The associated fungi had been grown in MEA with sea water for 7-14 days. MEA which grew the associated fungi were then cut cylindrical to be transferred to the Nutrient Agar (NA) inoculated bacteria MDR *S. haemolyticus* 1x24 hours with a density of 0.5 McFarland. Both were then incubated at room temperature for 1x24 hours. Inhibition zone showed that fungi produced a potential bioactive compound against MDR *S. haemolyticus*, furthermore only the most potential isolate was extracted of its secondary metabolites.

2.4. Secondary Metabolites Extraction of Potential Isolate

Extraction was initially started with growing the associated fungi into Malt Extract Broth (MEB) with sea water for 14 days [19]. Inoculum was separated between mycelia and medium with filter paper. Medium was added with ethyl acetate (EtOAc) and mycelia were added with methanol (MeOH). Both were then separated with separatory funnel, then samples were evaporated with rotary evaporator.

2.5. Antibacterial Susceptibility Test

Crude extract from potential isolate was prepared for antibacterial susceptibility test against MDR *S. haemolyticus*. Both crude extracts from medium and mycelia were tested in various concentrations from low to high. The concentration started from 50, 100, 150, 300, and 500 μg/ml. Inoculum of MDR *S. haemolyticus* with density 0.5 McFarland was swabbed into Mueller-Hinton Agar (MHA). Crude extracts were added on blank paper disc, then incubated for 1x24 hours at room temperature. Clear zone was measured by caliper.

2.6. Molecular Identification

DNA extraction was conducted using chelex [20]. Amplification was done using PCR with primer ITS region, ITS1 as forward primer and ITS4 as reverse primer. The process of denaturation was initially at 95 °C for 3 minutes, followed by 35 cycles (denaturation at 95 °C for 1 minute, annealing at 51.4 °C for 1 minute, and extension at 72 °C for 1 minute), then post cycling in 72 °C extension for 7 minutes and set 16 °C for finishing. Furthermore, the gel electrophoresis was conducted to see the DNA bands formed and it was visualized using Geldoc. The PCR products were then sequenced and analyzed using MEGA 6.06. The sequence then compared with another sequences with the ability as antibacterial agent in phylogenetic tree using Neighbor-Joining method.

3. Results and Discussions

The infection due to MDR-*S. haemolyticus* associated with implanted medical devices has a significant impact especially in expense for hospitalized patients. Traditional antimicrobial therapy was still used as a basic control, however, this infection continue to be a challenge [7] [21]. Therefore, to minimize this incident, the discovery of bioactive compounds that inhibit MDR-*S. haemolyticus* growth is an urgent need.

3.1. Collection of Sample

There were 7 soft corals collected from vicinity of Panjang Islands, encoded with PPSC-04, PPSC-11, PPSC-16, PPSC-21, PPSC-27, PPSC-41, and PPSC-42. In this study, 7 soft corals was obtained as a host for fungi selected as a sustainable sources. They are included in *Cladiella* sp., *Lobophytum* sp., *Sinularia* sp., and *Sarcophyton* sp. Soft corals have presented the majority of potential therapeutics including anticancer agents, immunomodulators, and useful antifouling agents. They also showed significant biological activity, including antimicrobial, Ca-antagonistic and anti-inflammatory properties [22]. The previous research explained that the genus mentioned above has been studied about their microorganisms symbiont as an antimicrobial agent [23][24][25][26].

3.2. Isolation and Purification

The isolation of fungal associated with 7 soft corals obtained 23 single colony isolates. Their morphological characteristics were shown in Table 1. Among 23 isolates, 18 isolates were including in filamentous fungi or molds and 5 others were including in yeasts. [27] explained that molds grow in compact structure called mycelia and yeasts grow in a single cell colony. Previous research about isolation fungi associated with *Sinularia* sp. was successfully isolated 15 fungi using MEA [24]. [17] stated that a successful isolation depends on substrate/ media, nutrition and environment conditions such as temperature and aeration.

Table 1. The Colony Morphology of Fungi associated with Soft Coral from Panjang Island Vicinity

Isolate Code	Colour	Shape	Size	Texture	Elevation
PPSC-04-A1	White	Circle	Large	Leather	Raised
PPSC-04-A2	Cream	Circle	Small	Creamy	Flat
PPSC-04-B1	White, yellow	Circle	Medium	Leather	Raised
PPSC-11-A1	White	Circle	Large	Cotton, smooth	Raised
PPSC-11-A2	Yellow greenish	Circle	Large	Granular	Flat
PPSC-11-A3	Dark Green	Pin, Circle	Medium	Rough	Raised
PPSC-11-A4	White	Circle	Small	Smooth	Flat
PPSC-11-A5	White	Circle	Large	Cotton, slim	Raised
PPSC-11-A6	White	Circle	Large	Cotton, soft	Raised
PPSC-11-B1	White, brown	Pin, Circle	Medium	Rough	Raised
PPSC-11-B2	White yellownish	Circle	Medium	Smooth	Flat
PPSC-16-A1	Cream	Circle	Small	Smooth	Flat
PPSC-21-B1	Green dust	Circle	Medium	Velvet	Raised
PPSC-21-B2	White	Circle	Medium	Cotton, soft	Raised
PPSC-21-B3	Cream	Circle	Small	Creamy	Flat
PPSC-21-B4	Dark green	Circle	Small	Filamentous	Raised
PPSC-27-A1	Yellow, Dark green, white	Circle	Large	Powdery	Raised
PPSC-27-A2	Yellow, white, dark green, yellow greenish	Circle	Large	Powdery	Raised
PPSC-27-A3	White	Circle	Small	Smooth	Flat
PPSC-27-A4	Green, bright yellow	Circle	Large	Powdery	Raised
PPSC-27-B1	White	Circle	Large	Cotton	Raised
PPSC-41-A1	Cream	Circle	Small	Creamy	Flat
PPSC-42-A1	Cream	Circle	Small	Creamy	Flat

3.3. Antibacterial Preliminary Test

Antibacterial preliminary test was conducted to minimize time, expense, and attempt. Using agar plug method, there were 4 isolates successfully combat MDR-*S. haemolyticus*, including PPSC-11-A2, PPSC-27-A1, PPSC-27-A2, and PPSC-27-A4. This potential isolates origin from soft coral PPSC-11 and PPSC-27. According to sclerites identification [28], this soft coral was identified as *Cladiella* sp. and *Lobophytum* sp. However, only the highest antibacterial activity isolate would be extracted its bioactive compound. The antibacterial activity from 3 isolates was performed in Figure 1. The most potential isolate encoded as PPSC-27-A4 with diameter 23±9.6 mm.

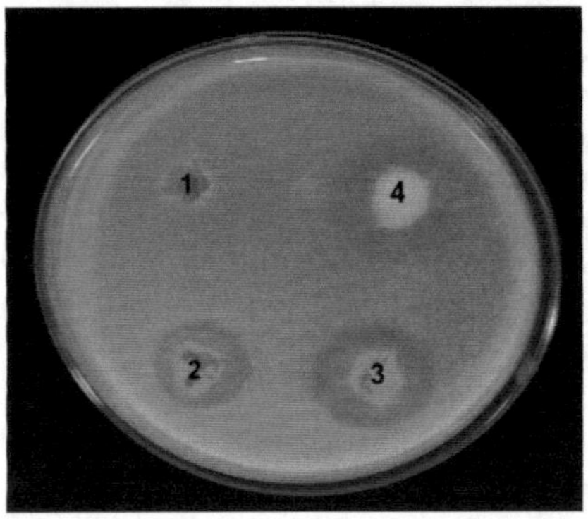

Figure 1. Antibacterial Activity from Fungi associated with Soft Coral against MDR-*S. haemolyticus*
Note: 1=PPSC-21-B4; 2=PPSC-27-A1; 3=PPSC-27-A4; 4=PPSC-27-A2

3.4. Secondary Metabolites Extraction of Potential Isolate

The purpose of secondary metabolites extraction was to find out bioactive compound from extracellular (released in medium) and intracellular (mycelia). This process was completed using polar and semi-polar solvent. [19] mentioned that organic solvent to extract bioactive compound from medium using ethyl acetate (EtOAc) and methanol (MeOH) to extract mycelia. The weight of crude extract from fermentation in 100 ml Malt Extract Broth (MEB) with sea water has been presented in Table 2. The most potential isolate encoded with PPSC-27-A4 has been extracted and tested for its antibacterial activity.

Table 2. Weight of crude extract from potential isolate encoded with PPSC-27-A4

Isolate Code	Source		Texture	Colour
	Medium	Mycelia		
PPSC-27-A4	0.0238 g	0.0395 g	Pasta	Yellow

3.5. Antibacterial Susceptibility Test

The crude extract of PPSC-27-A4 was tested for antibacterial susceptibility test in a various concentrations from low to high. Control was used to ensure the antibacterial activity from crude extract. Control positive used Tetracycline 30 µg and control negative used organic solvent. The diameter of inhibition has been shown in Table 3.

Table 3. Anti-MDR *S. haemolyticus* activity from crude extract of PPSC-27-A4

Source	Control (mm)		Concentration (mm)				
	Positive	Negative	50 µg/ml	100 µg/ml	150 µg/ml	300 µg/ml	500 µg/ml
Medium	21.5	-	10.8	11.6	11.4	11.8	11.5
Mycelia	21.5	-	11.7	10.9	-	12.2	10.3

Anti-MDR *S. haemolyticus* activity from crude extract of PPSC-27-A4 has the highest activity in 12.2 mm with concentration 300µg/ml from mycelia extract (Table 3). It means from this study, the intracellular metabolites were more effective than extracellular. Based on the literature, marine fungi majority produced hydrophobic compounds, through organic solvent extraction [29][30].

3.6. Molecular Identification

Molecular identification was used to preserve the potential isolate, thus the potential isolate could be reproduced and investigated in many cases in the future. Table 4 explained about homology analysis from sequence of PPSC-27-A4.

Table 4. Homology Analysis from Sequence of PPSC-27-A4 based on BLAST

Isolate Code	Close Relative	Homology	Acc. Number NCBI
PPSC-27-A4	*Trichoderma longibrachiatum* isolate A3S1-D3	99%	KJ767089.1

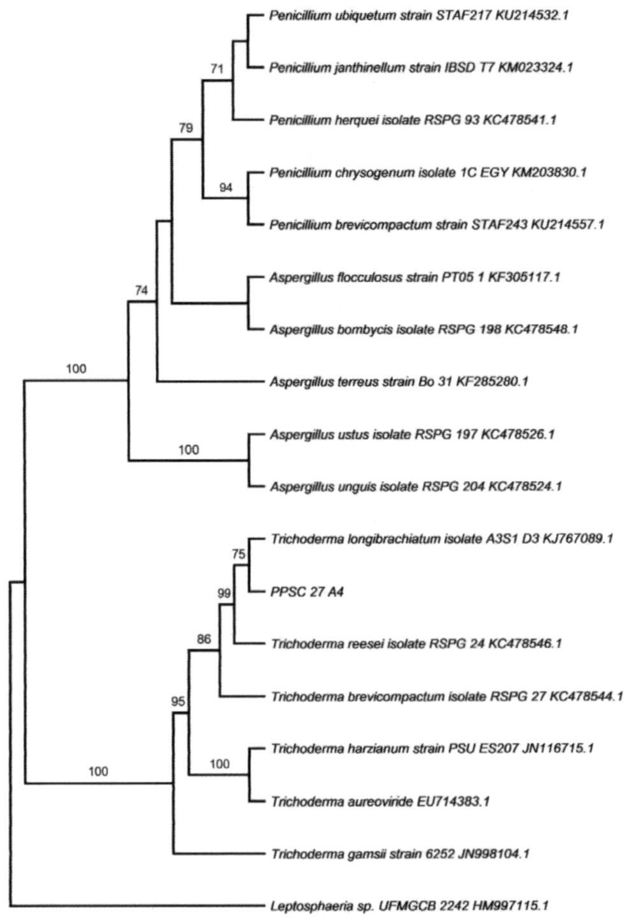

Figure 2. Phylogeny Tree based on comparison of 18S rRNA sequence using Neighbor-Joinig analysis with bootsrap 100

As mentioned by [13], the identification of potential isolate was very important, therefore, a further research is needed, so it could be reproduced. The result of sequences analysis with BLAST, this species had 99% similarity with *Trichoderma longibrachiatum* isolate A3S1-D3 (Table 4). *T. longibrachiatum* has been studied its chemical compound named ergokonin A with antifungal activity against Candida and Aspergillus species. This compound specifically could be *inhibitor of glucan synthesis in A. fumigatus* [31]. Phylogenetic tree was constructed using Neighbor-Joining to describe the relationship between this isolate and the other fungi (Fig. 2). PPSC-27-A4 was the same clade with *T. reseei* which has important metabolites in industrial products to produce second generations biofuels from cellulosic waste, that was cellulase enzyme [32].

4. Conclusion

Marine-derived fungi associated with soft corals have the ability to inhibit the growth of MDR-*S. haemolyticus*. It overcomes the supply of bioactive compound which is environmental friendly. Further research is still needed to discover chemical compound which inhibit the growth of MDR-*S. haemolyticus*.

Acknowledgment

This research was funded by the PMDSU (Program of Master Degree Leading to Doctoral Degree for Excellent Graduates) Scholarship from Ministry of Research and Higher Education, Indonesia with contract number 325/SP2H/LT/DRPM/IX/2016.

References

[1] Mack D 1996 The intercellular adhesion involved in biofilm accumulation of Staphylococcus epidermidis is a linear [beta]-1,6-linked glycosaminoglycan: purification and structural analysis *J. Bacteriol.* **178** 175–83
[2] Arciola C R, Campoccia D, Speziale P, Montanaro L and Costerton J W 2012 Biofilm formation in Staphylococcus implant infections. A review of molecular mechanisms and implications for biofilm-resistant materials *Biomaterials* **33** 5967–82
[3] Schleifer K-H 2009 Phylum XIII.Firmicutes Gibbons and Murray 1978, 5 (Firmacutes [sic] Gibbons and Murray 1978, 5) *Bergey's Manual® Syst. Bacteriol. SE - 3* **5** 19–1317
[4] Falcone M, Campanile F, Giannella M, Borbone S, Stefani S and Venditti M 2007 Staphylococcus haemolyticus endocarditis: clinical and microbiologic analysis of 4 cases *Diagn. Microbiol. Infect. Dis.* **57** 325–31
[5] Kumar Trivedi M 2015 Antibiogram Typing of Biofield Treated Multidrug Resistant Strains of Staphylococcus Species *Am. J. Life Sci.* **3** 369
[6] Valle J, Gomez-Lucia E, Piriz S, Goyache J, Orden J A and Vadillo S 1990 Enterotoxin production by staphylococci isolated from healthy goats *Appl. Environ. Microbiol.* **56** 1323–6
[7] Daniel B, Saleem M, Naseer G and Fida A 2014 Significance of Staphylococcus Haemolyticus in Hospital Acquired Infections *J. Pioneer Med. Sci.* **4** 119–25
[8] Fredheim E G A, Klingenberg C, Rohde H, Frankenberger S, Gaustad P, Flægstad T and Sollid J E 2009 Biofilm formation by staphylococcus haemolyticus *J. Clin. Microbiol.* **47** 1172–80
[9] Sidhu M S, Oppegaard H, Devor T P and Sørum H 2007 Persistence of multidrug-resistant Staphylococcus haemolyticus in an animal veterinary teaching hospital clinic. *Microb. Drug Resist.* **13** 271–80
[10] Frogatt J W, Johnston J L, Galetto D W and Archer G L 1989 Antimicrobial resistance in nosocomial isolates of {IStaphlococcus haemolyticus} *Antimicrob. Agents Chemother.* **33** 460–6
[11] Abad M J, Bedoya L M and Bermejo P 2011 Marine Compounds and their Antimicrobial Activities *Fortamex* 1293–306
[12] Wojnar J M 2008 Isolation of New Secondary Metabolites from New Zealand Marine Invertebrates
[13] Bugni T S and Ireland C M 2004 Marine-derived fungi: a chemically and biologically diverse group of microorganisms. *Nat. Prod. Rep.* **21** 143–63
[14] Ata A, Ackerman J, Bayoud A and Radhika P 2004 Bioactive Chemical Constituents of Cladiella Species *Helv. Chim. Acta* **87** 592–7
[15] Kelecom A 2002 Secondary metabolites from marine microorganisms* *Ann. Brazilian Acad. Sci.* **74** 151–70
[16] Proksch P, Edrada-Ebel R and Ebel R 2003 Drugs from the Sea - Opportunities and Obstacles *Mar. Drugs* **1** 5–17
[17] Kjer J, Debbab A, Aly A H and Proksch P 2010 Methods for isolation of marine-derived endophytic fungi and their bioactive secondary products **5** 479–90
[18] Balouiri M, Sadiki M and Ibnsouda S K 2016 Methods for in vitro evaluating antimicrobial activity: A review *J. Pharm. Anal.* **6** 71–9
[19] Rhoden S A, Garcia A, Bongiorno V A, Azevedo J L and Pamphile J A 2012 Antimicrobial activity of crude extracts of endophytic fungi isolated from medicinal plant trichilia elegans a. Juss *J. Appl. Pharm.*

Sci. **2** 57–9

[20] Walsh P S, Metzger D A and Higuchi R 1991 Chelex 100 as a medium for simple extraction of DNA for PCR-based typing from forensic material. *Biotechniques* **10** 506–13

[21] Brzychczy-Wloch M, Borszewska-Kornacka M, Gulczynska E, Wojkowska-Mach J, Sulik M, Grzebyk M, Luchter M, Heczko P B and Bulanda M 2013 Prevalence of antibiotic resistance in multi-drug resistant coagulase-negative staphylococci isolated from invasive infection in very low birth weight neonates in two Polish NICUs *Ann. Clin. Microbiol. Antimicrob.* **12** 41

[22] Kim (Editor) 2015 *Springer Handbook of Marine Biotechnology* (Heidelbergh : Springer)

[23] Nugraheni S A, Khoeri M M, Sabdono A and Radjasa O K 2010 ANTIBACTERIAL ACTIVITY OF BACTERIAL SYMBIONTS OF SOFTCORAL Sinularia sp . AGAINST PATHOGENIC RESISTANT Bacteria *J. Coast. Dev.* **13** 113–8

[24] Putri D A and Pringgenies D 2015 Effectiveness of Marine Fungal Symbiont Isolated from Soft Coral Sinularia sp. from Panjang Island as Antifungal *Procedia Environ. Sci.* **23** 351–7

[25] Elahwany A M D, Ghozlan H A, Elsharif H A and Sabry S A 2015 Phylogenetic diversity and antimicrobial activity of marine bacteria associated with the soft coral Sarcophyton glaucum *J. Basic Microbiol.* **55** 2–10

[26] Murti P D B and O K R 2012 Antibacterial Activity of Bacterial Symbiont of Soft **15** 297–302

[27] Madigan T M, Martinko J M, Stahl D A, and Clark D P 2013 *Biology of Microorganisms* 13th Ed (San Fransisco : Benjamin Cummings)

[28] Janes M P 2008 Laboratory methods for the identification of soft corals (Octocorallia : Alcyonacea) *Adv. Coral Husb. Public Aquariums. Public Aquarium Husb. Ser. vol.2.* **2** 413–26

[29] Blunt J W, Copp B R, Hu W P, Munro M H, Northcote P T and Prinsep M R 2009 Marine natural products *Nat Prod Rep* **26** 170–244

[30] Le Ker C, Petit K E, Biard J F and Fleurence J 2011 Search for hydrophilic marine fungal metabolites: A rational approach for their production and extraction in a bioactivity screening context *Mar. Drugs* **9** 82–97

[31] Vicente M F, Cabello A, Platas G, Basilio A, Díez M T, Dreikorn S, Giacobbe R A, Onishi J C, Meinz M, Kurtz M B, Rosenbach M, Thompson J, Abruzzo G, Flattery A, Kong L, Tsipouras A, Wilson K E and Peláez F 2001 Antimicrobial activity of ergokonin A from Trichoderma longibrachiatum *J. Appl. Microbiol.* **91** 806–13

[32] Schmoll M and Schuster A 2010 Biology and biotechnology of Trichoderma *Appl. Microbiol. Biotechnol.* **87** 787–99

Exploration of Fungal Association from Hard Coral against Pathogen MDR *Staphylococcus haemolyticus*

O Cristianawati[1], O K Radjasa[1,3], A Sabdono[3], A Trianto[3], A Sabdaningsih[1], MT Sibero[1], H Nuryadi[4]

[1]Coastal Resources Management, Diponegoro University Semarang 50241, Indonesia
[2]Director of Research and Public Services at Ministry of Research, Technology and Higher Education, Indonesia
[3]Department of Marine Science, Diponegoro University, Semarang 50241,Indonesia
[4]UPT Laboratorium Terpadu, Diponegoro University, Semarang 50241,Indonesia

E-mail: olvi.cristiana@yahoo.com

Abstract. *Staphylococcus haemolyticus* are opportunistic bacteria and as the second leading cause of nosocomial infections. It is a disease causing septicemia, peritonitis, otitis, and urinary tract infections and infections of the eye. It also a phenotype resistant to multiple antibiotics commercial. There is now an urgency to find an alternative antibiotics to combat this bacteria. It has been widely reported that many bioactive marine natural products from marine invertebrate have striking similarities to metabolites of their associated microorganisms including fungi. Hard coral associated microorganisms are among of the most interesting and promising marine natural product sources, which produce with various biological activities. The proposed work focused on the discovery of bioactive compounds and also estimated the phylogenetic diversity from fungal association of hard coral against pathogen MDR *Staphylococcus haemolyticus*. A total of 32 fungal association, FHP 7 which were isolated from *Favia* sp. capable of inhibiting the growth MDR. Molecular identification based on 18S rRNA gene sequences revealed that the active fungal association belonged 100% to the members from one of the genera *Trichoderma longibrachiatum*. Accession Number LC185084.1.

Keywords: Fungal association, secondary metabolites, *Staphylococcus haemolyticus, Trichoderma longibrachiatum*

1. Introduction

Staphylococcus haemolyticus which were included in the negative staphylococcus members colonized from the human skin as an opportunistic pathogen. This bacterium as the second leading cause of nosocomial infections are isolated in conjunction with S. *epidermidis*, for example through the insertion of medical devices [1]. They are a disease causing septicemia, peritonitis, otitis, and urinary tract infections and infections of the eye [2]. S. *haemolyticus* [3], [4] that the organisms known to show resistant to multiple antibiotics commercial in a phenotype scale [1], this was reinforced their resistance to several antibiotics including heteroresistance keglikopeptida. Resistance of antibiotic become a very serious problem that caused decreasing the number of recovery, the cost increasing rapidly for medical treatment extension and also increasing mortality of patient [5], [6], [7]. Using antibiotics against pathogenic bacteria has been increasing since the improper and uncontrolled treatment and become for Multi Drug Resistant (MDR) bacteria.

It has been widely reported many bioactive marine natural products have striking similarities with associated microorganisms from marine including fungi [8]. Marine fungal come out as an alternative solution to find potential compound from marine environment. Many potential marine-derived natural compounds were produced by marine organisms themselves or their associated microorganisms. reported that sustainability of marine organisms as the main problem of marine natural product exploration [9]. It has been almost impossible to harvest large numbers of organisms from nature without destroying the habitats.

A potential source of associated bacteria from marine invertebrates could be used for the isolation of bioactive metabolites which is collected from inhabiting the coastal waters [10]. Secondary metabolites can be found from association bacteria with soft coral because they have a similar characterization [11]. It can be possible that fungal association from hard coral can become an alternative to produce antibacterial compound which are against to pathogen MDR that fungal association from marine have potential as an natural drugs because they have an active secondary metabolites [12]. Secondary metabolites of marine fungal association are more specific than terrestrial fungal. However, the biodiversity of marine microorganisms is barely known. The biological diversity has been found in marine ecosystems which is higher than in tropical rain forests [13]. The reef invertebrates sources such as sponges, tunicates, bryozoans, soft corals and molluscs are accumulated as a large group of structurally unique natural products from the oceans. This diversity has been declared that the source of unique chemical compounds with the potential for industrial already develop as pharmaceuticals, cosmetics, nutritional supplements, molecular probes, enzymes and agrichemicals [14].

Coral reef ecosystem in Indonesia about 85.000 km^2 and represented 14 % world coral reef [15]. Panjang island is located in 2 mil from west jepara beach with lasge about 19,7 ha. It was reported that Panjang island surrounded by coral reef [16]. Coral reef is a productive ecosystem which can produce a useful bioactive compound with high biodiversity in the sea [14]. It has been very well established that terrestrial microorganism are sources of valuable bioactive metabolites for more than half a century [17]. Hard coral are included as a phylum Cnidaria which has potential biomedic compound for 21% species [15]. Many bioprospeksi activity of natural compound from marine has been resulted a thousand new compound [18]. Coral is the targeted source for natural bioactive product which known have a secondary metabolites [14]. Exploration of fungal association that produced a bioactive compound has been found in marine invertebrate which have a secondary metabolites [19].

Successfuly exploration of new compound was founded more than 330 compound from fungal association in 2002-2006. Some related research of bioactive compound produced by coral such as *Anthomastus bathyproctus* produced 3-Oxo sterols for cytotoxic [20], montiporyne A from *Montipora spp.* for anticancer [21] and montiporic acids from *M. digitata* as antimicrobial [22], reported that abundance and diversity of natural chemistry with some potential bioactivity can be gained into drugs discovery [18] . It is important to highlight an anti pathogen compounds that produced by marine fungi associated with invertebrates is providing the possible role an alternative to the commercial. This phenomenon has raised the importance of microbial symbionts of invertebrates as profilic sources of secondary metabolites with diverse biological activities applicable for health and industrial purposes. Therefore, it is reasonable to expect new classes of secondary metabolites are discovered which are exploration of untapped marine microbial diversity and resources will improve the rates.

2. Materials and Methods

2.1. Sampling of hard coral
Colonies of hard coral were collected from Panjang island, North Java Sea, Jepara, Central Java, Indonesia by scuba diving. Upon collection hard coral colonies were put into sterile plastic bags (Whirl-Pak, Nasco, USA).

2.2. Isolation of fungal symbionts from hard coral
The coral tissues were tapped on MEA medium containing chloramphernicol antibiotics, incubated at room temperature for 7 days. On the basis morphological features, colonies were randomly picked and purified by streak plates [23].

2.3. Inhibitory interaction tests
Antibacterial tests of hard coral-fungal symbiont against bacterial pathogen were performed by using plug method [24] and [25]. Bacterial pathogen (*S. haemolyticus*) used in this study was obtained from culture collection from Kariyadi Hospital. Culture of bacterial pathogen in the logarithmic phase was mixed with Mueller Hinton Agar (MHA) (1% v/v), which were then be poured on to the respective agar surface previously inoculated with hard coral fungal association and incubated for a day. The plates will be incubated at room temperature for 1 or 2 days. MDR activity was defined by the formation of inhibition zones around the fungal colonies.

2.4. PCR amplification and sequencing of 18S rRNA gene fragments
DNA extraction and amplification were also carried out at the Trophical Marine Biotechnology UPT Diponegoro University and sequencing at PT. Genetika Science. The DNA genomes from misselia fungal isolates were extracted using cheelex. The DNA concentrations were quantified by using Nanodrop. 18S rRNA fragments were generated by using universal primer ITS1 - ITS4 with base sequence ITS1: 5'- CTT GGT CAT TTA GAG GAA GTAA-3' and ITS4 : 5'- TCC TCC GCT TAT TGA TAT GC-3' [26]. These primers were used to obtain 400- 800 bp rDNA fragments for sequencing purposes. The DNA amplification was performed by Taq polimerase master mix with approximately 2 μg of total genome as a template. PCR was conducted as modification from: initial denaturation (5 min at 95^{0}C), 32 cycles of primer denaturation (1 min at 95^{0}C), annealing (1 min at 56,4^{0}C), and elongation (1 min at 72^{0}C) followed by a final elongation step (7 min at 72^{0}C) [27].. Purification of PCR products and subsequent sequencing analysis was conducted according to the method. The determined DNA sequence of strains were compared for homology to the BLAST database.

2.5. Phylogenetic analysis
All sequences were used at least 400-800 bp long. A phylogenetic tree was constructed using maximum-likelihood analysis. Only sequences of type strains was included in tree calculation. Alignment positions at which less than 50 % of sequences of the entire set of data have the same residues were excluded from the calculations to prevent uncertain alignments within highly variable positions of the 18S rDNA. Phylogenetic analysis was performed with MEGA software packages.

3. Results

3.1. Sample collection
Samples of Hard coral were collected from Panjang Island. Based on *purposive* method, 8 different samples already collected by scuba diving. The samples was then be isolated to produce the fungal association with specific media.

3.2. Isolation and Purification of Hard Coral
The isolations were used for knowing and selecting of coral fungal association, which have an antibacterial potential. Fungal association were planted on Malt Extracts Agar Media (MEA). The result of fungal asosiation was about 32 isolates. The isolates were then be identified based on morfology such as colony, texture and colour [28]. Moreover, every single isolate was continued for antibacterial test to combat pathogen MDR S. *haemolyticus*.

3.3. Antibacterial Test

Table 1. Fungal Association against MDR *S. haemolyticus*

Code	Bacteria	Code	Bacteria
	Staphylococcus haemolyticus		*Staphylococcus haemolyticus*
FHP 1	-	FHP 34	+
FHP 2	-	FHP 37	+
FHP 3	+	FHP 6	-
FHP 4	-	FHP 10	-
FHP 5	+	FHP 15	-
FHP 7	+	FHP 16	-
FHP 8	-	FHP 17	-
FHP 9	-	FHP 18	-
FHP 11	+	FHP 19	-
FHP 12	-	FHP 20	-
FHP 13	-	FHP 24	-
FHP 21	+	FHP 25	-
FHP 30	-	FHP 29	-
FHP 31	-	FHP 35	-
FHP 32	-	FHP 36	-
FHP 33	-	FHP 38	+

(+) success; (-) unsuccessful

Antibacterial test was showed 8 inhibition zone surrounding the plug miselia which was active combat with pathogen MDR. The growth rate of inhibition zone was observed until 9 days incubation (9x24 hours). The active isolates were signed by code FHP 3, FHP 5, FHP 7, FHP 11, FHP 21, FHP 34, FHP 37 and FHP 38. In 1x24 hours incubation, the inhibition zone was resulted 2,6 mm; 10,7 mm; 14 mm; 13 c mm; 22,3 mm; 14,6 mm; 18,7 mm dan 13,4 mm respectively. In the other hand, 9x24 hours incubation was resulted 22,3 mm; 10,7 mm; 26, 4 mm; 22,4 mm; 20,6 mm; 22 mm; 18,7 mm dan 15 mm. FHP 7 was picked up as the best isolate candidate based on growth inhibition zone.

3.4. Molecular identification
1. Polymerase Chain Reaction (PCR)
The DNA which was amplified by using PCR 18S rRna . The PCR product was visualized by using electrophoresis and gel documentation. It was showed that FHP 7 in 610 bp long. The result of amplification was presented in Figure 1.

IOP Conf. Series: Earth and Environmental Science **55** (2017) 012027 doi:10.1088/1755-1315/55/1/012027

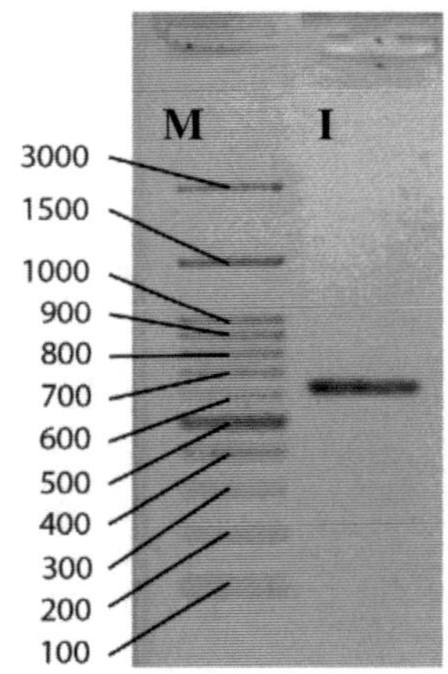

Figure 1. Visualization of PCR product by using electrophoresis.

where: M = *Marker Low DNA Mass Ladder*; I = FHP 7

3.5. Phylogenetic tree

Blast homology of FHP 7 was closely related 100 % with *Trichoderma longibrachiatum* strain JF694937.1. Analysis of phylogenetic by using Clustal W and MEGA 5 showed that phylogenetic tree of FHP 7 also closely related with genera from *Trichoderma* sp. The result of phylogenetic analysis was represented in Figure 2.

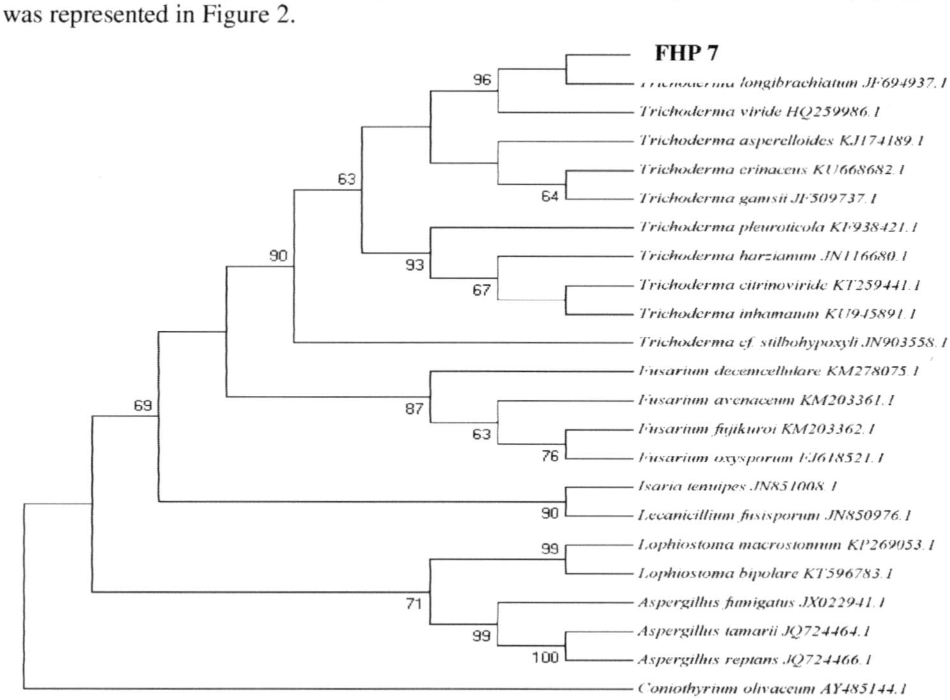

Figure 2. Phylogenetic tree of fungal association from hard coral

4. Discussion

The samples which were collected from Panjang island was classified by using *Coral of The World* book [29] and software Coral ID [30], [31]. Genera of 8 different coral was indentified, such as Genus *Acropora* (PP-HC-16-45) (Suharsono, 2008), genus *Pachyseris* (PP-HC-16-48) [29], genus *Favia* (PP-HC-16-35 [32]. Johan (2003), Genus *Favites* (PP-HC-16-15) [29], Genus *Goniastrea* (PP-HC-16-49) [34]. Genus *Isopora* (PP-HC-16-47) [32], Genus *Pavona* (PP-HC-16-44) and Genus *Caulastrea* (PP-HC-16-43). Coral *Favia* sp. was identified through the morphology using a microscope and molecular method based on the isolate candidate which was FHP 7.

The isolation process was used specific media for fungi which was MEA (*Malt Extract Agar*). From the isolation process, it was purified 32 isolate association Fungal association that already single colony was then be identified such as colony, texture and colour [28]. The fungal association from hard coral which then be continued for antibacterial test toward pathogen S. *haemolyticus* by using plug miselia method. It was representing that 8 isolates was built an inhibition zone. Association of microorganism can produce secondary metabolites which can be used for pharmacological activity [35], [36]. Fungal association from marine invertebrate which have a bioactive compound can give the benefit for antibacteri, antifungal, citotoxic, anticancer, antioxidant, antimalaria, nematicidal, antitumor, antineoplastic, antifouling, larvasida, neurotoksik, antimycobacterial, anti HIV, antiprotozoal, antivirus, antiulcer, antiparasite, antiinflamation, and immunosuppressant [37].

FHP 7 was choosed as the best candidate of fungal association from hard coral with code PP-HC-16-35 based on the inhibition zone activity until incubation in day 9. The characteristic FHP 7 was as the bacteriocidal which is killing the bacteria toward pathogen MDR. Microscopic test by using a microcope was conducted for checking conidiophore, phialides and conidia and supported by using identification book from [38], [39]. Moreover, FHP 7 was identified as a *Trichoderma longibrachiatum* with characteristic, such as short and tight phialides, soft green conidia.

Sequencing identification of FHP 7 was used BLAST (*Basic Local Aligment Search Tool*) system that was showed that fungal association belonged 100% to the members from one of the genera *Trichoderma longibrachiatum*. The isolates can be one level species for 99-100% homology condition. [40]. Identification based on molecular identification can be used for species level analysis of fungal.

5. Conclusion

A total of 32 fungal associations, FHP 7 which were isolated from *Favia* sp. capable of inhibiting the growth MDR S. *haemolyticus*. Molecular identification based on 18S rRNA gene sequences revealed that the active fungal association belonged 100% to the members from one of the genera *Trichoderma longibrachiatum*. Accession Number LC185084.1.

Acknowledgements
This research was partly funded by a research grant from the Ministry of Research, Technology and Higher Education, Indonesia for PMDSU (Program of Magister Leading to Doctoral Degree for Excellent Under Graduates) scholarship.

References

[1] B. Daniel, S. Mehvish, N. Gousia and F. Alsaba. 2014. Significance of Staphylococcus Haemolyticus in Hospital Acquired Infections. J PIONEER MED SCI. www.jpmsonline.com. Volume 4, Issue 3. July – September.
[2] A.R. Makki, S. Sharma, A. Duggirala, K. Prashanth, P. Garg, T. Das. 2011. Phenotypic and genotypic characterization of coagulase-negative staphylo- cocci (CoNS) other than Staphylococcus epidermidis

isolated from ocular infections. Invest Ophthalmol Vis Sci 52:9018–9022. http://dx.doi.org/10.1167/iovs.11-7777.

[3] J.W. Froggatt, J.L. Johnston, D.W. Galetto and G.L. Archer. 1989. Antimicrobial resistance in nosocomial isolates of Staphylococcus haemolyticus. Antimicrob Agents Chemother 33:460–466. http://dx.doi.org/10.1128/AAC.33.4.460.

[4] K. Hiramatsu. 1998. Vancomycin resistance in staphylococci. Drug Res Updates 1:135–150. http://dx.doi.org/10.1016/S1368-7646(98)80029-0.

[5] E. Leung, E. W. Diana, R. Mario, N. Hiroki. 2011. The WHO policy package to combat antimicrobial resistance. *Bulletine of World Health Organization* 89: 390-392.

[6] H. Gelband, M. H. Molly, P. Suraj, G. Sumanth, L. Jordan, B. Devra, W. Andrea and L. Ramanan. 2015. *The State of The World's Antibiotic 2015*. Washington, D. C (US): Center for Disease Dynamics, Economy and Policy.

[7] S. T. Odonkor and K. A. Kennedy. 2011. Bacteria resistance to antibiotics: recent trends and challenges. *International Journal of Biological & Medical Research* 2(4): 1204-2010.

[8] P. Proksch, R.A. Edrada and R. Ebel. 2002. Drugs from the seas-current status and microbiological implications. Applied Microbial. Biotechnol., 59: 125-134.

[9] O.K. Radjasa, M.M. Khoeri, C.C. Darusallam, H. Sudoyo. 2013. Bacterial symbionts of reef invertebra tes: screen ing for anti-patho genic bacteria activity. *J. Biodiversity*. 1-7.

[10] C.V.S Shankar, S.Sathianeson, V. Nadarajan and M. J. P. Stanislaus. 2015. Antibacterial and biofilm inhibitory activities of bacteria associated with polychaetes. Journal of Coastal Life Medicine 2015; 3(6): 495-502.

[11] J.G. Burgess, K.G. Boyd, E. Armstrong, Z. Jiang, L.Yan, M. Berggre, U. May, T. Pisacane, A. Granmo and D. R. Adams. 2003. The Development of a marinenatural product-based antifouling paint. Biofouling, Suppl. 197 -205.

[12] A.M. Mabrouk, Z.H. Kheiralla, E.R. Hamed, A.A. Youssry and A.E. Aty. 2008. Production of some biologicallyactive secondary metabolites from marine-derived. Malay. J. Micro. 4(1):14–24.

[13] T.O. Larsen, J. Smedsgaard, K.F. Nielsen, M.E. Hansen and J.C. Frisvad. 2005. Phenotypic taxonomy and metabolite profiling in microbial drug discovery. Nat Prod Rep 2005;22: 672–95. Le Crom S.

[14] O. K. Radjasa, A. Sabdono., J. Wiese and J. F. Imhoff. 2008. Coral as source of bacteria with antimicrobial activity. J.Coast. Dev: (11)(3):121-130.

[15] C. Reid, J. Marshall, D. Logan and D. Kleine. 2011. Coral reef and climate changes.Terumbu Translated from Krisantini dan L. Badib from book "Coral Reefs and Climate Change". Coralwatch, The University of Queensland, Australia. 272 hlm.

[16] Munasik, Ambariyanto, A. Sabdono, P. W. Diah, O. K. Radjasa and R. Pribadi. 2012. Spasial of hard coral (*Scleractinia*) In panjang Island, Central Java. *Buletin Oseanografi Marina* 1: 16-24.

[17] A. Kelecom. 2002. Secondary metabolites from marine microorganisms. An Acad Bras Cienc 74: 151-170

[18] M. C. Leal, J. Puga, J. Serodio, N.C.M. Gomes and R. Calado. 2012. Trends in the discovery of new marine natural products from invertebrates over the last two decades?Where and what are we bioprospecting. Plos One. 1 (1): 1?16.

[19] J. Kjer, A. Debbab, H.A. Aly, , and P. Proksch,. 2010. Methods for isolation of marine-derived endophytic fungi and their bioactive secondary products. Nature Protocols. 5(3): 479–490.

[20] G. Mellado, G.E. Zubía, M.J. Ortega, P.J. 2005. López- González: Steroids from the Antarctic octocoral An- thomastus bathyproctus,J.Nat.Prod. 68, 1111–1115.

[21] N. Alam, B.H. Bae, J. Hong, C.O. Lee, K.S. Im, J.H. 2001. Jung: Cytotoxic diacetylenes from the stony coral Montipora species, J. Nat. Prod. 64(8), 1059–1063.

[22] C.P. Marquis, A.H. Baird, R. de Nys, C. Holmström, N. 2005. Koziumi: An evaluation of the antimicrobial properties of the eggs of 11 species of scleractinian corals, Coral Reefs 24,248–253.

[23] G. Strobel and B. Daisy. 2003. Biosprospecting for Microbial Endophytes and Their Natural Products. Microbiol and Mol. Biol. 67:491-502.

[24] I.P.D. Santos , L.C.N.D Silva, M.V.D. Silva, J.M.D. Araújo, M.D.S. Cavalcanti and V.L.D.M. Lima. 2015. Antibacterial activity of endophytic fungi from leaves of Indigofera suffruticosa Miller (Fabaceae). Frontiers in Microbiology www.frontiersin.org 1. May 2015Volume 6 Article 350

[25] M. Balouiri, S. Moulay and K. I. Saad, 2016. Methods for *in vitro* evaluating antimicrobial activity: A rivew. *Journal of Pharmaceutical Analysis* 6(2016): 71-76.

[26] Y. Zhang, Mu, J., Feng, Y., Kang, Y., Zhang, J., Gu, P.J.,Wang, Y., Ma, L.F., and Zhu, Y.H. Broadspectrum antimicrobial epiphytic and endophytic fungi from marine organisms: Isolation, bioassay and taxonomy. Marine Drugs. 2009. (7): 97–112

[27] J. Wiese, B. Ohlendorf, M. Blumel, R. Schmaljohann and J.F. Imhoff. Phylogenetic identification of fungi isolated from the marine sponge Tethya aurantium and identification of their secondary metabolites. Mar . Drugs. 2011 (9). : 561 – 58

[28] L. Waluyo. 2008. Mikrobiologi Umum. Universitas Muhammadiyah Malang. Press.

[29] J.E.N. Veron, 2000. Coral of the world. Australian Institute of Marine Science, PMB 3, Townsville MC, Qld 4810 Australia, Vol. 1: 463 p.

[30] A. N. Budd, H. Fukami, N. D. Smith and N. Knowlton. 2012. Taxonomic Classification of The Reef Coral Family Mussidae (Cnidaria : Anthozoa : Scleractinia). Zoological Journal of The Linnean Society, 166 : 465-529.

[31] D. Huang, F. Benzoni, R. Arrigoni, A. H. Baird, M. L. Berumen, J. Bouwmeester, L. M. Chou, H. Fukami, W. Y. Licuanan, E. R. Lovell, R. Meier, P. A. Todd, A. F. Budd. 2014. Towards a Phylogenetic Classification of Reef Coral: The Indo-Pacific Genera *Merulina, Goniastrea,* and *Scapophyllia* (Scleractinia, Merulinidae). Royal Swedish Academy of Sciences: 1-18.

[32] Suharsono. 2008. Diversity of common coral in Indonesia. Research project of LIPI-P3O and growing up in beach area, Jakarta.

[33] O. Johan. 2003. Biology characteristic of coral. *Training Course* 7-12 Juli 2003. Funded by PSK-UI and Yayasan TERANGI.

[34] J.E.N. Veron. 1993. Corals of Australia and the Indo-Pacifi c. Univ. Hawaii. Press. Honolulu. 643.

[35] T. S. Bugni and C.M. Ireland. 2004. Marine derived fungi: a chemically and biologically diverse group of microorganisms. Nat. Prod. Rep. 21:143-163.

[36] M.E. Rateb, and R. Ebel. 2011. Secondary metabolites of fungi from marine habitats. Nat Prod Rep. 28:290–344

[37] J. Swathi, K. Narendra, K.M. Sowjanya and S. A. Krishna. 2013. Biological Characterisation of Secondary Metabolites from Marine Fungi Microascus sps. Int. J. Res. Pharm. Biomed. Sci. 4(3):754-759

[38] T. Watanabe. 2002. Pictorial atlas of soil and seed fungi morphologies of cultured fungi and key to species. CRC Press LLC. U.S.A.

[39] K.H. Domsch, W. Gams and T.H. Anderson. 1980. Compendium of Soil Fungi. Volume 1. Academic Press, London

[40] T. Sugita and A. Nishikawa, 2003. Fungal Identification Method based on DNA Sequence Analysis: Reassessment of the Method of teh Pharmaceutical Society of Japan and The Japanese Pharmapoeia. Health Science. 46 (6):531-543.

Isolation, Identification And Screening Antibacterial Activity from Marine Sponge-Associated Fungi Against Multidrug-Resistant (MDR) *Escherichia coli*

Mada Triandala Sibero [1,2,*], Aninditia Sabdaningsih[1,2], Olvi Cristianawati[1,2], Handung Nuryadi[2], Ocky Karna Radjasa[3,4], Agus Sabdono[3], Agus Trianto[3]

[1] Department of Coastal Resources Management, Diponegoro University, St. Imam Bardjo, SH, No. 15, Semarang 50241, Indonesia
[2]Laboratory of Tropical Marine Biotechnology, St. Prof. Soedarto,S.H, Tembalang, Semarang 50241, Indonesia
[3] Department of Marine Science, Faculty of Fisheries and Marine Science, Diponegoro University, St. Prof. H. Soedharto,SH, Tembalang, Semarang 50275, Indonesia
[4] Director of Research and Public Services at Ministry of Research, Technology and Higher Education, Indonesia

E-mail: madatriandala@hotmail.com

Abstract. Irrational used of antibiotic in several decades ago causing resistant in bacteria and decreasing the cure rate of infectious diseases. Multidrug-resistant (MDR) *Escherichia coli* is known to cause several infectious diseases such as urinary tract infection, nosocomial bloodstream infection, meningitis, bacteraemia, and gastrointestinal disease. Marine sponge-associated fungi have potential as source of new compound to combat MDR *E. coli*. The aims of this research were to isolate marine sponge-assosiated fungi, to screen potential fungi against MDR *E. coli*, to identify the potential fungi and its host sponge. There were 29 marine sponge-associated fungi successfully isolated from 9 sponges. Among 29 sponge-associated fungi screened, there were 7 isolates showed antibacterial activity against MDR *E. coli*. The best inhibition zone produced by MPS 14.1/MT 02 and MPS 14.3/MT 04 from sponge PP.SP.16.14. According to fungi identification result fungus MPS 14.1/MT 02 was identified as *Trichoderma asperellum* while MPS 14.3/MT 04 was identified as *Trichoderma reesei*. Sponge identification leaded the PP.SP.16.14 as *Cinachyrella* sp.

Key words: Antibacterial, Multi-drug resistant, Sponge associated-fungi, *Trichoderma*

1. Introduction

Irrational used of antibiotic for these several decades leads us to multidrug-resistant (MDR) bacteria era. This problem become a big issue for causing the decrease of cure rate thus allowing the increasing of the death rate [1,2,3]. One of bacteria which very common to infect human is *Escherichia coli*. These bacteria causing urinary tract infection, nosocomial bloodstream infection, meningitis, bacteraemia, and gastrointestinal disease [4,5,6,7]. For several cases showed these bacteria resistant to antibiotics such as *amoxilin-clavunalic acid*, *amikacin*, *aztreonam*, *cefazolin*, *cefuroxime*, *ceftazidime*, *cefotaxime*, *cephalosporin*, *ertapenem*, *imipenem*, *fluoroquinolones*, *fosfomycin*, *gentamycin*, *meropenem* and *piperacillin-tazobactam* [6,8,9]. This condition is known as multi-drug resistant. The resistance ability of *E. coli* influenced by the mutation or the presence of particular genes such as *mcr-1* and *tetA* [10,11]. Exploration of new antibiotic could be an answer of large number of ineffective antibiotics to cure *E. coli* infections. Indonesia's marine resources have the potential as source of new antibiotic. Sponge was reported to has antibacterial activity to against several pathogenic bacteria [12,13]. But the exploration of antibiotic compounds from sponge was stunted by conservation issue. To solve this problem, the exploration of new antibiotic from sponge-associated fungi is important. The aims of this research were to isolate marine sponge-associated fungi, to screen potential fungi against MDR *E. coli*, to identify the potential fungi and its host sponge.

2. Methodology

2.1 Sponges sampling

Sampling was done at February, 12th-13th 2016 in Teluk Awur (S 06°36'57.4" E 110°38'19.4") and Panjang Island (S 06°34'35.2" E 110°37'52.7") Jepara, Central of Java, Indonesia. The sponges were taken at depths between 1-5 m. Approximately 5 cm^2 of sponges were taken and put into ziplock plastic contains marine water for the fungi isolation.

2.2 Fungi isolation

Fungi isolation was done according to [14]. The sponges were cut with size 1x1 cm^2 then washed with sterile sea water for three times and then rinsed in ethanol (EtOH) 70% for 60 s for surface sterilization and then sprayed with sterile marine water and dried by the wind. After that put the sponge segments on the petri contained marine *Malt Extract Agar* (MEA) medium then incubated at room temperature for 7 days. For this step, an empty petri was opened as the environmental control. Each fungus which growth from the segment was compared to the control, if the morphology was different, the fungus was moved into new petri as single colony.

2.3 Screening antibacterial activity against MDR E. coli

This step performed using plug method [15,16]. The fungi were refreshed for 7 days and then the mycelia with the MEA medium were cut with circular shape with a diameter 1 cm^2. Then put on the petri contained *Nutrient Agar* (NA) that has been inoculated MDR *E. coli*. Petri were incubated for 24 h at room temperature. The clear zone was observed as the antibacterial activity from the fungi. Fungus with biggest clear zone were identified through morphology and molecular approaches.

2.4 Fungus identification through molecular approach

Potential fungus were cultivated in 7 days for DNA extraction. The DNA were obtained according to Turan *et al.* [17]. ITS 1 (5′-TCC GTA GGT GAA CCT GCG G-3′) was used as forward primer while ITS 4 (5′-TCC TCC GCT TAT TGA TAT GC-3′) as reverse primer. PCR mix consist of 1μL of ITS 1, 1μL of ITS 4, 12.5 μL of GoTaq Green Master mix, 10 μL of ddH_2O and 0.5 μL of DNA template. Amplification was performed with cDNA preheat at 95°C for 3 min, 30 cycles of denaturation at 95°C for 1 min, annealing at 51,88°C for 1 min and extension at 72°C for 1 min. The post cycling extension was done at 72°C for 7 min. The PCR products were loaded in agarose gel (1%) for electrophoresis then visualized by UVI Doc HD5 (UVITEC Cambridge).

2.5 Fungus morphology observation

Potential fungus was cultivated using slide culture method [18] for 7 days on MEA. Then the mycelia were observed under microscope with addition of lactophenol cotton blue (LPCB). The morphology characteristics were used for fungus identification.

2.6 Sponge identification

Sponge was identified through morphology and spicules identification. Shape, pit and colors were observed for the early identification. The spicules were obtained by immersed the sponge with chlorine for 30 min the centrifuged for 5 min. The supernatant were discard gently the added with chlorine again. This step were repeated until the transparent spicules seen clearly and then added ethanol (EtOH) 96% for cleaned the residue of the chlorine. After that the spicule were observed under a microscope.

3. Result and Discussion

3.1 Sponges collection

In total, 9 sponges were successfully collected from two locations. Of these, seven sponges collected from Panjang Island and two sponges collected from Teluk Awur. Mostly the sponges which collected had dark colours such as black, grey with deep yellow colour inside, and deep blue or purple. The colour of sponge usually affected by the light intensity [19]. The shape and characteristic of the sponges could be seen at Tabel 1.

Tabel 1. Sponges collection.

Sample code	Sampling location	Pictures
PP.SP.16.01	Panjang island	
PP.SP.16.13	Panjang island	
PP.SP.16.14	Panjang island	
PP.SP.16.20	Panjang island	
PP.SP.16.36	Panjang island	
PP.SP.16.39	Panjang island	
PP.SP.16.40	Panjang island	
TA.SP.16.01	Teluk Awur	
TA.SP.16.03	Teluk Awur	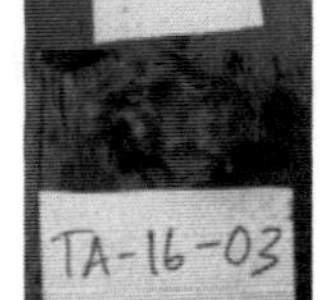

3.2 Sponge-associated fungi

Sponge-associated fungi were defined as fungi which isolated from sponge where the symbiotic mechanisms between fungi and its host still unknown. In addition, these fungi were very potential for pharmaceutical industry. Sponge-associated fungi such as *Aspergillus similanensis*, *Emericella variecolor, Hypocrea koningii* and, *Trichoderma hazarium* produced new chemical compounds with various biological activities such as antibacteria, antifungi, antioxidant, antitumor, and antihyperlipidemic [20,21,22,23]. There were 29 sponge-associated fungi isolated from 9 sponge collections that shown by Tabel 2.

Table 2. Sponge-associated fungi

No.	Sponge origin	Fungi isolates	No.	Sponge Origin	Fungi isolates
1.	TA. SP. 16. 03	MTS 3.1/MT 01	16.	PP. SP. 16. 01	MPS 1.1/ MT 15
2.	PP. SP. 16. 14	MPS 14.1/MT 02	17.	TA. SP. 16. 01	MTS 1.1/ MT 17
3.	PP. SP. 16. 14	MPS 14.2/ MT 03	18.	TA. SP. 16. 01	MTS 1.2/ MT 18
4.	PP. SP. 16. 14	MPS 14.3/ MT 04	19.	TA. SP. 16. 01	MTS 1.3/ MT 19
5.	PP. SP. 16. 14	MPS 14.4/ MT 05	20.	PP. SP. 16. 40	MPS 40.1/ MT 20
6.	PP. SP. 16. 14	MPS 14.5/ MT 06	21.	PP. SP. 16. 40	MPS 40.2/ MT 21
7.	PP. SP. 16. 14	MPS 14.6/ MT 07	22.	PP. SP. 16. 40	MPS 40.3/ MT 22
8.	PP. SP. 16. 14	MPS 14.7/ MT 08	23.	TA. SP. 16. 03	MTS 3.5/ MT 23
9.	PP. SP. 16. 14	MPS 14.8/ MT 09	24.	PP. SP. 16. 39	MPS 39.1/ MT 24
10.	PP. SP. 16. 14	MPS 14.9/ MT 10	25.	PP. SP. 16. 36	MPS 36.1/ MT 25
11.	PP. SP. 16. 14	MPS 14.10/ MT 11	26.	PP. SP. 16. 36	MPS 36.2/ MT 26
12.	TA. SP. 16. 03	MPS 3.2/ MT 12	27.	PP. SP. 16. 36	MPS 36.3/ MT 27
13.	TA. SP. 16. 03	MPS 3.3/ MT 13A	28.	PP. SP. 16. 40	MPS 40.4/ MT 32
14.	TA. SP. 16. 03	MPS 3.4/ MT 13B	29.	PP. SP. 16. 20	MPS 20.1/ MT 33
15.	PP. SP. 16. 13	MPS 13.1/ MT 14			

The biodiversity of sponge-associated fungi were influenced by the agar media in isolation step. Suitable nutrient in media will give higher biodiversity of isolated associated fungi [24,25]. This research used MEA because this agar medium is the common medium for environmental fungi isolation [14,26].

3.3 Screening of antibacterial activity

This research tried to find potential sponge-associated fungi to against pathogenic multi-drug resistant (MDR) *Escherichia coli*. The bacteria was clinical collection from Dr. Kariadi General Hospital Medical Center, Semarang, Central Java. According to antibiotic susceptibility test, this bacteria were resistant to 10 comercial antibiotics such as *Amoxicillin + Clavulanic acid*, *Ceftazidime*, *Ciprofloxacine*, *Cefepime*, *Gentamicin*, *Ampicillin sulbactam*, *Sulfamethoxazol+Trimetropi*, *Tetraciclin*, *Tigeciclin*, and *Piperacillin tazobactam*. Screening of antibacterial activity of 29 sponge-associated fungi was done by plug method. The presence of clear zone around the pluged agar was indicated the antibacterial activity of the fungi [15,16,27]. The result of antibacterial screening is showed by Tabel 3 and Figure 1.

The result of the screening antibacterial activity against MDR *E. coli* showed there were 7 isolates which performed antibacteria activity. Two biggest clear zones were produced by fungi MPS 14.1/MT 02 with inhibition zone was 17.2 mm^2 and MPS 14.3/MT 04 with inhibition zone was 16.9 mm^2. Furthermore, fungus MPS 14.1/MT 02 and MPS 14.3/MT 04 were identified by macro-microscopic and molecular approaches.

Table 3. Screening of antibacterial activity from sponge-associated fungi against MDR *E.* coli

Fungi isolate	Antibacterial activity	Diameter of clear zone (mm^2)
MTS 3.1/ MT 01	+	16.5
MPS 14.1/MT 02	+	17.2
MPS 14.2/ MT 03	-	-
MPS 14.3/ MT 04	+	16.9
MPS 14.4/ MT 05	+	12.2
MPS 14.5/ MT 06	-	-
MPS 14.6/ MT 07	-	-
MPS 14.7/ MT 08	+	12.1
MPS 14.8/ MT 09	-	-
MPS 14.9/ MT 10	-	-
MPS 14.10/ MT 11	-	-
MPS 3.2/ MT 12	-	-
MPS 3.3/ MT 13A	-	-
MPS 3.4/ MT 13B	-	-
MPS 13.1/ MT 14	-	-
MPS 1.1/ MT 15	-	-
MTS 1.1/ MT 17	-	-
MTS 1.2/ MT 18	-	-
MTS 1.3/ MT 19	-	-
MPS 40.1/ MT 20	-	-
MPS 40.2/ MT 21	-	-
MPS 40.3/ MT 22	-	-
MTS 3.5/ MT 23	+	16.1
MPS 39.1/ MT 24	-	-
MPS 36.1/ MT 25	-	-
MPS 36.2/ MT 26	-	-
MPS 36.3/ MT 27	-	-
MPS 40.4/ MT 32	-	-
MPS 20.1/ MT 33	+	10.9

\+ : performed antibacterial activity
\- : not performed antibacterial activity

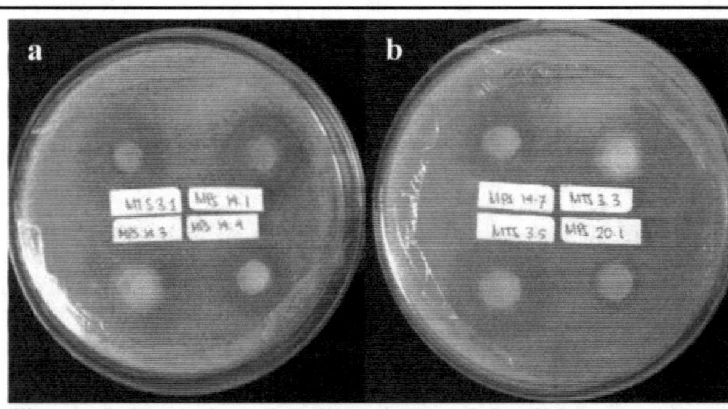

Figure 1. Screening of antibacterial activity from sponge-associated fungi.

3.4 Identification of Potential Fungi

Fungus MPS 14.1/MT 02 and MPS 14.3/MT 05 had colony distinct. The characteristics of the fungus MPS 14.1/MT 02 colony is shown by Figure 2.

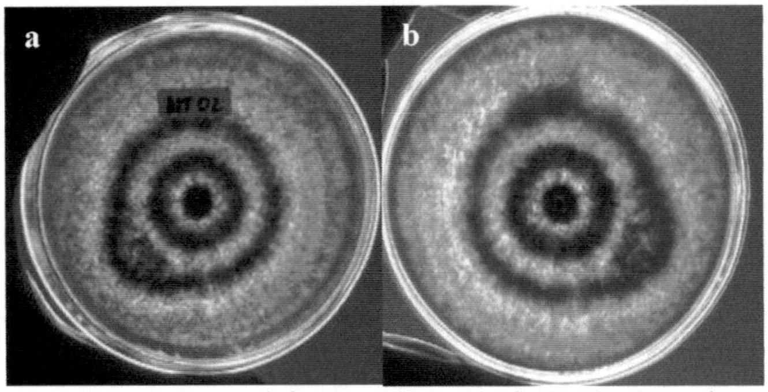

Figure 2. Colony of fungus MPS 14.1/MT 02; (a) colony characteristic and (b) reverse side.

Fungus MPS 14.1/MT 02 grew well at room temperature and had green colony with radial growth. This fungus didn't produce pigment in agar media. The microscopic morphology of MPS 14.1/MT 02 could at Figure 3.

Figure 3 shows that fungus MPS 14.1/MT 02 had septa in the hyphae, hyaline conidiophore with primary and secondary branches. The fungus produced sub-globose conidia and chlamydospore. According to the form of colony and its morphology under microscope, fungus MPS 14.1/MT 02 was judged as member of genus *Trichoderma* [28,29]. The colony of fungus MPS 14.3/MT 04 quite different from fungus MPS 14.1/MT 02. This fungus produced white mycelia with greenish yellow radial growth. In addition, fungus MPS 14.3/MT 04 released yellow colour at MEA media (Figure 4).

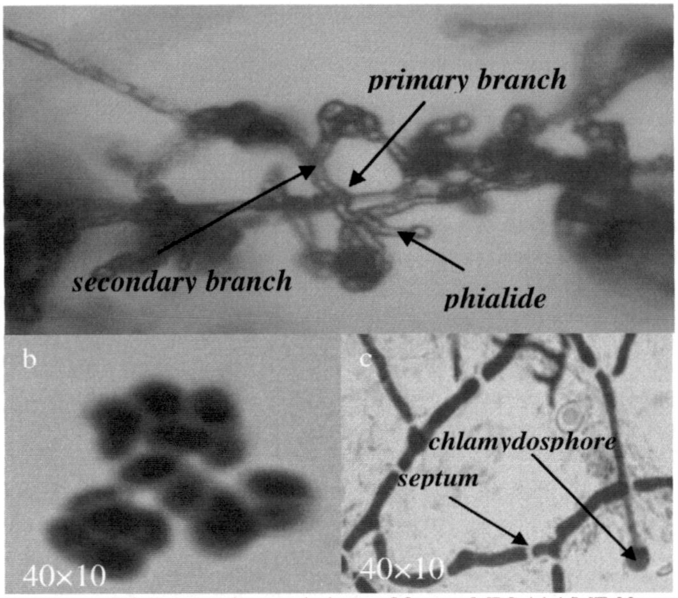

Figure 3. Microscopic morphology of fungus MPS 14.1/MT 02; (a) conidiophore, (b). conidia; (c) hyphae

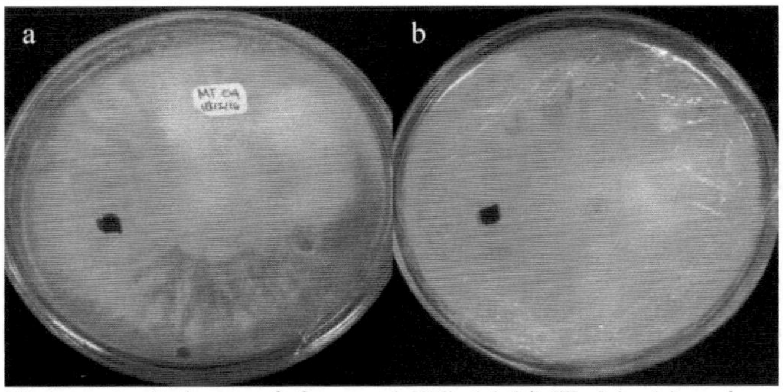

Figure 4. Colony of fungus MPS 14.3/MT 04; (a) colony morphology and (b) reverse side.

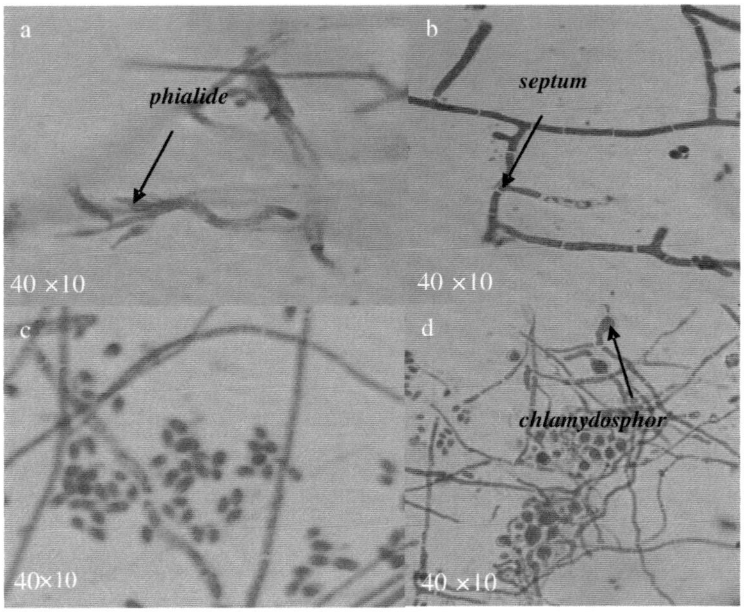

Figure 5. Microscopic morphology of fungus MPS 14.5/MT 04; (a) conidiophore, (b) hyphae, (c) conidia, (d) chlamydosphore

Fungus MPS 14.5/MT 04 also had septum and hyaline conidiophore, but braches of this fungus were rarely found. The conidia shape were sub globose and produced chlamydosphore. The characters of MPS 14.3/MT 04 are similar to 14.1/ MT 02 so that this fungus also judge as *Trichoderma*. This judgment were checked using molecular identification. For molecular identification, amplification was done in *internal transcribed space* (ITS) region because it is conserve region for fungi and usually give best result for fungi identification [30,31,32] The result of BLAST homology could be seen at Tabel 5 while the phylogenetic tree is shown by Figure 6.

Table 4. Homology of potential sponge associated-fungi

Fungi isolate	Molecular identification (BLAST closest relatives)	Similarity
MPS 14.1/MT 02	*Trichoderma asperellum* BK261A	99%
MPS 14.3/MT 04	*Trichoderma reesei* KU377472.1	99%

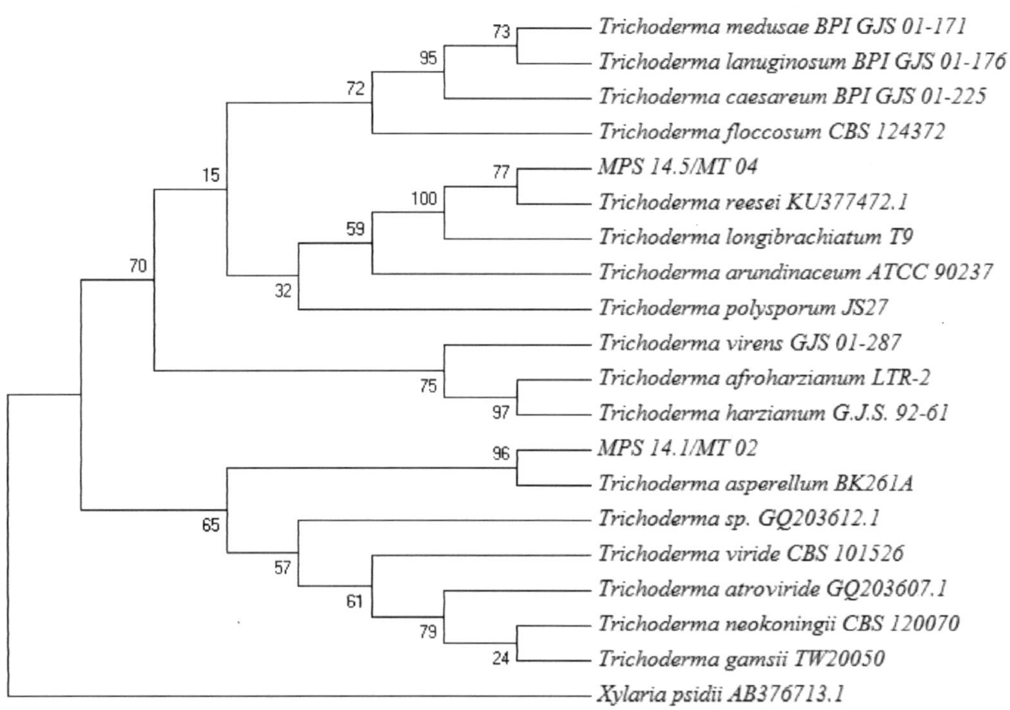

Figure 6. Phylogenetic tree of fungi MPS 14.1/MT 02 and MPS 14.5/MT 04

Blast homology search result of the potential sponge-associated fungi identified MPS 14.1/MT 02 as *Trichoderma asperellum* and MPS 14.5/MT 04 as *Trichoderma reesei*. Fungi from genus *Trichoderma* are usually found in soil and terrestrial plant as growth promotor and biocontrol agent to against plant disease [33,34,35,36]. The presence of *Trichoderma* in marine sponge demonstrate the tolerance of this fungi to salinity. Terrestrial fungi with salinity tolerance ability is known as facultative or marine-derived fungi. Several member of this genus were known as marine sponge-associated fungi such as *T. atroviride* and *T. cerinum* [37,38]. Marine-derived *Trichoderma* sp. isolated from Greenland seas produced new antibiotic *pyridones* while *T. longibrachiatum* which isolated from blue mussels produced long-chain peptaibols with antibacterial and antifungal activities [39,40]. *T. asperellum* from marine environment have been studied to produce *asperelines A-F, pertabiols* with antibacterial and antifungi activities while *trichodermatides A-D* with cytotoxic activity were isolated from marine-derived *T. reesei* [41,42]. Fungi *T. asperellum* MPS 14.1/MT 02 and *T. reesei* MPS 14.3/MT 04 isolated from sponge PP.SP.16.14 so that identification for this sponge was done through morphological and spicules identification.

3.5. Identification of sponge PP.SP.16.14

Sponge PP.SP.16.14 was collected from Panjang Island, found in depth less than 5 m and associated with corals. This sponge had characteristics such as globose shape with yellow with green line inside, had pit called as porocalices. This chacteristics owned by family tetillidae [43,44]. The pitures of sponge PP.SP.16.14 under water and outside water could be seen at Figure 7.

Figure 7. Sponge PP. SP. 16.14 (a. under water; b. outside of the water)

For sponge identification, spicules is very important as the specific character. Sponge PP.SP.16.14 had several forms of megascleres which showed by Figure 8.

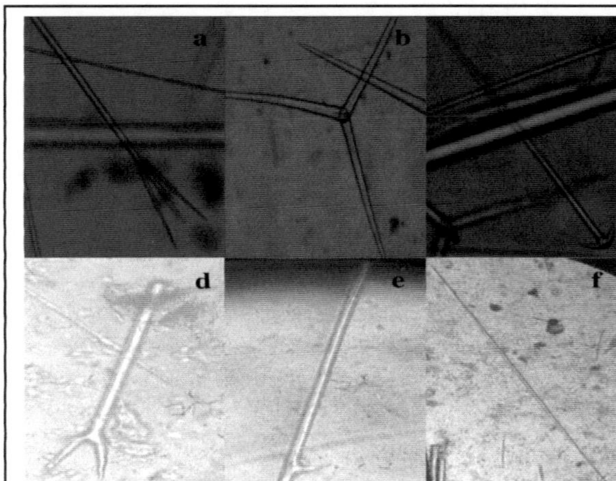

Figure 8. Spicule identification; (a) *protriaenes-spicule,* (b) *plagiotriaen-spicule,* (c) *anatriaen-spicule,* (d) *prodiaen-spicule,* (e) *promonoaen-spicule,* (f) *monaxon-spicule.*

In identification guide of sponge from family Tetillidae metion that sponge with specialized pore-bearing pits (porocalices), absence of meganthoxeas and calthrops with no discernible peripheral layer different from that of interior in cross section is member of genus *Cinachyrella* [43,44]. These characteristics are suitable to the characteristics of sponge PP.SP.16.14. So that this sponge was judged as *Cinachyrella* sp. Sponge *Cinachyrella* sp. is known as shallow water sponge and usually founded in tropical countries [45,46].

Acknowledgements
This research was supported by Directorate of Higher Education, The Ministries of Research, Technology and Higher Education, Republic of Indonesia. The authors thank to Dr. Jariya Sakayaroj, Researcher in Mycology Laboratory, NSTDA, Thailand for advices, support and supervision.

References
[1] Odonkor ST and Addo KK.2011. Resistance to Antibiotics: Recent Trends and Challenges. *Int. J. Biol. Med. Res.* 2(4): 1204–1210.
[2] Leung E, Weil DE., Raviglione M, and Nakatani H. 2011. The WHO policy package to combat antimicrobial resistance. *Bull. World Health Organ.* 89(5): 390–392.
[3] Center for Disease Dynamics, Economics and Policy. 2015. *State of The World's Antibiotics, 2015.* Washington, D.C: CDDEP.
[4] Edwards LA, Bajaj-Elliott M, Klein NJ, Murch SH, and Phillips AD. 2011. Bacterial-epithelial contact is a key determinant of host innate immune responses to enteropathogenic and enteroaggregative *Escherichia coli. PLoS One.* 6(10).
[5] Janny S, Bert F, Dondero F, Chanoine MH, Belghiti J, Mantz J, and Paugam-Burtz C. 2013. Fatal *Escherichia coli* skin and soft tissue infections in liver transplant recipients: Report of three cases. *Transpl. Infect. Dis.* 15(2): 49–54.
[6] Moore CE., Sona S, Poda S, Putchhat H, Kumar V, Sopheary S, Stoesser N, Bousfield R, Day N, and Parry CM. 2016. Antimicrobial susceptibility of uropathogens isolated from Cambodian children. *Paediatr. Int. Child Health.* 36(2): 113-117.
[7] Xu A, Zheng B, Xu YC, Huang ZG, Zhong NS, and Zhuo C.2016. National epidemiology of carbapenem-resistant and extensively drug-resistant gram-negative bacteria isolated from blood samples in China in 2013. *Clin. Microbiol. Infect.* 22(1), S1-S8.
[8] World Health Organization. 2014. Antimicrobial resistance: global reoport surveillance. *Bull. World Health Organ.* 61(3): 383–94.
[9] Yılmaz N, Ağuş N, Bayram A, Şamlıoğlu P, Cem Şirin M, Derici YK, and Hancı SY. 2016. Antimicrobial susceptibilities of *Escherichia coli* isolates as agents of community-acquired urinary tract infection (2008-2014). *Turk Urol. Derg.* 42(1): 32–36.
[10] Šeputienė V, Povilonis J, Ružauskas M, Žlabys P. 2006. Quinolone resistance among *Salmonella enterica* and *Escherichia coli* in Lithuania," *Biologija.* 3: 74–78.
[11] Rodríguez-Verdugo A, Gaut BS, and Tenaillon O. 2013. Evolution of *Escherichia coli* rifampicin resistance in an antibiotic-free environment during thermal stress. *BMC Evol. Biol.* 13(1): 50.
[12] Govinden-Soulange J, Marie D, Beesoo R, and Ramanjooloo A. 2014. Antibacterial properties of marine sponges from Mauritius. *Trop. J. Pharm. Res.* 13: 249–254.
[13] Sun S, Canning CB, Bhargava K, Sun X, Zhu W, Zhou N, Zhang Y, and Zhou K. Polybrominated diphenyl ethers with potent and broad spectrum antimicrobial activity from the marine sponge *Dysidea. Bioorg. Med. Chem. Lett.* 25(10): 2181–2183.
[14] Kjer J, Debbab A, Aly AH, and Proksch P. 2010. Methods for isolation of marine-derived endophytic fungi and their bioactive secondary products. *Nat. Protoc.* 5(3): 479–90.
[15] dos Santos IP, da Silva LCN, da Silva MV, de Araújo JM, Cavalcanti MS, and Lima VLM. 2015. Antibacterial activity of endophytic fungi from leaves of *Indigofera suffruticosa* Miller (Fabaceae). *Front. Microbiol.* 6: 1–7.
[16] Balouiri M, Sadiki M, and Ibnsouda SK. 2016. Methods for in vitro evaluating antimicrobial activity: a review. *J. Pharm. Anal.* 6(2): 71–79.
[17] Qiu WY, Yao YF, Zhu YF, Zhang YM, Zhou P, Jin YQ, and Zhang B. 2005. Fungal spectrum identified by a new slide culture and in vitro drug susceptibility using etest in fungal keratitis. *Curr. Eye Res.* 30(12): 1113–1120.

[18] Turan C, Nanni IM, Brunelli A, and Collina M. 2015. New rapid DNA extraction method with Chelex from *Venturia inaequalis* spores. *J. Microbiol. Methods.* 115: 139-143.

[19] Xue L and Zhang W. 2009. Growth and survival of early juveniles of the marine sponge *Hymeniacidon perlevis* (demospongiae) under controlled conditions. *Mar. Biotechnol.* 11(5): 640–649.

[20] Prompanya C, Dethoup T, Bessa LJ, Pinto MMM, Gales L, Costa PM, Silva AMS, and Kijjoa A. 2014. New isocoumarin derivatives and meroterpenoids from the marine sponge-associated fungus *Aspergillus similanensis* sp. nov. KUFA 0013. *Mar. Drugs.* 12(10): 5160–5173.

[21] Wu Q, Long HL, Liu D, Proksch P, and Lin WH. Varioxiranols I-L, new lactones from a sponge-associated *Emericella variecolor* fungus. *J. Asian Nat. Prod. Res.* 17(12): 1137–1145.

[22] Ding LJ, Gu BB, Jiao WH, Yuan W, Li YX, Tang WZ, Yu HB, Liao XJ, Han BN, Li ZY, Xu SH, and Lin HW. 2015. New furan and cyclopentenone derivatives from the sponge-associated fungus *Hypocrea koningii* PF04. *Mar. Drugs.* 13(9): 5579–5592.

[23] Li B, Huang QX, Gao D, Liu D, Ji YB, Liu HG, and Lin WH. 2015. New C 13 lipids from the marine-derived fungus *Trichoderma harzianum*. *J. Asian Nat. Prod. Res.* 17(5): 468–474.

[24] Kossuga MH, Romminger S, Xavier C, Milanetto MC, do Valle MZ, Pimenta EF, Morais RP, de Carvalho E, Mizuno CM, Coradello LFC, Barroso VM, Vacondio B, Javaroti DCD, Seleghim MHR, Cavalcanti BC, Pessoa C, Moraes MO, Lima BA, Gonçalves R, Bonugli-Santos RC, Sette LD, and Berlinck RGS. 2012. Evaluating methods for the isolation of marine-derived fungal strains and production of bioactive secondary metabolites. *Brazilian J. Pharmacogn.* 22(2): 257–267.

[25] Zhang XY, Bao J, Wang GH, He F, Xu XY, and Qi SH. 2012. Diversity and antimicrobial activity of culturable fungi isolated from six species of the South China Sea *Gorgonians. Microb. Ecol.* 64(3): 617–627.

[26] Gebala B and Sandle T. 2013. Comparison of different fungal agar for the environmental monitoring of pharmaceutical-grade cleanrooms. *PDA J. Pharm. Sci. Technol.* 67(6): 621–633.

[27] Rahaweman AC, Pamungkas J, Madduppa H, Thoms C, Tarman K. 2016. Screening of endophytic fungi from Chlorophyta and Phaeophyta for antibacterial actvity. *IOP Conference Series: Earth and Environmental Science.* 31: 1-9.

[28] Sriram S, Savitha MJ, Rohini HS, Jalali SK. 2013. The most widely used fungal antagonist for plant disease management in India, *Trichoderma viride* is *Trichoderma asperellum* as confirmed by oligonucleotide barcode and morphological characters. *Current Science* 104(10): 1332-1340.

[29] Chaverri P and Gary JS. 2003. *Hypocrea/Trichoderma (Ascomycota, Hypocreales, Hypocreaceae): Species with Green Ascospores.* (Netherlands: Centraalbureau voor Schimmelcultures) pp 1-35.

[30] White TJ, Bruns S, Lee S, and Taylor J. 1990. Amplification and direct sequencing of fungal ribosomal RNA genes for phylogenetics. *PCR Protocols: A Guide to Methods and Applications.* (Cambridge: Academic Press) pp 315–322.

[31] Larena I, Salazar O, González V, Julián MC, and Rubio V. 1999. Design of a primer for ribosomal DNA internal transcribed spacer with enhanced specificity for ascomycetes. *J. Biotechnol.* 75(2–3):187–194.

[32] Anderson IC, Campbell CD, and Prosser JI. 2003. Potential bias of fungal 18s rDNA and internal transcribed spacer polymerase chain reaction primers for estimating fungal biodiversity in soil. *Environ. Microbiol.* 5(1): 36–47.

[33] Rinu K, Sati P, and Pandey A. 2014. *Trichoderma gamsii* (NFCCI 2177): a newly isolated endophytic, psychrotolerant, plant growth promoting, and antagonistic fungal strain. *J. Basic Microbiol.* 54(5): 408–417.

[34] Saravanakumar K, Arasu VS, and Kathiresan K. 2013. Effect of *Trichoderma* on soil phosphate solubilization and growth improvement of *Avicennia marina*. *Aquat. Bot.* 104(2013):101–105.

[35] El-Komy MH, Saleh AA, Eranthodi A, and Molan YY. Characterization of novel *Trichoderma asperellum* isolates to select effective biocontrol agents against tomato *Fusarium* wilt. *Plant Pathol. J.* 31(1): 50–60.

[36] Hamed ER, Awad HM, Ghazi EA, El-Gamal NG, and Shehata HS. 2015. *Trichoderma asperellum* isolated from salinity soil using rice straw waste as biocontrol agent for cowpea plant pathogens. *J. Appl. Pharm. Sci.* 5(2): 91–98.

[37] Panizel I, Yarden O, Ilan M, and Carmeli S. 2013. Eight new peptaibols from sponge-associated *Trichoderma atroviride*. *Mar. Drugs.* 11(12): 4937–4960.

[38] Wiese J, Ohlendorf B, Blümel M, Schmaljohann R, and Imhoff JF. 2011. Phylogenetic identification of fungi isolated from the marine sponge *Tethya aurantium* and identification of their secondary metabolites. *Mar. Drugs*, 9(4): 561–585.

[39] Wu B, Oesker V, Wiese J, Schmaljohann R, and Imhoff JF. 2014. Two new antibiotic pyridones produced by a marine fungus, *Trichoderma* sp. strain MF106. *Mar. Drugs.* (12)3: 1208–1219.
[40] Mohamed-Benkada M, Pouchus YF, Vérité P, Pagniez F, Caroff N and Ruiz N. 2016. Identification and biological activities of long-chain peptaibols produced by a marine-derived strain of *Trichoderma longibrachiatum. Chem. Biodivers* 13(5): 521–530.
[41] Sun Y, Tian L, Huang J, Ma H, Zheng Z, and Lv A. 2008. Trichodermatides A – D , novel polyketides from the marine-derived fungus *Trichoderma reesei. Org. Lett.* 10(3): 393–396.
[42] Ren J, Xue C, Tian L, Xu M, Chen J, Deng Z, Proksch P, and Lin W. 2009. Asperelines A – F, peptaibols from the marine-derived fungus *Trichoderma asperellum.* 93(1): 1036–1044.
[43] van Soest RWM and Beglinger EJ. 2008. Tetractinellid and hadromerid sponges of the sultanate of Oman. *Zool. Meded.* 82(53): 749–790.
[44] van Soest RWM, Boury-Esnault N, Vacelet J, Dohrmann M,Erpenbeck D, de Voogd NJ, Santodomingo N, Vanhoorne B, Kelly M, and Hooper JNA. 2012. Global diversity of sponges (Porifera). *PLoS One.* 7(4).
[45] Freeman CJand Easson CG. 2016. Sponge distribution and the presence of photosymbionts in Moorea, French Polynesia. *PeerJ.* 4
[46] Nurhayati APD, Pratiwi R, Wahyuono S, Istriyati, Purnomo H, and Abdillah S. 2015. In vitro test and molecular docking of alkaloid compound in marine sponge *Cinachyrella anomala* against T47D cell cycle. *J. Mar. Sci. Res. Dev.* 5(2): 1–3.

Isolation, Characterisation and Antagonistic Activity of Bacteria Symbionts Hardcoral *Pavona* sp. Isolated from Panjang Island, Jepara Against Infectious Multi-drug Resistant (MDR) Bacteria

D. Ayuningrum[1], R. Kristiana [1], M.A. Asagabaldan[1], A. Sabdono[2], O.K. Radjasa [2,3], H. Nuryadi[4], A. Trianto[2]

[1]Student of Coastal Resources Management, Diponegoro University, Semarang, 50241, Indonesia
[2] Department of Marine Sciences, Diponegoro University, Semarang, 50241, Indonesia
[3]Director of Research and Public Services at Ministry of Research, Technology and Higher Education, Indonesia
[4]UPT Laboratorium Terpadu, Diponegoro University, Semarang, 50241, Indonesia

E-mail: diah.ayuningrum@student.undip.ac.id

Abstract. *Pavona* sp. is highly spread over Indonesian waters including Panjang Island. Several studies showed that bacteria symbionts hardcoral were the big source of antibiotic product, but there was limited research of the bacteria symbionts with hardcoral *Pavona* sp. In this research bacteria symbionts from hardcoral *Pavona* sp. had been collected from Panjang Island, Jepara. Marine bacteria symbionts were isolated by serial dillution method, while antibacterial activity was performed by using overlay and agar block method. The total of 2 from 5 isolates were active to MDR bacteria such as *Enterobacter aerogenes* and *Acinetobacter baumanii,* the code were PHC 44/04 and PHC 44/05. Then both of them were identified by morphological and molecular DNA characterization using 16 S rRNA gene sequence. The result of 16 S rRNA identification shows PHC 44/04 has 99% similarities with *Virgibacillus salarius* strain sa-Vb 1, while PHC 44/05 shows 99% similarities with *Pseudoalteromonas flavipulchra* strain NCIMB 2033.

Keywords : MDR bacteria, Antagonistic activity, *Virgibacillus salarius, Pseudoalteromonas flavipulchra.*

1. Introduction

Multi-drug Resistant (MDR) bacteria are bacteria that acquired non-susceptibility to at least one agent in three or more antibiotic categories [1][2]. To date, the emergence of MDR bacteria has spreaded worldwide and become a major public health problem. Data from Centers for Disease Control and Prevention (CDC) showed that the minimum number of illness and deaths caused by MDR are 2,049,442 illness and 23,000 deaths per year [3]. Furthermore, the estimated data of cases and death annual number for carbapenem-resistant *Enterobacteriaceae* are 9,300 case and 610 death; MDR *Acinetobacter* 7,300 and 500 death; Extended Spectrum Beta-Lactamases (ESBLs) 26,000 cases and 1,700 death; MDR *Pseudomonas aeruginosa* 6,700 cases and 440 death, and many more drug resistance bacteria infection cases and death [3]. Meanwhile, the case of drug resistency also occur in Asia, including Indonesia [4].

Antibiotic resistancy has been encompass any categories of antibiotics. CDC reported sixteen antibiotics resistance identified since 1940, those were Methicilin-resistant *Staphylococcus aureus* (MRSA), Penicillin-resistant *Pneumococcus* (PRP), Extended Spectrum Beta-Lactamases (ESBLs), Vancomycin-resistant *Enterococci* (VRE), and the latest is Ceftaroline-resistant *Staphylococcus* which found in 2011 [3]. Based on previous research ESBLs prevalency in Indonesia is the highest among ASEAN country, about 71.0% compared to Thailand 59.4% and *Asia Pacific Association Country* (APAC) 48,0% [5]. ESBLs *Enterobacteriaceae* resistency had covered the third generation of cephalosporin. In 2008 the percentage of ESBLs *E. aerogenes* resistant imipenem was 89.7% [6], cefoxitin resistant was 50.3% - 74.2%, ciprofloxacin resitant 51.4% and levofloxacin resistant 54.4% [7]. The resistency of the leading nosocomial agent (*Acinetobacter baumanni* and *Enterobacter aerogens*) to antibiotics such as imipenem, cefoxitin, ciprofloxacin, levofloxacin and many more are the reason to look for another new compound of antibiotics.

This condition compounded by the void development of antibiotics. Silver stated that there is a discovery void of antibiotics rather than gap since the fisrt antibiotic announce in year 1908 until 1987 [8]. The latest registered antibiotics after 1990's such as linezolid and daptomycin [3] were first found (patented) in 1978 and 1987, thus there have been no novel antibiotics discoveries since 1987[8][9]. This, lacked for the development of new novel antibiotics compound. Marine natural products have been studied as antibiotics candidat [10][11]. One of the marine natural product source is coral which have many activities as antimicrobial, antivirus, antifungal, anti-plasmodial, cytotoxic, and anti-inflamatory [12][13][14].

Pavona is a genus from family Agariciidae which commonly found in Indopacific region including Indonesia. According to previous research [15] one genus from Agariciidae which found in Panjang Island vicinity, Jepara from depth 0.5 – 7 m is *Pavona.* Although widely spread in tropical regions, the exploration of bioactive compound from *Pavona* sp is still limited. Corals which have been widely explored are Plexauridae, Clavulariidae, Xeniidae and Gorgoniidae [10]. However the main barriers for secondary metabolites from corals entering the clinical trial is the issue of availability of raw materials. The low concentration of secondary metabolites is less than 10^{-6}% by weight of the wet invertebrates [16]. It is certainly not conservative and disturb the balance of aquatic ecosystems. Therefore coral microsymbiont is the answer for the challenge. Coral symbionts bacteria have the similar compound with their host which can be source of new antibiotics compound[17]. Moreover bacteria have the characteristics of easy to breed, takes relatively shorter culture and do not require much space.

The purpose of this research were to isolate bacteria symbiont from coral *Pavona* sp, to screen antagonistic activity against human patogenic MDR bacteria, and to characterize active isolate morphologically and molecularly using PCR amplifications 16S rRNA gene sequences.

2. Materials and Methods

2.1. Sampling and isolation of symbiont bacteria

Hardcoral *Pavona* sp was collected from Panjang island vicinity, Jepara, Central Java, Indonesia by skin diving to a depth approximately 3m. During collection, *Pavona* sp was put into sterile ziplock plastic bag and placed inside the cool box.

The isolation and purification of symbionts bacteria were conducted in Tropical Marine Biotechnology Laboratory, Diponegoro University. The isolation process was conducted by serial dillution method, furthermore the purification process was carried out by streak plate method [18].

2.2. Screening of antagonistic activity
The screening of antagonistic activity was also conducted in Tropical Marine Biotechnology Laboratory, Diponegoro University. The purpose of this process was to screen the antagonistic activity between isolate and tested bacteria (MDR). The screening method against pathogenic MDR bacteria (*Pseudomonas aeruginosa, Klebsiella pneumonia, Enterobacter aerogenes,* MRSA, *Escherischia coli, Enterobacter cloacae complex, Staphylococcus haemolyticus* and *Acinetobacter baumannii*) was performed by overlay method [17]. Each 1% of the test-bacteria cultures at logarithmic phase for as much as 0,5 Mc Farland was mixed into soft agar medium which then poured on to the agar surface which was previously inoculated with isolates for seven days incubation in room temperature. The isolates which have antagonistic activity will form clear zone surround it.

The result of antagonistic activity was confirmed using agar block method [19]. Each of the isolates was cultured on the agar media for 7 days. Cylindrical pieces were cut out from each cultures. The blocks were placed on to agar medium which have been inoculated with a fix amount of test-bacteria (0,5 Mc Farland). Cultures stayed for 14 – 18 hours at 2 – 8 ^{0}C and then incubated at 37^{0}C for 24 hours. The antagonistic activity was measured in milimeter (mm) clear zone on the 24th until 48th hours. The strongest potential isolate with the greatest clear zone was characterized by morphological and molecular characterization using 16S rRNA gene sequence.

2.3. Morphological characterization
Morphological characterization was conducted by observing colony morphology and gram's staining. Observation of the colony include the colony colour, size, form, elevation and margin [18]. Gram's staining method was conducted by using gram A (crystal violet), gram B (lugol iodine), gram C (alcohol) and gram D(Safranin) [20].

2.4. DNA extraction
Chelex method with several modification was used for DNA extraction [21]. Selected colonies from the best isolate were put in to 50 – 100 µl ddH$_2$O and 1 ml of 0,5 % saponin in PBS 1x (stayed overnight at 4^{0}C). The incubated mixture was centrifuged at 12000 RPM for 10 minutes and discard the supernatant. As much as 1ml PBL 1x was added to the mixture and centrifuged again at 12000 RPM for 5 minutes and discard the supernatant. Then 100 µl ddH$_2$O and 50 µl of 20% Chelex 100 (shake well and ensure the chelex crystal mix well to the sample) were added to the final mixture. The mixture was boiled for 10 minutes and vortex once after 5 minutes. Sentrifuge the mixture with 12000 RPM for 10 minutes. The supernatan was moved in to steriled microtube and saved in -20^{0}C. The examination of DNA concentration was conducted using NanoDrop 2000 spectrophotometer (Thermo Scientific). DNA purity and concentration was calculated using 260/280 and 260/230 ratios by NanoDrop 2000 spectrophotometer (Thermo Scientific).

2.5. PCR amplification 16S rRNA gene sequence
Primers used for PCR amplification 16S rRNA were universal primer 27 F ((5'-AGAGTTTGATCMTGGCTCAG-3) and 1492 R (5'- GGTTACCTTGTTACGACTT-3'). The PCR mixture contain GoTaq®Green Master Mix Promega (12.5 µl), primer 27 F (1 µL), primer 1492 R (1 µL), template DNA (1 µl) and ddH$_2$O (9.5 µl), so that the total volume was 25 µl.

PCR reaction was conducted in a MJ Mini Personal Thermal Cycler (BIO RAD) using optimation consist of initial denaturation at 95^{0}C for 3 minutes, then followed by 30 cycles of denaturation at 95^{0}C for 1 min each cycle, annealing at 53.9^{0}C for 1 minutes, extension at 72^{0}C for 1 minutes, and followed by final extension at 72^{0}C for 7 minutes. PCR products were examined using agarose 1 % gel electrophoresis and the result was visualized by using UVIDoc HD5 (UVITEC Cambridge).

2.6. DNA sequencing

DNA template was sent to PT. Genetika Science (Jakarta, Indonesia) for 16S rRNA sequence determination. The allignment of sequences result were done using MEGA 5.05, then allignment result was inserted to the BLAST program to create homology of any closely related bacteria in the gene bank database (NCBI, http://www.ncbi.nlm.nih.gov). The result of BLAST Homology was deposite to DNA Data Bank of Japan (DDBJ, www.ddbj.nig.ac.jp) in orger to obtain the accession number.

2.7. Phylogenetic analysis

Phylogenetic tree was constructed using neighbor-joining analysis inside the program MEGA 5.05. The result of phylogenetic tree construction was examined using bootstrap method.

3. Result

3.1. Isolation and purification of symbionts bacteria

The total of symbionts bacteria from one sample *Pavona* sp. obtained 5 pure isolates with different morphology each other. Morphological characterization of each isolates was performed in table 1 as follow.

Table 1. Morphological characteristic of bacteria symbionts *Pavona* sp.

Isolate code	Colour	Size	Shape	Margin	Elevation
PHC-44/01	Pinkish white	*Pin point*	Round	Entire	Convex
PHC-44/02	White turbid	Medium	Round	Entire	Convex
PHC-44/03	Thin white turbid	Small	Round	Entire	Convex
PHC-44/04	White	Medium	Round	Entire	Convex
PHC-44/05	Orange	Medium	Round	Entire	Convex

3.2. Screening of antagonistic activity

Screening of antagonistic activity from isolates of bacteria symbionts *Pavona* sp. against MDR bacteria obtained 3 active isolates. The result of antagonistic activity screening was presented in table 2 as follow.

Table 2. Result of antagonistic activity screening from bacteria symbionts*Pavona* sp.

Isolate Code	MDR								Antagonistic Activity
	P. aeruginosa	*K. pneumonia*	*E. aerogenes*	*A. baumanni*	MRSA	*E. coli*	*E. cloacae complex*	*S. hemolyticus*	
PHC 44/01	-	-	-	-	-	-	-	-	Not active
PHC 44/02	+	-	+	-	-	-	-	-	Active
PHC 44/03	-	-	-	-	-	-	-	-	Not active
PHC 44/04	-	-	++	-	+++	-	-	-	Active
PHC 44/05	-	-	+	+++	+++	++	-	+	Active

Confirmation for the best isolates was conducted by using agar block method. The confirmation result was shown in table 3.

Table 3. Confirmation of antagonistic activity from isolates of bacteria symbionts *Pavona* sp.

Isolate code	***Enterobacter aerogenes* (mm)**		***Acinetobacter baumannii* (mm)**	
	Replication 1	Replication 2	Replication 1	Replication 2
PHC-44/04	29.60	30,23	-	-
PHC-44/05	-	-	15	16

3.3. Morphological characterization

The result of morphological characterization using gram staining showed that isolate PHC-44/04 was gram positive bacteria, meanwhile isolate PHC-44/05 was gram negative bacteria (fig. 1).

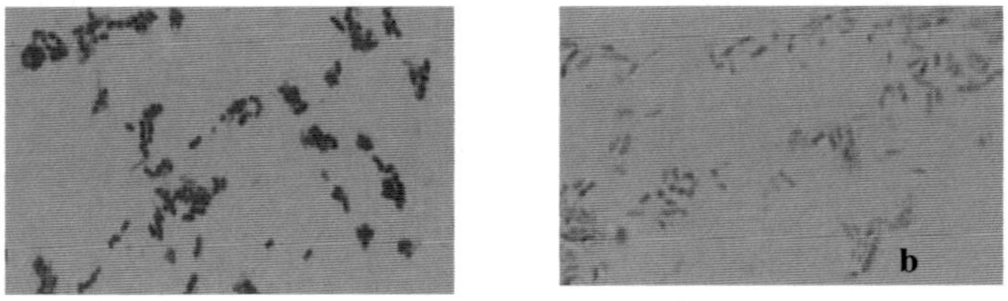

Figure 1. (a) gram staining for isolate PHC-44/04 showed gram positive bacteria, (b) gram staining for isolate PHC-44/05 showed gram negative bacteria (magnification 1000 times)

3.4. Molecular characterization and Phylogenetic analysis

The result of isolate PHC-44/04 BLAST homology showed that this bacterium was closely related to the family Bacillaceae and genus *Virgibacillus*. Meanwhile BLAST homology result for isolate PHC-44/05 showed correlation with *Pseudoalteromonas flavipulchra* as much as 99%. Phylogenetic tree showed that isolate PHC-44/04 was closely related 99% to *Virgibacillus salarius* strain Sa-Vb-1 (Fig. 2)

4. Discussion

The result of isolation obtained 5 bacteria isolate symbiont *Pavona* sp. Coral microbiota store various symbionts such as algae (dinoflagellates), viruses, bacteria and arkhaea which known as coral holobiont. Bacteria symbionts corals provide carbon, nitrogen, sulfur biogeochemical cycles and nourish also protect the coral from pathogen infection [22]. A wide variety of microbiota live in the mucus layer, tissue and skeletal calcium carbonate. The highest composition and abundant of bacteria can be found in mucus layer [23]. Mucus layer contain *mucopolysaccharida* which is nutrition for bacteria consortium. While, the amount of bacteria consortium increase it will make competition to each other. Therefore some bacteria have ability to produce secondary metabolites to maintain their existence.

The ability to inhibit the growth of MDR bacteria was different in each isolate. Isolate PHC-44/02 and PHC-44/05 inhibit in the first replication only, while isolate PHC-44/04 inhibits in the second replication. Furthermore, the antagonistic activity of isolate PHC-44/04 was confirmed by agar block method as shown in the table 3. The diameter of the inhibition zone obtained 29.60 mm and 30.23 mm. This is supported by Radjasa et al [17] which showed that the bacteria symbionts coral *Pavona* have antagonist activity against gram positive and negative bacteria.

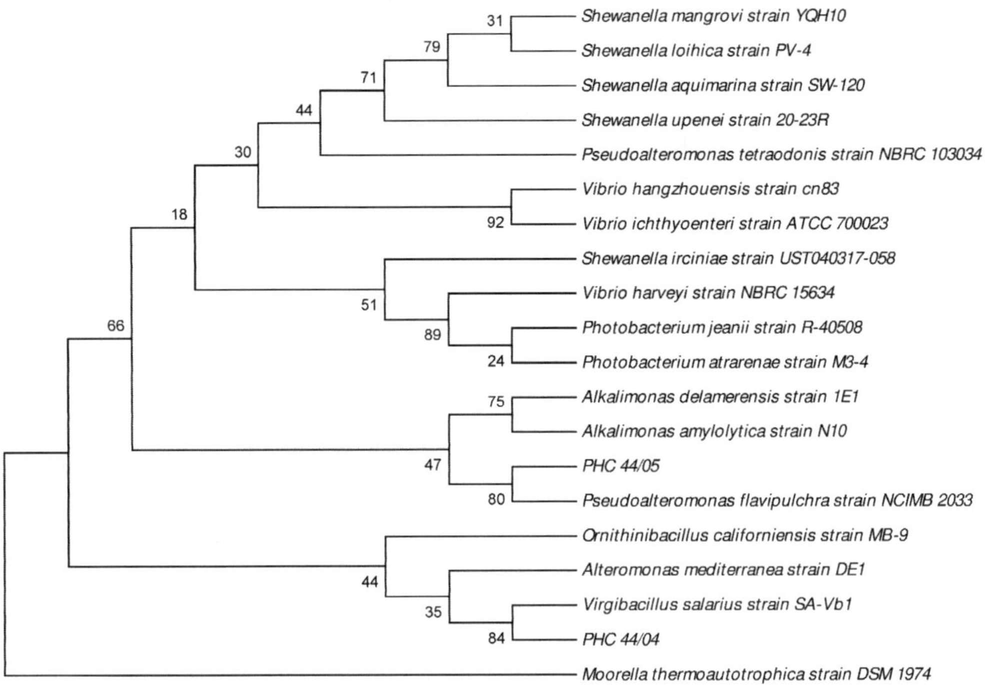

Figure 2. Phylogenetic tree from isolate PHC-44/04 and PHC-44/05

The results of microscopic identification showed that isolate PHC-44/04 stained purple (gram-positive bacteria). Meanwhile isolate PHC-44/05 showed red colour (gram negative) (fig. 1). This is due to gram-positive bacteria cell wall was thicker so it is more powerful to tie the color of the crystal violet (gram A). Thick peptidoglycan layer also makes crystal violet color does not fade when washed with gram C or alcohol. So when colored with dyes gram D (counterstain) bacteria remain purple. This mechanism was contrast with the gram negative bacteria (PHC-44/05). Moreover, microscopically isolates PHC-44/04 looks short bacillus shaped with spores inside, singly or in pairs. Those characteristics of is closely similar to the characteristics possessed by family Bacillaceae [24]. Meanwhile isolate PHC-44/05 looks bacillus shaped without spore and singly. The results of sequence homology 16S rRNA of isolate PHC-44/04 and PHC-44/05 showed similarity with *Virgibacillus salarius* strain Sa-Vb 1 with accesion number LC 190934 and *Pseudoalteromonas flavipulchra* strain NCIMB 2033 as much as 99%.

The phylogenetic tree showed that isolate PHC-44/04 is at the same clade with *Virgibacillus salarius* strain SA-Vb 1, which is halofil bacteria isolated from the Saharan Salt Lake [25]. Many strains of the genus Virgibacillus obtained from Brazilian sponge have antimicrobial activity [26]. This, suggested that many species of the genus were potential source of antimicrobial compounds. Moreover strain *Virgibacillus* sp. TS2A5 also has antibacterial activity against *Staphylococcus aureus* and *Escherichia coli* were also found in the same waters with *Virgibacillus salarius* strain SA-Vb 1 [27]. On the other hand, isolate PHC-44/05 is in the same clade with *Pseudoalteromonas flavipulchra* strain NCIMB 2033 (fig. 2) with characteristic a marine, gram negative bacteria, lemon-yellow colour and has antibacterial activity against pathogenic bacteria [28] [29] also cytotoxic activity against sea urchin eggs [30]. This is proven that marine bacteria symbionts have potential ability to be source of new antibiotics compound from marine environment.

5. Conclusion

Isolation and purification process result 5 pure isolate of bacteria symbionts *Pavona* sp, which are 3 or them have antagonistic activity against MDR bacteria. The strongest isolates against MDR *Enterobacter aerogenes* and *Acinetobacter baumannii* are isolate PHC-44/04 and PHC-44/05 which has 99% similarities with *Virgibacillus salarius* strain Sa-Vb 1 and *Pseudoalteromonas flavipulchra* strain NCIMB 2033.

Acknowledgement

This research is supported by the grant from Ministry of Research, Technology and Higher Education, Indonesia under the Master Program of Education Leading to Doctoral Degree for Excellent Graduates (PMDSU) scheme No. 2993-07/UN7.5.1/PG/2016.

References

[1] Magiorakos A, Srinivasan A, Carey R B, Carmeli Y, Falagas M E, Giske C G, Harbarth S, Hindler J F, Kahimeter G, Olsson-Lijequist B, Paterson D L, Rice L B, Stelling J, Struelens M J, Vatopoulos A, Weber J T, and Monnet D L. 2012. Multi-drug resistant, extensively drug-resistant and pandrug-resistant bacteria: an international expert proposal for interim standard definitions for acquired resistance. *Clin Microbiol Infect* **18:** 268 - 281.

[2] Nikaido H. 2010. Multidrug Resistance in Bacteria. *Annu Rev Biochem.* 119–46

[3] Centers for Disease Control and Prevention [CDC]. 2013. *Antibiotic resistance threats in the United States, 2013.* United States.

[4] Lestari E S and Severin J A. 2009. *Antimicrobial Resistance in Indonesia Prevalence, determinants and genetic basis.* Rotterdam: Erasmus Universiteit

[5] Mendes R E, Mendoza M, Banga Singh K K, Castanheira M, Bell J M, Turnidge J D, Lin S S F and Jones R N. 2013. Regional resistance surveillance program results for 12 Asia-Pacific nations (2011) *Antimicrob. Agents Chemother.* **57** 5721–6

[6] Biendo M, Canarelli B, Thomas D, Rousseau F, Hamdad F, Adjide C, Laurans G and Eb F. 2008. Successive emergence of extended-spectrum β-lactamase-producing and carbapenemase-producing *Enterobacter aerogenes* isolates in a university hospital. *J. Clin. Microbiol.* **46** 1037–44

[7] Lu P L, Liu Y C, Toh H S, Lee Y L, Liu Y M, Ho C M, Huang C C, Liu C E, Ko W C, Wang J H, Tang H J, Yu K W, Chen Y S, Chuang Y C, Xu Y, Ni Y, Chen Y H and Hsueh P R. 2012. Epidemiology and antimicrobial susceptibility profiles of Gram-negative bacteria causing urinary tract infections in the Asia-Pacific region: 2009-2010 results from the Study for Monitoring Antimicrobial Resistance Trends (SMART). *Int. J. Antimicrob. Agents* **40** S37–43

[8] Silver L. L. 2011. Challenges of Antibacterial Discovery. *Clinical Microbiology Reviews* **24** 71–109

[9] World Health Organization [WHO]. 2014. *Antimicrobial resistance Global Report on Surveillance.* Switzerland

[10] Kim S (ed). 2015. *Handbook of Marine Biotechnology*. London: Springer

[11] Blunt J W, Copp B R, Keyzers R A, Munro M H G and Prinsep M R. 2012. Review Marine natural products. *Nat. Prod. Rep.* **29** 144–222

[12] Raina J, Tapiolas D, Motti C, Foret S and Seemann T. 2016. Isolation of an antimicrobial compound produced by bacteria associated with reef-building corals. *PeerJ* 4:e2275 https://doi.org/10.7717/peerj.2275.

[13] Mohammadizadeh F, Afkhami M, Ehsanpour M, Bahri A and Taziani H M. 2014. Screening for antibacterial, antifungal and cytotoxic agents in three hard coral species from Persian Gulf. *Bangladesh J. Pharmacol.* **9** 171–5

[14] Meyer M, Delberghe F, Liron F, Guillaume M, Valentin A and Guyot M. 2009. An antiplasmodial new (bis)indole alkaloid from the hard coral *Tubastraea* sp. *Nat. Prod. Res.* **23** 178–82

[15] Munasik, Ambariyanto, Sabdono A, Wijayanti D P, Radjasa O K and Pribadi R. 2012. Sebaran spasial karang keras (Scleractinia) di Pulau Panjang , Jawa Tengah. *Bul. Oseanografi Mar.* **1** 16–24

[16] Proksch P, Edrada R A and Ebel R. 2002. Drugs from the seas - Current status and microbiological implications. *Appl. Microbiol. Biotechnol.* **59** 125–34

[17] Radjasa O, Wiese J, Sabdono A and Imhoff J F. 2008. Corals As Source of Bacteria With Antimicrobial Activity *J. Coast. Dev.* **11** 121–30

[18] Benson H J. 2002. *Microbiological Applications A Laboratory Manual in General Microbiology 8th Edition.* New York: The McGraw-Hill Companies

[19] Nedialkova D and Naidenova M. 2005. Screening The Antimicrobial Activity of Actinomycetes Strains Isolated From Antartica. *J. Cult. Collect.* **4** 29–35

[20] Madigan M T, Martinko J M, Stahl D A and and Clark D. 2013. *Brock Biology of Microorganisms 3th Ed.*San Francisco: Benjamin Cummings.

[21] de Lamballerie X, Zandotti C, Vignoli C, Bollet C and de Micco P. 1992. A one-step microbial DNA extraction method using "Chelex 100" suitable for gene amplification. *Res. Microbiol.* **143** 785–90

[22] Kuang W, Li J, Zhang S and Long L. 2015. Diversity and distribution of Actinobacteria associated with reef coral *Porites lutea*. *Front. Microbiol.* **6** 1–13

[23] Sweet M J, Croquer A and Bythell J C. 2011. Bacterial assemblages differ between compartments within the coral holobiont. *Coral Reefs* **30** 39–52

[24] Rosenberg E (ed). 2006. *The Prokaryotes Firmicutes and Tenericutes* 4th edition. Springer

[25] Hua N P, Hamza-Chaffai A, Vreeland R H, Isoda H and Naganuma T. 2008. *Virgibacillus salarius* sp. nov., a halophilic bacterium isolated from a Saharan salt lake. *Int. J. Syst. Evol. Microbiol.* **58** 2409–14

[26] Santos O C S, Pontes P V M L, Santos J F M, Muricy G, Giambiagi-deMarval M and Laport M S. 2010. Isolation, characterization and phylogeny of sponge-associated bacteria with antimicrobial activities from Brazil. *Res. Microbiol.* **161** 604–12

[27] Khoeri M M, Radjasa O K, Sabdono A and Sudoyo H. 2011 Bioprospecting of Bacterial Symbiont of Tunicate *Didemnum Molle* from Sambangan, Karimunjawa Islands. *J. Coast. Dev.* **14** 255–61

[28] Yu M, Wang J, Tang K, Shi X, Wang S, Zhu W M and Zhang X H. 2012. Purification and characterization of antibacterial compounds of *Pseudoalteromonas flavipulchra* JG1. *Microbiology* **158** 835–42

[29] Yoghiapiscessa D, Batubara I and Wahyudi A T. 2016. Antimicrobial and Antioxidant Activities of Bacterial Extracts from Marine Bacteria Associated with Sponge *Stylotella* sp. *Am. J. Biochem. Biotechnol.* **12** 36–46

[30] Ivanova E P, Shevchenko L S, Sawabe T, Lysenko A M, Svetashev V I, Gorshkova N M, Satomi M, Christen R and Mikhailov V V. 2002. *Pseudoalteromonas maricaloris* sp. nov., isolated from an Australian sponge, and reclassification of [*Pseudoalteromonas aurantia*] NCIMB 2033 as *Pseudoalteromonas flavipulchra* sp. nov. *Int. J. Syst. Evol. Microbiol.* **52** 263–71

IOP Conf. Series: Earth and Environmental Science **55** (2017) 012030 doi:10.1088/1755-1315/55/1/012030

Encapsulation of Phycocyanin-Alginate for High Stability and Antioxidant Activity

Hadiyanto[1*], Meiny Suzery[2], Deny Setyawan[2], Dian Majid[2], Heri Sutanto[3]

[1] Center of Biomass and Renewable Energy, Department of Chemical Engineering, Faculty of Engineering, Diponegoro University, Jl. Prof Soedarto, SH, Tembalang, Semarang 50275

[2] Department of Chemistry, Faculty of Science and Mathematics, Diponegoro University, Jl. Prof Soedarto, SH, Tembalang, Semarang 50275
[3] Department of Physics, Faculty of Science and Mathematic, Diponegoro University, Jl. Prof Soedarto, SH, Tembalang, Semarang 50275

*Corresponding author E-mail: hadiyanto@live.undip.ac.id

Abstract. The aim of this study was to obtain optimal condition of phycocyanin-alginate encapsulation, encapsulation efficiency and phycocyanin load, physicochemical properties of beads, in vitro release study, stability and antioxidant activity. The result product with alginate content 1,5% (w/v) and 2% (w/v) produced were in spherical shape than product with alginate content 2,5% (w/v) by ratio of phycocyanin 1:1. Increasing alginate content on encapsulation process will increase of encapsulation efficiency and phycocyanin load. In vitro released study showed that phycocyanin-alginate beads were more resistant in simulated gastric fluid, while rapidly release in simulated intestinal fluid. The antioxidant activity showed that phycocyanin antioxidant activity decreased after encapsulation process due to duration of storage and the possibility of a cracking which will cause reduced stability of phycocyanin.

Keywords: encapsulation, phycocyanin, alginate, extrusion, antioxidant

1. Introduction

Phycocyanin is a blue-light pigment found in cyanobacteria and two eukaryotics algae such as *Rhodophyta* and *Crytophyta*. Phycocyanin is soluble in water and has a strong fluorescent properties as an antioxidant and it is normally used in food, cosmetics, biotechnology and drugs. Phycocyanin is a blue phycobiliprotein composed of two homologous units: α-chain and β-chain attached to the cysteine cluster [1]. Moreover, besides its use of antioxidant, phycocyanin has also properties as an anti-cancer, anti-inflammatory [2], neuroprotection [3], and prevention of hypertension [4].

However application of phycocyanin as active compounds in functional foods is often accompanied by problems of stability caused by elevation of moisture content, light and temperature which cause degradation of phycocyanin [5]. Several researchers showed that microencapsulation is an effective and economical method to maintain the stability of phycocyanin [6]. Coating in encapsulated product will protect the active compound from external influences and also regulates the release of the compound capsules [7].

Alginate is a biodegradable polymer that is mostly used as coating material because it can easily form a gel beads in a solution without the use of organic solvents and at room temperature [4]. The current study is carried out to evaluate the use of alginate as coating material for encapsulation of phycocyanin under variation of concentration of alginate, while $CaCl_2$ was used as linking agent. Ca^{2+} was chosen because not toxic and is frequently used as a fitting cross formation with a polymer alginate microspheres. Moreover $CaCl_2$ easily connected with alginate because Ca^{2+} ions bound to guluronic acid residue which is a component of alginate[8].

2. Material and Methods

2.1. Material

Phycocyanin was extracted from microalgae *Spirulina platensis*. Alginate and $CaCl_2$ used in this study were food grade quality obtained from the local market. All other reagents used were analytical grade.

2.2. Encapsulation of phycocyanin – alginate

Alginate solution was prepared with concentration of 1,5% (w/v), 2% (w/v), 2,5% (w/v) referring to previous research of Yan et al[9]. Phycocyanin added to the alginate with the ratio of 1: 1 (w/w) and then stirred using a magnetic stirrer until the solution is homogeneous. The homogenized suspension was inserted into the syringe and injected into a solution of $CaCl_2$ 2,5% (w/v) using a 23G gauge needle and stirred by a magnetic stirrer. The beads were collected and washed using distilled water to remove Cl^- ions. The $CaCl_2$ and distilled water were collected to calculate the encapsulation efficiency and beads were stored in the refrigerator.

2.3. Determination of encapsulation efficiency and phycocyanin load

Encapsulation efficiency (EE) was calculated from the mass phycocyanin, which was not encapsulated compared to the mass of phycocyanin added at the beginning of process [10]:

$$\mathrm{EE}(\%) = \frac{\text{initial mass of phycocyanin} - \text{mass of uncoated phycocyanin}}{\text{Initial mas of phycocyanin}} \times 100 \qquad (1)$$

Phycocyanin load was calculated by dissolving 0.025 gr dried encapsulated products in Na phosphate buffer pH 7.4. After 20 min then it was diluted using 25 mL flask and then allowed to stand for 30 minutes in order to precipitate alginate. The filtrate was separated from the solution and absorbance was measured. The phycocyanin load was calculated using equation 2:

$$\text{Phycocyanin load}(\%) = \frac{\text{mass of phycocyanin in beads}}{\text{mass of coating material in beads}} \times 100 \qquad (2)$$

2.4. Characterization of encapsulation

Beads were characterized using SEM instrument to determine the morphology and FT-IR was used to determine the chemical interactions that occur after the encapsulation.

2.5. In vitro release study

Dissolution test was conducted by weighting 253.8 mg beads under concentration variation of 2.5% (equivalent to 116.5125 mg pure phycocyanin). The beads then was put into HCl buffer at pH 1.2 as simulated gastric fluid(SGF) medium after 2 hours and then transferred to Na phosphate buffer pH 7.4 as simulated intestinal fluid(SIF) medium for 6 hours. A 5 mL sample was measured every 60 minutes and was measured for its absorbance by using Uv-Vis spectrophotometer.

2.6. Stability and antioxidant activity assay

Antioxidant test using DPPH method was referred to Molyneux[6]. A 9 mg phycocyanin was weighed for a control and 25 mg of the encapsulated samples (equivalent to 11.5 mg of pure phycocyanin) were then dissolved in Na phosphate buffer pH 7.4 using 25 mL volumetric flask to obtain a concentration of 360 ppm for control and 460 ppm for sample encapsulation. The samples were prepared by varying concentration of 18 ppm, 36 ppm, 54 ppm, 72 ppm and 90 ppm for control and 23 ppm, 46 ppm, 69 ppm, 92 ppm and 115 ppm for encapsulation. A 2 mL of sample was added to 2 mL of 50 μM DPPH (in methanol) and the mixture was shaken and then incubated in the dark for 30 minutes. Blank solution consisted of 2 ml Na phosphate buffer pH 7.4 and 2 mL of DPPH 50 μM. The solution was measured for its absorbance at the maximum λ. The percentage of inhibition was calculated by equation 3:

$$\%\ \text{Inhibition} = \frac{A_{blank} - A_{sampel}}{A_{blanko}} \quad (3)$$

Storage stability was performed under temperature variation of 35, 45, 55^{0}C for 3 days. Samples were measured for their antioxidant activities after incubation for 3 days.

3. Result and Discussion

3.1. Preparation of phycocyanin – alginate encapsulation

Figure 1 shows the product of encapsulation with variation of concentration. Most of the products showed the homogenous sizes however, at concentration of 2.5%, the shape of encapsulated products was tailed beads. This is because at concentration of 2.5%, the alginate solution has a high viscosity which affecting the extrusion.

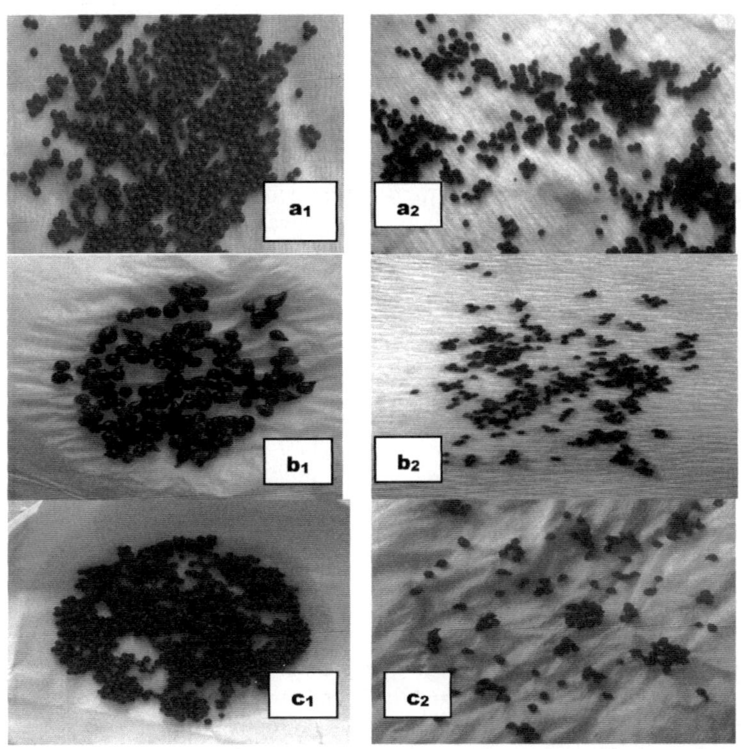

Figure 1. Morphology of encapsulated phycocyanin at various concentrations. Note: Concentration of (a) 1,5%; (b) 2%; (c) 2,5% and (1) Before dried; (2) After dried

At a concentration of 1.5% the beads produced are more lenient than the concentration of 2% and 2.5% where the beads produced are harder. Increase the levels of alginate used in the producing beads increased the numbers of polymer chains that bind to the cross-linking material [11]. An increase amount of polymer chains bound to the coating material lead to harder beads produced at a concentration of 2% and 2.5% than the concentration of 1.5%. The drying process have shrunk the bead sizes due to loss of water evaporation [9]. The diameter of beads at a concentration of 1.5% is 0.756 ± 0.0532 mm, 2% concentration is 0.872 ± 0.0622 mm and concentration of 2.5% is 1.032 ± 0.0795 mm.

In this study the ionic cross-linking agent used is Ca^{2+} derived from $CaCl_2$. Ca^{2+} has been selected because it is not toxic and is frequently used as cross-linking of microsphere formation with Na alginate polymer [12]. Alginate can form a formation of egg - box by using ionic cross-linking agents, such as divalent cations (i.e., Ca^{2+}). The divalent cation binding group or guluronic blocks of alginate chains, as the structure of guluronic blocks that enable high-level coordination from the divalent ions. Guluronic blocks of one polymer then forms a bond with guluronic blocks of other adjacent polymer forming egg-box model of cross-linking, producing a gel structure [12].

3.2. Encapsulation efficiency and phycocyanin load

The encapsulation efficiency was calculated from the free phycocyanin and from the mass of initial phycocyanin added to encapsulation process [10]. The result of encapsulation efficiency and phycocyanin load presented in Table 1.

Table 1. Encapsulation efficiency and phycocyanin load.

Alginate Concentration	Encapsulation Efficiency (%)	Phycocyanin Load (%)
1,5%	53.5334 ± 0.6132	38.7856 ± 0.6589
2%	65.7334 ± 2.7106	42.3856 ± 1.7379
2.5%	71.7599 ± 0.1319	45.9095 ± 1.1788

The results of encapsulation efficiency and phycocyanin load increase by increasing concentrations of alginate. The results obtained are in agreement with Yan et al., [9] where both the encapsulation efficiency and phycocyanin load increase with increased concentrations of coating material. However, the research conducted by Yan et al., [9] resulted lower encapsulation efficiency (53.84 ± 0.81%) than our experiments (71.7599 ± 0.1319%).

Increased levels of alginate used in the production of microspheres causing the increase number of cross-linking material bonded to the polymer chains that lead to increasing the amount of trapped protein. $CaCl_2$ concentration did not have a significant effect on the encapsulation efficiency and the phycocyanin load due to phycocyanin has a high molecular weight such that it cannot emanated from alginate gel [13].

3.3. Beads characterization using FT-IR

FTIR was used to detect of new interactions marked by the advent of a new wave number shift (Figure 2). Peak which appears at wave numbers of 3400.04 cm^{-1} (1.5%), 3401.09 cm^{-1} (2%) and 3421.41 cm^{-1} (2.5%) caused by *stretching* O-H group, while wave numbers of 1624.13 cm^{-1} (1.5%), 1616.33 cm^{-1} (2%) and 1630.34 cm^{-1} (2.5%) caused by *stretching* -COO group, peak which appears at wave numbers 1424.69 cm^{-1} (1.5%), 1419.84 cm^{-1} (2%) and 1412.58 cm^{-1} (2.5%) caused by *bending* C-H group. In addition, peak appears at wave numbers of 1032.25 cm-^{1} (1.5%), 1032.5 cm^{-1} (2%) and 1033 cm^{-1} (2.5%) caused by C-O alkoxyl group and peak at wave numbers 671.04 cm^{-1} (1.5%), 689.96 cm^{-1} (2%) and 687.9 cm^{-1} (2.5%) caused by *bending* C-H groups [14].

The spectra of FT-IR of phycocyanin encapsulated with alginate showed that the spectra of phycocyanin-alginate have similarity to the spectra of alginate, some peaks were not shifted in wave numbers significantly indicating only occurs electrostatic interaction between phycocyanin with alginate in the beads [9].

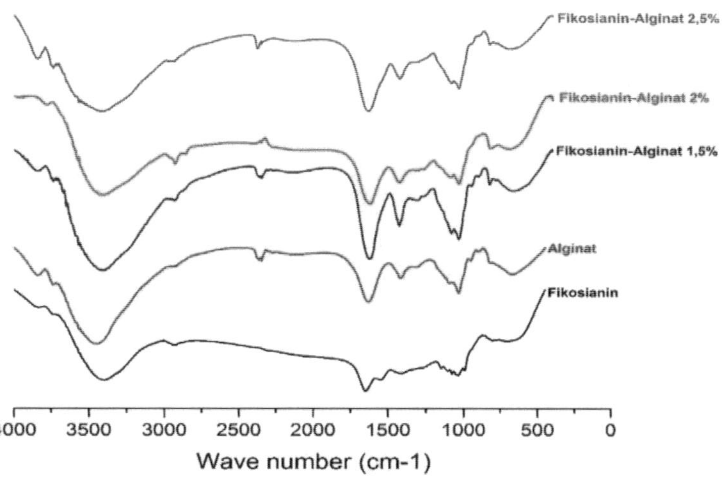

Figure 2. FT-IR analysis of phycocyanin and coating materials

3.4. Beads characterization using scanning electron microscopy

Characterization using scanning electron microscopy was aimed to determine the morphology of the beads resulting from the encapsulation process. The results of morphological analysis encapsulation phycocyanin-alginate at a concentration of 2% and 2.5% shown in Figure 3.

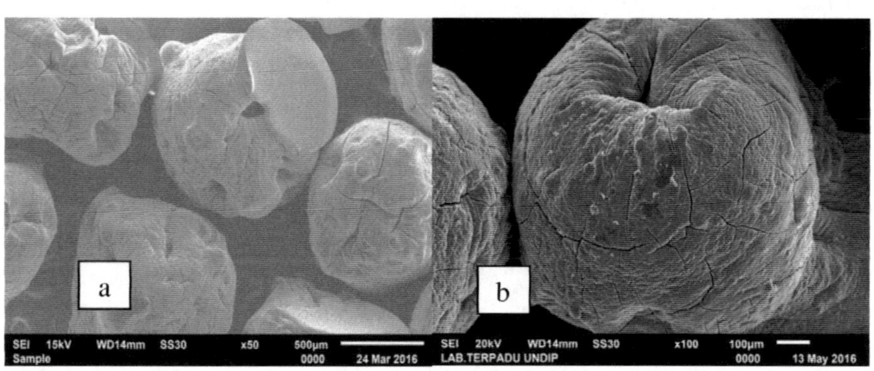

Figure 3. SEM images of phycocyanin – alginate beads (a) concentration of 2%, (b) concentration of 2,5%

Figure 3 shows that the morphology of beads at concentration of 2% was more spherical than beads concentration of 2.5% which have a "tail" (tailings) due to greater viscosity alginate [14]. The surface morphology of phycocyanin-alginate beads was not smooth due process of rapid evaporation of water at cold temperatures [14]. The particles tend to agglomerate which is used by a particular space of the polymer and the electrostatic force [14].

3.5. *In vitro release study*

One objective of the phycocyanin encapsulation is to protect the phycocyanin from acidic gastric fluid. Phycocyanin was stable at pH 5 – 7.5 at room temperature (25 ± 2^{0}C) but easily decolorized at pH above or below 5 – 7.5[15]. Phycocyanin release test results shown in Figure 4.

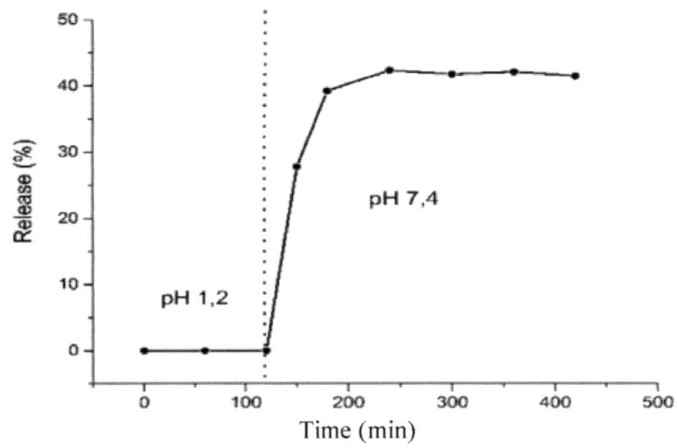

Figure 4. In vitro release study

After 2 hr of incubation, there was no release observed in the simulated gastric fluid medium but it released rapidly in the simulated intestinal fluid medium. The test of phycocyanin release shows that the alginate can hold phycocyanin release rate in simulated gastric fluid medium (acidic condition) and expected to avoid phycocyanin damage at pH below and above 5 - 7.5[15]. At pH 1.2, swelling of alginate beads was longer due to the strong hydrogen bonding that occurs between -COOH and -OH of the polymer chain [16] and increasing the attraction between the amine group on phycocyanin protonated and the carboxyl group of the alginate[17]. At pH 7.4, swelling of alginate beads more rapidly due to the counter ion is neutralized by the ionized carboxyl group at phycocyanin-alginate beads and the electrostatic repulsion between the carboxylate groups are ionized[18]. This can be explained by the presence of ion exchange between Ca^{2+} ion in the hydrogel and Na^+ ions in phosphate buffer[19].

3.6. *Antioxidant activity assay*

The principle of antioxidant activity assay is by measuring the antioxidant activity quantitatively through the DPPH radical arrest by a compound as free radical activity value reduction (IC_{50} value). IC_{50} (Inhibitory Concentration) is defined as the concentration of test compounds that can reduce free radicals by 50%. The smaller IC_{50} value, activity of reduction of free radicals are higher. Antioxidant activity assay is presented in Table 2.

Table 2. Antioxidant activity assay

Sample	IC_{50} (ppm)
Control	197,6738 ± 9,3296
Concentration of 2,5%	280,2862 ± 0,7481
Concentration of 2%	382,1489 ± 132,2927

IC_{50} results in this study was smaller than the previous study conducted by Thangam et al.[2] , however they used different source of phycocyanin. The results of IC_{50} in the level of 200 ppm was due to the possibility of Na phosphate buffer that has a high polarity causing interference on the antioxidant test [3]. The encapsulated phycocyanin showed greater antioxidant activity than the controls. This is possibly due to the storage time of the making encapsulation to test the antioxidant activity long enough to affect the value of IC_{50}.

3.7. Antioxidant stability assay

Antioxidant stability test performed by incubating the dried beads at a temperature of 35⁰C, 45⁰C and 55⁰C for three days. The antioxidant activity was calculated on the third day by using DPPH assay. The results of stability testing of antioxidant show in Figure 5.

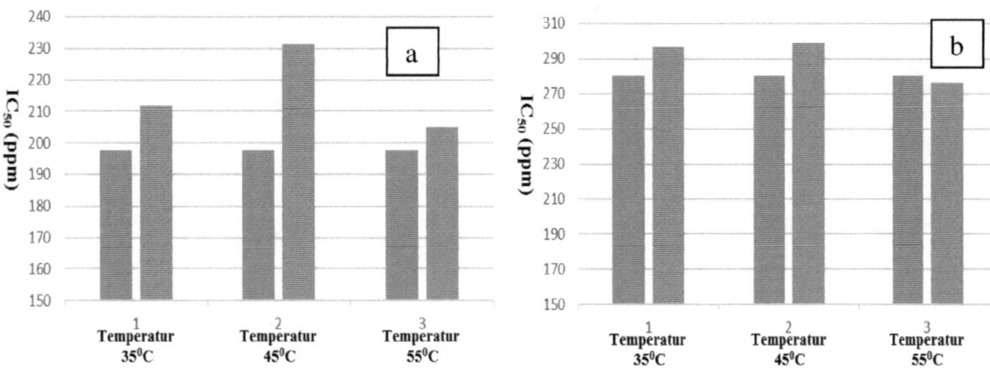

Figure 5. Antioxidant stability (a) Control; (b) Encapsulation at alginate concentration of 2,5%

The results show that encapsulation phycocyanin-alginate has a margin increase of IC_{50} less than the control. This indicates that encapsulated phycocyanin is more stable [20]. The increase of temperature has increased the antioxidant activity [21].

4. Conclusion

The encapsulation at a concentration of 2.5% has highest efficiency, phycocyanin load and antioxidant activity than the other concentrations. FTIR characterization results showed that there was the emergence of a new wave numbers, but there was a shift wave numbers are small due to the electrostatic interaction after encapsulation. In the results obtained morphology using SEM surface morphology showed physical cracking after being dried at a temperature of 4⁰C. From the in vitro release study using encapsulation concentration of 2.5% showed that encapsulation phycocyanin - alginate can hold phycocyanin release rate in simulated gastric fluid medium and release rapidly in the small intestine fluid medium. Encapsulation phycocyanin - alginate is more stable at higher temperatures compared to phycocyanin without encapsulation.

References

[1]. Eriksen N T, 2008 Production of phycocyanin—a pigment with applications in biology, biotechnology, foods and medicine.*Applied Microbiology and Biotechnology*, **80**, 1–14.

[2]. Thangam R, Suresh V, Asenath Princy W, Rajkumar M, Senthilkumar N, Gunasekaran P,Kannan S 2013 C-Phycocyanin from Oscillatoria tenuis exhibited an antioxidant and in vitro antiproliferative activity through induction of apoptosis and G0/G1 cell cycle arrest. *Food Chemistry*, **140**, 262–272.

[3]. Mitra S, Siddiqui W A, and Khandelwal S 2015 C-Phycocyanin protects against acute tributyltin chloride neurotoxicity by modulating glial cell activity along with its anti-oxidant and anti-inflammatory property: A comparative efficacy evaluation with N-acetyl cysteine in adult rat brain,*Chemico-Biological*

Interactions, **238**, 138–150.

[4]. Ichimura M, Kato S, Tsuneyama K, Matsutake S, Kamogawa M, Hirao E, Omagari K, 2013 Phycocyanin prevents hypertension and low serum adiponectin level in a rat model of metabolic syndrome, *Nutrition Research* **33**, 397–405.

[5]. Chaiklahan R, Chirasuwan N, and Bunnag B 2012 Stability of phycocyanin extracted from Spirulina sp.: Influence of temperature, pH and preservatives. *Process Biochemistry*, **47**, 659–664.

[6]. Molyneux P 2003 The use of the stable radical Diphenylpicrylhydrazyl (DPPH) for estimating antioxidant activity. *Songklanakarin Journal of Science and Technology*, **26**.

[7]. Frascareli E C, Silva V M, Tonon R V, and Hubinger M D 2012 Effect of process conditions on the microencapsulation of coffee oil by spray drying. *Food And Bioproducts Processing*, **90**, 413

[8]. Kikuchi A, Kawabuchi M, Watanabe A, Sugihara M, Sakurai Y and Okano T 1999 Effect of Ca2+-alginate gel dissolution on release of dextran with different molecular weights. *Journal of Controlled Release*, **58**, 21–28.

[9]. Yan M, Liu B, Jiao X, and Qin S 2014 Preparation of phycocyanin microcapsules and its properties. *Food and Bioproducts Processing*, **92**, 89–97.

[10]. Ge X, Wan Z, Song N, Fan A, Wu R 2009 Efficient methodsfor the extraction and microencapsulation of red pigmentsfrom a hybrid rose. *J. Food Eng.* **10**, 122–128

[11]. Manjanna K M, Pramod Kumar T M, and Shivakumar B 2010 Calcium alginate cross-linked polymeric microbeads for oral sustained drug delivery in arthritis. *Drug Discoveries & Therapeutics*, **4**, 109–122.

[12]. Lee K Y, and Mooney D J 2012 Alginate: Properties and biomedical applications. *Progress in Polymer Science*. **37**, 106–126.

[13]. Yoo S H, Song Y B, Chang P S, & Lee, H G 2006 Microencapsulation of α-tocopherol using sodium alginate and its controlled release properties. *International Journal of Biological Macromolecules*, **38**, 25–30.

[14]. Mladenovska K, Cruaud O, Richomme P, Belamie E, Raicki R S, Venier-Julienne M C, Goracinova K 2007 5-ASA loaded chitosan–Ca–alginate microparticles: Preparation and physicochemical characterization. *International Journal of Pharmaceutics*, **345**, 59–69.

[15]. Sarada R, Pillai M G & Ravishankar G A 1999 Phycocyanin from Spirulina sp: influence of processing of biomass on phycocyanin yield, analysis of efficacy of extraction methods and stability studies on phycocyanin. *Process Biochemistry*, **34**, 795–801.

[16]. Tavakol M, Vasheghani-Farahani E, Dolatabadi-Farahani T, & Hashemi-Najafabadi S 2009 Sulfasalazine release from alginate-N,O-carboxymethyl chitosan gel beads coated by chitosan. *Carbohydrate Polymers*, **77**, 326–330.

[17]. Zhang L, Guo J, Peng X, & Jin Y 2004 Preparation and release behavior of carboxymethylated chitosan/alginate microspheres encapsulating bovine serum albumin. *Journal of Applied Polymer Science*, **92**, 878–882.

[18]. Lin YH, Liang H F, Chung C K, Chen M C & Sung H W 2005 Physically crosslinked alginate/N,O-carboxymethyl chitosan hydrogels with calcium for oral delivery of protein drugs. *Biomaterials*, **26**, 2105–2113.

[19]. Bajpai S K and Sharma S 2004 Investigation of swelling/degradation behaviour of alginate beads crosslinked with Ca 2+ and Ba 2+ ions. *Reactive and Functional Polymers*, **59**, 129–140.

[20]. Suzery M, Hadiyanto H, Sutanto H, Soetrisnanto D, Majid D, Setyawan D, & Azizah N 2015 The improvement of phycocyanin stability extracted from Spirulina sp using extrusion encapsulation technique. *AIP Conf. Proceeding*, **1699**.

[21]. Hadiyanto H., Sutanto A A and Sudarto Y. 2014. Ultrasound assisted extraction of antioxidant from Coleus tuberosus peels. Carpathian Journal of Food Science and Technology **6(1),**58-65

The Correlation of Upwelling Phenomena and Ocean Sunfish Occurrences in Nusa Penida, Bali

C K Tito[1,2]* and E Susilo[2]

[1] Master of Marine Science, Faculty of Fisheries and Marine Science, Diponegoro University, Jl. Prof. H. Sudarto, SH Tembalang, Semarang, 50275

[2] Institute for Marine Research and Observation (IMRO), Ministry of Marine Affairs and Fisheries, Jl. Baru Perancak, Negara, Jembrana, Bali, 82251

E-mail: camellia.tito@gmail.com

Abstract. Sea surface Temperature (SST) is an important oceanographic variable that can figure the upwelling phenomena. This study aims to determine the variability of SST in relation to upwelling phenomena in the Indian Ocean Southern of Bali Island and the Ocean Sunfish occurrences in the southern of Nusa Penida. Data loggers and remote sensing approach that record temperature was used. An Onset HOBO U20 Water Level Logger U20-001-02 was deployed in Crystal Bay (08°42'S and 115°27'E) at 8 meters depth. The daily field SST data were available from June 2011 to December 2014 with 30 minutes time interval. The monthly satellite images obtained from MODIS on board the Aqua satellite. While the ocean sunfish occurrences were based on rate of encounter (ROE) of previous works by Putra (2015) on July to October 2014. It was found that field data and MODIS have a high correlation (r=0.89) with Root Mean Square Error (RMSE=1.64°C). The upwelling phenomena characterized by the evidence of the colder water mass (SST < 25°C) and a higher concentration of chlorophyll-a, reach its peak in August to September. This phenomena coincidence with the high occurrences of Ocean Sunfish in Crystal Bay on August to October.

1. Introduction

The sea surface temperature (SST) is one of the important physical properties of the marine ecosystem. It's variability influence on various marine processes including physical, chemical, and biological process above [1]. SST gradients can be used to describe the dynamics of oceanographic condition associated with current systems, eddies, fronts, jets, and upwelling phenomena that can be linked to one process to another process [2]. Physically, the upwelling process is characterized by the colder water mass at the surface ocean and higher salinity than the surrounding area. The upwelling area supplies phytoplankton as the feed for the zooplankton, fish larvae, and adult pelagic fishes [3]. The higher phytoplankton abundance triggers the fishing activities at surrounding area [4-6].

In addition to fishing activities, these conditions also had a positive impact on the marine tourism activities. Diving is one of the activities are undertaken by tourists in the waters of Nusa Penida, one of the diving spots which became the main attraction is Crystal Bay. This dive site is known as a inhabiting of Ocean Sunfish, which can generally be seen between July and September [7, 8], mostly at peak season July to October [9] each year, while the SST of this area decreased [10].

Ocean Sunfish is distinguished by their role in the marine food web, its primarily consume gelatinous zooplankton [11-13], and also bizarre appearance and strange basking behavior. Recent studies suggested some hypotheses to explain this basking behaviour related to SST: thermal recharging [14-17] and parasite elimination [18]. Another tagging studies have revealed the ocean sunfish dive several hundred meters deep, possibly for forage [19]. Thus, it has been hypothesized that Ocean Sunfish might bask at the sea surface to re-warm the body after deep dives into cold water [20].

Generally, SST dataset can be obtained from 2 kinds of methods, field measurements and remote sensing approach [21]. SST can be measured with a wide synoptic coverage at fine spatial resolution and repeated time series by using remote sensing satellite [22]. Both infrared and microwave bands have highlighted the global SST map very well [23] but are not the case with coastal areas [24]. These sensors were not able to provide a good quality data for coastal and estuaries zone which were less than 2 km from the coastline [25]. The higher resolution (spatial and temporal) of SST data, especially from field measurements, has a great potential to monitor the oceanography dynamics in the estuarine and coastal areas. Unfortunately, there is a distinct gap in the space and time of SST data obtained from field measurement.

This study aims to determine the variability of SST from field measurements and satellite data associated with upwelling phenomena in the Indian Ocean Southern of Bali Island and the Ocean Sunfish occurrences in the southern of Nusa Penida. First, we will present the logger configurations and measurement method. Then, we discuss the diurnal, monthly and seasonally variability of SST from both field measurements and satellite data.

2. Materials and Method

2.1. Study Area

Geographically, Nusa Penida waters are dealing directly with the Lombok Strait in the northern part and the Indian Ocean in the southern part. Lombok Strait is one of the major Indonesian Through Flow (ITF) pathway through to the Indian Ocean with the volume transport reach 2.6 Sv (1Sv = 106 m3 s-1). This ITF relatively brings the cool and fresh water from the Pacific Ocean to the Indian Ocean [26, 27]. In the other hand, the South Java Current (SJC) and upwelling process in the Indian Ocean have a significant role in maintaining the SST variability along the Java and Sumatera coastal area [28].

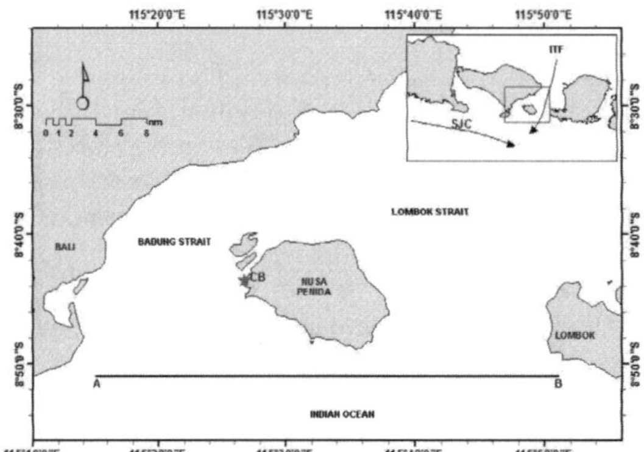

Figure 1. Nusa Penida waters and surrounding (★: Onset HOBO deploy location in Crystal Bay/CB, AB: horizontal line transect of SST and Chl-a MODIS; ITF: Indonesian Through Flow pathway; SJC: South Java Current)

2.2. In situ SST Measurements

SST measured used data loggers at a user defined time interval were used, with the date, time and sampling interval set by software through a specific interface. An Onset HOBO U20 Water Level Logger U20-001-02 deployed in Crystal Bay (08°42'S and 115°27'E), attached to branching coral at approximately 8 meters below the surface (Figure 1). The loggers have autonomy of about one year, at a sampling interval of one record by 30 minutes, which has been selected for the whole period of study. This logger can record temperatures in range -200 to 500°C with the accuracy about ± 0.37° @ 20°C and ± 0.5° over -5° to 50°C. Each logger includes calibration certificate of accuracy in accordance with NIST-traceable standards at three pressure points distributed throughout the range. This logger can record 21,700 combined pressure and temperature measurements up to 30 meters water depth at sea level (Onset Computer Corporation, 2010). The data logger was generally replaced every 10 to 12 months and downloaded using the HOBOware Pro with an Onset Optic USB Base Station (BASE-U-4) and a coupler (COUPLER2-B). The daily SST data were available from June 2011 to December 2014 with 30 minutes time interval. Furthermore, the HOBO SST data were analyzed to look at the diurnal and monthly variation in Crystal Bay, Nusa Penida.

2.3. Remote Sensing Data

The monthly average of SST and Chl-a images obtained from MODIS (Moderate Resolution Imaging Spectroradiometer) on board the Aqua satellite was used in this study. The 3 level images with a 4-km spatial resolution from 2011 to 2014 were downloaded from Giovanni website (http://gdata1.sci.gsfc.nasa.gov/). A horizontal line transects of the monthly average of SST and Chl-a was used to know the spatial and temporal evolution of the upwelling process in the Indian Ocean Southern of Bali Island.

3. Results and Discussion

3.1. Diurnal Sea Surface Temperature Cycle

The complete data series of daily SST variability is presented in Figure 2. The low temperature was observed twice a day, it's in the midnight (12 to 4 am) and mid-afternoon (11 am to 5 pm) (Figure 2a). Equally, with the low temperature recorded, the warm temperature was observed twice a day as well, it's in the morning (6 to 10 am) and (6 to 10 pm). The lowest temperatures were observed in the boreal winter (December – February), with the temperature variability was 0.39°C approximately. The moderate temperature variability observed in the transition period (March – May) and the boreal summer (June – August), it's 0.81°C and 0.72°C respectively. Meanwhile, the strongest temperature variability of daily temperatures (> 1.75°C) observed at the end of boreal summer (September – November), while the coolest temperature (± 24°C) also observed during this period.

The hovmoller diagrams figure out the temporary diurnal temperature variability for each monsoon period very clearly (Figure 2b). During the boreal winter period, the waters tend to be warmer with average 27.30 ± 0.10°C and 26.85 ± 0.23°C, respectively. During this period, the daily variation of SST was very small. The waters being colder (less than 26°C) at the boreal summer period and continued drop until September. The temperature tends to be cooler with average 25.84 ± 0.19°C and 25.11 ± 0.53°C, respectively. During this period, the daily variation of SST was bigger than ever. The low and warm temperatures were clearly observed twice a day in the peak of this time, especially in the mid-evening.

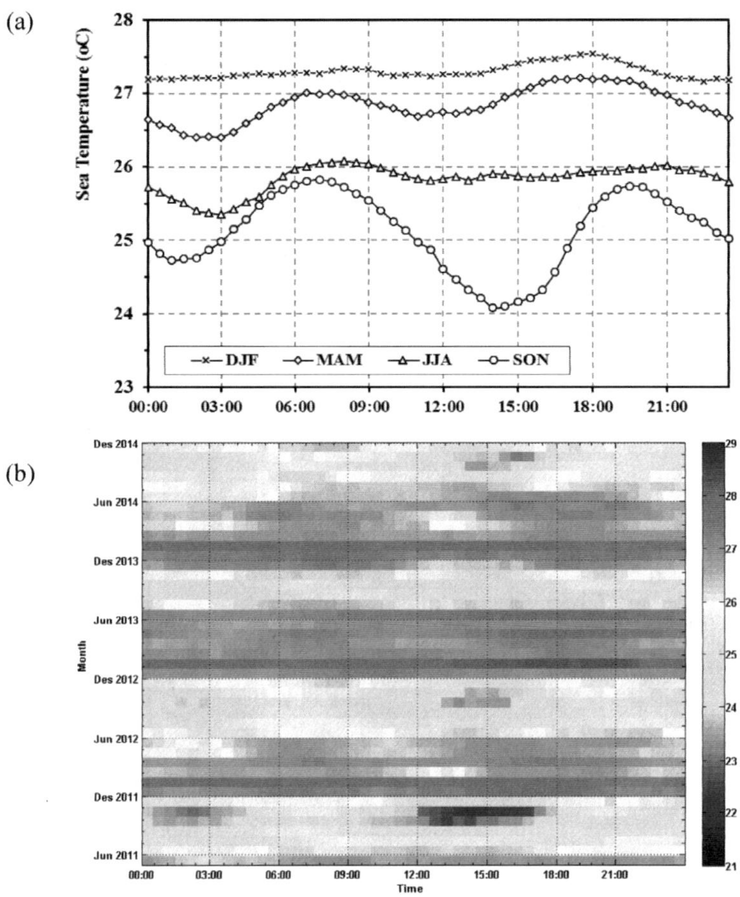

Figure 2. Diurnal cycles of temperature in Crystal Bay, Nusa Penida from HOBO (a) Daily SST average based on monsoon period; (b) Time – monthly average of SST hovmoller diagram.

Diurnal cycle and variation of SST mostly caused by solar radiation and the earth's rotation. Near the equator, the seasonal variation generally weaker rather than in the higher latitude. Equator regions get a longer exposure more than other areas [29]. Evaporation process, wind and wave mixing also touted influence the SST variabilities. In weak winds condition and strong solar surface heating would trigger a warm water mass in the subsurface layer [21].

However, it is still unclear what main caused this diurnal cycle in Crystal Bay. The SST distribution in the Indonesia region is complex. It is difficult to understand the processes that control the SST variability [28]. Local monsoon, tidal mixing, and ITF probbaly become the most influence factor [28, 30]. During the boreal summer that associated with the rains season, the deep convective clouds will reduce the diurnal shortwave radiation forcing caused the diurnal SST variability is weaker. While in the boreal winter, when the wind blows stronger, causing the vertical mixing bringing the cold water to the surface layer. This process is reinforced by the tidal mixing that caused stronger vertical mixing in this region. The role of ITF on the diurnal SST cycle in this area also still unclear.

3.2. Monthly Cycles of Sea Surface Temperature

The monthly cycle of SST observed from data loggers (HOBO) and remote sensing data (MODIS) showed Figure 3. The monthly average SST from HOBO varied between 24.56°C - 27.77°C with the mean is 26.24°C and the standard deviation is 1.0°C. The highest temperatures recorded in January and the lowest were in September. The monthly SST variability for each year (2011-2014) was less than 3.5°C. The highest monthly temperature cycle observed at 2011 with the variability reached 3.40°C. Meanwhile, for the next three year after (2012, 2013, 2014), the monthly temperatures variability were 3.20°C, 2.99°C, and 2.87°C, respectively (Figure 3a).

Comparing the monthly cycle of SST from HOBO and MODIS showed that the two datasets have high correlation (r=88.68 and RMSE=1.64°C). During boreal summer, SST was likely to be more uniform. The upwelling process causing the colder water lifted to surface layer from the bottom. This is evidenced by SST from both HOBO and MODIS. While in boreal winter, MODIS give higher SST than HOBO. The SST difference between this dataset reaches up to -0.84°C until 2.99°C. MODIS using infrared wavelength to measure the SST and only in the top surface layer (we call it's 'skin' layer). The skin SST is the radiometric temperature of the sea surface and usually measured by the infrared wavelengths (3.7–12 µm) that are used by the satellite instruments. Whereas HOBO deployed at 8-meter depth which is often termed the 'bulk' SST. Bulk SST is defined to be the temperature within the upper few layer and convective processes. There will differ from the skin temperature by values as large as 1-3 K in the tropics [31].

Generally the spatial and temporal distribution of SST in the southern waters of Nusa Penida influenced by monsoon (Figure 3b). During boreal summer, SST decreased significantly compared to other months. Strong winds caused the Ekman transport, which causes the surface water mass moved away from the coastal area to offshore. This process leads "an empty space" on the coastal side and the pressure gradient. The water masses from the sub-surface layer was moving towards the coastal and follow the slope of the seafloor moving towards the surface and cause upwelling. The upwelling process in the Indian Ocean south of Java and Bali was characterized by the evidence of the cold water mass (SST < 28°C) and high concentration of chlorophyll_a (Chl-a) more than 0.5 m/m^3 [2]. This upwelling reached its peak in August or September [32]. This cold-water mass pushed westward to the west coast of Sumatra [33-35]. This upwelling also affects on SST variability in Crystal Bay. The cold-water masses began to be seen since the beginning of boreal summer and persist until October.

MODIS showing the spatial distribution of SST in the southern part of Nusa Penida tends to be cooler than the east side (Lombok Strait) and western side (Badung Straits). The warm water is visible on both the straits are a mass of water that tends to ITF pathway. In the early boreal summer of 2013 and 2014, there was warm water masses appearance in the southern waters of Nusa Penida. This condition does not occur in the previous 2 years. This water masses lead the intensity of upwelling weaker which is characterized by a narrow size of the area and of short duration. The size of the area and duration of upwelling in the southern part of Nusa Penida can be used as a benchmark upwelling strength. Based on MODIS data in 2012, showing the intensity upwelling tend to be stronger than other years (Figure 3b). SST measured colder, spread in a wider area (115.325 - 115.575°E) with a longer duration. Spatial and temporal distribution of primary production seen very clearly. The Chl-a concentration reaches up to 2.5 mg m^{-3} with an area spreading from 115.425 - 115.575°E. Likewise, in the western part of Nusa Penida (Badung Straits).

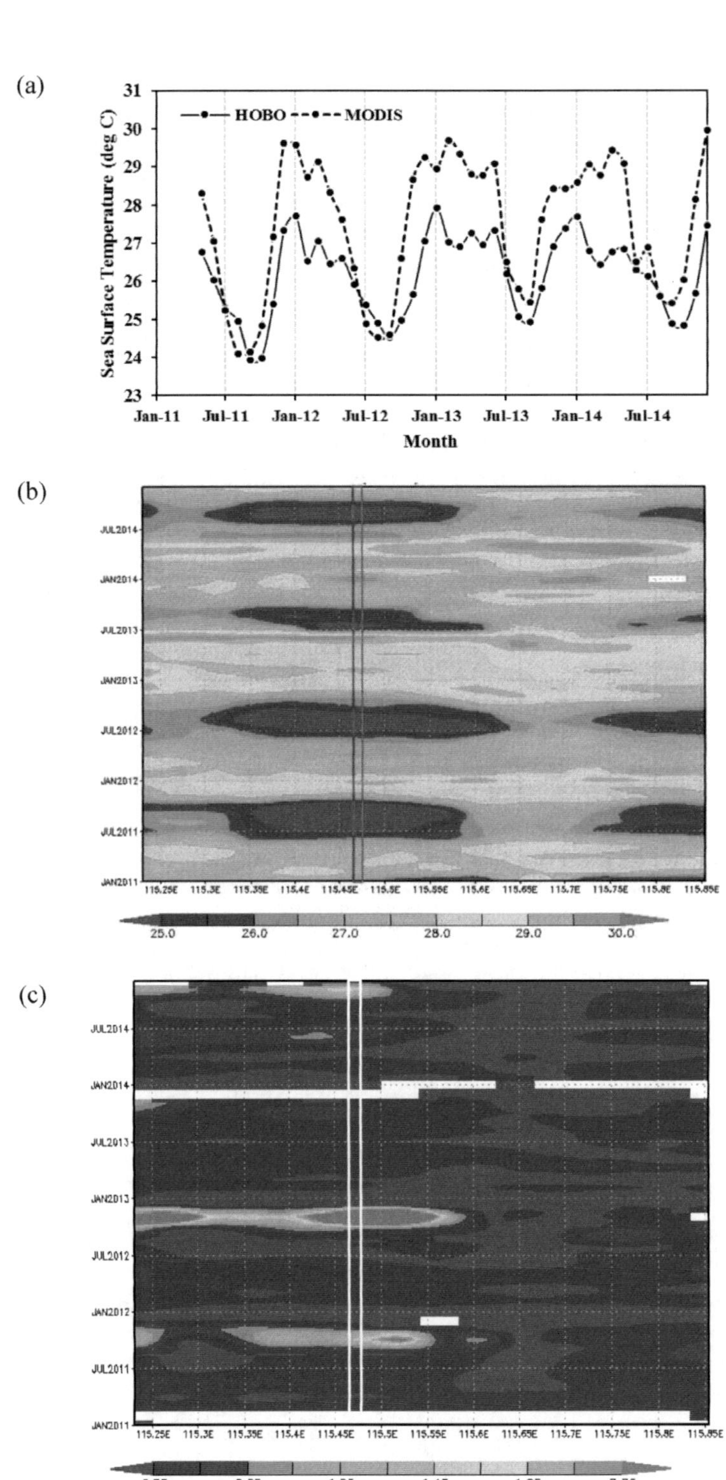

Figure 3. Monthly SST variability in Crystal Bay, Nusa Penida, (a) HOBO and MODIS, monthly climatology 2011 – 2014; Longitude – monthly average hovmoller diagram of SST (b) and Chl-a (c) from MODIS. The red and white box represents the Crystal Bay site.

3.3. Relation of Upwelling and Ocean Sunfish Occurrences

Nusa Penida comprises a group of islands in the southeast of Bali. Based on ecological surveys since 2002 – 2009, in Nusa Penida can found 1419 hectares of coral reef, 230 hectares of mangrove with 13 species and 108 hectares of seagrass bed with 8 species. Furthermore a marine rapid ecological assessment (REA) in Nusa Penida's water found 296 species of coral reef [36]. There is also tremendous fish biodiversity; the REA found 576 species of fish, five of which are new species in Nusa Penida [37]. This island contains a high level of marine biodiversity and has significant tourism potential, visited by more than 200,000 tourists each year. One of unique and charismatic fish that occur predictably annually in Nusa Penida's waters is Ocean Sunfish. This marine megafauna became one of main attraction that undertaken by tourists through diving activities [7]. The Ocean Sunfish season is between July-September each year [6, 7]

Some research works already initiated by few scientists as a preliminary study to documented Indonesian Ocean Sunfish. On going works were initiates by Marianne Nyegaard, PhD research at Murdoch University (Perth, Western Australia) from 2011 and now on. She spent three consecutive sunfish seasons diving with the Ocean Sunfish in Nusa Penida waters. During the 2014 she undertook field research, specifically to look at the movement of the Ocean Sunfish, and the potential for them to be resident in the Bali area. Recent studies by Putra [8], under Marianne Nyegaard's supervision, documented the sighting of Ocean Sunfish in the waters of Nusa Penida. From this work, approximately 70 sighting of Ocean Sunfish from about 400 diving activity recorded in Nusa Penida waters during August to October 2014. The analysis shows that, the most frequently sighting and the highest rate of encounters (ROE) of the Ocean Sunfish were in Crystal Bay (Figure 4) [8].

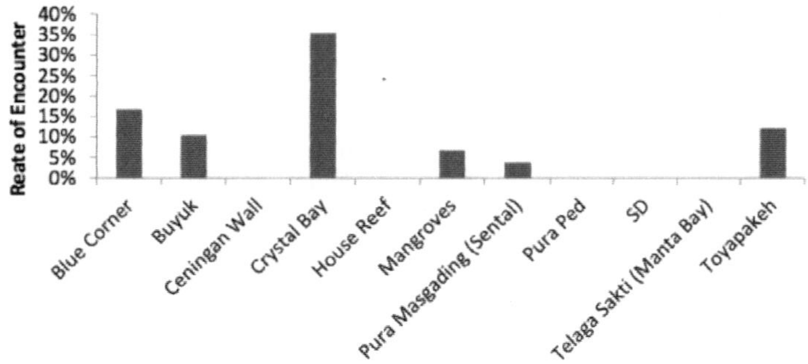

Figure 4. Rate of encounters (ROE) of the Ocean Sunfish in Nusa Penida waters during August to October 2014 [8].

This higher occurrence of the Ocean Sunfish in Nusa Penida waters coincided with the lower SST recorded from data loggers (HOBO) and remote sensing data (MODIS) at the early of boreal summer (September-November) (Figure 3a and b). In this period, the distribution of Chl-a concentration reaches up to 2.5 mg m^{-3} in surrounding Nusa Penida waters (Figure 3c). This lower SST and higher abundance of Chl-a characterized that there are upwelling process at surrounding area [3-6]. Possibly, this area supplies phytoplankton as the feed for the gelatinous zooplankton that triggers the Ocean Sunfish activities at surrounding area. Previous works found that the Ocean Sunfish is distinguished by their role in the marine food web, its primarily consume gelatinous zooplankton [11-13].

4. Conclusion

The variability of SST in the southern waters of Nusa Penida influenced by monsoon and upwelling phenomena that lifted cold-water mass and also high concentration of Chl-a. In boreal summer 2014, the higher occurrences of the Ocean Sunfish in Nusa Penida waters coincided with the lower temperatures and the higher Chl-a concentration.

Acknowledgments

This work was funded by Ministry of Marine Affairs and Fisheries, Indonesia under Institute for Marine Research and Observation since 2011 to 2014. The Authors would like to express appreciations to Eghbert Elvan Ampou, Suciadi Catur Nugroho and Nuryani Widagti for their field support. This work also supported by the Indonesian Endowment Fund for Education (LPDP) within Ministry of Finance, Indonesia.

References

1. Lalli, C.M. and Parsons, T.R., *Biological Oceanography, An Introduction, 2nd Edition*, Elsevier Butterworth-Heinemann.
2. Robinson, I.S., *Discovering the Ocean from Space. The Unique Applications of Satellite Oceanography*. 2010: Springer-Verlag Berlin Heidelberg.
3. Hendiarti, N., H. Siegel, and T. Ohde, *Investigation of different coastal processes in Indonesian waters using SeaWiFS data.* Deep Sea Research Part II: Topical Studies in Oceanography, 2004. **51**(1–3): p. 85-97.
4. Hendiarti, N., et al., *Seasonal Variation of Pelagic Fish Catch Around Java.* Oceanography, 2005. **18**(4): p. 112-123.
5. Rintaka, W.E., et al. *Variasi Sebaran Suhu, Salinitas dan Klorofil Terhadap Jumlah Tangkapan Lemuru di Perairan Selat Bali saat Muson Tenggara*. in *Pertemuan Ilmiah Nasional Tahunan X ISOI 2013*. 2014. Jakarta.
6. Sartimbul, A., et al., *Variations in chlorophyll-a concentration and the impact on Sardinella lemuru catches in Bali Strait, Indonesia.* Progress in Oceanography, 2010. **87**(1–4): p. 168-174.
7. Carter, E., M. Welly, and W. Sanjaya, *Lesson Learned in the Development and Establishment of the Nusa Penida Marine Protected Area 2008-2014*. 2014: Coral Triangle Center.
8. Ruchimat, T., R. Basuki, and M. Welly, *Nusa Penida Marine Protected Area (MPA) Bali - Indonesia: Why Need to be Protected?* Transylvanian Review of Systematical and Ecological Research, 2013. **15**(1).
9. Putra, M.I.H., E. Indrayanti, and M. Zainuri, *Variabilitas Suhu dan Kecepatan Arus terhadap Keberadaan Ikan Matahari (Mola Ramsayi) di Perairan Kepulauan Nusa Penida.* Jurnal Oseanografi, 2015. **4**(3): p. 545-555.
10. Welly, M., et al., *Profil Wisata Bahari Nusa Penida, Kabupaten Klungkung, Propinsi Bali*. 2011, Coral Triangle Center: Sanur. p. 22.
11. Cardona, L., et al., *Massive Consumption of Gelatinous Plankton by Mediterranean Apex Predators.* PLoS ONE, 2012. **7**(3): p. e31329.
12. Harrod, C., et al., *Reply to Logan & Dodge: 'Stable isotopes challenge the perception of ocean sunfishOcean Sunfishas obligate jellyfish predators'.* Journal of Fish Biology, 2013. **82**(1): p. 10-16.
13. Nakamura, I. and K. Sato, *Ontogenetic shift in foraging habit of ocean sunfish Ocean Sunfish from dietary and behavioral studies.* Marine Biology, 2014. **161**(6): p. 1263-1273.
14. Pope, E.C., et al., *The biology and ecology of the ocean sunfish Ocean Sunfish: a review of current knowledge and future research perspectives.* Reviews in Fish Biology and Fisheries, 2010. **20**(4): p. 471-487.
15. Cartamil, D.P. and C.G. Lowe, *Diel movement patterns of ocean sunfish Mola mola of southern California*. Mar Ecol Prog Ser, 2004. 266:245–253
16. Dewar, H., et al., *Satellite tracking the world's largest jelly predator, the ocean sunfish, Ocean Sunfish, in the Western Pacific.* Journal of Experimental Marine Biology and Ecology, 2010. **393**(1-2): p. 32-42.
17. Nakamura, I., Y. Goto, and K. Sato, *Ocean sunfish rewarm at the surface after deep excursions to forage for siphonophores*. Journal of Animal Ecology. 2015. doi: 10.1111/1365-2656.12346.
18. Hixon, M.A., *The halfmoon, Medialuna californiensis, as a cleaner fish*. Calif Fish Game, 1979. 65:117–118.
19. Sims, D.W., et al., *Satellite tracking of the World's largest bony fish, the ocean sunfish (Ocean Sunfish L.) in the North East Atlantic.* Journal of Experimental Marine Biology and Ecology, 2009. **370**(1-2): p. 127-133.

20. Daniel, P.C. and G.L. Christopher, *Diel movement patterns of ocean sunfish Ocean Sunfish off southern California.* Marine Ecology Progress Series, 2004. **266**: p. 245-253.
21. Martin, S., *An Introduction to Ocean Remote Sensing.* 2 ed. 2014, United Kingdom: Cambridge University Press.
22. Rajeesh, R. and G.S. Dwarakish, *Satellite Oceanography– A review.* Aquatic Procedia, 2015. **4**: p. 165-172.
23. Reynolds, R.W., et al., *An Improved In Situ and Satellite SST Analysis for Climate.* Journal of Climate, 2002. **15**(13): p. 1609-1625.
24. Thomas, A., D. Byrne, and R. Weatherbee, *Coastal sea surface temperature variability from Landsat infrared data.* Remote Sensing of Environment, 2002. **81**(2–3): p. 262-272.
25. Fisher, J.I. and J.F. Mustard, *High spatial resolution sea surface climatology from Landsat thermal infrared data.* Remote Sensing of Environment, 2004. **90**(3): p. 293-307.
26. Sprintall, J., et al., *Direct estimates of the Indonesian Throughflow entering the Indian Ocean: 2004–2006.* Journal of Geophysical Research: Oceans, 2009. **114**(C7): p. n/a-n/a.
27. Tillinger, D., *Physical oceanography of the present day Indonesian Throughflow.* Geological Society, London, Special Publications, 2011. **355**(1): p. 267-281.
28. Qu, T., et al., *Sea Surface Temperature and its Variability in the Indonesian Region.* Oceanography, 2005. **18**.
29. Yang, Y., et al., *What Controls Seasonal Variations of the Diurnal Cycle of Sea Surface Temperature in the Eastern Tropical Indian Ocean?* Journal of Climate, 2015. **28**(21): p. 8466-8485.
30. Kida, S. and S. Wijffels, *The impact of the Indonesian Throughflow and tidal mixing on the summertime sea surface temperature in the western Indonesian Seas.* Journal of Geophysical Research: Oceans, 2012. **117**(C9): p. n/a-n/a.
31. Donlon, C., et al., *The Global Ocean Data Assimilation Experiment High-resolution Sea Surface Temperature Pilot Project.* Bulletin of the American Meteorological Society, 2007. **88**(8): p. 1197-1213.
32. Kunarso, et al., *Variabilitas Suhu dan Klorofil-a di Daerah Upwelling pada Variasi Kejadian ENSO dan IOD di Perairan Selatan Jawa sampai Timor.* Ilmu Kelautan, 2011. **16**(3): p. 171-180.
33. Susanto, D. and J. Marra, *Eff ect of the 1997/98 El Niño on Chlorophyll a Variability Along the Southern Coasts of Java and Sumatra.* Oceanography, 2005. **18**(4): p. 124-127.
34. Susanto, R.D., A.L. Gordon, and Q. Zheng, *Upwelling along the coasts of Java and Sumatra and its relation to ENSO.* Geophysical Research Letters, 2001. **28**(8): p. 1599-1602.
35. Ling, Z., et al., *Different effects of tropical cyclones generated in the South China Sea and the northwest Pacific on the summer South China Sea circulation.* Journal of Oceanography, 2011. **67**(3): p. 347-355.
36. Turak, E. and L. De Vantier, *Biodiversity and Conservation Priorities of Reef-Building Corals in Nusa Penida*. Final Report to Conservation International, 2009. 66 pp.
36. Allen, G.R. and M.V. Erdman, *Reef fish of Nusa Penida, Indonesia*. Final Report to Conservation International, 2008. 22 pp.

Initial Study Of Potency Thermal Energy Using OTEC (Ocean Thermal Energy Conversion) As A Renewable Energy For Halmahera Indonesia

Y O Andrawina [1], D N Sugianto [2], I Alifdini[1]

[1]Oceanography Department, Faculty of Fisheries and Marine Sciences, Diponegoro University Indonesia, Tembalang Semarang, Indonesia 50275

[2] Center for Coastal Disaster Mitigation and Rehabilitation Studies, Diponegoro University, Jl. Prof. Soedarto SH Tembalang, Semarang 50275 Indonesia

Email: yochiokta@gmail.com, inovasita@gmail.com, dennysugianto@yahoo.com

Abstract. The Blue Energy is the one of renewable energy. Indonesia as a maritime country has many potention to implement this Blue energy, but it still an advance technology to convert this energy into the electricity. The concept of OTEC with Land Based System is to boost the cost effectiveness of ocean technology. OTEC Land based system which is integrated with the cooling system, water desalination systems, aquaculture, and agriculture. Geographically Halmahera district is located in the Eastern Province of North Maluku. Halmahera is the one of remote island in Indonesia that still lack of electricity. By processing the Argo Float data in 2013-2015 using ODV (Ocean Data View) software, shows the difference of ocean temperature in Halmahera sea surface the range is 29°C. In depth of 1000 m, the sea temperature ranging from 2^{O}C. The Electrical potential of OTEC a long 55 km Halmahera beach are around 93,1% as potential source of OTEC technology. The electricity OTEC prediction is 5,12 MW. Capacity factor of OTEC is 0.857, means that Halmahera have electrical potential for applying OTEC for each year is 35,88 GWh/year.
Keyword: OTEC, Renewable Energy, Thermal energy, Blue Energy, Halmahera

1. Introduction

The demand of energy in Indonesia is increase each year. The distribution of electricity in Indonesia is still worst. There are many of renewable energy that can be applied in Indonesia to solve this problem. Since Indonesia is a vast archipelago country (> 18,000 islands), The Blue Energy technologies should play an important role in the future. The potential of The Blue Energy ratified by ASELI version (Association Energy of Indonesian Seas) in 2011; the potential of tidal current theoretically is 160 GW but technically 22,5 GW, and the potential practical 4,8 GW; the tide has the potential theoretical 510 GW, technical potentially is 1,2 GW, the potential practical 1,2 GW, and the latter is the sea temperature have the potential theoretically is 57 GW, technical potential is 52 GW, and the potential practical is 43 GW [1]. From this data indicate that ocean thermal energy temperature is suitable to be implemented in the tropical areas such as Indonesia, but it still need an advanced technology to convert the ocean thermal energy into the electrical energy.

OTEC (Ocean Thermal Energy Conversion) in Indonesia have not been developed in the laboratory or in the field. Whereas OTEC technology itself has a variety of useful byproduct such as water

desalination, hydrogen, lithium, ice sea water, greenhouse crops, and cultivation. Today, the Association Energy of Indonesian Seas (ASELI) still promote the implementation of OTEC technology through the cooperation with the government of Japan and Indonesia [2].
Halmahera is a remote island in Indonesia that is potential to built OTEC technology. By processing the Argo Float data using ODV (Ocean Data View) software, shows the difference of ocean temperature in Halmahera sea surface range is 24° to 29°C. In depth of 1000 m, the sea temperature ranging from 2^{O} to 3 ^{O}C. From these data indicate that the area in Halmahera has the potential to set up power plants using OTEC, so it can raise the government sector in Maluku Islands [3]. The aim of this research is initial study of OTEC potention in Halmahera area, and do the feasibility study of OTEC in Halamahera area.

2. Methodology

Indonesia as an archipelago island has a vast area of ocean and Indonesia is a tropical island that is make indonesia has big potention on thermal energy. Research and analisys is needed to esthablish the idea of renewable energy from themal energy by using OTEC (Ocean Thermal Energy Conversion) technology.

2.1. Literature Study

2.1.1. Halmahera. Geographically Halmahera district is located in the Eastern Province of North Maluku. East Halmahera located at 1° 4 '- 0° 40' LS and 126° 45 '130° 30' east longitude. Halmahera is the one of remote island in Indonesia that still lack of electricity, because the electric will be turn off at 6.00- 18.00 WITA [4 -5]. The need for electricity in the province of North Maluku supplied by several isolated systems. such as Systems Ternate, Soa Siu, the Vedas, Bacan, Dofa, Jailolo, Sanana, Tobelo, Subaim, Maba-Buli, Patani, and Sofifi. From 12 systems that supply electricity in the province of North Maluku, 4 system (System Soa Siu, the Vedas, the Maba-Buli, and Patani) are in a state of "Surplus", and eight other system (System Ternate, Bacan, Dofa, Jailolo, Sanana Tobelo, Subaim, and Sofifi) are in a state of "deficit". The electrification ratio in North Maluku province reached 49.44% and the ratio of electrified villages amounted to 93.06%. The waiting list has reached 2,812 PLN or demand of 7.1 MVA [10].

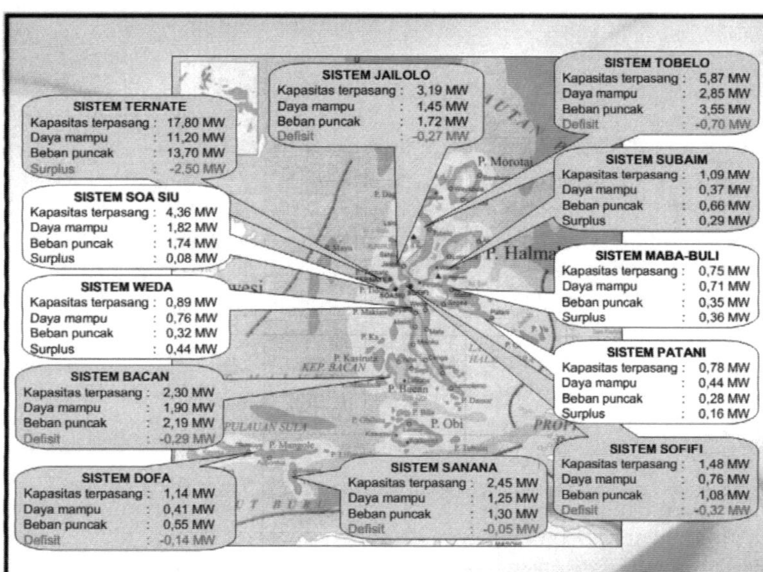

Figure 1. Electricity system in North Maluku Province.

Table 1. The Balance of power system in Maluku and North Maluku 2010- 2014

		2010	2011	2012	2013	2014
Requirement						
Household	GWH	327	360	393	427	462
Comersial	GWH	83	87	90	93	97
Public	GWH	71	76	81	87	94
Industry	GWH	7	8	8	9	10
Total Requirement	GWH	489	530	572	616	663
Growth	%	7,5	8,4	8,0	7,7	7,5
Decrease & loses	%	7,8	7,7	7,6	7,5	7,5
Decrease own use	%	2,0	2,0	2,0	2,0	2,0
Total decrease & Loses	%	9,8	9,7	9,6	9,5	9,5
Peak load	MW	108	116	126	135	145
Installed power	MW	43	106	123	128	204
Additional power	MW	64	20	9	80	47
Backup power	MW	42	10	6	73	106

The table 1 shows the distribution of fuel to North Maluku province served by Pertamina UPMS VIII Jayapura [10]. Pertamina UPMS VIII serve the supply and distribution of fuel that includes four provinces, There are the province of Papua, West Papua, Maluku and North Maluku [10]. There are some obstacles encountered in the provision of fuel in this region. These constraints, such as geographical position is an archipelago with many areas still remote (remote areas). These condition makes Halmahera lack of electricity.

2.1.2. OTEC. The operating system of OTEC seems like a steam engine, with evaporation and condensation [3-4]. Differences pressure caused turbine spun to generate electricity. OTEC technology can be used one of the hybrid system cycle. Hybrid system advantage are open and closed cycles. Hybrid cycles used seawater laid low pressure in the tank (vacuum chamber) to give steam. Steam are used to vaporize fluid low dotted boiling (ammonia or another) will move turbine to generate electricity. Sea water vapor and condensed to produce freshwater desalination . Using ammonia anhydrous as a working fluid, modeled with Rankine cycle saturated the principle of OTEC. Mechanism working will be explained by figure 2 [5] .

2.2. Collecting Data and Data Processing

Data that will be use is origin from Argo Float Data (2013-2015). Argo Float is array global more than 3,000 who have a profile for measuring temperature and salinity at 2,000 m altitude from the sea. This allows for the first time, monitoring continuous of the temperature, salinity, and the velocity from the upper sea. The data delivered and available for public in hours after collection. The data can be download from www.argodatamgt.org. Data processed by ODV (Ocean Data View). The difference temperature of Halmahera ocean and can be displayed in diagram that can easily to understand[5].

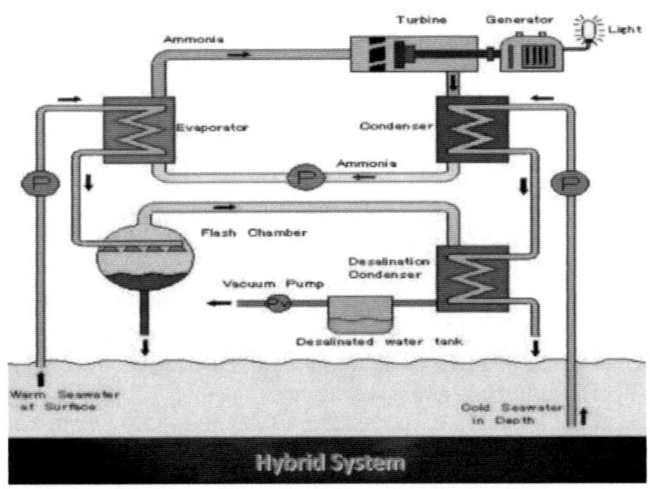

Figure 2. Flow diagram and schematic Hybrid System

2.3. Data Analysis
Data analysis from literature study and processing Argo Float data with ODV should be analysed[7].. The analysis that will do in this initial study are analysis of potential area, eficiency of the technology, electrical analysis, and the pay back period. This analysis aimed to explore all opportunities, advantage, and loss of OTEC potential in Halmahera. There are many theoritical equation and term that are used to do this analysis, such as:

2.3.1. Analysis maximum efficiency of an OTEC system
There is a theoretical limit, up to a maximum efficiency of an OTEC system by converting heat stored in the warm surface water of tropical oceans into mechanical work[11].

$$\eta max = (Tw - Tc)/Tw$$

where:

ηmax	: maximum efficiency
Tw	: absolute temperature from warm water
Tc	: absolute temperature from cold water

2.3.2. Analysis Payback Period
Adopting a formula based on zero net present value is given by: [8] :

$$X = d\text{-}i / A \quad (2)$$

Equation 1 explain how to get result of X .

$$Y = (1\text{-}CC).X \quad (3)$$

Input equation 1 to equation 2 so we can produce result of Y

$$O = (1 + i) / (1 + d) \quad (4)$$

After that arithmetic equation 3 to get result of O

$$N = \text{Log}\ \{ Y / \text{Log}\ O\ [year] \} \quad (5)$$

Finally input equation 2 and equation 3 to equation 4 to get result of pay back period of OTEC.

IOP Conf. Series: Earth and Environmental Science **55** (2017) 012032 doi:10.1088/1755-1315/55/1/012032

Where:

CC : Estimated costs
A : Income and expenses per year
N : Payback Period
d : Discount Rate
i : Inflation

3. Result and Disscussion

3.1. Analysis of Potential Source in Halamahera

The location of a commercial OTEC plant has to be in an environment that is stable enough for an efficient system operation. The aim of these analysis is to know the potential place for developing OTEC in Halmahera. By processing the Argo Float data using ODV (Ocean Data View) software, shows the difference of ocean temperature in Halmahera sea surface range is 24° to 29°C. In depth of 1000 m, the sea temperature ranging from 2^{O} to 3 ^{O}C. The relation of depth and temperature shows in the figure 3 that has been processing using ODV softwere and MS.Excel.

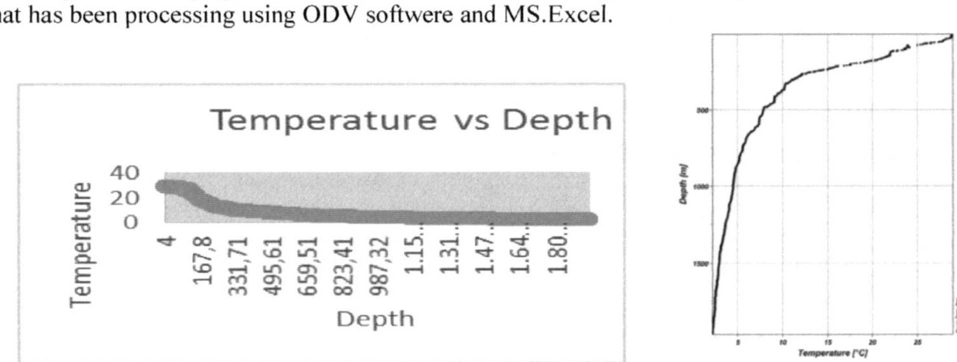

Figure 3 Large-scale distribution of resources OTEC thermal Halmahera

After processing data with ODV (Ocean Data View) it shows in figure 5 that the longitude and latitude affect the distribution of surface temperature. The range of temperature shows with the gradient of the colour. The figure 4 shows that the North Halmahera esspecially Akelamo is a potential place to built the OTEC, because the difference temperature in this area is very significant.

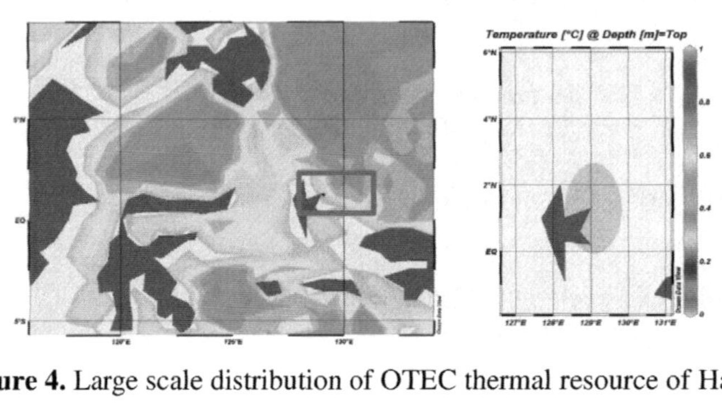

Figure 4. Large scale distribution of OTEC thermal resource of Halmahera

3.2. Analysis The Efficiency of OTEC in Halmahera

This analysis is the result of the calculation by using maximum efficiency of an OTEC system equation [term 2-5] in the methodology.

$$\eta max = \frac{29\,°C - 2\,°C}{29\,°C}$$
$$= 0{,}931$$
$$= 93{,}1\%$$

The result of the efficiency is adequate, so the north sea of Halmahera is the most potential in the development of sea surface water temperatures for OTEC. If it take 100 KW for input power and carnot efficiency 93,10344 , so the estimation result of the net power become 93,103 KW.

3.3. Analysis of Electrical Potential

Electrical potential of OTEC technology a long 55 km Halmahera beach, around 95 % as potential source of OTEC at more than a 1000 m depth. The long of coastline is 55 km. The Source of OTEC potential is 93,1 %: 0.931 x 55 km = 51,2 km. The distance between 100 MW OTEC is 10 km, means that potential power for each kilometer is {100/10} MW. Capacity factor of OTEC is 0.8 [5,12]. The assumed of Electrical Potential represented in table 2.

Table 2.Asumption for the calculation electrical potential

No	Variabel	Value	Unit
1	Energy Source	51,2	Km
2	OTEC Energy Potential	98,304	MWh/day
3	Total Generator	10	/km
4	Effeciency	93,1	%
5	Depth	1000	Meter
6	Power capacity for each generator	20	KWh
7	OTEC electrical potential	35,88	GWh/year

3.4. Analysis of Pay Back Period

The assumed value used to estimate the cost of payback time N for a 5250 KW OTEC plant, which may represent targeted the limit of OTEC. Based on equation number 4 from on estimation cost by L. A. Vega, Ph.D. of Pacific International Center for High Technology Research (PICHTR), Honolulu, Hawaii Capital Investment [7-9]. The result of assumption represented in table 2, with revenue from electricity sales at the end of the first year of the project. With CF (capacity factor) 80% and electricity price US$ 10/kWh, A = US$ 3,675,000. , favourable compares with the projected life OTEC plan, typically 20 – 30 years.

Table 3.Assumption for the calculation of the payback period

No	Variabel	Value	Unit
1	N	11	Year
2	A	3.675.000	US $ / kW
3	CC	7.000	US $ /kW
4	D	20	%
5	I	10	%

3.5. Future Plan of OTEC in Halmahera

In this project the OTEC system with Land-based system. The Land-based system do not require a sophisticated mooring system, lengthy power cables and more extensive maintenance as required with open ocean environment. Land-based system is more easier and cheapest than using the off-shore. In addition,for the future plan the land-based sites allow OTEC to be associated with industries such as agriculture and those needing cooling and desalinated water [Fig.5]. Below is desaign for future plan of OTEC in halmahera:

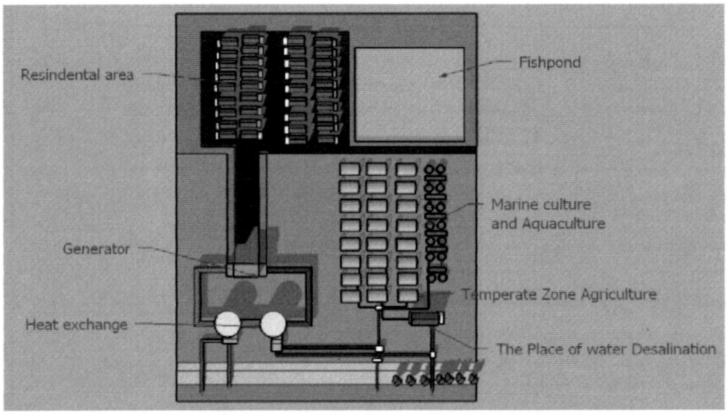

Figure 5. Layout desaign OTEC in Halmahera

Figure 5. shows the OTEC that will be implemented in Halamahera not only for generate the electricity, but it associated with the water desalination, temperate zone agriculture, aquaculture and fish pond to encourage the local prosperity economically.

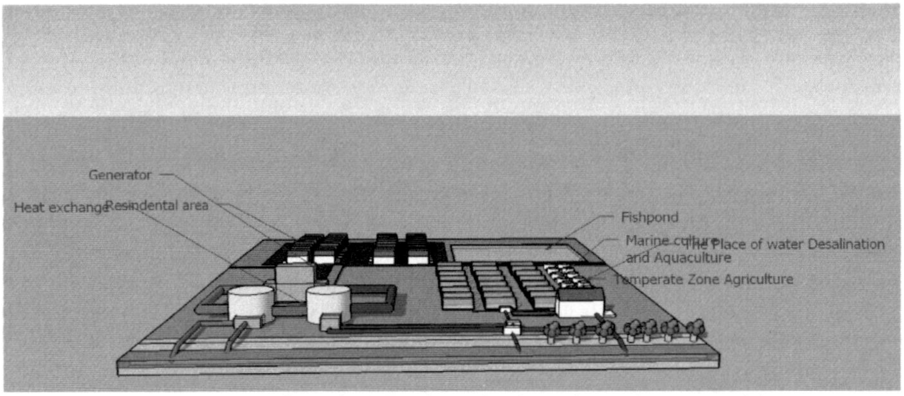

Figure 6. Desaign of OTEC in Halmahera from the ocean side

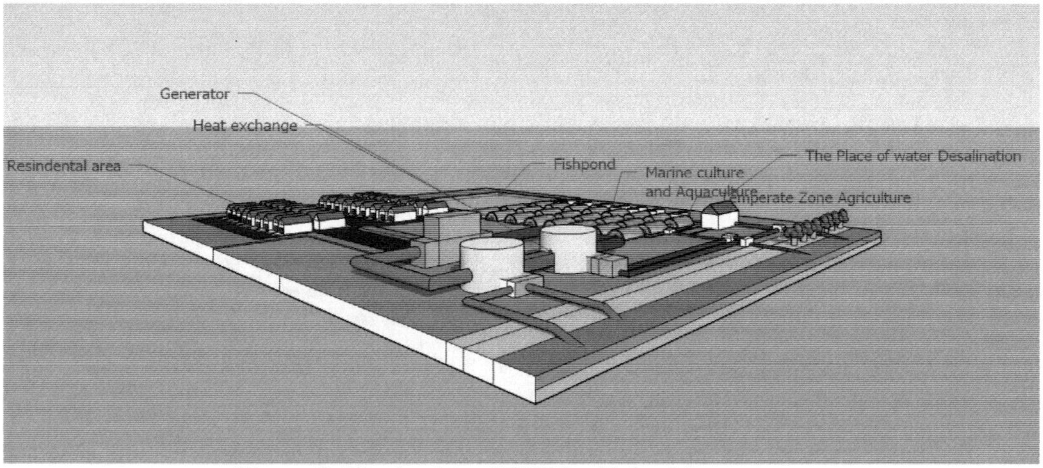

Figure 7. The desaign of OTEC with land-base system

Those upon figure are shows the future [Fig.5, Fig.6, and Fig.7] design of OTEC plant, wich is using Land- base system concept. OTEC has some benefit, such as:

- Helps produce fuels such as hydrogen, and lithium
- Produces base load electrical energy.
- Produces desalinated water for industrial, agricultural, and residential uses.
- As a resource for on-shore and near-shore marine culture operations.
- Provides air-conditioning for buildings.
- Provides moderate-temperature refrigeration.
- Has significant potential to provide clean, cost-effective electricity for the future.

4. CONCLUSION

OTEC by using the concept of land based system which is integrated with the cooling system, water desalination systems, aquaculture, and agriculture. By using the Land-base system is more cheaper, safer and easier than using the Off-shore system. This renewable energy can produce not only the electricity energy, but also for agriculture, and fisheries to develop the remote island in Indonesia such as Halmahera. Halmahera is a remote island in Indonesia that is potential to built OTEC technology. By processing the Argo Float data using ODV (Ocean Data View) software, shows the difference of ocean temperature in Halmahera sea surface range is 29°C. In depth of 1000 m, the sea temperature ranging from 2^{O}C. The Electrical potential of OTEC a long 55 km Halmahera beach are around 93,1% as potential source of OTEC technology. The electricity OTEC prediction is 5,12 MW. Capacity factor of OTEC is 0.857, means that Halmahera have electrical potential for applying OTEC for each year is 35,88 GWh/year. More comprehensive research and study are needed to know the possibility location for OTEC to built and estimated cost.

5. Acknowledgement

We would thank to Oceanography Department Diponegoro University, Center for Coastal Disaster Mitigation and Rehabilitation Studies (PKMBRP) Diponegoro University, Center of Excellence Science and Technology (PUI) Diponegoro University, Community of Ocean Modelling Diponegoro Univerity (COMED) and Ministry of Research Technology and Higher Education (Ristek Dikti) Republic of Indonesia for funding our research.

References

[1] Vega, LA 2012 *Ocean thermal energy conversion* Encyclopedia of Sustainability Science and Technology(van den Berg :Springer) July pp.7296-7328.J.

[2] Mamahit, C 2011 *Development of ocean thermal energy conversion* Elektromatika Journal.H. Poor, An Introduction to Signal Detection and Estimation (New York: Springer-Verlag 1985) , vol. 1 no. 1 pp. 55-64 ch. 4.

[3] B. Abdul, and M.R. Putri *Water Masscharacteristics Ofweda Bay, Halmahera Island, North Maluku* Jurnal Ilmu dan Teknologi kelautan Tropis vol. 5 no. 2 pp. 365-376 Desember 2013.

[4] Siahaya, Y & Lydia, S 2010 *Ocean thermal energy conversion (OTEC) power plant and it's by products yield for small islands in Indonesia sea water* (Bandung: ICCHT-5th International Conference on Cooling and Heating Technologies Bandung, Indonesia, 9 - 11 December 2010, Bandung) pp. 1-9

[5] Nugraha, AS & Estiko, R 2010 *Ocean current energy conversion system in wallacea region using variable speed control approach* Journal of Mechatronics Electrical Power and Vehicular Technology **1** no.1 pp 27-34

[6] Achiruddin, D 2010 *Ocean energy* (Jakarta: METI)

[7] Schlitzer, R. 2015. Ocean Data View. odv.awi.de

[8] Steinbach, R.B. 1982 *Mini-OTEC*: A Hardware Perspective Society of Naval Architects and Marine Engineers Spring Meeting Honolulu Hawaii pp 289-306 April 1982

[9] Quinby - Hunt M.S., Wilde P. and Dengler A.T 1986 *Potential Environmental Impacts of OpencycleOcean Thermal Energy Conversion* in *Environ Impact Assess Rev*(New York: Elsevier Science Pub. Co.,Inc., New York, NY) pp. 77-93

[10] The Ministry of energy and mineral resources 2009 *Master Plan Pembangunan Ketenagalistrikan 2010 S.D. 2014* (Jakarta : The Ministry of energy and mineral resources) pp 144-151

[11] Adrian Rizki Sinuhaji 2015 *Potential Ocean Thermal Energy Conversion (OTEC) in Bali* ISSN 2413-5453 **1** 5-12

Effect of Reaction Temperature on Biodiesel Production from *Chlorella vulgaris* using CuO/Zeolite as Heterogeneous Catalyst

Dianursanti[1], M Delaamira[1], S Bismo and Y Muharam[1]

[1]Chemical Engineering Department, Universitas Indonesia, Depok, Indonesia

email : danti@che.ui.ac.id ; muthia.delaamira@yahoo.com

Abstract. Human needs for fossil energy increase every year. Biodiesel is the main way to resolve this world problem. Biodiesel produces from vegetable oil. But then, the alternative way came from the uses of microalgae in Chlorella vulgaris type causes by its simplicity of growing. In the other hand, this microalgae known for its high lipid content by considering several parameter such as light intensity, medium nutrition, pH and also salinity. Lipid content will be extracted by using Bligh-Dryer method which will be reacted with methanol along transesterification. Beside, there come another matter which is the utilization of homogeny catalyst. The difficulty of separation is the main matter so then biodiesel need to be washed in case normalizing the pH and this process will decrease the quality of biodiesel. To resolve this problem, we'll be using a heterogeneous catalyst, zeolite, with ability to catalyst the process. Zeolite is easier to separate from the biodiesel so there will not be needed washing process. Heterogeneous catalyst work as well as homogeneous. Variation implemented on transesterification included reaction temperature of 40°C, 60°C, and 80°C. Reaction time, catalyst percentage and the solvent amount remain steady on 4 hours, 3% and 1:400. Complete best result obtained at 60°C with the yield of 36,78%. Through this, heterogeneous catalyst CuO/Zeolite proved to have a capability for replacing homogeneous catalyst and simplify the production of biodiesel particularly in separation step.

Keywords : Biodiesel, Chlorella vulgaris, CuO, Zeolite, Temperature

1. Introduction

Declining of world's crude oil reserve and increasing of green fuel needs led to the developing of renewable fuel production [1]. The movement of world economy requires development of technology as well as energy section. Green diesel demand predicted to reach 900 million tons in 2020 [2]. Green fuel demand was not balanced by the supply, managing the fossil fuel for being the main fuel sources.

Greater amount of green fuel production will maintain the situation and replaced the fossil fuel production. Green fuel as well as green diesel and biodiesel produced mostly form plant oil which needs time to grow and grant the oil ready. The alternative sources would better be the microalgae. Essential composition of microalgae will be the positive value rather than its oil. Lipid's content of microalgae reaches the percentage of 34% w/w, which on the other side, palm oil only reach 17-23,8% [3]. Moreover, photosynthesis efficiency of microalgae (6-8%) way higher than the other usual plant with 1,2-2,2% [4].

Microalga will be used as the sources for biodiesel by the research using the transesterification process. *Chlorella vulgaris* would be the microalgae species used because of its high lipid content [5]. The catalyst which dominate the reaction of transesterification is the homogeny catalyst such as KOH and NaOH because of its solubility with methanol and helping the reaction to reach methyl ester [6]. The disadvantage of using the homogeny catalyst would be the needed of biodiesel washing to neutralize the pH which cause the adding of water in system. Water need to be dried and it cost one more process. Also, homogeny catalyst could not be recovered, recycle nor reuse from the system. However, heterogeneous catalyst has the advantage of replacing the disadvantage from the homogeny catalyst.

Heterogeneous catalyst that could be used for the transesterification process will be zeolite. Zeolite which impregnated with metal oxide will result a high yield of biodiesel. This statement supported by result obtained by Sani in 2013 whose Hierarchical H-Beta Zeolite uses as a catalyst for algae's oil transesterification with the conversion of 99%. The effect of big pore structure (729 m^2/g) led to the diffusion of microalgae's oil to the active site of catalyst. On the other side, BET surface area should be greater too, which cause the amount of acid site to increase. Si/Al ratio will effect on acidity surface [7]. Heterogeneous catalyst for transesterification process usually needs the temperature up until >350°C [8], but the zeolite uses decrease the temperature process because of its channels and greater pore [7].

In this paper, zeolite will be used as heterogeneous catalyst. To improve the capability of conversion, metal oxide should be inserted in between zeolite, which in this case CuO will be used. Jack Vincent and Dheeban Chakravarthi on 2009 shows that transesterification process using heterogeneous CuO might reach 90%. In this research, CuO/Zeolite will be used as a catalyst in transesterification process, and temperature will be the variation to acknowledge the ideal condition needed to reach highest biodiesel conversion.

2. Experimental Method

2.1 Pre-Treatment of Zeolite
First, zeolite washed by the water. Right after, washed zeolite immersed in HF 1% for about 30 minutes. Make sure that all the zeolite being gently mixed to enhanced the purification from polluter. Dry zeolite about 2 hours in 110°C microwave, then immersed again in the 1M of NaOH for 24 hours and dry zeolite for 2 hours in 110°C.

2.2 Zeolite Impregnation
Mix zeolite in 0,001M $Cu(NO_3)_2$ for 4 hours. Right after, do the decantation for a day and separate the zeolite (at the bottom) with the residue. Mix impregnated zeolite with NaOH solution (pH of 8) for 4 hours. Do the decantation for a day and separate it as mentioned before. Dry zeolite in 110°C for 2 hours and then do the calcination occur in 350°C for 2 hours.

2.3 Lipid Extraction
1ml of cake (biomass) being centrifuge for 10minutes (4500 rpm). Every 1 ml of cake react with 2ml of methanol and 1ml of chloroform using sonicator for 5 minutes. Right after, add up 1 ml chloroform and 1 ml of demin water. React using sonicator for 10 minutes. Centrifuge the tube for around 5 minutes to distinguish the phase. Throw the upper part (methanol, water and biomass) and uses the bottom part (lipid and chloroform).

2.4 Transesterification
Extracted lipid react with methanol and zeolite catalyst in three neck flash and condenser system. Ratio of lipid : methanol uses is 1:400 and catalyst uses is 3% of w/w. Do the reaction in 4 hours with the variation of temperature (40°C, 60°C, and 80°C). Biodiesel purification using decantation need in total of 24 hours and then purification between biodiesel and glycerol could be done manually. Biodiesel

phase then being mixed with boiled water up until water phase being clear. Right after, biodiesel phase then being heated until 100°C to evaporate water. Biodiesel percentage calculated using equation below.

$$Y_{px} / Y_{bx} = \frac{m_{biodiesel}}{m_{sel}} x100\% \ldots(1)$$

$$Y_{ps} / Y_{bs} = \frac{m_{biodiesel}}{m_{lipid}} x100\% \ldots(2)$$

where,

Y_{px} or Y_{bx}	= Yield of Biodiesel per Biomass Weight (g.g^{-1})
Y_{ps} atau Y_{bs}	= Yield of Biodiesel per Lipid Weight (g.g^{-1})
$m_{biodiesel}$	= Biodiesel Weight (gr)
m_{cell}	= Biomass Weight (gr)
m_{lipid}	= Microalgae's Lipid Weight (gr)

2.5 GC-MS Analysis

Sample which being analyzed is the most optimum biodiesel obtained. This analysis is done to figure the methyl ester existence. Composition of methyl ester obtained analyze using Gas Chromatography and Mass Spectrophotometry (GCMS) in Pusat Laboratorium Forensik, Markas Besar Polisi Republik Indonesia.

3. Result and Discussion

3.1 Reaction Temperature Variation

Based on research done by Kim in 2014, effect of reaction temperature in transesterification reach the number of 47%, which beat the other factor such as solvent amount, reaction time and catalyst percentage (each of it has the percentage of 26,7%, 17,5% and 8,3%). To observe the effect of the variation, the other parameter remained consistent in 4 hours and 3% w/w of catalyst with the variation of 40°C, 60°C, and 80°C. Following figure shows the result of the reaction variation.

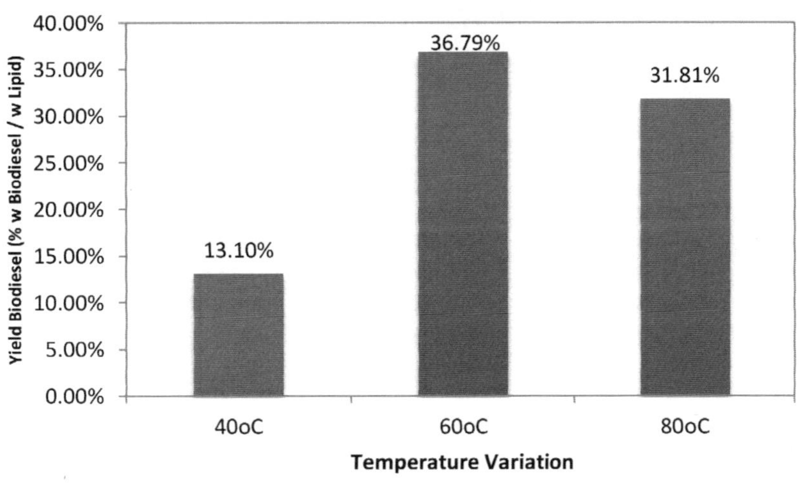

Fig1. Reaction Temperature Variation for Transesterification Process

Consecutively, conversion obtained is 23,91%, 28,66% and 23,97% (Y_{ps}), and on the other side it result an amount of 2,28%, 6,41% and 5,54% for Y_{xs} (from biomass to biodiesel percentage) and 13,1%, 36,79%, and 31,81% for Y_{ps} (from lipid to biodiesel percentage). Effect for the reaction temperature more likely gave a big effect for the yield produced. Biodiesel produced gain the highest point at 60°C, while at 80°C it decreased. The 80°C for the system temperature excess the boiling point of the methanol (64,7°C). Methanol easily evaporate into gases for they are a volatile compound and produced the bubble in a great amount. The gases conduce a three-phase reaction system inside the three neck flash [8]. The bubble will prevent the contact of each reactant [9]. Higher temperature should be effective to escalate the catalyst activity, accelerate movement of molecule so that the frequency of collision from one and another molecule is bigger and reaction will go even faster [10]. On the other hand, 60oC reach the maximum point for producing biodiesel because there is a limitation of methanol boiling point. It reduces the amount of methanol so that the methanol become insufficient [11].

Teo et.al. in 2014 reacted the *Nanochloropsis sp.* oil with heterogeneous base catalyst such as CaMgO and $CaMgO/Al_2O_3$ in variant reaction temperature. It shows that in 60°C, it reach the maximum yield while reaction temperature up until 100°C shows the decrease level of biodiesel produced cause by the methanol loss (evaporated). However, in 40°C, catalyst is not yet activated. Reaction rate of the molecule and frequency of the collision reduce so that reaction itself became slow. Reaction constant in low temperature show a small level [12]. N the other hand, low reaction temperature result the small amount of copper methoxide form. Copper methoxide is a mid reactant whose function as a bond breaker of triglyceride. In low temperature, this process will took a lot more time.

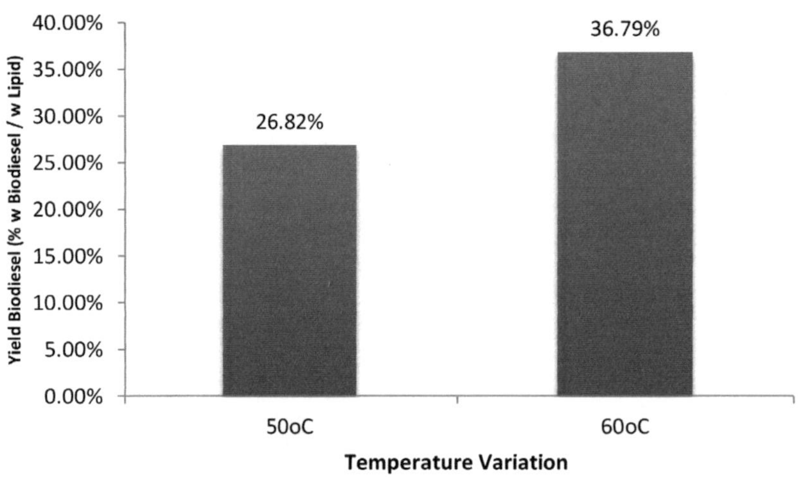

Fig2. Reaction Temperature Variation for Transesterification Process (50°C and 60°C)

Optimum variation temperature checked by further research operate in 50°C condition. New variation (50°C) produce only 4,61% of Y_{px}, as well as the Y_{ps} of 26,82%, smaller than the 60°C reaction. according to research by Teo in 2014, this variation seems to have a parallel result where the maximum point reached on 60°C. This maximum yield do not exceed the boiling point of methanol. apart from Teo, Kumar in 2015 and Kim in 2014 also prove that the best reaction temperature is 60oC.

3.2. GC-MS Analysis

Methyl ester obtained by the reaction of microalgae lipid and methanol catalyzed by CuO/Zeolite being analyzed using GC-MS. This result is being compared with methyl ester obtained by lipid tranesterification catalyzed by homogeneous catalyst such as KOH and NaOH.

Compared to other kind of oil or lipid, microalgae lipid is kind of different. It does not rich of polyunsaturated fatty acid with four or more double bond. Eicosapentaenoic acid (EPA, C20:5n-3) and docosahexaenoic acid (CHA, C22:6n-3) is one of the most common composition of microalga oil. Fatty acid and FAME with 4 or more double bonded will easily being oxidize within storage phase and will decreasing the ability of commercialization. Generally, biodiesel produced by C. vulgaris will produce more unsaturated methyl ester. On the other hand, more saturated FAME will cause higher melting point, better oxidative stabilization, and lower NOx emission [14]. Greater unsaturated FAME in microalgae oil lead to a lower melting point of biodiesel, therefore it suits more to country with lower ambient temperature and need more additive.

Table1. FAME analyzed by GC-MS compared to Previous Research

Fatty Acid Methyl Ester (FAME)		**CuO/Zeolit *C. vulgaris***	**Comparison** KOH *C. vulgaris* (Anggraini, 2013)	**KOH *C. vulgaris* (Religia, 2014)**
Methyl Myristate	C14:0	Nd	4,48	Nd
Methyl Palmitate	C16:0	58,42	28,3	31,22
Methyl Palmitoleate	C16:1	Nd	11,1	4,45
Methyl 7,10-hexadekdienoate	C16:2	Nd	Nd	2,39
Methyl 7,10,13-heksadektrienoat	C16:3	Nd	Nd	5,52
Methyl Stearate	C18:0	Nd	17,1	11,55
Methyl Oleate	C18:1	41,58	39,1	30,83
Methyl Linoleate	C18:2	Nd	Nd	14,05
Total of Saturated FAME		49,88	42,77	58,42
Total of Unsaturated FAME		50,12	57,23	41,58

GC-MS analysis in Table1. shows the methyl ester variant produced. Similar research done by Anggraini and Pijar in 2013 and 2014 using homogeneous catalyst which show that Methyl Palmitate and Methyl Oleate commonly and mostly produced by C. vulgaris rather than other methyl ester. On the other hand, in consecutive 4 years, it also show that ratio of saturated methyl ester and unsaturated methyl ester is more than 1.

Transesterification process with CuO/Zeolite as the catalyst only result Palmitate and Oleate as its methyl ester. This outcome lead to 2 main reasons. First reason is because of the newly come strain of Chlorella vulgaris, and second reason is because of the pH from the catalyst. pH from homogeneous catalyst will lead to a higher pH in reaction (between 10-12) and heterogeneous catalyst will only result a lower pH (between 8-9). The pH itself could result the minimum amount of methyl ester produced by the reaction. From GC-MS analysis on Table2., it can be seen that there are several adverse reaction. Most adverse reaction produce siloxane with amount of 17,66% (*tetrasiloxane, cyclotetrasiloxane, cyclohexasiloxane, cycloheptasiloxane, cyclooctasiloxane*, serta *cyclononasiloxane*). Siloxane is an organosilicon compound which has Si-O-Si chain. Siloxane lead to a Si-O bond with organic compound, which in this research, the organic compound assumed as methanol or the other fatty acid. Si-O which being trapped inside zeolite after activation process react with methanol excess in system. Si-O itself is a destructed framework from HF cleaning phase.

CuO as a catalyst still need to compete with free Si-O chain inside system. Siloxane in system will lead to "shine" on biodiesel. On the other hand, Ala-beta-ala trimethylsilyl ester (4,95%) is a methyl ester compound whose structure being distracted by siloxane. It explain why methyl ester produce being decreased even more. Siloxane and nitrogen mostly attack unsaturated methyl ester, forming ala-beta-ala trimethylsilyl ester, and decrease the amount of unsaturated methyl ester.

Reverse reaction produced the product written in Table2. Recombination between ester group in methyl ester form another ester compound such as 1-Alanine, N-valeryl -, hecadecyl ester. On the other hand, alcohol group formed by the hydroxyl (OH^-) which react with another component. Hydroxyl itself constructed by reaction of methanol and base catalyst (CuO/Zeolite) and it recombine with another compound so the alcohol group compound formed. Aldehyde and ketone group form by dehydration and further oxidation reaction from alcohol group. Eicosane is the rest free fatty acid which not reacted with methanol. (C:16).

Table2. GC-MS Analysis for Transesterification Product

No	Component	Percentage
1	FAME	33,7%
2	Siloxane	17,66%
3	14-BETA-H-PREGNA	9,94%
4	Ala-beta-ala trimethylsilyl ester	4,95%
5	Estra-1,3,5(10)-trien-17.beta.-ol	4,77%
6	4-and 5-(1,1,3-Trimethyl-2-butenyl)-2-cyclopenten-1-ol (isomer a)	3,63%
7	Cyclohexyl 2-Methylenebutanyl Ketone	3,17%
8	(tetrahydroxycyclopentadienone)tri carbonyliron	2,99%
9	1H-Pyrrole-2,5-dione, 1-(hydroxymethyl)	2,28%
10	l-Alanine, N-valeryl-, hexadecyl ester	2,25%
11	Eicosane	2,18%
12	Benzil Alkohol	1,90%
13	Benzenemethanol	1,30%
14	(trans)-1,2-Dihydroxy-1,2-dihydro- acronycine	1,28%
15	N-methyl-N-n-propylamine	0,36%

4. Summary

Best variation reaction shown in 60°C reaction condition with Y_{ps} of 36,79% and Y_{px} of 6,41%. Further temperature will cause an evaporated methanol to form and inhibit the reaction, on the other hand, reaction temperature below 60oC will cause the slower reaction rate and production.

References

[1] Demirbas A 2009 *Biofuels. In: Demirbas A, editor. Biofuels: Securing the planet's future energy needs* **336** (London (England): Springer)

[2] Bold, H.C. and Wayne, M.J. 1978 *Introduction to the Algae Second Edition (*New York: Pretice-Hall Mc. Engelwood Cliffs)

[3] Santos, C 2013 *Enhanced lipidic algae biomass production using gas transfer from a fermentative Rhodosporidium toruloides culture to an autotrophic Chlorella protothecoides culture. Bioresource technology*, **138**, pp.48–54.

[4] Aresta, M., Dibenedetto, A. and Barberio, G 2005 Uti*lization of macro-algae for enhanced CO_2 fixation and biofuels production: Development of a computing software for an LCA study* **86(14-15), pp.1679–1693**

[5] Li, Y 2008 *Effects of nitrogen sources on cell growth and lipid accumulation of green alga Neochloris oleoabundans.* 81(4), **pp.629– 36.**

[6] Sharma C., Singh B., Korstad J 2011 *Latest Developments on Application of Heterogenous Basic Catalyst for an Efficient and Eco Friendly Synthesis of Biodiesel: A Review* **90** 1309-1324.

[7] Carrero A 2011 *Hierarchical Zeolites as Catalysts for Biodiesel Production from Nannochloropsis Microalgal Oil* **148-153**

[8] Islam, Taufiq-Yap, Chu, Ravindra, and Chan 2013 Transes*terification of palm oil using KF and $NaNO_3$ Catalyst Supported on Spherical Milimetric gamma-Al_2O_3.* **59** 23-29

[9] Naik R, Radhika, Sravani, Hareesha, Mohanakumari, Bhavanashindu 2015 *Optimized Parameters for Production of biodiesel from Fried Oil* **2**

[10] Salamah, Siti 2014 *Kinetika Reaksi Esterifikasi Minyak Biji Kapuk pada Pembuatan Biodiesel*. **11-18**

[11] SabirUsta, Mathur, dan Samad 2016 *Optimization of Transesterification Process for Production of Thumba Methyl Ester* **5**

[12] Pal, Krupal 2011 *Investigations of Transesterifications of Canola Oil With Methanol and Ethanol for A New Efficient Method of Biodiesel Production*

[13] Teo, Yap, and Ng. 2014 *Alumina Supported.Unsupported Mixed Oxides of Ca and Mg as heterogeneous Catalysts for Transesterification of Nanochloropsis sp. Microalga's Oil*

[14] Chisti, Yusuf 2007 *Biodiesel from Microalgae* **294-306**

IOP Conf. Series: Earth and Environmental Science **55** (2017) 012034 doi:10.1088/1755-1315/55/1/012034

Annona muricata modulate brain-CXCL10 expression during cerebral malaria phase

Kis Djamiatun* [1,2], Sumia M A Matug [2], Awal Prasetyo [2,3], Noor Wijayahadi [4], Djoko Nugroho [5]

[1]Parasitology Department Medical Faculty Diponegoro University
[2]Biomedical Science Post Graduate Program Medical Faculty Diponegoro University
[3]Pathology Anatomy Department Medical Faculty Diponegoro University
[4]Farmacology Department Medical Faculty Diponegoro University
[5]Biostatistic Department Public Health Faculty Diponegoro University

Corresponding author: Email: ramus64@yahoo.com

Abstract. Cerebral malaria (CM) contributes in malaria mortality. People in endemic region get benefices by using *A. muricata*-leaf extract (AME) before qualified for receiving standard anti-malaria, because AME restrains malaria infection and modulate immune responses. CXCL10 expressed by astrocytes limit brain inflammation. Vascular leakage was found in the brain of experimental CM. Additionally, biomarker related with vascular leakage, angiopoietin-2 (Ang-2) levels increase in CM-patients. Objectives of this study were to determine the efficacy of ethanolic-AME in regulating brain-CXCL10-expression and Ang-2 levels during CM-phase. The study was post-test-only-control-group design. Thirty Swiss-mice were randomly divided in 6 groups. C+ and C- groups were PbA-inoculated and healthy-mice, respectively. X1 and X2 groups were healthy-mice treated with AME 100 and 150 mg/Kg BW/day, respectively. X3 and X4 groups were PbA-inoculated and received either dose mentioned above. CXCL10 was stained by IHC, and determined by Allred score. Plasma-Ang-2 was measured by elisa-method. Kruskal-Wallis-test showed the difference of CXCL10-expression among the studied groups ($p=0.003$). CXCL10-expression of C+ group was lower than healthy-mice which were C-, X1 and X2 groups ($p=0.008$, $p=0.045$, and $p=0.012$). CXCL10-expression of X3 was comparable to healthy mice (C-, X1 and X2), and was higher than C+ and X4 groups ($p=0.012$ and $p=0.028$). CXCL10-expression of X4 group was lower than C- and X2 groups ($p=0.011$ and $p=0.016$). Kruskal-Wallis-test showed no difference of Ang-2-levels among 6 groups ($p = 0.175$). The conclusion is *A. muricata* influences brain-CXCL10 expression during CM phase, but has no association with Ang-2 levels during CM phase.

Key words: *Annona muricata*, malaria, CXCL10, Angiopoietin-2

1. Introduction

Malaria is a fatal disease in tropical and subtropical countries. WHO estimated 214 million malaria cases globally in 2015, and 438,000 of them were died [1, 2]. *Plasmodium falciparum* infection is the most common cause for severe malaria including cerebral malaria (CM). The increase risk for severe malaria might occur when the early malaria treatment for uncomplicated malaria is delayed.

Uncomplicated *P. falciparum* malaria has clinical manifestation alike many other diseases including influenza and other common diseases related with fever. This, therefore, is difficult to do early diagnose with subsequently rapid malaria treatment of uncomplicated malaria which may avoid severe condition [3]. Malaria patients can get benefices by using traditionally malaria treatment before they are eligible for standard-malaria therapy. Anti-malaria plants increase worldwide interest, and are further studied for their capability as anti-plasmodial [4-6]. *Annonaceae* which comprises approximately of 123 genera and 2,100 species, is noticeable due to their broad traditional uses against malaria [7]. Anti-malaria of *Annona muricata* leaf extract (AME) treatment was shown *in vivo* by using imprinting control region (ICR) mice inoculated with 10^7 *P. berghei* ANKA (PbA) [8]. The CM studies showed that tumour necrosis factor-alpha (TNF-α), a pro-inflammatory cytokine, contributed to CM pathogenesis [9]. Ethanolic-AME treatment was associated with reduce TNF-α and nitric oxide (NO) produced by splenocytes of PbA-inoculated Swiss mice as a model for experimental cerebral malaria (ECM). The doses used were of 100 and 200 mg/Kg BW/day given orally before and after 10^7 PbA-inoculation. Two out of six mice was died in the group of PbA inoculated and treated with 200 mg/Kg BW/day, therefore the use of lower doses was recommended [10].

The increase levels of many chemokines were associated with CM. The chemokines most strongly implicated in CM are those bind to (C-X-C motif) receptor 3 (CXCR3), including CXCL9, and CXCL10. CXCL10 which binds to CXCR3 with a high-affinity, is the first known chemokines which is able to direct the trafficking of activated effector CD4+ and CD8+ T lymphocytes, natural killer (NK) and natural killer T (NKT) cells [11-13]. The importance of CXCL10 had been suggested by several studies in malaria. A study of CM in Ghanaian children demonstrated that out of 36 biomarkers, only CXCL10 was serum marker independently associated with CM mortality. In addition, study of cerebral spinal fluid (CSF) of CM patients showed that CXCL10 was one of eight biomarkers significantly up regulated in the CM group. [14]. The ECM study showed that CXCL10 was highly induced in the brain especially at the neuron of PbA infected mice, and CXCL10 deficient mice were protected partially from CM mortality [11]. The recent study showed that CXCL10 provoked the apoptosis of both human brain micro vascular endothelial cells and neuroglia cells *in vitro* [15]. In contrary to those found in CM and ECM studies, an *in vitro* study on astrocyte showed a reduce CXCL10 expression during hypoglycaemia and hypoxia which both were two condition found in severe malaria [16]. Astrocytes are abundance in the brain, and this cells are important in maintaining the brain homeostasis and prevent the brain inflammation induced by T cells [17]. The aim of the study was to determine whether ethanol-AME capable of modulating CXCL10 expression in the brain of Swiss mice during CM phase. A significantly lower CXCL10 expression than normal was found during CM phase of positive control mice without AME treatment. Dose of 100 mg/Kg BW/day AME was able to normalize CXCL10 expressed in the brain.

Vascular dysfunction is one of many other CM pathogenesis, and Angiopoietin-2 (Ang-2) is biomarker contribute in it. Protective immune responses develop in malaria patients targeted to combat the infection without inducing any immunopathology that induced severe malaria including CM and severe malaria anaemia [18, 19]. CM is characterized by the present of inflammation that contributes in vascular dysfunction. Based on that point of view, it is proposed to integrate adjuvant therapy improving vascular dysfunction in CM management [20]. The vascular endothelial dysfunction in malaria indicates by many biomarkers including increase of Ang-2 due to increase vascular endothelial cell activity [21, 22]. Anti-malaria therapy affected the improve vascular endothelial function and better thrombocyte count that both were indicated by the decline of Ang-2 levels [23]. The increase of Ang-2 in CM patients is a better indicator for disease severity and mortality than Ang-1 which released by thrombocytes. This fact suggested that vascular endothelial dysfunction in CM might preferably due to the increase of Ang-2 rather than the reduced Ang-1 released by thrombocytes [24-26]. Based on those studies, therefore, Ang-2 level can be used to prove the benefice particular treatment which alleviate CM. The sequestered parasitized red blood cells and accumulation of leukocytes, both have a strong relation with CM [27]. The treatment used may traditional herbs having anti-malaria and anti-inflammatory effects. The herb used is expected to lessen Ang-2 level which indicates the ease of vascular dysfunction in CM. Natural product including traditional plants are used for malaria treatment in malaria endemic area [28]. Natural product having both anti-malaria and immunomodulatory effects had been extensively studied in *in vitro* and *in vivo* [29-34]. The advantages of leaf of *Annona muricata* toward malaria have been reported. The AM leaf extract (AME) showed anti-plasmodial effect and low toxicity in *in vitro* studies by using *P. falciparum* and normal kidney cells, respectively [35, 36]. Methanol and water based AME showed low toxicity, and both of them have anti-inflammatory and anti-oxidant activity in *in vitro* and *in vivo* studies [37, 38]. Recent published data showed that ethanolic-AME reduce tumor necrosis factor-alpha (TNF-α) and nitric oxide (NO) produced by stimulated spleen of Swiss mice during CM phase [39]. The aim of the study was to determine whether ethanol-AME capable of reducing Ang-2 in the blood of Swiss mice during CM phase. No significant lower of Ang-2 levels in those receiving AME treatment was found during CM phase. It is conclude that AME might not influence Ang-2 levels of Swiss mice during CM phase.

2. Material and methods

2.1. Preparation of ethanol extract leaves of Annona muricata

Ethanolic extract of *A. muricata* leaf was kindly provided by experts from SidoMuncul company, Indonesia. Raw material was sorted, washed, dried and weighed, and milled coarse. The materials then further processed until the condensed-extracts was obtained.

2.2. Experimental Mice and Animals handling

This study was post test only control group design. Thirty adult female Swiss mice used were 8-10 weeks old, and 30 – 35 grams body weight. The strain and health of mice were certified by agriculture department of Indonesian Republic. They were housed in six plastic cages under hygienic conditions. The animal care room were maintain in temperature at 25 ± 4° C with 12 hours alternate exposure to light and dark. The mouse adaptation for 7 days was done before experiment. The experiment procedures done in the mouse were approved by ethical committee of health researches, Medical Faculty of Diponegoro University and Karyadi Hospital Semarang. The mice divided in 2 control and 4 treated groups. All mice received hygienic standard diet and water. C(-) group was healthy mice without AME treatment, while X_1 and X_2 groups were healthy mice treated with AME 100 and 150 mg/kg BW/ day, respectively. AME treatment was given orally for period of 14 days. C(+) was group of PbA-inoculated mice. X_3 and X_4 groups were PbA-inoculated mice which received AME either dose mention above in period of 7 days before and 7 days after PbA inoculation.

2.3. Malaria Parasite.

Plasmodium berghei ANKA (PbA) used for inoculation was provided by Parasitology department of Medical Faculty UGM,Yogyakarta, Indonesia. PbA was passage twice in swiss mice before experiments. The blood was collected from the second passage when the mice had parasitemia level of 15 – 20 %. Dose of 10^7 iRBC collected from the second passage was injected peritoneally to swiss mice used for the AME study.

2.4. Parasitemia levels

Parasitemia level was observed on thin blood slide stained by Giemsa. Parasitemia was determined by technician of MFDU-Parasitology Department. This was done at day 3, 5 and 7 p.i. Parasitemia level was percentage of parasitized Red Blood Cells (pRBCs) out of 1000 RBCs in each blood slide.

2.5. Immuno Histochemestry of CXCL10 and Allred score

Starr Trek Universal HRP detection system (Catalog number: STUHRP700 H, L10; Biocare Medical, Concord, California, USA) and primary antibody which was anti-CXCL10 polyclonal antibody for mouse/ rat (Catalog number: bs-1502R; Bioss Antibodies, USA) was used to stain CXCL10 expressed in the mouse brain. Parafine block of brain was cut as thick as 3 – 4 µm, and attached to the object glass. This was followed by deparaffinization by using xylene and rehydrated by using alkohol. The next step was decloaking proccess which followed by cooling at room temperature for 1 hours, and washed the section by using aquadest, and put PBS on it for 5 minutes. The normal wine serum was put on the section for 10 minutes, and then washed the section by using PBS. The IHC staining by using CXCL10 primary antibody and HRP detection system were then done on the section. The samples, then, were observed under light microscope. Allred score was used for evaluating CXCL10 expression.

IOP Conf. Series: Earth and Environmental Science **55** (2017) 012034 doi:10.1088/1755-1315/55/1/012034

2.6. Ang-2 measurement

Plasma used for Ang-2 measurement was isolated from whole blood collected from *plexus retro-orbital*. Ang-2 levels were measured by Elisa method.

2.7. Statistical analysis

Kappa test used was used to evaluate the agrreement beween two pathology anatomy expert who determine the Alred score. Variable immunohistochemestry examinations (CXCL10 levels) were measured by two (2) blindly anatomical pathologist, and test the proportion of responses to 'deal' wear kappa test. Kappa values <0.40 means poor agreement, values between 0.41 to 0.60 indicates moderate agreement, values between 0.61 to 0.80 indicates a good agreement, and values ≥0.81 means very good agreement. Kruskal-Wallis test followed by Mann-Whitney U test was used to analyses CXCL10 expressed in the brain. The Ang-2 levels then analyzed using Kruskal-Wallis test.

3. Result and Discussion

3.1. CXCL10 expression

CXCL10 expression was determined by Allred score obtained by calculating intensity and proportional score (Figure 1). Kappa test result showed Kappa value of 0.547 meaning moderate agreement between 2 experts determines the Allred score on CXCL10 expression (Table 1).

The CXCL10 expressions were normally distributed in all studied groups. Analysis of variance, however, showed there were data in the studied groups were not homogen (p = 0.001). Kruskal Wallis Test, a nonparametric test, therefore, was performed, and it showed significant difference among the studied groups ($p = 0.003$). C(+) group was significantly expressed lower CXCL10 than those of healthy groups without or with *A. muricata* treatment (C(-), $p = 0.008$; X_1, $p = 0.045$; or X_2, $p = 0.012$) (Figure 2A, Table 2.). CXCL10 expression of X_3 group was comparable with those of healthy mice (C(-), X_1 and X_2), meanwhile the expression was significantly higher than C(+) and X_4 groups (p = 0.012 and p = 0.028, respectively). X_4 group showed CXCL10 expression significantly lower than healthy groups which were negative control and X_2 groups ($p = 0.011$ and $p = 0.016$, respectively) (Figure 2; Table 2).

3.2. Parasitemia

Parasitemia levels were low at day 3 p.i. in all PbA-inoculated groups which were C(+), X_3 and X_4 groups. A sharp increase of parasitemia levels were observed on day 7 p.i. The mean ± SD % parasitemia levels of C(+), X_3 and X_4 groups were 23.91 ± 12.89 %, 21.59 ± 10.15 %, and 20.91 ± 13.98 %, respectively. Shapiro-Wilk test showed that parasitemia levels were normally distributed in each group ($p > 0.05$). Test of homogeneity of variance revealed the homogeneity of parasitemia levels ($p > 0.05$). These test then followed by one-way anova test which showed no difference of parasitemia levels among those three groups ($p = 0.916$).

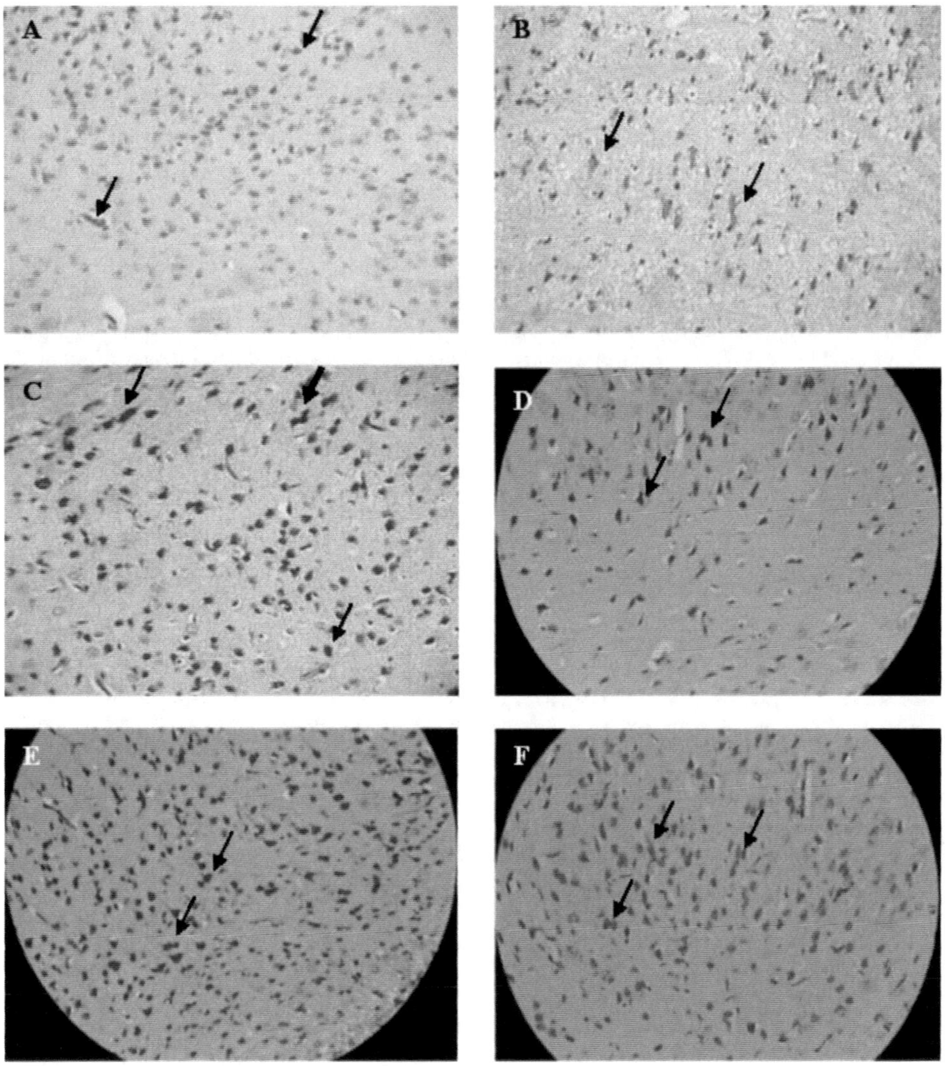

Figure 1 showed IHC of brain-CXCL10 expression used to determine the Allred score. Allred score was sum of proportion score and intensity score observed using light microscope with magnification 400 x. Fig 1.A, B and C were CXCL10 score 1, 2 and 3. The score was determined by the proportion of cells expressed CXCL10. Score 1 = one cell was positive for CXCL10; score 2 = 1 – 10% positive; score 3 = 10 - 33% positive; score 4 = 34 – 66% positive; and score 5 > 66% positive. Fig 1.D, E, F were CXCL10 intensity 1, 2 and 3.

Table 1. *Kappa* test for examination CXCL10 level
Observer 1 * observer 2 Cross tabulation

Count

		Observer 2							Total
		2	3	4	5	6	7	8	
Observer 1	2	3	2	0	0	0	0	0	5
	3	0	13	2	1	0	0	0	16
	4	1	3	8	5	1	0	0	18
	5	1	0	11	28	22	2	0	64
	6	0	1	2	8	83	12	3	109
	7	0	0	0	1	6	36	4	47
	8	0	0	0	0	0	1	5	6
Total		5	19	23	43	112	51	12	265

Kappa = 0.547 (*moderate agreement*)

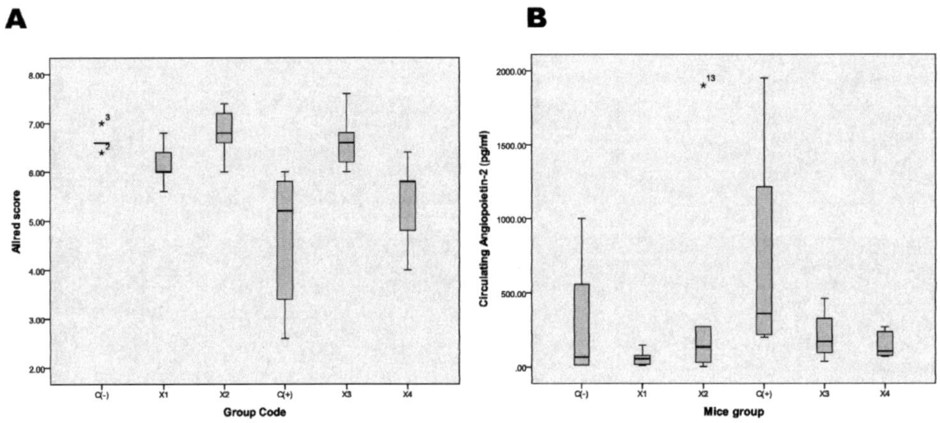

Figure 2. was box plot of Swiss mice in ethanolic-AME study. Fig 2 A. was CXCL10-expression in the brain. **Fig 2 B.** was circulating angiopoietin-2 levels.
These groups were C(-) group which was healthy mice without AME treatment; $X_1$and X_2 groups were healthy mice which received 100 and 150 mg AME/kg BW/day, respectively; C(+) group was group of PbA-inoculated mice; X_3 and X_4 groups were groups of PbA-inoculated swiss mice which received 100 and 150 mg AME/kg BW/day, respectively.

Table 2. Statistical analysis of CXCL10 expressed in the brain of healthy or PbA-inoculated mice with or without *A. muricata* treatment

Groups	C(-)	X_1	X_2	Control (+)	X_3	X_4
C(-)	(-)	(p=0.089)	(p=0.389)	(p=0.008)*	(p=0.829)	(p=0.011)*
X_1		(-)	(p=0.090)	(p=0.045)*	(p = 0.203)	(p=0.140)
X_2			(-)	(p=0.012)*	(p=0.673)	(p=0.016)*
C(+)				(-)	(p=0.012)*	(p=0.459)
X_3					(-)	(p=0.028)*
X_4						(-)

Mann-Whitney U test with significant difference ($p < 0.05$)

This recent IHC study was comparing the CXCL10 expression in the brain of PbA-infected mice with and without AME treatment (Fig 2 A). No significant difference was found among positive control and those received AME 150 mg/kg BW/day (Table 2). The difference however, was observed between positive control and X_3 ($p = 0.013$). CXCL10-expressions of X_3 group were comparable to those of negative control. AME treatment might normalize CXCL10 expression in the brain during CM phase of Swiss mice. CXCL10 is expressed by astrocytes, the abundant cells in the brain which function to maintain CNS homeostasis [9]. Astrocytes are able to induce T-cell apoptosis and subsequently avoid inflammation [17, 40]. Since CXCL10 is chemokine which attract Th1 cells, thus the reduce CXCL10 expressed by astrocytes is speculated to reduce the Th1-cell apoptosis in the brain of PbA-inoculated swiss mice. This then, maintains inflammation in the brain of PbA-infected Swiss mice. More recent study found that inflammatory activated astrocytes had no role in Th1 polarization [41]. Whether this occure in PbA-infecd swiss mice has not been studied. All together support futher studies including the use of CXCL10-deficient Swiss mice in order to strengthen the notion that CXCL10 expressed by astrocytes has important role in ECM. The CXCL10 generation is induced by IFN-γ, a cytokine signature of Th1 cells. Interestingly, IFN-γ is also needed for the release of IL-10, an anti-inflammatory cytokine, by microglia/ macrophage and Th1 cells [42, 43]. The relation between those cytokines and astrocyte-CXCL10 expression necessitates to be clarified in PbA-infected mice treated with AME.

Preventing neuronal and endothelial apoptosis alone may not sufficient to improve survival of those suffer from CM, it need immune modulator treatment [44, 45]. This therefore, AME treatment is remain possible providing any protection on CM. Study on C57BL/6 mice inoculated with 5 x 10^5 PbA-pRBC showed that two distinct anti-apoptosis treatment successfully prolonged survival in sepsis animal model, were not successful prevent neuronal apoptosis and improve survival of PbA-inoculated mice [44]. C57BL/6 mice inoculated with 10^6 pRBC was used to evaluate the protective effect of Neuregulin-1. Serum CXCL10 levels of PbA-inoculated C57BL/6 mice with and without any Neuregulin-1 and arthemeter treatment were compared. Neuregulin-1 significantly reduce serum CXCL10 and other inflammatory cytokines (TNF-α, IL-1β, IL-6 and G-CSF) while significantly increase anti-inflammatory cytokines including IL-5 and IL-13 at day 11 than positive control group, PbA-inoculated mice without treatment. Neuregulin-1 prevented mortality of ECM mice, and other beneficial effects including preserved BBB stability and reduced leukocyte accumulation in the brain [45]. Whether any of these occur in PbA-inoculated mice treated with AME need to be elucidated.

Brain CXCL10 expression increased in most brain infection [46, 47]. CXCL10 produced by monocytes and netrophils was responsible for CM developed in PbA-infected mice [48]. CXCL10 KO mice or CXCL10 neutralization reduced inflammation in the brain of ECM mice model [49]. Swiss mice used in the recent study are CM susceptible mice, therefore it is expected that CXCL10 expression increases in those PbA-inoculated mice particularly on day 7 p.i. which is CM phase. The CXCL10 expressions were statistically different among the six-studied groups ($p = 0.003$). The CXCL10 expression of PbA-infected Swiss mice (the positive control group) was significantly lower than healthy groups which were without or with AME treatment (negative control, $p = 0.008$; X_1, $p = 0.045$; or X_2,

$p = 0.012$). The difference was also observed by comparing X_4 and X_2 groups ($p = 0.016$). This study was the first observed CXCL10 in the brain of Swiss mice during CM phase. The recent finding demonstrated that Swiss mice might modulate the immune response to ameliorate inflammation in the brain during CM phase. The increase number of large astrocyte clusters and astrocyte dilatation were found in PbA-infected CBA/CaH mice [50]. Neuronal and astrocyte apoptosis was found in PbA-inoculated C57BL/6 mice [51]. Astrocyte dilatation and apoptosis may modulate the astrocyte-CXCL10 expression in PbA-inoculated Swiss mice and explained the reduce CXCL10 expression in PbA-inoculated Swiss mice. Hipoxia and hypoglicemia is possibly reduce astrocyte-CXCL10 expression during malaria. *In vitro* study showed that the hypoxia and hypoglycemia result in reduce astrocyte-CXCL10 expression [16]. Whether all of these influencing brain-CXCL10 expression during CM phase of Swiss mice need to be elucidated.

The C57BL/6 mice study showed that astrocytes expressed CRG2 (CXCL10) mRNA [52]. By comparing C57BL/6 and Balb/c mice, it was found that genetic factors had strongly influence on intracranial chemokines. Astrocyte-CXCL10 expression in the brain of C57BL/6 mice was reaching peak earlier than Balb/c mice during *Toxoplasma* encephalitis [53]. Most CXCL10 studies on PbA used C57BL/6 mice, while this study used Swiss mice inoculated with 10^7 PbA-pRBC. C57BL/6 mice inoculated with 10^6 pRBC showed that CXCL10 mRNA in the brain was higher in PbA-inoculated C57BL/6 mice than those inoculated mice received vitamin D. The treated group showed a reduce adhesion molecules result in reduce pathologic T cell accumulation in the brain. Spleens of treated mice were also showed reduce differentiation, activation and maturation of DC, while increase T regulatory cell and IL-10 expression. Survival of PbA-inoculated mice was significantly longer in those received vitamin D [54]. Additionally, other study using C57BL/6 mice inoculated with 10^6 PbA-pRBC showed the increase CXCL10 mRNA in the brain and parasite biomass in PbA-inoculated mice with reticulocytosis than those without reticulocytosis [55]. The two studies compared CXCL10 expression between PbA-inoculated groups with and without particular treatment, while the trend of reduce brain-CXCL10 expression in our study was found by comparing PbA-inoculated than healthy Swiss mice. Additionally, similar finding was also observed by comparing healthy and PbA-inoculated group which both of them received similar treatment dose.

The similar degree of CXCL10 expression were found in those negative control, healthy Swiss mice receiving 100 or 150 mg/kg BW/day for period of 14 days. CXCL10 expressed by those healthy groups in our study has not yet been explained, therefore further studies warrant to be performed to confirm and find mechanism involve. Astrocytes cultured without any stimulation result in detectable level of CXCL10 in the culture supernatant [56]. This might in contribute in explaining that astrocyte of healthy Swiss mice expressed CXCL10 in our study.

Parasitemia levels were not different among positive control and two groups receiving AME treatment. Parasitemia might be not a sensitive marker for evaluating the benefice of AME treatment in this recent study. The pRBCs was sequestered in the brain blood vessels and other tissues of PbA-infected mice [57-59]. This may cause no association between parasitaemia levels and the development of CM. However, pRBCs sequestered in the brain blood vessels during ECM developed during PbA infection is remain a matter of debate, and even if there seems to be happening to a small extent [60]. PbA-infected mouse used and treatment used resulted in an obvious pRBC sequestered in the brain. CXCL10 KO mice reconstituted with bone marrow of wild type mice showed an obvious pRBCs and leukocytes sequestered in the brain [48].

Ang-2 levels were not normally distributed, and Kruskal-Wallis test showed no significant different among six studied group ($p = 0.175$). The interesting finding showed that the median value of Ang-2 levels of positive control group tended to be higher than negative control (Fig 2B). This study was the first showing circulating Ang-2 levels of Swiss mice during CM phase. A trend of higher Ang-2 levels found in positive control group during CM phase was not entirely supported by previous finding in CM patients. CM patients showed significantly higher circulating Ang-2 levels [24]. The study compared CM patients with retinopathy (CM-R) and CM without retinopathy (CM-N). The study recruited CM patients based on WHO criteria. Retinopathy considered positive when one or more of retinal finding including hemorrhage, vessel change, and papilloedema. Ang-2 levels of CM-R patients were significantly higher than CM-N patients. Vascular dysfunction, indicated by the retinopathy, was obviously present in CM-R. The study defined CM-R patients as confirmed CM patients. The study, however, remained needed to be validated. It is open possibility that CM occur without obvious vascular dysfunction. Increase mRNA Ang-2 expression in the brain was observed in day 5 after PbA-inoculation of C57/BL/6 mice [21]. Circulating Ang-2 levels had been observed in PbA inoculated C57BL/6 mice as a model for placental malaria [61]. This recent study done, therefore, was the first study proved that circulating Ang-2 levels is not associated with CM phase of Swiss mice. The use of circulating Ang-2 level as biomarker in CM phase of Swiss mice should be taken with caution. No different found among the six studied group indicates that ethanolic-AME might have no relation with circulating Ang-2 levels of Swiss mice during CM phase. The finding, however warrant to be further confirmed by evaluating Ang-2 expression in the brain of Swiss mice during CM phase.

Acknowledgement
KD and SMAM contributed equal work in this article and share first authorship. KD received PNBP grant 2015 from Medical faculty Diponegoro University, which partly supported this research. SMAM received post graduate schoolarship and research funding from Libya Goverment. We thank to P.T. SidoMuncul Herbal Medicine Industry, especially to Mrs. Wahyu Widayani, Apt who kindly provided ethanolic-*A. muricata* leaf extract and its analysis.

References

[1]. WHO. *Malaria*. Malaria 2016 April 2016; Available from: http://www.who.int/mediacentre/factsheets/fs094/en/.
[2]. WHO, *World Malaria Report 2015*. 2015, WHO: Geneva.
[3]. WHO, *Management of severe malaria* 2012, WHO Press: Geneva.
[4]. Batista R, Silva Ade J, Jr., and de Oliveira A B 2009 *Molecules* **14** 3037.
[5]. Kaur K, Jain M, Kaur T, and Jain R 2009 *Bioorganic & Medicinal Chemistry* **17** 3229.
[6]. Oliveira A B, Dolabela M F, Braga F C, Jacome R L, Varotti F P, and Povoa M M 2009 *Anais da Academia Brasileira de Ciencias* **81** 715.
[7]. Mabberley D J, *Plant-book*. 3rd ed ed. 2008, Cambridge: Cambridge University Press.
[8]. Somsak V, Polwiang N, and Chachiyo S 2016 *Journal of pathogens* **2016** 3264070.
[9]. Zhu X, Liu J, Feng Y, Pang W, Qi Z, Jiang Y, Shang H, and Cao Y 2015 *Experimental parasitology* **156** 1.
[10]. Karolina M E, Fransisca Prameshinta H, Dharmana E, and Djamiatun K 2016 *Jurnal Kedokteran Brawijaya* **29** 4.
[11]. Campanella G S, Tager A M, El Khoury J K, Thomas S Y, Abrazinski T A, Manice L A, Colvin R A, and Luster A D 2008 *Proceedings of the National Academy of Sciences of the United States of America* **105** 4814.
[12]. Hansen D S, Bernard N J, Nie C Q, and Schofield L 2007 *Journal of immunology* **178** 5779.
[13]. Luster A D, Alon R, and von Andrian U H 2005 *Nature immunology* **6** 1182.
[14]. Armah H B, Wilson N O, Sarfo B Y, Powell M D, Bond V C, Anderson W, Adjei A A, Gyasi R K, Tettey Y, Wiredu E K, Tongren J E, Udhayakumar V, and Stiles J K 2007 *Malar Journal* **6** 147.
[15]. Wilson N O, Solomon W, Anderson L, Patrickson J, Pitts S, Bond V, Liu M, and Stiles J K 2013 *PloS one* **8** e60898.
[16]. Bakmiwewa, *The Astrocyte: a Crossroads in Cerebral Malaria Pathogenesis*, in *Pathology School of Medical Sciences Faculty of Medicine* 2015, University of Sydney.
[17]. Gimsa U, Mitchison N A, and Brunner-Weinzierl M C 2013 *Mediators of inflammation* **2013** 320519.
[18]. Bakir H Y, Tomiyama C, and Abo T 2011 *Biomedical research* **32** 203.
[19]. Perkins D J, Were T, Davenport G C, Kempaiah P, Hittner J B, and Ong'echa J M 2011 *International journal of biological sciences* **7** 1427.

[20]. Carvalho L J, Moreira Ada S, Daniel-Ribeiro C T, and Martins Y C 2014 *Memorias do Instituto Oswaldo Cruz* **109** 577.

[21]. Kim H, Erdman L K, Lu Z, Serghides L, Zhong K, Dhabangi A, Musoke C, Gerard C, Cserti-Gazdewich C, Liles W C, and Kain K C 2014 *Infection and immunity* **82** 371.

[22]. Gomes L T, Alves-Junior E R, Rodrigues-Jesus C, Nery A F, Gasquez-Martin T O, and Fontes C J 2014 *PloS one* **9** e109246.

[23]. Brouwers J, Noviyanti R, Fijnheer R, de Groot P G, Trianty L, Mudaliana S, Roest M, Syafruddin D, van der Ven A, and de Mast Q 2013 *PloS one* **8** e64850.

[24]. Conroy A L, Phiri H, Hawkes M, Glover S, Mallewa M, Seydel K B, Taylor T E, Molyneux M E, and Kain K C 2010 *PloS one* **5** e15291.

[25]. Conroy A L, Lafferty E I, Lovegrove F E, Krudsood S, Tangpukdee N, Liles W C, and Kain K C 2009 *Malaria journal* **8** 295.

[26]. Conroy A L, Glover S J, Hawkes M, Erdman L K, Seydel K B, Taylor T E, Molyneux M E, and Kain K C 2012 *Critical care medicine* **40** 952.

[27]. Ioannidis L J, Nie C Q, and Hansen D S 2014 *Parasitology* **141** 602.

[28]. Ntie-Kang F, Onguene P A, Lifongo L L, Ndom J C, Sippl W, and Mbaze L M 2014 *Malaria journal* **13** 81.

[29]. Esmaeili S, Naghibi F, Mosaddegh M, Sahranavard S, Ghafari S, and Abdullah N R 2009 *Journal of ethnopharmacol* **121** 400.

[30]. Bassey A S, Okokon J E, Etim E I, Umoh F U, and Bassey E 2009 *Indian journal of pharmacology* **41** 258.

[31]. Bero J, Ganfon H, Jonville M C, Frederich M, Gbaguidi F, DeMol P, Moudachirou M, and Quetin-Leclercq J 2009 *Journal of ethnopharmacol* **122** 439.

[32]. Innocent E, Moshi M J, Masimba P J, Mbwambo Z H, Kapingu M C, and Kamuhabwa A 2009 *African Journal of Traditional, Complementary and Alternative Medicines* **6** 163.

[33]. Dell'agli M, Galli G V, Bulgari M, Basilico N, Romeo S, Bhattacharya D, Taramelli D, and Bosisio E 2010 *Malaria journal* **9** 208.

[34]. Waknine-Grinberg J H, El-On J, Barak V, Barenholz Y, and Golenser J 2009 *Planta medica* **75** 581.

[35]. Osorio E, Arango G J, Jimenez N, Alzate F, Ruiz G, Gutierrez D, Paco M A, Gimenez A, and Robledo S 2007 *Journal of ethnopharmacology* **111** 630.

[36]. Mohd Abd Razak M R, Afzan A, Ali R, Amir Jalaluddin N F, Wasiman M I, Shiekh Zahari S H, Abdullah N R, and Ismail Z 2014 *BMC complementary and alternative medicine* **14** 492.

[37]. Ishola I O, Awodele O, Olusayero A M, and Ochieng C O 2014 *Journal of medicinal food* **17** 1375.

[38]. Moghadamtousi S Z, Rouhollahi E, Karimian H, Fadaeinasab M, Abdulla M A, and Kadir H A 2014 *Drug design, development and therapy* **8** 2099.

[39]. Maria Estela K, Fransisca Prameshinta H, Dharmana E, and Djamiatun K 2016 *Jurnal Kedokteran Brawijaya* **29** 4.

[40]. Xie L and Yang S H 2015 *Brain research* **1623** 63.

[41]. Yang J F, Tao H Q, Liu Y M, Zhan X X, Liu Y, Wang X Y, Wang J H, Mu L L, Yang L L, Gao Z M, Kong Q F, Wang G Y, Han J H, Sun B, and Li H L 2012 *Clinical and experimental immunology* **170** 254.

[42]. Beurel E, Harrington L E, Buchser W, Lemmon V, and Jope R S 2014 *PLoS one* **9** e86257.

[43]. Ishii H, Tanabe S, Ueno M, Kubo T, Kayama H, Serada S, Fujimoto M, Takeda K, Naka T, and Yamashita T 2013 *Cell death & disease* **4** e710.

[44]. Helmers A J, Lovegrove F E, Harlan J M, Kain K C, and Liles W C 2008 *The American journal of tropical medicine and hygiene* **79** 823.

[45]. Solomon W, Wilson N O, Anderson L, Pitts S, Patrickson J, Liu M, Ford B D, and Stiles J K 2014 *Journal of neuroinflammation* **11** 9.

[46]. Lokensgard J R, Mutnal M B, Prasad S, Sheng W, and Hu S 2016 *Journal of neuroinflammation* **13** 114.
[47]. Liu Y, Chen L, Zou Z, Zhu B, Hu Z, Zeng P, Wu L, and Xiong J 2016 *Journal of medical virology* **88** 1596.
[48]. Ioannidis L J, Nie C Q, Ly A, Ryg-Cornejo V, Chiu C Y, and Hansen D S 2016 *Journal of immunology* **196** 1227.
[49]. Nie C Q, Bernard N J, Norman M U, Amante F H, Lundie R J, Crabb B S, Heath W R, Engwerda C R, Hickey M J, Schofield L, and Hansen D S 2009 *PLoS pathogens* **5** e1000369.
[50]. Ampawong S, Chaisri U, Viriyavejakul P, Nontprasert A, Grau G E, and Pongponratn E 2014 *International journal of clinical and experimental pathology* **7** 2056.
[51]. Wiese L, Kurtzhals J A, and Penkowa M 2006 *Experimental neurology* **200** 216.
[52]. Morrell C N, Srivastava K, Swaim A, Lee M T, Chen J, Nagineni C, Hooks J J, and Detrick B 2011 *Infection and immunity* **79** 1750.
[53]. Lane T E, Asensio V C, Yu N, Paoletti A D, Campbell I L, and Buchmeier M J 1998 *Journal of immunology* **160** 970.
[54]. Strack A, Schluter D, Asensio V C, Campbell I L, and Deckert M 2002 *Glia* **40** 372.
[55]. Lo U, Selvaraj V, Plane J M, Chechneva O V, Otsu K, and Deng W 2014 *Scientific reports* **4** 7405.
[56]. Krauthausen M, Kummer M P, Zimmermann J, Reyes-Irisarri E, Terwel D, Bulic B, Heneka M T, and Muller M 2015 *The Journal of clinical investigation* **125** 365.
[57]. Hearn J, Rayment N, Landon D N, Katz D R, and de Souza J B 2000 *Infection and immunity* **68** 5364.
[58]. Franke-Fayard B, Janse C J, Cunha-Rodrigues M, Ramesar J, Buscher P, Que I, Lowik C, Voshol P J, den Boer M A, van Duinen S G, Febbraio M, Mota M M, and Waters A P 2005 *Proceedings of the National Academy of Sciences of the United States of America* **102** 11468.
[59]. Amante F H, Stanley A C, Randall L M, Zhou Y, Haque A, McSweeney K, Waters A P, Janse C J, Good M F, Hill G R, and Engwerda C R 2007 *The American journal of pathology* **171** 548.
[60]. Lou J, Donati Y R, Juillard P, Giroud C, Vesin C, Mili N, and Grau G E 1997 *The American journal of pathology* **151** 1397.
[61]. Silver K L, Zhong K, Leke R G, Taylor D W, and Kain K C 2010 *PLoS one* **5** e9481.

Socio-Economic Spatial for the Sustainability of the Estuary Ecosystem in Pelabuhan Ratu Coastal West Java

Supriatna, L.[1,3], Supriatna, J.[2], Harmantyo, D.[3]

[1] Program Study of Enviromental Science, Universitas Indonesia.
Jl. Salemba Raya No. 4, Jakarta 10430, Indonesia
[2] Departmen of Biology, Faculty of Matematics & Natural Science, Universitas Indonesia.
Kampus UI Depok, Depok 16424, Indonesia
[3] Departmen of Geography, Faculty of Matematics & Natural Science, Universitas Indonesia.
Kampus UI Depok, Depok 16424, Indonesia

E-mail: ysupris@yahoo.com;jatna.supriatna@gmail.com;djokoharmantyo@yahoo.com

Abstract. An estuary area is a typical ecosystem where a vast array of fish breed to enlarge populations. These regions are made productive by organic material in the form of foodstuff, while receiving sunlight sufficient enough to illuminate the brackish waters. These zones must be protected due to their fertile waters and surrounding fertile land. Estuary areas are threatened by nearby landfill and waste debris along and upstream of the river, which consequently contaminate the estuary zone. Socio-economic conditions of the community also affect the sustainability of the estuary. In this case, the Socio-Economic Spatial Model, based on Geographical Information System (GIS) and trade off analysis, were used to elaborate the ecosystem sustainability in the Cimandiri estuary, West Java. This research also uses the temporal analysis of land use change upstream and monitors the community activity around the estuary. The research showed a change in the spatial and temporal land use consequently altering the watershed and the socio-economic analysis showed the community use of the estuary as unsustainable for the region and ecosystems within.

Keywords : estuary, socio-economic, spatial, trade off, sustainability.

1. Introduction

Currently, the understanding of an *estuary* among society is limited, in particular the ways that a community directly uses an estuary. The activity of the community in the region can affect the estuary [1]. Population growth will pressure land area encroaching on the estuary and can affect the estuary ecosystems, which will ultimately lead to environmental degradation.

Regional planning usually emphasizes economic factors for the growth of its region rather than the balance of the environment factors, including the ecosystem balance [2]. The ecosystem, including the estuary ecosystem, can maintain the balance of surrounding environments. As it is said in the report of the World Commission on Environment and Development (WCED) about Our Common Future, sustainable development emphasizes the balance of social, economic and environmental spheres for future generations [3].

In order to properly conserve the estuary region and existing ecosystem, a holistic approach to consider social, economic and political factors is needed. The economic background of a community that uses an estuary as a main source of livelihood is a determinant of the degree of degradation in the region [4]. Community behaviour can directly or indirectly affect the surrounding ecosystem, in particular when the area is used irresponsibly for economic benefit, farmed incorrectly or waste is incorrectly disposed.

In their research, Fortini *et al.* [5] suggest that economic analysis can be used to focus on some of the advantages and disadvantages of timber production and *Acai oil* in the Amazon estuary and the phenomenons that arises from production activities. The result of the analysis was to draw up an alternative strategy to design better production, with consideration of economic, ecological and social impacts of production activities.

In a study area with a diverse range of estuaries, there is a trade off approach related to the social, economic and ecological factors affecting the estuary. The management of the fisheries sector as a whole becomes a big challenge, especially the conflicts of interest between socio-economic factors and the existing ecology in the estuarine area. An integrated model is needed with the approach of measuring the exchange (trade-off) between the ecological system and the socio-economic estuary systems [6]. Application of the trade-off method calculation in ecosystems (socio-ecological) depends on the results from assessment of the existing calculations [7]. The use of such methods reveal the fact that the values obtained are actual values. The use of the trade off method calculation and synergies for the ecosystem, in an industrial-oriented estuary region, reveals a relationship and mutual interest in sustainable estuary management measures [8].

The purpose of this study is to first examine the spatial use of estuary directly by the community, assess the existence of the estuary within the social and economic conditions of its community, and then evaluate the relationship of the estuary with the social and economic characteristics of the community. This is facilitated by the trade off analysis of the community by selecting the most important economic, social and ecological factors.

The simple and effective method to integrate spatial data from various sources can be quantified to make ecology and economy data area (zoning) in an estuary [9]. This model is integrating the spatial and tabular data to use traditional ecological and economic analysis. Similar research revealed the impact of the decline in production and consumption of Japanese eels (glass eel) in the Sagami estuary [10]. The decline had raised concern on the state of ecological decline and the factors of economic interests of the Sagami community.

An ecosystem-based model (economic and ecological factors) in the area of an estuary can be applied on an ongoing basis and act to majorly benefit the local community and surrounding areas. Not only in theory but in fruition, this model can create an investment in biodiversity and is not for certain specialties [11]. A more holistic and balanced approach needs to be taken on the policy framework of the beaches and coastal area in Indonesia [12]. The policy framework should be able to overcome the problem of vulnerability when addressing coastal communities on availability of marine resources. This research utilizes social, economic and government policies.

An economic-based model will give evidence that an interdisciplinary approach to a particular area can be done with management-based coastal area planning [11]. Cost approach methods allow integration of the most basic model of connectivity and activities to make certain planning areas in the city. The potential for this type of local planning in local decision-making process depends on the individual interests of planners rather than being systematic in nature [13].

A strong knowledge of management (production, mobilization, integration) is the key to achieving a sustainable socio-economic field to address the many challenges in the coastal area. For example, balancing the ecology, conservation and economic development of a region. The relationship between theory and practice with the socio-economic integrative approach requires involvement, participation, communication, and implementation between the community and the stakeholders [14]. The results of the social networks analysis have the potential to be a valuable tool in support of decision-making in conservation planning [15].

A collaboration is needed between the stakeholders and the government to protect (active conservation measures) and develop (development and investment) beach and estuary areas [16], [17], as well as benefiting the surrounding community [18]. In order to obtain a large amount of revenue through economic activity and to inspire people to be environmentally conscious, the collaboration between stakeholders and governments is a necessity.

Carvalho & Fidelis [19] stated that the institutional features of estuary management involve various agencies, stakeholders and users, as well as the principles of governance that are relevant. The result is a model of governance that is capable of ensuring the implementation of estuary planning, with a strong contribution to the involvement of all stakeholders and users in the planning of development. Thus, the conciliation of interests and active participation in decision-making need a firm place within the framework of collaborative governance.

In their research, Chen *et al.* [20] stated that the value of ecological functions is at risk and under pressure from estuary degradation and declining biodiversity. Also, Ottinger *et al.* [21] stated that the impact of anthropogenic activities on many estuaries has led to substantial changes in the surrounding environment.

The objective this research was socio-economic spatial and trade off analysis of the community activity around the estuary.

2. Methods

The following study of the socio-economic relationship between the community and the estuary used spatial analysis and trade off analysis. This research was carried out from June to December 2015. The study area is the estuary located in the area of Cimandiri [22] and the surrounding areas, which are on the South coast of West Java in the subdistrict of Simpenan and the Sukabumi Region.

The social and economic data on population was collected from the community living within and utilizing the resources of the Cimandiri estuary area. Population data sampling was arranged using the simple random sampling technique. Sampling is based on interviews with the community that directly utilize the Cimandiri estuary. This community (population) has approximately 1,200 people (Simpenan Sub-district, 2015). The majority of community members earn their living as sand miners and fish larvae fishers. Samples were taken from 200 respondents (which equates to 16.67% of the people who use the resources within the Cimandiri estuary). Out of the sample within the community 100 respondents were fish larvae fishers (eel) and 100 respondents were sand miners. The study was also carried out on a sample of stakeholders (community leaders and agencies related to the research theme) who were around the Cimandiri estuary area. The sample method was purposive sampling, utilizing interview techniques.

The variables of this study were determined by considering the presence of an estuary in conjunction with socio-economic aspects of the community, such as: education, employment, income, housing, knowledge and attitudes. The data processed included information regarding individuals living between Cimandiri Estuary and the community's shelters, all of whom directly utilize Cimandiri estuary. Socio-economic data is the knowledge and attitudes of the community towards the existence of the estuary. Socio-economic data is obtained by processing information to calculate social, economic and ecological trade off (profits and losses).

The relationship between the community who utilize the existence of the Cimandiri estuary and the estuary itself was analyzed using socio-economic analysis of the surrounding area. First, how far the influence of the Cimandiri estuary reaches in the surrounding area was analyzed. Second, the socio-economic conditions of local communities was analyzed and third, the social, economic and ecological trade offs were determined.

Spatial analysis was used to determine the region for study, that is, the region influenced by the estuary and the surrounding community. The radius for analysis was hence found to be the distance from the mouth of the River Cimandiri to the houses of the community members who directly use the Cimandiri estuary.

The study of tradeoffs (exchange) is one of the main principles of economics. People face a trade off between desires and the source of their desires. There is no issue if the source is unlimited and infinite, however when the source of their desire is limited a sacrifice is needed. This sacrifice is a trade off; giving up something that is liked to get something that is liked more. Analysis of trade-offs in managing various interests and desires should be done wisely [7], [8].

3. Result and discussion

The boundary of the Cimandiri estuary region for study was the area from the river mouth to the houses of community members who directly use the Cimandiri estuary. Figure 1 and Figure 2 show that fish larvae fishermen and sand miners live around the estuary of the Cimandiri River. Almost all of them reside in the five administrative regions of: Pelabuhanratu village, Jayanti village and Citarik village in the Pelabuhanratu district; Cidadap village and Loji village in the Simpenan district. The distance of community member's houses from the Cimandiri river estuary was less than 5 km.

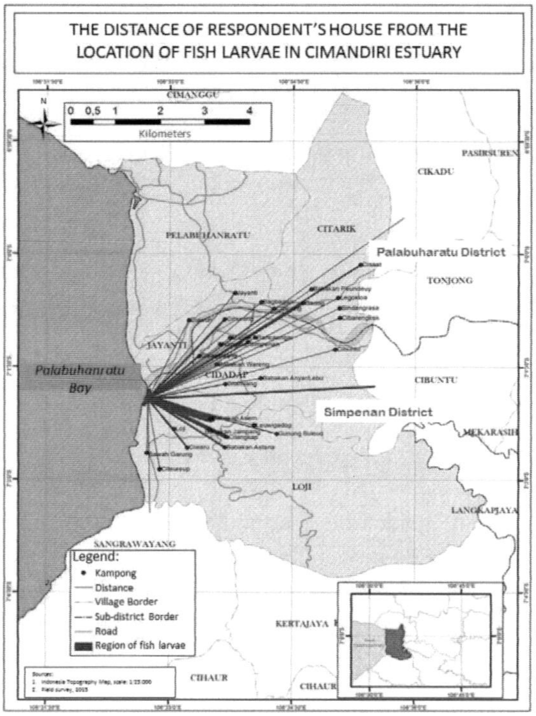

Figure 1: Map of the distance of respondent's' house from the location of fish larvae in Cimandiri estuary

Spatially, community member residences of the sand miners were an average distance of about 2.1 km from the Cimandiri estuary. There are nine hamlets in Cidadap village, four hamlets in Jayanti village, three hamlets in Loji village, two hamlets in Citarik village and two hamlets in Pelabuhanratu village.

Analysis of knowledge and behavior towards the Cimandiri estuary found the following:

a. All sand miner respondents (100%) are aware of the Cimandiri estuary, while the awareness of the fish larvae fisher respondents of the Cimandiri estuary is only 97%.
b. The importance of Cimandiri estuary for the sand miner respondents is 100% important while for fish fisher respondents is only 96% important.
c. 100% of sand miner respondents and 94% of fish larvae farmer respondents were aware that Cimandiri estuary is a habitat of fish larvae as a food for small fish.
d. The estuary of Cimandiri River is constantly deformed by the sediments carried by the river flow and big ocean currents (waves) that affects the formation of the estuary. 98% of sand miner respondents knew this, while 96% of fish larvae fisher respondents makers were aware.

e. The estuary of Cimandiri River must be protected from pollution, especially from the trash carried from the river upstream.98% of sand miner respondents and 85% of fish larvae fisher respondents were aware of this.
f. The changes in form of the estuary and pollution are due to human activity. 95% of sand miner respondents were aware of this and 72% of fish larvae fisherman respondents.
g. The river stream and Cimandiri estuary are always dirty, especially in rainy season. Respondents who knew this were 100% of sand miner respondents and 98% of the fish larvae fisher respondents.
h. Cimandiri estuary region is sometimes used for beach tourism activities (fishing in the estuary, taking fish larvae, and other tourist activities). Sand miner respondents who acknowledge this numbered 49%, while fish larvae farmer respondents numbered 66%.
i. The respondents are aware that the land around the estuary Cimandiri is fertile, which 98% of both sand miner respondents and fish larvae fisher respondents were aware of.
j. Estuary and beaches around Cimandiri are places for fishing. Both sand miner respondents and fish larvae fisher respondents are equally aware of this at 98% awareness.
k. The respondents who have ever utilized Cimandiri estuary are 100% of sand miner respondents and 98% of fish larvae fishermen.

From the results of the questionnaire, it appears that sand miner respondents are more aware of the Cimandiri estuary, showing that sand miner respondents were better educated about estuaries than respondents who farm fish larvae. More details can be seen in Table 1.

Table 1. The respondent percentage of the existence Cimandiri estuary

No	Respondent's knowledge and behavior	Respondent(%)	
		Fish Larvae	**Sand**
1	Perceiving the existence of Cimandiri estuary	100	97
2	The importance of Cimandiri estuary to the respondent	100	96
3	Respondents aware that the Cimandiri is a habitat for fish larvae as a food for small fishes	100	94
4	Cimandiri estuary constantly deformed by the sediments carried by the river flow and big ocean currents (waves) that also affect the formation of the estuary	98	96
5	Cimandiri estuary should be protected from pollution, especially the trash carried from upstream	98	95
6	Changes in the form estuary and the pollution occurred due to human activity	95	72
7	The stream and estuary are always dirty, especially the rainy season	100	98
8	Estuary area is sometimes used for beach tourism activities	49	66
9	The respondents aware that the surrounding land of the estuary is fertile	98	98
10	Estuary and beaches around the Cimandiri estuary is a place for fishing	98	98
11	The respondents have ever used the Cimandiri estuary	100	98

Source: Field survey and data processing (2015)

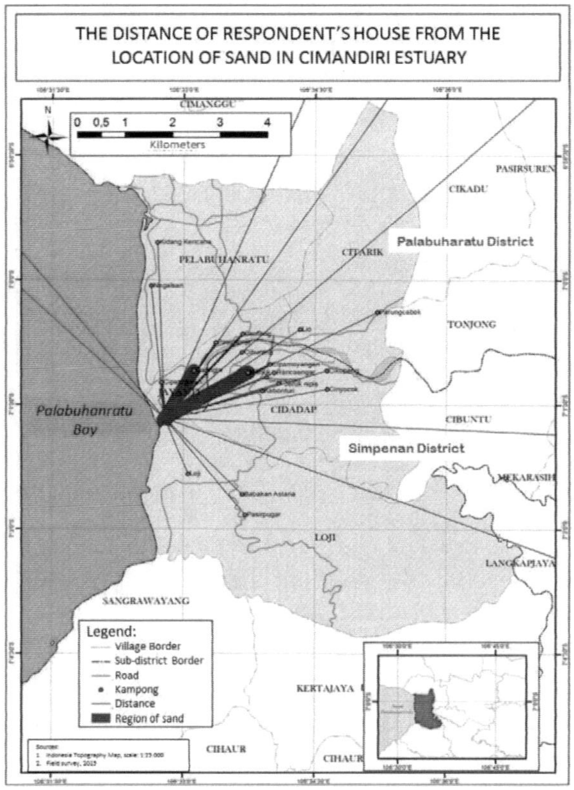

Figure 2: Map of the distance of respondents' house from the location of the sand in Cimandiri estuary

The results of the analysis show that all of the public concern about the estuary in Cimandiri is mainly related to the conservation and protection of fish larvae. In most cases, public figures did not know about the regulations that related to the estuary, although some of them answer that they do know.

Socio-economic trade off analysis revealed that fish larvae fishers and sand miners with an average age of 40 years old think that taking fish larvae is not desirable, while respondents with an average age under 40 years old believe that taking fish larvae (eel) is a desirable job. Economic trade off analysis in the sand miner community shows that sand mining is a job that is desireable by respondents.

The next step was to synthesize the trade off approach to the social, economic and ecological analysis to derive policy alternatives in the model of sustainable landscape in the Cimandiri estuary. The stakeholder respondents comprised of the community that directly use the estuary, public figures around the estuary, and authorities that deal with estuary problems. Community as the main stakeholders (primary stakeholders), have a high level of concern for the estuary. The public figures around the estuary which is a second-tier stakeholder (secondary stakeholders) have a medium level of interest and influence on the decision-making process. Finally, government officials or authorities have the lowest interest levels but a high influence in the decision-making process [23].

Table 2. The criteria in determining the importance of estuaries.

Criteria	Sub criteria	Community	Public figure	Authorities
Social	Importance / benefit of estuary on surrounding communities.	98% answered that it is important.	100% answered that it is important.	100% answered that it is important.
	Estuary as a place to make a living in the surrounding communities.	98% answered as a place to make a living.	83% answered as a place to make a living.	90% answered as a place to make a living.
Economic	Community income from coastal tourism around the estuary	47% answered that the income is from tourism activities.	60% answered that the income is from tourism activities.	40% answered that the income is from tourism activities.
	Community income is from the estuary (fish larvae and sand).	97% answered that their income is from the fish larvae and sand.	80% answered that their income comes from the fish larvae and sand.	70% answered that their income comes from fish larvae and sand.
Ecological	Perceiving the estuary as a habitat for small fish	96% answered that estuary is a habitat for small fish	100% answered that estuary is a habitat for small fish	100% answered that estuary is a habitat for small fish
	Estuary must be protected from pollution	97% answered that it is must be protected.	100% answered that it is must be protected.	100% answered that it is must be protected.

Source: Data processing (2015)

Table 2 shown that 98% of community answered that they saw the estuary is a place to make a living and so be beneficial for the community. Even public figures (100%) answered that the estuary is important to the community for them to take the advantage of it, and 83% said that the estuary is the livelihood for the community.

Of the economic criteria, community's income from tourism activities apparently doesn't provide a significant income compared to larvae of fish and sand. The questionnaires result shown that 97% (community), 80% (the public figures) and 70% (authorities) answered the income of local communities comes from taking fish larvae and sand mining. Ecological criteria on the habitat of small fish and the protected estuary reveалled that all of the stakeholders answered above 95%. Ecological criteria were the highest criteria, where the stakeholders answered that the estuary must be protected. The synthesis is available in Table 3.

Table 3. The social economic relationship of community to the existence of estuary

• Fish larvae community.	• 70% of fishermen and farmers take fish larvae every day.
• Sand miners community.	• 95% of sand miners take the sand every day.
• socio-economic trade off in the community	• The job of fish larvae fishers is to take eels larvae (30% of the answer), while the job of sand miners is mining the sand (93% of the answer).
• ecological socio-economic trade off in the community	• Ecological criteria is the highest criteria, where the stakeholders selected protecting estuaries.

Source: Data processing (2015)

The community and their socio-economic relationship in this study emerged by using trade off analysis. Trade off analysis of livelihoods (employment) is performed by the community who directly use the Cimandiri estuary and trade off analysis by the stakeholder is performed with the social, economic and ecological aspects in Cimandiri estuary. The result of this research is contrary to the results of the research with the community behavior approach on the social, economic and political aspects within the estuaries [4]. The economic analysis was calculated by considering gains and losses [5], estimates of the potential costs of land use patterns, and scenarios [6]. The research by Kismartini [7] and Lopeza *et al.* [8] also used the trade off method but the study was not in the area of estuaries. Meanwhile, Jacobs *et al.* [8] and Wang *et al.* [6] have done research using the trade off method that took place in estuary areas but the research was in the sub-tropical area.

4. Conclusion

There is a very close relationship between the socio-economic life of the community and the Cimandiri estuary. The bond that the community has made results from either the direct or indirect use of the environment through practises such as fishing, sand mining and tourism. The results of the trade off analysis considering social, economic and ecological factors has shown that ecological criteria are the most important for stakeholders, compared to social, economic and the other criteria.

Rerefences

[1] Clarke, D.W., Boyle, J.F., Chiverrell, R.C., Lariob, J., & Plater, A.J. 2014. A sediment record of barrier estuary behaviour at the mesoscale: Interpreting high-resolution particle size analysis. *Journal of Geomorphology,* 221, 51-68.

[2] Guneroglu, N., Acar, C., Guneroglu, A., Dihkan, M., Karsli, F. 2015. Coastal land degradation and character assessmen.t of Southern Black Sea landscape. *Journal of Ocean & Coastal Management xxx* (2015), 1-8.

[3] Musacchio, L.R. 2009. The scientific basis for the design of landscape sustainability: A conceptual framework for translational landscape research and practice of designed landscapes and the six Es of landscape sustainability. *Journal of Landscape Ecology*, 24, 993-1013.

[4] McAuliffe, S. Potts, J. Canessa, R. 2014. Establishing attitude sand perceptions of recreational boat users based in the River Hamble Estuary, UK, towards Marine Conservation Zones. *Journal of Marine Policy,* 45, 98-107.

[5] Fortini, L.B. & Carter, D.R. 2014. The economic viability of smallholder timber production under expanding acaí palm production in the Amazon Estuary. *Journal of Forest Economics* 20 223-235.

[6] Wang, Y., Hua, J., Pan, H., Li, S., Failler, P. 2016. An integrated model for marine fishery management in the Pearl River Estuary: Linking socio-economic systems and ecosystems. *Journal of Marine Policy,* 64, 135-147.

[7] Lópeza, B.M., Baggethun, E.G., Llorentea, M.G., Montesaa, C. 2014. Trade-offs across value-domains in ecosystem services assessment. *Journal of Social Ecological Indicators,* 37, 220- 228.

[8] Jacobs, S., Wolfstein, K., Vandenbruwaene, W., Vrebos, D., Beauchard, O., Maris, T., Meire, P. 2015. Detecting ecosystem service trade-off sand synergies: A practice- oriented application in four industrialized estuaries . *Journal of Ecosystem Services,*16, 378-389.

[9] Souza, F.E.S. & Silva, C.A.R. 2011. Ecological and economic valuation of the Potengi estuary mangrove wetlands (NE, Brazil) using ancillary spatial data. *Journal of Coast Conservation,* 15, 195-206.

[10] Aoyama, J., Shinoda, A., Yoshinaga, T., Tsukamoto, K. 2012. Late arrival of Anguilla japonica glass eels at the Sagami River estuary in two recent consecutive year classes: ecology and socio-economic impacts. *Journal of Fish Science* 78: 1195-1204.

[11] Guimaraes, M.H., Sousa, C., Dentinho, T., Boski, T. 2014. Economic base model for the Guadiana estuary, Portugal an application for Integrated. *Journal of Coastal Zone Management Marine Policy,* 43, 63-70.

[12] Schulte, D.F., Gorris, P., Baitoningsih, W., Adhuri, D.S., Ferse, S.C.A. 2015. Coastal livelihood vulnerability to marine resource degradation: A review of the Indonesian national coastal and marine policy framework. *Journal Marine Policy,* 52, 163-171.

[13] Huber, P.R., Shilling, F., Thorne, J.H., Greco, S.E. 2012. Municipal and regional habitat connectivity planning. *Journal of Landscape and Urban Planning,* 105, 15-26.

[14] Rodríguez, D.P., Giebels, D., de Jonge, V.N. 2014. Strengthening coastal zone management in the Wadden Sea by applying 'knowledge-practice interfaces. *Journal of Ocean & Coastal Management,* xxx, 1-12.

[15] Mills, M., Romero, J.G.A., Borland, K.V., Cohen, P., Pressey, R.L., Guerrero, A.M., Ernstson, H. 2014. Linking regional planning and local action: Towards using social network analysis in systematic conservation planning. *Journal of Biological Conservation,* 169, 6-13.

[16] Cayeiro, M.L.P., & Ruiz, J.A.C. 2015. Evaluation of a programme of integrated coastal zone management: The Ecoplata Programme (Uruguay). *Journal of Marine Policy,* 51, 527-535.

[17] Ernoul, L., & Johnson, A.W. 2015. Environmental discourses: Understanding the implications on ICZM protocol implementation in two Mediterranean deltas. *Journal of Ocean & Coastal Management,* 103, 97-108.

[18] Beeharry, Y., Chadee, T.M., Bokhoree, C. 2014. Policy Analysis for Performance Assessment of Integrated Coastal Zone Management Initiatives for Coastal Sustainability. *APCBEE Procedia,* 9, 30-35.

[19] Carvalho, T.M., & Fidélis, T. 2013. The relevance of governance models for estuary management plans. *Journal of Land Use Policy,* 34, 134- 145.

[20] Chen, Q., Liu, J. , Ho, K.C., Yang, Z. 2012. Development of a relative risk model for evaluating ecological risk of water environment in the Haihe River Basin estuary area. *Journal of Science of the Total Environment,* 420, 79-89.

[21] Ottinger, M., Kuenzer, C., Liu, G., Wang, S., Dech, S. 2013. Monitoring land cover dynamics in the Yellow River Delta from 1995 to 2010 based on Landsat 5 TM. *Journal of Applied Geography,* 44, 53-68.

[22] Supriatna, L., Supriatna, J., Koestoer, R.H., Takarina, N.D. (2016). Algorithm model for determination of Cimandiri Estuarine boundary using remote sensing. *AIP Conference Proceedings* 1729, 020079.

[23] Kismartini. 2004. Analisis trade-off sebagai alat analisis kebijakan publik. *Jurnal Dialogue,* 1 (3), 1-19.

The Effect of One-to-One Counseling to Pregnant Women's Knowledge about Anemia in Semarang

N P R Egryani[1], F Saktini[2], N Susilaningsih[2], V D Puspitasari[3], A R Gumay[4]

[1] Student of Medical Faculty Diponegoro University. Prof.Soedarto Street, Semarang, Indonesia.
[2] Histology Departement of Medical Faculty Diponegoro University. Prof.Soedarto Street, Semarang, Indonesia.
[3] Obstetric Gynecologist Departement of Diponegoro National Hospital. Prof.Soedarto Street, Semarang, Indonesia.
[4] Physiology Departement of Medical Faculty Diponegoro University. Prof.Soedarto Street, Semarang, Indonesia.

Abstract. The incidence of anemia in Semarang was constantly increasing years by years. The promotive and preventive activities through counseling of anemia were essential to increase the knowledge of pregnant women for reducing the incidence of anemia. Given the fact that one-to-one counseling using pamphlet was rarely implemented, so that this research about one-to-one method needs to be done. The aim are to determine the difference between participant's pre and post-intervention knowledge about anemia and to determine the correlation among age, education, occupation, income, history of anemia (confounding variables) with the participant's post-intervention knowledge. The study used quasi experimental design without control group. The research subjects were pregnant women from Puskesmas Gayamsari Semarang which were chosen by cluster random sampling method. There were 48 research subjects recruited from antenatal care (ANC) and home visits. The pre-intervention knowledge scores were 10 (5-21), while the post-intervention scores were 24 (9-33) ($p<0,05$) so that, there was significant difference between pregnant women's pre and post-intervention knowledge about anemia. Confounding variables were insignificantly correlated with post-intervention knowledge. The conclusion is one-to-one counseling effects the pregnant women's knowledge about anemia in Semarang.

Keywords: one-to-one counseling, knowledge about anemia, pregnant women.

1. Introduction

The prevalence of anemia in pregnant women was increased from 17,93% (2011) to 19,14% (2012) in Semarang[1]. Pregnant women are more vulnerable to have anemia than others because of the elevated iron demand to produce more red blood cells, expand the plasma volume and feto-maternal growth[2].
The number of anemia in pregnant woman can be decreased by some actions, start from promotive, preventive, and curative actions. Iron and folic acid supplements, B12 vitamin and healthy foods are essential for pregnant women to prevent anemia[3]. The preventive and curative action cannot run smoothly without the promotive action. Therefore, counseling is chosen as the main action in promoting healthy life which can directly increase participant's knowledge[4], so that the participant especially pregnant women become more concern about their health and supposed to consume nutritious meals within the pregnancy

IOP Conf. Series: Earth and Environmental Science **55** (2017) 012036 doi:10.1088/1755-1315/55/1/012036

period. But, education, mass media, social culture, economic, environment, past experiences and age give big influences for someone's knowledge[5].
This far, the healthcare providers tend to conduct the counseling using group approach (one way communication) which involve many pregnant women in once counseling than individual approach using one-to-one communication. It made the participants absorb the information ineffectively so there is no enough knowledge's enhancement. There was no research about the effect of one-to-one counseling using pamphlet to pregnant women's knowledge about anemia. So that the research which determines the difference between participant's pre and post-intervention knowledge about anemia (definition, normal amount of Hemoglobin (Hb) in pregnant woman, vulnerable group, causes, sign and symptom, effect, comorbid factor and prevention) as well as the correlation among age, education, occupation, income, history of anemia (confounding variables) with the participant's post-intervention knowledge need to be done as multicenter research in Indonesia. The research conducted in Semarang is a part of Multicenter Research held by National Analysis and Research Development Institution of Indonesian Medical Student Senate Association or *Badan Analisis dan Pengembangan Ilmiah Nasional Ikatan Senat Mahasiswa Kedokteran Indonesia (BAPIN ISMKI)* which is headed by Medical Faculty Indonesia University Team as the Main Center of this research after winning multicenter research proposal in Temu Ilmiah Nasional (TEMILNAS) 2015.

2. Research Method

2.1 Research plan
The study used *quasi experimental* design without control group to determine the difference between pregnant women's pre and post-intervention knowledge about anemia from one-to-one counseling using pamphlet about anemia [Fig 1]. The primary data were collected by fulfilling the questionnaire and interview during January to February 2016.

2.2 Sample
The participants are pregnant women who were fit into the inclusions criteria such as Indonesian citizen, reproductive ages, have been diagnosed as pregnant woman by physician, approve the informed consent and regularly join antenatal care (ANC) in Puskesmas while automatically exclude if they belong to the criteria such as no longer visiting the Puskesmas, have physical and mental disorder, and cannot face to face during interview. The research center that was chosen in Semarang City region. Samples collection was done by *cluster random sampling* method using free randomized software on www.random.org . This method was chosen because it is one of the best method of *probability sampling* to represent the population of the sample. The number of the samples are 660 in entire Indonesia, divided into 14 Institution Center who joined this multicenter research so each Institution Center (include Semarang) got 48 samples to be collected.

2.3 Research plot
The research carried out one-to-one and face-to-face counseling between the counselor and participant. It began by giving the participants pre-intervention questionnaire, counseling about anemia using pamphlet [Fig 1] and it ended by giving the same post-intervention questionnaire. The pamphlet consists some aspects, such as the definition of anemia, normal amount of Hb in pregnant women, vulnerable group of having anemia, causes, signs and symptoms, effects of anemia, comorbid factors and how to prevent anemia [Fig 1]. The questionnaire consists of sample's bio-data (age, level of education, occupation, income, history of anemia, and history of having counseling about anemia) and six questions about anemia aspects with different scores, they are vulnerable groups of anemia (maximum score is 4), vulnerable group of having anemia (maximum score is 6), causes of anemia (maximum score is 6), signs and symptoms (maximum score is 7), effects of anemia (maximum score is 6), how to prevent anemia (maximum score is 7). The maximum score for all aspects is 36.

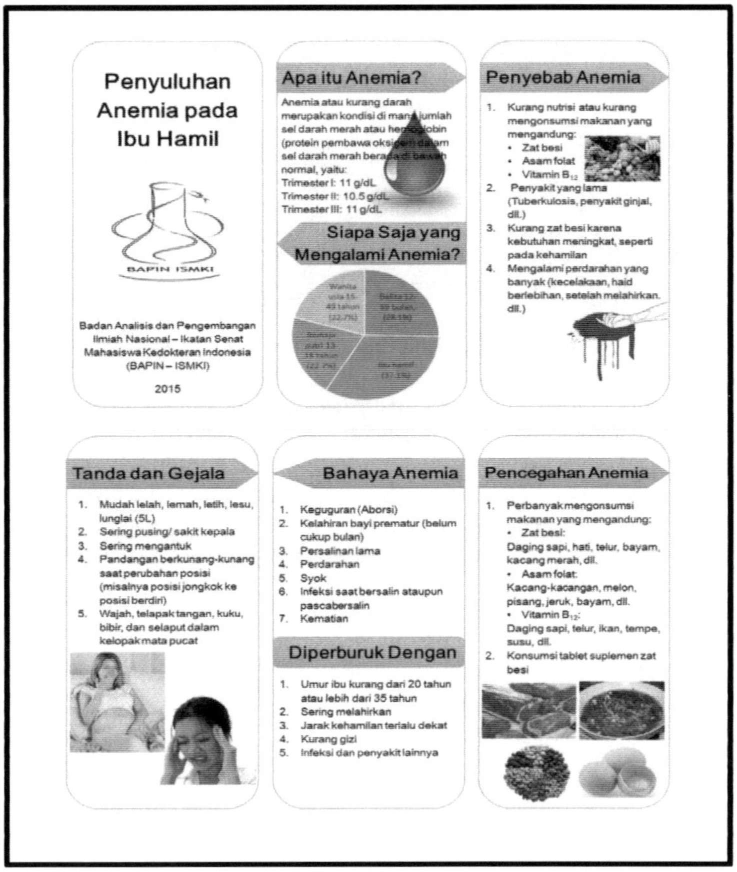

Figure 1. Pamphlet about anemia

2.4 Data Analysis

Wilcoxon test was used to analyze the difference between participant's pre and post-intervention knowledge about anemia, because there was no normal data distribution found by seeing the result from *Test of Normality Shapiro-Wilk* ($p < 0{,}05$), and there was no normal data distribution after transforming the data. Confounding variables (age, education, occupation, income, and history of anemia) were analyzed in bivariate with post-intervention knowledge to find the correlations among those variables.

3. Result and Discussion

Puskesmas Gayamsari Semarang was chosen as the research center using *cluster random sampling* by dividing Semarang region into four regions, they were North Semarang, East Semarang, South Semarang, and West Semarang. 48 participants who fit into inclusions criteria were recruited by visiting the Puskesmas on antenatal care schedules. They came from eight different villages. The description of participant's characteristics are available on Table 1.

The result of Wilcoxon test showed the value $p = 0,000$ ($p < 0,05$) which means that there was significant difference between participant's pre and post-intervention knowledge about anemia as the effect of one-to-one counseling about anemia using pamphlet [Table 2].

Meanwhile, the result of bivariate analysis between each confounding variables to participant's post-intervention knowledge were not correlated. *Spearman* correlation test showed the *p* value of correlations between age, education, income (ordinal scale) to participant's post-intervention knowledge, they were 0,720; 0,276; and 0,499 ($p > 0,05$). Those three confounding variables have very low correlation power (r) (age = 0,053; education = 0,160; income = 0,100) with positive correlation direction [Table 3]. The result of biserial correlation test between occupation and history of anemia which categorized into nominal scale to participant's post intervention knowledge were not correlated too, with the value $p = 0,541$ for occupation and $p = 0,591$ for history of anemia [Table 4]. Table 3 and 4 show that there were no correlations among each confounding variables and participant's post-intervention knowledge. Therefore, the multivariate analysis is not necessary to be done to find the most correlated variables.

This intervention through one-to-one counseling using pamphlet as one of many methods used to promote healthy lifestyle especially prevention of anemia in pregnant women was successfully increased participant's knowledge about anemia. It is appropriate with the definition of counseling to spread information and embed some believes to everyone. This study have reached a step named "know" by analyzing the answers of the questionnaire pre and post-intervention[4]. There was similar study by giving counseling in a form of pregnant women class which show significant result between pre and post knowledge ($p = 0.001$) about Maternal and Baby Health book[6]. There was also significant difference between pregnant women's pre and post knowledge about iron tablet ($p = 0,000$)[7].

Reflect to the result of correlations between confounding variables and participant's post-intervention knowledge (Table 3 and 4), seems there were no correlations among that variables which means that the one-to-one counseling was the one and only variable influencing the difference between pre and post intervention knowledge. It was caused by the homogeneity of participant's characteristic. However, this study contradict with the result of Ghimire *et al* who mentioned that pregnant women's knowledge were correlated with high level education ($p = 0.002$) and ANC visits more than four times ($p = 0.007$)[8]. Onyeneho *et al* even said there was significant correlation between anemia prevention's knowledge and health service nearby house ($p = 0.031$), secondary level of education ($p = 0.001$), well payed occupation ($p = 0.017$) and ages ($p = 0.027$)[9].

This counseling method was very useful. Every single pregnant woman had a chance to increase and improve their knowledge through this one-to-one counseling about anemia, no matter what their education levels, being a housewife or career woman, different numbers of incomes, have a history of anemia or not, and have received counseling before or not. The success of knowledge's improvement was born from the effective communication carried out by communicator and respondent which emphasize individual approach through two-ways feedback communication and face-to-face counseling[10]. Besides, the role of pamphlet was very important as the instrument used to help the participants in absorbing information, because of its attractive figures and clear explanations. Dwiatmoko *et al* also used pamphlet in his study as the instrument for two ways verbal-nonverbal communication[11]. On the other half, full concentration significantly affects the elevating of knowledge through active listening and active reading activities. Good concentration starts from effective communication where there are many chances for the communicant to speak and the communicator can appreciate the others as the God's creature so that the trustworthy relationship can be created, then the motivation and interest of certain information appears that help respondents to focus on one thing without easily distracted by anything around them[12],[13]. The other benefits are the flexibility of this method with no limitation in place and time, easily conduct in formal or informal way.

The limitations of this study are showed by the type of questionnaire which gave the same scale to each multiple choices and allowed the participants to choose more than one correct answers that can create enough bias because the respondent could think on their own without knowing whether it is true or false.

Tabel 1. Description of participant's characteristics

Characteristics	n	%
Age		
< 21 years old	3	6,3
21 – 35 years old	40	83,3
> 35 years old	5	10,4
Education		
Primary	2	4,2
Secondary	36	75
High level	10	20,8
Occupation		
No (housewife)	25	52,1
Yes	23	47,9
Income		
< Rp 1.5 million	15	31,3
Rp 1.5 – 3 million	25	52,1
Rp 3 – 5 million	5	10,4
> Rp 5 million	3	6,3
Diagnosis of anemia		
Have anemia	10	20,8
No anemia	38	79,2
Counseling about anemia		
Have counseling	5	10,4
Never have	43	89,6

Tabel 2. The difference between participant's pre and post-intervention knowledge about anemia

Knowledge	n	Median (minimum-maximum)	Mean ± SD	*p*
Pre-intervention	48	10 (5-21)	9,92±4,547	0,000
Post-intervention	48	24 (9-33)	22,27±6,229	

Table 3. *Spearman* correlation test between age, education, and income to participant's post intervention knowledge

Variables	*p*	r	Note
Age	0,720	0,053	Not correlated
Education	0,276	0,160	Not correlated
Income	0,499	0,100	Not correlated

Table 4. Bi-serial correlation test between occupation and history of anemia to participant's post intervention knowledge

Variables	Mean ± SD	Median (min – max)	*p*
Occupation			
No (housewife)	22,64 ± 6,23	24 (10 – 33)	0,541
Yes	21,87 ± 6,34	24 (9 – 30)	
History of anemia			
Have anemia	22,5 ± 7,71	26 (10 – 33)	0,591
No anemia	22,21 ± 5,9	24 (9 – 30)	

4. Conclusion and Advise

Based on the result and discussion above, it can be concluded that the pre-intervention knowledge scores were 10 (5-21), while the post-intervention scores were 24 (9-33) ($p<0.05$). There was a difference between participant's pre and post-intervention knowledge ($p = 0.000$) so that the one-to-one counseling effects the pregnant women knowledge about anemia in Semarang. Meanwhile, there were no significant correlations among confounding variables and post-intervention knowledge.

Health Care Providers are expected to directly apply the one-to-one counseling using pamphlet to every single pregnant woman (with or without anemia) who regularly come to ANC in every health care services around Indonesia in order to increase the pregnant women's knowledge about anemia. Not only "know" about anemia but also constantly apply the preventive action of anemia. It is also good to change the questionnaire type into essay or open-ended question to get and assess more critical, clear and specific answers in case on avoiding the bias.

5. Acknowledgement

The author thanks to Medical Faculty of Indonesia University Team as the Main Center of this multicenter study, head and maternal-child policlinic's staffs of Puskesmas Gayamsari Semarang, and the health surveillance providers. The author also appreciate the hard work of Medical Faculty of Diponegoro University Team and the pregnant women in Puskesmas Gayamsari region who sincerely participate in this study, and everyone who cannot be mentioned one by one for every help and spirit that are given directly and indirectly.

References

1. Semarang City Health Office 2013 Nutrition Programs Report 2012 (Semarang: Semarang City Health Office)
2. Bencaiova G, Burkhardt T, Breymann C 2012 Anemia-prevalence and risk factor in pregnancy Eur. J. Intern. Med. **23**(1):529-533
3. World Health Organization 2001 Iron deficiency anemia: assessment, prevention, and control: a guide for program managers (Geneva: WHO) p 3-9
4. Maulana, HDJ 2012 Health promotion (Jakarta: EGC)
5. Notoatmodjo S 2007 Education and health behavior ed. 2 (Jakarta: PT. Rineka Cipta)
6. Harsanto AN 2015 The influence of antenatal class toward mother's knowledge of maternal and child health handbook [undergraduated thesis] (Semarang: Universitas Diponegoro)
7. Puspitasari R, Suprapti, Istiana S 2013 The counseling effect against knowledge of pregnant women in first semester about Fe tablet Jurnal Kebidanan **2**(2)

8. Ghimire N, Pandey N 2013 Knowledge and practice of mothers regarding the prevention of anemia during pregnancy, in teaching hospital, Kathmandu JCMC. **3**(5):14-7
9. Onyeneho N, Subramanian S 2016 Anemia in pregnancy: factor influencing knowledges and attitudes among mother in Southeastern Nigeria Springer Link.
10. Open Learn Works Health education, advocacy and community mobilization module: 7. Introduction to health communication Available from: http://www.open.edu/openlearnworks/mod/oucontent/view.php?id=164
11. Dwiatmoko S, Kristiana D 2011 Influence of verbal and written communication on knowledge, attitude and denture hygiene of the removable denture user Dentika Dental Jurnal **16**(1):14-17
12. Niven N 2002 Health psychology guidance for nurse and healthcare professional ed 2 (Jakarta: EGC) p 35
13. Rusbult C 2011 Effective learning skills Available from: http://www.asa3.org/ASA/education/learn/203.htm

Nutrient Content And Acceptability Of Snakehead-Fish (*Ophiocephalus Striatus*) And Pumpkin (*Cucurbita Moschata*) Based Complementary Foods

Etika Ratna Noer[1], Aryu Candra [1], Binar Panunggal [1]

[1]Department of Nutrition Science, Faculty of Medicine, Diponegoro University, Indonesia.

Email: etikaratna@undip.ac.id

Abstract. Poor nutrient-dense complementary foods is one of the common factors contributed for decline growth pattern in children. Snakehead-fish and Pumpkin Complementary Feeding (SPCF) base on locally food can help to reduce child malnutrition. Specifically, high protein and vitamin A in SPCF may improve immunity and nutrition status of malnutrition children. This study aimed to formulate low-cost, nutritive value and acceptable of SPCF on malnutrition children in coastal area. Carbohydrate content was determined by difference, protein by Kjeldahl, betacaroten by spectofotometri and sensory evaluation using a five point hedonic scale. Fe and zinc was determined by AAS. There is an effect of the substitution of snake-head fish flour and yellow pumpkin flour toward the nutrient content and the acceptability

Keywords: snake-head fish, pumpkin, complementary feeding, infant

1. Introduction

The increasing malnutrition of children becomes serious and urgent, so it has to be solved. Malnutrition is a condition in which our body is deficiency of energy, protein and other nutrient. In Semarang, the prevalence of malnutrition on 2013 was 0.9% or 801 cases [1]. Therefore, a good strategy is required to solve malnutrition problems.

Complementary Foods (CF) can be given based on the criteria of the infant, and usually it is served in soft or mushy. Instant porridge is considered more practical and hygienic for the consumption. In infant's malnutrition cases, the energy deficiency, protein and vitamin A, is associated each other. According to the Ministry of Health, CF instant porridge should meet the requirements of 100 gram nutrient of infant food. Those are minimum energy 400 Kcal, protein 15-22 gram, and vitamin A 250-350 μg [2]. The main sources to produce CF porridge generally made of mixture of rice flour, skim milk, refined sugar and palm oil. To enhance the nutritional content of the CF porridge, those sources can be substituted by high protein and vitamin A food source.

Protein, plays important role in infant's growth, cells maintenance, damage tissue replacement, and muscles tissue deposit [3]. Intake of dietary protein may stimulate the synthesis of albumin serum which play role in the protein regulation of the body [4]. Animal protein has a high quality because it contains the amino acid essential with the complete arrangement in accordance of the body needs and high digestibility [5]. One of the sources of animal protein can be retrieved from snake-head fish

(*Ophiocephalus striatus*). It contains 25,5% protein and the value of acceptability in digest system is good [6]. One of the products that can be produced from snake-head fish, which is suitable for CF, is by flouring. The snake-head fish flour, 76.9% per 100g flour, while it only contains 30g protein of 100g skim milk [7].

Vitamin A roles in immune system, protect the epithelial cell integrity, eyes surface, mouth and digest and respiration [3]. Vitamin A deficiency can reduce the immune of the body and it will increase the risk of infectious diseases that could aggravate the malnutrition of the infants [8]. The rich vitamin A food source is also needed to meet the requirements of vitamin A content in malnutrition infant, especially in their CF. Pumpkin (*Cucurbita moschata)* is one of the local food which contain a high beta-carotene at 1.569 μg/100g. Flouring can be an alternative way to optimize of food consumption because it is flexible and practical in the processing, especially in the porridge processing.

The SPCF, influence the content of the nutrient which also can affect the preference. The processing of the composition is aimed can produce the instant porridge of SPCF which can be accepted by the infant.

2. Materials and Methods

2.1 Formulation

The production of SPCF with the substation of snake-head flour and pumpkin flour were made by dry mixing method, in which all of the source were already process and mixed in the dry state. The initial composition of CF , before substitute could be called F0, 35% of rice flour, 46% skim milk, 14% vegetable oil and 5% of refined sugar. The formulation ofF1 which pumpkin 5% and snake-head 10%, F2 which pumpkin 8% and snake-head 10%, F3 which pumpkin 5% and 13% snake-head, F4 which pumpkin 8% and snake-head 13%. What is F0-F4

2.2 Preparation of samples

Snake-head flour which was used in this experiment is the result of flouring it, fresh snake head-fish (*Ophiocephalus striatus*), it weight was about 200-250g/fish and made by steam it, fillet, dried, milled and sieved. The result of yellow pumpkin flour was by flouring the pumpkin. In thin sliced, dried, milled and sieved with the 80 mesh sieve. The drying process of the snake-head fish was done with hot display ±50^{0}C for about ±4 hours, while the pumpkin dried at ±50^{0}C for about ±12 hour.

2.3 Analysis of samples

In the study, the data collected from dependent variable was the content of the nutrients and the level of the substitute MP-ASI instant porridge preference. The content of the nutrients which was analyzed are the content of protein by using Kjeldahl method, fat content by using *soxhlet*. The next was the analysis of the carbohydrate content by using difference method. Spectrophotometry was used to analyze the content of beta-carotene, zinc and Ferum (Fe).

To determine the preferences of this product, the method used was hedonic test. The effect of the variation in this product is tested by *One Ways Anova* and then continued by the *Wilcoxon Signed Rank Test*.

3. Results and Discussion

The result of the nutrients content in SPCF porridge, which had been substitute by snake-head flour and yellow pumpkin flour is briefly presented in Table 2 and 3.

Table 2. Mean Nutrient Content of SPCF **Porridge**

Formula	Protein (%)	Fat (%)	Carbohidrate (%)	Energy (kkal/100g)	Betakaroten (μg/100g)	Fe (mg/100g)	Zinc (mg/100g)
F0	2.23±0.22[e]	0.17±0.06[d]	86.03±0.90[a]	354.57±2.39[a]	0.76±0.13[e]	0.56±0.03[e]	0.16±0.02[e]
F1	9.74±0.96[d]	0.40±0.03[c]	75.78±1.52[b]	345.72±6.31[ab]	12.77±0.71[d]	1.27±0.21[d]	0.46±0.61[d]
F2	13.64±0.67[c]	0.52±0.53[bc]	69.94±1.58[c]	338.96±7.54[bc]	15.99±0.56[c]	1.55±0.26[c]	0.67±0.46[c]
F3	16.07±0.79[b]	0.66±0.10[ab]	65.65±0.83[d]	332.86±6.08[bc]	19.95±0.91[b]	1.66±0.31[b]	0.87±0.10[b]
F4	19.88±0.76[a]	0.77±0.09[a]	59.47±0.67[e]	324.29±3.56[c]	26.08±1.27[a]	1.72±0.27[a]	0.89±0.07[a]
	p= 0.001	p= 0.001	p= 0.001	p= 0.001	p= 0.001	p= 0.001	p= 0.001

Note : *superscript* (a,b,c, d,e) shows significantly different.

There were differences of protein level between MP-ASI instant porridge (p=0.001). According to the further test result, it showed that the level of the protein of treatment group is significantly different from the control group. Based on the Table 2, the protein highest level in SPCF instant porridge is in the group of 13% substitution of snake-head flour and 8% of pumpkin flour (Formula F4).

Table 3. Delta *Mean* Nutrient contain in SPCF

Formula	Delta *Mean*						
	Δ Protein (%)	Δ Fat (%)	Δ Carbohidrate (%)	Δ Energy (kkal/100g)	ΔBetacarotene (μg/100g)	Δ Fe (mg/100g)	Δ Zink (mg/100g)
F1	7.51±0.96[d]	0.23±0.03[c]	-10.25±1.52[b]	-8.85±6.31[ab]	12.01±0.71[d]	2.16±1.48[bc]	0.34±0.07[c]
F2	11.41±0.67[c]	0.35±0.53[bc]	-16.09±1.58[c]	-15.61±7.54[bc]	15.23±0.56[c]	3.83±1.78[b]	0.51±0.04[b]
F3	13.84±0.79[b]	0.49±0.10[ab]	-20.38±0.83[d]	-21.71±6.08[bc]	19.19±0.91[b]	5.41±1.37[ab]	0.64±0.06[ab]
F4	17.65±0.76[a]	0.60±0.09[a]	-26.56±0.67[e]	-30.28±3.56[c]	25.32±1.27[a]	7.55±0.76[a]	0.76±0.02[a]

Note : *superscript* (a,b,c, d,e) shows significantly different.

Based on the Table 3, the increase of the protein level is happened after the substitution of snake head-flour and pumpkin flour. The more it is substituted, the higher level of protein level we can get.

Snake-head fish flour contains 76.9g/100g of protein while in the skim milk we only get 30g/100g. The substitution of snake-head fish flour in the SPCF instant porridge increases the protein level. It is because the protein from snake-head fish is higher than skim milk.

Pumpkin flour contains 7.81g/100g protein. Although it is not as high as the snake-head fish flour, the yellow pumpkin flour substitutions can increase the protein of SPCF porridge at 5% or 8%.

The high quality of protein, play important role in infant growth. Infant at the age 6-12 months, the growth is rapidly increase and the baby more dependent on food complementary [8]. The protein source which is used is snake-head fish flour and skim milk. Fish have biological protein value of 75, while the protein of cow milk is 93. Protein which has more 70 of biological protein value can be considered as the support for infant's growth process [9].

There were differences between the fat content of SPCF instant porridge formula (p=0.001). According to the further test result, it showed that the fat content among the treatment group was significantly different from the control group. The highest fat content in SPCF instant porridge were at the group of 13% of substation of snake-head fish flour and 8% of yellow pumpkin flour (Formula F4). Based on the Table 3, there were the increasing of fat content in MP-ASI instant porridge after substitute with the snake-head fish flour and yellow pumpkin flour.

The increasing level of the fat content in SPCF instant porridge can be affected by the fat content of the flour at 0.55 % of 100g flour [7]. The fat of the fish contains of fat acid long chain of omega 3 (EPA and DHA) which good for the growth process. Although the substitution of the snake-head fish flour can increase the fat level at 0.23g-0,6g, yet, it is not fulfill the requirements of the Technical Specification BMS Food Companions (Infant Complementery Foods). The Ministry of Health (2007), decided the content of fat for 6-12 months baby instant porridge should meet the requirements of 10-15g/100gr. It is because the fat content from SPCF material limited [2].

There are differences between the carbohydrate content of SPCF instant porridge formula (p=0.001). According to the further test result, it showed that the carbohydrate content among the treatment group was significantly different from the control group. The lowest carbohydrate content is in the group on 13% substitution of snake-head fish and 8% of pumpkin flour (Formula F4). Based on the Table 3, there was the decreased level of carbohydrate content after the substitution. The more it is substituted; we will get lower carbohydrate content.

The decreasing of carbohydrate contents is caused when the main material in instant porridge substitute with the yellow pumpkin flour which has high beta-carotene and the substitutions of the skim milk with the snake-head fish flour with lower carbohydrate content. The more we substitute the main material with the snake head fish and yellow pumpkin flour, the lower we will get the carbohydrate content. Carbohydrate is important for the toddlers' growth, but it is better if takes at least 52-54% of the energy needed and Formula F4 has already fulfill the requirements of baby's carbohydrate needed.

Based on the *Anova one Ways* test, it showed that there was a difference of energy level between the SPCF formula (p=0.001). The substitution of the snake-head fish and yellow pumpkin flour, decrease the energy level of SPCF instant porridge. Based on Table 3, there was energy level decreasing of SPCF instant porridge after the substitutions.

Based on the calculation result, the energy from the substituted biscuit of snake head fish flour and pumpkin flour is at the range of 324.29-345.72 kcal/100g. Among all of the substituted SPCF instants porridge, is below the minimum requirements of energy content (400kcal/100g). If the baby got the medium quantity and quality of breast feeding, the energy intake can be fulfill through the breast feeding and the shortage of 356 kcal is expected can also be fulfilled through the breast feeding, especially SPCF [10].

There are differences between the beta-carotene level of SPCF instant porridge formula (p=0.001). According to the further test result, it showed that the beta-carotene level among the treatment group was significantly different from the control group. The highest level of beta-carotene was fond on the group of 13% substitution of snake head fish flour and 8% substitution of pumpkin flour (Formula F4). Based on Table 3, there was improvement of beta-carotene level of substituted MP-ASI instant porridge.

The level of beta-carotene n baby's porridge produced more than the specification after it was converted into vitamin A, the level were at 12.77-26,08mg/100g or below 30 mg, so it will not the cause of carotenoids accumulations. Beta-carotene is an antioxidant which has role in immune system, maintenance the integrity of epithelial cell, visions and as well as for the growth [3]. Beta-carotene will be converted into vitamin A in the body and it will become the precursor of vitamin A. Vitamin A is essential for visions, growth, differentiation and proliferation, reproduction, and immune system [8].

There are differences between the Fe of MP-ASI instant porridge formula (p=0.001). Based on the Table 3, it showed that the highest level of Fe were at the group of 13% substitution of snake-head fish flour and 8% substitution of yellow pumpkin flour (Formula F4). The more we substitute the flour with both of the snake head fish and yellow pumpkin flour, the higher we will get the level of powder.

There are differences between the zinc level of MP-ASI instant porridge formula (p=0.001). The level of zinc in the instant porridge was at the group of 13% of snake-head fish and 8% of pumpkin flour (Formula F4). Based on the Table 3, there was an improvement of zinc level after the substitution. The more we substitute the flour, the more we will get the higher zinc level on it.

The level of preferences of color, senses, texture and taste of the MP-ASI instant porridge were presented in Table 4.

Table 4. Hedonic test of SPCF Instant Porridge

Formula	Color		Flavour		Texture		Taste	
	Rerata	Ket	Rerata	Ket	Rerata	Ket	Rerata	Ket
F0	3.08±0.95[a]	Netral	3.44±0.92	Netral	2.96±0.84	Netral	3.44±0.92[a]	Netral
F1	3.6 ± 0.71[a]	Like	3.16 ± 0.89	Netral	2.72±0.74	Netral	2.8 ± 1.00[b]	Netral
F2	3.44 ±0.71[a]	Netral	3.08 ±1.04	Netral	2.88±0.78	Netral	2.96 ±0.98[a]	Netral
F3	3.88 ±0.73[b]	Like	3.24 ±1.05	Netral	2.80±0.64	Netral	3.04 ± 0.98[a]	Netral
F4	3.52 ±0.77[a]	Like	3.08 ±1.04	Netral	2.72±0.89	Netral	2.72 ±0.94[b]	Netral
	p= 0.01		p=0.58		p=0.49		p=0.02	

Note : *superscript* (a,b,c, d,e) shows significantly different.

The highest preferences of SPCF instant porridge were at the 10% substitution of snake head-flour and 8% of yellow pumpkin flour (F3). Among all of the substitution (F1,F2,F3and F4) had a neutral assessment from the panelist.

Based on the hedonic test, the substitution of the flour with the snake head fish flour and yellow pumpkin flour could increase the preferences of the color in MP-ASI instant porridge. Statistically, there were some effects that could change the level of color preferences of the panelist toward the substitution of the flour.

Formulation of SPCF instant porridge which was produced, has a sweet and savory taste but it also has a fishy, strong flour taste and bitter after taste. The sweetness of the taste came from the sugar and yellow pumpkin flour, while the skim milk and the sanke-head fish flour gave a savory taste. The used of sugar would strengthen the taste of porridge, but remember, its use should be controlled.

Before the steaming process in making the snake-head flour, the fish had already soaked into the lime water for about an hour, but then, the fishy taste from still could be tasted on the porridge. Bitter after taste was caused by the hydrolysis of the amino acid that happened on the *Maillard* reaction in the flouring process.

The highest the protein content were at the substitution 13% of snake-head fish and 8% pumpkin flour (Formula 4). The highest levels of beta-carotene also were at F4 formula. Meanwhile, the energy content from the fourth formula was at the lower level compare to the requirements. The preferences level, did not show any significantly differences so any formula can be selected. Based on the consideration, the formula which can be used for infant is formula F4.

Determining the serving dosage is important for the sake of food labeling. The best amount of nutrient in SPCF, for the 7-11 months baby was presented in Table 5. The amount of the dose in SPCF was determined based on the calculation of the needed in energy level for 7-11 months baby.

Tabel 5. Contribution SCPF instant porridge per serving to Dietary Adequacy Index for infant (7-11 months)

Nutrient	Nutrient/100 g	Nutrient/servings	RDA	%RDA
Energy (kkal)	324.29	97.29	725	13
Protein (g)	19.88	5.96	18	33
Fat (g)	0.77	0.23	36	0.6
Carbohidrate (g)	59.47	17.84	82	22
Zinc (mg)	1.02	0,205	3	6,7
Fe (mg)	0.89	0.18	7	2.6
Vitamin A (μg)	2.173.3	652.00	400	163

RDA : Recommended Dietary Allowances

At the age of 7-11 months, the baby requires 30-45% additional energy which is divided into 2-3 of meal times. One serving of SPCF instant porridge was determined at 30 g which contribute 15% of sufficient energy for 7-11 months baby, based on the Recommended Dietary Allowance (RDA) 2013. Based on the contribution nutrients analysis, SPCF instant porridge can give the best protein level at 33,11% and vitamin A at 163% on the nutrients RDA for 7-11 months baby, 9 kg weight. Further research needs the use of MCT (*Medium Chain Triglycerides*) can be used to increase the fat level of SPCF it is also increase the energy level of the porridge so it can fulfill the requirements of energy level needed.

4. Conclusion

There is an effect of the substitution of snake-head fish flour and yellow pumpkin flour toward the nutrient content and the acceptability. The substitution could improve the level of protein, fat, beta-carotene, zinc, Fe and also the level of preference toward the SPCF instant porridge.

Based on the nutrient content value and the acceptability, MP-ASI instant porridge which is recommended, substitution SPCF instant porridge with 13% of snake-head fish and 8 % of pumpkin flour (Formula F4).

5. References

[1]. Ministry of Health 2013 *Semarang City Health Office in Nutrition Program Reports* (Semarang)
[2]. Ministry of Health 2007 Decision of the Ministry of Health Indonesia Number: 224/Menkes/SK/II/2007 Technical spesification of complementary infant feeding (Jakarta)
[3]. Trahms CM, McKean KN 2008 Nutrition during infancy in: Mahan LK, Escott-Stump S. Krause's Food and Nutrition Therapy 12th ed (Canada: Elsevier)
[4]. Caso G, Scalfi L, Marra M, Covino A, Muscaritoli M, Mc Nurian M, Garlick PJ and Contaldo F 2000 Albumin synthesis is diminished in men consuming a predominantly vegetarian diet Journal Nutrition 130:528-533
[5]. Muchtadi D 2010 Evaluation technique of protein value (Bandung:Alfabeta)
[6]. Santosa AH 2001 Extraction albumin of snake head fish (*Ophiocephalus striatus*) *[undergraduated thesis]* (Malang: Brawijaya University)
[7]. Sari DK, Marliyati, Sri A, Kustiyah, Lilik, Khomsan A, Gantohe, Tommy M 2014 Organoleptic test formulation biscuit based snake head fish *(Ophiocephalus Striatus)* Agritech 2(34)
[8]. Parízková J 2010 Nutrition, physical activity, and health in early life 2nd edition (USA: CRC Press)

[9]. Rolfes SR, Pinna K, and Whitney EN 2008 Understanding Normal and Clinical Nutrition (USA: Wadsworth Cengage Learning)
[10]. World Health Organization 2000 Complementary Feeding: Family Foods for Breastfed Children. Department of Nutrition and Development (Geneva: WHO)

Isolation and Identification of *Aeromonas hydrophila* and *Saprolegnia* sp. on Catfish (*Clarias gariepinus*) in Floating cages in Bozem Moro Krembangan Surabaya

Rahayu Kusdarwati[1], Kismiyati[2], Sudarno[3], Hendi Kurniawan[4] and Yudha Teguh Prayogi[5]

[1]Department of Fish Health Management and Aquaculture Faculty of Fisheries and Marine, Airlangga University, 031-5914151,
[2,3]Department of Fish Health Management and Aquaculture, Faculty of Fisheries and Marine
[4,5]Study Pograme Aquaculture Faculty of Fisheries and Marine, Airlangga University

rahayukusdar@gmail.com

Abstract. Catfish (*Clarias gariepinus*) is one of the familiar freshwater fish cultured in Indonesia farmer. One of the reason is the high mortality of the catfish infected by *Aeromonas hydrophila* and *Saprolegnia* sp. *Motile Aeromonas septicemia (*MAS) is a common bacterial disease, caused by *Aeromonas hydrophila*, which affects freshwater fish. In Southeast Asia, the outbreak of this disease was firstly reported from West Java in 1980, when a total of 82.5 tons a month of catfish were lost, while in Central Java in 1984, the total loss was 1.6 tons. Saprolegniosis can cause economic loss due to high mortality from its case reaching 10% to 50% . This research aimed to identify and determine the percentage of *A. hydrophila* and *Saprolegnia* sp. the catfish farmed in bozem Moro Krembangan, Surabaya, East Java. Meanwhile, a supporting parameter in this research is the value of water quality parameter including pH, temperature, ammonia and dissolved oxygen that were measured during sampling. The results showed that of the 20 samples taken from the two cages, 19 fish were positively infected by *A. drophila*. percentage of infections of *A. hydrophila* that infect umbo catfish in Moro Krembangan, was 95%, while the percentage *Saprolegnia* sp. was 90%.

1. Introduction

Surabaya has a system of outdoor water reservoirs that have the same functionality with reservoirs namely Bozem. These Bozems include: Bozem Moro Krembangan, Kalidami, Bratang, Rungkut industrial, Wonorejo, Kedurus and Broken Gorge. Ecosystem of Bozem is built as Surabaya is a city in flood-prone with lack of water catchment areas. This is due to the development in the city of Surabaya that develops rapidly. In General Bozem is built without cutting off rivers. In general Bozem serves as a controller and water storage, and the society also utilizes it as a place to fish farming and irrigation of field (planning and development agency of the city of Surabaya, 2010). One of the farmed fish in water of Bozem was catfish. Catfish is one type of fish consumption that are quite popular in Indonesia. Some advantages of the fish are to have meat that are thick and tasty as well as easy maintenance, its relatively quick growth, and high selling prices, so that catfish arelargley farmed in to involves several factors: the surroundings (environment), the farmed fish (host), and the disease-causing organisms (pathogens). *A. hydrophila* can affect fish farming and often lead to outbreaks of disease with mortalities occasionally approaching 80-100 percent in a short period of 1-2 weeks [1]. While from the fungus genus, many *Saprolegnia* sp. infect eggs and freshwater fish which can lead to massive death of fish in fish farms [2]. *Saprolegnia* sp. quickly transmitted to other catfish that were in one pool. Thus, it spread more quickly and potentially inflicts considerable losses for farmers [3].

Fish farming in Bozem Moro Krembangan having some problems such as the water environmental problems that is less support, because the waters such a reservoir of water from various sources and disposal of waste. Based on this research the reason they will do. The purpose of this research is to determine The presence of infection Aeromonas hydrophila and Saprolegnia sp. on catfish (*Clarias gariepinus*) cultured in Bozem Moro Krembangan Surabaya.

2. Research Methodology

2.1. Research Materials

Research materials that are used in this research are *Clarias gariepinus* obtained from Bozem Moro Krembangan, Surabaya, aquades sterile, spirtus, alcohol 70%, selective medium Triptic Soy Agar (TSA), Saboraud Dextrose Agar (SDA) and safranin. The instruments used are a petri dish, needle ose, mikropipet, ring, light microscopy, object glass, hot plate stirer, thermometer, pH meters, DO the test kits and ammonia test kit.

2.2. Research Methods

This study used a survey method. The determination of the sampling method used purpossive sampling [4]. Sampling is to retrieve data of only a portion of the population who are expected to describe the nature of the population from the object of research.

2.3. Work Procedures

2.3.1. Sampling

Catfish taken for the sample are consuming catfish with a size of 10-15 cm age 2-3 months with a weight of 100-110 grams. The number of cages used for sampling amounted to two cages, the total population is 400 catfish, in one population cage is 200 catfish, and taken a sample of 5% of the population that is as many as 20 catfish out of a total population or 10 for each cage.

2.3.2. Sterilization of Tools

Equipment such as needles of ose before used first were sprayed alcohol 70% then conducted burning directly. Petri dish was sterilized by autoclave at a temperature of 121 ° C with pressure 1 atm for 10-15 minutes.

2.3.3. Isolation of Aeromonas hydrophila and Saprolegnia sp.

The medium used for the isolation of *A. hydrophila* was Trypticase Soy Agar (TSA), while the medium for isolation of *Saprolegnia* sp. was Saboraud Dextrose Agar (SDA). The isolation was carried out using streak methods at medium of isolation. Furthermore, purification was done by moving the isolates that grew into new media until pure isolate was obtained.

2.3.4. Identification of Aeromonas hydrophila and Saprolegnia sp.

Bacterial identification was conducted with the macroscopic test, Gram staining and biochemistry test. Observations of macroscopic samples test of catfish included size, shape, and pigmentation seen from the inside, beside and top. Microscopic tests were done by staining Gram by observing the color and shape of bacteria as well as biochemical test was done. Biochemical test involved Motile test, indol test, O/F test, and H2S test.

Identification of the fungus *Saprolegnia* sp. with macroscopic test and microscopic test. Observations of macroscopic samples test catfish includde size, pigmentation (the colour of the colony) and shape seen from the inside, beside and top. While the observations of the microscopic test were conducted by observing the color, the shape of the hypha, forms spores, Spore and form layout cells of fungi by using inverted microscope [5].

2.3.5. Calculating the percentage of Fish infected by Aeromonas hydrophila and Saprolegnia sp.

The percentage of infections was the number of fish infected by the bacteria/fungi that are divided with many samples taken multiplied by one hundred. Resulting in the percentage of fish infected with the disease.

$$\text{The percentage of catfish infected bacteria/fungi (\%)} = \frac{\text{Number of infected fish x 100}}{\text{The number of fish samples}}$$

3. Results and Discussion

3.1 Identification and Isolation of Aeromonas hydrophila and Saprolegnia sp.

Isolation of *A. hydrophila* and *Saprolegnia* sp. was conducted on 20 samples of fish that were taken that showed pathologic conditions including hemorrhagic septicemia and tail or fin rot. Isolation of a fungus was conducted by taking body parts catfish that showed clinical symptoms of fungus infection. The fish part isolated was part of the surface of the body, fins, gills, and tail, because the parts that would change as result of infection of fungus *Saprolegnia* sp. On the skin of infected catfish there was smooth yarn resembling cotton and spreading to the surface of the skin to another. This is in accordance with statement [6], stating that fish infected with *Saprolegnia* sp. Have objects that resemble cotton on the surface of their body, tail, fins or gills.

The results of the identification of the bacterial isolate showed negative staining Gram, motile, form rod, while biochemical test results are fermentative, positive oxides, positive glucose and negative indol.

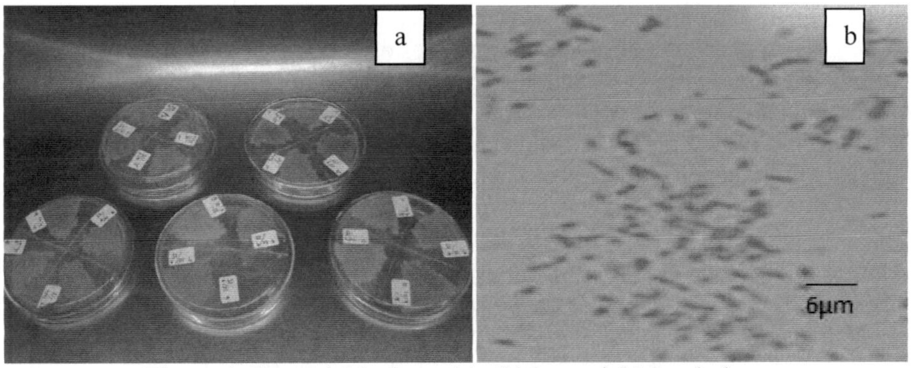

Figure 1. Bacterial isolates (a), (b) bacterial Morphology

Identification of fungus isolate in macroscopic colony characteristics can be seen with a yellowish white with forms such as cotton. This is in accordance with [6], who states that the genus *Saprolegnia* have characterized the colony resembling cotton, colored white and the colony can be found on the surface of the skin, fins and gills.

Microscopically *Saprolegnia* sp. morphology appear to have long Hypha and aseptat (not (not septad), tube-shaped, and has the zoospora on its corner. Sporangia widens and the short as well as at the end of the sporangia looks a bit dark. The zoospora *Saprolegnia* sp was resulted from the tip of the long Hypha . Zoosporangia Saprolegnia SP.that are long and cylindrical has a length of 180-350 μm in width 20-24 μm [6]. Macroscopic and microscopically Observation can be seen in Figure 2.

Figure 2. a colony of *Saprolegnia* sp.

One of the types of diseases that are often found in cultivated organisms is bacterial disease caused by the bacterium *Aeromonas hydrophilla*, which is the pathogenic bacteria cause disease "Motile Aeromonas Septicemia" (MAS), especially for freshwater fish species in tropical waters [7]. The bacterium *Aeromonas hydrophila* is one of the dangerous disease-causing bacteria in freshwater fish farming. The bacteria attack the Catfish that is one of the leading commodities in freshwater and can infect fish in all sizes which can lead to death until it reaches 80%, resulting in huge losses in the freshwater fish farming ventures [8].

Aeromonas hydrophila is a bacteria that can grow in the waters of the optimum operating temperature of 20-30 ° C [9]. *Aeromonas hydrophila* is able to grow at temperatures up to 37 ° C [10]. *Aeromonas hydrophila* extracellular products of consist of the α and β hemolisin, protease, elastase, lipase, cytotoksin, enterotoxin, gelatinase, caseinase, lecithinase, and leucocidin. The liver is the organ target of Aeromonas hydrophila and disruption can have an effect on metabolic processes [11]. *Aeromonas hydrophila* is pathogens because it only can cause disease in fish populations that are weak or as secondary infections when fish are infected with other diseases [12].

The process of when *Aeromonas hydrophila* bacteria infects into the body of the fish is preceded by that bacteria stick on the skin surface by utilizing the flagella to move towards the fish and then use hooks to attach to strong on the outermost layer of a body of fish scales. Scales are covered by the substance chitin. During the process of *Aeromonas hydrophila* bacteria takes plac, it produces enzymes that play a role in kitinase layer of chitin degrades, so that bacteria can easily fit into the body of the fish. In addition to utilizing the *Aeromonas hydrophila* kitinase, the bacteria also secrete enzymes such as lesitinase in an attempt to get into the blood stream and go directly to the kidney to breed [13].

Saprolegnia sp. are fungi with yarn shape, resembling cotton, white to gray and Brown. *Saprolegnia* sp. fungi belongs to the class of Oomycetes, also known as fungi with algae because its nature is similar to that of algae but do not contain chlorophyll [14]. Compiled by the threads of a Hypha, it has no bulkhead separators (aseptat), but the branched into misselium. The walls of *Saprolegnia* sp. Hypha are composed of glycogen and cellulose [9]. Hypha branches *Saprolegnia* sp. form woven wool like yarn. *Saprolegnia* sp. has a Hypha which is larger, tapered shape and able to spread faster than the other fungi.

3.2 The percentage of Clarias gariepinus Infected by Aeromonas hydrophila and Saprolegnia sp.

The data results of the counting of the *Aeromonas hydrophila* percentage of fungi farmed in floating net cage in Bozem Moro Krembangan Surabaya can be seen in table 1.

Table 1. The results of the counting of the percentage of the Catfish infected *Aeromonas hydrophila*.

Cages	**The number of Fish Samples (fish)**	**The number of fish infected by *Aeromonas hydrophila* (fish)**		**The percentage (%)**
		+	-	
A	10	9	1	90
B	10	10	0	100
The total number of	20	19	1	95

Results data of counting of percentage of fungi *Saprolegnia* sp. observed in cages floating in Bozem Moro Krembangan Surabaya (seen in table 2)

Table 2. The results of the counting of the percentage of the Catfish infected with *Saprolegnia* sp.

Cages of floating	**The number of fish sample (fish)**	**The The number of fish infected by *Aeromonas hydrophila* (fish)**		**The percentage (%)**
		+	-	
A	10	10	0	100
B	10	8	2	80
Total	20	18	2	90

On this study, the estimated value of water quality is obtained : temperature 30°C, DO 4 ppm, pH 6 - 7 and ammonia 1.0 ppm.

Of the 20 fish samples taken, all were experiencing clinical symptoms such as wounds to the skin also damaged fins and tail. Then at 19 fish's kidney organ were suffered swelling and only 1 fish had not experienced swelling in his kidney. The high percentage of fish infected by *Aeromonas hydrophila* and *Saprolegnia* sp. was 95% and 90% indicating that the waters in Morokrembangan Surabaya Bozem were less support for fish farming activities. Remember that waters had the function as shelter of water, such as rain water and waste water, both household waste as well as industrial waste, with the aim to control the occurrence of floods.

According to [15] *Aeromonas hydrophila* bacteria infection can happen in 4 levels, namely Acute septicemia which is fatal, infection quickly with marked swelling of the internal organs. Sub acute, can be seen with symptoms such as wounds and bleeding on the scales. Chronic can be seen with symptoms such as ulcers and sores of the skin damage whose development lasts a long time. Latent, it can happen with not showed symptoms of the disease, but on the internal organs contained disease-causing bacteria. While infection of *Saprolegnia* sp. on fish generally are secondary infections that preced the onset of bacterial infection, the aquatic environment are not heeded, the parasitic infestations, post-harvest handling, dense stocking high causing easy fish infected with *Saprolegnia* sp. [16]. Bad water quality, low dissolved oxygen, high organic matter content and the presence of bacteria will create an environment that is less good for fish and cause stress. The low oxygen levels in the waters can cause fish become stressed out so that the immune system can decline [17] . At that time, the pathogen will easily fit into the body of the fish, either in the form of bacteria, fungi or parasites. While according to [18], pH changes can cause fish stressed so that they are easily affected by desease, and automatically law pH can cause the damage on skin, so pathogen can easily infect it. the value of optimal water quality for fish growth, namely: water temperature 28-32 °C. In addition to temperature, DO and pH, the content of ammonia is a factor to consider. The content of DO is in the water so that fish growth is ideal not less than 2 ppm, pH 6.5 to 8.5 and an ammonia content of less than 1 ppm.

4 Conclusion

Catfish (Clarias gariepinus) cultured in Moro Krembangan Surabaya, East Java, are mostly infected with Aeromonas hydrophila and *Saprolegnia* sp., It is caused by the environmental conditions that are unfavorable. The percentage of catfish (Clarias gariepinus) infected with *Aeromonas hydrophila* is 95% and infected with *Saprolegnia* sp. is 90%.

References

[1] Lukistyowati, I. dan Kurniasih. 2012. Pelacakan Gen Aerolysin dari *Aeromonas hydrophila* pada Ikan Mas yang diberi Pakan Ekstrak Bawang Putih. Jurnal Veteriner 13 (1) : 43-50.

[2] Akbar, J. 2008. Pemanfaatan Ekstrak Bunga Kecombrang (*Nicolaia speciosa horan*) Terhadap Penyembuhan Infeksi Jamur *Saprolegnia* sp. Pada Ikan Nila Merah. Fakultas Perikanan Universitas Lambung Mangkurat.

[3] Khairyah U., Kismiyati., dan Kusdarwati R. 2013. Identifikasi dan Prevalensi Jamur pada Ikan Gurami (*Osphronemus gouramy*) di Desa Ngrajek, Kecamatan Mungkid, Kabupaten Mungkid, Jawa Tengah.

[4] Singarimbun dan Effendi. 1995. *Metode Penelitian Survei*. Jakarta : PT Pustaka LP3ES Indonesia.

[5] Yuasa K., Novita P., Meliya P., dan Edy B.K, 2003. Teknik Diagnosis Penyakit Ikan Budidaya Air Tawar di Indonesia. Japan International Cooperation Agency dan Panduan Diagnosis Penyakit Ikan. Balai Budidaya Air Tawar Jambi. Jambi.Stoskopf, M. K. 1993. Fish Medicine. WB Saunders Company. Mexico.

[6] Rahmaningsih, S. 2012. Penagruh Ekstrak Sidawayah dengan Konsentrasi yang Berbeda untuk Mengatasi Infeksi Bakteri *Aeromonas hydrophyla* pada Ikan Nila (*Oreochromis niloticus*). Jurnal Ilmu Perikanan dan Sumberdaya Perairan.

[7] Sanoesi, E. 2008. Penggunaan Ekstrak Daun Pepaya (*Carica papaya Linn*) terhadap Jumlah Sel Makrofag pada Ikan Mas (*Cyprinus carpio* L) yang Terinfeksi Bakteri *Aeromonas hydrophila*. Jurnal Penelitian Perikanan, Vol 11, No. 2.

[8] Indrawati, I. 2007. Potensi Ekstrak Rimpang Lengkuas (*Alpinia galanga*) terhadap jamur *Saprolegnia* sp. dari telur ikan mas (*Cyprinus carpio*) Secara in Vitro. Skripsi. Fakultas Kedokteran Hewan. Universitas Airlangga. Surabaya.

[9] Austin, B. and D. Austin. 1987. Bacterial Fish Pathogen. Ellis Hoorwood. London. pp. 250-262.

[10] Cipriano RC, GL Bullock and SW Pyle. 1984. *Aeromonas hydrophila* and *Motile Aeromonad Septicemia* of Fish. Fish Diseases Leaflet 68, US. Fish and Wildlife Service. West Virginia.

[11] Kabata, Z. 1985. Parasites and Disease of Fish Cultureed in Tropics. Taylor and Francis, London.

[12] Mangunwardoyo, W., R. Ismayasari., E. Riani. 2010. Uji Patogenisitas dan Virulensi *Aeromonas hydrophila* Stanier pada Ikan Nila (*Oreochromis niloticus* Lin.) melalui Postulat Koch. J. Ris. Akuakultur Vol. 5 Tahun 2010: 245-255.

[13] Nuryati, S., F.B.P. Sari., dan Taukhid. 2009. Identifikasi dan Uji Postulat Koch Cendawan Penyebab Penyakit pada Ikan Gurami. Program Studi Budidaya Perairan. Fakultas Perikanan dan Ilmu Kelautan. Institute Pertanian Bogor. Bogor.

[14] Sniezko SF and HR Axelrod. 1971. Diseases of Fishese. TFH Publication Ltd., Hongkong.

[15] Bruno D. W and Woo, P. T. K.. 1999, Fish Disease and Disorders vol. 3 (Viral, Bacterial and Fungal Infection). CAB International Publishing. Wallingford, UK. Pp 366-432.

[16] Shafrudin, D., Yuniarti dan Setiawati, M., 2006. Pengaruh Kepadatan Benih Ikan Lele Dumbo (Clarias sp.) Terhadap Produksi Pada Sistem Budidaya Dengan Pengendalian Nitrogen Melalui Penambahan Tepung Terigu. Jurnal Akuakultur Indonesia 5(2) : 137-147.

[17] Taufik,P. 1984. Faktor Kualitas Air Dapat Mempengaruhi Timbulnya Suatu Penyakit Pada Ikan. Majalah Pertanian No 3. Departemen Pertanian. Jakarta.

The Thermocline Layer and Chlorophyll-a Concentration Variability during Southeast Monsoon in the Banda Sea

Nikita Pusparini[1,3], Budi Prasetyo[3], Ambariyanto[1,2], Ita Widowati[1]

[1]Department of Marine Sciences, Faculty of Fisheries and Marine Sciences, Diponegoro University, Jl. Prof. H. Soedharto, SH, Tembalang Semarang. 50275
[2]Coastal Disaster Rehabilitation and Mitigation Center, Integrated Laboratory, Diponegoro University, Jl. Prof. H. Soedharto, SH, Tembalang, Semarang 50275
[3]Indonesian Agency for Meteorology, Climatology, and Geophysics (BMKG)

Email address: nikitapusparini@gmail.com

Abstract. Thermocline layer and chlorophyll-a concentration can be used to investigate the upwelling region. This investigation is focused in the Banda Sea because the upwelling event in this area is quite large and has a longer upwelling duration than other waters in Indonesia. In addition, Banda Sea is also influenced by climatic factors such as monsoon. The aim of this research is to determine the validation of secondary data (from satellite imagery data and model) and in situ observation data (from research cruise) and to determine the variability of thermocline layer and chlorophyll-a concentration during Southeast Monsoon in the Banda Sea. The data used in this study were chlorophyll-a concentration, seawater vertical temperature at depths 0-400 meters, and sea surface temperature from remote sensing and in situ data. Spatial and temporal analysis of all parameters was conducted by quantitative descriptive method. The results showed that the variability of thermocline layer and the chlorophyll-a distribution were strongly related to seasonal pattern. In most cases, the estimates of thermocline layer and chlorophyll-a concentration using remote sensing algorithm were higher than in situ measured values. The greatest variability occurred in the eastern Banda Sea during the Southeast Monsoon with shallower thermocline layer, more abundance of chlorophyll-a concentration, and lower sea surface temperature.

Keywords: thermocline layer, chlorophyll-a, monsoon, upwelling

1. Introduction

The Indonesian Archipelago is so unique that it represents the only low latitude inter-ocean connection on the earth, with the Indonesian Trough Flow (ITF) communicating between the Pacific Oceans and Indian Oceans [1,2]. In relation to oceanography parameters such as sea surface temperature (SST), it has been suggested that complex feedback mechanisms may exist between the monsoons and the physical dynamics within the Indonesian Seas [3].

During the Northwest Monsoon, winds force the surface waters into the eastern archipelago, depressing the thermocline, while in the Southeast Monsoon, when dry air is brought from the Australian continent, the eastern seas is empty and there is evidence of large-scale upwelling events in the Banda Sea and elsewhere in the archipelago [1,4,5,6,7]. These upwelling events are one of the processes to control the productivity dynamics of the region. Whatever the mechanism for the upwelling event, there is an observable decrease of sea surface temperatures in the Banda Sea during the Southeast Monsoon [8]. For the eastern Indonesian waters, the southeastern wind prevails from April to November, with surface currents in the Flores Sea and Banda Sea directed westwards, whereas from December to March the northwestern wind reverses these currents to the east [30]. The seasonally changing hydrographic conditions in this area greatly influence the pelagic ecosystem, as indicated by monsoon-dependent changes in chlorophyll-a concentration and primary production and in zooplankton biomass. In the Banda Sea, the monsoonal periods coincide with upwelling and downwelling [1,4]. The Southeast Monsoon-induced westerly surface currents are partly fed by upwelling in the eastern Banda Sea, which results in nutrient enrichment of the photic layer. The northwest monsoon-induced easterly surface currents results in downwelling and oligotrophic conditions in the upper water layers. These changes in vertical movement in the water column chiefly concerned the upper 150 meters, while deeper layers remain relatively undisturbed [9].

In determining the oceanographic phenomenon, it's widely used data source from remote sensing or from in situ observation. Each data source has plus and minus. Remote sensing data are subject to several sources of error that affect their accuracy, for example calibration, atmospheric correction algorithm error, and problems in deriving parameter from radiances [10]. In situ data typically provide high quality and accurate data, but can only produce extremely limited spatial observations because of the expense of sea operation and the large area of the ocean. In situ data provide high quality and real information of parameter that satellite cannot, and satellite provide spatial and temporal observation that in situ cannot.

The analysis of time series is an important and valuable approach adopted in several studies for its ability to improve the spatial and temporal resolution of seasonal and inter annual pattern in biological and oceanographic data [11]. Moreover, information of the seasonal factor affecting the upwelling events in the Banda Sea is still limited. Therefore, a study on the variability of thermocline layer and chlorophyll-a concentration spatially over a prolonged period of time was urgently needed. This should be done in order to determine the validation of secondary data (from satellite imagery data and model) and in situ observation data (from IJEP research cruise). The final objective is to determine the variability of thermocline layer and chlorophyll-a concentration as upwelling indicator in the Banda Sea. This research focused in the condition during the Southeast Monsoon (June-July-August) until September.

2. Material and Methods

2.1. Study Area

The Banda Sea, lying in between the Flores Sea and the Arafura Sea, is a semi-closed waters in Indonesia. It comprises 3 major areas, the Northern Banda Basin with maximum depth 5801 meters in the northwest, the Southern Banda Sea Basin with maximum depth 5417 meters in the center, and the Weber Deep with maximum depth 7439 meters in the east. The study area was restricted to the area of 3^0S – 9^0S and 123^0E – 132^0 E (Figure1a). In this study, it was also used data from in situ observation using Baruna Jaya VIII Research Vessel as part of IJEP (Indeso Joint Expedition Program) cruise. This expedition was conducted in September 2016 across the northern area of Banda Sea, the Molucca Sea, and the Celebes Sea. The two survey stations for this study were arranged in 4.13^0 S 127.63^0 E and 3.38^0 S 127.34^0 E (Figure1b).

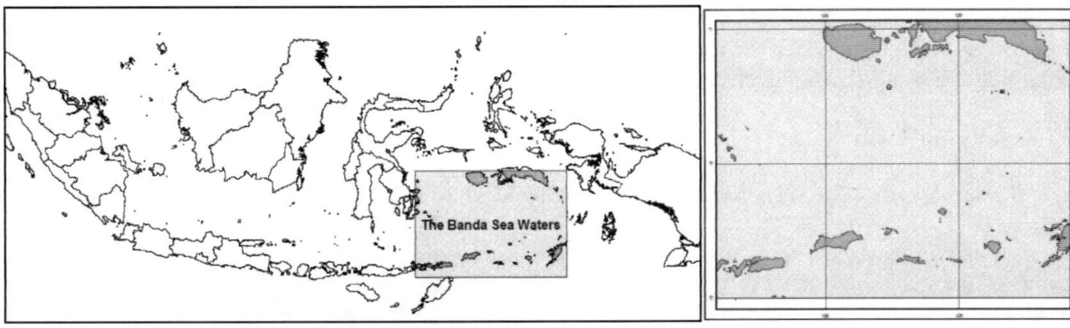

Figure 1. (a) Research study location of the Banda Sea, (b) The two red dots in the north indicate the stations of in situ observation as part of IJEP cruise

2.2. Data and Methods

Remote sensing data used in this research were monthly mean vertical sea water temperature, monthly mean chlorophyll-a concentration, and sea surface temperature as additional data. In order to make better description from time series data, this research used minimum 10-year-average of parameters observed. Monthly mean vertical sea water temperature data of 0-400 meters depth were extracted from SODA (Simple Ocean Data Assimilation) reanalysis for the period of January 1981 – December 2010. It now uses GFDL MOM5/SIS numerics at finer 0.25° x 0.25°x50 lev resolution, similar to the ocean component of the GFDL CM2.5 coupled climate model, and includes an active sea ice component. The Optimal Interpolation filter has also been augmented (relative to previous releases) with bias correction in order to allow for reduced bias in estimates of long term trends. Monthly mean Chlorophyll-a data were extracted from Aqua MODIS satellite data from July 2001 until December 2011. All data were in NetCDF format and were extracted using GrADS (Grid Analysis and Display System) software. We also delineated upwelling events based on the values of monthly mean sea surface temperature anomaly following the previous research method with 0^0 C > SST anomaly > -0.5^0 C as borders for upwelling initiation stage or termination stage and SST anomaly < -0.5^0 C as a border for fully developed stage of an upwelling event [12]. Monthly mean sea surface temperature data were extracted from SST Pathfinder from January 1981 – December 2010. To determine the bathymetry condition, we used Etopo1 bathymetry data from NGDC (National Geophysical Data Centre).

In situ observation was carried out during research cruise using Baruna Jaya VIII Research Vessel on September 2016. Profiles of sea water vertical temperature and chlorophyll-a concentration were obtained using CTD (conductivity, temperature, depth). Surface was defined by 0-10 meter depth. The water samples were collected at each observation station in various depths, filtered through Whatmann Cellulose Nitrate Membrane Filters 0.45μm and stored in cold storage room until further analysis in water quality laboratory. The chlorophyll-a concentration from all in situ observation station were quantified spectrophotometrically in absorbance at 630, 647,664, and 750 nm using formula (1) [13]:

$$C_a = 11.85\,(e_{664} - e_{750}) - 1.54(e_{664} - e_{750}) - 0.08\,(e_{630} - e_{750}) \qquad (1)$$

with C_a is concentrations of chlorophyll-a, respectively, mg/L; absorbance e_{630} ,e_{647} ,e_{664} are corrected optical densities at the respective wavelength (nm), and absorbance e_{750} is correction of turbidity (nm). After determining the concentration of pigment in the extract, then the amount of pigment per unit volume was calculated using formula as follows:

$$Chlo\ a, mg/m^3 = \frac{C_a\ x\ extract\ volume, L}{volume\ of\ sample, m^3} \qquad (2)$$

The variability of thermocline layer and chlorophyll-a concentration from remote sensing data and in situ observation were compared and determined spatially and temporally, especially during the

Southeast Monsoon (June-July-August) and second Transitional Monsoon (September-October-November). Generally, when the thermocline layer is shallower, sea surface temperature is lower, and chlorophyll-a condition is higher in abundance, so the upwelling condition is said to be happened.

3. Results and Discussion

3.1. Variability of the Thermocline Layer and Chlorophyll-a Concentration in the Banda Sea

3.1.1. Variability of the Thermocline Layer in the Banda Sea. The thermocline depth is defined as the depth of the maximum vertical temperature gradient. The depth of 20°C isotherm is widely used to represent the thermocline depth [14]. Moreover, the previous research found that the upper limit of thermocline layer in Indonesian waters were slightly shallower than it in equatorial Pacific [15], so the upper limit of thermocline layers were identified by 22°C isotherm. Based on the 30-year-average SODA model data (1981-2010), the thermocline layer in the Banda Sea vary among region with the depth ranges between 85 until 130 meters (Figure 2).

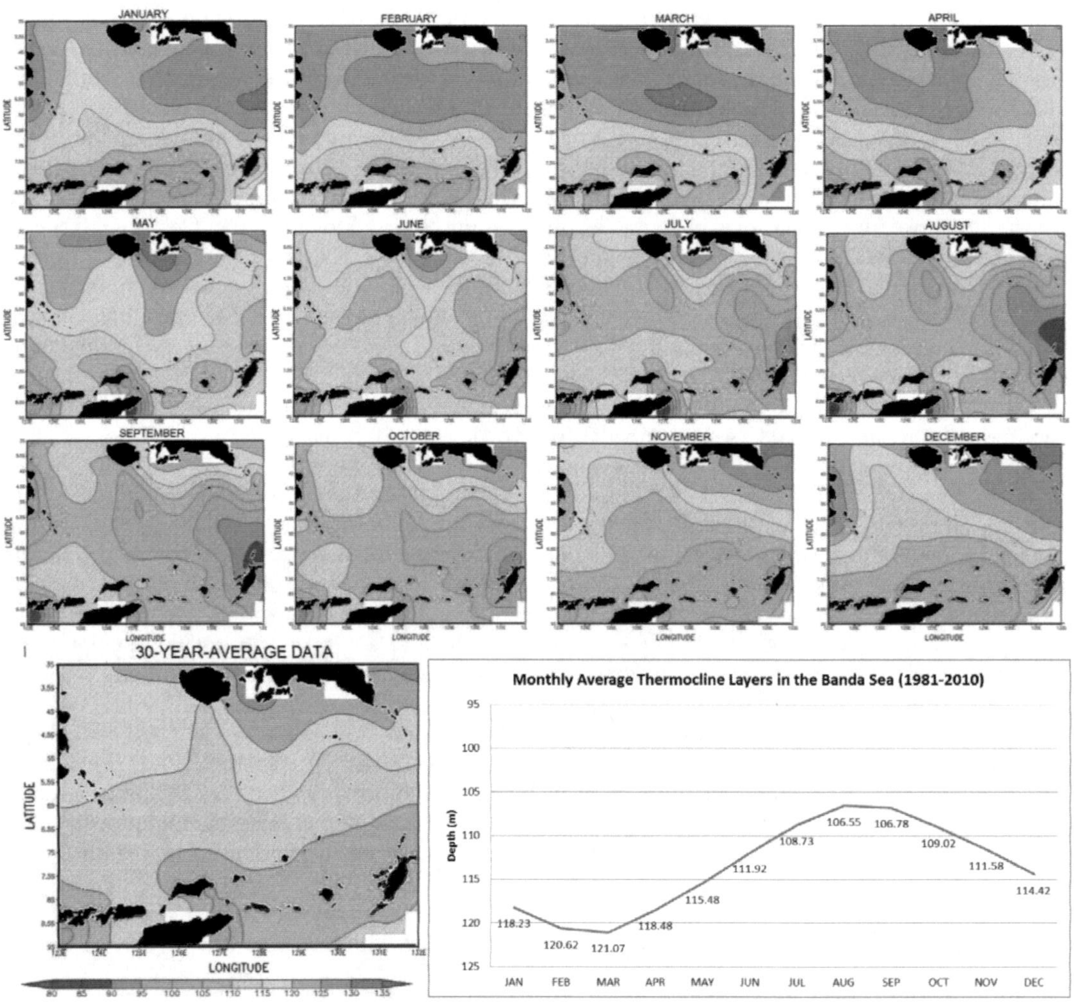

Figure 2. Spatial and temporal variability of Thermocline Layer in the Banda Sea

The variability of the thermocline layer is affected by monsoonal system as well as local and regional environment. In the Northwest Monsoon (wet season) occurred during December to February, the thermocline layer was seen only in the southern Banda Sea and deeper layer was seen in the most area of Banda Sea. The deepest thermocline layer was mostly reached in February until March. In the First Transitional Monsoon occurred during March to May, shallower thermocline layer was seen in the southern Banda Sea and extended to the eastern Banda Sea. In the Southeast Monsoon occurred during June to August, shallower thermocline layer was seen in the most area of the Banda Sea and reached the shallowest peak in the eastern Banda Sea. In the Second Transitional Monsoon occurred during September to November, shallower thermocline layer was seen in the southern and eastern Banda Sea but the area of shallow thermocline layer was narrower as the depth of thermocline began to rise. In general, the fluctuation of thermocline layer depth seemed to follow the seasonal pattern.

The upwelling period in the Banda Sea occurred from April through December. Moreover, they calculated the Ekman pumping upwelling was 1,27Sv. [7]. Wind stress curl induces Ekman upwelling from April to December, with weaker downwelling for the rest of the year [4]. This condition is followed by the shallow thermocline layer in the Banda Sea [7]. The thermocline layer temporal variability in Figure 2 showed that the initial of thermocline shoal was in April and tended to be shallower until September. The highest variability in this research was seen in the eastern Banda Sea. Easterly wind occurred during the Southeast Monsoon and brought water mass from the eastern Banda Sea caused the water mass gap in that area. The water mass from deeper layer uplifted to the surface through the upwelling event [4,16]. Therefore, the eastern Banda Sea tends to be the location of the greatest upwelling event in the Banda Sea.

3.1.2. The Variability of Chlorophyll-a Concentration in the Banda Sea. The variability of chlorophyll-a concentration in the Banda Sea was described by the comparison of remote sensing data (from Aqua MODIS satellite data) and in situ data. The time series of monthly composites Aqua MODIS data was from 2001 until 2011. The sea surface chlorophyll-a concentration vary among the region of the Banda Sea, so it will be easy to determine the variability of the parameter using spatial and temporal analysis.

A relatively high concentration and wide distribution of sea surface chlorophyll-a was obviously seen during April to August. During this period, high concentration of chlorophyll-a in the eastern area tends extend to the center and western area of the Banda Sea. The spatial and temporal distribution of chlorophyll-a concentration (Figure 3) also showed the evidence of a difference between the character and dynamics of the eastern and western Banda Sea. The eastern Banda Sea exhibit the highest chlorophyll-a concentration up to 1 mg/m^3. The area near the island showed high concentration of chlorophyll-a. This situation indicated that marine productivity within this area was governed by nutrient load from the river and coastal discharge. Moreover, that condition also indicated that the upwelling event is controlled by monsoonal system. The chlorophyll-a distribution pattern indicates the gradation value that is high in coastal areas, especially estuary and getting lower toward the open sea. High concentrations in the area near of the islands were caused by the accumulation of nutrients that come from rivers and inputs from land runoff [17,18,19].

The monthly average chlorophyll-a concentration in the Banda Sea ranged from 0.11 to 0.43 mg/m^3. The maximum peak is in June, meanwhile the minimum peak is in February, March, September, and October (Figure 3). This condition is consistent with upwelling event that occurred in April to December [20]. In general, the abundance of chlorophyll-a concentration datasets didn't follow the seasonal pattern. This meant that the local effect gave stronger influence than monsoonal effect. But, the chlorophyll-a distribution still follow the monsoonal effect. The patterns of chlorophyll-a distribution was usually controlled by the seasonal monsoon [12,21,22,23]. The datasets from remote sensing data (SeaWiFS data) and field data support the idea that upwelling during the Southeast Monsoon dominates the east/west distribution of chlorophyll-a, and supports higher phytoplankton biomass (and therefore, shallower euphotic depths) in the eastern Banda Sea [8].

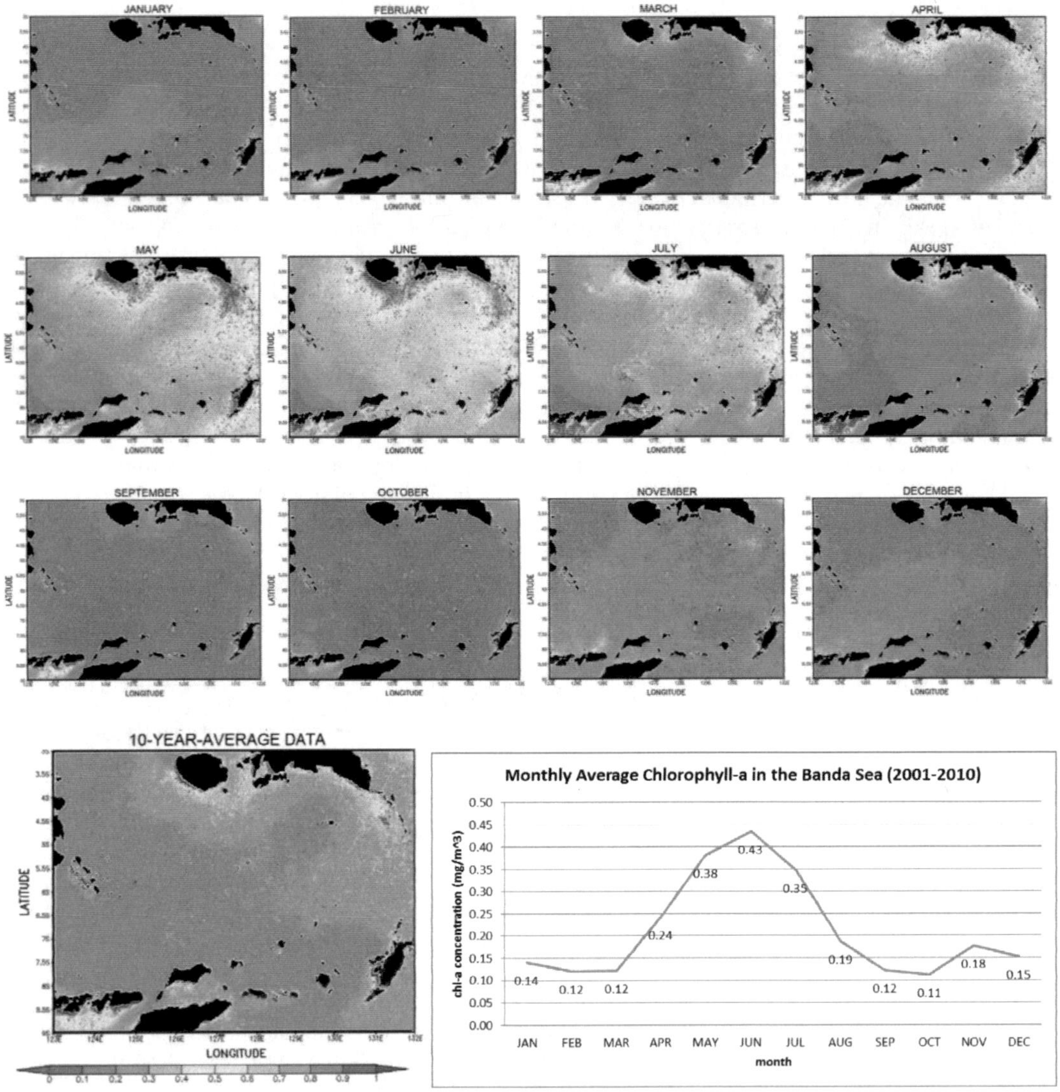

Figure 3. Spatial and temporal variability of Chlorophyll-a concentration in the Banda Sea, from 10-year-average Aqua MODIS data (2001-2010)

3.2. The Comparison of In Situ and Remote Sensing Data

The IJEP (INDESO Joint Expedition Program) cruise was carried out aboard the Indonesian Research Vessel, RV. Baruna Jaya VIII. The northern area of Banda Sea was identified by Station 1 and Station 2. We used CTD data to determine the sea vertical temperature and chlorophyll-a concentration.

Based on the in situ observation using CTD Rosettes, the 22^{0}C isotherm existed in various deep, depend on the bathymetry condition. The thermocline layer in station 1 was shallower than that of in station 2. The comparison of SODA model data and in situ data from CTD showed that SODA model data algorithm over-estimated the thermocline layer up to 1.4 times (Figure 4). Based on this result, the thermocline layer in 2016 was shallower for about 34 meters depth than the average condition (1981-2010). Although there are difference between the thermocline layer from SODA model data and CTD data (in situ), but the value seems to be linear (Figure 4).

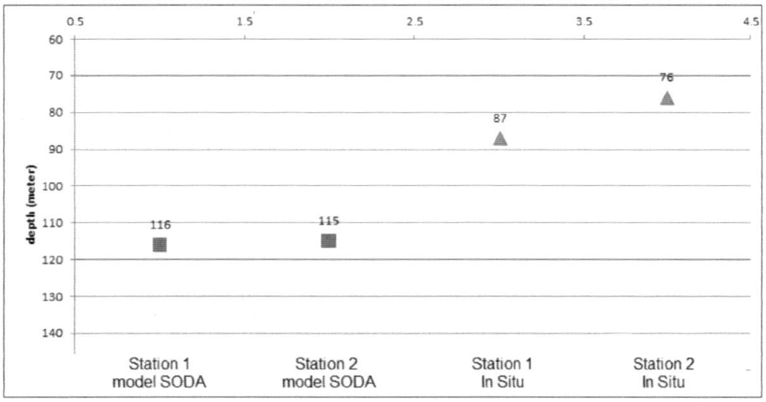

Figure 4. The comparison of thermocline layer from in situ and remote sensing data (SODA model data)

The good performance was also shown by the comparison of chlorophyll-a concentration. The comparison of Chlorophyll-a concentration from in situ data and remote sensing data (Figure 5) showed that the value of in situ data was lower than the remote sensing data, but the value seems to be linear. This indicates overall good performance of Aqua MODIS data compared with in situ data. In previous research, Aqua MODIS data also has a strong relation with in situ data up to 80% [18].

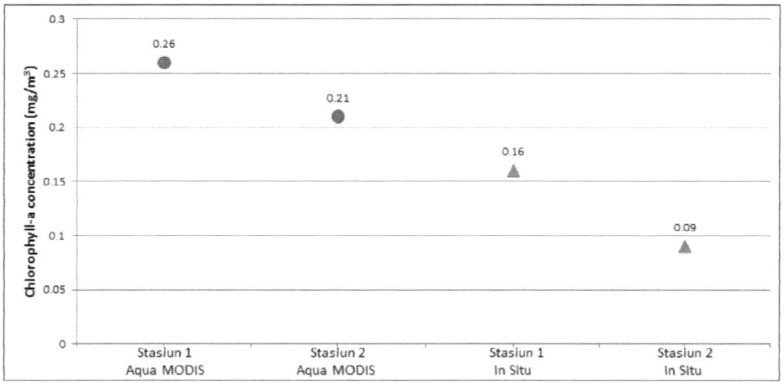

Figure 5. The comparison of chlorophyll-a concentration from in situ and remote sensing data (Aqua MODIS data)

According to the result shown, the chlorophyll-a concentration from Aqua MODIS data seemed to over-estimate the in situ data up to twice. Based on this result, the chlorophyll-a concentration in 2016 was less abundant for about 0.11 mg/m^3 than the average condition (2001-2011). In situ data used the 10 meter depth chlorophyll-a data, meanwhile the remote sensing data used the sea surface chlorophyll-a data. Supposedly, this was the reason why there was difference between in situ and remote sensing data in this research.

Many researchers exhibit the same problems. Remote sensing data may over-estimate coastal chlorophyll, with values 30%–77% higher than the next closest climatology [21]. Remote sensing data of chlorophyll-a SeaWiFS over-estimated chlorophyll-a (up to six times) at stations near the turbidity front and in the Rıo de la Plata Estuary waters, but provided adequate estimates (within 10%) in the subtropical waters of the Brazil Current [22]. Moreover, although remote sensing data were always higher than the other datasets in the open ocean, the relatively small differences could be due to natural variability.

In general, when comparing the in situ and satellite derived data, one has to remember that both kinds of estimates are subject to errors. In situ water determinations are subject to several potential sources of errors and need for further improvement in the methodology. More importantly, on a qualitative level, the remote sensing data for all parameters showed similar patterns to those seen in the in situ observation (Table 1).

Table 1. The comparison of all parameters from In Situ and remote sensing data

Observation Parameters in September		Station 1	Station 2
Chlorophyll-a (mg/m^3)	In Situ Data	0.156	0.088
	Aqua MODIS data	0.262	0.209
Thermocline Layer (meter)	In Situ Data	93	108
	SODA data model	115	116

3.3. Variability of the Thermocline Layer and Chlorophyll-a Concentration as the Upwelling Indicator in the Banda Sea

To determine the variability of the parameters, we used the anomaly value. The negative anomalies of monthly average thermocline layer describe the thermocline shoals, meanwhile the positive anomalies describe the thermocline sinking. The thermocline layer anomaly showed that the thermocline shoal ranged 10 – 20 meters. The greatest anomaly occurred in the eastern Banda Sea during Southeast Monsoon. The thermocline started to be deeper during the second Transitional Monsoon (Figure 6). The maximum variability of thermocline layer depth showed in the center until eastern area of the Banda Sea during Southeast Monsoon (Figure 6a) and mostly showed in the eastern area of Banda Sea during Second Transitional Monsoon (Figure 6b). The southern and northern area of Banda Sea showed minimum variability since the thermocline layer depth remains shallow along the year.

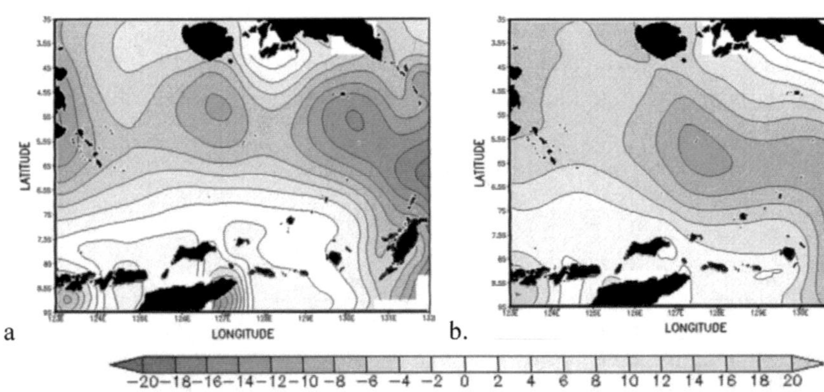

Figure 6. The thermocline layer anomalies in the Banda Sea, derived from SODA model data (1981 – 2010) during (a) Southeast Monsoon (JJA) (b) Second Transitional Monsoon (SON)

The variability of thermocline layer is affected by monsoonal system. Shallower thermocline layer that mostly occurred in the southern and northern area of Banda Sea (Figure 6) was supposedly caused by the difference in bathymetry condition. The southern area is surrounded by islands with 2000 meters of the maximum depth. Meanwhile, the other area of Banda Sea has higher depth up to more than 7000 meters (Figure 7). The maximum variability of thermocline layer mostly occurred in the deepest area of the Banda Sea (the eastern Banda Sea).

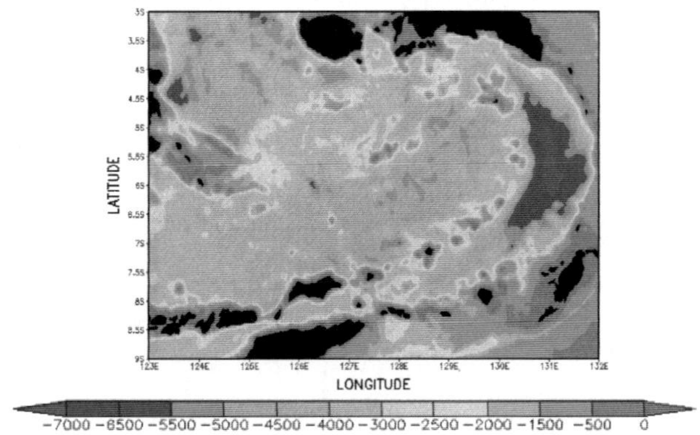

Figure 7. Bathymetry condition in the Banda Sea

The positive anomalies of monthly average Aqua MODIS data describe the more abundance of chlorophyll-a concentration, meanwhile the negative anomalies describe the less abundance. The range of chlorophyll-a concentration was -1 until 1 mg/m^3. The chlorophyll-a concentration seemed to be higher during Southeast Monsoon (Figure 8a) than during Second Transitional Monsoon (Figure 8b). This condition is consistent with monthly average chlorophyll-a concentration (Figure 3).The highest variability of chlorophyll-a concentration occurred mostly in the eastern Banda Sea and extended to the center area during Southeast Monsoon. During the Second Transitional Monsoon, the chlorophyll-a concentration variability started to decrease.

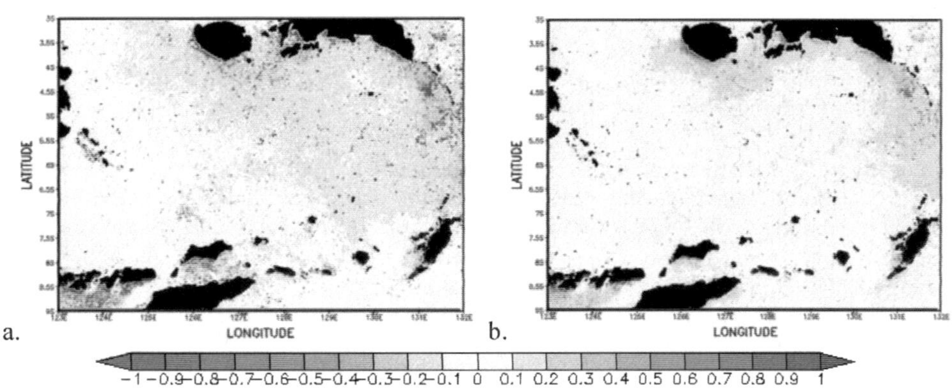

Figure 8. The chlorophyll-a anomalies in the Banda Sea, derived from Aqua MODIS data (2001 – 2010) during (a) Southeast Monsoon (JJA) (b) Second Transitional Monsoon (SON)

We also used sea surface temperature from SST Pathfinder (1981-2010) as an additional data. Sea surface temperature has the close relationship with layer of sea water [24]. Accordingly, sea surface temperature data can be used to interpret the phenomenon (such as thermal front, currents, upwelling, and biological activity) that occurred in the ocean [25,26]. According to previous research, the high value of sea surface temperature mostly occurred during Northwest Monsoon, and the low value of sea surface temperature mostly occurred during Southeast Monsoon. The temperature difference of west area minus east area of the Banda Sea calculated was 2^{0}C [27]. Moreover, the upwelling event can be determined from the sea surface temperature anomalies. Based on the data processing of SST Pathfinder (1981-2010), the upwelling initiation in Banda Sea typically occurred in May and fully developed upwelling event occurred in June and the largest upwelling event occurred in August. The average sea surface temperature in the Banda Sea is 28,34^0 C.

Based on the anomaly of all parameters in the Banda Sea, it can be seen that in general, upwelling event supposedly occurred in May until August. During those periods, the chlorophyll-a concentration was increasing, meanwhile the thermocline layer and sea surface temperature was decreasing (Figure 9). This condition can be inferred as upwelling event. This value might be different in other cases, depend on the time series used in the data processing and the number of in situ stations observed. In general, this result was consistent with previous research that in the south of East Java and Banda Sea have upwelling with longer duration (3–4 months) and higher intensity (decrease in temperature reached >2°C below average), the primary productivity was higher than other locations [28].

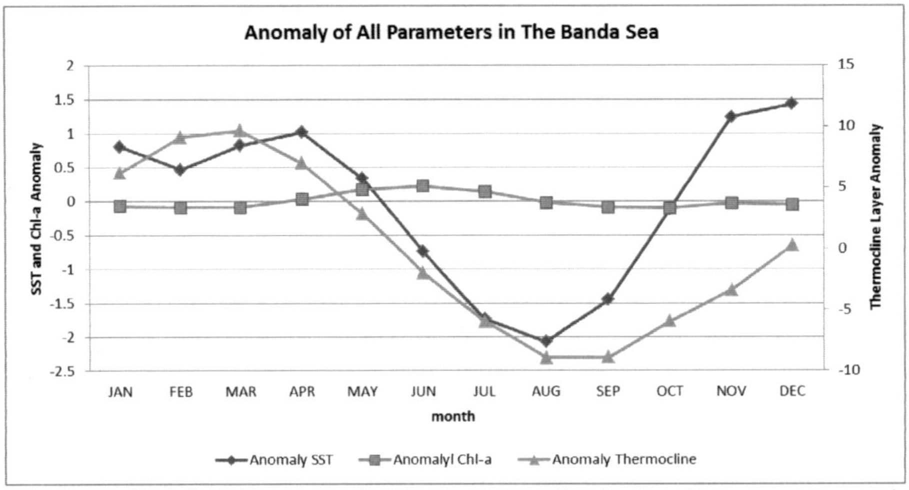

Figure 9. Anomalies of All Parameters in The Banda Sea, Sea Surface Temperature were derived from SST Pathfinder data, Chlorophyll-a concentration was derived from Aqua MODIS data, Thermocline layer (22^{0}C isotherm) is derived from SODA model data.

It is also the evidence that upwelling did occurred during the Southeast Monsoon. If the upwelling is strong or continues for a sufficient length of time, the thermocline eventually breaks the surface and is then seen as a front along the coast which moves back and forth depending upon the strength of the upwelling. The upwelling is described with the shallower thermocline upper limit, the lower sea surface temperature, and the higher chlorophyll-a concentration compared with the monthly average value. Moreover, the sea surface chlorophyll-a concentration has a strong correlation with fish productivity [17,29] and other biological parameters, for example the micro nekton of the Banda Sea showed in general larger spatial variability in numbers and biomass in August than in February [30]. To determine this case, it is needed an advance and further analysis.

4. Conclusion

The remote sensing data seemed to be overestimated nearly double the value of in situ data, but they showed similar patterns (linear) with in situ data. The variability of thermocline layer, sea surface temperature, and chlorophyll-a distribution was strongly related to seasonal pattern, although the local effect gave more influence in the abundance of chlorophyll-a concentration. In general, the upwelling event occurred in May until August (during Southeast Monsoon) in The Banda Sea. The Southeast Monsoon dominates the east/west distribution of chlorophyll-a, shallower thermocline layers and lower sea surface temperature, especially in the eastern Banda Sea.

5. Acknowledgement

This research was supported by Indonesian Agency for Meteorological, Climatological, and Geophysics. The authors would like to thank to the institution produced SODA model data, SST Pathfinder data, and Aqua MODIS data for providing open source data for analysis to be used in this paper. We also thank to INDESO (Infrastructure Development of Space Oceanography) and Indonesian Ministry of Marine Affairs and Fisheries who has been carried out a research cruise as a part of IJEP (INDESO Joint Expedition Program). We also thank to all of crews aboard Baruna Jaya VIII Research Vessel, for realizing the sampling program.

References

[1] Wyrtki, Klaus. 1961. Physical Oceanography of The Southeast Asian Waters.--Naga Report. 2: p 195.

[2] Gordon, A.L., Susanto, R.D. and Vranes, K., 2003. Cool Indonesian Throughflow as A Consequence of Restricted Surface Layer Flow. Nature, 425(6960) pp 824-828.

[3] Godfrey JS. 1996. The Effect of The Indonesian Throughflow on Ocean Circulation and Heat Exchange with The Atmosphere: A Review. J Geophys Res 101 pp 12217–12237

[4] Wyrtki, Klaus. 1957. The Water Exchange Between The Pacific and The Indian Oceans in Relation to Upwelling Processes.--Proc. 9th Pac. Sci. Congr. 16 pp 61-66.

[5] Meyers G. 1996. Variation of Indonesian Through Flow and the El Niño-Southern Oscillation. J Geophys Res 101 pp 12255–12263

[6] Ilahude AG. 1998. Three and a half decades of oceanographic surveys in the Indonesian waters (1960–1995). In: Cresswell G, Wells G (eds) The ASEAN-Australia Regional Ocean Dynamics Expeditions 1993–1995. Proc Symp, Lombok Indonesia, June 1995. AMSAT, Canberra pp 19–28

[7] Susanto RD, Gordon AL, Zheng Q (2001) Upwelling Along The Coasts of Java and Sumatra and Its Relation to ENSO. Geophys Res Lett 28(8) pp1599–1602

[8] Moore TS, John Marra, Ali Alkatiri. 2003. Response of the Banda Sea to the Southeast Monsoon. Mar Ecol Prog Ser Vol.261 pp 41-49.

[9] Zijlstra, JJ, M.A. Baars, S.B. Tijssen, F.J. Wetsteyn, J.I. Witte, A.G. Ilahude, Hadikusumah. 1990. Monsoonal Effects on The Hydrography of The Upper Waters (< 300 M) Of The Eastern Banda Sea and Northern Arafura Sea, with Special Reference to Vertical Transport Processes. Proc. Snellius-tl Symp., Neth. J. Sea Res. 25 pp 431-447.

[10] Gregg, W.W. and M.E.Conkright.2001. Global Seasonal Climatologies of Ocean Chlorophyll: Blending In Situ and Satellite Data for the Coastal Zone Color Scanner Era. Journal of Geoph. Res. Vol.106 (C2) pp 2499-2515.

[11] Vontepotte, V and Melin F.2011. Interannual Variations in The SeaWiFS Global Chlorophyll-a Concentration (1997-2007). Deep Sea Res. Oceanography Res.Pap.58(4) pp 429-441.

[12] Nababan, B, N.Rosyadi, D Manurung, NM Natih, and R.Hakim. 2016. The Seasonal Variability of Sea Surface Temperature and Chlorophyll-a Concentration in The South of Makassar Strait. The 2nd International Symposium on LAPAN-IPB Satellite for Food Security and Environmental Monitoring. LISAT-FSEM.

[13] Jeffrey, S.W and G.F.Humphrey. 1975. New Spectrophotometric Equations for Determining Chlorophylls a, b, and c, in Higher Plants, Algae, and Natural Phytoplankton. *Biochem.Physiol. Pflanzen, 167:191.*

[14] Tomzack, M. dan J.S.Godfrey. 2001. Regional Oceanography: An Introduction, Pergamon, New York. P 422.

[15] Bray, Nancy A., S. Hautala, J.Chong, John P. 1996. Large-scale sea level, thermocline, and wind variations in the Indonesian throughflow region. Journal of geophysical research, vol. 101, (C5) pp12239-12254.

[16] Wyrtki, K., 1987. Indonesian Through Flow and The Associated Pressure Gradient. Journal of Geophysical Research: Oceans, 92(C12), pp 12941-12946.

[17] Sachoemar, S.I, T.Yanagi, R.S.Aliah. 2012. Variability of Sea Surface Chlorophyll, Temperature, and Fish Catch Within Indonesian Region Revealed by Satellite Data. Mar.Res.Indonesia.Vol.37 (2) pp 75-87.

[18] Semedi, B and N.M. Safitri.2015. The Estimation of Chlorophyll-a Distribution in Madura Strait Using MODIS Satellite Imagery Data and In Situ Observation During Southeast Monsoon. Res.J.of Life Science Vol.2(1) pp 40-49.

[19] Wirasatriya, Anindya. 2011. Distribution Patterns of Chlorophyll-a concentration and Total Suspended Solid (TSS) in Toli Toli Bay, Celebes. Oseanografi Marina Bulletin October Vol.1 pp 137-149.

[20] Gordon, A.L and R.D. Susanto. 2001. Banda Sea Surface Layer Divergence. Ocean Dynamics 52 pp 2-10

[21] Conkright, ME and WW Greg.2003. Comparison of global chlorophyll climatologies: In situ, CZCS, Blended in situ-CZCS and SeaWiFS. Int.J.Remote Sensing Vol.24, 5 pp 969-991. DOI: 10.1080/01431160110115573

[22] Armstrong, Roy A, F.Gilbes, R. Guerrero, C.Lasta, H.Benavidez, and H.Mianzan. 2004. Validation of SeaWiFS-derived chlorophyll for the Rio de la Plata Estuary and Adjacent Waters. Int.J.Remote Sensing Vol.25, 7-8 pp 1501-1505. DOI: 10.1080/014311603 10001592517

[23] Nontji, Anugerah. 2005. The Nusantara Sea. Jakarta: Djambatan. p 368.

[24] Tajduddah, Muslim. 2016. Observation of Sea Surface Temperature on Spatial and Temporal Using Aqua MODIS Satellite in West Banda Sea. The 2nd International Symposium on LAPAN-IPB Satellite for Food Security and Environmental Monitoring. Procedia Environmental Sciences 33 pp 568 -573.DOI: 10.1016/j.proenv.2016.03.109

[25] Reynold, R.Smith, T.Liu, C.Chelton, D. Casey, K. Schlar. 2007. Daily High-resoulution Blended Analysis for Sea Surface Temperature. J.Clim.20 pp 5473-5496.

[26] Robinson, IS. 1991. Satellite Oceanography: An Introduction for Oceanographers and Remote Sensing Scientist. England: Ellis Ltd. pp 89-100.

[27] Boely T, Gastellu-Etchegorry J, Potier M, Nurhakim S .1990. Seasonal and Interannual Variations of The Sea Surface Temperatures (SST) in the Banda and Arafura Sea Area. Neth J Sea Res 25 pp 425–429

[28] Kemili, P. and M.R. Putri.2012. Influences of Upwelling Duration and Intensity Based on Sea Surface Temperature Anomaly Toward Primary Productivity Variability in Indonesian Waters. Marine Tropical Science and Tech. Journal Vol.4 (1) pp 66-79.

[29] Wardani, R.T.2012.Comparison Analysis of Chlorophyll-a Between Terra and Aqua MODIS Satellite Imagery from Sea Surface Temperature and Total Suspended Solid Data (Case Study in Madura Strait and Adjacent waters.

[30] Schalk, PH, JIJ Witte, S.Budiharjo, AM Hatta.1990. Monsoon Influences on Mincronecton of The Banda Sea (Indonesia). Netherlands Journal of Sea Research 25(4) pp 601-610.

IOP Conf. Series: Earth and Environmental Science **55** (2017) 012040 doi:10.1088/1755-1315/55/1/012040

Identification of wave energy potential with floating oscillating water column technology in Pulau Baai Beach, Bengkulu

I Alifdini[1], D N Sugianto[2], Y O Andrawina[1] and A B Widodo[3]

[1] Oceanography Department, Diponegoro University, Jl. Prof. Soedarto SH Tembalang, Semarang 50275 Indonesia
[2] Center for Coastal Disaster Mitigation and Rehabilitation Studies, Diponegoro University, Jl. Prof. Soedarto SH Tembalang, Semarang 50275 Indonesia
[3] Electrical Engineering Department, Diponegoro University, Jl. Prof. Soedarto SH Tembalang, Semarang 50275 Indonesia

Email: inovasita@gmail.com, dennysugianto@yahoo.com, yochiokta@gmail.com, adrianbelaw@gmail.com

Abstract. Pulau Baai is a beach, which is located in Bengkulu, Indonesia. This location has swell waves, which is beneficial for wave energy, because it directly faces the Indian Ocean. Floating Oscillating Water Column (OWC) is a prototype used to generate electricity from wave energy. The objective of this research is to identify how much electricity can be generated from floating OWC. This research used a quantitative method by processing wind data (speed and direction) from ogimet.com in 2000-2016. The wind speed rate for wave energy potential of this location is above 5.14 m/s. Wind data is converted to significant wave height and periods data by Sverdrup, Munk, and Bretschneider (SMB) method. Significant wave height rate of this location is 0.06 – 5.33 meters. Assuming that this power plant uses 3 chambers of floating OWC, the power output of OWC is 1.9 GW/year. Thus, suppose each residents' house uses 1300 watt, this power plant can be used for 1,461,538 residents per year.

Keywords: Wave Energy, Electricity, Floating OWC, Pulau Baai Beach, Bengkulu.

1. Introduction

The demand for energy in Indonesia increases every year. The energy crisis also has been a crucial issue in recent years. This energy crisis occurs due to the depletion of fossil fuel reserves. Because of the excessive human consumption and dependency towards fossil fuels, the fuel reserves become depleted, while it takes thousands and even millions of years for its renewal. [1]. Total energy needs final in Sumatra from 2013 to 2050 is estimated to grow at a rate of average growth 5% each year from 35 million TOE in 2013 increased to 68 million TOE in 2025 and to 212 million TOE in 2050 [2]. Indonesia will certainly meet an energy crisis if only rely on energy from fossil (non-renewable energy). Therefore, we can do the mitigation energy crisis by using the potential of wave energy as a renewable energy.

Amongst the numerous devices that have been suggested for wave energy conversion, the oscillating water column (OWC) wave energy device is probably the most extensively studied type [3]. The oscillating water column (OWC) is a type of wave energy converter (WEC) that has been object of study since the 1940s, when Yoshio Masuda started developing a navigation buoy powered by this device. It consists of a pneumatic chamber open at the bottom to the sea water and at the top to the atmosphere through an air turbine [4].

An Oscillating Water Column (OWC) is basically a device that transforms the mechanical energy of the waves into electric power. The waves entering the chamber to compress and decompress the air around this SWL so that an oscillating airflow is created. This airflow is passed through a power take-off (PTO) system consisting of a turbine and an induction generator that transforms this motion into electrical power [5]. The oscillating motion of the water column acted upon by waves produces a bidirectional flow through the turbine that drives an electrical generator [4]. An offshore OWC will have to float, which uniquely requires that both the wave activated body and the OWC are modeled in a coupled fashion as each absorbs power from the waves [6].

Floating Oscillating Water Column is a particular type of OWC device known as the backward bent duct buoy (BBDB) first proposed by Masuda [7]. It has a submerged opening aligned downstream of the incident wave propagation direction. It has a single air chamber and is free to move in six degrees of freedom. The device is constructed of thin walls enclosing the water column. The PTO system is provided by means of an air turbine connected to an electric generator. The motion of the water column relative to the OWC body creates oscillating pressure in the chamber and air flow through the turbine. A relief valve provides a way to keep the pressure in the air chamber within acceptable limits to prevent the turbine from stalling [8].

Oscillating water column (OWC) wave energy conversion devices consist of a partially submerged chamber open to wave forces at the base as illustrated in Figure 1. The wave forces cause the water column within the chamber to rise and fall, driving the air in and out (inhalation and exhalation) of the chamber typically through a Wells or variable pitch type air turbine. An electrical generator is then utilized to convert the oscillatory airflow established into electrical energy. The pneumatic gearing provided by the air coupling facilitates the conversion of low frequency wave power into high frequency electrical power [9].

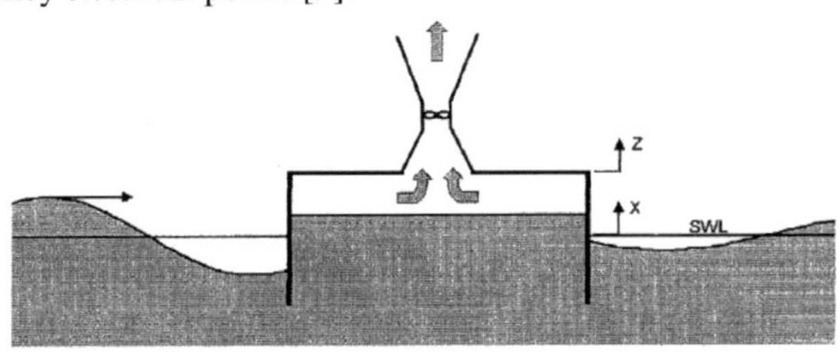

Figure 1. Floating Oscillating Water Column Wave energy device (air flow arrows indicate the exhalation phase) [9]

Pulau Baai is a beach which is located in Bengkulu, Indonesia which directly faces the Indian Ocean [10]. This location is very easy to access and some of its area is used for tourism and ports. Therefore, this location requires more source of electrical energy. The objective of this paper is calculating wave energy potential and converted to electrical energy in Pulau Baai Beach Bengkulu by using Floating Oscillating Water Column Technology.

2. Methodology

The study was conducted using a quantitative method. The data were collected from indirect observations. Based on its objectives, this study is categorized as an applied research methods [11].

2.1. Research Location

The study was conducted in the Pulau Baai Beach Bengkulu city with latitude coordinates 3 ° 48'38.31 "S to 4 ° 1'34.37" S and longitude 102 ° 2'35.36 "E to 102 ° 27'38.46" E. The identification of wave energy was done before the wave hit the coast.

IOP Conf. Series: Earth and Environmental Science **55** (2017) 012040 doi:10.1088/1755-1315/55/1/012040

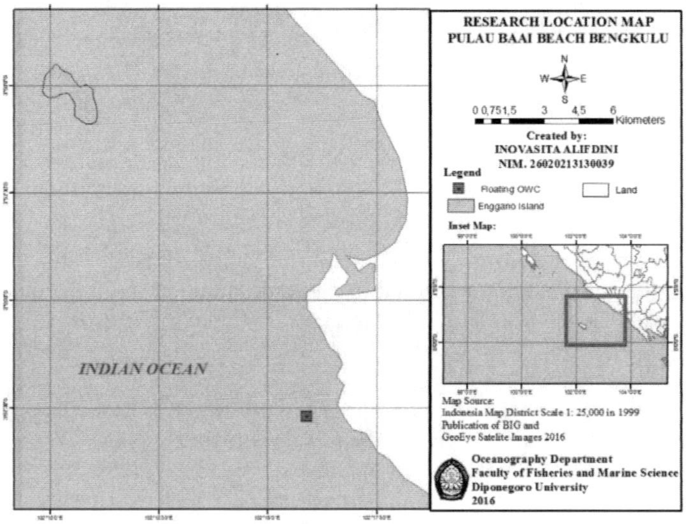

Figure 2. Research location map

2.2. *Wave Forecasting*

The data used in this study were the wind data from Ogimet (2000-2016). Then, the wind data were calculated using the SMB method for wave forecasting. Ocean winds were estimated using the wave height and period and Sverdrup-Munk-Bretschneider's SMB method. Wind speed is estimated from the following equation (1).

$$\frac{gH_{1/3}}{U_{10}^2} = 0.26 \tanh\left[\left(\frac{1}{f_m U_{10}}\right)^{3/2} \frac{(3.5g)^{3/2}}{10^2}\right] \qquad (1)$$

where, g: gravity (m/s^2), $H_{1/3}$: significant wave height (m), U_{10}: wind speed at 10m above the ground (m/s), f_m: peak frequency of ocean wave spectrum (1/s). However, the peak frequency of the ocean wave spectrum could not be taken; it was treated as *fm* by taking the reciprocal of the observed significant wave period at this time.

Wind speed at 10 m above the ground was assumed in the equation (1), it was corrected using the power law from the following equation (2).

$$U_z = U_{10}\left(\frac{Z}{10}\right)^{\alpha} \qquad (2)$$

where, U_Z: wind speed at measured station (m/s), U_{10}: wind speed at 10 meter height (m/s), Z: distance from the ground, α: power index (0.12) [12].

To determine the fetch length in the SMB method, cosine average method was used. Based on this method, 15 radials were extend in the upwind direction ± 45 degree (at 6 degree intervals). Then the fetch length (X) was calculated as:

$$X = \frac{\sum_{i=1}^{15} Xi \cos\alpha_i}{\sum_{i=1}^{15} \cos\alpha_i} \qquad (3)$$

where, Xi : fetch length which measured from waves observation to the fetch ends. α: deviation on both sides of the wind direction, using a 6-degree increase, up to an angle of 42 degrees on either side of the wind direction (degrees).

According to this method, to accomplish the fetch-limited condition, the wind duration must be greater than t_{min} that is:

$$\frac{gt_{min}}{U} = 6.5882 \exp\left\{\left[0.0161\left(\ln\left(\frac{gX}{U^2}\right)\right)^2 - 0.3692\left(ln\left(\frac{gX}{U^2}\right)\right) + 2.2024\right]^{0.5} + 0.8798\left(ln\left(\frac{gX}{U^2}\right)\right)\right\} \quad (4)$$

where U: corrected wind speed (m/s)
In the fetch limited condition, the equation for predicting wave height (Hs) is:

$$\frac{gH_s}{U^2} = 0.283 \tanh\left[0.0125\left(\frac{gX}{U^2}\right)^{0.42}\right] \quad (5)$$

If the wind duration is smaller than $t_{min,}$ the condition is called duration limited. In this condition, equivalent fetch must be calculated by substituting the wind duration into the equation (4). Then, the wave heights are estimated by substituting the equivalent fetch into the equation (5) [13].

2.3. Energy and Power Calculation Analysis

The type of OWC used in this study was the floating type, which has mechanical oscillator model utilized in the present study, (Figure 3) was the fixed OWC model proposed by Szumko [9] with the inclusion of air compressibility. The lower-case variables *k,* band *m* are the owe water plane stiffness, radiation damping and mass respectively. The corresponding upper-case parameters for the floating structure are *K, Band M.* The mass terms include the hydrodynamic mass. It must be noted that for the floating structure, *K* also includes the mooring line stiffness. The turbine damping is modelled by the linear damping parameter A and the air compressibility by the linear stiffness /1. The *x* coordinate is the OWC mean free surface elevation and *z* is the floating structure displacement relative to the still-water level (see also Figure 1) [9].

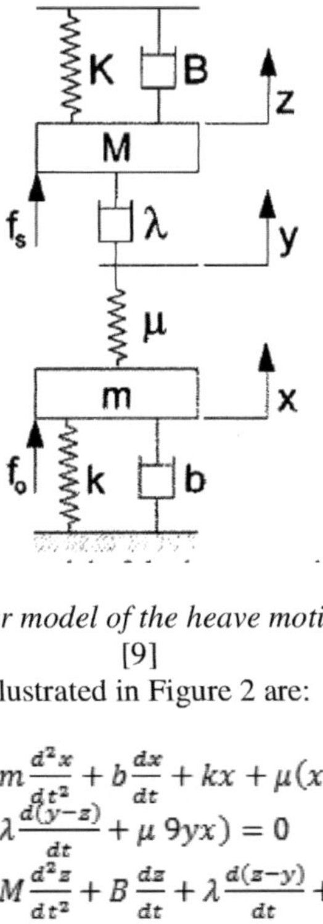

Figure 3. *Discrete mass-spring-damper model of the heave motions of a floating OWC WEC device* [9]

The equations of motion of the system illustrated in Figure 2 are:

$$m\frac{d^2x}{dt^2} + b\frac{dx}{dt} + kx + \mu(x-y) = f_0 \quad (6)$$
$$\lambda\frac{d(y-z)}{dt} + \mu\, 9yx) = 0 \quad (7)$$
$$M\frac{d^2z}{dt^2} + B\frac{dz}{dt} + \lambda\frac{d(z-y)}{dt} + Kz = f_s \quad (8)$$

The wave forces on the OWC, f_0, and the floating structure, f_s, are assumed to be related via the parameter *r* (see Eq. 4). In general, *r* is complex, allowing both a magnitude and phase difference between the forces. In the present analysis, the floating owe is assumed to be axisymmetric. In the limit of large wavelength, or small wave number, *r* can also be shown to be equivalent to the area ratio of the OWC opening to the total base area of the floating wave energy converter [9].

$$\begin{Bmatrix} f_0 = rf \\ f_s = (1-r)\,f \end{Bmatrix} \; 0 \leq r \leq 1 \tag{9}$$

$$f = Fe^{i\omega t} \qquad x = Xe^{i\omega t} \qquad y = Ye^{i\omega t} \qquad z = Ze^{i\omega t} \tag{10}$$

$$(k + \mu - m\omega^2 + ib\omega)X - \mu Y - Fr \tag{11}$$

$$\mu X - (\mu + i\lambda\omega)Y + i\lambda Z = 0 \tag{12}$$

$$\big(K - M\omega^2 + i\omega(B + \lambda)\big)Z - i\lambda\omega Y = F(1 - r) \tag{13}$$

Making the substitutions $\alpha = k - m\omega^2$, $\beta = b\omega$, $\gamma = K - M\omega^2$, $\delta = B\omega$ and $A = \lambda\omega$ and solving the set of simultaneous equations.
The fixed OWC solution [9] can be retrieved by setting S→ ∞ and T→ ∞ (Le. K→ ∞);

$$A^2_{opt\_f} = \frac{(1+Q^2)R^2\mu^2}{R^2+Q^2(1+R)^2} \text{ and} \tag{14}$$

$$P_{\max\_f} = \frac{|Fr|^2 Q^2 \omega}{4\mu(QR+\sqrt{R^2+Q^2(1+R)^2}\sqrt{(1+Q^2)R^2})} \tag{15}$$

The forcing term *Fr* is the force on the oscillating water column. The limiting case of incompressible air for a fixed OWC may be obtained by setting R→ 0 ($\mu \to \infty$).

$$P_{\max\_f\_incomp} = \frac{|Fr|^2}{4b(1+\sqrt{(1+Q^2)})} \tag{16}$$

The ratio of maximum fixed OWC power capture ratio for the compressible and incompressible flow cases is then

$$\frac{P_{\max\_f}}{P_{\max\_f\_incomp}} = \frac{QR(1+\sqrt{1+Q^2}}{QR+\sqrt{R^2+Q^2(1+R)^2}\sqrt{(1+Q^2)R^2}} \tag{17}$$

In interpreting Eq. 23, the relationship between the expressions for *Q* and *R*, stated previously, must be considered. Solving these equations yields

$$Q = \frac{\mu}{\beta} R = \frac{\mu}{b\omega} R \tag{18}$$

The parameters *Q* and *R* may be represented as a function of the ratio of the wave frequency to the incompressible system natural frequency, Q_o, as

$$\begin{Bmatrix} Q = \frac{\alpha}{\beta} = \frac{k-m\omega^2}{b\omega} = \frac{1}{2\zeta_0}\left(\frac{1}{\Omega_0} - \Omega_0\right) \\ R = \frac{\alpha}{\mu} = \frac{k-m\omega^2}{\mu} = \kappa_0(1-\Omega_0^2) \end{Bmatrix} \tag{19}$$

The air compressibility spring rate expression may be estimated assuming isentropic compression with only small changes in volume (relative to the total chamber volume). It is expressed in terms of the ratio of specific heats of air, c_p/c_v, atmospheric pressure, *p*, the owe water surface area, *A*, and the chamber height, *h*, as

$$\mu = \frac{\left(\frac{c_p}{c_v}\right)pA}{h} \tag{20}$$

The power capture presented is normalized as [9]:

$$P = \frac{P_{max}}{(\frac{|F|^2 \omega_{00}}{\mu})} \tag{21}$$

where *Pmax* is maximum power.

2.4. *Data Analysis*

Data analysis was conducted to identify the potential development of Floating OWC (Oscillating Water Column) in Pulau Baai Beach Bengkulu. It included the review of wave energy potential for electricity. Our study focused on calculating the electrical energy from wave using Floating OWC prototype to be used in Pulau Baai beach, and calculated the residents who could use this technology.

3. Results and Discussion

3.1. *Forecasting Wave*

Based on Sverdrup, Munk, and Bretschneider (SMB) method, the highest significant wave height in the period from 2000 to 2016, is 5.33 meters, with a period of 9.94 second. The minimal significant wave height is 0.062 meters with a period of 0.062 second. Meanwhile, the average significant wave height is 2:01 meters and the average significant wave period is 6.78 second. The data distribution is shown below in Figure 4 (a & b):

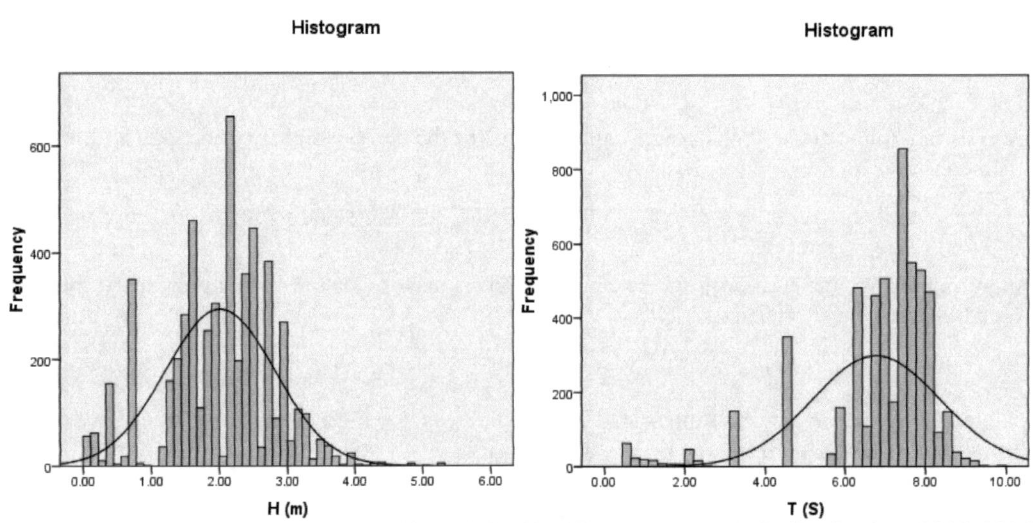

Figure 4. (a) Significant wave height and **(b)** Significant wave periods distribution (2000-2016)

.Based on the Beaufort scale, the wind speed that has the potential to produce waves that can be used as an energy plant is 10 knots (equivalent to the wave height of 2 meters and above). Based on table 1 and 2, the significant wave height value is above 2 meters, there are as much as 53.8%.

Table 1. Significant waves height distribution (2000-2016).

Range (m)	Frequency (amount of data)	Percentage (%)
0-1	660	12.34
1.1-2	1809	33.84
2.1-3	2455	45.93
3.1-4	394	7.37
4.1-5	21	0.39
5.1-6	6	0.11

Table 2. Significant waves periods distribution (2000-2016).

Range (s)	Frequency (amount of data)	Percentage (%)
0-1	92	1.72
1.1-2	50	0.93
2.1-3	63	1.18
3.1-4	148	2.77
4.1-5	348	6.52
5.1-6	191	3.57
6.1-7	1352	25.32
7.1-8	2308	43.22
8.1-9	754	14.11
9.1-10	34	0.64

In addition, this research analyzed the daily height and period of significant wave as shown in Figure 5 (a & b):

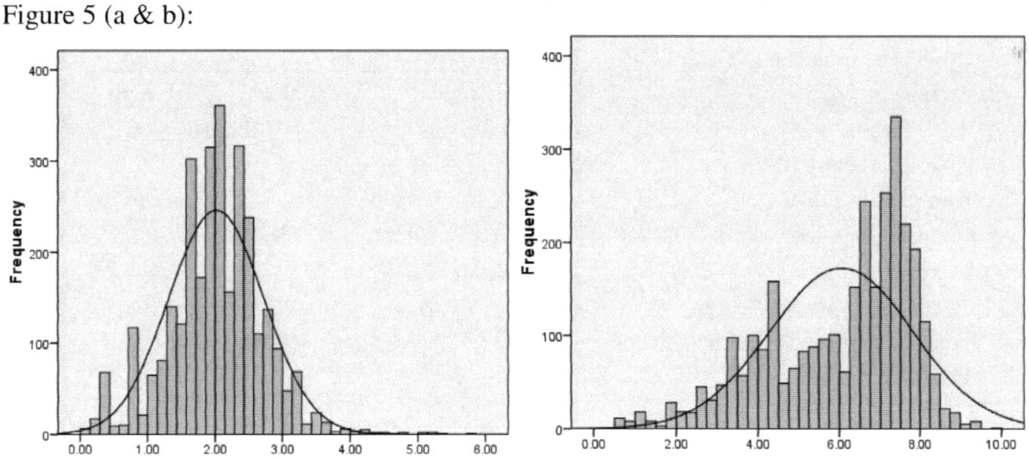

Figure 5. (a) Daily significant wave height and **(b)** Daily significant wave period distribution

The data of daily significant wave height and daily significant wave period distribution is presented on table 4 and 5, the significant wave height values above 2 meters are 52.65%. Therefore, Pulau Baai Beach is suitable for developing renewable energy from ocean waves.

Table 3. Daily significant wave height

Range (m)	Frequency	Percentage (%)
0-1	247	8.11
1.1-2	1195	39.23
2.1-3	1411	46.32
3.1-4	173	5.68
4.1-5	14	0.46
5.1-6	6	0.19

Table 4. Daily significant wave periods

Range (s)	Frequency	Percentage (%)
0-1	18	0.59
1.1-2	56	1.84
2.1-3	110	3.61
3.1-4	301	9.89
4.1-5	356	11.70
5.1-6	367	12.06
6.1-7	608	19.98
7.1-8	1000	32.87
8.1-9	212	6.96
9.1-10	14	0.46

Table 5. Maximum power

Parameters	Symbol	Value	Unit
Wave force	f_0	1.42×10^4	N
Floating stucture force	f_s	$3,57 \times 10^3$	N
Total area of OWC	A	76.18	m^2
Optimal Damping	$A^2_{opt\_f}$	9.22×10^8	m
Maximal Power	P_{max_f}	4.28	J/s
Power incompressibel air in OWC (R=0)	$P_{max\_f\_incomp}$	0.04	J/s
Time duration each year	t	3.15×10^7	s
The Power	P	0.04	J/s
The Energy Working	E_w	$1,16 \times 10^6$	watt

3.2. Energy and Power Generating

Generating energy and power from wave energy have been calculated based on the methods in section (2.3) before. Some parameters of this calculation were assumed in some sizes. Cross-sectional area of OWC were assumed 100 m^2, height of OWC's chamber to be 25 meters, and OWC operated to be 15 meters depth. The results are as follows:

3.2.1. Maximum Power

Maximum power was assumed as the power which can be generated by maximum wave height and period, maximum Sea Surface Temperature (SST), and maximum wave force. The maximum significant wave height value is 5.33 meters, the maximum significant wave period value is 9.95 second, the maximum Sea Surface Temperature (SST) value is 31 degrees, and the maximum of wave force is 17858.41 N. The power calculation in maximum condition are presented in the Table 5.

3.2.2. Minimum Power

Minimum power is assumed as the power which can be generated by the minimum wave height and period, the minimum Sea Surface Temperature (SST), and the minimum of wave force. The minimum significant wave height value is 0.062 meters, the minimum significant wave period value is 0.062 second, the minimum Sea Surface Temperature (SST) value is 29 degrees, and the minimum of wave force is 2.44 N. The power calculation in minimum condition are presented in the Table 6 below:

Table 6. Minimum power

Parameters	Symbol	Value	Unit
Wave force	f_0	1.95	N
Floating stucture force	f_s	0.49	N
Total area of OWC	A	76.18	m^2
Optimal Damping	$A^2_{opt\_f}$	8.06×10^8	m
Maximal Power	P_{max_f}	4.27	J/s
Power in compressible air in OWC (R=0)	$P_{max\_f\_incomp}$	0.69	J/s
Time duration each year	t	3.15×10^7	s
The Power	P	0.002	J/s
The Energy Working	E_w	6.83×10^4	watt

3.2.3. Average power

Average power is assumed as the power which can be generated by the average wave height and period, the average Sea Surface Temperature (SST), and the average of wave force. The average significant wave height value is 2.01 meters, the average significant wave period value is 6.78 second, the average Sea Surface Temperature (SST) value is 29.5 degrees, and the average of wave force is 2956.33 N. The power calculation in average condition are presented in the Table 7 below:

Table 7. Average power

Parameters	Symbol	Value	Unit
Wave force	f_0	2,365.55	N
Floating structure force	f_s	591.39	N
Total area of OWC	A	76.18	m^2
Optimal Damping	$A^2_{opt\_f}$	8.35×10^4	m
Maximal Power	P_{max_f}	4.28	J/s
Power in compressible air in OWC (R=0)	$P_{max\_f\_incomp}$	0.05	J/s
Time duration each year	t	3.15×10^7	s
The Power	P	0.02	J/s
The Energy Working	E_w	7.53×10^5	watt

3.3. Recommendations Based on Power Output

The results of the calculation in section (3.2.1), (3.2.2) and (3.2.3), are summarized in Table 8. From the table, the average power output of floating OWC (one chamber) is 6.60 x 10^5 watt. The power plant generates at least 1 GW as a standard. So, to implement this power plant in Pulau Baai beach, at least 3 chambers of floating OWC are needed. The total energy of 3 chambers of OWC is 1.9 GW per year. If we assume each residents' house used 1300 watt, this power plant can be used for 1,461,538 residents per year.

Table 8. Summary of power output

Average Energy Calculation Each Year			Unit
Energy of OWC	Average	7.53 x 10^5	Watt
	Minimal	6.83 x 10^4	Watt
	Maximal	1,16 x 10^6	Watt
Average		6.60 x 10^5	Watt

4. Conclusion

In this study, we overview the identification of wave energy potential in Pulau Baai beach, Bengkulu. Based on the results of the study, this power plant has a great potential to be implemented and developed, although its establishment will cost a larger amount of money. However, we still have to develop renewable energy before we run out of fossil energy in the future.

5. Acknowledgments

We would thank to Oceanography Department Diponegoro University, Center for Coastal Disaster Mitigation and Rehabilitation Studies (PKMBRP) Diponegoro University, Center of Excellence Science and Technology (PUI) Diponegoro University, and Ministry of Research Technology and Higher Education (Ristek Dikti) Republic of Indonesia for funding our research.

References

[1] Suryaningsih R and Irhas 2014 *Proc. Conf. and Exhibition Indonesia Renewable Energy & Energy Conservation [Indonesia EBTKE-CONEX] (Jakarta)* vol. 47 (Amsterdam: El-Sevier) p 211
[2] Farida Z, Yenny D S, Ainur R, Dwi H, Dian R, Dian R, Ervan M, Fitria S, Sadmoko H P, Cecilya M, Joko S and Agus N 2014 *Indonesian Energy Outlook* vol 1, ed F Rahman and S Abdurahman (Jakarta: National Energy Council Republic of Indonesia) p 85
[3] Utku Ş and Aydoğan Ö 2011 Modelling the interaction between water waves and the oscillating water column wave energy device *Mathematical and Computational Applications* **16** 631
[4] Gomez R P F, Henriques J C C, Gato L M C, and Falcão A F O 2012 *Proc. Int. Conf. on Ocean Energy (Dublin)* vol 1 (Paris: Ocean Energy System) p 1
[5] Garrido A J, Erlantz O, Izaskun G, Jon L, Francisco J M, Pedro L, and Julien M 2015 *Hindawi Publishing Corporation Mathematical Problems in Engineering* 11
[6] Bull D and Johnson E 2013 *Proc. 10th European Wave and Tidal Energy Conference (Aalborg) vol.* 6198C (Aalborg: EWTEC Technical Committee) p 1
[7] Masuda Y. Kimura H. Liang X. Gao X. Mogensen R M. and Anderson T 1996 *The Second European Wave Power Conference (Lisbon) vol* 1 (Brussels: European Commission) pp. 1–3
[8] Aur´elien B, Jorgen H, Adi K, Torgeir M, and Jorgen K 2011 *Proc. 30th Int. Conf. on Ocean, Offshore and Arctic Engineering (Rotterdam)* vol 1 (Rotterdam: OMAE) p 5
[9] Stappenbelt B and Cooper P 2010 Mechanical model of a floating oscillating water column wave energy conversion device *Annual Bulletin of the Australian Institute of High Energetic Materials* **1** 34
[10] Supiyati, Suwarsono and Kristina H 2012 The pattern of tidal currents in Pulau Baai Port Bengkulu using surface water modelling system (SMS) 8.1 software *J. Gradien* **8** 792
[11] Sugiono 2010 *Quantitative and Qualitative Research Methods* (Bandung: Alfabeta) p 334
[12] Watanabe K and Nomura K 2013 *Proc. of the 7th International Conference on Asian and Pacific Coasts (Nusa Dua)* vol 1 (Makassar: University of Hasanuddin) p 867

[13] Etemad S A, Kazeminezhad M H and Mousavi S J 2009 On the prediction of wave parameters using simplified methods *J. Coastal Research* **56** 506

Habitat characteristic of two selected locations for sea cucumber ranching purposes

Retno Hartati*, **Agus Trianto, and Widianingsih**

Marine Science Departemen, Faculty of Fisheries and Marine Science,. Diponegoro Univesity. Jl. Prof. Soedharto, SH, Tembalang, Semarang. Telephone & fax : 0247474698;

retnohartati.undip@yahoo.com

Abstract. Sea cucumbers face heavily overfished because of their high prices and very strong market demand. One effort suggested to overcome this problem is sea ranching. The objectives of present works were to determine biological, physical, and chemical characteristics of prospective location for sea ranching of sea cucumber *Holothuria atra*. Two location at Jepara Waters (Teluk Awur and Bandengan WateRs of Jepara Regency) were selected. The determination of chemical (salinity, temperature, dissolved oxygen of water, phosphate, nitrate, nitrite and ammonium of water and sediment, organic matters of sediment), physical (transparancy, sedimen grains size, water current direction and its velocity), biologycal characteristic (coverage of seagrass and its macroalgae associated, phytoplankton as well as chlorophyl-a and phaeopytin of water and sediment) ware determined. The result of present work showed that some characteristic were matched with requirement as sea ranching location of sea cucumber because the density of sea cucumber in the sea is a function of habitat features. For sediment feeding holothurians of the family Aspidochirotida, the biologycal characteristic act as very important considerations by providing sea cucumber food. High cholophyl-a and phaeopytin in sediment also represent a prosperous habitat for sea cucumber ranching.

Keywords : sea cucumber, habitat, characteristic, sea ranching

1. Introduction

Sea cucumbers, known as 'beche-de-mer' or 'trepang', are generally consumed in Asia, where it is regarded as traditional medicine, a delicacy and an aphrodisiac. The rising demand in these markets prompted declines of many holothurian populations worldwide. The stocks of many tropical species are reportedly over-harvested [1][2][3]. Sea cucumbers being defenceless animals offer no resistance at the time of capture and are indiscriminately fished out. They also do not make any attempts to move away like fish or prawns. This has resulted in large scale capture of sea cucumbers including small and immature forms. In Indonesia, the sign of depleted stock was also very prompted. It showed by decreased production, reduced size of individual catch, farther and deeper fishing area, and more new species introduced in the market [4][5][6].

Sea ranching could be proposed when natural populations in the sea have been depleted due to overfishing. Sea cucumber lend themselves well for sea ranching since they immediately settle down to the bottom and also remain at the same place where they are sea ranched since sea cucumbers have limited movements [7]. Also young forms are known to live among coral reefs and seagrass bed for protection. This habit also helps them to survive better and contribute to the fishery. As in the sea ranching the processes between release and harvest are largely left to nature and the level of care that can be offered to sea cucumber throughout the growth process is reduced, yet still able to produce marketable size of sea cucumber, therefore it is important first to select a suitable site where natural populations are distributed.

Holothuria atra commonly known as black sea cucumber, is one of the most abundant and widely distributed sea cucumber species in most parts of the Indo-Pacific region [8][9]. They inhabit a wide range of depths and a broad variety of habitat ranging from rocky reefs to mudflats [10]. For sea ranching purposes, the objectives of present works were to determine biological, physical, and chemical characteristics of prospective location for sea ranching of sea cucumber *Holothuria atra*.

2. Materials and Methods

Teluk Awur (06°37.437'S, 110° 38.317'E-06°37.142'S, 110° 38.290'E) and Bandengan Waters (06°33.905'S, 110° 39.315'E-06°33.762'S, 110°39.142'E-) of Jepara Regency were surveyed on dry season (June 2016) to assess their suitability as sea raching location of sea cucumber *H. atra*. There were 5 line layed on each location, and 3 transect were plotted in each line for sampling of biotic and abiotic factors.

Biotic environment factor in each proposed location were assesed include seaweed and sea grass associated communities, epiphytes on the seagrass and phytoplankton.

Five transect line and three transect was established at each site running perpendicular to the shore and ending at approximately 500 meters deep edge of the bed. The edge of the bed was defined as the furthest growing sea grass from shore. The quadrat consisted of 1-m^2 of PVC-pipe separated into 100 squares. If any seagrass was found in a square, then it was identified and counted for their shoot. Percent cover for each species was calculated from the shoot density according to [11]. The number colony of macroalgae found in each quadrat was also recorded. A sample of each species was saved for later identification in the lab. Ten or more blades of seagrass were selected at random, the epiphytes were carefully scraped from these blades, identified and counted.

Phytoplankton was collected from sea water samples which were taken from the surface waters (0.5-1 m depth) using a plankton net with a 100-µm mesh and a 20-cm opening. The net was pulled behind a boat at the end of a 10-m-long nylon rope for 10 minutes. The distance covered was about 500 m and the volume of seawater filtered for each plankton haul was 350 m^3. After each haul, the sample was taken to the laboratory where it was set in 5% buffered formalin. Phytoplankton identification based on [12]. Phytoplankton cell counting was done using a haemocytometer under a compound microscope.

Seawater and sediment samples were taken to measure chlorophyll a and total chloropigments to determine the phytobenthic biomass. Sediments were taken using a 60ml syringe with the top section cut off as a mini-corer, analysing only the top 1 cm of the sediment. Chlorophyll extraction used a standard ethanol extractions and pre- and post-acidification measurements [13]. A Perkin-Elmer Lambda 3BUV/VIS spectophotometerwith a 1 nm spectral bandwidth and optically matched 4 cm micro-cuvettes are used in the present work. Total chloropigments was calculated as the sum of chlorophyll a and phaeophytin.

Abiotic environmental factors, such as depth, salinity, pH, dissolved oxygen and temperature were measured in situ using a multi-probe meter. The depth of the light penetration was measured using the Secchi disc and was recorded to the nearest 5 cm.. The sample of sea water and sediment/substrates were taken for analyses of nitrate, nitrite, phosphat, ammonia, organic matter and carbon content, while sediment were also for physical characteristic (texture) analyses.

IOP Conf. Series: Earth and Environmental Science **55** (2017) 012041 doi:10.1088/1755-1315/55/1/012041

3. Results and Discussion

Selection of suitable site is the most important for the success of the sea ranching programme for *H. atra*. They prefer lives in coastal sea grass beds, soft and hard substrates of coral reefs as have been previously reported by [14][15][16][17]. Teluk Awur and Bandengan waters were chosen as proposed sea ranching of *H. atra* since they may have such characteristics.

3.1. Seagrass and Seaweed community

Seagrass are flowering plants growing in shallow coastal waters of tropical seas. They occupy a wide range of habitats from coarse sands and coral rubble to soft muddy bottroms, and from the intertidal zone down to 20 m water depth (or more). Altogather there are 12 species of seagrass inhabit Indonesia waters [18], while in present work, there were 5 species found in Bandengan waters (Table 1) namely *Thalassia hemprichii, Enhalus acoroides, Halodule pinifolia, Cymodocea serrulata, Cymodocea rotundata.* The range average shoot density were 2,8-13,6 shoot/m2 and the seagrass in line transect L2 and L3 were more dense. The cover persentage of seagrass was 23,4-56,3. The highest coverage was *C. serrulata* (75%).

There were found 7 species in Teluk Awur Waters (Table 2), i.e. *T. hemprichii, E. acoroides, Syngirodium isoetifolium, H pinifolia, C. serrulata, C. rotundata, and H. ovata.* The species composition is more varied, there were more line with more seagrass species. *E. acoroides* and *T. hemprichii* found in all line transect while *C. serrulata, C. rotundata* found in four line out of five. Average density was range from 13,2-39,9 shoot/m^2 with covarage of 21,3-39,9 %/m^2. The highest coverage was *Thalassia hemprichii* (44,3%).

There were also seaweed found in the same location of sea grasses. There were two species (*Halimeda opuntia* and *Udotea argentea)* lived in Bandengan waters and more varied species in Teluk Awur waters. *There were 9 species of seaweed in teluk Awur waters i.e. Padina australis, Halimeda opuntia, H. macroloba, Halimeda sp., Caulerpa racemosa C. Serullata, Dyctiota sp., Udotea argentea and Sargassum cristaefolium.*

Seagrass serves as diverse habitat for many marine species including fish and many invertebrates. It also serves as a protection for juvenile marine species and a source of nutrient that helps to sustain many complex food chains and recognised as one of the most productive marine ecosystems [19]. One group of marine invertebrates that inhabit seagrass beds is Echinoderms such as Holothuroidea (sea cucumbers) [20].

In shallow water and on intertidal shores, like in Teluk Awur and Bandengan Waters, seagrasses and microalgae growing directly on the sediment surface, whereas macroalgae are a relatively minor component due to the often turbid water and the difficulty of attaching in a fluid sedimentary environment. But too much of vegetation like algal beds are also not suitable as sea ranching sites [7].

Table 1. Seagrass and seaweed associated composition, shoot or colony density abundance (A = shoot or coloni/m^2) and coverage (B=%/m^2) in Bandengan Waters of Jepara

		L1		L2		L3		L4		L5	
No.	Seagrass species	A	B	A	B	A	B	A	B	A	B
1	*Enhalus acoroides*			12	44,5	8,8	21,9			1,6	9,4
2	*Thalassia hemprichii*	4	37,5	13,6	37,5	7,2	37,5	6,4	37,5	4	37,5
3	*Halodule pinifolia*			7,2	20,3						
4	*Cymodocea rotundata*	4	37,5								
5	*Cymodocea serrulata*			2,4	18,8	8	28,1	8,8	75		
	Average	4	37,5	8,8	30,3	8	29,2	7,6	56,3	2,8	23,4
	Seaweed species										
1	*Halimeda opuntia*			3,2		3				5	
2	*Udotea argentea*					2					
	Average			3,2		2,5				5	

Table 2. Seagrass and seaweed associated composition, , shoot or colony density abundance (A = shoot/m^2 or coloni/m^2) and coverage (B=%/m^2) in Teluk Awur Waters of Jepara

		L1		L2		L3		L4		L5	
No.	Seagrass species	A	B	A	B	A	B	A	B	A	B
1	*Enhalus acoroides*	16	16,4	21,6	39,1	19,2	17	16,8	16,2	33,6	31,8
2	*Thalassia hemprichii*	25,6	44,3	8,8	17,2	20	20,3	13,6	16,2	22,4	29,1
3	*Halophila ovata*	11,2	30,5	11,2	19,8	8	14,8				
4	*Halodule pinifolia*			10,4	19,8	11,2	14,1				
5	*Cymodocea rotundata*	12,8	35,2	10,4	24,2	10,4	37,5			8	37,5
6	*Cymodocea serrulata*	5,6	14,6	13,6	39,8	14,4	37,5	14,4	37,5		
7	*Syringidium isoetifolium*	8	75							10	75
	Average	13,2	36	12,7	26,7	15	23,5	13,2	23,3	21,3	39,9
No.	Seaweed species										
1	*Padina australis*			1		3		5			
2	*Halimeda opuntia*	2,5		2,5		3		2,3		3,3	
3	*Halimeda macroloba*			4		3		2,5			
4	*Halimeda sp.*	1								6	
5	*Caulerpa racemosa*			2		4,7		2		4,5	
6	*Caulerpa serullata*					8		2		3,5	
7	*Dyctiota sp*			4							
8	*Udotea argentea*					2,3		3			
9	*Sargassum cristaefolium*	1,5		1		5		4		7	
	Average	5		14,5		29		21,8		24,3	

3.2. Epiphytes

Many tropical seagrass species have high primary production rates [21] and provide a substantial proportion of the primary productivity for associated ecosystems. Epiphytes density and number of genera found in each species of sea grasses were presented in Table 3 and 4.

Table 3. Average epiphytes density (cell/cm^2) and number of genera (n) in seagrass blades in Bandengan Waters

No.	Sea grass species	Chrysophyta	Chlorophyta	Cyanophyta	Total Density (No. of genera)
1	*Enhalus acoroides*	3.727 (23)	443 (3)	1.489 (5)	5.659 (31)
2	*Thalassia hemprichii*	1.097 (16)	125 (3)	646 (5)	1.867 (24)
3	*Halodule pinifolia*	670 (11)	43 (2)	818 (5)	1.531 (18)
4	*Cymodocea rotundata*	810 (15)	28 (3)	470 (4)	1.307 (22)
5	*Cymodocea serrulata*	1.261 (25)	88 (2)	913 (5)	2.262 (25)

Table 4. Average epiphytes density (cell/cm^2) and number of genera (n) in seagrass blades in Teluk Awur Waters

No.	Sea grass species	Chrysophyta	Chlorophyta	Cyanophyta	Total Density (No. of genera)
1	*Enhalus acoroides*	2.946 (23)	234 (4)	1.227 (5)	4.407 (32)
2	*Thalassia hemprichii*	1.012 (14)	111 (2)	643 (2)	1.765 (20)
3	*Halophila ovata*	1.131 (16)	48 (1)	446 (5)	1.625 (22)
4	*Halodule pinifolia*	670 (11)	64 (2)	818 (5)	1.552 (18)
5	*Cymodocea rotundata*	1.011 (15)	33 (3)	601 (4)	1.645 (22)
6	*Cymodocea serrulata*	1.131 (16)	48 (3)	678 (4)	1.857 (23)
7	*Syringidium isoetifolium*	739 (6)	-	537 (4)	1.275 (10)

3.3. Phytoplankton

Phytoplankton was determined for their composition and abundance. Therewere 12 and 9 genera found in Bandengan and Teluk Awur waters respectively, all belong to Division of Chromophyta and Cyanophyta. Total of 5,86 x 10^5 cell/L in Bandengan waters was quite higher compare to 3,76 x 10^5 cell/L in Teluk Awur waters (Table 5).

Tabel 5. The average abundance of phytoplakton (cell/L) number of genera (n) found in Bandengan and Teluk Awur waters

Division	Bandengan Waters	Teluk Awur Waters
Chromophyta	325.000 (32)	554.000 (34)
Cyanophyta	45.000 (4)	56.000 (4)
Abundance (cell/L)	370.000 (36)	610.000 (38)

For deposit feeder such as *H. atra*, phytoplankton do not have direct effect but the increasing temperature, as well as of the extent and intensity of daylight, implying an increase in phytoplankton biomass and, in turn, a larger supply of food for sea cucumbers [22]. In the other hand, the phytoplankton plays very important role on feeding of holothurian planktonic larva.

3.4. Cholophyll-a and phaeophytin

Estimating the concentration of chlorophyll-*a* remains the most common method for assessing algal biomass. The concentration of chlorophyll-*a* has also been shown to relate to primary productivity and can be used to assess the physiological health of algae by examining its degradation product, phaeophytin. This degradation product has been shown to contribute 16-60% of the chlorophyll-*a* content in seawater and freshwater. Chlorophyll-a concentration in Bandengan sea water was slightly higher than Teluk Awur waters (Table 6) but the consentration of phaeophytin and total chloropigments of both location was almost the same.

Tabel 6. The content of chlorophyll-a, Phaeophytin and Total chloropigments in the seawater and sediment of Bandengan and Teluk Awur waters

	Bandengan waters		Teluk Awur Waters	
Pigment	Seawater	Sediment	Seawater	Sediment
Klorofil-a (μg/l)	0,002421	0,005160	0,001783	0,018556
Phaeophytin (μg/l)	0,059220	0,123888	0,060140	0,143882
Total chloropigments (μg/l)	0,061641	0,129048	0,060923	0,162438

Chlorophyll-a in sediment could be representative of microphytobenthos, i.e. benthic microalgae which are important and resilient ecosystem modifiers and are also important in coastal food webs because of their high accessibility to consumers, such as sea cucumber. The chlorophyll-a, phaeophytin and total chloropigment in sediment of Teluk Awur was higher than Bandengan waters. It may due to high abundance of microphytobenthic organisms.

Both chemosynthetic and photosynthetic bacteria in the sediment contribute substantially to primary production [23]. In addition to their role in primary production, the microphytobenthos (photosynthetic algae and bacteria) structure the physical environment by mediating oxygen and nutrient flux and stabilizing the sediment[24][25].

3.5. Water quality and nutrient content of the sea water and sediment

Water quality such as temperature, salinity, pH, dissolved oxygen, light transparancy were measured in situ and the value was presented in Tabel 7. Sea cucumbers being stenohaline cannot tolerate wide ranges is salinity. Therefore river mouth, estuaries and other bays where the salinity goes down below 10 ppt during the monsoon season have to be avoided [7]. The salinity both location (Bandengan and Teluk Awur waters) were 22-32 and 23-33(‰) and suitable for sea cucumber life.

The seawater temperature of Bandengan and Teluk Awur waters ranged of 27-34 and 28-35 °C. H. *atra* is unusually tolerant of heat, and is almost the only macroscopic organism living in these warm pools at low tide on the warmest days. H. *atra* face high temperature of the environment by coating of light-colored sand with which it habitually covers its body, enabling it to retain a slightly lower body temperature than it does when not sand-coated. H. *atra,* are negatively phototropic since protect themselves from strong light by its sand coating. They usually react to quick changes in light intensity, as by withdrawal of tentacles when covered by a shadow during feeding.

Tabel 7. Water quality parameters of Bandengan and Teluk Awur Waters

Locations	Station	Temperature (°C)	Salinity (‰)	pH	Dissolved oxygen (ppm)	Light intensity (Cm)	Depth (Cm)
Bandengan Waters	L1	28-30	25-32	7-8	4-6	102-115	102-115
	L2	27-33	26-31	7-8	3,54-6	98-110	98-110
	L3	28-34	28-30	7-8	4-6	107-109	107-109
	L4	28-33	26-30	7-8	2,5-6	110-116	110-116
	L5	28-30	22-29	7-8	4-6	100-120	100-120
Teluk Awur Waters	L1	28-33	28-30	7-8	4-6	115-120	115-120
	L2	29-30	28-30	7-8	4-6	110-120	110-120
	L3	29,5-35	29-33	7-8	4-6	104-125	104-125
	L4	29-32	23-32	7-8	4-6	115-130	115-130
	L5	30-32,5	24-32	7-8	4-6	110-150	110-150

Nitrogen (N), one of the primary nutrients that mediate bottom-water hypoxia, can be removed from beach through biological uptake, denitrification, or permanent burial in sediments. Denitrification is the bacterial conversion of nitrates (NO_3^-) in watersheds to gases (N_2 and N_2O) that can enter the atmosphere. The concentration of NH_3-N, NO_3-N, N0$_2$-N, and PO4-P presented in Tabel 8 and 9.

Tabel 8. The nutrient content in the sedimen of Bandengan and Teluk Awur Waters

Locations	Station	Nutrients (mg/gr) NH3-N	NO3-N	N02-N	PO4-P
Bandengan Waters	L1	0,004	0,6638	0,0080	0,2402
	L2	0,002	1,0170	0,0019	0,2042
	L3	0,002	0,2894	0,0008	0,1197
	L4	0,002	0,0468	0,0020	0,1241
	L5	0,002	0,2145	0,0065	0,9234
Teluk Awur Waters	L1	0,098	0,1660	0,0115	2,1580
	L2	0,002	0,1702	0,0069	1,7942
	L3	0,002	0,3319	0,0054	0,8937
	L4	0,002	0,1787	0,0020	0,3046
	L5	0,002	0,2340	0,0026	0,3621

Tabel 9. The nutrient content in the sea water of Bandengan and Teluk Awur Waters

Locations	Station	Nutrients (mg/gr)			
		NH3-N	NO3-N	N02-N	PO4-P
Bandengan Waters	L1	0,033	0,0145	0,0013	0,0136
	L2	0,007	0,0123	0,0009	0,0132
	L3	<0,002	0,0158	0,0007	0,0112
	L4	<0,002	0,0026	0,0017	0,0054
	L5	0,006	0,0111	0,0012	0,0012
Teluk Awur Waters	L1	<0,002	0,0004	0,0018	0,0136
	L2	<0,002	0,0023	0,0012	0,0172
	L3	<0,002	0,0013	0,0014	0,0116
	L4	<0,002	0,0026	0,0014	0,0078
	L5	0,0022	0,00045	0,0021	0,0023

3.6. Sediment characteristics

According to [26], bottom sediment characteristics are one of the crucial components affecting the habitat preference of sea cucumbers and hence studies on ecology of sea cucumber beds are essential. The sediment character of Bandengan and Teluk Awur Waters is presented in Tabel 10. The presence of sediment reduces the stress impacts as limited growth rate caused by the handling and fluctuation in environmental parameters by allowing sea cucumbers to bury in accordance with their diurnal cycle and in response to stress conditions.

Tabel 10. The sediment character of Bandengan and Teluk Awur Waters

Location	Station	Percentage (%)					Organic matter (%)
		Gravel	Coarse sand	Fine sand	Silt	Clay	
Bandengan Waters	L1	10,15	21,15	66,85	1,85	0,00	3,34
	L2	2,20	6,60	90,90	0,30	0,00	4,03
	L3	2,15	7,65	89,75	0,45	0,00	4,02
	L4	2,30	7,60	89,85	0,25	0,00	4,03
	L5	0,00	36,10	60,20	3,70	0,00	2,77
Teluk Awur Waters	L1	2,00	32,20	62,20	3,60	0,00	4,93
	L2	11,40	21,20	65,00	2,40	0,00	4,71
	L3	11,00	24,10	61,35	3,55	0,00	4,20
	L4	10,20	20,20	67,90	1,70	0,00	4,24
	L5	24,20	21,00	42,70	12,10	0,00	4,95

Holothurians can assimilate low content of organic matter as live diatoms, bacteria and detritus by passing sediment through their gut system. By feeding and movement characteristics, they grind organic material and sediment into finer particles, mix up the substrate and recycle the detrital material on the top layers of the sediment. Rocky beds have to be avoided since the sea cucumbers live on the mud or sand and fixed on the organic matter present in the same. *H. atra* are deposit-feeders that ingest sediment with organic matter [28][29][30] and their gut contents are dominated by decaying materials from macroalgae and seaweed, shell fragments from mollusks, crustaceans and barnacles, echinoderm ossicles, many pelagic and benthic foraminifera and diatoms [31][32]. Some studies have also shown that sea cucumbers ingest feces of marine animals[33][34], even their own feces [35][36]. Sediment is an absolutely necessary material for not only feeding of sea cucumbers but also for burrowing and wellbeing of these benthic organisms. Robinson *et al.* (2013) suggested that the presence of sand improves the growth due to the increased surface area for the development of natural food as bacteria and microphytobenthic primary production. Also, diet digestion facilitates by the presence of sand for consumption

According to [15] in the shallow water (<10 m) *H. atra* prefer seagrass habitat with sediments characterized by 2–3.5% organic content, 15–25% of gravel and coarse sand (0.7–1.2 mm) but not silt–mud. Study of [37] also found that *H. atra* primarily in sand-dominated habitats. The preference towards the specific habitat characteristics seems to be associated with their feeding and protection. An understanding of habitat preference would be useful to improve the management of these sea cucumber populations and enable more precise stock assessment.

The present work is to conduct survey for natural beds in purpose to sea ranch the wild stock of sea cucumber. The other most important aspect in this programme is that the area where sea cucumber is sea ranched should be a protected area and free from activities of the fishermen and fishing operations [7]. If trawl net is operated over the area the stocked sea ranched will enter the net resulting in total failure of the programme. The fishing activity in Bandengan are more than in Teluk Awur Waters, since there were many fishing boat harbour in the beach. Another aspect to be borne in mind is that the area where the young or wild *H. atra* are left should be free from strong currents which will sweep away all the seed sea ranched. Teluk Awur can be more suitable for biological reserve for *H. atra.*

4. Conclusion

Based on biological, chemical, physical characteristic of natural beds of Bandengan and Teluk Awur Waters revealed that some characteristic were matched with requirement as sea ranching location of sea cucumber because the density of sea cucumber in the sea is a function of habitat features. For sediment feeding holothurians of the family Aspidochirotida, the biologycal characteristic act as very important considerations by providing sea cucumber food. High cholophyl-a and phaeopytin in sediment also represent a prosperous habitat for sea cucumber ranching

Acknowledgement

The author wish to thank to Rector of Diponegoro University for funding this research through scheme of International Research Publication of PNBP DIPA Universitas Diponegoro No. : SP DIPA-042.01.2.400898/2016, Date of 7 December 2015, Bugdet year of 2016

Referrences

[1]. Lovatelli A, Conand C, Purcell S, Uthicke S, Hamel J F, and Mercier A editors 2004 *Advances in sea cucumber aquaculture and management (FAO fisheries technical paper 463)* (Rome United Nations Food and Agriculture Organisation).

[2]. Bruckner A (editor) 2005 *The Proc. of the Technical workshop on the conservation of sea cucumbers in the families Holothuridae and Stichopodidae* (*NOAA Technical Memorandum NMFSOPR 44*) Silver Spring MD 239 pp

[3]. Uthicke S and Conand C 2005. *SPC Beche-de-mer Inf. Bull.,* **21** 9–14.

[4]. Hartati R, Widianingsih, Purwati P 2009a *Teripang (Holothuroidea : Echinodermata) di Indonesia : Biologi, Budidaya dan Konservasinya.* (Yogyakarta Penerbit Navila Idea) 72 hal
[5]. Hartati R, Widianingsih, Pringgenies D 2009b *Pembenihan dan pembesaran Teripang Pasir (Echinodermata: Holothuroidea)* (Semarang Badan penerbit Universitas Diponegoro) 128 hal.
[6]. Purwati P, Hartati R dan Widianingsih 2012. *Mar. Res. Indonesia* **35(2)** 23-30
[7]. James D B 1993 Sea ranching of sea cucumber. Marine Fiseheries Information (*Tehnical and Extension Series 124*) (India Central marine Fisheries Research Institute Cochin, India) pp 15-17
[8]. Uthicke S 2001 *Mar Ecol Prog Ser* **210** 125–138.
[9]. Conand C 2008 Population status, fisheries and trade of sea cucumbers in Africa and the Indian Ocean. Sea Cucumbers. *A Global Review of Fisheries and Trade (FAO Fisheries Technical Paper 516)* ed Toral-Granda V, Lovatelli A, and Vasconcellos M (Rome FAO) pp 143–193
[10]. Purcell S W, Gossuin H and Agudo N S 2009. *SPC Beche-de-mer Infor Bull* **29** 3-6
[11]. English S, Wilkinson C, and Baker V 1994 *Survey manual for tropical marine resources* (Townsville North Queensland Australia Australian Institute of Marine Sciences) 390 pp.
[12]. Tomas C 1997 *Identifying Marine Phytoplankton* 1st Edition (London Academic Press) 858 p
[13]. Nusch E A 1980 *Arch Hydrobiol Beih Ergeb Limnol* 14 14–30
[14]. Kinch J, Purcell S, Uthicke S, and Friedman K 2008 Population status, fisheries and trade of sea cucumbers in the Western Central Pacific *A Global Review of Fisheries and Trade (FAO Fisheries Technical Paper 516)* ed Toral-Granda V, Lovatelli A, and Vasconcellos M (Rome FAO) pp 7–55
[15]. Dissanayake D C T, and Stefansson G 2012 *J. of the Mar. Biol. Ass. of the UK* **92(3)** 581-590
[16]. Setyastuti A 2014 *J. Ilmu Teknol Kelaut Trop* **6(1)** 31-39
[17]. Asha P S, Diwakar K, Santhanavalli G and Manissery M K 2015 *Indian J. Fish.* **62(1)** 52-57
[18]. Nienhuis P H, Coosen J and Kiswara W 1979 *Netherlands J of Sea Res* **23(2)** 197-214
[19]. Rasheed M A, Dew K R, Kerville S P, McKenzie L J and Coles R G 2006 Seagrass distribution, community structure and productivity for Orman Reefs, Torres Strait – March and November 2004. (*DPI Information Series QI06088)* (Cairns DPI) 38 pp
[20]. Tuapattinaja M A, Pattikawa J A, and Natan Y 2014 *AACL Bioflux* **7(5)** 351-356
[21]. Mateo M A, Cebrián J, Dunton K and Mutchler T 2006 Carbon flux in seagrass ecosystems. (*Seagrasses: Biology, Ecology and Conservation)* ed Larkum A WD, Orth RJ, and Duarte CM (Dordrecht The Netherlands Springer) pp 159-192
[22]. Benítez-Villalobos F, Avila-Poveda O H, and Gutiérrez-Méndez I S 2013. *Sex Early Dev Aquat Org.* **1** 13–24 doi: 10.3354/sedao00003
[23]. Valiela I 1984 *Marine ecologycal process* (New York Springer Science + Bussines Media LLC) 559 p.
[24]. Yallop M L, de Winder B, Paterson D M, and Stal L J 1994 *Est Coast Shelf Sci* **39** 333–344
[25]. MacIntyre H L, Geider R J, and Miller D C 1996 *Estuaries* **19(2A)** 186-201
[26]. Slater M J and Jeffs A G 2010 *J. Sea Res* **64** 241–249
[27]. Moriarty D J W 1982 *Aust J Mar Freshw Res* **33** 255–263
[28]. Uthicke S 1999 *B Mar Sci* **64** 129–141.
[29]. Michio K, Kengo K, Yasunori K, Hitoshi M, Takayuki Y, Hideaki Y, and Hiroshi S 2003 *Mar Pollut Bul.* **47** 118–125
[30]. Hauksson E 1979 *Sarsia* **64** 155-160
[31]. Zhang B L, Sun D L, and Wu Y Q 1995 *Mar Sci* **3** 11–13 (in Chinese with English abstract)
[32]. Yuan X T, Yang H S, Zhou Y, Mao Y Z, Zhang T, and Liu Y 2006. *Aquaculture* **256** 457–467
[33]. Zhou Y, Yang H S, Liu S L, Yuan X T, Mao Y Z, Liu Y, Xu X L, and Zhang F S 2006 *Aquaculture* **256** 510–520
[34]. Goshima S, Fujiyoshi Y, Ide N, Gamboa R U, and Nakao S 1994 *Suisanzoshoku* **42** 261–266 (in Japanese, with English abstract)
[35]. Kang K H, Kwon J Y, and Kim Y M 2003 *Aquaculture* 216 87–93
[36]. Bellchambers L M, Meeuwig J J, Evans S N, and Legendre P 2011 *Aquat Biol* **14** 57–66

Social Relation between Businessman and Community in Management of Intensive Shrimp Pond

Indra Gumay Febryano[1], James Sinurat[2] and Messalina Lovinia Salampessy[3]

[1] Department of Forestry, Faculty of Agriculture, University of Lampung, Bandar Lampung, Indonesia
[2] Graduate Programme, University of Nusa Bangsa, Bogor, Indonesia
[3] Faculty of Forestry, University of Nusa Bangsa, Bogor, Indonesia

indragumay@yahoo.com, sinurat.james@gmail.com, meis_forester@yahoo.com

Abstract. Expansion of aquaculture, especially shrimp culture, is the primary cause of deforestation of mangrove along coastal zone. This phenomenon is pretty much related to social relation between businessman of intensive shrimp pond and community around coastal zone. The objective of this research is to explain social relation between businessman and community in managing intensive shrimp pond. This research is a kind of qualitative research and the method used is a case study. The result of this research shows that the behaviour of the majority of businessman of intensive shrimp pond is not accordingly with environmental concerns as they compelled conversion of mangrove and they disposed waste of shrimp pond into the sea. Such kind of behaviour caused degradation of water ecosystem and marginalizing local community. Corporate Social Responsibility (CSR) which was implemented by businessman of intensive shrimp pond in the area of social, religion, and education can downgrade the coming up of social turbulence. Otherwise, CSR in enabling economic community and environmental management was not conducted yet. CSR in environmental management can be conducted by businessman of intensive shrimp pond by considering the existence of mangrove and pond management and waste in a better way, so that environment around ponds is not polluted and the sustainability of shrimp pond business as well as income of community can be guaranteed. Accordingly with the result of this research, CSR is not only involving businessman of intensive shrimp pond and community, but also involving local government in terms of right and responsibility of citizen as well as management and development of community.

Keywords: social relation, intensive shrimp pond, mangrove, coastal management

1. Introduction

Mangroves have function and benefit which are very important in supporting life within coastal regions [1, 2]. However, the majority of the mangroves have been exploited commercially and mangrove areas as land resources have also been converted into other uses, such as agriculture, fisheries, urbanization, mining, and salt ponds [3]. Aquaculture is a major cause of mangrove deforestation globally [4, 5]. The World Bank, Asian Development Bank, and FAO have actively promoted the shrimp farming industry as a means to create jobs, bring foreign exchange, and reduce poverty in developing countries [6].

Shrimp exports provide major contribution to the economies of producer countries, but these countries are often less clear system of governance in ensuring fair use of resources. Often, the external costs of the industry are not paid by those who get benefit, but the impact go to the poorest and most vulnerable [6]. The case in Sinaloa, Mexico showed that the negative impact of the conversion of mangroves and shrimp farms waste has led to social conflicts between businessman and rural communities driven by a cooperative (economic enterprise) of fishermen [7].

Conflicts that arise in the expansion of shrimp farming industry in the Gulf of Fonseca, South Honduras, shows how individuals and companies that have power were able to gain access to coastal land concessions, thereby depriving poor access [8]. Resistance of NGOs, environmental groups and farmers toward increase of the activity of shrimp farms also appeared in Thailand [9]. Activities of shrimp farms in Thailand has encouraged the formation of networks oriented social and ecological relationships as a way to unify the commercial interests, the environment, economic growth and social welfare [10].

A phenomenon that occurs in the management of mangrove, particularly converted into intensive shrimp ponds, strongly associated with social relations arising between businessman and the surrounding community. The social relationships is a picture of the businessman of intensive shrimp ponds in doing a variety of ways that can reduce social unrest arising from the negative impact of the their activity, like degradation of coastal ecosystems and the marginalization of local communities. Therefore, the research conducted was to explore social relation between businessman and community in management of intensive shrimp ponds.

2. Methods

The location of the research is Pesawaran Regency, Lampung Province, Indonesia (Figure 1). Geographically, Pesawaran Regency is located at coordinates 104.92°–105.34° E and 5.12°–5,84° S. Pesawaran Regency is an area of expansion of South Lampung Regency in 2007 with an area of 1173.77 km square and is administratively divided into eleven districts, where four districts are located in the coastal region, namely: Padang Cermin District, Teluk Pandan District, Punduh Pidada District and Marga Punduh District [11].

This study is a qualitative research and the method used is case study method. In general, the case study is a strategy that is more suitable when the principal question of a study with regard to how or why, when researchers have little opportunity to control events that will be investigated, and where the focus of the research lies on contemporary phenomenon in the context of real life [12]. The case studies provide broad access and opportunities to researchers to examine in depth, detailed, intensive and thorough study of the social unit.

The data collection was conducted in several ways, namely: in-depth interviews, participant observation and document analysis. The key informants made up of businessmen intensive shrimp ponds (14 persons), organization's management of Shrimp Club Indonesia (SCI) of Lampung Province (3 person), community (46 persons), Mitra Bentala NGO (3 persons), and government officials of Pesawaran Regency (15 persons). The data collected then were further analyzed descriptively in relation that occurred between businessman and community in the management of intensive shrimp ponds.

3. Results and Discussion

3.1. Impact Exploitation Intensive Shrimp Farming

In the 1980s, when Pesawaran Regency was still part of South Lampung Regency, exploitation of shrimp farms in the region was still conducted traditionally. In that time, businessman still rely on land extensification; where the ponds size were between 0.5-2 hectare and types of shrimps were black tiger shrimp (*Penaeus monodon*). Along with the decline in production due to increased disease that attacked the tiger shrimp around the end of the 1990s, cultivation of shrimp farming was conducted in semi-intensive and intensive way; where the ponds size were between 0.25 to 0.3 hectare and cultivated species of shrimp are white shrimp (*Litopenaeus vannamae*). Type of white shrimp were considered by businessman were more resistant to disease than black tiger shrimp.

Data [14] showed in the time period between 2007 to 2011 alone, there had been a vast increase in shrimp ponds of 32.23%. Most of the shrimp farms in the region, or 77.96% were intensively cultivated. Construction of shrimp farms by businessman were to convert area of mangrove, either in part or as a whole. According to [15] Pesawaran Regency has 838.653 hectare of mangrove. The mangrove area was degraded from year to year due to the conversion of mangrove into shrimp ponds.

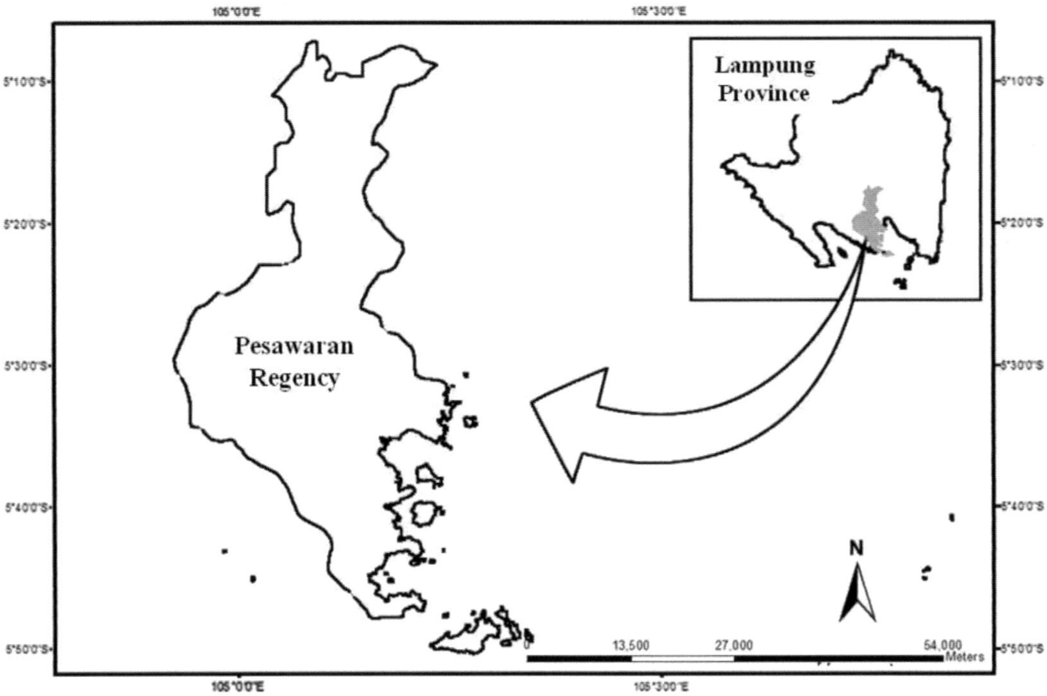

Figure 1 The Location of the Research Source: [13]

Mangrove degradation made community becoming marginal, because the community are increasingly difficult to make ends meet, such as firewood and wood buildings from the mangrove. Also, the community who work as fishermen, due to the destruction of foraging, spawning and breeding of various species of fish, shrimp, oysters and other marine creatures. Disposal of waste most of intensive shrimp ponds without treatment processes increased water pollution. As a result, fishermen are forced to fish up to the middle of the sea, as a result that their catch has declined.

In addition to polluting the water, seawater originating from the most intensive shrimp ponds experiencing seepage into lands surrounding communities. Because of that agricultural land can not be cultivated and well water became salty. Construction of partly intensive shrimp ponds also resulted partly buried irrigation channels which made a number of the fields experiencing water shortages. Some of farmland were also submerged as a result of the closing of the drain by the construction portion intensive shrimp ponds. Development of intensive shrimp farms have exhausted eroded hilly areas adjacent to mangrove land hoarding. The use of heavy equipment and road drainage damage as a result of these activities had damaged public facilities. The existence of the ponds also make community access to the sea became more difficult.

Mangrove in the Pesawaran Regency grow on lands that are outside the state forest. Based on the narrative community leaders and village elders, in the past there were no one claimed ownership of mangrove lands in the region; because most community have profession as farmer working the fields or fields and fishermen who fish in the sea. Shrimp farming began to develop in the 1980s prompted some people to own mangrove land to be turned into ponds which were traditionally managed. Other people owned mangrove control land for the purpose of sale to investors from outside the region. In some locations, it had also occurred forcible takeover of mangrove lands ownership to the community by investors with inadequate compensation; so could lead to violent conflict. With the rapid development of shrimp farming cultivated semi-intensive and intensive around the end of the 1990s, most of the land was in the coastal mangrove Pesawaran Regency has shifted ownership to investors from outside the region.

Pesawaran Regency does not have local regulations that specifically regulate mangrove management in the region, but there are some policies related to the management of mangrove and shrimp ponds as well as its exploitation. The policy is permission for exploitation shrimp farms, the establishment of Ponds Monitoring and Evaluation Team, Pesawaran Regency Decree No. 162.B/III.06/HK/2009 regarding the Protection of Mangrove Management Agency of Punduh Pedada District Pesawaran Regency, Pesawaran Regent Decree No. 175/III.06/HK/2009 regarding Mangrove Working Group and Local Regulation of Pesawaran Regency No. 4 of 2012 regarding Spatial and Territory Planning of Pesawaran Regency 2011-2031. In the implementation, the above policies are more supportive on intensification of shrimp farming than mangrove conservation.

3.2. Relationships between businessman of Intensive Shrimp Farming and Community

The negative impacts of intensive shrimp ponds on community's lives and the environment has the potential of a great conflict. Community in some villages in the coastal areas several times blocked access road leading to the pond locations. But it was done only limited impact caused by damage to the road, not because of the activity in converting mangrove ponds and contaminate the water. Trucks activities carrying shrimp feed and crop yields, as well as the activity of heavy equipment used in construction of intensive shrimp ponds have allegedly accelerate the process of destruction of the road in their region. Issues of road damage is usually settled amicably between the businessman and the community. Businessman will contribute, in cash or material way, while people who do roadwork worked together.

Some communities have started realizing the environmental impact caused by the activity of intensive shrimp ponds. The increasing of public awareness can not be separated from the facilitation undertaken by one of the local NGOs in Lampung Province which was named the Mitra Bentala NGO. Exposes which were quite intensively conducted by the NGO at the local and national mass media in the period 2008-2011 related to the conversion of mangrove performed by intensive shrimp ponds in Pesawaran Regency. In 2010, Mitra Bentala NGO had facilitated a multi-stakeholder workshop in the two districts, Padang Cermin District and Punduh Pedada District. The workshop was attended by representatives of coastal communities and relevant stakeholders to establish a rescue deal of coastal mangrove in Pesawaran Regency.

The workshop resulted in several recommendations, namely: stop all kinds of land conversion, especially coastal mangroves; synergy between the relevant agencies in determining the allotment of coastal areas; supports community participation in the management of mangrove resources; urged relevant agencies to not issue a permit to use that can lead to conversion of mangroves; provides strict sanctions against the perpetrators of environmental destruction (farmers, tourism and public) or other parties; encourage the participation of entrepreneurs in order to contribute to environmental conservation, social awareness to the community, and support the improvement of facilities and infrastructure in the region; companies must perform sewage treatment ponds; the Pesawaran Regency Government facilitate coastal villages in the manufacture of village regulations on the management of natural resources; asks the Pesawaran Regency Government and Council of Local People Representative of Pesawaran Regency to quickly create Local Regulation with regard to mangrove protection; and cracked down/penalizing businessman and community that caused demolition of the coastal environment. The above recommendations then used as a policy negotiations at the district level to encourage the existence local regulation of mangrove management. However, legal drafting of local regulation is apparently encountered many obstacles in the process.

Social and environmental impacts as a result of massive conversion of mangrove can be mitigated by the businessman. Various social activities carried entrepreneurs to communities around the intensive shrimp farms. Businessman participating in social activities at different levels in each location. Social activities are generally done to contribute to the celebrations of religious holidays and the day of independence of the Republic of Indonesia, as well as holiday allowances to people who were in the vicinity of their business. In addition, there also are on scholarships, mass circumcision, etc. There are also entrepreneurs who participated in the construction of facilities and infrastructure, such as mosques, schools, village hall, village office, roads and others. Various social activities above have made one of the businessmen appointed traditional leaders by the local community. Traditional institutions are still to exist in several locations.

Position as traditional leaders make people become reluctant to businessman. This of course also affect how people behave towards social and environmental impacts caused by business activities run by entrepreneur intensive shrimp ponds. Personal approach was also conducted with local community leaders. Similarly, the local authorities, such as the village and district heads. The relation that exists with village chiefs and district heads facilitate businessman in administrative proceedings pond licensing, manufacture statement on land of mangrove land it has, and the support of local government officials to the presence of shrimp farms in the region.

The existence of intensive shrimp ponds can empower the surrounding community, so businessman involving some people by hired them in cultivation, harvesting, post-harvest pool cleaning, security personnel, cooks, and others. Recruitment of security that comes from the surrounding community pond is one of the ways in which businessman by utilizing the identity and social relations which owned the local community. Usually the security personnel recruited a community leader who has a strong influence as respected and honored in his village, so that the security problems of the local communities for activities intensive shrimp ponds can be minimized.

Of the many social activities that have been undertaken by businessman intensive shrimp ponds to communities around the location of his business, it can be stated that businessman are already doing most of corporate social responsibility (CSR). [16] explained that CSR has become the concern of various circles, such as the government, politicians, academics, and community. From the point of government, CSR can be seen as part of the company's participation in regional development financing sources. From the political point of view, CSR is a means of companies to obtain government support. From the view point the community, CSR is the right of local people to benefit from the company's presence on the improvement of their living conditions. From the point of companies, CSR is a process of internalization of external factors, namely: people, planet, and profit; where a good company not only hunt mere economic profit (profit), but has a concern also to the preservation of the environment (planet) and social welfare (people).

CSR activities conducted by each businessman of intensive shrimp ponds at different levels, depending on the willingness and ability of businessman. CSR conducted thus far was able to reduce the social and environmental impacts arising from the massive conversion of mangrove into shrimp ponds intensive. However, CSR is merely the religious social activities, development of infrastructure, education, and others. Businessman of intensive shrimp ponds have not been doing CSR in economic empowerment, such as the provision of a revolving fund for community development efforts in the field of agriculture, fisheries, livestock, trade, and so forth. In addition, the most important thing is that CSR in environmental management up to now has not been done by the businessman.

CSR has not been done by businessman in the management of the environment is closely related to the lack of insight and knowledge of businessman about the importance of good environmental quality. It can be seen from the massive conversion of mangrove lands into intensive shrimp ponds. In addition, spacious green belt of mangrove there has not been adequate. This condition is exacerbated by the unavailability of good sewage treatment, where the majority of intensive shrimp farms dispose of waste directly into the water. Studies conducted [17] in the Gulf of Hurun, Padang Cermin District, Pesawaran Regency show Harmful Algal Blooms phenomenon (HABs). The phenomenon is one indication of the decline in water quality that is allegedly due to the increase in nutrients (nitrogen and phosphorus) from wastewater aquaculture, hatchery (hatchery), and fish farming in floating net. Input of nitrogen and phosphorus cause water eutrophication that can trigger phytoplankton population explosion that is harmful to aquatic organisms. This is in line with research [18] that shows how coastal eutrophication in Hainan Island, China is driven by the introduction of aquaculture waste untreated threatens aquatic ecosystems in the region.

Behavior of businessman shrimp farms that dump waste into surrounding waters has resulted in water quality to decrease; whereas sea water used by businessman as a medium for their aquaculture. As a result, these behaviors trigger the onset of a disease affecting shrimp cultivated by businessman. Disease attack has caused a sharp decline in shrimp production in almost all ponds in the Pesawaran Regency since the late 2000s. Mass death of fish grouper was also experienced by owners of floating net cages. [19] explains that the principles underlying ecological shrimp farm management becomes important, because it can help to understand and contribute to solving problems faced diseases in shrimp farming. Aquaculture is basically a natural ecological processes, although in the case of intensive shrimp farming is done to achieve the proportion of the industry; so that the shrimp farms rely heavily on environmental services provided by nature and should consider the carrying capacity of the shrimp farming ecosystem perspective, including aspects such as the intensity of cultivation, density ponds and sustainability.

Massive conversion of mangrove into shrimp ponds has eliminated the function of the mangrove ecosystem as natural biofilter; consequently reusable waste pond, causing pollution to the pond system itself and affecting mangrove and coastal habitat remaining, especially the harvesting of the sea product by the local community. According to [20], management of intensive shrimp ponds should be done in an environmental friendly by businessman to the existence of mangroves as well as the management of shrimp ponds and better waste, so that the environment around the pond is not polluted and sustainability of shrimp ponds and the community's livelihood be guaranteed.

4. Conclusion

Behavior of the majority businessman of intensive shrimp ponds are not environmental friendly by converting mangroves and dispose waste into the sea This behavior have resulted in the degradation of coastal ecosystems and marginalizing local communities. Although the policy of Pesawaran Regency Government has accommodated function protection of mangrove, but its implementation further supports the intensification of intensive shrimp ponds. The relation between the businessman of intensive shrimp ponds and the surrounding communities and CSR in the areas of social, religious, educational, and others were able to dampen social unrest arising from the presence of the pond activity in the region. However, businessman have not been doing CSR in the economic empowerment and environmental management.

CSR in environmental management can be performed businessman of intensive shrimp ponds to the existence of mangroves as well as the management of ponds and better waste, so that the environment around the pond is not polluted and sustainability of shrimp farms and the community's livelihood be guaranteed. Therefore, CSR does not only involve businessman and community, but also involves Local Government of Pesawaran Regency in terms of rights and obligations of citizens as well as management and development of community. Assertiveness of Pesawaran Regency Government in implementing policies related to sustainable management of mangrove can be encouraged by providing pressure through a network involving national and local NGOs, local institutions, universities, research institutions, and others, so the Pesawaran Regency Government expected to be more in favor of conservation of mangroves.

References

[1] Walters B B, Ronnback P, Kovacs J M, Crona B, Hussain S A, Badola R, Pimavera J H, Barbier E and Dahdouh-Guebas F 2008 *Aquat. Bot.* **89** 220–36
[2] Bosire J O, Dahdouh-Guebas F, Walton M, Crona B I, Lewis III R R, Field C, Kairo J G and Koedam N 2008 *Aquat. Bot.* **89** 251–59
[3] Kusmana C, Wilarso S, Hilwan I, Pamoengkas P, Wibowo C, Tiryana T, Triswanto A, Yunasfi and Hamzah 2005 *Teknik Rehabilitasi Mangrove* (Bogor: Fakultas Kehutanan Institut Pertanian Bogor)
[4] Barbier E B and Cox M 2003 *Contemp Econ Policy* **21** 418–32
[5] Valiela I, Bowen J L and York J K 2001 *BioScience* **51** 807–15
[6] Environmental Justice Foundation 2003 *Smash & Grab: Conflict, Corruption and Human Rights Abuses in the Shrimp Farming Industry* (London: Environmental Justice Foundation)
[7] Cruz-Torres M L 2000 *J. Pol. Ecol* **7** 63-90
[8] Dewalt B R, Vergne P and Hardin M 1996 *World Dev.* **24(7)** 1193–208
[9] Vandergeest P, Flaherty M and Miller P 1999 *Rural Sociol.* **64** 573–96
[10] Vandergeest P 2007 *World Dev.* **35(7)** 1152–71
[11] Pemerintah Kabupaten Pesawaran 2016 *Sejarah Terbentuknya Kabupaten Pesawaran* http://pesawarankab.go.id/profil/sejarah-pesawaran/
[12] Yin R K. 2003. *Case study research: design and methods* (Thousand Oaks, Calif: Sage Publications)
[13] Febryano I G, Suharjito D, Darusman D and Hidayat A 2015 *J. Analisis Kebijakan Kehutanan* **12(2)** 125–42
[14] Dinas Kelautan dan Perikanan Kabupaten Pesawaran 2012 *Profil Kelautan dan Perikanan Kabupaten Pesawaran Provinsi Lampung* (Gedong Tataan: Dinas Kelautan dan Perikanan Kabupaten Pesawaran)
[15] Saputro G B, Hartini S, Sukardjo S, Susanto A dan Poniman A 2009 *Peta Mangroves Indonesia* (Jakarta: Pusat Survei Sumber Daya Alam Laut, Badan Koordinasi Survey dan Pemetaan Nasional)
[16] Nasdian F T 2014 *Pengembangan Masyarakat* (Jakarta: Yayasan Pustaka Obor Indonesia)

[17] Hasani Q, Adiwilaga E M and Pratiwi N T M 2012 *Makara Journal of Science* **16(3)** 183–91
[18] Herbeck L S, Unger D, Wu Y and Jennerjahn T C 2013 **57** 92–104
[19] Kautsky N, Ronnback P, Tedengren M and Troell M 2000 *Aquaculture* **191(1–3)** 145–61
[20] Febryano I G, Suharjito D, Darusman D and Hidayat A 2015 *J. Analisis Kebijakan Kehutanan* **12(2)** 125–42

IOP Conf. Series: Earth and Environmental Science **55** (2017) 012043 doi:10.1088/1755-1315/55/1/012043

Designing Cognitive Intervention to Improve the Awareness Index of the Residents in the Landslide Area

N. Susanto[1], T. T. Putranto[2], E. A. Ulfa[1]

[1] Industrial Engineering Department, Faculty of Engineering, Diponegoro University
[2] Geological Engineering Department, Faculty of Engineering, Diponegoro University

E-mail: nophie.susanto@gmail.com

Abstract. Considering Semarang as a city with a high potential of landslides occurrences in its almost area, human as the part of the system should be played as a centre of the disaster management system to reduce the natural disaster risk. The study area is located in Manyaran district (the west of Semarang) which is categorised as a vulnerable of landslide area. This study aims at establishing a cognitive intervention based on a cause analysis (Fault Tree Analysis/FTA) to find the cause of low value and improve the awareness index of residents as the implementation of human-based disaster management model. The FTA result was then combined with the demographical data to generate the design of the cognitive intervention. The FTA result conducted that the preparedness of emergency planning had the lowest value (18.2%) which was caused by the lack individual preparation including lack of residents knowledge, and the absence of observation facilities as well as the lack of evacuation planning. Analysis of demographical data resulted in a situation of lack socialisation and knowledge of the residents regarding the landslide occurrence. The model of cognitive intervention then utilised some tools such as video, module and discussion to improve the awareness index.

1. Introduction

Semarang city has a quite rapid development characterised by the implementation of development in all fields. This development must be balanced with a more specific handling of the disaster control to protect the assets and development outcomes that have been achieved and to provide a sense of security from disasters (before/pre or during disaster/emergency response) as well as post-disaster.

Semarang city has the potential landslides were quite high in almost all regions [1]. Recently, the studies focused only on the technical aspects of disaster management. The studies of disaster management largely focused on supply chain management of the logistics system [2] [3] [4] [5] and the technical problems of disaster from the point of view of geology, especially in the city of Semarang [6] [7]. Public awareness of landslides should be increased through people's behaviour toward disaster prevention, an understanding of the equipment factors as well as the establishment of the community's commitment to the security plan prepared by the Government of Semarang together with the community. The policies and programs undertaken by the Government of Semarang for this matter are still focused on the institutions whose responsible for the emergency response and post-disaster when residents in disaster-prone areas need an adequate training and education regarding the condition of the area. Human factors should be considered because humans are the most disadvantaged side in the event of disasters even they are also suspected as one cause of the disaster. Human empowerment as individuals and community part is a critical point for the success of disaster management program.

Based on the previous research [8], analysis of awareness of Semarang residents to landslides has been conducted by calculating the index on the awareness of residents in 19 areas. From this study, it was known that Manyaran district (33.36) had the lowest level of individual and family awareness index and classified as "not ready" (0-39) on a scale of 0-100. Based on administrative data on 2016, the population in Manyaran includes 4899 households and a total population is 15,020 people with an area of 150 ha, consisting of dry ground (grounds/building/emplacement and orchards) amounted to 145.25 ha and ground purposes of public facilities (sports fields, recreational parks and cemeteries) of 4.75 ha.

Manyaran district is a densely populated area and it requires empowering people to improve the success of disaster management and to prevent material losses and casualties in the event of landslides. Analysis of the causes of low index of residents' awareness needs to be done as a community empowerment approach using Fault Tree Analysis.

The human factor in empowering the community is needed, especially regarding the cognitive and physical readiness and awareness of residents on preventing and overcoming disasters. An ergonomics of cognitive interventions in disaster management should be highly considered as the key to successful mental aspects of disaster management. With these cognitive factors, human cognition during the critical phase was not bothered by stress or emotional aspect of decision makers and agents collaborated [9]. Recent studies of human factors in disaster management can be further analysed based on the research result of [10] about improving human awareness index in children. Lachland et al. [11] investigated elderly people in disaster management. Other studies analysed the role of government [12], strategy to overcome the trauma of the disaster [13] and disaster risk management [14] [15].

A new novel model of intervention is purposed in this study. This model utilises a cognitive aspect of human being in disaster management. Cognitive intervention is a form of psychological interventions, techniques and exercise therapy in counselling [16]. Cognitive interventions can be used for the conceptual design of cognitive interventions to reduce the depressive level of breast cancer patients after mastectomy [17]. The function of cognitive interventions in addition to reducing depressive also to change individual behaviour to fit the purpose of the intervention. Cognitive interventions in disaster management should be highly considered as the key to successful mental aspects of disaster management is applied. The results of the Fault Tree Analysis approach used as the basis for the design of cognitive interventions on the human condition combined with typography and vulnerabilities that exist in the area. Thus, this study aims at analysing the cause of the low index of awareness of residents to landslides using Fault Tree Analysis as well as designing the cognitive interventions for the communities in landslide areas.

2. Research Methods

2.1. Research Area and Sample

This research was conducted in Manyaran district of Semarang city from March to June 2016. The sample in this study were 40 people aged ≥ 17 years, illiteracy and the board of neighbourhood (RT) or Hamlet (RW), and Family Welfare Programmed (PKK) cadres thereby contributing to disseminate information to the other resident. The sampling technique in this research is purposive sampling or judgmental sampling.

2.2. Research Variables

This study has four independent variables; they were Knowledge and Attitude (KA), Emergency Planning (EP), Warning System (WS), and Resources Mobilization (MR) from LIPI-UNESCO/ISDR [18]. The dependent variable in this study is the index of awareness of residents to landslides. Measurements were made using a Likert scale questionnaire which is then weighted for each parameter with the weight of each parameter of 25%. After weighting, calculation of Awareness Index can be taken into account using following formula:

$$Preparedness\ Index\ =\ \frac{Total\ Riil\ Score\ of\ Parameter}{Maximum\ Score\ of\ Parameter}x10 \quad (1)$$

From the calculation of the index, it will be obtained awareness index which then can be categorised based on the detail in Table 1.

Table 1. Category of awareness index (Sopaheluwakan, 2006)

Nr.	Index score	Category
1	80-100	Very ready
2	65-79	Ready
3	55-64	Almost ready
4	40-54	Less ready
5	0-39	Not ready

The determination of cognitive intervention was varied according to the characteristic of the residents in the each of district that obtained from the result of the survey.

2.3. Research Procedure

The study was begun with a preliminary study through field surveys by spreading the questionnaire according to the specified parameters. From the results of the questionnaire Likert scale in the form of resident awareness index, a parameter with the lowest value can also be identified. 5 point Likert scale was selected considering the diversity of the respondents' education level and to reduce charging errors generated by respondents caused by too many options. From the lowest parameter, a Fault Tree Analysis was conducted to find the cause of low index.

Along with the spread of the questionnaire, characteristics of the population in the study area was obtained, as well. The design of cognitive interventions carried out based on the combination of the Fault Tree Analysis and data characteristics of the population.

3. Result and Discussion

3.1. Population Characteristic

There were in total 40 people in Manyaran district from RW V, RW VI, and RW IX. The selection of RW and RT location was conducted based on data of landslide occurrences, the advice and recommendation of the Head of the Manyaran district as well as the result of survey research. District Disaster Awareness (KSB) of Manyaran district was less active and do not run well. In addition, most of the resident in the Manyaran district were not realised that their area is prone to landslides (based on a survey on April 2016).

Pretest phase was conducted to determine the awareness index of the residents in the Manyaran district before obtaining cognitive interventions. The result showed that the awareness index awareness residents to landslides was 32.20 and categorised as "Not Ready". Index for the parameters of knowledge and attitudes was 44.25%, parameter of self-preparedness plan amounted to 18.21%, the warning system parameter was 22.17%, and the parameters of resource mobilisation was 25.83%. The parameter of preparedness plans was a parameter with the lowest percentage. This parameter was served as the basis of Fault Tree Analysis.

3.2. Fault Tree Analysis

The information contained in the Fault Tree Analysis was obtained through interviews with residents, the district government and the KSB (District Disaster Awareness). Figure 1 showed the Fault Tree Analysis of the awareness parameter in the Manyaran district.

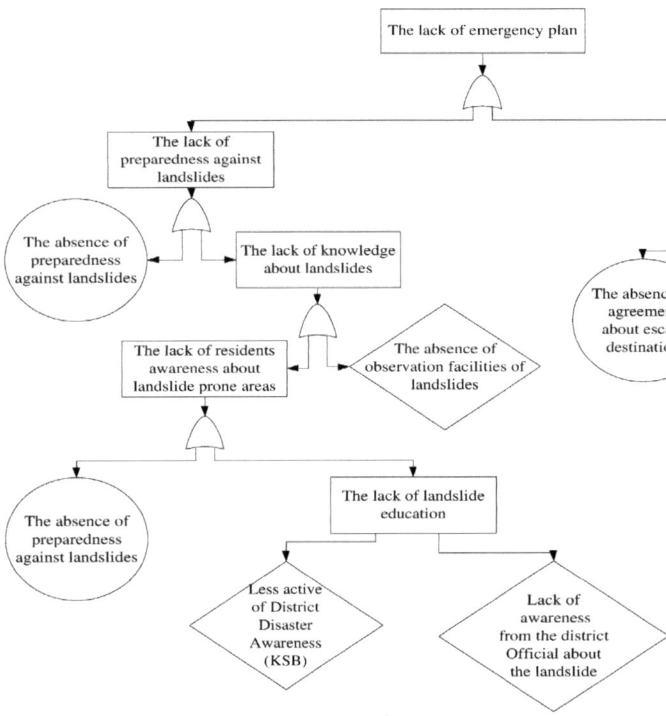

Figure 1. *Fault Tree Analysis* of awareness parameter in Manyaran district

Based on Figure 1, it can be explained that the basic event of awareness parameter was the lack individual preparation, the absence of landslide hazard map, the absence of agreement about escape destination, the absence landslide signs as well as the absence of evacuation route. That basic event was then used as the basis for cognitive interventions while did not ignore the other three parameters that expected to improve the awareness index in Manyaran district. In addition to a basic event, there was undevelopment event including the absence of observation facilities of landslides and less active District Disaster Awareness (KSB). The undevelopment event were not considered as the basis of the design of cognitive interventions because these undevelopment events were certain failure event that was not sufficiently related to the research conducted. By doing so, further cause analysis was not taken into account.

To analyse FTA qualitatively, it was found firstly the minimal cut set using Boolean algebra. The next step was to analyze the failure that led directly to the occurrence of a top event and the quantitative analysis using probability theory to get the top probability event occurred. A cut set is the set of the basic event which happens when all the basic event are occurred, and then there will be a top event. The Minimal cut set is the set of the smallest combination of basic event where if the event occurs, it will cause the top event occurs [19]. For example:

T = Top event

P = Primary Event (Basic event)

G = intermediate event
S = conditioning event
So, the study can be translated into:

T = Lack of plan for emergencies in the Manyaran district (Top event)
G1 = Lack of the preparedness against landslides
G2 = The absence of escape destination
G3 = Lack of the knowledge about landslide
G4 = Residents are not aware of the area prone to landslides
G5 = Lack of education about landslides
S1 = The absence of observation facilities for landslides
S2 = Less active of District Disaster Awareness (KSB)
S3 = Lack of awareness from the district official about the landslide
P1 = Lack of preparation against landslides
P2 = The absence of agreement about the escape destination
P3 = The absence of the signs of landslides prone pins
P4 = The absence of evacuation route
P5 = The absence of a landslide hazard map

So the Fault Tree Analysis can be drawn as seen in Figure 2.

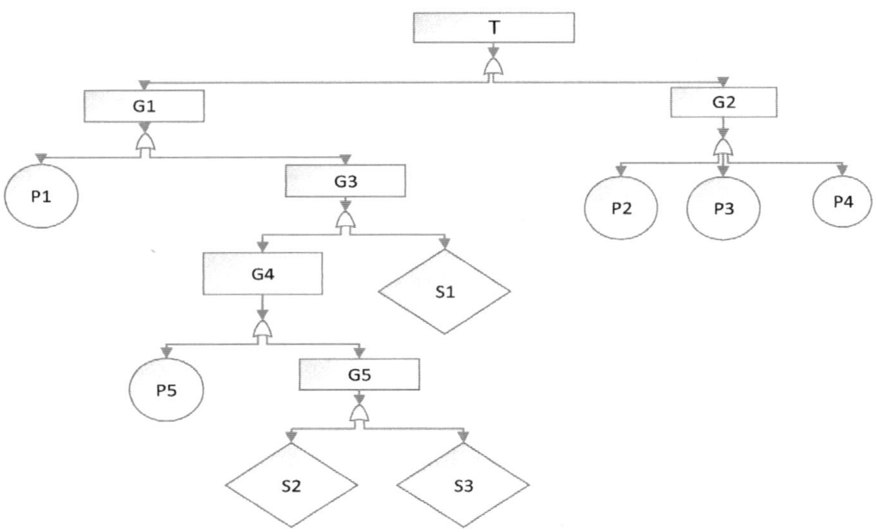

Figure 2. Fault Tree Analysis for Manyaran district with code

The Boolean equation was:

T = G1 + G2
G1 = G3 + P1
G2 = P1 + P2 + P3
G3 = G4 + S1
G4 = G5 + P5
G5 = S2 + S3

Using a top-down approach, it can be found:

T = G1 + G2 (because G1= G3 + P1 and G2 = P2 + P3 + P4)
= G3 + P1 + P2 + P3 + P4 (because G3 = G4 + S1)
= G4 + S1+ P1 + P2 + P3 + P4 (because G4 = G5 + P5)
= G5 + P5+ S1+ P1 + P2 + P3 + P4 (because G5 = S2 + S3)

= S2 + S3 + P5+ S1+ P1 + P2 + P3 + P4

So the minimal cut set were {P1}, {P2}, {P3}, {P4}, {P5}, {S1}, {S2}, {S3}. As aforementioned, the top event was the low score of emergencies plan. The results of the qualitative analysis showed that the causes of low score of plan for emergencies in Manyaran district were under these events:

- Lack of the preparedness against landslides
- The absence of a landslide hazard map
- The absence of an agreement about the escape destination
- The absence of the signs of landslides prone pins
- The absence of evacuation route
- The absence of observation facilities for landslides
- Lack of awareness from the district official about the landslide
- Less active of District Disaster Awareness (KSB).

The design of cognitive interventions was emphasised the basic event because the development event was a specified failure that was indirectly related to the research conducted. For example, observation facilities of landslides and lack of awareness about the landslides by the district official were the responsibility of local government and the Regional Disaster Management Agency (BPBD) Semarang. They have a responsibility to undertake the procurement of equipment and increase the government awareness regarding the landslide. Less active of District Disaster Awareness (KSB) is the responsibility of BPBD Semarang.

3.3. Design of Cognitive Intervention

The establishment of cognitive interventions in this study carried out the modules and videos about landslides, as well as the discussion. These instruments were selected based on the result of the survey regarding the characteristics of the population in the study area. They were indicated have a diverse educational level and rarely get the socialisation of the government regarding the landslide disaster that may occur in their area. Based on the results of the survey in April 2016, it was found that residents in Manyaran district did not know that the area was prone of the landslide. The knowledge level was inadequacy, as well. There were also an insufficient preparedness plans against landslides and the warning system when landslides occurred, while the mobilisation of resources in the form of disaster preparedness activities carried level RT/RW/district was not well occupied.

These conditions affected the awareness index of the resident, and it prominently needed an effort to improve index. The cognitive intervention was utilised video, modules and discussion media. The media for delivering the information in the form of the module was used because the residents in Manyaran district insufficiently obtained the information on landslides and environmental conditions.

3.3.1. Video

The use of video in this study was based on research conducted by Gloria [20]. Selection of video as a medium of intervention given to the Manyaran district based on research that video is the most effective media of information and it expected can draw enthusiasm of the residents. The selection of cognitive intervention media by using video is also based on the recommendations of the Regional Disaster Management Agency (BPBD) Semarang. BPDP was suggested to make a video about landslides as media dissemination and delivery of information to the residents in landslide-prone area (Survey result on May 2016). The video was included some material which is established based on the basic events of Fault Tree Analysis and a parameter of awareness index for landslide areas.

3.3.2. Module

The module was included the basic information based on the result of the basic events in Fault Tree Analysis, materials about the parameter of awareness index as well as a short comics about the importance of residents' awareness to landslides. The resident characteristic in Manyaran district was identified as a population with insufficient knowledge and information of landslide area caused by due to the lack of information on landslides from the official and District Disaster Awareness (KSB). So, the module was designed to detail to deliver understandable basic information and to improve the resident intention about landslide occurrences.

The video and module material used in the cognitive intervention consisted of four parameters as seen in Table 2.

Table 2. Material of video and module for cognitive intervention

PARAMETER	VARIABLE	INDICATOR	MATERIAL
Knowledge and attitudes	Natural events and disasters (type, source, quantity, location)	Explaining the condition of disaster-prone areas	Understanding of knowledge of regional situation
		Explaining the impact	The impact of landslides
		Explaining the signs	Signs of landslides
		Explaining the source of the landslide	The causes of landslides
	Physical vulnerability	Explains the vulnerability of communities to natural disasters	The rate of occurrence of landslides urban village
	Attitude when the landslide occurred	Describes the self-rescue measures in time of disaster	Attitude when the landslide occurred by Major Disaster Scenario
Plan for emergencies	Disaster preparedness	Preparedness against disasters	Individual Preparedness Plan *(Basic Event)*
	Evacuation plans	The availability of escape destination	Meeting point in the landslide occurrence *(Basic Event)*
		Availability of hazard maps	Landslide hazard map of the district *(Basic Event)*
		Availability of evacuation maps, route / lane	Evacuation route map *(Basic Event)*
		Availability of signs alarm	Signs prone to landslides *(Basic Event)*
Disaster warning system	Disaster warning systems that traditionally evolved/ heredity applied and/or local agreements	Warning system agreed and the level of understanding and the effectiveness of the warning system	Warning system agreed
Resource Mobilization	Resource Mobilization	Resource mobilization activities and effectiveness of conducted resource mobilization done	Explanation on resource mobilization

3.3.3. Discussion

The discussion was started with material transfer via the slide show (power point) medium. The utilisation of the slide show was considered with the aim to deliver a large scope of information and to optimise the intervention implementation. The material of the presentation was designed in the form of points, comprehensive and understandable statement. By doing so, it was expected the resident can achieve a higher knowledge about the presentation material. Besides. Animation and picture were used to attract the resident attention. The material was consisted of four parameters in awareness index; they were knowledge and attitude (KA), emergency planning (EP), warning system (WS), and resource mobilisation (RM). After the presentation, the discussion session was conducted to check the resident comprehension and to explain further the unclear material presented in the first session.

4. Conclusion and Further Study

The result of the study conducted the causes of low awareness index toward landslides in Manyaran district (index score = 32.30) were the absence of preparedness to the landslide, the absence of landslide hazard map, the absence of the agreement about escape destination, the absence of landslide signs as well as the absence of evacuation route.

The design of cognitive intervention can be different for each area depends on the characteristic of the population and the result of root cause analysis. Manyaran district was identified as an area with insufficient knowledge and information of landslide occurrences and by doing so, module, video and discussion were utilised to improve the awareness index.

The further studies should be implemented the design of cognitive intervention as well as analysed the role of the other stakeholders in disaster management, such as government/local official and non-departmental institution (for example, BPBD). Besides, an advanced study regarding collaboration design between the stakeholders should be carried out to improve the total awareness index.

Acknowledgement

The authors thank Diponegoro University who funded the project "Desain intervensi kognitif untuk meningkatkan indeks kesiapsiagaan individu di daerah rawan longsor" within "Riset Pengembangan dan Penerapan (RPP)" scheme, Contract Number: SP DIPA-042.01.2.400909/2016, December 7th, 2015, with funding resource: DIPA PNBP of Diponegoro University 2016.

References

[1] Putranto T T, Susanto N, Dwiyanto J S, Anatoly N and Rifqi A 2015 *Proc. Seminar Nasional Kebumian Ke-8 (*Graha Sabha Pramana: Jakarta) pp 87-97

[2] Hale T and Moberg C R 2005. Improving supply chain disaster preparedness: A decision process for secure site location. *International Journal of Physical Distribution and Logistic Management.* **35 (3)** 195-207

[3] Hagani A and Afshar A M 2009 Supply Chain Management in Disaster Response. *Final Project Report.* Mid-Atlantic Universities Transportation Center. Department of Civil and Environmental Engineering, University of Maryland

[4] Cozzolino A 2012 Humanitarian Logistics and Supply Chain Management. *Humanitarian Logistic,* SpringerBriefs in Businness

[5] Hong J, Jeong K, and Xie Y 2015 A Multi-Objective Approach to Planning in Emergency Logistic Network Design. *International Journal of Industrial Engineering: Theory, Applications and Practice* **22(4)** 4-12

[6] Purba J O 2014 Pembuatan Peta Zona Rawan Tanah Longsor Di Kota Semarang Dengan Melakukan Pembobotan Parameter. *Bachelor report in Geodetic Engineering Department Diponegoro University* http://eprints.undip.ac.id/42838/ [Online accessed Dec 20th, 2015]

[7] Sriyono A 2012 Identifikasi Kawasan Rawan Bencana Longsor Kecamatan Banyubiru, Kabupaten Semarang *Bachelor report in Geographic Department*, Semarang State University

[8] Susanto N and Putranto T T 2015 How to develop a human-centered design of disaster management based on the human factor toolkits?: A case study of landslides area in Semarang City. *Proc. Int. Conf on Industrial Engineering Theory, Methodology, and Application* (Yogyakarta, October 27-28th, 2015)

[9] Coelho D A 2013 *Int. J. of Human Factors and Ergonomics* **2(1)** 1-10

[10] Short S S, Nurke R V, Jensen A R, Berg B M, Goodhue C J, Barthel E R and Upperman J S 2013 *Management Strategies for Children during Natural Disaster. Natural Disaster: Prevention, Risk Factors and Management.* Raskovic, B and Mrdja, S., (eds), Nova Publishers New York

[11] Lachlan K A, Spence P R, and Lin X 2013 *Sepf-efficacy and Learning Processes Associated with the Elderly during Disasters and Crises.* Natural Disaster: Prevention, Risk Factors and Management. Raskovic, B and Mrdja, S., (eds), Nova Publishers New York.

[12] Molk P 2013 The Unintended Consequences of Government Involvement: Connecticut and Tropical Storm Irene. *Natural Disaster: Prevention, Risk Factors and Management.* Raskovic, B and Mrdja, S.,

(eds), Nova Publishers New York.
[13] Kar N 2013 *Coping with Disaster Trauma: Observations from Around the World. Natural Disaster*: Prevention, Risk Factors and Management. Raskovic, B and Mrdja, S., (eds), Nova Publishers New York
[14] Gonzales C G F 2006 Risk Management of Natural Disasters: The Example of Mexico. *Dissertation* Universitaetsverlag Karlsruhe
[15] Freeman P K, Martin L A, Linnerooth-Bayer J, Mechler R, Pflug R, and Warner K 2003 *Disaster Risk Management: national system for the comprehensive management of disaster risk and financial strategies for natural disaster reconstruction. Environment Division, Sustainable Development Department, Inter-American development Bank* (Washington, D.C. Felipe Herrera Library)
[16] Mary B 1995 *Psychological interventions: a guide to strategies* Greenwood Publishing Group
[17] Sari E 2010 *Rancangan Konseptual Intervensi Kognitif Untuk Menurunkan Tingkat Depresivitas* (Bandung: Universitas Padjajaran)
[18] Sopaheluwakan J 2006 Kajian Kesiapsiagaan Masyarakat Dalam Mengatisipasi Bencana Gempa Bumi & Tsunami.Jakarta : LIPI-Unesco/ISDR
[19] Vesely W E 1981 *Fault Tree Handbook* Washington D.C. : U.S. Nuclear Regulatory Commision.
[20] Gloria R 2013 *Efektivitas Media Penyampaian Pesan pada Kegiatan Literasi Media (Studi pada SMA Negeri 2 Bandar Lampung)* (Bandar Lampung : Universitas Lampung)

Genetic diversity of the causative agent of ice-ice disease of the seaweed *Kappaphycus alvarezii* from Karimunjawa island, Indonesia

E Syafitri[1,2*], S B Prayitno[3], and W F Ma'ruf[3] O K Radjasa[4]

[1]Doctoral of Coastal Resource Management, Diponegoro University, Semarang 50241, Indonesia
[2] Department of Fishery, Dharmawangsa University, Medan, North Sumatra, Indonesia
[3] Department of Fisheries, Diponegoro University, Semarang 50275, Indonesia
[4] Department of Marine Science, Diponegoro University, Semarang 50275, Indonesia

E-mail : esyafitri@yahoo.com

Abstract. An essential step in investigating the bacterial role in the occurrence of diseases in *Kappaphycus alvarezii* is the characterization of bacteria associated with this seaweed. A molecular characterization was conducted on the genetic diversity of the causative agents of ice-ice disease associated with *K. alvarezii* widely known as the main source of kappa carrageenan. *K. alvrezii* infected with ice-ice were collected from the Karimunjawa island, North Java Sea, Indonesia. Using Zobell 2216E marine agar medium, nine bacterial species were isolated from the infected seaweed. The molecular characterizations revealed that the isolated bacteria causing ice-ice disease were closely related to the genera of *Alteromonas, Bacillus, Pseudomonas, Pseudoalteromonas, Glaciecola, Aurantimonas*, and *Rhodococcus*. In order to identify the symptoms causative organisms, the isolated bacterial species were cultured and were evaluated for their pathogenity. Out of 9 species, only 3 isolates were able to cause the ice-ice symptoms and consisted of *Alteromonas macleodii, Pseudoalteromonas issachenkonii* and *Aurantimonas coralicida. A. macleodii* showed the highest pathogenity.

Keywords : Genetic diversity, causative agents, ice-ice, *Kappaphycus alvarezii*, Karimunjawa island Indonesia

1. Introduction

Genus *Kappaphycus* is farmed as raw material for the extraction of the phycocolloid, carrageenan, and as food which is considered as an important commercial commodity for many developing countries such as Indonesia, Malaysia, Philippines, and Tanzania [1]. The production of farmed seaweed in Indonesia steadily increased from 1,728,475 tonnes in 2007 to 9,298,475 tonnes in 2013 [2]. The significant increased in seaweed production consequently stimulated carrageenan industry that opens job creation, reducing unemployment and contributing to national economic growth.

Microorganisms have been identified as the colonizers of seaweeds, and bacteria have been regarded as the primary colonizer [3]. This bacterial seaweed association, however, often disturbed the seaweed production by the occurrence of disease so-called ice-ice disease. According to [4], ice-ice disease led to a significant decrease in seaweed production and decrease in carrageenan yield compared to the healthy crop ranging from 25 to 40 %.

Bacteria are known to be the main component of seaweed epibionts with higher diversity and had implications in the development of the disease [5]. It has been further found that non-pigmented bacterium and a yellow bacterium, *Cytophaga-Flavobacterium* complex and *Vibrio-Aeromonas* complex were identified as causative microbes of the whitening of *Kappaphycus alvarezii.* [6] reported a bacterium, *Pseudoalteromonas bacteriolytica* to be the causative agent of the red spots disease of the cultured brown seaweed species, *Laminaria japonica.* In Oshoro Bay, Hokkaido, Japan, a bacterium, *Alteromonas* sp. was identified as the symptom-causative agent of *Laminaria religiosa* [7]. Further research conducted by [8] reported that *Pseudoalteromonas issachenkonii* was a bacterium that degraded algal polysaccharides, thallus of the brown alga *Fucus evanescens.*

However, very little is known about seaweed disease and its status in Indonesia. First report on seaweed disease in Indonesia was published in early 2000, 75% of seaweed farming in Pari Island, Thousand Island was collapsed largely due to emerging diseases [9]. This phenomenon was still exists in present days. The disease was spread along with the extension of cultivation area and environmental degradation. Due to the regular outbreaks and economic loss caused by bacterial community, several strategies have been attempted to stop spreading and progressing of seaweed diseases. However, those approaches were demonstrated ineffective and there were very little success on a large spatial scale. Two alternatives approaches to identify seaweed diseases are by understanding the causative agents of particular disease and the search of anti-seaweed disease obtained from natural products [10,11].

Based on the information above, in the effort to fully understand the causative agen of *K. alvarezii* ice-ice disease in Karimunjawa island, North Java Sea, Indonesia, molecular based-approach was conducted to estimate the diversity of the causative agents and investigated the symptom causative bacteria using pathogenity test [7]. The objective of this research were to determine the diversity of the causative agents from bacterial community associated with ice-ice diseased on *K. alvarezii* and investigated the symptom causative bacteria using pathogenity test.

2. Material and Methods

2.1. Sampling site

Infected seaweed samples of *K. alvarezii* were collected from cultivation farms at Karimunjawa island, North Java Sea, Indonesia (Figure 1), in May 2014 (05°47'22.4"S - 110°26'57"E). After collection, the infected seaweed materials were placed onto sterile borosilicate bottles containing autoclaved seawater and brought in chilled condition, to the Tropical Marine Biotechnology Laboratory, Diponegoro University for bacterial isolation

2.2. Isolation of bacteria from infected seaweed

One gram of infected seaweed sample was grinded using mortar and diluted in 10 mL of sterilized seawater. The sample was vortexed to homogenize the solution. Afterwards, one mL of the solution was diluted into 9 mL of sterilized seawater to make 10^{-1} seaweed dilution. The processes were repeated until 10^{-6} seaweed dilution. The 0,1 aliquots from 10^{-1} to 10^{-7} dilution were spread onto Zobell 2216E marine agar medium. Then, they were incubated upside down at room temperature for 4x24 hours. Based on morphological features, colonies were randomly picked and purified by making streak plates [12].

2.3. Molecular Identification

Genomic DNA of each bacterial isolate from diseased seaweed was extracted by using Chelex method [13], the DNA concentrations and purity were quantified and qualified by using NanoDrop 2000 spectrophotometer. DNA extracts were amplified by PCR using universal primers 27F (5'AGAGTTTGATCMTGGCTCAG-3') and 1492 R (5'TACGGTTAACCTTGTTACGACTT-3') and temperature cycle of amplification was as follows: initial denaturation at 95^0 C for 3 min, and then successive denaturation (95^0 C for 1 min), annealing (55^0 C for 1 min), and extension (72^0 C for 1 minutes). Series of denaturation, annealing and extension were repeated until 30 cycles. A final extension was performed at 72^0 C for 7 min [14]. The PCR products were analyzed by 1 % agarose gel electrophoresis and the result was observed using UVIDoc HD5 (UVITEC Cambridge). DNA sequencing was conducted at 1st BASE DNASequencing, Malaysia. Purification of PCR products and subsequent sequencing analysis were performed according to [15]. The nearly complete 16S rDNA gene sequences were used to search the genetic profile similarity in the GenBank database with the Blast algorithm to determine relative phylogenetic positions. Sequences then were aligned by using ClustalW software and the phylogenetic analyses were performed by using MEGA 6. The phylogenetic trees were determined using the neighbour-joining method with Kimura's two-parameter. The resultant tree topology was evaluated by bootstrap analyses of the neighbour-joining method based on 1.000 resampling [15].

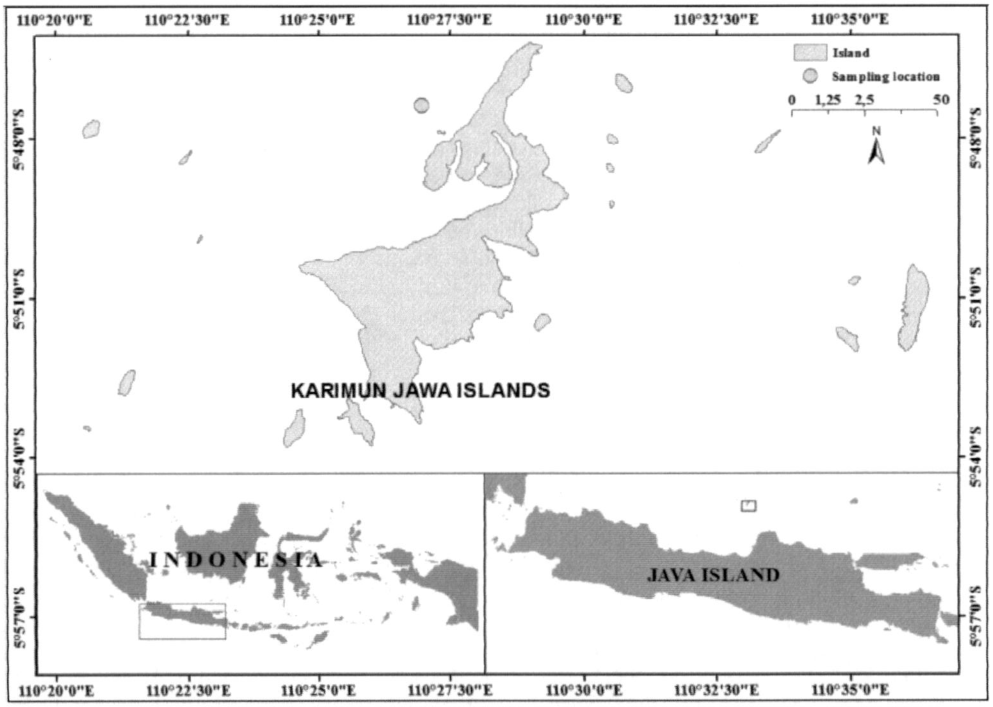

Figure 1. Sampling site for the collection of *K. alvarezii* from seaweed farm

2.4. DNA Sequences
DNA sequences of all bacterial isolates have been deposited in the DNA Database Bank of Japan (DDBJ) in the following accession numbers: LC085337-LC085345.

2.5. Pathogenity test
All strains that previously isolated from diseased seaweed were cultured in 10 ml of ZoBell 2216E broth medium at 37^0 C until cell density reached 10^8 ml for 2 days. Healthy, epiphyte-free part of main branches of *K. alvarezii* were then cut into 5 cm long pieces and distributed individually into 300 mL flask containing 200 mL autoclaved seawater and were acclimated for 10 days. The cells were then

harvested by centrifugation at 5000 rpm for 10 minutes, washed twice and re-suspended with PBS solution, and then two mL subsamples of each bacterial suspension were inoculated into 200 mL SSW in 300 mL erlenmeyer flask. A set of control experiments (without bacterial inoculum) was also carried out simultaneously. All transfers of algal materials and media were done under aseptic conditions in clean bench using pre-sterilized glass wares, forceps and blades. Flasks with samples were placed on a rotary shaker at 100 rpm for 7 days under approximately 50 μmol photons $m^{-2}s^{-1}$, 28^0 C, and 12:12 L:D cycle. Observations were conducted daily until visible sign of ice-ice by whitening of thallus was occurred up to a maximum of 1 week

3. Results and Discussion

3.1. Isolation and Molecular Identification of Associated Bacteria from Infected Seaweed

Management of healthy seaweed aquaculture and control of ice ice disease are important component in seaweed production. The problem of ice-ice disease has not only affected the seaweed farmers but also the nation as a whole. Farmings in many producing countries suffered the drastic decline in aquaculture production with negative growth due to ice-ice disease. [4], showed that ice-ice also leads to the decrease of carrageenan yield, viscosity and gel strength of infected thalli.

To support the integrated prevention of ice ice disease, information about genetic variation of bacterial pathogens and the availability of fast and accurate detection are required. Detection method that has been developed at this time is molecular technique through analysis of genomic DNA. Here, the utilization of the 16S rDNA has been used as a systematic parameter as a marker for the identification of causative agent of ice-ice disease of seaweed *Kappaphycus alvarezii* from seaweed farming in Karimunjawa islands, North Java Sea, Indonesia.

A total of 9 bacterial strains have successfully been isolated from diseased seaweed, *K. alvarezii*. Based on the analysis of 16S rDNA sequences and phylogenetic tree, all isolates were closely related with the following bacterial genera of *Alteromonas, Bacillus, Pseudomonas, Pseudoalteromonas, Aurantimonas,* and *Rhodococcus* (Table 1 and Figure 2).

Table 1. Molecular characterization of bacterial causative agents

Isolate	Length (bp)	Closest Species	Similarity (%)	Accession Number	Group
KAKJ1	1187	*Alteromonas macleodii* 107	96%	NR_037127.1	Gammaproteobacteria
KAKJ2	1258	*Bacillus oceanisediminis* H2	91%	NR_117285.1	Bacilli
KAKJ3	1237	*Pseudomonas stutzeri* ATCC 17588	95%	NR_041715.1	Gammaproteobacteria
KAKJ4	1107	*Pseudoalteromonas issachenkonii* KMM 3549	95%	NR_025139.1	Gammaproteobacteria
KAKJ5	1060	*Bacillus hunanensis* JSM 081003	97%	NR_108948.1	Bacilli
KAKJ6	1239	*Bacillus megaterium* ATCC 14581	97%	NR_116873.1	Bacilli
KAKJ7	916	*Alteromonas marina* SW-47	96%	NR_025260.1	Gammaproteobacteria
KAKJ8	1192	*Aurantimonas coralicida* WP1	94%	NR_042319.1	Alphaproteobacteria
KAKJ9	1311	*Rhodococcus rhodochrous* DSM 43241	95%	NR_116689.1	Actinobacteria

It is predicted that one of the factors associated with high ice ice occurrence in the field is the high incidence of epiphytes. Moreover, [16] found that the combined effect of stress and biotic agents, such as opportunistic bacteria are primary factors of the ice-ice disease. Compared to previous research, [5] found that bacterial genera of *Cytophage, Aeromonas, Pseudomonas, Alteromonas, Flavobacteriummm., Arthrobacter* and *Vibrio* have been associated with the diseased seaweed *K. alvarezii*. This study showed that molecular identification of bacteria isolates have not been reported from previous study. The difference leads to an assumption that the bacteria have site specific properties in expressing their pathogenity [1]. The 16S rRNA gene sequencing assist resolved the exact taxonomic position of seaweed bacteria, and provided more detail information on their phylogenetic position among their closest relatives (Table. 1 and Figure. 2).

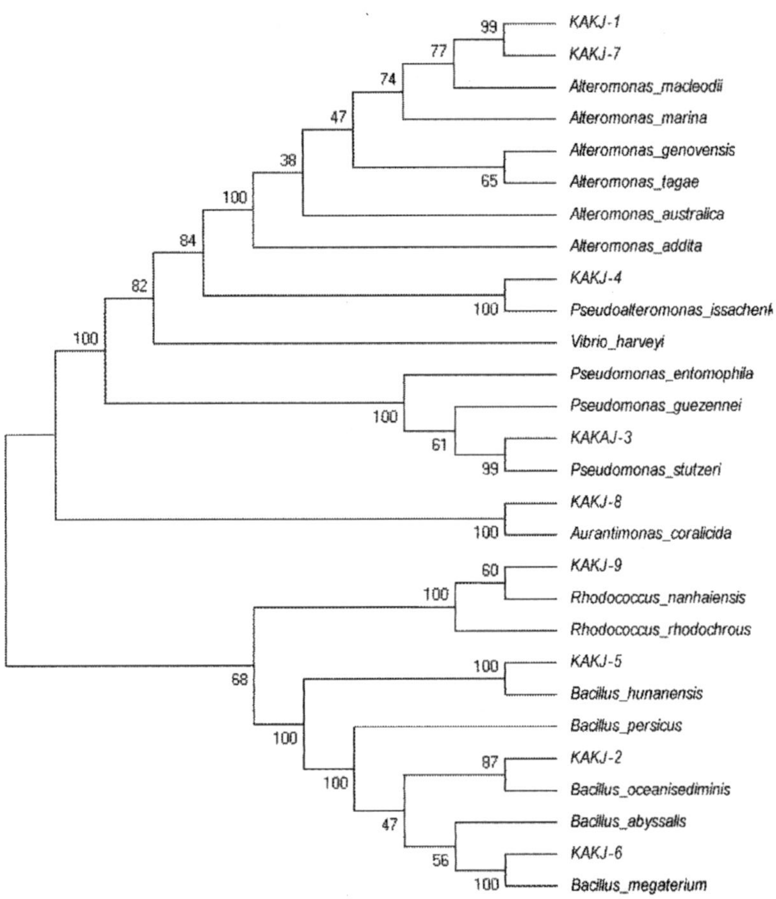

Figure 2. Phylogenetic tree of the causative agent of ice-ice disease based on the 16S rDNA sequence.

The similarity analysis showed that the isolates were between 91% and 97%. Among the isolates, strain KAKJ 5 (NR_108948.1) was similar to *Bacillus hunanensis* JSM 081003, while KAKJ 6 (NR_116873.1) similar to *Bacillus megaterium* ATCC 14581 had the higher similiraty (97%) and the lowest similarity strain was found in KAKJ2, which had 91% similarity to *Basillus oseanisediminis* H2. To estimate genetic affiliation of the pathogenic bacteria cause ice-ice disease on seaweed, a neighbor-joining tree including identified isolates and representatives marine microorganisms was constructed. A phylogenetic analysis of the 16S rDNA data for selected strains belonging to the group of the Gammaproteobacteria, Bacilli, Alphaproteobacteria, and Actinobacteria produced the dendogram shown in Figure 2. This comparison was made to determine the species to which the nine selected isolates are most closely related and to determine how closely the four taxa are related to each other.

3.2. Pathogenity test

Among the nine strains screened for pathogenity ability on healthy thallus seaweed, strains designated as KAKJ1, KAKJ4 and KAKJ8 indicated initial pathogenic activity, the thallus were distinct and showed obvious thallus deterioration as compared to the control and other bacterial species (Figure 3). The thallus seaweed inoculated with *A. macleodii* 107 (KAKJ1) showed ice-ice symptoms after 1 day post infection (1 in score), where as the thallus seaweed inoculated with *P. issachenkonii* KMM 3549 (KAKJ4) and *A. coralicida* WP1 (KAKJ8) showed symptoms after 3 days post infection (1 in score), respectively (Figure 3). After 7 days post incubation, thalllus inoculated with *A. macleodii* 107 (KAKJ1) had white margins of 8,4 mm in length (3,67 in score), higher compared to other isolates (Figure 3). Bleaching and fragmentation intensified as incubation time progressed. The experimental controls remained without any symptoms for the entire duration of the experiment. These results revealed that the thallus seaweed with *A. Macleodii* 107 showed symptoms earlier as compared to other remaining pathogenic bacteria.

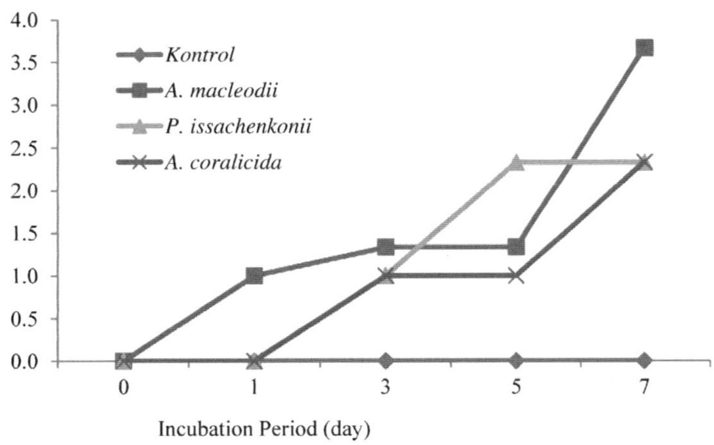

Figure 3. The average indicator of ice-ice disease symptoms on seaweed *K. alvarezii* in score during the 7 days of infection.

The pathogenity test is the main criterion for the identification of bacteria suspected of being the aetiological agents of a plant disease. In the present pathogenity studies, three bacterial isolates of this study were found to be pathogenic to the seaweed *K. alvarezii* as compared by inoculation method, where *A. macleodii* was found to be the worst, followed by *P. issachenkonii* KMM 3549 and *A. Coralicida* WP1. Hence, based on pathogenity test, *A. macleodii* 107 qualifies itself as the symptom-causing bacterium. This genus was defined as comprising heterotrophic, catalase and oxidase-positive, gram-negative rods, motile by a single polar flagellum, which produce buds, grows at 15-37 ^{0}C and pH 6-9 and require NaCl for growth and positive for hydrolysis of gelatin and strach, can utilize cellobiose, D-fructose, D-glucose, D-mannose, L-lactose and maltose [17].

The ability of *A. macleodii* 107 to produce algal component-degrading enzymes and utilize various algal components may possibly be contributory factors to the development of ice-ice disease in cultivated red seaweeds. Carrageenan and cellulose is a polysaccharide cell wall component of *Kappaphycus* species and constitutes the bulk of the cell's interstitial matrices [18]. According to [4], ice-ice disease led to a significant decrease in seaweed production and decrease in carrageenan yield and average molecular weight of carrageenan extract compared to the healthy crop ranging from 25 to 40 %. This depolymerization was attributed to the carrageenolytic activity produced and secreted by occasional bacteria. Cellulase and carrageenase isolated from bacteria are able to release epidermal and medullary protoplasm in *K. alvarezii* [19]. According to [16], these hydrolytic enzymes might be a factor in the whitening/bleaching observed in seaweed thalli during ice-ice infection. During infection the thalli at high concentration and start to utilize carrageenan, penetrate into the medulla of the seaweed thallus could cause epidermal degradation and destruction of the cell's pigment-containing plastids, resulting in the initial bleaching of the infected part [16,20].

The development of ice-ice disease in *K. alvarezii* depends on several factors to which the seaweed was exposed. [16] found that the combined effect of stress and biotic agents, such as opportunistic pathogenic bacteria, are primary factors of the ice-ice disease. The infection of the seaweed by these pathogenic bacteria may depends on their ability to successfully attach to the seaweed surface. The attachment of *A. macleodii* 107 to the seaweed thalli was required for the bacterium to promote the ice-ice disease and it is enhanced by the flagellum of the microorganism. Previous studies on the behavior of motile and non motile bacteria of *Pseudomonas fluorescens* showed that motile strains colonized the substrate more successfully than non-motile ones, due to their active flagella [21]. This may explain the possible success of higher number of *A. macleodii* 107 colonised onto the seaweed surface and start hydrolytic activity inducing the symptomatic of ice-ice disease.

Another factor shown to affect the attachment of *A. macleodii* 107 was water movement. [16] observed that continuous stirring during cultivation diminishes bacterial fixation. In the present study, water motion (continuous stirring) was not introduced during disease induction, when stirring was not used, some of the inoculated *A. macleodii* 107 cells were able to colonise the seaweed surface.

4. Conclusion

Based on the findings presented in this study, it can be concluded that among isolates which mainly appeared on ice-ice branches, three strains, designated as *A. macleodii* 107, *P. issachenkonii* KMM 3549 and *A. coralicida* WP1 were found to be pathogenic to the seaweed *K. alvarezii* as compared by inoculation period, where *A. macleodii* 107 was found to be the worst.

References

[1] Vairappan C S, Chung C S, Hurtado A Q, Soya F E, Lhonneur G B and Critchley A 2008 Distribution and symptoms of epiphyte infection in major carrageenophyte-producing farms *J. Appl. Phycol.* **20** 477–83

[2] Zamroni A and Yamao M 2012 An assessment of farm-to-market link of Indonesian dried seaweeds: Contribution of middlemen toward sustainable livelihood of small-scale fishermen in Laikang Bay *Afr. J. Agric. Res.* **7** 1709–18

[3] Lachnit T, Blümel M, Imhoff J f and Wahl M 2009 Specific epibacterial communities on macroalgae: phylogeny matters more than habitat *Aquat. Biol.* **5** 181–6

[4] Mendoza W G, Montano N E, Ganzon-Fortes E T and Villanueva R D 2002 Chemical and gelling profile of ice-ice Kappaphycus striatum (Schmitz) Doty Gigartinales, Rhodophyta *J. Appl. Phycol.* **14** 409–18

[5] Largo D B, Fukami K and Nishijima T 1995 Occasional pathogenic bacteria promoting ice-ice disease in the carrageenan-producing red algae Kappaphycus alvarezii and Eucheuma denticulatum (Solieriaceae, Gigartinales, Rhodophyta) *J. Appl. Phycol.* **7** 545–554

[6] Sawabe T, Makino H, Tatsumi M, Nakano K, Tajima K, Iqbal M M, Yumoto I, Ezura Y and Christen R 1998 Pseudoalteromonas bacteriolytica sp. nov., a marine bacterium that is the causative agent of red spot disease of Laminaria japonica *Int. J. Syst. Bacteriol.* **48** 769–74
[7] Vairappan C S, Suzuki M, Motomura T and Ichimura T 2001 Pathogenic bacteria associated with lesions and thallus bleaching symptoms in the Japanese kelp Laminaria religiosa Miyabe (Laminariales, Phaeophyceae) *Hydrobiologia* **445** 183–191
[8] Ivanova E P, Sawabe T, Alexeeva Y V, Lysenko A M, Gorshkova N M, Hayashi K, Zukova N V, Christen R and Mikhailov V V 2002 Pseudoalteromonas issachenkonii sp. nov., a bacterium that degrades the thallus of the brown alga Fucus evanescens. *Int. J. Syst. Evol. Microbiol.* **52** 229–234
[9] Darmayati Y 2010 Pathogenic bacteria caused ice-ice disease on seaweed Kappaphycus alvarezii (Doty, 1986) (Eucheuma cotonii) *Sixth Annual Scientific Meeting ISOI 2009* ed B Nababan et al (Jakarta: Ikatan Sarjana Oseanologi Indonesia) p 553
[10] Egan S, Fernandes N D, Kumar V, Gardiner M and Thomas T 2014 Bacterial pathogens, virulence mechanism and host defence in marine macroalgae: Bacterial pathogens of macroalgae *Environ. Microbiol.* **16** 925–38
[11] Sabdono A, Sawonua P H, Kartika A G D, Amelia J M and Radjasa O K 2015 Coral Diseases in Panjang Island, Java Sea: Diversity of Anti–Pathogenic Bacterial Coral Symbionts *Procedia Chem.* **14** 15–21
[12] Radjasa O K, Martens T, Grossart H-P G, Brinkhoff T, Sabdono A and Simon M 2007 Antagonistic Activity of a Marine Bacterium Pseudoalteromonas luteoviolacea TAB4.2 Associated with Coral Acropora sp. *J. Biol. Sci.* **7** 239–46
[13] Walsh P S, Metzger D A and Higuchi R 2013 Chelex 100 as a medium for simple extraction of DNA for PCR-based typing from forensic material *Biotechniques* **54** 134–139
[14] Lee Y K, Jung H J and Lee H K 2006 Marine bacteria associated with the Korean brown alga, Undaria pinnatifida *J. Microbiol.* **44** 694–8
[15] Susilowati R, Sabdono A and Widowati I 2015 Isolation and Characterization of Bacteria Associated with Brown Algae Sargassum spp. from Panjang Island and their Antibacterial Activities *Procedia Environ. Sci.* **23** 240–6
[16] Largo D B, Fukami K and Nishijima T 1999 Time-dependent attachment mechanism of bacterial pathogen during ice-ice infection in Kappaphycus alvarezii (Gigartinales, Rhodophyta) *Sixteenth International Seaweed Symposium* 1 vol 11, ed J M Kain (Jones), M T Brown and M Lahaye (Springer) pp 643–650
[17] Chen Y-G, Xiao H-D, Tang S-K, Zhang Y-Q, Borrathybay E, Cui X-L, Li W-J and Liu Y-Q 2009 Alteromonas halophila sp. nov., a new moderately halophilic bacterium isolated from a sea anemone *Antonie Van Leeuwenhoek* **96** 259–66
[18] Lechat H, Amat M, Mazoyer J, Gallant D J, Buléon A and Lahaye M 1997 Cell wall composition of the carrageenophyte Kappaphycus alvarezii (Gigartinales, Rhodophyta) partitioned by wet sieving *J. Appl. Phycol.* **9** 565–572
[19] Zablackis E, West J A, Liao M-L and Bacic A 1993 Reproductive Biology and Polysaccharide Chemistry of the Red Alga Catenella (Caulacanthaceae, Gigartinales) *Bot. Mar.* **36** 195–202
[20] Solis M J L, Draeger S and dela Cruz T E E 2010 Marine-derived fungi from Kappaphycus alvarezii and K. striatum as potential causative agents of ice-ice disease in farmed seaweeds *Bot. Mar.* **53** 587–94
[21] Korber D R, Lawrence J R and Caldwell D E 1994 Effect of motility on surface colonization and reproductive success of Pseudomonas fluorescens in dual-dilution continuous culture and batch culture systems *Appl. Environ. Microbiol.* **60** 1421–1429

Capacity Building Resource Management of Coastal Areas To Improve the Local Economic Based by Cross-Cutting Partnerships: Case Study on Panjang Beach Bengkulu City

Titi Darmi
Politics and Social Science Department
Muhammadiyah University Bengkulu, Indonesia
Jl. Bali Bengkulu City, +06285234332266, +062736-22765, 38119
titi.harmadi@gmail.com

Abstract. The western part of Bengkulu city is a coastal region approximately 525 km, along the coast lies the tourism object which become the flagship of Bengkulu province. Coastal tourism is made as the pre-eminent is Panjang beach. Management of two beaches has not been managed optimally, so that is not yet on the economy have implications for communities, the people majority are fishermen, the highest poverty levels of Bengkulu city. The purpose of this research is to examine the development capacity of resource management and coastal areas of Panjang beach should be able to increase income community along the coast. Research method is using qualitative research types with approach case studies, which can investigate and phenomena identify that occur, and explain how and why about the income community arround the beaches, using the instrumental case study. Determination of samples is done based on the purposes and research problems, primary data sources are collected by in-depth interviews, FGD, and secondary data source are collected by observation and documentation. Data relating to be interpreted set forth in proposition because the result of the organized with matching pattern competition. The study results indicate there has been no integrated and suitability management between province and the city; stakeholders participation in community empowerment on the coast has been optimal; leadership commitment related to enhance institutional capacity is too weak, and there has not been a policy of managing the Panjang beach Bengkulu based on information technology.

Keywords: Capacity building the coastal region, Stakeholder, Participation, Public Management

1. Introduction

Indonesian as a maritime state whose has broadly ocean. It can be understood by Indonesia President Mr. Jokowi so that he will decides into his Nawacita program aimed for becoming Indonesia as World Maritime Axis abbreviated (PMD). To reach successful of PMD program, Indonesia needs management policy effectively and efficiently, in order to these regions potential able to advantageous for societies prosperity. Indonesia perfectly on the front line position, which able to influence and formulate strategies of management development in coastal area around overseas. Indonesia has shoreline source and Economy Zone Exclusively abbreviated as (ZEE) as wide as 5,8 million, that means Indonesia ia important internationally[1].

In order to, all stakeholders able to develop their own capacities, organization capacity and system capacity [2] while the shoreline development able to place, able to increase its society economy life, stakeholders should have ability capacity and skill in shoreline management and able to integrate so does capitalizing all of potentiates. Developing individual capacity is a way of establishing and strengthening powers of organization management shoreline[3]. In developing shoreline establishment, it needs collaboration between all stakeholders so that able to increase authority capacities of local management into shoreline development[4]. Authority party can take partnership cooperation in each sector, especially for private partner.

The first thing that should be done into shoreline resource management is setting down human resources as priority aiming for managing a shoreline optimally[5]. In managing a shoreline is done by bottom-up approach and executing rearranges regulation, by involving a local government official through determining society aspirations, so that they will take a role in deciding development of their own environment[6]. The tourism development implicates to the society aspirations, economy, social and culture[7], it contributes to the local society life quality[8]and[9]. The shoreline development to be born a tourism area will produce economy impact for region that is visited. It will influence business sector and creating job opportunities for societies around, for example: tour guide, waiter, bell boy, etc.

A shoreline arranging to be born as tourism area will impact on the facility building and infrastructures well, for example: building new tourism destination, mall, restaurant, hotel, street, etc, this is indirectly will influence economy life for society around so does motivating people for creating business opportunity, for example: local café known as *warteg*, tube rent, bicycle rent, handicrafts, etc. This shoreline management can impact into society economy life, it should be managed maximally by Co-management model by management together assume continuity to solve existence problem, with partnership approach[10].

Because of the complexity of Indonesian government systems, there is regulation that imbalance, individual sector between one dignitary to others that represent relating institution, it impact on the shoreline management of Bengkulu city (Panjang beach) doesn't be managed optimally[11] or neglected because Panjang beach is an icon of Bengkulu city tourism[12] This case causes disappointing stated by Ministry of State Apparatus and Bureaucracy Reform of Indonesian Republic abbreviated as MENPAN RB. This will influence Panjang beach management not being optimal because there is an inconsistence authority between Bengkulu province and city into Panjang beach management[13].

Panjang beach potency is very good to be icon of tourism mainstay and stimulating economy growth in Bengkulu province. It locates in the Bengkulu city area, confronting Indian Ocean. While, the poverty rates in Bengkulu city ranked into four from ten district or city in Bengkulu province, the poverty rates in 2013 reached 21,51 %, that mean it had high rates nationally in position 11,47% [14]. The literature above had been shown what link real happen and how did the policy of Panjang beach shoreline as tourism area could be managed by involving various sectors (government, private and societies) so that it could increase society economy as long as this area. This research also contributes into local policy of shoreline study especially for shoreline tourism development by involving every sector so that they will take an action according to their role.

2. Method

The research method uses qualitative approach by case study approach, that able to investigate and identify phenomenon held, and explaining how and why, *instrumental case study* is used. The informant determinations are done bases on goal, research problem and determining main stakeholder, stakeholder who involve into tourism management in Panjang beach. These informants are classified into three components as follows (1) government component (Tourism Agency, Marine and Fisheries Agency, Department of Public Work, Small-Medium Enterprise Agency); (2) privates' components (Travel Agency, Hotel Owner as long as around Panjang beach, Restaurant owner, Fisheries Owner); (3) societies components (Public Interest Group known as LSM, surrounding societies along this beach, Community selling along the coast, fishing communities).

The secondary and primary data sources, the secondary data that relate to tourism object of Panjang beach has been published both of internet or tourism Agency so are rules that relevant with this research topic. The primary data collecting is done such as; (1) researcher as main source in this

IOP Conf. Series: Earth and Environmental Science **55** (2017) 012045 doi:10.1088/1755-1315/55/1/012045

research do interviewing with three sectors that involve into shoreline management of Panjang beach; (2) accommodating society aspiration along this shoreline through deep interviewing; (3) recognizing physical condition, social economy along this beach, data that is taken includes; geographical area data, communities data, economy data, facilities and infrastructures readiness data along this beach, public access data, relating data organization, authorizing data or land utilizing along this beach.

Then the valid data is interpreted and arranged into cause effect propositions that are arranged by matching competition pattern, comparing pattern bases on empirical data with pattern predicted, if this pattern has similar result then it will strengthen validity. Pattern matching of this research is "the development of shoreline management capacity that is involving each sector aiming for increasing community economy life along this area". Then identifying potentiate, problem, opportunity, and threat for arranged development strategy of shoreline management capacity in Panjang beach by applying SWOT analysis, the strategy is arranged to maximize potency, minimize problem, by opportunity utilizing and avoiding threat of shoreline management.

3. Existing of Shoreline Area in Bengkulu

Bengkulu city is capital city of Bengkulu Province geographically located on the 10 20' 14" 10 20' 22" east longitudes and 3 34' – 3 59 south latitude. This city has area as wide as 151,7 Km^2. Panjang beach has other name called Puteri Gading Cempaka beach and has 3 tourism destinations such as Jakat, Tapak Padri dan Pasir Putih beaches, they have area as wide as ± 178,50 Ha, Panjang beach belonging to three sub-district area such as Ratu Agung, Ratu Samban and Teluk Segara sub districts, while Panjang beach belonging to three districts such as Sumur Meleleh, Pasar Pantai Malabro and Lempuing districts. It has as wide as 7 km by 50 meter of street wide, the centre distance from Bengkulu city is about ± 2 Km, consist of ± 500 meter of ebb flow and straight lines, when the ebb flow water able to reach wider because it doesn't has coral. It has strategic position, has a white sand uniquely there is no palm trees but it is only proliferated by casuarinas and pine trees, this place also has various facilities, inn, restaurant, café, mall, children playground and athletics area, Bengkulu communities usually run nothing shoes, it is consider as therapy for accelerating blood circulation. (Tourism Agency Profile of Bengkulu City, interviewing with Beach management subsection chief.

Figure 1. Existing Panjang Beach

In the management of this area based on the Government Regulation number 50 in 2011 about National Tourism Development master plan, specifying the length of the Beach and surrounding area as a National tourism development. Regulatory Area of Bengkulu province number 02 in 2012 about the Spatial Plan of Bengkulu province emphasizing Panjang Beach was designated as a strategic area. Long Beach coastal development reflected on the Mayor's Regulation number 14 in 2012 about the Spatial Plan of Bengkulu city in 2012 – 2032 made Panjang beach as a strategic area of nature tourism. The above regulation is already strengthening the perpetrator can execute policy management optimally, doing a sequence of planning activities, utilization, control and evaluation. The Mayor of Bengkulu commissioned the Tourism management of Panjang Beach based on the decision of the Mayor of Bengkulu Number 2008of the designation of the Tourism and Culture Department of Bengkulu city as in charge of the Arrangements, management and maintenance of Panjang Beach Tourism Area, the task given them are doing cross-sector coordination in order to perform the tasks settings, Setup, management and maintenance of the coastal area of long beach.

Department of tourism and culture has to be partnership with institutional or service agencies involved for land use of the coastal area including Panjang Beach (1) public works of providing infrastructure services in the form of making roads, public toilet; (2) the marine and Fisheries Agency, foster, help facilitate Community fishermen in the coastal area of Panjang Beach; (3) Office of small and medium enterprises, coordinating and facilitating sellers (PKL) and small medium enterprises in Panjang Beach; (4) Besides Tourism partners with private parties who use the land around hotel, restaurant and cafe. For the Setup, management and development of coastal areas of long Beach based on master plan development of Long Beach Attractions and contained in the draft of the construction Term Mengah abbreviated as (RPJM) for 2005-2010 development focus include recreational Beach, tourism, urban tourism, water tourism, ecotourism tours and tours of the harbor.

The condition of the population of Panjang Beach in the coastal region of three districts include Ratu Samban, Ratu Agung and Teluk Segara districts, communities had a population of 291,540 inhabitants[15]. The number of individuals or the people who live in the coastal area of Panjang beach, taken from the welfare status of the lowest 30% in very district in Bengkulu city. To the age of 5 years until the age of 18 years who do not continue school, in Ratu Agung district individuals totaled 302 inhabitants, as many as 152 in souls and in society that do not Immediately Teluk Segara district continue school as much as 211 inhabitants[16]. This means that there are still many people in a coastline in the age which should be a time of school, but not "school. Based on interviews with fishing communities on the coast cause they don't continue the school due to economic factors and low motivation for children to learn.

Portrait Head of family households with women's in coastline areas in three districts, subdistricts Ratu Samban family female head count as much as 137 families, Ratu Agung district has 253 family and in Teluk Segara Immediately as much as 204 family. The kind of work women as heads of households nearly 60% work in commercial, petty trade sekor as street vendor, and trading with warteg offers food that is sold from one to the other visitors to visitors along the coast, other 40% with type of work in the field of fishery, farming, breeding and services (being a washerwoman in other households). In the coastal area of Panjang Beach, there is still a lot of amount of individual categories of productive age (15 years – 60 years) but does not have a job or are waiting for jobs, Ratu Agung District individuals working as much as 3.580 life wile who does not work as many as 5,395 Ratu Samban population who worked as much as 1,819 individuals and not working as much as individual 2,778, Teluk Segara District community who worked a number of 2,515 people and not working as much as 3,423 people. Interview with Sumur Meleleh Head factors of high society which does not work due to the low capacity of the human resource community, many citizens still unconfirmed at the age of 5 years old to 18 years old are not coastal regions in school. Whereas, the individuals who work or who are waiting for a job at the age of 18 up to 60 years old with the status of well-being of a lowest in every district as many as 30% of Bengkulu city, with working type such as catch fish, industrial management, trading, services, work in hotels or restaurants and others as in table 1. Trading activity was the highest type of work undertaken by the community, Ratu Samban district recorded 897 people, Ratu Agung district as much as 837 people and Teluk Segara district as much as 854, while the kind of work services include educational services, health care, civic and governmental fields with the second highest order of interest by the public. While this type of work in the field of buildings, gas and electricity, in the highest Ratu Agung with a population of 1,011 inhabitants. According to an interview with high society types of work in the field of development, electricity and gas for all coastal areas of long beach in the last 3 years much of the development process, for example the construction of hotels, cafes, restaurants and the mall. Unconfirmed in Long Beach have been many standing hotel, cafe, and restaurant. That is, of some type of work performed, the types of jobs dominated trade, services and buildings.

Table 1. Jobs Types of the Welfare Status of An Individual With 30% the lowest in 3 Districts

Job Types	Ratu Samban District	Ratu Agung District	Teluk Segara District
Fisheries	53	7	245
Industrial Management	56	22	26
Trading	897	837	854
Hotel and Restaurant	71	23	19
Mining	5	23	15
Service	827	277	753
Agriculture and Horticulture	21	199	7
Animal Husbandry	2	4	2
Electricity and Gas and Building	41	1.011	335
Aquaculture	2	6	5
Financial and Insurance	3	8	18
Information and Communication	9	12	12
Transportation and Warehousing	103	257	109
Others	188	156	112

Source: [16]

The level of well-being of the community by 30% the lowest in 3 coastline sub regions there are still 172 households that do not have toilet facilities, while individuals who do not have a place end of the stool as much 1,626 inhabitants. This means that the level of feasibility of housing inhabited celebrated coast very anxious and yet livable. Residential housing coupled with conditions there is no flow of electric lighting by as much as 55 rumah[16]. These data indicate that there are still citizens have a place to live that is not yet feasible, of impropriety this residence contributes to the poverty level of the coastal region of Panjang beach. The results of the interviews with officials of the society other than the condition of the sub factors of poverty, because the culture of the community in coastal areas do defecate into the beach.

On the Panjang coast there is a area named Jakat beaches, have potential resource diversity. With the condition of the beach ramps, small as well as having wavy beach sand white there is a wealth of fish ecosystems, each morning or afternoon fishing boats going and returning with fish catching results carry 'see Figure 3'. Results catch fish other than selling fresh fish most citizens manage fish with salted fish caramengelola and sale into the beach as in 'Figure 4'. The Government is already providing support at the fishermen in the coastal area of Panjang Beach for example provide assistance capture equipment, machines, nets and other group gatherings and already have fishermen and has a cooperaton named Bina Masyarakat Pesisir to strengthen the organization of the fishermen in the coastal area of Panjang Beach. Every day especially the holidays many residents bathe and play into the coast see 'Figure 5'. There are a few private parties rent facility with giving the tire in order to swim in the ocean is more fun. The above potential have not been made by the community opportunity about the beach, should this potential impact of locals to open businesses around Jakat coastal areas so as to increase income family.

Figure 2. Fisheries Activity

Figure 3. Sales of Salted Fish in the Coastal Area of Panjang Beach

Figure 4. The Activity of the Public Bathing and Playing on the Beach on the Afternoon

4. Results and Discussions

4.1 Coastal Area Management Capacity Development of Panjang Beach through Cross-Sector Partnerships

The publication of the Law Number 23 in 2014 about local government describes the Authority Delegation Central Government to local governments into coastal areas management. Based on government regulation Number 50 in 2011 this coastal area designated as a National Tourism development area. This regulation provides a great opportunity for the local Government of Bengkulu city province and Bengkulu together involving the public, private and cross-partnership in the coastal areas management. [17] Public participation and the private will have an impact both in the management of coastal areas, the community included determining management policies one of the strategies in order to make the management of the coastal region in accordance with the needs of local residents, thus the public will feel to have and maintain and preserve the environment, e.g. the community keep clean by removing litter in its place. Management of coastal areas with cross-sector collaboration and determine the plans, standards, indicators, monitoring the use of new management approaches, followed by action, evaluate whether is in compliance with what is planned, will impact significantly on the development of coastal areas.

Panjang Beach coastal management is the responsibility of the Department of tourism and culture of Bengkulu city by carrying out coordination and cooperation to other parties that use the resources of the coastal region[18]. The concepts of management of coastal regions are conducted by the collaboration of the Government, the public and private sector. The planning and provision of infrastructure is the responsibility of the provincial government and the City Government, besides serving to bring in tourists and oversee the performance of Panjang Beach. While the surrounding communities authorized to manage parking lot, and take advantage of the area around the beach to sell (street vendors). The involvement of private parties provides a means of hospitality, dining, children's playground, outdoor swimming pool. Conforming local availability of infrastructure in the coastal area of long Beach was already pretty good compared to 8 years ago, it is now urgent that the repair was done Setup merchants of times five (street vendors), so unsightly, such as the street vendors are provided a decent facility, placed at one point and are styled properly in order not to impressed slums. In addition, problems encountered is littering, the number of rubbish along this Beach impressed not clean.

On the other site of the coastal areas is no society which daily activities of fishing, tourism and Culture has yet to do the synergy with the Department of marine and Fisheries in Bengkulu City or province, as well as work program between the Department of marine and Fisheries Department and the marine City and provincial fisheries often make the development program did not fit the needs of fishing communities, based on the observation that there are several buildings into the beach allowed faded and not enabled Unfortunately, because of the State budget be worth it – nothing. According to residents, the Department of marine and fisheries of the province nor the city either in making coastal area development programme length, does not capture the aspirations of the community. [19]Yet its optimal management of the coastal areas of long beach caused no synergy and coordination, there are still sectors of ego among the Regent's authority from both the province and the city as well as from the Service Department-coupled in the management of coastal areas of Panjang beach.

4.2 The Potential Problems, Opportunities and Threats of Coastal Area Management Capacity Development of Panjang Beach.

Based on the existing condition of the coastal region of Panjang beach, it can be known to potential, problems, opportunities, and threats. The potential of existing conditions of the coastal area is supports capacity building for coastal areas through a cross-sector partnerships. While the problem is the weakness of coastline area became obstacles for capacity development of the coastal areas with cross-sector partnerships. Opportunities is the State or condition of the coastal area that is not part of the activity or the society activity as long as Panjang beach that could be developed. Although it does not include activities that will be developed, these conditions can support the economy of communities in the coastal areas. Next is the threat, which is also a condition that can inhibit the development of the capacity of coastal areas of long beach. The potential, problems, opportunities, and threats of a long coastal area development can be seen in Table 2 below:

Table 3. SWOT Matrix Development Capacity of the Panjang Beach Coastline area's of Bengkulu City

	Strength		**Weakness**
1	Panjang Beach entry requirement of a good beach, besides having white sand, overgrown pine-tree, has a length about 7 km	1	The level of poverty of people in the coastal area are still high, the human resource capacity is still low
2	Already equipped infrastructure, roads, sports venues, malls, hotels, restaurants, public toilets, a children playground area	2	Yet there is a synergy and guidance of management authorities and stakeholer involved Panjang beach management
3	Its strategic location close to the city centre	3	The cleanliness of the beaches have not been managed optimally Arrangement of street vendors haven't done well
	Oppotunity		**Threat**
1	Panjang Beach has been known in the neighboring province	1	Security-prone of Panjang Beach visitors
2	Become a local mainstay attractions of Bengkulu city	2	The low tourist visits from abroad and outside the province
3	There are some location in Panjang beach are settlement	3	Trash littered the roadside as well as both coasts
		4	There is no public transport in the region

4.3 Management of Capacity Development Strategy for Panjang Beach Coastal Areas through Cross-Sector Partnerships

See the real conditions of the coastal areas of Panjang Beach found potential problems, opportunities, and threats in managing coastal areas of Panjang Beach. With the SWOT matrix that has been analyzed can determine the strategy using the potential and capitalize on opportunities (S-O) to achieve the management of coastal areas that will contribute to the income of the local people. So too with the discovery of weaknesses and capitalize on opportunities (W-O) will create the strategy so that the purpose of the development of the coastal area can be achieved. The SWOT analysis has identified power, this power can overcome the threats (S-T) in the management of coastal areas of long beach, with weaknesses/problems encountered can be mengantispasi the threat of (W-T). Based on the above strategy formulation analysis can determine the strategies in the development of the capacity of coastal areas of Panjang beach can be formulated as follows:

4.3.1 Cooperation between Regions, the Synergy between the Province and the City

In the development of the resource capacity of the coastline of things that ought to be taken into account is how the region can manage and make use of existing resources. In Law Number 23 of 2014 about local government, cooperation is required in administering the Government. Cooperation or mutual could have meant together find a solution any problems encountered. The study said that cooperation between the regions to develop the economic potential of daerah[20] however, the sustainability of the cooperation among the area subject to the commitment of the head of the region as well as support from the Central Government, both financially as well as regulation as the legal umbrella. In the context of capacity development of Panjang beach each coastal Rainforests in order

to optimize the potential of regional cooperation between regions can be one alternative innovations with consideration of efficiency and effectiveness, synergistic and mutually beneficial. For example, by forming community or forum tourism between provinces or between cities on Sumatra island, could also make up the poverty reduction forum coastal areas, with the forum can share experiences, adopt a policy or share the coastal area management strategies. The cooperation could be in the form of profit sharing in managing coastal areas of long beach, for example in making tourist fishing port, in order to make the program work in accordance with Bengkulu City RPJM from 2010 to 2015.
In order for sustainable tourism activities and impact the growth of the need to adopt the principles of tourism. Tourism Office of party can supervise and coordinate all activities of tourism. It can also form the organizational management with the participation of the local community.

Table 3. Capacity Building Strategy Matrix of the Coastal Region -Based on Cross-Sector Partnerships

	Strategy S-O		**Strategy S-T**
1	Local governments and the city of Bengkulu province conducting cooperation between regions, for example, make the MoU with the neighboring province to undertake activities in the coastal area of Panjang beach	1	The Setup is performed or a grouping of street vendors (PKL) in several places in the coastal areas of Panjang Beach
2	Leveraging iconic long beach as a tourist attraction of Bengkulu province, set a target of making the region a national scale length of coastline	2	Do integrated management waste management
3	Government partners with communities living Inn beside the beach, to support tourism activities	3	Promoting tourism by leveraging the position of provincial capital
	Strategy W-O		**Strategy W-T**
1	Increase the human resource capacity of, organize your environment, and requiring the public in making program development in the coastal areas of long beach	1	Provides a community that will be heading to the coastal areas of Panjang Beach
2	Increased institutional coordination involved in the land use in coastal areas of Panjang beach, as well as involve stakeholders in development programs	2	Ask for help financing as well as promotion to the Central Government as the implementation of a national tourism development
		3	Providing security personnel along the coast

4.3.2 *Improving Human Resource Capacity*

In the management of the coastal region of the most important elements of long beach is the human resources (HR) Reform of the civil State (ASN). HR ASN is an assets the Government so that these assets can benefit most out then to do capacity building of HUMAN RESOURCES for her important ASN, conducted capacity building individual/SDM. Tourism was assigned to manage the long coastal area, is expected to make planning, coaching and HR capacity ASN. HR placement ASN adapted to the competence, the estimated number of employees assigned appropriately refers to: 1) condition of the problem; 2) population and the vast coastal areas and; 3) levels of economic activity.

In order for employees assigned to understand and can perform the task given, authorities can make improvements the human resource capacity of the ASN as an effort of human resource in order to maximize the potential of existing resources, through several strategies, for example by way of always learning and understanding as well as giving a solution to the problem at hand. The purpose of improving human resource capacity of the ASN as an alternative to help Governments, member organizations and organizations in developing the expertise and skills needed to realize the objectives that have been determined[21].

In, HR ASN tourism agency who were given a mandate in the management of coastal areas have the capacity and capability, they can mentranformasikan what is owned to the community, e.g., HR can build the ASN and facilitates local community in order to overcome the problems encountered and can play an active role in doing development, employees who have high integrity, can make a valuable economic empowerment. The Office given authority in the management of coastal areas of long beach is expected to have a capacity of agarbisa coordinate and synergize the other parties involved with the development of coastal areas, for example, coordinating with the Office of small and Medium Businesses and the Department of marine and fisheries, because in the long coastal location there is a settlement of fishermen.

4.3.3 Improving Synergy and Coordination between Institutional and Related Stakeholders

The mandate of the regional autonomy laws of development of the coastal area is the responsibility of the local authorities, because of the numerous parties involved either from among the Government itself as well as from the local community i.e. governmental, private parties and non-governmental organizations. The success of development programs will not run optimally when implementing elements do not collaborate, coordinate and synergize with other work units or partnering with all stakholder involved. The involvement of stakeholders and the various parties can enhance the development of the coastal areas became more komprehensip good in economics, the environment, the provision of infrastructure as well as socio-cultural fields. Develop and implement a range of cross-sector partnerships between the alternative local government, private and community colleges by making the regulation governs the role and responsibility of the task against awab respectively.

In the management of the coastal region of the structural approach is more effective, because this approach emphasizes the role of the related service in order to involve all stakeholders, at least in terms of structuring the political system and social structure. The important approach[22] to social intraksi and the existence of a coordination unit significant effect when there is a competence of the internal resources. Competencies between units will be the motivation to do the management of coastal areas for the better. Whereas the non structural approach can be done with social intraksi on local communities with the hope of knowledge, skill and awareness can enhance its role for the development environment. The assigned service have an obligation to guide, nurture and be the catalyst for the communities that lived along the coast. Related service sometimes forget the villages as stakeholders to engage in the planning and management of Long Beach. To the front of the provincial government and municipalities in the management of coastal areas of long beach need to enhance partnership of cross-sector, especially at college because the natural processes of the coastal environment and ecological changes can only be understood, researched by scientists and then understanding the base consideration for City Government to carry out a development program that puts coastal communities as perpetrators and the purpose of improving the socio-economic area.

4.3.4 Integrated Management in the Management of Coastal Areas of Panjang Beach

Planning and management of coastal areas integrated development should pay attention to three items so that sustainable development can occur[23]: (1) regulatory/policy to boost the economy need instruments as environmental parameters to see the cost benefit analysis. For example need to waste concern hotels, restaurants and garbage from the street vendors in order not to pollute the environment, the need for waste management in the four fish fish auction, so as not to cause air pollution and others; (2) the quality of life of the community of concern, provide a means of education for coastal communities, the provision of health facilities and adequate sanitation, and disaster mitigation; (3) the policy should pay attention to environmental issues. Integrated area management will likely negatively impact mengantispasi desired. For example, the involvement of small businesses of high pengeloaan waste due to the large number of street vendors highly contributed to the trash along the beach. Management of coastal areas of long beach can also ensures economic growth which can be enjoyed fairly and proportionately by all the parties involved (stakeholders), and keep the power support as well as the environmental quality of coastal, so that development can take place in a sustainable. Integrated management can identify analyze potential, opportunity, problem or threat, because every sector can engage appropriate fields of science and its role. Approach to the alignment of the management and utilization of coastal area is very important, so is expected to materialize one plan and one management and achieved sustainable development and welfare of society as a whole.

5. Conclusion

Management of coastal areas of Panjang beach as a National Tourist attraction has not been managed optimally may impact because not to increased local community economy. Whereas the policy of tourism development as a strategic value National. The provincial government and the city has the opportunity to be able to manage the Panjang beach area that can be contributed to improve the welfare of society. Thus required a prudent management by removing egosektoral and puts the interests of the community, the environment, social and culture, both in the short and long term. Management with cross-sector partnership and will be maximum with alternatives; (1) increase the capacity of the human resource community capacity and ASN; (2) conducting cooperation between areas that have a single vision and vision to enhance the development of coastal areas of long beach community; (3) improving institutional coordination, synergy and stakholder; (4) should the management of Panjang beach area is carried out by integrated management, alignment done by involving the public, private and community colleges.

References

[1] USAID 2003 Contoh Rencana Strategis pengelolan terpadu Teluk Balikpapan in *Pengelolaan Wilayah Pesisir Terpadu* (T Knight M & SEd USA).

[2] L Brown A Lafond and K. Macintyre 2001 Measuring Capacity Building (University of Noth Carolina at Chapel Hill 23 W Frankli Street Suit 304) March, 919–966.

[3] Senay Habtezion and Jonathan Padgham 2012 *Integrated Coastal Zone Management Lessons In Capacity Building & good governance for Coastal Adaptation* (University of Ghana: START) vol 1.

[4] Jeremy P 2015 Governance fit for climate change in a Caribbean coastal-marine context (*Mar Policy)* **51**, 486–498.

[5] R H Harahap 1994 Pengembangan Kapasitas Perencanaan Daerah dalam Pengelolaan Pesisir Secara Terpadu di Pantai Timur (Provinsi Sumatera Utara Indonesia).

[6] Giovanni D Anna et al 2015 Governance analysis in the Egadi Islands Marine Protected Area: A Mediterranean case study (*Mar. Policy*) 61.

[7] S A Eshliki and M. Kaboudi 2012 Community Perception of Tourism Impacts and Their Participation in Tourism Planning: A Case Study of Ramsar (Iran, *Procedia - Soc. Behav. Sci)* **36**, 333–341.

[8] M H Hanafiah I Azman M R Jamaluddin and N Aminuddin 2016 Responsible Tourism Practices and Quality of Life: Perspective of Langkawi Island communities (*Procedia - Soc. Behav Sci)* 222, 406–413.

[9] S R Lakshmi and T L Shaji 2016 Transformation of Coastal Settlements Due to Tourism (*Procedia Technol)* ,**24**, 1668–1680.

[10] L Carlsson and F Berkes 2005 Co-management: Concepts and methodological implications, (*J. Environ. Manage)* ,**75**(1), 65–76.

[11] Arie Saputra Wijaya 2015Tarik Ulur Pengelolaan Pantai Panjang *Rakyat Bengkulu*, Bengkulu.
[12] Pedoman Bengkulu 2016 Lihat Pantai Bengkulu Menpan Kecewa *Pedomanbengkulu.com*, (Bengkulu, Indonesia) p 21 .
[13] Aditya 2013 Kumuh dan Rawan, Pantai Panjang Akan Diserahkan ke Kota? *Bengkulu Ekspress* (Bengkulu, Indonesia).
[14] B. K. Bengkulu 2016 Tim Koordinasi Penanggulangan Kemiskinan Daerah (TKPKD) Kota Bengkulu.
[15] Badan Statistik Provinsi Bengkulu 2013 Bengkulu Dalam Angka (BPS Provinsi Bengkulu, Bengkulu).
[16] TNP2K 2015 Basis Data Terpadu untuk Program Perlindungan Sosial.
[17] L Havard L Brigand M Cari M. Cariño M Cari and M Cariño 2015 Ocean & Coastal Management Stakeholder participation in decision-making processes for marine and coastal protected areas : Case studies of the south-western Gulf of *(Ocean Coast. Manag)* vol 116 pp 116–131.
[18] S Kang Liang S Pearson W Wu Y jie Ma L lu Qiao X H Wang J Mei Li and X lin Wang 2015 Research and integrated coastal zone management in rapidly developing estuarine harbours: A review to inform sustainment of functions in Jiaozhou Bay (China: *Ocean Coast Manag)* **116**, 470–477.
[19] Pedoman Bengkulu 2015 Kelola Pantai Panjang, Pemprov Ajak Pemkot Bersinergi," *Pedomanbengkulu.com* (Kota Bengkulu, Indonesia) Apr.
[20] B T Harsanto and S Rosyadi 2015 Format Kelembagaan Kerjasama Antar Daerah untuk Pembangunan Ekonomi Kawasan Berkelanjutan (Bandung: UNISBA MIMBAR) **31**(1), 211–220.
[21] D Horton A Alexaki S Bennett-lartey K N Brice D Campilan F Carden J D S Silva and L T Duong 2002 *Organizations around the World.*
[22] W Tsai 2002 Social Structure of Coopetition Within a Multiunit Organization: Coordination, Competition, and Intraorganizational Knowledge Sharing (*Organ. Sci)* **13**(2), 179–190.
[23] Muttaqiena et at 2004 Pengelolaan Wilayah Pesisir Secara Berkelanjutan Pasca Tsunami Desember 2004 *http://slideshare.net/abida/pengelolaan-pesisir.*

IOP Conf. Series: Earth and Environmental Science **55** (2017) 012046 doi:10.1088/1755-1315/55/1/012046

The Dynamics of Sea Surface Height and Geostrophic Current in the Arafura Sea

Umaroh[1,2], Sutrisno Anggoro[1], Muslim[1]
[1]Department of Marine Sciences, Faculty of Fisheries and Marine Sciences, Diponegoro University, Jl. Prof. H. Soedharto, SH, Tembalang Semarang. 50275
[2]Indonesian Agency for Meteorological, Climatological, and Geophysics

Email address: umaroh@bmkg.go.id

Abstract. The aim of this research is to analyze the dynamics of sea surface height and geostrophic current in relation to upwelling and downwelling phenomenon. The climatic factor such as monsoon are also calculated to understand dynamic of all parameters. The location of this research is in the waters of Arafura Sea waters. The data used in this research include sea surface height anomaly and geostrophic current. The method used in this research is done by analyzing the relation of sea surface height and geostrophic current with the eddy current flow. Spatial and temporal analysis using Surfer will be used in this research to visualize the parameters. The result showed that the upwelling (downwelling) phenomenon occurs when the geostrophic current rotate with the cyclonic (anticyclonic) flow followed by higher (lower) sea surface height. The upwelling event mostly occurred during the Southeast Monsoon and downwelling event mostly occurred during Second Transitional Monsoon until Northwest Monsoon.

Keywords: sea surface height, geostrophic current, eddy, upwelling, downwelling

1. Introduction

Arafura Sea is located in the eastern part of Indonesia, bordered with the Banda Sea in the north, the Torres Strait in the east, the northern waters of Australia in the south, and the Banda Sea and the Timor Sea in the west. The Arafura sea is part of Sahul Shelf-shaped and semi-closed waters with an average depth of about 30-90 meters and small part of the area has a depth of >3000 meters [1].

The circulation and dynamic of sea water always occur continuously, the circulation exists both on the surface and in the deep. One of the circulation is sea current flow [2]. The general circulation of surface currents in Arafura waters is affected by the northwest and southeast monsoon cycle which changes periodically every year. This causes seasonal changes in oceanographic parameters including density, temperature, and salinity were also influenced by the flow of river water [3]. These waters also be the pathways of Indonesian Through Flow (ITF) which is the water mass flow from the Pacific Ocean to the Indian Ocean passing through the Indonesian seas seasonally in the Arafura Sea causing changes in the sea surface temperature.

Geostrophic current is current that occur due to geostrophic balance. This current is the result of the difference between the horizontal pressure gradient on the moving water mass and is balanced by the Coriolis force [4]. This current is related to the slope of (slope) of higher sea surface with lower one formed by the Ekman transport and is offset by the Coriolis force [5]. Therefore, this current movement is related to sea surface height. Other factor which play role in geostrophic current circulation is

Indonesian Through Flow [6,7]. The characteristic of the current is significantly needed to be known, especially for the fishery sector (vertical current or upwelling) and for planning sector, such as tackling of sea contamination (horizontal current surface) [8].

Research on the geostrophic flow variability has widely determined, but for the Arafura Sea region is still relatively rare. The sea surface and surface geostrophic variations were determined in the Sunda Strait. They found that seasonal variations in the Sunda Strait sea surface height caused a monsoon [9]. The research of geostrophic currents seasonal variation in the waters of the Arafura -Timor found that the dynamics of the surface geostrophic currents in the waters of the Arafura -Timor occur due to changes sea surface height due to the monsoon each season [10]. The results of these two studies stated that the monsoon (monsoon) greatly affects the variability of geostrophic currents. The anomaly characteristic of sea surface high in special terms and temporal on the northern waters and southern of Java island [11].

According to previous research, by observe to the course of current geostrophic resultant could be established the eddy current phenomenon observed from current vortex which formed [12]. This things is appropriate to previous research that in the southern hemisphere, eddy current with cyclonic rotation is dealing with upwelling event and anti-cyclonic rotation is dealing with downwelling event [12,13]. Mesoscale eddies is current of motion circling cause of heavy current, held more than a week to a couple of months, the condition held cause an eddies rotation could happen for 10-30 days [14]. The research of potential fishing zone can executed by combine the analysis between eddies current with anomaly high current surface.

Because the location of the waters is strategically influenced by the monsoon which blows periodically throughout the year, it is necessary to conduct a research on the dynamics of sea levels and currents geostrophic in the Arafura Sea to understand the dynamics or characteristics of the distribution of high sea surface height anomaly. This research aimed to determine the parameter condition spatially and temporally to finally analyze the upwelling / downwelling event in the Arafura Sea.

2. Material and Methods

2.1. Study Area

The Arafura Sea, lies in the west of the Pacific Ocean and overlies the continental shelf between Australia and Indonesia (Papua). The study area was restricted to the area of 5°S – 15°S and 124°E – 141°E (Figure 1).

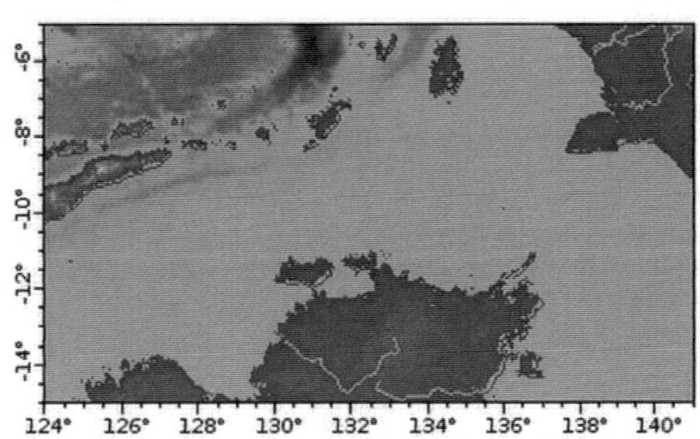

Figure 1. Research study location of the Arafura Sea

2.2. Data and Methods

Data used in this research were monthly geostrophic currents and sea surface height anomaly along 20-year period from January 1996 to December 2015. Geostrophic currents data and sea surface height anomaly used in this research is an altimetry satellite data. All data were in NetCDF Format (.nc) and extracted using ODV (Ocean Data View) then plotted using Surfer Software.

Data processing procedure begins by calculating the monthly average of sea level anomaly parameters as well as the direction and speed of geostrophic currents are used in accordance with the data period to look at the general condition (normal) of these parameters in the Arafura Sea. After that, we calculated the average value of those parameters during monsoonal periods, namely: the Northwest Monsoon (December to February / DJF), the First Transitional Monsoon (March to May / MAM), the Southeast Monsoon (June to August / JJA), and the Second Transitional Monsoon (September to November / SON). The results of the calculation will be plotted in the form of two-dimensional (longitude-latitude) with Surfer software.

Furthermore, based on the results of the mapping, geostrophic flow pattern overlaid with sea surface height anomaly. Both of these parameters are combined into a single image (composite data) to make it easier to analyze the eddy currents as well as an indication of upwelling and downwelling event in the Arafura Sea.

3. Results and Discussion

The monthly composite of sea surface height and geostrophic current in the Arafura Sea have been determined. The upwelling event supposedly occurred in the area with cyclonic eddy and lower sea surface height. The downwelling event supposedly occurred in the area with anti-cyclonic eddy and higher sea surface height. In this research, we focused only on the eddy-induced upwelling and downwelling event. The upwelling event in the figures is shown in the red circle area and the downwelling event in the black circle area. We used the anomaly of sea surface height and geostrophic current to determine their variability during monsoonal periods.

3.1. Variability of Sea Surface Height and Geostrophic Current during Northwest Monsoon (DJF)

During Northwest Monsoon, the anomaly of sea surface height ranged 0-0.32 meters. The area with lower sea surface height occurred in the western Arafura Sea, with the lowest area was in the south-western Arafura Sea. This low area mostly exhibited the cyclonic eddy event. Thus, in this research, those conditions represent the upwelling event (red circles). The area with the higher sea surface height occurred in the eastern Arafura Sea, with the highest area was along the eastern Arafura Sea during January until February. This highest area mostly exhibited the anticyclonic event. Thus, in this research, those conditions represent the downwelling event (black circles).

In December, the upwelling event occurred in the centre area of Arafura Sea and in the north Timor Sea, while the downwelling event occurred in the northeast and southeast Arafura Sea (Figure 2). These conditions were occurred until January. In February, the upwelling event was decreasing but the upwelling event was more than the downwelling event. In general, during the Northwest Monsoon, the upwelling event occurred mostly in the west Arafura Sea.

IOP Conf. Series: Earth and Environmental Science **55** (2017) 012046 doi:10.1088/1755-1315/55/1/012046

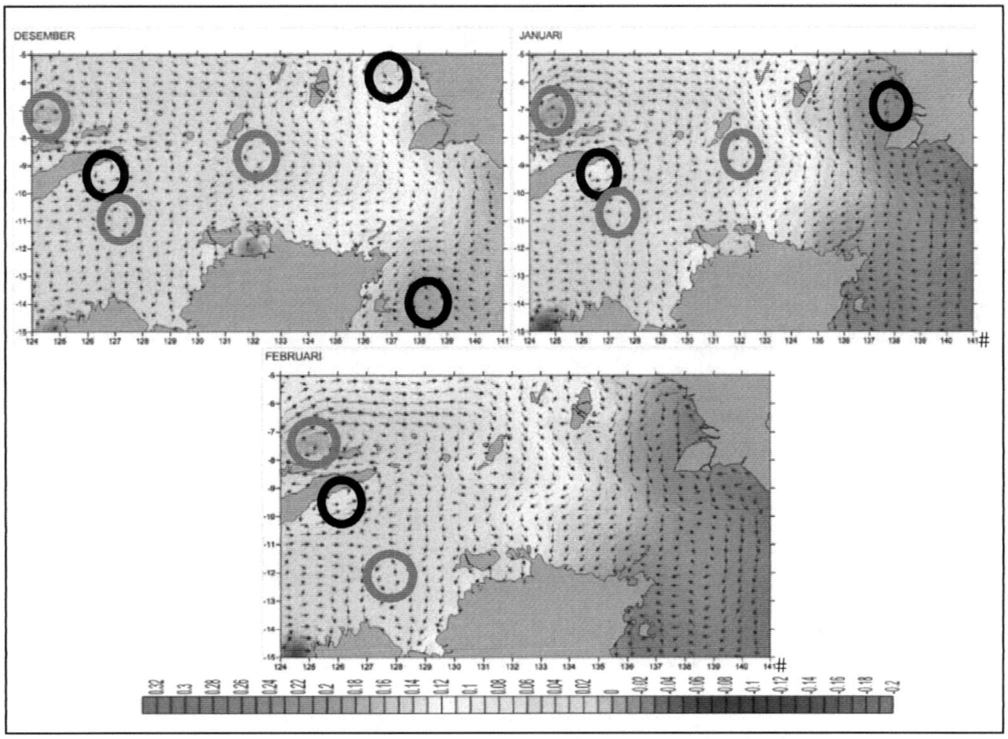

Figure 2. The composite of sea surface height and geostrophic current anomaly during the Northwest Monsoon (December-January-February). Red circles represent cyclonic eddy event (upwelling event) and black circles represent anti cyclonic eddy event (downwelling event).

3.2. Variability of Sea Surface Height and Geostrophic Current during First Transitional Monsoon (MAM)

During First Transitional Monsoon, the anomaly of sea surface height ranged -0.1 - 0.32 meters. This sea surface height anomaly value was lower than the value during the Northwest Monsoon. The area with lower sea surface height mostly occurred in the western Arafura Sea, but the lowest area occurred in the south-eastern area in May (Figure 3). The area with the higher sea surface height mostly occurred in the north-eastern Arafura Sea, with the highest area was along the eastern Arafura Sea in March.

The unique condition was occurred during April-May (Figure 3). In April, the upwelling event tended to occur in the south Arafura Sea. This condition was a proof that upwelling event moved from north area to the south area during this First Transitional Monsoon. In May, the lowest area was in the eastern Arafura Sea, and the highest area was in the western Arafura Sea. In general, during the First Transitional Monsoon, the upwelling event occurred mostly in the southern Arafura Sea.

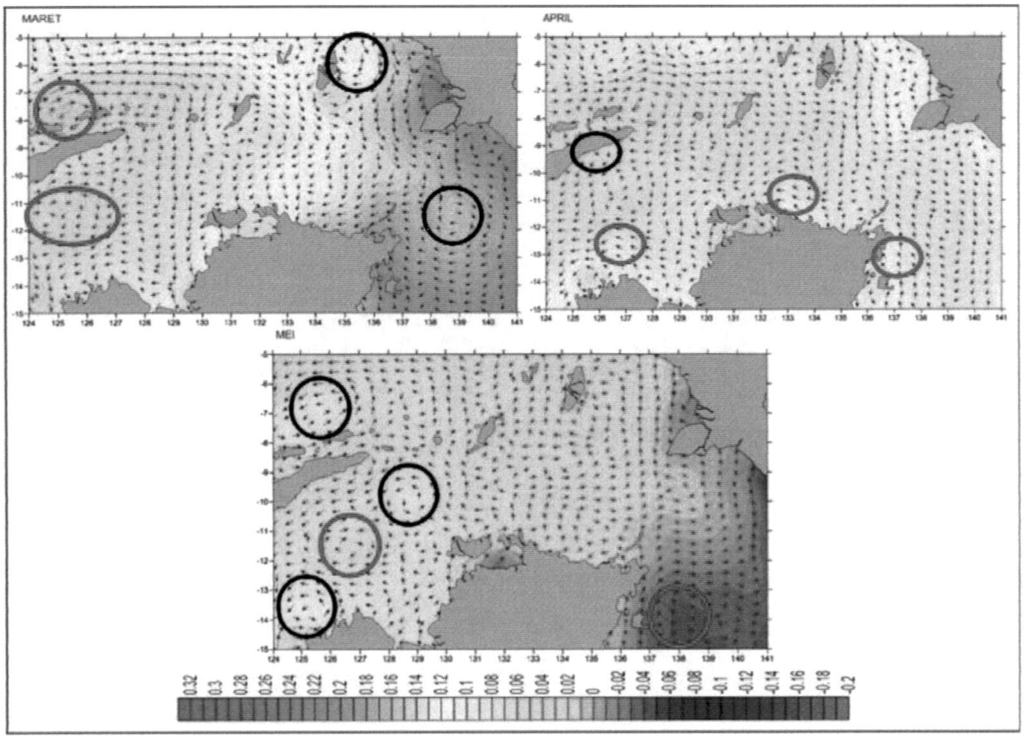

Figure 3. The composite of sea surface height and geostrophic current anomaly during the First Transitional Monsoon (March-April-May). Red circles represent cyclonic eddy event (upwelling event) and black circles represent anti cyclonic eddy event (downwelling event).

3.3. Variability of Sea Surface Height and Geostrophic Current during Southeast Monsoon (JJA)

During the Southeast Monsoon, the anomaly of sea surface height ranged -0.2 - 0.14 meters. This sea surface height anomaly value was the lowest compared by the other Monsoonal periods. The area with shallower sea surface height mostly occurred in the eastern Arafura Sea, with the lowest area was in eastern Arafura Sea. This was the opposite of the condition during Northwest Monsoon. The area with the higher sea surface height mostly occurred in the western Arafura Sea, with the deepest area was in the western Arafura Sea in June.

The sea surface height anomaly in the western Arafura Sea exhibit lower value than the other areas. This condition was caused by the water mass flow from east to the west during Southeast Monsoon. Based on Figure 4, the upwelling event mostly occurred during this Southeast Monsoon, compared with other monsoonal periods. The upwelling event occurred in the centre area and extended to the southern Arafura Sea. The downwelling event occurred in some area beside the upwelling area.

IOP Conf. Series: Earth and Environmental Science **55** (2017) 012046 doi:10.1088/1755-1315/55/1/012046

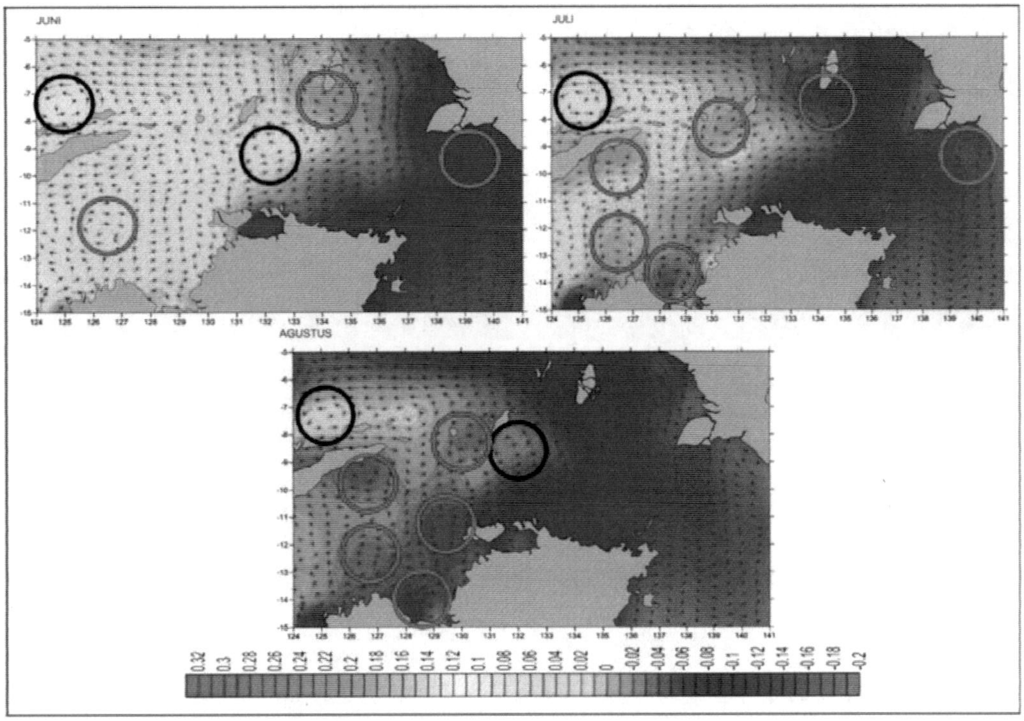

Figure 4. The composite of sea surface height and geostrophic current anomaly during the Southeast Monsoon (June – July- August). Red circles represent cyclonic eddy event (upwelling event) and black circles represent anti cyclonic eddy event (downwelling event).

3.4. Variability of Sea Surface Height and Geostrophic Current during the Second Transitional Monsoon (SON)

During the Second Transitional Monsoon, the anomaly of sea surface height seemed to increase, compared with the Southeast Monsoon. The lowest sea surface height area mostly occurred in the eastern Arafura Sea, with the lowest area was in eastern Arafura Sea. The area with the higher sea surface height mostly occurred in the western Arafura Sea during October and in the southeastern Arafura Sea in November.

Based on Figure 4, in September, the area with shallower sea surface height was almost the same with the condition during the Southeast Monsoon, but the sea surface height anomaly was in higher value. In October, the sea surface height anomaly was in the higher value and the upwelling event only occurred in the closed area near the island Melville of Australia. The downwelling event mostly occurred during this time.

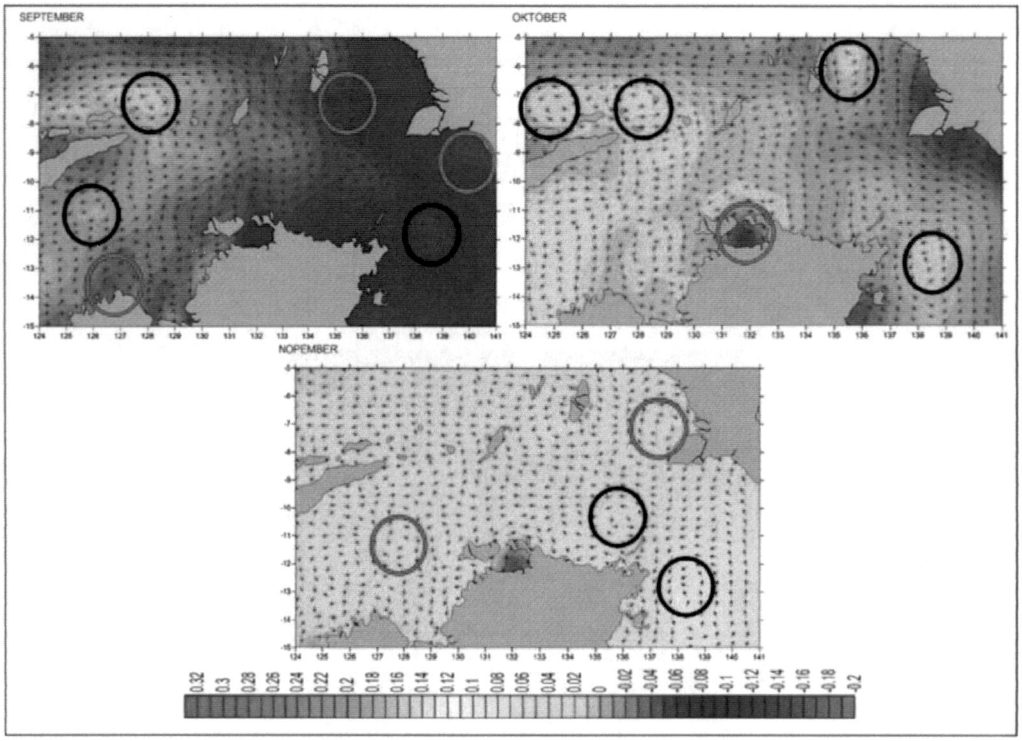

Figure 5. The composite of sea surface height and geostrophic current anomaly during the Second Transitional Monsoon (September – October- November). Red circles represent cyclonic eddy event (upwelling event) and black circles represent anti cyclonic eddy event (downwelling event).

3.5. Temporal Analysis of Sea Surface Height and Geostrophic Current in the Arafura Sea

Based on the monthly sea surface height anomaly in the Arafura Sea (Figure 6), it can be seen that the negative anomaly occurred from June until October, meanwhile the positive anomaly occurred from November until May. The negative sea surface height anomaly characterized the upwelling event, and the positive sea surface height anomaly characterized the downwelling event. This condition is dealing with the upwelling and downwelling event described spatially.

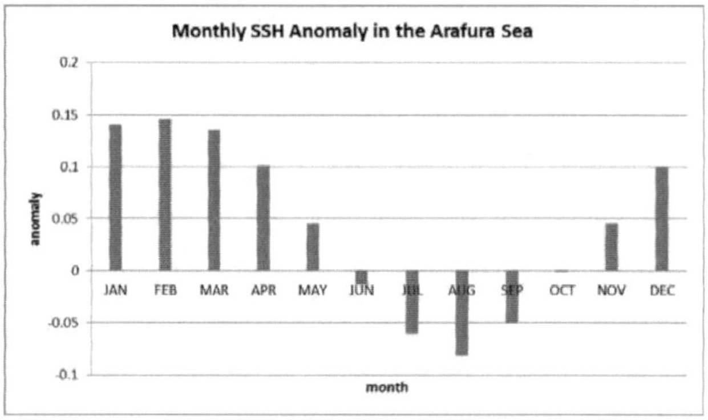

Figure 6. Monthly sea surface height anomaly in the Arafura sea

4. Conclusion

The sea surface height and geostrophic current anomaly can describe the upwelling and downwelling event. The upwelling event supposedly occurred in the area with low sea surface height with cyclonic eddy. The downwelling event supposedly occurred in the area with high sea surface height with anti-cyclonic eddy. The upwelling event mostly occurred during the Southeast Monsoon and downwelling event mostly occurred during Second Transitional Monsoon until Northwest Monsoon.

5. Acknowledgement

This research was supported by Indonesian Agency for Meteorological, Climatological, and Geophysics. The authors would like to thank to the institutions that provide open source data for analysis to be used in this paper.

References

[1] Alongi, D.M., K. Edyvane, do Ceu Guterres, W.S. Pranowo, S. Wirasantosa, and R. Wasson. 2011. *Biophysical profile of the Arafura and Timor Seas*. Report prepared for the Arafura Timor Seas Ecosystem Action (ATSEA) Program. 32p.

[2] Marpaung, S. and Prayogo, T. 2014. *The Analysis of Surface Geostrophic Current Based On Satellite Altimetry Data*. National Seminar on Remote Sensing.

[3] Gordon, A.L. and R.A. Fine. 1996. *Pathways of water between the Pacific and Indian Oceans in the Indonesian Seas. Nature*, 379 (6561):145-149.

[4] Brown, J, A. Colling, D. Park, J.Philips, D. Rothery and J. Wright. 1989. *Ocean Circulation*. New York. Pergamon Press.

[5] Hadi, S. and I.M. Radjawane. 2009. *Sea Currents*. College Book. ITB: Department of Oceanography.

[6] Gordon, A.L. 2005. *Oceanography of the Indonesian Seas and Their Through Flow. Oceanography*, 18(4):14-27.

[7] Susanto. D, Amy Ffield, Arnold L. Gordon. 2012. *Variability of Indonesian Through Flow within Makassar Strait, 2004–2009*. Journal of Geophysical Research JGR – Oceans.

[8] Pranowo, W.S. 2012. *Upwelling – Downwelling* Dynamics of Arafura and Timor Seas. Widyariset, Vol. 15 (2) : 415-423.

[9] Oktavia, R, J.I. Pariwono and P. Manurung. 2011. *Variation of Sea Surface and Surface Geostrophic Current in Sunda Strait waters based on Tidal Wave and Wind Data year 2008*. Journal of Topical Marine Science and Technology Vol.3 (2):127-152.

[10] Ramadyan, Fachri and Ivonne M.Radjawane. 2013. *Seasonal Surface Geostrophic Current in the Arafura – Timor Sea*. Journal of Tropical Marine Science and Technology Vol.5 (2):261-271.

[11] Marpaung, S. and Harsanugraha, K. W. 2014. *The Characteristic of Sea Surface Height Anomaly Distribution in the southern and northern Java Sea*. National Seminar on Remote Sensing.

[12] Martono. 2009. *The Monthly Characteristic and Variability of Surface Wind in Indonesian Ocean*. Makara Sains Vol. 13 (2): 157-162.

[13] Stewart, R.H. 2008. *Introduction to Physical Oceanografi*. Dept. of Oceanography. Texas: A & M University.

[14] Bell J, Johnson J, Hobday A. 2011.*Vulnerability of Tropical Pacific Fisheries and Aquaculture to Climate Change*. Secretariat of the pacific community edn. Noumea, New Caledonia. 927pp.

Vulnerability Assessment: The Role of Coastal Informal Settlement Growth to Social Vulnerability in Genuk Sub-District, Semarang City

Sariffuddin[1*], Khristiana Dwi Astuti[1], Gustika Farhaeni[2], Lutfiyatul Wahdah[3]

[1] Department of Urban and Regional Planning, University of Diponegoro,
[2] Perum Perumnas – Semarang Jl. Wilis No. 23 Candisari, Semarang
[3] Master of Environmental Science, University of Diponegoro

*Email: sariffuddin@live.undip.ac.id

Abstract. Tidal, land subsidence and flooding in Semarang City brought many consequences to the city's economic and social conditions. In the economic sector, the largest contributor to GDP derives from industries that are located in the coastal area. The environmental problems also affect the ability of social adaptation and vulnerability. This issue encourages the efforts of the public and the government. City conditions are constantly changing force people to adapt for survival, through endogenous. At the same time, the government is trying to help for the same goal, through a variety of urban development. Although it has the same goal, many found gaps, which, if not followed could increase the vulnerability of society and lead social dysfunction. According to of the phenomenon, this article is based on questionnaire surveys, observation, and previous studies in Semarang City. Quantitative data obtained from questionnaire surveys of 133 respondents of the household. The questionnaire distributed in 3 villages in Genuk sub-district, i.e. Terboyo Kulon, Terboyo Wetan, and Trimulyo. Industrialisation in the coastal area is main caused informal settlement hyper growth in the coastal area. For the present study sought to find out how the process of adaptation that naturally affects the vulnerability of communities using qualitative methods. The results showed the social, economic condition is strongly influenced adaptability of society.

Keywords: adaptive capacity, social vulnerability

1. Introduction

The rise of the resilient city idea is an attempt to answer the uncertainty of the city in the future [1, 2]. This debate is often associated with disaster issue [3, 4], climate change [5-7] and urban economic security [8, 9]. Seen from climate change and disasters perspective, the city is considered resilient if the urban system can absorb the pressure and continue to function normally. To understand the metaphor of resilient city, urban ecology, science is the closest discipline for the scope of city planning [10]. A city is an ecosystem where organisms (include man) living inside that influence each other [11]. The ecosystem itself consists of 'structure' and 'function'. The structure indicates the interaction system while the function indicates the role of each element in the system [10, 11]. Any change emerging from inside and outside the system will have an impact on the stability of the ecosystems [12, 13]. In the scope of climate change and urban growth, vulnerability and risk assessment are useful to support the policy development that the city can adapt in facing climate change. Rosenzweig [6] argue that vulnerability and risk cities assessment for climate change are based on (1) climate hazards faced by the city, (2) vulnerabilities due to a city's social, economic, or physical attributes, and (3) adaptive capacity aspects.

There are many definitions of vulnerability; there are even more than ten definitions mostly debated in the world. Cutter [14] gathers the definition of vulnerability into 18 definitions according to the scientists' proposed perspectives. The debate on vulnerability terminology began in the 1980s, where vulnerability research focused on three basic concepts: (1) related to the exposure (biophysical/technology), and (2) the consequences of the loss of social vulnerability) or (3) a combination of both. The vulnerability research falls at least under three alternative categories: (1) vulnerability to the risk of hazard exposure, (2) social vulnerability and (3) the vulnerability of the place/location. However, the vulnerability definition proposed by IPCC [15] is often used as a reference that defines vulnerability as the "the degree to which geophysical, biological and socioeconomic systems are susceptible to and unable to cope with the adverse impacts of climate change". There are three components forming the vulnerability that is the level of sensitivity, the level of exposure, and adaptive capacity [16]. The level of sensitivity shows the internal state of the system that shows the degree of vulnerability to interference. Meanwhile, the adaptive capacity shows the ability of a system to adjust to environmental change or change in policy so that the potential negative impacts can be mitigated and positive impacts can be maximised or, in other words, it is the ability to cope with the consequences of climate change [16].

Furthermore, the vulnerability is a phenomenon that is influenced by the social construction of institutions and economic dynamics [17]. Social construction is a long-term social value of human interaction with the natural surroundings in the sphere of social-ecological systems which are mutually influential [18]. This social construction affects the adaptation ability as one of the aspects of social vulnerability research. Adaptability plays an important role in the vulnerability assessment of the location/ place. In this perspective, the vulnerability is formed from the risks biophysical and social responses [19]. In a general perspective, adaptability is very dependent on the economic ability, in fact, it is more influenced by the knowledge, experience, and community response to the weather [20].

Semarang towards the resilient city: a brief review?

One of the efforts to realise Semarang City to become resilient is through an ACCCRN program (Asian Cities Climate Change Resilience Network) which is funded by the Rockefeller Foundation and is managed by Mercy Corps Indonesia [21]. ACCCRN program aims to realise the resilient city against climate change through three outcomes: (1) capacity building, (2) network development for knowledge and learning, and (3) the expansion and scaling up [12]

One of the biggest concerns is that the largest contributors to GDP Semarang that are the industrial sector and trade as well as hotel and restaurant will be threatened since they would not be able to continue their business due to sea levels rising. In 2011, these two economic sectors which were mostly located on the coast were accounted for not less than 52.28% of the total GDP of Semarang. The industrial sector contributed for 27.36% increase compared to the GDP in 2010 which was 24.16% to a growth rate of 12.60%. Trade, hotels, and restaurants accounted for 28.01% of GDP, which increased to 12.03% from the previous year that reached 27.92% [22].

Despite the economic growth of the city, the increasing sea levels widen the tidal inundation area and soak the infrastructures, factory buildings, warehouses, and the residence. In 2012, the 20.76% of tidal inundation area was predicted to extend into 22.32% of the total area of Semarang in 2015 [23]. The extension of the inundated area is influenced by the increase in sea level of 5.43 cm/year due to global sea-level rise of 2.65 mm/year and a reduction in the land of 5.165 cm/year [24]. The direct result is the damaged infrastructures as well as the high cost of development and maintenance of city infrastructure.

These issues become the concern of Semarang City to realise a resilient city in response to the global climate change [25]. Semarang coastal area becomes the most affected area due to climate change which is characterised by the increase of tidal inundation area, soil degradation, and abrasion [26, 27]. The local communities are 'forced' to endure and accept the environmental condition which continues to deteriorate. With limited capabilities, they have to live 'in harmony' with the environment by elevating the building, the floor of the house, and closing the water channel during high tide [27, 28]. Therefore, this study tries to find out how the process of informal settlement hypergrowth influences on the social vulnerability of the community?.

2. Methodology

This article applies descriptive statistics and spatial analysis. Descriptive statistics are used to assess the vulnerability of Semarang coastal settlements while superimpose/overlay is used to determine the pattern of growth of the coastal settlements. The research variables refer to the statement mentioned by Adger [29], consisting of exposure, sensitivity, and adaptive capacity. Table 1 below shows the variables, indicators, and weight to assess the vulnerability.

Table 1 Variables, Indicators, and Weight

Variables	Indicators	Weight
Exposure	Houses along the river	0,25
	Houses affected by flood	0,25
	Houses affected by flood and tide	0,25
	Land subsidence	0,25
Sensitivity	Electrical service	0,25
	Drainage quality	0,25
	Drinking water quality	0,25
	House quality	0,25
Adaptive Capacity	Education	0,25
	Social welfare	0,25
	Family insurance	0,25

Vulnerability assessment is based on the formula presented by Turner and Kasperson [30]:

$$\textbf{Vulnerability} = \frac{\textbf{Exposure x Sensitivity}}{\textbf{Adaptive Capacity}}$$

Based on the results of the vulnerability assessment, it is then grouped into five groups based on the index of adaptive capacity (IAC) and the index of exposure and sensitivity (IES). IAC represents the X axis while the Y axis is represented by the IES vulnerability matrix as shown in Figure 1.

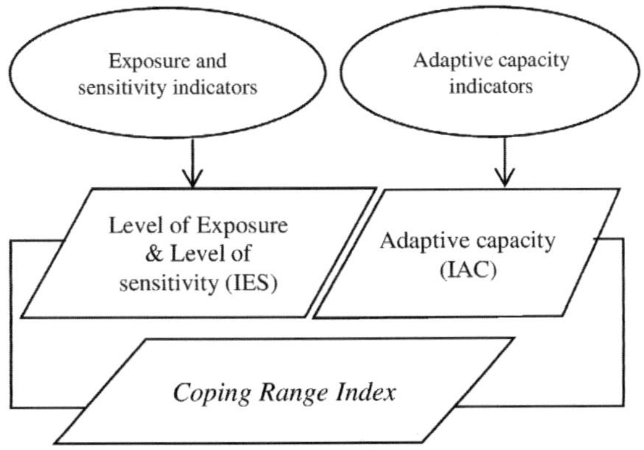

Figure 1. The analytical procedure [31]

To obtain the relative position of the village towards the other village in tolerating a disaster (coping range) is by looking at the value of IES and IAC position in the quadrant system. If the value of the index of exposure and sensitivity (IES) and the index of adaptability (IAC) is in quadrant 5, the tolerance against catastrophic events is low (narrow coping range) or very vulnerable. The level and position of the vulnerability of each area (village) can be grouped according to the following chart:

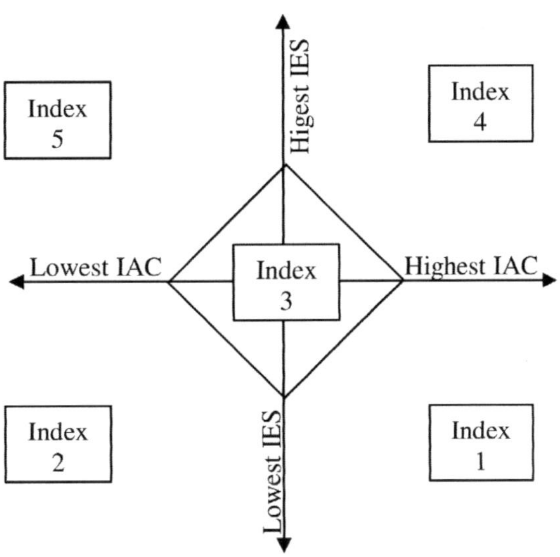

Note:
Index 1: Very invulnerable
Index 2: Invulnerable
Index 3: Almost vulnerable
Index 4: Vulnerable
Index 5: Very vulnerable

Figure 2 Vulnerability classification according to coping capacity index [31]

The data is collected based on questionnaires distributed to 113 households with a level of accuracy of 90%. Table 2 presents a sample according to the village.

Table 2 Sample in each village

Village	Number of households	Sample
Trimulyo	1.018	33
Terboyo Wetan	1.106	37
Terboyo Kulon	1.549	43
Total		113

3. Data and Analysis

The terminology of vulnerability in this article refers to the notion proposed by Cutter [14] which focuses on the area vulnerability which is affected by vulnerabilities as a result of disaster risk and vulnerability due to social conditions response. The discussion in this article will be divided into four parts that are the explanation of the potential hazard of tidal inundation, the pattern of informal settlements growth, and the social vulnerability assessment and the description of the community adaptation.

3.1 Tidal Inundation

The entire study area is the area prone to tidal inundation. Semarang City Regional Regulation No. 7/2014 mentions about the drainage system master plan of Semarang 2011-2013, of which three villages in the research area that are Trimulyo, Terboyo Wetan, and Terboyo Kulon, are included in the drainage system of East Semarang that are the sub-drainage system of Tenggang river (Terboyo Kulon), Sringin River (Terboyo Wetan), and Babon river (Trimulyo). The main problem is the tidal flooding that occurs almost twice a week. The settlements located along the river are often flooded by tidal inundation for about 10-30 cm high. Tidal inundation is getting worse from year to year. Before the development of industrial zones in Terboyo Wetan and Trimulyo (in the 1980s), the tidal inundation could be accommodated by rice fields or ponds located on the waterfront. However, once they were converted into an industrial area, the tidal inundation extended to the settlements.

Tidal inundation also becomes an unfinished problem for the coastal community in Semarang [26, 27]. The community tries to resolve these ecological problems according to their economic ability by elevating the house floor, constructing dams/gutter in front of the house to retain the tide [32], or closing the drainage during tide [28]. The study of city resilience strategy undertaken by Tahta [33] shows the economic loss due to the damage of the coastal areas of Semarang which reached no less than Rp. 354.991.473.017,-. Compared to the local revenue, such losses spent half of Semarang total revenue which averagely was Rp780 billion each year [22]. This indicates that the achievement of the city's economy is incomparable with the losses due to the damage of the coastal environment [28].

3.2 Informal Settlement Growth

Industrialisation is a major cause of informal settlement growth in Genuk, Semarang. The Ministry of Public Work [34] showed that the increase of the building from 2000-2006-2012 based on satellite imagery and primary survey. The most rapid growth occurred in 2000-2006 which was marked by some new industries and warehouses. In 2000, space covering an area of 70.091 ha increased to be 24.112 Ha in 2006 and increased into 22.719 Ha in 2012. By the end of 2013, 116.922 Ha in Genuk sub-district was built.

Massive development in Genuk's coastal areas extends and deepens the tidal inundation areas since the open space as a catchment area is decreased. There are 64% of neighbourhoods located on the river banks and 72% of the houses that are inundated. The residents who work as industrial workers do not have sufficient capability to cope with the threat of the tidal inundation.

Houses in Genuk industrial zone are categorised as informal settlements. The residents build their houses without the intervention of private agent (developer) or the city government. The house construction by the community was started along with the development of industrial zones in Small Industrial Environment) and Terboyo Megah Industrial Area. The growing number of the houses over the years has increased along with the growth of the population. The Industrial Zone of Genuk was previously paddy fields, ponds, garden/fields, and only a small area that was built houses.

The increasing number of the houses is the implications of high population growth both natural growth and migration. Urbanisation contributes the most impact compared to natural population growth which increases twice higher than the natural growth. Based on the statistical data, Genuk District population growth in 2014 was 2.79%. Almost every year the rate of migration tends to increase. The impact is the increased need for houses for the new residents. Within the past 20 years, the population is centralised in Terboyo Wetan and Trimulyo Village with Ngilir River as the centre of growth. It results in high density and crowds in which one house is occupied by more than two families, especially in the urban village. Moreover, the distance between houses is no more than one meter. Similar problems also occur in the other four villages (Terboyo Kulon, Genuksari, Gebangsari, Banjardowo and Muktiharjo Lor). The total area of the five villages is still sufficient to accommodate the newcomers. Uncontrolled housing density widely impacts the physical and socio-economic conditions of the population.

Many houses are built along the river bank even on the river. The houses of the immigrants who work in the informal sector such as pedicab drivers, street vendors, market porters or construction workers are built on the river bank. The river bank is located in the north of Terboyo Megah Industrial Zone in Terboyo Wetan Village.

Figure 3 The form of Informal settlement growth

Obviously, these houses are built on illegal land. The houses in Terboyo Kulon are also built on the land and river by attaching pillars to the houses. The existence of these houses poses a major problem to the conditions of settlement environment. The pillars hamper the water flow and the garbage which results in the rainwater overflow and inundates the roads and houses.

3.3 Social vulnerability assessment

Environmental problems are worsened and become a social problem. The community assumes that the current environmental management is a useless effort because the environment that has been maintained and repaired can certainly be damaged again. Principally, therefore, the environmental problems are caused by the behaviour that is the public ignorance in managing the housing environment, the inability to manage and to restore the environmental conditions to be better, and the community unwillingness because environmental management is a useless act. The overlapping forms of adaptation designed by the government and public institutions indicate the institutional influences on social vulnerability. The absence of integrated planning between the real conditions in the field carried out by the government would be a separate issue that impact on the vulnerability of coastal communities. The vulnerability that occurs in the community is more influenced by the economic dynamics than institutions. The absence of collective adaptability in force in the group of citizens has also become a new problem to the degree of vulnerability of the community.

From the assessment of the level of vulnerability, there are five categories: index 1, index 3, and index 4 out of 5 vulnerability indexes. Index 1 indicates the lowest level of vulnerability (only two respondents out of 113 respondents). The low vulnerability level is caused by the ability to adapt (adaptive capacity) especially the level of income and education level of the community. Index 3 is a vulnerability in quadrant five which is right in the middle of the Cartesian quadrant. This level shows that the vulnerability level of the community is on the threshold (coping range), and if it does not receive any intervention on its adaptive capacity, it will move to vulnerable quadrant. In the last level, that is the index 4 or vulnerable; there are 67 respondents out of 113 respondents. Most people whose livelihood is industrial workers fall into this category.

Table 3 Level of Vulnerability

No	Vulnerability Index	Number	Percentage
1	Index 1	8	7.08%
2	Index 2	3	2.65%
3	Index 3	52	46.02%
4	Index 4	14	12.39%
5	Index 5	36	31.86%

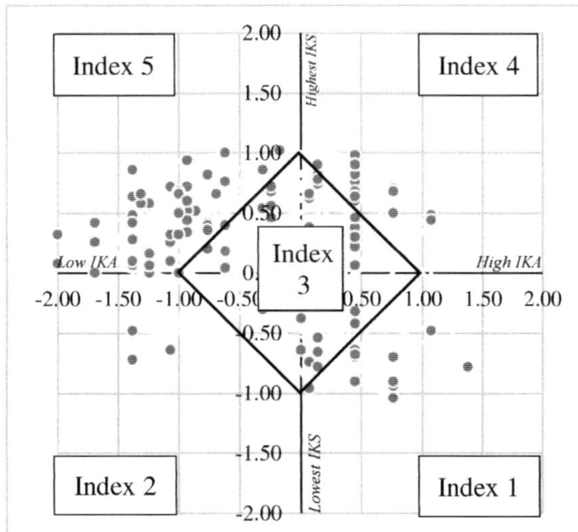

Figure 4 Vulnerability Matrix

Community ignorance regarding the management and improvement of the environment has caused the lack of ethics and wisdom to the environment. This worsens the settlement environment since it has been practised for years by many people. Due to this ignorance, the community also utilises the river as waste canal and toilet. Public awareness of waste management is also low. There are a lot of people who throw their garbage in the river or another vacant area. Also, the community does not separate the drainage with household waste. Besides, public participation in the environmental management is also low. The community frequently participates in cooperation (*gotong-royong*) whenever there are damaged infrastructures such as damaged road or bridge. The participation in managing environmental hygiene, river cleanliness, drainage, and others is still low.

Semarang coastal areas become vulnerable because the mitigation and adaptation capability is 'burdened' to the ability of the people who live in self-made settlements. There is no procedure and clear standard to overcome the ecological problem. This vulnerability is predicted to grow higher due to the urbanisation drive of Kedungsepur area which is 1.36% per annum [34]. Rapid urbanisation has led to environmental degradation, congestion, unemployment, crime, and socio-political conflict [35] which increases the informal sector amounted to 1.47% per annum [22]. Satterthwaite [36] suggests that the greatest vulnerability of the city comes derive from the informal settlement which has now reached 49.86% [22] in Semarang City.

As previously discussed, the collaborative adaptation is not carried out by the government, and at the same time, the community forms new problems. Socio-ecological systems that run circularly in the study area is a short walk, which starts from the 1980s characterised by urbanisation as a result of mega industrial projects [28]. The ability of people to adapt is highly affected by the economic capacity and does not explain the existence of local knowledge and local technologies that affect their adaptation efforts. It became one picture that opinions Adger and Vincent [20] who said that adaptability is influenced by experience and knowledge of citizens does not apply for two reasons: (1) people who stay are migrant communities, (2) The range of experience of social ecology is relatively short and does not run for generations.

3.4 Community Adaptation Form

The adaptation process becomes highly dependent on the ability of the community which has no standards. The economic ability of the coastal communities defines their adaptation strategy and capability. High-income community is capable of elevating their homes but the low-income community is only able to elevate the floor of the house. Therefore, it is not difficult to find houses which are no less than3 meters high. Adger and Vincent [20] mention that vulnerability is a socially constructed phenomenon influenced by institutional and economic dynamics. In the study area, both components influencing this vulnerability to go hand in hand. It is even prone to minor 'conflict' as perceived by the public. The forms of adaptation are likely more physical that is by elevating the physical buildings such as roads, bridges, river embankments, floors, and community houses [28, 32]. The adaptation made by the local community is a survival process and is based on their ability (self-help) or from government aids. The informal process done by the community is to reduce the impacts of the flood (both tidal inundation and flood) and to meet their basic needs. This adaptation is done by elevating the building, constructing a dam in front of the house, and closing the drainage during the tide [26, 27]. Meanwhile, the adaptation strategy by the community to meet the basic needs is by building artesian wells. The absence of the tap water service in Terboyo Kulon, Terboyo Wetan and Trimulyo Village has forced the people to build a well which is then distributed among the community informally. In fact, the PDAM pipes for the port are installed in the three villages.

Residents who live in Genuk coastal areas are mostly not natives of Semarang City. The native referred in this article are citizens who were born and grew up in the same place. This becomes important as the methods (social learning process) for the community in facing their environment within the scope of the social-ecological system [18]. They come for working in the industrial sector, but due to the limited education and skills, the factories cannot accept them. The houses which are built do not meet the health requirements. There are many houses which are only 21 m^2 high and are not equipped with a window, built with the incomplete wall, and have a unpaved floor which is inundated during the flood. This condition occurs in a neighbourhood located in industrial areas and riverbanks. The slum houses have long emerged that was started to spread in 1997 [37].

The community ability to adapt is associated with class society. There are three classes of people in the Genuk coastal area which affect the adaptation ability. The first class is informal workers and communities living in illegal areas or marginal land. Most people work as porters and parking attendants with low or even no education background. They consider that a poor environment is a logical consequence of coastal settlements. Also, a community in this group does not have a significant contribution in environmental management. Second is the industrial worker community. They have middle to the high educational background and give a high contribution to environmental management. Many community forums are led by this group. The third is a wealthy community which has a major asset in their environment. Their concern for the environment is quite low, and they consider the environmental degradation is caused by the lower-class community who is not aware of the environment [38].

This adaptation is not only conducted by the community but also by the government through the elevation of the main road (Kaligawe Street). The road is routinely elevated to avoid flooding. However, in reality, this form of adaptation creates new problems faced by the community. Being elevated, the road also serves as an 'embankment' retaining the water. As a result, the inundation in Terboyo Kulon, Terboyo Wetan, and Trimulyo can not flow as the road obstructs it. What becomes a new phenomenon now is the adaptation efforts designed by the government and carried out by the community which turns to be ecological and social conflicts. The government implements integrated and formal development planning within the scope of community's informal development planning (bottom up). This indicates that the implemented environmental planning is unable to design a better environment for the community.

4. Conclusion

Social vulnerability is influenced by the ability of adaptation. The adaptation strategy which is not collaborative between the government and the community will create a new problem. The community's capability to adapt highly depends on their welfare which in fact is not wealthy enough. There are still basic needs (water) that have not been fulfilled, and the community has to fulfil it by building artesian wells. The experience and local knowledge have not yet fully supported community adaptability. This is caused by the local communities who are not native residents. They are newcomers who work as labours in the factory industry. Local knowledge is not constructed because there is no ecological experience that the community forms from their birth until the present days. This condition is contradictory to the opinion mentioned by Adger and Vincent [20] that the ability of social adaptation is more influenced by their life experiences and knowledge in residence. Traditional knowledge that is running in the community grows and develops through generations and forms the adaptive values which are used. The principle of circular relationship [39] and social-ecological system [18] have to be explained clearly. However, the condition of 'hereditary' occurred in Semarang is not the same with the hereditary condition in rural areas in which village communities are composed of unchanged various generations. In contrast to the Genuk coastal communities, most migrant communities are not from Semarang but from around Semarang [28]. The reason is what makes the difference in the findings of the study with a good understanding of the literature presented by Adger and Vincent [20], Soemarwoto [39], Cutter, Barnes [40] and Folke [18]. Collaborative planning that combines form with the adaptation of government agencies that have been made by the public within the scope of the new research that should be studied more in depth in the future.

5. Acknowledgment

The author would like to thank the University of Diponegoro which has funded the research in the development of the scheme Penerimaan Negara Bukan Pajak (PNBP) Fiscal year 2015 through the Budget Implementation List (DIPA) Universitas Diponegoro No. DIPA– 313-17UN7.5.1/PG//2015, 22 Avril 2015.

References

[1] Leinchenko R 2011 Climate change and urban resilience (Current Opinion in Environmental Sustainability vol 3) p 164-168.
[2] Jordan Y C G, Abduwasit, Hartling Sean 2014 Traits of surface water pollution under climate and land use changes: A remote sensing and hydrological modelling approach (Earth-Science Reviews vol 128) p 181-195.
[3] Pelling M 2003 The Vulnerability of Cities: Natural Disasters and Social Resilience (London: EARTHSCAN Publications Ltd)
[4] Shaw R, H Srinivas and A Sharma 2009 Urban risk reduction: a perspective (United Kingdom: Emerald)
[5] Yuen B and A Kumssa 2011 Climate Change and Sustainable Urban Development in Africa and Asia (New York: Springer)
[6] Rosenzweig C et al. 2011 Urban Climate Change in Context, in Climate Change and Cities: First Assessment Report of the Urban Climate Change Research Network ed C. Rosenzweig et al. (Cambridge: Cambridge University Press)
[7] Calthorpe P 2011 Urbanism in the age of climate change (Washington: ISLAND PRESS)
[8] Capello R and P Nijkamp 2004 Urban dynamics and growth, in Advances in Urban Economics, ed B Baltagi, E Sadka and D Wildasin (Amsterdam: Elsevier)
[9] Simmie J and R Martin 2010 The economic resilience of regions: towards an evolutionary approach (Cambridge Journal of Regions, Economy and Society Vol 3 no 1) p. 27-43.
[10] Pickett S T A C, 2002 The Ecosystem as a Multidimensional Concept: Meaning, Model, and Metaphor. (Ecosystems vol 5) p. 1 - 10.
[11] Tansley A G 1935 The use and abuse of vegetational concepts and terms (Ecology vol 16) p. 284-307.
[12] Brown A, A Dayal and C R D Rie 2012 From practice to theory: emerging lessons from Asia for building urban climate change resilience (Environment and Urbanization vol 24 no 2) p. 531-556.

[13] Benjaminsen TA et al. 2012 Does climate change drive land-use conflicts in the Sahel? (Journal of Peace Research Vol 49 No 1) p. 97-111.
[14] Cutter S L 1996 Vulnerability to environmental hazards (Progress in Human Geography Vol 20 No 4) p. 529-539.
[15] IPCC 2012 Summary for Policymakers. In: Managing the Risks of Extreme Events and Disasters to Advance Climate Change Adaptation, in A Special Report of Working Groups I and II of the Intergovernmental Panel on Climate Change ed C B Field, V Barros, T F Stocker, D Qin, D J Dokken, K L Ebi, M D Mastrandrea, K J Mach, G-K Plattner, S K Allen, M Tignor and P M Midgley (Cambridge: Cambridge University Press)
[16] Gallopín G C 2006 Linkages between vulnerability, resilience, and adaptive capacity (Global Environmental Change Vol 16 No 3) p. 293-303.
[17] Adger W N et al 2003 Adaptation to climate change in the developing world (Progress in Development Studies Vol 3 No 3) p. 179-195.
[18] Folke C 2006 Resilience: The emergence of a perspective for social–ecological systems analyses (Global Environmental Change Vol 16 No 3) p. 253-267.
[19] Cutter S L, B J Boruff and W L Shirley 2003 Social Vulnerability to Environmental Hazards (Social Science Quarterly Vol 84 No 2) p. 242-261.
[20] Adger, W.N. and K. Vincent 2005 Uncertainty in adaptive capacity (Comptes Rendus Geoscience Vol 337 No 4) p. 399-410.
[21] Mulyana W et al 2013 Urbanisation, Demographics and Adaptation to Climate Change in Semarang, Indonesia, in Technical Briefing (Jakarta: UNFPA)
[22] Indonesia Statistic 2012 Kota Semarang dalam angka 2012 (Semarang: Indonesia Statistic and Planning and Development Board of Semarang)
[23] Nugraha A L, Purnama B S and T Adity 2012 Pemetaan risiko bencana banjir rob Kota Semarang, in The 1st Conference on Geospatial Information Science and Engineering (Yogyakarta: Jurusan Teknik Geodesi UGM)
[24] Wirasatriya A 2005 Kajian kenaikan muka laut sebagai landasan penanggulangan rob di pesisir Kota Samarang (Semarang: Diponegoro University)
[25] Azis A et al 2010 City resilience strategy: Semarang's adaptation plan in responding to climate change (Semarang TAHTA Undip: Semarang)
[26] Dewi A 2007 Community-based analysis of coping with urban flooding: a case study in Semarang, Indonesia, in International institute for geo-information science and earth observation enschede (Netherlands: ITC)
[27] Marfai M A K et al 2008 The impact of tidal flooding on a coastal community in Semarang, (Indonesia. Environmentalist vol 28) p. 237-248.
[28] Sariffuddin, S and R. Susanti 2011 Penilaian kesejahteraan masyarakat untuk mendukung permukiman berkelanjutan di Kelurahan Terboyo Wetan, Semarang (Makara seri sosial humaniora vol 15 No 1) p. 29 - 42.
[29] Adger WN 2006 Vulnerability (Global Environmental Change Vol 16 No 3) p. 268-281.
[30] Turner B L et al 2003 A framework for vulnerability analysis in sustainability science (Proceedings of the National Academy of Sciences Vol 100 No 14) p. 8074-8079.
[31] Boer R 2002 Analisis penilaian tingkat kerentanan, M.P.K.K. (VA), (Bogor: Center for Climate Risk and Opportunity Management-Bogor Agriculture University).
[32] Marfai M A and D R Hizbaron 2011 Community's adaptive capacity due to coastal flooding in Semarang coastal city, Indonesia (Analele UniversităŃii din Oradea – Seria Geografie Vol 2) p. 209 - 221.
[33] Tahta 2010 Strategi ketahanan kota rencana adaptasi Kota Semarang hadapi perubahan iklim (Semarang: Asian Cities Climate Change Resilience Network)
[34] Kementerian Pekerjaan Umum 2012 Draft Technical Report on Semarang Metropolitan Area (Kedungsepur) (Jakarta: Ministry of Public Work)
[35] Wilonoyudho S 2011 Overurbanization in Semarang City (Jurnal Teknik Sipil dan Perencanaan Vol 13 No 2) p. 191-200.
[36] Satterthwaite D 2011Editorial: Why is urban health so poor even in many successful cities? (Environment & Urbanization Vol 23 No 1)
[37] Sariffuddin S, Pradoto Wisnu, Damayanti Maya 2006 Quality of Life and The Perception of Community in Semarang (Case Study: Settlement Area in Genuk, Semarang). in 2nd International Conference on Environment and Urban Management (Science, Nature and Justice) (Semarang: Soegijapranata Catholic University)

[38] Sariffuddin S 2010 Pengaruh kesejahteraan masyarakat terhadap kondisi lingkungan permukiman di sepanjang Sungai Ngilir, Semarang. in Perubahan Iklim di Indonesia Mitigasi dan Strategi Adaptasi dari Tinjauan Multidisiplin (Yogyakarta: Sekolah Pascasarjana UGM)
[39] Soemarwoto O 1983 Ekologi, Lingkungan Hidup dan Pembangunan (Jakarta: Djambatan)
[40] Cutter S L et al 2008 A place-based model for understanding community resilience to natural disasters (Global Environmental Change Vol 18 No 4) p. 598-606.

Waves Induce Sediment Transport at Coastal Region of Timbulsloko Demak

Purnomo Hawati[1,2], Denny Nugroho Sugianto[1], Sutrisno Anggoro[1], Anindya Wirasatriya[1], Sugeng Widada[1]

[1] Faculty of Fisheries and Marine Diponegoro University, Prof. Soedharto, SH Road, Semarang, Indonesia
[2]Meteorology Climatology and Geophysics Agency, Kemayoran, Jakarta Pusat 10720, Indonesia
phawati.undip@gmail.com

Abstract. Waves have variation of length and period. This influenced by seasonal variation. Impacts of seasonal variation related to wave is playing important roles in sediment transport mechanism and coastal changes. This condition not only treats coastal area but also decreases economic capacity of community. This research aim to determine different seasonal based on Indonesia Northwest-Northeast monsoon and Southeast-Southwest monsoon and theirs relation to sediment transport. Area of study is suffered by 938.73 H of abrasion. Conducted by CERC methods, transforming BMKG Ahmad Yani wind database (2005-2015) into wave's data of seasonal variation and using field measurement recorded by ADCP. Definitely, maximum speed of wind is reached at 23 knots from December to February, wind direction predominantly from North West direction. Range of significant Wave Height (Hs) all season is 24.66-30.32 cm and Significant Wave period (Ts) count at 3.64-3.78 sec. Net of sediment volume annual 72,353.40 m^3year^{-1} and sediment movement pattern is forwarding from west to east direction. Correlation between breaking waves and sediment transport is linier.

Keywords: Wind, Waves, Sediment Transport, Timbulsloko, Demak

1. Introduction

The changing of coastal area has influenced by several factors. It has regularly affected by natural mechanism and has rapidly received human activities. One of natural factor which has influencing coastal area is sediment transport. In fact, measuring the mechanism of displacement some sediment from one to the other area is possible and reliable [1]. The movement of sediment has been basically triggered by breaking wave. Breaking wave and decreasing of bathymetries will be followed by unstable peak of wave and turbulence of current has been delivered sediment along coastal area [2]. The breaking wave which has been existed in surf zone will be transporting sediment along coastal area. On the other hand, location near equator or low latitude area is depending on seasonal aspect because this factor has been affected of waves [3]. Indonesia has two major seasons depend on rainfall. There are four part condition regularly which known as northwest-northeast monsoon, first transitional season, southeast-southwest monsoon and the second transitional season [4]. The differences of wavelength and wave height are influenced by seasonal variation. Studying of characteristics of waves is basically important because it is dealing with understanding the foundation and determining the amount of wave. The sophisticated information is compulsory because it can estimate the changed coastal area in present frame and predict its evolution in the future [5]. This research had been aimed to determine sediment transport relied on different seasonal condition because of Indonesia monsoon.

2. Methodology

2.1. Area of Study

The massive of coastal change mainly in abrasion of Demak Resident reach 938.73 H furthermore it considered as secondly ranking of abrasion in coastal area along Pantai Laut Utara (PANTURA) after Brebes[6]. The highest abrasion has been deconstructed village area of Timbulsloko and the community not only suffered by this condition but also loosed their ponds as major income [7]. Area of study had focused in coastal area of Demak area which has borderline with Semarang at west side and Jepara at east side. There are directly in front of Java Sea. All situations of area which had been discussed depict on Figure 1.

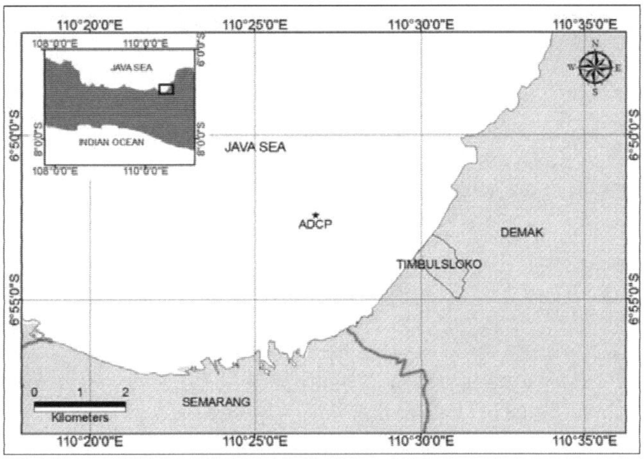

Figure 1. Area of Study

2.2. Collected Waves Field Data

The precedent procedures of data collecting was conducted by Wave Recorder Sontek Argonaut Type Xr 750 MHz particularly this instrument is highly recommended because this instrument is used more than two decades to measure waves and current [8]. The instrument was mounted at 6.89°S and 110.45°E position and staggered in 9.6 m water depth. Data of waves and current had been collected during 3-10 June 2016, sampling rate per 10 minutes were recording height and periods of waves, speed and direction of current.

2.3. Collected Wind Data

The obligatory in this research is assessment wind data which has information of speed and direction. These had been retrieving from BMKG Ahmad Yani 2005-2015 synoptic data which has enclosed by OGIMET. Wind data produce windrose graph. Transforming wind to wave in further processing of its value can generate measured longshore current and sediment transport known as methods of SMB [9].

2.4. Assessment of transforming Wind to Wave
Wind database ten years was divided into four season classification West monsoon (December, January, February); transitional season I (March, April, May) East monsoon (June, July, August) and transitional season II (September, October and November).

2.5. Measuring of Generated Wave and Sediment transport
Method to determine wind data convey into transport sediment speed along breaker zone and its volume using empirical equation [10]. The resuming formulas are followed by equation through (1) until (10).

Wind data transform to waves (Height and Period) are calculating sequentially wind stress factor (UA), wind speed (UW).

$$U_A = 0.71 U_W^{1.23} \quad (1)$$

$$U_W = R_L U_L \quad (2)$$

Waves which have happened at offshore determine wavelength (L_0), when they approach shoreline the refraction (Kr) and shoaling (Ks) are reforming of new height of waves (H). These formulas present at (3) and (4) equation.

$$L_0 = 1.56 T^2 \quad (3)$$

$$H = H_0 K_r K_s \quad (4)$$

Determining of breaking wave height and depth of breaking wave followed by equation (5) and (6)

$$\frac{H_b}{H_0'} = \frac{1}{3{,}3(H_0'/L_0)^{1/3}} \quad (5)$$

$$\frac{d_b}{H_b} = 1.28 \quad (6)$$

Mainly discussed on this study is sediment transport which has speed of longshore current along coastline (7) and volume of annual transport sediment (9) and daily (10) complete equation formulated are shown bellow.

$$V = 1{,}17(gHb)^{1/2} \sin\alpha_b \cos\alpha_b \quad (7)$$

$$P_l^n = \frac{\rho g}{8} H_b^2 C_b \sin\alpha_b \cos\alpha_b \quad (8)$$

$$Qs = 1290 P1 (m^3/year) \quad (9)$$

$$Qs = 3{,}543 P1 (m^3/day) \quad (10)$$

3. Result and Discussion

3.1. Waves Height and Period among field measurement
Simultaneous recorded data of waves during seven days at offshore area by instrument ADCP are presented on Figure 2. Wave's height fluctuation generates differentiation between highly and lowest height of its. The highest peak which reached 31.5 cm had happened after more than 24 hours of deployment and slowly decreased. On the day three until the end of time period of data collection wave's height more stable than before. Waves Height Significant is 13.7 cm. Waves period represented by Tmin; Tmax and Ts were 3.6 sec, 6.3 sec and 4.04 sec. Long waves is 25.46 m assisted by equation (3). Wave's classification is transitional and ratio of depth and long wave (d/L) is 0.38.

IOP Conf. Series: Earth and Environmental Science **55** (2017) 012048 doi:10.1088/1755-1315/55/1/012048

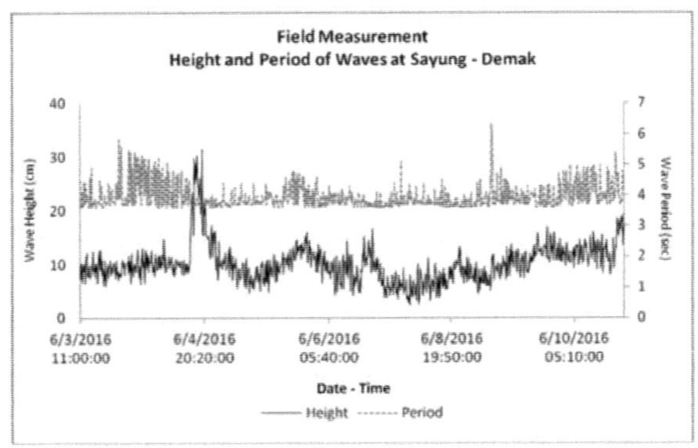

Figure 2. Field measurement of wave height and period recorded by ADCP

3.2. Wind Speed and Direction Plot

As mention above (sec 3.3) stated that there are four divided seasonal in ten years wind data. Figure 3 is represented all of seasonal variation and dominant of wave direction. At season of Northwest-northeast monsoon wind mostly originated at northwest contrary to Southeast-southwest monsoon wind directly opposite its directions. In transitional I and II wind were relatively coming from all direction.

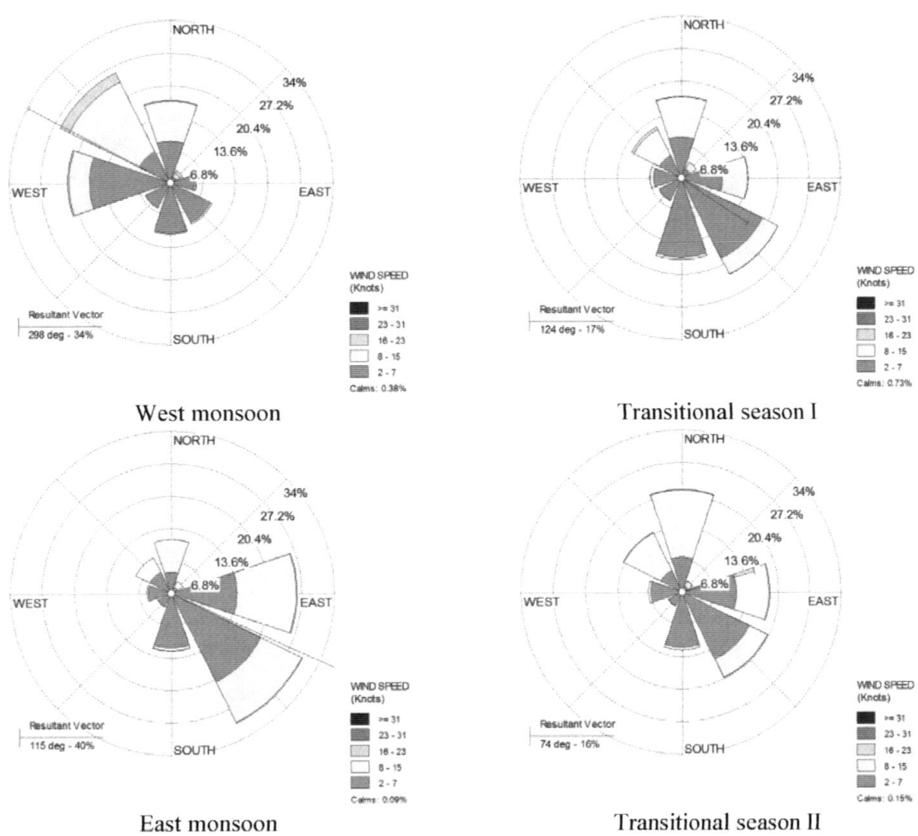

Figure 3. Wind speed and direction ten years (2005-2015) relied on varieties of seasonal condition

Plotting of wind called windrose proves that the value of speed in December, January and February is higher than other months, dominant wind force had been received from northwest direction and maximum wind speed velocity is 23 knot. Wind blowing on March until May come from two directions even though at northwest wind speed more than 16 knot still exists. It highly considerable at east monsoon wind forced from east direction with dominant velocity range is 8-15 knot. Uniformly speed at class 8-15 knot has happened at all season. Blowing wind circulate from north direction is highly occurring from September to November with velocity of wind at class 8-15 knot.

3.3. Significant Wave Height (Hs) and Significant Wave Period (Ts)

Transforming wind-wave data was connected into parameter itself within empirical methods. Based on that experienced, Table 1 present number of waves height and waves period among seasonal time and related to field measurement. Recording ADCP instrument had settled in between east monsoon season (June, July and August). Significant wave height at all season circumstances was compared between one and each others. The value of Hs is ranging from 24.66 cm to 30.32 cm. The highest predominant Hs had happened in period West monsoon and circulating begins December. Non monsoon season appointed lowest value of significant wave height (Hs) than monsoon[11]. In fact, at East period Hs number is below than Transition II. Hs value decreased after West monsoon because the direction of wind suddenly predominant came from 124 degree or relative opposite with primarily condition. In reality, study area coastline onset 50^0 orientations from north direction.

The number of significant wave's period (Ts) ranging from 3.64 sec to 3.78 sec. Periods of wave among seasonal event proved that the highest Ts also had happened from December to January season linier with Hs number. Interesting value of significant wave's period was rarely different value. Figure 4 and Figure 5 depicted Hs and Ts value.

Table 1. Summary of wave height and wave period.

Time	Waves Height (cm)			Waves Period (sec)		
	Hmin	Hmax	Hs	Tmin	Tmax	Ts
West monsoon (December, January, February)	2.13	134.90	30.32	2.97	5.74	3.78
Transition I (March, April, May)	6.65	151.53	24.66	3.12	5.98	3.64
East monsoon (June, July, August)	2.13	110.36	26.01	2.97	5.36	3.68
Transition II (September, October, November)	2.13	140.73	28.34	2.97	5.83	3.74
ADCP 6/3/2016-6/10/2016	2.4	31.5	13.7	3.6	6.3	4.04

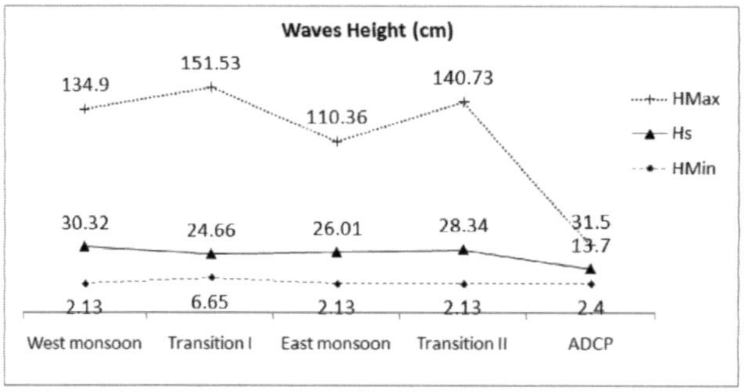

Figure 4. The number of wave's height (cm) among seasonal variation and field measurement

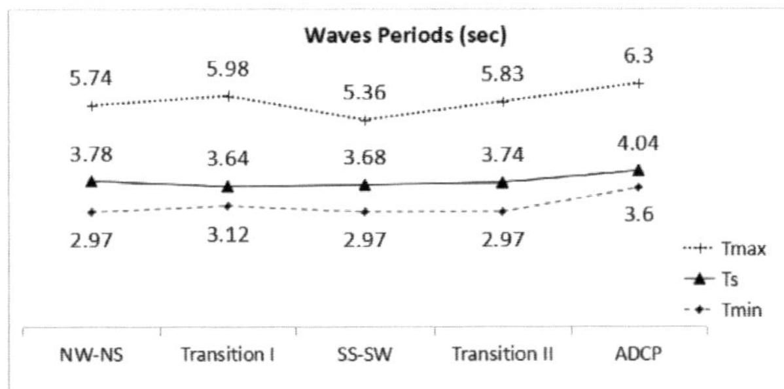

Figure 5. The number of wave's period (sec) among seasonal variation and field measurement

West monsoon period has the highest number of significant wave's based on calculation. Wave conveys the number of transport sediment because it influences breaking wave height and periods. Transported sediment is more massive when storming wind has happened and it will potential bring suspended sediment and reform area of coastal [12].

3.4. Sediment transport related to seasonal variation

Transforming wind-wave data was produced several parameter which play important roles to determine and describe the value of sediment transport. Table 2 shows different number among season. Arranging from west predominantly monsoon the number of annual sediment transport is outstanding because of its breaking wave deeper than others. Speed of longshore transport during December until two months later correlate to sediment transport.

During June until August the value of breaking wave height (Hb) and breaking wave depth (db) decreased because direction of wind come from relative south east to northwest must be consider at area of study is bordered by mainland. This condition influence of magnify of wind speed.

Table 2 Summary of breaking wave number on surfzone.

Season	α_0 (^{0})	Hb (m)	db (m)	αb (^{0})	Notation
West monsoon (December, January, February)	298	1.17	1.29	57.15	(-)
Transition I (March, April, May)	22.5	1.00	1.11	19.73	(+)
East monsoon (June, July, August)	75	0.77	0.85	48.56	(+)
Transition II (September, October, November)	320	0.92	1.02	33.01	(-)

Generally, all of seasonal variations prove their breaking wave height (Hb) and breaking wave depth (db) linier with speed of transport (V) and volume (Qs). During September-November phase the velocity of longshore current higher than next period but its high of breaking wave influence volume of sediment. If the breaking wave height and its depth increase, sediment transport number will be magnify. This kind of situation present on Table 3.

Sediment which has displaced based on calculated ten years data of wind shows the Nett of sediment volume is 72,353.40m3year-1 and predominantly moving forward from west to east direction. Timbulsloko located in near of Semarang Bay. Wave and current influence of stress of displacement particularly in coastal area with bay morphology [13].

Table 3 Summary of velocity of longshore current and sediment transport Net.

Season	V (ms^{-1})	Qs daily (m^3day^{-1})	Qs annual (m^3year^{-1})	Direction of displacement
West monsoon (December, January, February)	1.80	311.50	113,697.69	West-East
Transition I (March, April, May)	1.16	160.10	58,436.69	East-West
East monsoon (June, July, August)	1.60	149.21	54,461.12	East-West
Transition II (September, October, November)	1.61	196.04	71,553.51	West-East
Sediment Net		198.23	72,353.40	

On the other hand dominant factor which generate sediment transport is surf zone area. At this point breaking waves has been occurred and mixed some materials. Although there are several factors of properties of sediment, bathymetry, shape of wave and its properties, angle of coastal area [14]. Hence vegetation are play important roles in sediment net because it can support of increasing sediment when the thick of vegetation higher. Flux of sediment is related to its velocity [15].

4. Conclusion

Studies of sediment transport has been founded that during December until February in area of study held significant wave height and period based on condition which has speed of wind maximum velocity is 23 knot. Sediment transport which has transformed from wind to wave data and its connection with different seasonal variation was considering number of waves when it has slowly destructed. Based on this study, correlation between breaking waves and sediment transport is linier. Breaking wave is triggering sediment transport and it will be displaced sediment from one area depend on speed of transport along coastal area. Based on the study, different wave forming which had generated by wind results the variation number of longshore current and volume of sediment.

Necessity of studying sediment transport plays important roles because changing of coastal area will destruct not only morphological but also economical factors. Sediment transport as one of natural factor has rapidly effect in coastal area either abrasion or accretion. Focusing in area of study, after it has reliable database of sediment transport and its wave's properties, this study is recommend for the next management of coastal area must be prepare well plan and consider natural factors and their effects for coastal region and particularly communities who have been living and working in this area.

5. Acknowledgments

This acknowledge present for Centre of Education and Training Meteorology, Climatology and Geophysics (BMKG) which has encouraging scholarship at Master of Science Marine of Diponegoro University. Dr. Ita Widowati and Dr. Diah Permata Widjayanti in research collaboration with Wageningen University, Netherland. Lembaga Penelitian dan Pengabdian kepada Masyarakat (LPPM) and Pusat Unggulan Iptek (PUI). All collaborated colleagues in field studies and laboratory activity.

6. References

[1] Zheng J, Li R, Yu Y, and Suo A 2014 Influence of Wave and Current Flow on Sediment-Carrying Capacity and Sediment Flux at The Water – Sediment. Water Science and Technology. 338,1090–1099

[2] Baldock T E, Manoonvoravong P, and Pham K S 2010 Sediment Transport and Beach Morphodynamics Induced by Free Long Waves, Bound Long Waves and Wave Groups. Coastal Engineering 57, 898-916

[3] Chempalayil S P, Kumar V S, Dora G U, and Johnson G 2015 Near Shore Waves, Longshore Currents and Sediment Transport along Micro-tidal Beaches, Central West Coast of India. International Journal of Sediment Research 29 (3), 402–413

[4] Tjasyono H B 2007 Meteorology in Indonesian Equatorial Region. ASEAN Regional Climate Model, Meteorological and Geophysical Agency Symposium , 1-8

[5] Splinter K D, Davidson M A, Golshani A and Tomlinson R 2012 Climate Controls on Longshore Sediment Transport. Continental Shelf Research 48, 146–156

[6] Ministry of Marine and Fisheries 2016 Restoration of Coastal Area of Pantura Jawa (Demak Resident). Interim Report Montlhy Edition. 03, 2

[7] Wetland International 2016 Building With Nature. News of Wetland Conservation. 24(1), 9

[8] Merckelbach L M 2006 A Model for High-Frequency Acoustic Doppler Current Profiler Backscatter from Suspended Sediment in Strong Currents. Continental Shelf Research 26(11), 1316–1335

[9] Siswanto A D 2015 Longshore Current Characteristics in Madura Strait. Procedia Environmental Sciences (ICTCRED 2014). 23 , 34–38.

[10] CERC 1984 Shore Protection Manual. US Army Corps of Engineers.I,1-143

[11] Shanas P R and Kumar V S 2014 Coastal Processes and Longshore Sediment Transport along Kundapura Coast, Central West Coast of India. Geomorphology. 214. ,436–451

[12] Aagaard T and Jensen S G 2013 Sediment Concentration and Vertical Mixing under Breaking Waves. Marine Geology. 336, 146–159

[13] Carlin J A, Lee G, Dellapenna T M, and Laverty P 2016 Estuarine, Coastal and Shelf Science Sediment Resuspension by Wind, Waves and Currents during Meteorological Frontal Passages in A Micro-tidal Lagoon. Estuarine, Coastal and Shelf Science. 172, 24–33

[14] Wang P, Yuan W, and Min L 2012 Measuring and Modeling Suspended Sediment Concentration Profiles in The Surf Zone. Journal of Palaeogeography.1(2), 172–192

[15] Ma G, Han Y, Niroomandi A, Lou S and Liu S 2015 Numerical Study of Sediment Transport on A Tidal Flat with A Patch of Vegetation. Ocean Dynamics. 65, 203–222

Potential impact of climate variability on respiratory diseases in infant and children in Semarang

Budiyono[1], **Rismawati**[2], **S P Jati**[3] and **P Ginandjar**[4]
[1,2,3,4] Faculty of Public Health, Diponegoro University – Indonesia
Jl. Prof. Soedarto SH, Tembalang, Semarang – Indonesia

E-mail: kenang92@yahoo.com

Abstract. Temperature, humidity, and rainfall may influence respiratory disease, including acute respiratory infection (ARI) and pneumonia. In Semarang, the temperature and humidity has increased 0.1°C and 1.6% respectively during 2002-2011. ARI and pneumonia in children under 5 years had increased during 2012-2014. This study aimed to analyze the relationship of climate variability and ARI and pneumonia incidence. It was an ecological study. Subject consisted of patients visited primary health care of Bandarharjo from 2011 to 2015. Pneumonia was related to infants (<1-year-old) and children (1-4 years old), while ARI was related to children (≥5 years old). Data of climate was obtained from Agency for Meteorology, Climatology and Geophysics (BMKG) Semarang. Pearson correlation (α=0.05) was used to analyse the correlation of the 60 samples. Mean of temperature was 27.96° C, relative humidity was 74.73%, and rainfall was 179.98 mm/month. The total of ARI was 38523 cases and pneumonia was 1558 cases. Temperature, humidity, and rainfall had no correlation to pneumonia. Humidity had a significant correlation to ARI on female children and total ARI (r=0.3 and r=0.26; p–value=0.02 and 0.04 respectively). Rainfall and temperature had no correlation to total ARI. This study concluded humidity has potential impact to ARI.

Keywords: climate variability, acute respiratory disease, pneumonia, infant, children

1. Introduction

Climate change refers to any change in climate over time due to natural variability or as a result of human activity [1]. The change of climate influences regional weather change, such as heat wave, extreme weather, temperature and precipitation [2]. Average temperature of earth surface rises to exceed the safe threshold of 2°C above preindustrial average temperature during the 21st century [3].

Climate change harms human health, both directly and indirectly. Direct effects include changes of earth system, such as rising temperatures, increasing climate variability, rainfall or snowfall in some areas, and drought. Indirectly, climate-related ecosystem changes increases the range, seasonality, and infectivity of some vector-borne diseases [4]. Drought conditions produce pollution that creates multiple health challenges such as more dust and particulates [5].

Some studies showed the relationship between climate and respiratory diseases. The proportion of primary care visits due to respiratory diseases varied according to seasonal changes [6], [7]. Ambient temperature and absolute humidity correlate to the occurrence of upper respiratory tract infection (URTI) [8].

IOP Conf. Series: Earth and Environmental Science **55** (2017) 012049 doi:10.1088/1755-1315/55/1/012049

Acute respiratory disease caused of death among children aged <5 years and it was contributed 15% globally and 17% in South-East Asia Region in 2015 [9]. Pneumonia and diarrheal are among the leading causes of child mortality in the world [10]. One of the populations most at risk for pneumonia is children under five years [11].

The mean minimum temperature, rainfall and the incidence of respiratory syncytial virus (RSV) and pneumonia infections was the significant correlation [12]–[14]. The number of hospitalizations due to pneumonia on children, infants, and adults, was in accordance with the increase or decrease in temperature, humidity, precipitation, wind speed, and thermal comfort index [15]. Temperature, sunshine duration, relative humidity, and concentration of pollutants were significantly correlated to severe acute respiratory infection [16]. Climate change directly threaten respiratory health by promoting or aggravating respiratory diseases or indirectly by increasing exposure of risk factors for respiratory diseases [17].

The epidemiological outcome of climate change on disease patterns worldwide will intense, especially in developing countries [3]. Temperature is related to the prevalence of upper and lower respiratory tracts infection in developing country [18]. Climate change will affect the health of urban populations [19]. Higher temperatures occur in urban areas than in rural area [20], and have a different effect on hospitalizations for respiratory disease [21].

The evidence of climate change in Semarang is monthly increase of surface temperatures over the last 100 years [22]. Over the next 100 years, the temperature may rise up to 1°C [23]. According to Minanda et al, during 2002-2011, the mean temperature was 27.7°C and relative humidity 76%. The climate, however, has changed during 2002-2011, the mean temperature has increased 0.1°C and the mean of relative humidity has increased 1.6% [24].

Based on primary health care reports in 2013 and 2014, acute respiratory infection (ARI) in the upper tract system on children was the main health problem in Semarang. The cases were 85125 (in 2013) and 56376 (in 2014). According to the hospital report in Semarang, cases of ARI were 2290 cases in 2013 and 1696 cases in 2014. Pnemonia in infants have also increased during 2012-2014. The average of pneumonia cases in children under five years old was 3079 [25], [26]. The ARI and pneumonia in Semarang are potentially influenced by climate variability. Objective of this study was to analyse the climate variability and respiratory disease in infant and children who visited in PHC Bandarharjo-Semarang.

2. Methodology

The research was an ecological study [27]–[29] and carried out in coverage area of the primary health care (PHC) Bandarharjo, Semarang. The coverage area consists of 4 villages, i.e. Bandarharjo, Tanjung Mas, Kuningan, and Dadapsari. The population of children under-five years from January 2011 to December 2015 estimated 7530 to 7841. The total study subject were 38993 cases of acute respiratory infection (ARI) in children ≥ 5 years old and 1588 cases of pneumonia in children aged <1 year and 1-4 years old who visited the PHC of Bandarharjo during January 2011 to December 2015.

Acute respiratory diseases included acute respiratory infection (ARI) in the upper tract system ("ISPA") and pneumonia. ARI and pneumonia were diagnosed by the medical doctor and paramedic used standard diagnostic (guideline) for ARI and pneumonia [30]. ARI located above the epiglottis were considered as upper airway diseases, while those at the epiglottis and below were considered as diseases of the lower respiratory tract. Symptoms related to the respiratory tract (such as coughing), but without diagnosis, were defined as a non-specific respiratory disease [6]. It was excluded from the samples. Pneumonia is a form of acute respiratory tract infection (ARTI) that affects the lungs. When an individual has pneumonia, the alveoli in the lungs are filled with pus and fluid, which makes breathing painful and limits oxygen intake [11].

Reliability and validity of data were ensured by cross check to registered diseases at PHC of the Bandarharjo. ARI and pneumonia were old and new cases.

Climate refers to the average state of the atmosphere for a given time scale (hour, day, month, season, year, the decade and so forth) and for a specified geographical region. The average-state of atmosphere involves temperature, precipitation, wind, cloudiness and sunshine, pressure, visibility, humidity and elements with noteworthy human impacts [31]. Monitored data of temperature, relative humidity and rainfall at Semarang level and area of PHC Bandarharjo were obtained from Agency for Meteorology, Climatology and Geophysics (BMKG) of Semarang. Semarang has 9 climate monitoring stations and distributed in Ngaliyan, Tanjung Mas, Siliwangi, Genuk, Tlogosari, Candi, Klipang, Mijen, Gunungpati. The Tanjung Mas station is the one to monitor the variability of temperature, relative humidity and rainfall at surrounding area included area PHC of Bandarharjo.

The climate variability used mean of temperature (°C), mean of relative humidity (%) and mean of rainfall (mm) per month both at Semarang level and area PHC of Bandarharjo. The temperature measured by wet and dry bulb thermometer. Relative humidity measured by the wet and dry hygrometer. Rainfall measured by rainfall instrument type Hillman with an auto record. The BMKG maintained the climate equipment at least a week to ensure sensitivity of equipment and it was calibrated every 5 years or when replaced by new equipment. Monitoring of climate (temperature, relative humidity) conducted every day on 07.00, 13.00, and 18.00 at local time and recorded in form, it is called Fklim71. The rainfall is measured on 07.00 at local time and recorded in rainfall card. The climate data at Semarang level and area PHC of Bandarharjo reported to BMKG of Semarang every month [32]. Rainfall is classified into four levels: 0-50 mm/month (low), >50-150 mm/month (medium), >150-300 mm/month (high), and > 300 mm/month (very high) [33]. The climate data at Semarang level for picturing and comparing the temperature, relative humidity and rainfall to climate at area PHC of Bandarharjo.

The data of climate at area PHC of Bandarharjo as independent variables were obtained from daily monitoring at the Tanjung Mas station, and it was transformed to mean of temperature, relative humidity and rainfall per month. As dependent variables were ARI and pneumonia cases per month. Total of samples were 60 climate, 60 ARI and 60 pneumonia cases. It was calculated from average climate data and diseases (ARI and Pneumonia) per month during 2011-2015. Spearman correlation was used to analyze the correlation between climate and ARI and pneumonia cases with α=5% [34].

3. Result and Discussion

3.1. City of Semarang

Semarang is located between the 6°50 '- 7°10' south latitude and 109°35 '- 110°50' east longitude [35]. Restricted to the west with Kendal, at East with Demak, at Semarang regency in the south and the north is limited by the length of the Java Sea coastline which covers 13.6 km [36]. Altitude of Semarang lies between 0.75 to 348 m above the sea level [35]. The total area of Semarang is 373.7 km^2, composed of 16 districts and 177 villages [37]. The population of the Semarang amounted to 1575068 people, consisting of 787705 men and 797176 women [38]. The population density of Semarang was 4172 per km^2 [39].

The climate condition at Semarang level during 2011-2015 consisted of temperature 27.97°C±0.83°C, relative humidity 74.83%±7.88%, and rainfall 180.75 mm±149.34mm per month. Rainfall was classified as very high [33]. Mean of temperature during five years has increased 0.58°C, but mean of relative humidity and mean of rainfall had fluctuated. Mean of temperature in 2011 was 27.72°C, in 2012 and 2013 were 27.95°C, and in 2014 was 27.98°C. The highest mean temperature was in 2015 (28.30°C). Relative humidity during five years ranged 58-90% per month, while rainfall ranged 0-736 mm per month.

3.2. Area of primary health care of the Bandarharjo

PHC Bandarharjo covers an area that lies 0.75 meters above sea level. The area is located at shore line and most of it is covered by tidal inundation (local term: rob) [35]. Total area PHC of Bandarharjo is 761.1 km^2 and it consists of Bandarharjo village (342.7 hectare), Tanjung Mas village (330 hectare), Kuningan village (46.9 hectare), and Dadapsari village (41.5 hectare). The total population of four villages in 2014 was 78394 [37]. The distribution of population: Bandarharjo village (20967 people), Tanjung Mas village (30678), Kuningan village (15427 people), and Dadapsari village (11322 people).

Temperature in area of PHC Bandarharjo during 2011-2015 (table 1) was 28.15°C±0.72°C, relative humidity was 76.13%±6.45%, and rainfall was 180.27 mm/month±167.34mm (very high).

Mean of temperature for the last four years (2012-2015) has increased 0.23°C. Mean of temperature in 2011 was 28.13°C, in 2012 was 28.05°C, in 2013 was 28.12°C, in 2014 was 28.15°C and in 2015 was 28.28°C. The highest mean of temperature was in 2015. The mean of relative humidity and rainfall per year fluctuated. Relative humidity ranged 63-86%/month and rainfall ranged 0-992 mm/month.

Based on mean rainfall per month, type of month in PHC Bandarharjo can classify as wet month (rainfall>100 mm), humid month (60-100 mm), dry month (<60 mm) [40]. The wet month is more frequent than the dry month and humid month. Cases of pneumonia and ARI in PHC Bandarharjo were higher in the wet month than in the dry and humid months (data not showed). It was clarified that during the dry season there were 21% fewer visits for respiratory disease in health facility [6].

Climate change influences the regional weather changes in accordance with temperature, sea level, precipitation, and extreme weather events. Those will cause downstream effects on the environment that lead to adverse health effects [3]. Coastal communities, in low-income countries, are vulnerable to a range of health effects due to climate variability and long-term climate change [41]. Geographically, the area of PHC Bandarharjo lies in coastal area and the most of the population are categorized as low income.

During rainfall or wet month the temperature will decrease [42]. The pattern of peak of the pneumonia cases was in January-February-March and November-December. The patterns of pneumonia cases was similar to rainfall and humidity, particularly in high rainfall (figure 1), and the correlation was not significant (p=0.08). High transmission of pneumonia occurs during high rainfall (June-October) and regular surface wetness [43]. Pneumonia is affected by colder temperature. A temperature decrease from one day to the next had an adverse impact on childhood pneumonia [44]. The increased incidence of respiratory disease (pneumonia) during the colder periods of the year is due to the low temperatures [15]. Associated with cold days in Central Australia climate zone, there was an increased in acute respiratory admissions for children [21]. The conceptual mechanism of infection may illustrate that a decrease in air temperature causes a decrease in the temperature of the nasal airway. Cooling of the nasal airway compromises respiratory defend against infection by slowing mucociliary clearance and inhibiting leukocyte phagocytosis. The sign of warming of the body and airway associated with fever and nasal congestion that are natural way to defend against infection [45]. The most common pathogens that caused pneumonia are *Streptococcus pneumoniae*, *Haemophilus influenza* type b (Hib), and respiratory syncytial virus (RSV) [11], [46]. Contradictive with Kim's finding that the risk of pneumonia was higher in the dry season than in the rainy season. There was variability in the relationship between climate factors and pneumonia [14].

Mean of temperature (0.17°C) and mean of relative humidity (1.3%) were higher in area PHC of Bandarharjo than Semarang in general, but mean of rainfall in area PHC of Bandarharjo was lower (-0.48 mm/month). According to the statistical test, there were significant different between mean temperature, relative humidity, and rainfall in Semarang and in area PHC of Bandarharjo (p-value=0.0001) (data not showed). Bandarharjo had higher temperature because it lies in the coastal area. It was confirmed that global mean sea surface temperatures (SST) have risen, with associated warming temperature in the coastal area [47].

Total patients visited (old and new visits) to PHC of Bandarharjo was 63222. Total pneumonia (table 1) in Bandarharjo village was 1072 (68%), Tanjung Mas 363 (22%), Kuningan was 107 (7%) and Dadapsari village was 50 (3%) out of 1592 cases. The average of pneumonia cases was higher in children aged 1-4 years than in infant (<1 year). In Kuningan village, Pneumonia cases in children under five years both in male and female in 2012 to 2015 tended to increase. It had the same trend with the temperature, although the causal mechanisms were not well understood.

Table 1. The number of Pneumonia cases on infants and children under five years and climate during 2011-2015 in area PHC of Bandarharjo.

Year	Bandarharjo village		Tanjung Mas village		Kuningan village		Dadapsari village		Mean temperature	Mean relative humidity	Mean rainfall
	Male	Female	Male	Female	Male	Female	Male	Female			
	Pneumonia on infant (<1 year old)										
2011	6	10	5	3	1	1	2	1	28.13	76.92	194.5
2012	50	30	14	13	2	3	0	0	28.05	75.42	177.17
2013	60	40	9	17	4	3	2	3	28.12	77.17	202.58
2014	26	31	11	11	3	3	0	2	28.15	76.33	199.42
2015	44	44	19	12	7	12	4	4	28.28	74.83	127.67
Mean	37.2	31	11.6	11.2	3.4	4.4	1.6	2	28.15	76.13	180.27
	Pneumonia on children (1-4 years old)										
2011	24	21	9	6	7	0	0	0	28.13	76.92	194.5
2012	75	61	23	16	1	1	2	2	28.05	75.42	177.17
2013	125	106	35	34	5	4	1	1	28.12	77.17	202.58
2014	85	66	37	24	11	14	3	3	28.15	76.33	199.42
2015	69	97	35	30	18	16	5	15	28.28	74.83	127.67
Mean	75.6	70.2	27.8	22	8.4	7	2.2	4.2	28.15	76.13	180.27

The highest cases of ARI in children ≥5 years (Table 2) were in Bandarharjo village (31.53%) and the lowest cases was in Dadapsari village (20.75%). Cases of ARI in Tanjung Mas and Kuningan were 8356 cases (21.24%) and 10243 cases (26.28%) respectively. ARI cases declined in all villages since 2011, but rose in 2015. Based on gender, ARI in children ≥5 years found higher proportion in female than male.

Table 2. The number ofARI cases on children ≥5 years and climate during 2011-2015 in area PHC of Bandarharjo.

Year	Bandarharjo village		Tanjung Mas village		Kuningan village		Dadapsari village		Mean temperature	Mean relative humidity	Mean rainfall
	Male	Female	Male	Female	Male	Female	Male	Female			
2011	2189	2216	1022	1021	1649	2115	1286	1413	28.13	76.92	194.5
2012	1220	1906	930	1348	1333	1694	1053	1386	28.05	75.42	177.17
2013	731	1154	665	1051	608	855	539	719	28.12	77.17	202.58
2014	433	715	367	618	316	595	290	535	28.15	76.33	199.42
2015	603	1123	463	871	372	706	301	564	28.28	74.83	127.67
Mean	1035.2	1422.8	689.4	981.8	855.6	1193	693.8	923.4	28.15	76.13	180.27

The proportion of total pneumonia by sex was higher in male than female in three villages (Bandarharjo, Tanjung Mas, and Kuningan) (table 1). As many as 82.1% children with pneumonia were under five years of age, and almost half of children with pneumonia (46.4%) were infants. The boys were leading in all age groups [48].

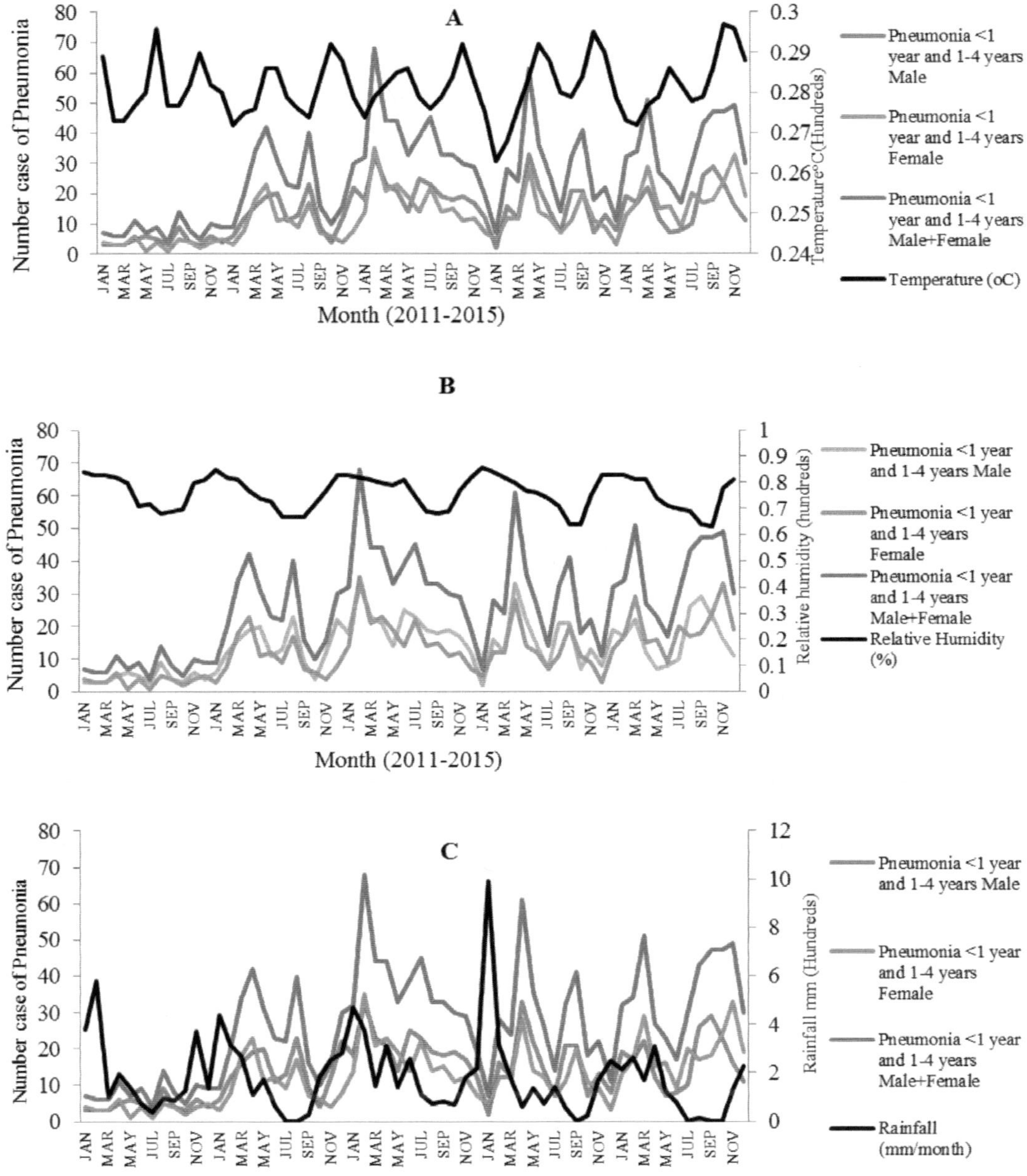

Figure 1. The pattern of monthly average of the temperature (A), humidity (B), rainfall (C) and Pneumonia (infant and children <5 years old)

IOP Conf. Series: Earth and Environmental Science **55** (2017) 012049 doi:10.1088/1755-1315/55/1/012049

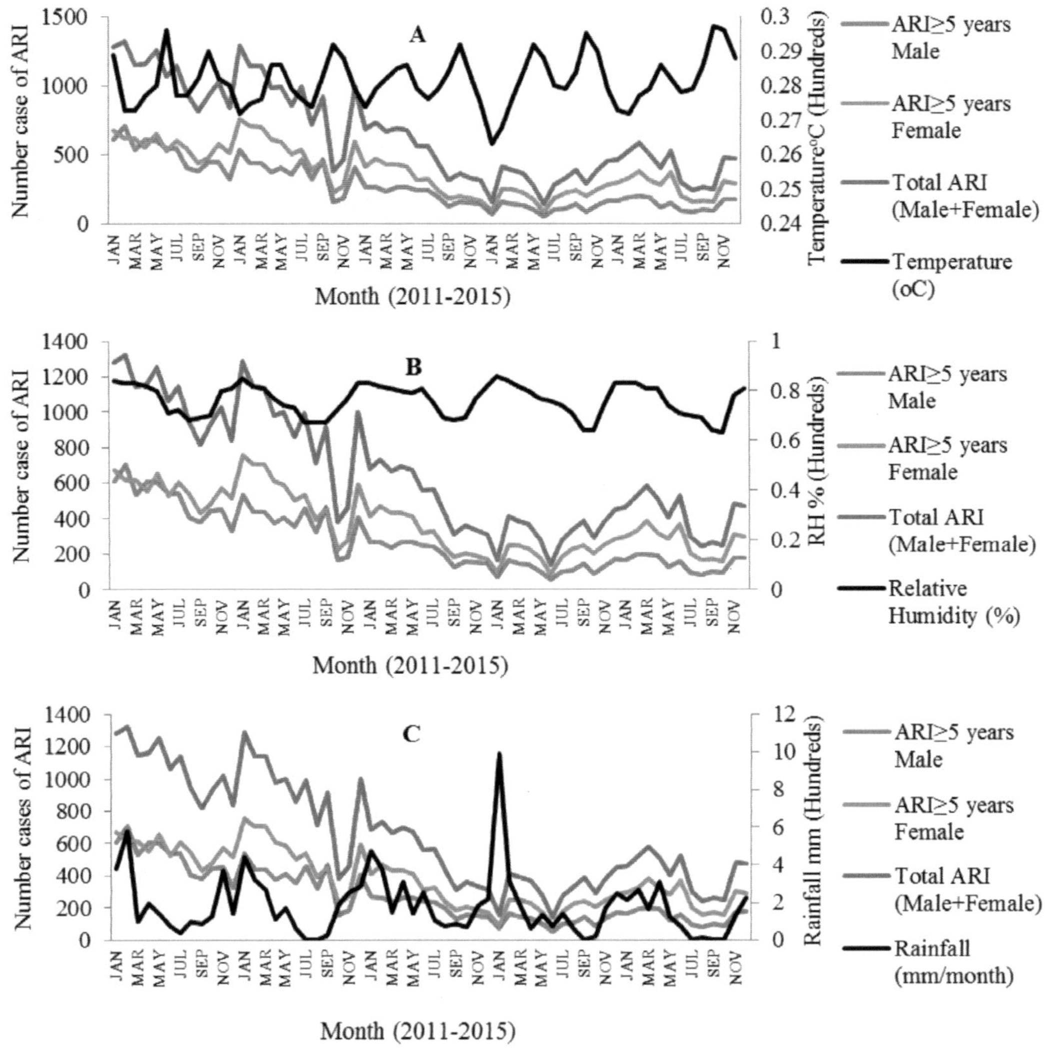

Figure 2. The patterns of mean temperature (A), humidity (B), rainfall (C) and total ARI (children ≥5 years old)

3.3. Correlation between climate and Pneumonia and ARI

Temperature, relative humidity, rainfall had no correlation to total pneumonia in infant and children. The average cases of Pneumonia per month tend to increase during 2011-2015, both of infant and children under five years old and also by gender. Relative humidity had a positive correlation to total ARI (r=0.26; p–value=0.04). Relative humidity had the positive correlation to ARI in female children (r=0.3; p–value=0.002), temperature and rainfall had no correlation to pneumonia (p=0.35, r=-0.12; p=0.08, r=-0.22 (table 3).

IOP Conf. Series: Earth and Environmental Science **55** (2017) 012049 doi:10.1088/1755-1315/55/1/012049

Table 3. The summary correlation between climate and pneumonia and ARI

Climate	Pneumonia <1 year		Pneumonia 1-4 years		Total Pneumonia (<1 year+1-4 years)	ARI≥5 years		Total ARI≥5 years
	Male (n-60)	Female (n-60)	Male (n-60)	Female (n-60)	(n-60)	Male (n-60)	Female (n-60)	(n-60)
Mean temperature	p=0.72 r=0.05	p=0.26 r=0.15	p=0.66 r=0,06	p=0.20 r=0.17	p=0.35 r=0.12	p=0.17 r=-0.18	p=0.12 r=-0.21	p=0.1 r=-0.19
Mean relative humidity	p=0. 17 r=-0.18	p=0.87 r=0.02	p=0.62 r=-0.07	p=0.59 r=-0.07	p=0.53 r=-0.08	p=0.10 r=0.21	p=0.02 r=0.30	p=0.04 r=0.26
Mean rainfall	p=0.08 r=-0.23	p=0.47 r=0.09	p=0.12 r=-0.21	p=0.10 r=-0.21	p=0.08 r=-0.22	p=0.22 r=0.16	p=0.21 r=0.17	p=0.20 r=0.17

Respiratory illness remains a major killer and cause of morbidity for children, causing almost 20% of all under-five deaths. Vulnerability of children to respiratory disease not all of which are directly affected by climate change [49]. Children more vulnerable than adults, because children of more rapid metabolism, the immaturity of the child's respiratory system, were the development of the lungs and thorax is progressive and continues until ten years of age, as well as the capacity to adapt to climate issue [49]-[50].

Our result was quite similar to Falagas that relative humidity had correlation to upper and lower respiratory tract infection [51]. Relative humidity causes both direct and indirect health effect. Very low or high relative humidity may cause physical discomfort, as the relative humidity of the air directly affects to temperature perception. Relative humidity as a determinant of the incidence of infections will depend on the relative strength of the settling rate of aerosols and the survival of airborne pathogens [52]. The incidences of respiratory infection were found to be lower among people who live in environments with a relative humidity level between 40% - 70% [53]. The recovery of the airborne pathogen (virus) was higher at a higher relative humidity and the stability of the aerosol was at the relative humidity of 60%. Humidity performs and important role and can affect the transmission of the virus [50]. When relative humidity moved from 1% to 6% there was a 39% increase in the cases of respiratory syncytial virus (RSV) [7]. Seasonality of certain ARI pathogens can be explained by meteorological influences, for example Rhinovirus correlated to relative humidity [54].

Rainfall had no correlation to total ARI, but had the same pattern with the total of ARI, whereas temperature and relative humidity had no same pattern with total ARI. Primary care visits for respiratory disease, especially those due to upper airway diseases, are related to the rainy season [6].

Children's vulnerability to respiratory disease may related to their proximity to traffic, the level of crowding in their homes, and the cooking or heating fuel burned within their homes. But it can also be related to ambient outdoor air quality [49]. Indoor aspect due to indoor air pollution resulted from fuel for cooking. The quality of housing contributed to respiratory disease of home's inhabitant. The tighter of the building made by brick will easy to spread indoor air pollution and influence the home's inhabitant [55]. The coverage of healthy housing proportion in coverage area PHC Bandarharjo in 2012-2014 tended to decline, i.e. 508 out of 582 inspected houses (87.29%), 510 out of 600 (85%), 277 out of 400 (69.25%) respectively. Lack of housing quality and the overcrowding found in temporary housing for the homeless contribute to morbidity from respiratory infections and activation of tuberculosis [56].

Respiratory infections are also influenced by outdoor air pollution. The outdoor air pollution affected by climate change. Change in climate globally will affect to regional weather and potential to produce air pollution through anthropogenic, natural emission and atmospheric process [2], [57]. If the climate becomes warmer and more variable, air quality is likely to be affected [57]. Climate change also affected ozone concentration [58] and particulate matter (PM) concentration in polluted environments by 0.1–1 mg/m^3 over the coming decades [59].

Air pollutions (PM, Ozone, sulfur dioxide) are environmental risk factors of respiratory diseases [60]. Ground-level ozone can exacerbate chronic respiratory diseases and cause short-term reductions in lung function. Exposure of PM can aggravate chronic respiratory [57]. Ozone inhalation induces epithelial damage and consequent inflammatory responses in the upper and lower airways [60]. Short-term exposure to $PM_{2.5}$ increases the risk for hospital admission for cardiovascular and respiratory diseases [61]. The weekly number of RSV positive cases was also correlated to the mean PM_{10} concentration [12]. Exposures to carbon monoxide, sulfur dioxide, and nitrogen dioxide can affect respiratory illnesses, lung irritation, and alterations in the lung's defense systems [57]. Particulate matter significant associated with emergency department visits due to asthma, wheezing, bronchitis, and lower respiratory tract symptoms. High ozone levels seem to be linked to asthma and asthma-like symptoms [60].

4. Conclusion

Pneumonia cases in children under five years both in male and female in 2012 to 2015 tended to increase and had same trend with temperature. Relative humidity had potential impact on total ARI incidence and female children. The effect of relative humidity on ARI varies between sexes. Rainfall had the same pattern with the total ARI and it has potential to respiratory infection. Cases of pneumonia and ARI in PHC Bandarharjo were higher in the wet month than in the dry and humid months. It is advisable to consider other factors that may contribute to the incidence of respiratory tract infections, such as indoor and outdoor air pollution, housing condition.

5. Acknowledgment

The authors would like to thank Agency for Meteorology, Climatology and Geophysics (BMKG) Semarang - Indonesia for serving climate data at Semarang level and area of PHC Bandarharjo, and to District Health Office of Semarang and Primary Health Care of Bandarharjo, Semarang for providing data of diseases and health profile.

References

[1] Parry M L, Canziani O F, Palutikof J P, van der Linden P J and Hanson C E 1997 Climate Change 2007 *Impact, Adaptation and Vulnerability* Eds Cramer W and Murdiyarso D (Cambridge:Cambridge University Press) p 27

[2] Patz J A, McGeehin G A, Bernard S M, Ebi K L, Epstein P R, Gubler D J, Reiter P, Romieu I, Rose J B and Samet J M The potential health impacts of climate variability and change for the United States: executive summary of the report of the health sector of the U.S. national assessment 2000 *Environ. Health Perspect.* **108** pp 367–76

[3] Costello A, Abbas M, Allen A, Ball S, Bell S, Bellamy R, Friel S, Groce N, Johnson A, Kett M, Lee M, Levy C, Maslin M, McCoy D, McGuire B, Montgomery H, Napier D, Pagel C, Patel J and Antonio J 2009 Managing the health effects of climate change *Lancet* **373** pp 1693–733

[4] Maibach E, Nisbet M and Weathers M 2011 Conveying the human implications of climate change *A Climate Change Communication Primer for Public Health Professionals* (Fairfax VA:George Mason University) pp 8-11

[5] Takaro T K, Knowlton K and Balmes J 2013 Climate change and respiratory health *Respir. Med.* **7** pp 349–361

[6] Rosa A M, Ignotti E and Botelho C 2008 Respiratory disease and climatic seasonality in children under 15 years old in a town in the Brazilian Amazon *J. Pediatr.* **84** pp 543–49

[7] Omer S B, Sutanto A, Sarwo H, Linehan M, Djelantik I G G, Mercer D, Moniaga V, Moulton L H, Widjaya A, Muljati P and Gessner B D 2008 Climatic, temporal, and geographic characteristics of respiratory syncytial virus disease in a tropical island population *Epidemiol. Infect.* **136** pp 1319–27

[8] Ma T M, Harju T H, Peitso A, Bloigu A, Silvennoinen-Kassinen S, Leinonen M and Hassi J 2009 Cold temperature and low humidity are associated with increased occurrence of respiratory tract *Respir. Med.* **103** pp 456–62

[9] WHO *World Health Statistic* 2015 First (Geneva: WHO Press) p 73

[10] WHO/UNICEF 2013 Ending Preventable Child Deaths from Pneumonia and Diarrhoea by 2025 *The integrated Global Action Plan for Pneumonia and Diarrhoea (GAPPD)* First (Geneva: WHO Press) p 5

[11] Tong B N 2013 Priority Medicines for Europe and the World *A Public Health Approach to Innovation Update on 2004 Background Paper Background Paper 6.22 Pneumonia* pp 7-15

[12] Vandini S, Corvaglia L, Alessandroni R, Aquilano G, Marsico C, Spinelli M, Lanari M and Faldella G 2013 Respiratory syncytial virus infection in infants and correlation with meteorological factors and air pollutants *Ital. J. Pediatr.* **39** pp 1–6

[13] Pica N and Bouvier N M 2014 Ambient Temperature and Respiratory Virus Infection *The Pediatric Infectious Disease Journal* **33** p 311-13

[14] Kim J, Kim J H, Cheong H, Kim H, Honda Y and Ha M 2016 Effect of Climate Factors on the Childhood Pneumonia in Papua New Guinea : A Time-Series Analysis *Int. J. Environ. Res. Public Heal.* **13** pp 1–16

[15] De Souza A, Fernandes W A and Lastoria G 2012 Potential impacts of climate variability on respiratory morbidity in children, infants, and adults *J. Bras. Pneumol.* **38** pp 708–15

[16] Silva D R, Viana P and Alice M 2013 Respiratory viral infections and effects of meteorological parameters and air pollution in adults with respiratory symptoms admitted to the emergency room *Influenza Other Respir. Viruses* **8** pp 42–52

[17] Ayres J G, Forsberg B, Annesi-Maesano I, Dey R, Ebi K L, Helms P J and Medina-Ramon M 2009 Climate change and respiratory disease : statement *Eur. Repir. J.* **34** pp 295–302

[18] Erling V, Jalil F and Lars A 1999 The impact of climate on the prevalence of respiratory tract infections in early childhood in Lahore, Pakistan *J. Public Health Med.* **21** pp 331–39

[19] Kovats S and Akhtar R 2008 Climate, climate change and human *Environ. Urban.* **20** pp 165–175

[20] Satterthwaite D, Huq S, Pelling M, Reid H and Lankao P R 2007 Adapting to Climate Change in Urban Areas: *The possibilities and constraints in low-and middle-income nations* (London: International Institute for Environment and Development (IIED)) pp 7-8

[21] Green D, Bambrick H, Tait P, Goldie J, Schultz R, Webb L, Alexander L and Pitman A 2015 Differential Effects of Temperature Extremes on Hospital Admission Rates for Respiratory Disease between Indigenous and Non-Indigenous Australians in the Northern Territory *Int. J. Environ. Res. Public Health* **12** pp 15352–65

[22] Mulyana W, Dodman D, Zhang S and SchensulD 2013 Technical Briefing:*Climate vulnerability and adaptation in the Semarang Metropolitan Area: a spatial and demographic analysis* (London and New York:UNFPA-IIED) pp 1-4

[23] PAKLIM 2014 *Integrated Climate Change Strategy in Semarang* (Jakarta) p 1

[24] Minanda R K, Budiyono, Winarni S, Jati S P and Ginandjar P 2013 Case Study: Correlation Between Climate Variability and Dengue Hemorhagic Fever (DHF) Incidence in Semarang City During 2002-2011 in *International Conference on Environment and Health: Integrating Research Community Outreach and Service Learning* pp 22–3

[25] District Health Office of Semarang 2013 *Health Profile of Semarang* (in Bahasa) (Semarang: District Health Office of Semarang) pp 33-4

[26] Widoyono, Pramudiyanto A A, Endang S, Pandu H, Prayitno G and Triatmi 2014 Health Profile of Semarang (in Bahasa) (Semarang: District Health Office of Semarang) pp 30-1

[27] Pinfold J V, Horan N J and Mara D 1991 Seasonal effects on the reported incidence of acute diarrhoeal disease in Northeast Thailand *Int. J. Epidemiol.* **20** pp 777–86

[28] Blumenthal U J, Fleisher J M, Esrey S A and Peasey A 2001 Epidemiology: a tool for the assessment of risk in *Water Quality: Guidelines, Standards and Health* Eds Fewtrell L and Bartram J (London: IWA Publishing) p 142

[29] Bonita R and Beaglehole R 2006 *Basic epidemiology* 2nd editio (Geneva: WHO Press) pp 41-3

[30] Ministry of Health Republic of Indonesia 2010 *Gudeline of Pneumonia management in children under five years old* (in Bahasa) 2nd edition (Jakarta: Ministry of Health Republic of Indonesia) pp 14-6

[31] Houghton DD 2002 Introduction to Climate Change *Lecture notes for meteorologists*, First. (Geneva: WMO) p 3

[32] Agency for Meteorology, Climatology and Geophysic (BMKG) 2006 *Mechanism of Climate Observation and Data Report of Climate and Agroclimate* (in Bahasa) (Jakarta: Agency for Meteorology, Climatology Geophysic) pp 1–18

[33] Agency for Meteorology, Climatology and Geophysic (BMKG) 2016 Analysis of Atmosphere Dynamic of Sea and Rainfall Forecasting (in Bahasa) (Jakarta: Agency for Meteorology, Climatology Geophysic) p 1
[34] Sugiyono 2002 *Statistic for research* (in Bahasa), Fourth edition (Jakarta: CV AlfaBeta) pp 11-5
[35] Central Agency on Statistics of Semarang (BPS) 2016 *The Latitude of Semarang* (in Bahasa) (Semarang:Central Agency on Statistics of Semarang (BPS)) p 1
[36] Central Agency on Statistics of Semarang (BPS) 2016 *The Administrative Border of Semarang* (in Bahasa) (Semarang:Central Agency on Statistics of Semarang (BPS)) p 1
[37] Central Agency on Statistics of Semarang (BPS) 2016 *The Area of Semarang* (in Bahasa) (Semarang:Central Agency on Statistics of Semarang (BPS)) p 1
[38] Central Agency on Statistics of Semarang (BPS) 2014 *The Population Number of Semarang Based on Age 2012-2014* (in Bahasa) (Semarang:Central Agency on Statistics of Semarang (BPS) p 1
[39] Central Agency on Statistics of Semarang (BPS) 2014 The Population Density of Semarang 2012-2014 (in Bahasa) (Semarang:Central Agency on Statistics of Semarang (BPS)) p 1
[40] Setiawan O 2012 Rainfall and temperature variability analysis in Bali (in Bahasa) *J. Anal. Kebijak. Kehutan.* **9** pp 66–79
[41] Nicholls R J, Wong P P, Burkett V, Codignotto J, Hay J, McLean R, Ragoonaden S, Woodroffe C D, Abuodha P, Arblaster J, Brown B, Forbes D, Hall J, Kovats S, Lowe J, McInnes K, Moser S, Rupp_Amstrong S and Saito Y 2007 Coastal systems and low-lying areas in *Climate Change 2007: Impacts, Adaptation and Vulnerability. Contribution of Working Group II to the Fourth Assessment Report of the Intergovernmental Panel on Climate Change* First Eds Parry M L, Canziani O F, Palutikof J , van der Linden P J and Hanson C E (Cambridge:Cambridge University Press) pp 315–356
[42] Katsaros K and Buettner K J K 1969 Influences of Rainfall on Temperature and Salinity of the Ocean Surface *J. Appl. Meteorol.* **8** pp 15–8
[43] Omonijo A G and Matzarakis A 2014 Pneumonia Occurrence in Relation to Population and Thermal Environment in Ondo State, Nigeria *African Rev. Phys.* **9** pp 511–25
[44] Xu Z, Hu W and and Tong S 2014 Temperature variability and childhood pneumonia : an ecological study *Environ. Heal.* **13** pp 1–8
[45] Eccles R 2002 An Explanation for the Seasonality of Acute Upper Respiratory Tract Viral Infections *Acta. Otolaryngol.* **122** pp 183–91
[46] Rudan I, Boschi-Pinto C, Biloglav Z and Campbell H 2008 Epidemiology and etiology of childhood pneumonia *Bull. World Health Organ.* **86** pp 408–16
[47] Hartmann D L, Klein Tank A M G, Rusticucci M, Alexander L V, Brönnimann S, Charabi Y, Dentener F J, Dlugokencky E J, Easterling D R, Kaplan A, Soden B J, Thorne P W, Wild M and Zhai P M 2013 Observations: Atmosphere and Surface. In: Climate Change 2013: *The Physical Science Basis.* Contribution of Working Group I to the Fifth Assessment Report of the Intergovernmental Panel on Climate Change Eds Stocker T F, Qin D, Plattner G K, Tignor M, Allen S K, Boschung J, Nauels A, Xia Y, Bex V and Midgley P M (Cambridge:Cambridge University Press) pp 190-194
[48] Devleta H and Evlijana Z 2015 Characteristics of Peneumonia Hospotalizations at Pediatric Clinic Tuzla *SANAMED* **10** pp 57–63
[49] Bartlett S 2008 Climate change and urban children *Impacts and implications for adaptation in low-and middle-income countries* (London: IIED) pp 6-8
[50] Passos SD, Gazeta R E, Felgueiras A P and Beneli P C 2014 Do pollution and climate influence respiratory tract infections in children? *Rev. Assoc. Med. Bras.* **60** pp 276–82
[51] Falagas M E, Theocharis G, Spanos A, Vlara L A, Issaris E A, Panos G, Lon H and Peppas G 2008 Effect of meteorological variables on the incidence of respiratory tract infections *Respir. Med.* **102** pp733–37
[52] Arundel A V, Sterling E M, Biggin J H and Sterling T D 1986 Indirect Health Effects of Relative Humidity in Indoor Environments *Environ. Health Perspect.* **65** pp 351–61
[53] Alsmo T and Alsmo C 2016 A Comparison of Relative Humidity between Two Swedish Buildings with Different Ventilation Solutions *J. Environ. Prot.* **7** pp 855–73
[54] du Prel J, Puppe W, Gro B, Knuf M, Weigl J A I and Schaaff F 2009 Are Meteorological Parameters Associated with Acute Respiratory Tract Infections ? *CID* **49** pp 861–68
[55] Barnes C S, Alexis N E, Bernstein J A and Cohn J R 2008 Climate Change and Our Environment : The Effect on Respiratory and Allergic Disease *J. Allergy Clin. Immunol. Pract.* **1** pp 137–41
[56] Krieger J and Higgins D L 2002 Housing and Health : Time Again for Public Health Action *Am. J. Public Health* **92** pp 758–68

[57] Bernard S M, Samet J M, Grambsch A, Ebi K L and Romieu I 2001 The Potential Impacts of Climate Variability and Change on Air Pollution-Related Health Effects in the United States *Environ. Health Perspect.* **109** pp 199–209
[58] Ordonez C, Mathis H, Furger M, Henne S, Huglin C, Staehelin J and Prevot A S H 2004 Changes of daily surface ozone maxima in Switzerland in all seasons from 1992 to 2002 and discussion of summer 2003 *Atmos. Chem. Phys. Discuss.* **4** pp 7047–88
[59] Daniel J and Winner D A 2009 Effect of Climate Change on Air Quality *Atmos. Environ.* **43** pp 51–63
[60] Amato G D, Cecchi L, Amato M D and Liccardi G 2010 Urban Air Pollution and Climate Change as Environmental Risk Factors of Respiratory Allergy : An Update *J. Investig. Allergol. Clin. Immunol.* **20** pp 95–102
[61] Dominici F, Peng R D, Bell M L, McDermott A, Zeger S L and Samet J M 2006 Fine Particulate Air Pollution and Hospital Admission for Cardiovascular and Respiratory Diseases *JAMA* **295** pp 1127–34

Early Childhood Environmental Education in Tropical and Coastal Areas: A Meta-Analysis

D R Sawitri [1*]

[1]Faculty of Psychology, Diponegoro University, Jl. Prof. Soedarto, S.H., Tembalang, Semarang 50275, Central Java, INDONESIA *Mobile: +628112798878, fax: +62247460051,

e-mail address: dian.r.sawitri@gmail.com

Abstract. Early childhood years are the period of the greatest and most significant developments in ones' life, and are generally regarded as the basis upon which the rest of their life is constructed. However, these early years are those that traditionally have received the least attention from environmental education. This paper was aimed to summarize several day-to-day activities that can be conducted to educate children in their early years about environment. Environmental education is an educational process that deals with the human interrelationships with the environment, and that uses an interdisciplinary problem solving approach with value clarification. Environmental education is aimed at producing a community that is knowledgeable about the biophysical environment and its associated problems, aware of how to solve these problems, and enthusiastic to work toward their solution. It highlights the progress of knowledge, understanding, attitudes, skills, and commitment for environmental problems and considerations. Further, environmental education can help children expand their ecological worldview, promote active care to the environment, and explain the relationship between modern life style and current environmental problems. Several types of environmental education have been identified from the literature, such as outdoor activities in natural outdoor setting, school gardening, play-based learning, and drawing activities. Each of these activities has its own characteristics and effects on children's environmental-related attitudes and behaviors. Through these activities, the unique characteristics of tropical and coastal areas can potentially be used to facilitate children to learn about nature and environment. Recommendations for childhood education practitioners and future researchers are discussed.

Keywords: environmental education, children, outdoor activities, play-based learning, classroom activities

1. Early Childhood Years and Environmental Education

The essential objective of environmental education includes individual's active involvement in working toward the ways to cope with environmental-related problems[1]. Environmental education is also conducted to help individuals acquire environmental awareness and sensitivity, and also understand of environmental problems, pro-environmental ideas, values, and concerns. In other words, environmental education helps them learn the skills to identify and resolve the environmental problems, as well as to provide opportunity to be actively involved in environmental-related actions [2].

In early childhood education, environmental education is becoming an increasingly essential learning area to study [3], as there is a strong need to educate children since their early years to learn how to act in a sustainable way [4]. Today, these children are born into a world with major environmental global issues, such as pollution, climate change, and biological diversity loss. These problems are widely discussed by most of the people on earth such as politicians, academics, and the media throughout the world. These issues are also problems which are articulated and used as educational content in schools. Early childhood education in the United States includes young children within the age range of birth through eight. However, in different countries, this term is more widely used to explain education for young children who have not yet begun kindergarten, generally before they reach five years old [5].

The early childhood years are the phase of the greatest and most significant developments in an individual's life. These years are generally recognized as the foundation upon which the rest of one's life is constructed [6,7]. The role of early childhood education in shaping key academics, social, and cognitive skills that have lifelong consequences for the individual and society has been shown by a significant empirical evidence [5]. Additionally, a large body of literature from different fields such as neuroscience, health, and economics demonstrates that early investments in human capital can offer significant returns both to individuals and to the larger community. Therefore, early childhood environmental education has the potential to make an important impact on young children development as well as on a sustainable world [5].

2. Informal Learning

Environmental education for young children will be more interesting when it is delivered through an informal learning. Informal learning is defined as learning that takes place outside of a classroom environment, such as museums, interactive science centers, forests, botanical gardens, zoos and wildlife parks [8]. Informal learning can improve children's understanding by having the learning process happen in an environment, where children may be unaware that they are learning and that they are naturally motivated by the environment. Compared to a formal classroom environment where information is often delivered orally, or using letters and numbers that need to be decoded, this approach is more exciting to follow [9].

As opposed to formal school, informal learning allows children more to display differing knowledge and understandings. This is because an informal setting is not structured to relate in simple ways to the formal curriculum that children have experienced. Different children have different access to different parts of the knowledge base through, for instance, their different interests, their exposure to different books, CDs, websites, and television, and their different leisure and holiday activities [8]. Additionally, informal learning settings are more likely to provide young children with a variety of free-choice activities to exploring things and construct new knowledge and beliefs [10]. This type of learning can serve as an opportunity to make a topic entertaining and enjoyable, and to forge links with everyday experiences and observations. This can make the learning experience more meaningful and not easy to forget [11]. Several informal learning that can be applied in children in tropical and coastal areas includes outdoor activities in natural outdoor settings, school gardening, play-based learning, and drawing activities.

3. Outdoor Activities in Natural Outdoor Settings

For young children, small-scale actions at the level of the school yard and the local environment might be the most appropriate settings to learn about their environment, as they should not be overwhelmed with indefinite environmental problems. Young children need a direct experience which does not go beyond their intellectual capacity [1].

Responsible environmental behavior is a learned response or action. For young children, outdoor activities can be a basis for developing an environmental responsibility in early childhood. The purpose of outdoor activities is to give children out-of-classroom educational experiences which includes direct contact with a variety of environment. Previous study showed that increased awareness and knowledge of environmental action strategies contribute to increased motivation to take action. Through personal experiences, diverse environmental education programs such as field trips, hiking, camps, and adventure activities can develop young children's environmental sensitivity, affective relationship to the natural environment, perform outdoor behavior and exercise their social relationship [12].

Natural outdoor settings are outdoor environments which range from relatively natural to relatively wild. These settings are in contradiction of maintained or built spaces, such as cut grassy areas, landscaped park settings, paved areas, or playgrounds [13,5]. The importance of natural spaces in education is not a new idea, as educational theorists emphasize the role of direct experiences in nature for young children's development and well-being [13, 5]. For example, in tropical areas or in coastal areas, trips to nearby forests or beach make it possible for young children to learn to take responsibility for themselves, their personal belongings, and the environment by not littering and by tiding up. Hiking and walking in the forests makes them learn about the terrain, animals, and plants. Farming can make the students learn about farm life. These trips can make them to take responsibility for their equipment and food [12].

Experiences, often with significant adults, that socialize children to view nature in positive and meaningful ways, were more likely to be beneficial. Nature activities in childhood and youth, as well as examples of parents, teachers, and other role models who show an interest in nature, are key 'entry-level variables' that predispose young children to take an interest in nature themselves and later act for its protection. An authoritative parenting style, combining high standards for behavior and clear rules with encouragement for children to express their thoughts and feelings, is associated with high levels of social responsibility in children[14]. Education programs and membership in environmental clubs and organizations can be seen as arena to gain increased knowledge about environmental issues and learn environmental action skills and capacities[1]. Parents and other family members are critical role models of interest in public issues and prosocial values such as environmental responsibilities. Young children are more likely to participate in community activities if their parents are also active in this way, or give them approval and encouragement to take part[15].

Early childhood educators' beliefs regarding the difficulty of using natural outdoor setting is also a key insight and integral in guiding future research and professional development efforts[5]. Natural outdoor settings hold endless possibilities for learning in all curricular domains. However, early childhood educators may not recognize the potential opportunities for learning in natural outdoor settings nor the alignment between early childhood pedagogy and the opportunities offered by nature experiences. Early childhood educators who were limited in their own childhood exploration of natural outdoor environments may lack the foundational experiences necessary to see the affordances in natural outdoor settings[16,5]. Teachers' beliefs as to benefits of outdoor settings may limit the opportunities that teachers provide for children in the outdoors. For example, early childhood educators tend to associate outdoor settings with physical and social development, as opposed to a wider range of developmental benefits, and view the role of natural elements in outdoor setting as having aesthetics rather than educational value[17,5]. Understanding early childhood educators' beliefs, practices, and perceived barriers in relation to natural outdoor settings can inform efforts to encourage and support the use of natural outdoor settings in early childhood education[5].

4. School Gardening

School gardening covers a range of activities to increase the horticultural complexity of the schoolyard. School gardening can involve having potted plants, raising beds on asphalt, doing indoor composting, and carrying-out in-ground plantings[18,19]. It also includes a systematic approach to redesign the outdoor space around schools into learning landscapes[20]. The purposes of gardening are academic, behavioral, recreational, social, such as to increase children's sense of belonging, self-esteem, and compassion. It can also be political, as the schoolyard can be a visible community asset, and this activity can also serve as an effort for environmental remediation.

Several rationales for school gardening to be an alternative of environmental education for children are that this activity can broaden children's experience of the complexity of the ecosystem and knowledge about the nature and culture continuum. For example, vegetable gardening can teach children about food systems ecology, and exposure to nature and gardening in can shape their attitudes and environmental values[18]. Gardens cover the principles of biodiversity and organic pest management which contains ponds or recycling streams, trees, and butterfly attractors. This place can be a safe place for a wide variety of flora and fauna beyond the crops, flowers, and bushes purposely grown, and this can also demonstrate ecosystem complexity. Gardens that children help to plan and grow allow a close and personal experience with the environment, repeated sensory contact, and an interaction with the processes of nature. These benefits are believed as necessary for a healthy human development[21].

5. Play-based Learning

Previous studies in the area of play-based learning and pedagogy generally seeks to focus on the ways in which play is related to and/or used in early childhood education as a foundation for pedagogy. It is important for researchers in this area to formulate the way in which "play" is understood, located, or identified in relation to pedagogy[22, 23].

Previous study investigated how teachers can create different types of pedagogical plays. These activities can provide opportunities for young children to develop knowledge through experiences about environmental education[24]. This work suggested three ways of thinking about children's play-based activities when engaging with environmental education as a learning area, including open-ended play, modeled play and purposefully framed play. First, open-ended play which includes play experiences where the teacher provides young children with materials suggestive of an environmental or sustainability concept. In this activity, teachers provide minimal engagement and interaction which allows children to examine and explore the materials as a basis for learning about the concept of environment or sustainability. For example, children are provided with materials and then asked to make dirty water "clean." There is little or no discussion or modeling of the activity, and the children are left to participate in open-ended play with the materials.

Second, modeled play which involves play experiences where the teacher illustrates, explains, and/or demonstrates the use of materials suggestive of an environmental or sustainability concept prior to allowing the children to use these materials with minimal adult interaction as basis for learning about the concept of environment or sustainability. For example, children are shown a model of how the materials could be used to make a filter for cleaning water. Children participate in the play activity following modeling by the teacher. This activity can involve a live demonstration or can also use a video-based model.

Third, purposefully framed play which includes play activities in which the teacher provides children with materials suggestive of the concept of environment or sustainability. The teachers then provide opportunities for the children to be involved in an open-ended play, followed by modeled play and then teacher-child interaction/engagement. For example, children are provided with materials and participate in open-ended play. Following the open-ended play activity, the teacher models how the material could be used, and the teachers engage in discussions with the children about the activity.

Additionally, there are several organizing themes in environmental-education related play which involves social facilitation, play places, spatial play, play content, and intermittent play[25]. First, social facilitation recognized that supportive parents or other immediate others or peers made play experiences possible in wild environments. Parents can guide the activity by choosing to live near nature, experiencing it directly and through discussion with their children. They can send their children to camp, and traveling with their children to wild places. Parents can also facilitate the play activity by saving items to use in collecting or allowing their children to come home dirty. Parents can provide books and magazines in their homes devoted to the scientific aspects of nature.

Second, play places which include the use of many types of play areas. Participants reported using many interstitial spaces, including wooded areas and ponds along golf course fairways, watersheds through urban areas, unfinished houses in construction sites, bridges, and traditional rural and wild landscapes.

The third organizing theme is spatial play, which distinguished exploration of large-scale environments from stationary play in one area. These environments included time spent away from home on vacation or with relatives, exploring by following creeks and ditches, and using forts or tree houses that were near-distant, that is, close to home but far enough away to require some travel and provide privacy from parents. The participants can spend several weeks away from home, either traveling or staying with relatives. During these activities, their relatives can provide access to new play areas.

Fourth, both play places and spatial play led to the broadest and most complex organizing theme, play content. The content of play describes in detail regarding activities occurred during childhood. One cluster of activities consisted of fantasy play based around story lines in movies and books. A second content cluster involved creative acts with clay and plants. A third grouping consisted of searching for, catching, trapping, or collecting animals, insects, and fossils. There were also games such as hide-and-seek and dam wars. By climbing trees, crossing creeks, and swinging on vines, participants demonstrated physical prowess. Last, many responses included creating structures such as forts and shelters. Play content can be child–child play in nature and child–nature play. Child–child play involved activities dominated by interactions between children in which they used natural objects as barriers, weapons, or tools. For example, throwing walnuts at each other during war games would be child–child play. In child–nature play, either solitary or with others, the focus was on observing, catching, exploring, or creating with natural objects.

Finally, another organizing theme was intermittent play through chore-like rehabilitation of wildlife, gardening, caring for pets, and picking berries. Although these activities resembled work, they sometimes provided novel and enjoyable experiences. Many respondents cared for a variety of pets during their childhood. Others mentioned finding sick or injured wildlife and having the opportunity to rehabilitate wild animals. Other responses included mowing the lawn, feeding animals, and spending time outside as a family.

The global theme that covers all the themes was environmental pleasures, acknowledging the strong perceptual and affect component of recalled experiences of playing outdoors. The difference that we identified as child–child and child–nature play merits further investigation. The style of interaction during play in and with nature may predict a generative interest in natural history. Young children whose play tended to involve interacting with nature (e.g., catching frogs in a creek) as opposed to interacting with each other (e.g., war games with walnuts) may be more likely to have pleasurable, overt, and implicit learning experiences with and about natural objects and creatures. That possibility may explain why many adults may recall play experiences in nature as a child are not interested in natural history or biodiversity issues as an adult. Childhood play provides opportunities for developing layers of skill sets: physical coordination, ancillary skills (e.g., tolerance for dirtiness, heat, and cold; wayfinding), social interactions, and creative acts, including art and construction. Most children develop these outcomes by playing outdoors. Children who engaged in child–nature play probably experienced an additional dimension of nature that they continued to value into adulthood.

Parents and peers played a role in access to wild and interstitial places through choosing home location and granting freedom to play and explore. Both access to natural areas and the content of childhood play in nature probably reflect to some extent the values of parents and peers. Parents appear to play a significant although indirect role in experientially preparing their young children to benefit from structured environmental education through unstructured, self-directed experiences in nature. Environmental educators should help parents to provide unstructured and child-directed experiences with natural environments.

Child play activities in nature are either child–child (e.g., playing war) or child–nature (e.g., exploring a creek) focused activities. Child–nature interactions are more developmentally important than child–child interactions. Parents, environmental educators, and nature centers can stimulate child–nature interactions that encourage exploring for animals by providing equipment ranging from empty mayonnaise jars to insect nets. Many parks loan fishing equipment and could also loan nature-study equipment. Environmental learning centers could offer programs for parents on how to purchase quality nature-study equipment, microscopes, and telescopes for their children. To help their children start effectively to use their new gift, the video-game generation of parents would also need instruction on what in nature their children can observe with these nature study gadgets.

6. Drawing Activities

Children's drawings may provide us with insights into their cognitive, affective and social development because they are unique, and have complex thoughts and feelings. Psychotherapists and clinical counsellors have analyzed children's drawings as projections of their personality traits. These experts are likely to use children's art as a powerful tool in communication[26].

For young children, drawings are believed to reflect their mental representations and conceptual knowledge about the objects they draw. When children's mental models of the world become more extensive and differentiated, their drawings become more precise and comprehensive[27]. Using drawings to probe understandings was a useful approach in researching children's learning. Drawings enabled them to visualize and reveal to the child and teacher 'qualities of understanding' that can be hidden through other research procedures such as interview, observation, or traditional cognitive paper-and-pencil tests. This is because individuals are more ready to think about images long before thinking in words and they expand upon this noting how humans tend to feel more satisfied if they can translate words into images. In science, 'complex notions are often represented by familiar, everyday images[28]. Previous research showed that drawings to be an effective method for analyzing 7 to 9 years old English and Mexican children's environmental perceptions[29]. Empirical material consisting of drawings can be used to visualize the way in which children think about the environment.[30]

Previous study also demonstrated that drawing activities can facilitate children to learn about tropical rainforest[30,31]. Drawing activity is one of the unique strategies to evaluate children's perceptions regarding the environment, and also their major expectations and concerns for the future in relation to environment[29]. Drawing techniques provide a relatively easy way to gather social information from and about children[32]. The use of drawings for evaluation purposes is a powerful tool, since most children tend to enjoy drawing activity without showing any sign of pressure.

For children who do not like answering questions, drawing tests can be completed quickly, easily and in an enjoyable way.[33] Drawings avoid linguistic and oral communication barriers. These fun activities also enable comparisons between groups of different languages and abilities.[34] The content of children's drawings may provide insight into their feelings and thoughts about the world.[35] Children's drawings provide a 'window' into their thoughts and feelings, mainly because they reflect an image of his/her own mind.[36] For example, a previous study used drawings from children to discover the variety and kinds of concerns children might have about the environmental crisis. Children were asked to draw a picture about what it means to them when someone says, 'You have to save the planet'. Findings showed that 87% of the children were very much aware of the environmental crisis. Nearly half (47%) described themselves or others taking personal action for positive social/environmental changes.[32]

Other study demonstrated that before and after the children visited the project and took part in a workshop, they were each asked to draw a picture of a 'tropical rainforest'—drawing[31]. At the first stage, young children were encouraged to express their knowledge about tropical rainforests. This was supervised by teachers in their school and the children were given the task in as relaxed an atmosphere as possible. They had as much time as they needed to complete their drawing and it was emphasized that it had to be their own work from their existing knowledge. No secondary sources of information or discussion were permitted. At the second stage, they were asked to do the same thing. However, before they entered the second stage, they visited the tropical rainforest.

One of the key differences between the set of children's first drawings i.e., before they went to see the rainforest and second drawings after they visited the rainforest, was the extent and depth of the trees and plants drawn. In the first drawing, they drew fewer trees and plants. Additionally, the trees were drawn very much in a stylized fashion such as they have lollipop shaped trees, or sometimes they have fruits such as apples or oranges or bananas. These things did not happen anymore after they

IOP Conf. Series: Earth and Environmental Science **55** (2017) 012050 doi:10.1088/1755-1315/55/1/012050

visited the real rainforest. In their second drawings, children drew many more recognizably different species of tropical rainforest trees and plants. The accuracy and details of their drawings had significantly improved. They drew more accurate and architecturally shaped rainforest trees and plant shapes. The shapes of the leaves were accurately drawn with details such as leaf veins and drip points included.

7. Conclusion

Scientists have started to be aware of the importance of children to learn about environment since their early years, especially through informal learning. Informal learning is identified as an appropriate method for children in their early years to learn about their environment. This includes outdoor activities in natural outdoor settings, school gardening, play-based learning, and drawing activities. Parents and educators are encouraged to use these range of activities to educate children about environment. Future researchers are encouraged to investigate a variety of methods to develop environmental education strategies for early childhood education in tropical and coastal areas.

References

1. Chawla L, Cushing DF 2007 Education for Strategic Environmental Behavior, Environmental Education Research; **13**: 437–452.
2. Carmi N, Arnon S, Orion N. 2015 Transforming Environmental Knowledge into Behavior: The Mediating Role of Environmental Emotions, The Journal of Environmental Education **46**: 183-201.
3. Pearson E, Degotardi S. 2009 Education for Sustainable Development in Early Childhood Education: A Global Solution to Local Concerns? International Journal of Early Childhood **41**: 97–111.
4. Hedefalk M, Almqvist J, Östman L 2015 Education for sustainable development in early childhood education: a review of the research literature, Environmental Education Research **21**: 975-990.
5. Ernst J 2014. Early Childhood Educators' Use of Natural Outdoor Settings as Learning Environments: An Exploratory Study of Beliefs, Practices, and Barriers, Environmental Education Research **20**: 735-752.
6. Davis J. 2009 Revealing the Research 'Hole' of Early Childhood Education for Sustainability: A Preliminary Survey of the Literature, Environmental Education Research;**15**: 227-241.
7. Rutter, M. 2002 The interplay of nature, nurture and developmental influences: The challenge ahead for mental health, Archives of General Psychiatry; **59**: 996–1000.
8. Gerber BL, Marek EA, Cavallo AML. 2001 Development of an informal learning opportunities assay, International Journal of Science Education; **23**, 569–583.
9. Ramey-Gassert L, Walberg III HJ, Walberg HJ. 1994 Re-examining connections: museums as science learning environments, Science Education; **78**, 345–363.
10. Falk JH, Dierking LD. 2000 Visitor experiences and the making of meaning. Walnut Creek: AltaMira Press).
11. Tofield S, Coll RK, Vyle B, Bolstad R. 2003 Zoos as a source of free choice learning, Research in Science and Technological Education; **21**, 67–99.
12. Palmberg IE, Kuru J. 2000 Outdoor Activities as a Basis for Environmental Responsibility, The Journal of Environmental Education, **31**:4, 32-36
13. Wilson R. Nature and Young Children: Encouraging Creative Play and Learning in Natural Environments 2012. London: Routledge.
14. Baumrind D. Current patterns of parental authority, Developmental Psychology Monographs 1971, 4, 1–103.
15. Fletcher AC, Elder G, Mekos D. Parental influences on adolescent involvement in community activities, Journal of Research on Adolescence 2000; **10**: 29–48.
16. Crim C, Desjean-Perrotta B, Moseley C., Fall. Partnerships Gone Wild Preparing Teachers of Young Children to Teach about the Natural World, Childhood Education 2008; **6:** 29-48
17. Davies M. Outdoors: An Important Context for Young Children's Development. Early Child Development and Care1996; **115**: 37–49
18. Blair B. The Child in the Garden: An Evaluative Review of the Benefits of School Gardening, The Journal of Environmental Education2009; **40**:15-38,

19. Graham H, Beall DL, Lussier M, McLaughlin P, Zidenberg-Cherr S. Use of school gardens in academic instruction, Journal of Nutrition Education Behavior, 2005, **37**: 147–151.
20. Brink L, Yoast B. Transforming inner-city school grounds: Lessons from learning landscapes. Children, Youth and Environments2004; **12**: 208–232.
21. Thorp L, Townsend C. (2001, December 12). Agricultural education in an elementary school: An ethnographic study of a school garden. Proceedings of the 28th Annual National Agricultural Education Research Conference in New Orleans, LA (pp. 347–360).
22. Langford R. Critiquing child-centred pedagogy to bring children and early childhood educators into the centre of a democratic pedagogy, Contemporary Issues in Early Childhood2010; **11**, 113–127.
23. Stephen C. Pedagogy: The silent partner in early years learning, Early Years: An International Journal of Research and Development2010; **30:** 15–28.
24. Mackenzie AC, Edwards S. Toward a Model for Early Childhood Environmental Education: Foregrounding, Developing, and Connecting Knowledge Through Play-Based Learning, The Journal of Environmental Education 2013; **44**: 195-213.
25. Vadala CE, Bixler RD, James JJ. Childhood Play and Environmental Interests: Panacea or Snake Oil? The Journal of Environmental Education 2007; **39:** 3-18
26. Malchiodi CA. 2003. Handbook of art therapy (New York, The Guilford Press). Reith, E. (1997) Drawing development, in: A. M. Kindler (Ed.) Child development in art, (Virginia, National Art Education Association), 59–79.
27. White R, Gunstone R. 2000. Probing understanding (London, Falmer Press).
28. Barazza L. Children's Drawings About the Environment, Environmental Education Research1999, **5**:1, 49-66.
29. Alerby E. A way of visualizing children's and young people's thoughts about the environment: a study of drawings, Environmental Education Research 2000; **6:** 205–222.
30. Bowker R. Children's Perceptions and Learning About Tropical Rainforests: An Analysis of Their Drawings, Environmental Education Research 2007;**13**: 75-96.
31. King, L.D. Doing their Share to Save the Planet. Children and Environmental Crisis1995. (New Jersey, Rutgers University Press).
32. Lewis D, Greene J. Your child's drawings… their hidden meaning1983. (London, Hutchinson and Co).
33. Chambers D.W. Stereotypic images of the scientist: the draw-a-scientist test, Science Education 1983, 67, pp. 255-265
34. Crook C. Knowledge and appearance, in: N. H. FREEMAN & M. V. Cox (Eds) Visual Order: the nature and development of pictorial representation1985. (Cambridge, Cambridge University Press
35. Thomas G, SILK A. An Introduction to the Psychology of Children's Drawings 1990. (London, Harvester Wheatsheaf)
36. Davis J. Educating for sustainability in the early years: Creating cultural change in a child care setting, Australian Journal of Environmental Education 2005; **21**: 47–55.

Diatoms and Water Quality of Telaga Warna Dieng, Java Indonesia

Tri Retnaningsih Soeprobowati[1,2,*], Sri Widodo Agung Suedy[1], Hadiyanto[2,3)]

[1]Department Biology, Faculty Sciences and Mathematics, Diponegoro University, Semarang, Indonesia
[2] School of Postgraduate Studies, Diponegoro University, Semarang, Indonesia
[3] Department of Chemical Engineering, Diponegoro University, Semarang, Indonesia
Telephone/Fax: +62-24-8442990 ;

trsoeprobowati@live.undip.ac.id

Abstract. Diatoms are popular name for Bacillariophyte, the microalgae with the specific characteristic of silicious cell walls that well preserved in the sediments. The different diatoms assemblage in the sediment layers indicate different environment at the time of the diatoms live and deposited. Telaga Warna is small lake in Dieng Plateau. Telaga means lake, Warna means colour. It is called Telaga Warna because previously have 4 colours i.e. red, white, blue, and yellow which was influenced by weather, time, and site of view. This study aims to analysis the diatoms communities and water quality of Telaga Warna Dieng, Java, Indonesia. Coring conducted at three different locations. Water sampling carried out on all three spots. Analysis of diatoms consist of three stages: digestion, preparation ,and identification of diatoms. There were 59 diatoms species found in Telaga Warna Dieng that were belong to 9 groups diatoms of centric, arafid, eunotoid, birafid, monorafid, birafid, epitemid, nitzschioid, and surirelloid. *Eunotia, Pinnularia*, and *Melosira* were the dominant genus from Telaga Warna. The water quality parameters that exceeded Indonesia Water Quality Standard were pH (2.2 – 5.4), Pb, Cd, Cr, Cu. Based on the abundance of species *Eunotia* and *Pinnularia* in Telaga Warna Dieng indicates that water tends to be acidic. The dominance of *Melosira* indicates waters rich in nitrogen. Based on the total concentration of nitrogen and phosphorous, Telaga Warna was in an eutrophic – hipereutrophic conditions with total nitrogen concentration > 1.9 mg/L and total phosphorus concentration > 0.1 mg/L.

Keywords: diatoms, water quality, Telaga Warna Dieng, pH, eutrophic

1. Introduction

The Dieng plateau is located in Central Java, Indonesia with a 63 km^2 concession area, approxiametly 25 km from north of the city Wonosobo. DiengPlateau, is the second highest plateau in the world after Nepal about 2,093m asl. The geothermal resource in Dieng was recognized since Dutch colonial period and in 1964/1965 the area was identified as one of the best geothermal prospect in Indonesia by UNESCO[1]. There are many lakes, creater (hydrothermal), valley, water springs, hills, and vulcan complex in one area makes Dieng has unique characteristic and very rare encountered in other countries[2].

Telaga Warna Lake is the famous creater lake and the most interesting and famous in the Dieng area, due to different color reflection, sometimes the color of the lake is red or white or blue and yellow depending on the weather, time, and place to see it. The color is a reflection of sunlight by sediment or rock at the bottom of the lake that supposedly once the volcanic crater. Red and yellow color is a reflection of the element sulphur deposition, the white color comes from the deposition of limestone rocks and quartz. But now, TelagaWarna only visible only blue indicating sediment mud. Its colorful appearance makes the lake an interesting tourist atraction. Gas bubbles can be seen rising to the lake surface and the air has sulphurous odour[3].

Diatoms also known as Bacillariophyte have uniquely ornamented and diversly shaped transparant silicate cell walls, microscopic unicellular algae abundant in almost all aquatic habitat, and part of base aquatic food webs. Diatoms community are responsive to the environmental change such as physical and chemical characteristics include pH, water deph, nutrient avability, salinity also current condition in the environment. Diatoms assemblages are often specific to particular habitats and also used to characterize those habitats. For that reason, characteristic diatoms examination have been widely used to investigate the status of the lake and can explain about pollution state, monitoring program and paleontology for lake environment[4]. Palaeolimnological studies offer an opportunity to improve knowledge about the past environmental conditions[5].

The study about diatoms as indicator for water quality of the lake have been conducted for long periode. Epipelic diatoms preserved in the sediment might tell the story about past climatic changes that can be inferred from changes in species abundant within a sediment core. When diatoms died their cell walls turn down in the bottom of the lake and preserved in the sediment. The use of diatoms to reconstruct past ecological conditions in Rawapening Lake, Java was well established[6]. The general objectives of these journal are to study subfossil diatoms assemblage to access natural and human activity induced environmental changes in Telaga warna Dieng and to study the water quality of Telaga Warna Dieng.

2. Material and Method

The survey was conducted in 31 August 2014 and 2 February 2015. There are 3 sampling sites with diferent characteristic, D1 located near sulphurous area, D2 located near riparian vegetation, and D3 located near Pengilon lake. At each site, dissolve oxygen, electrical conductivity, pH, temperature, and turbidity were measured with three times replication. The water samples were also collected at each site at the same time.

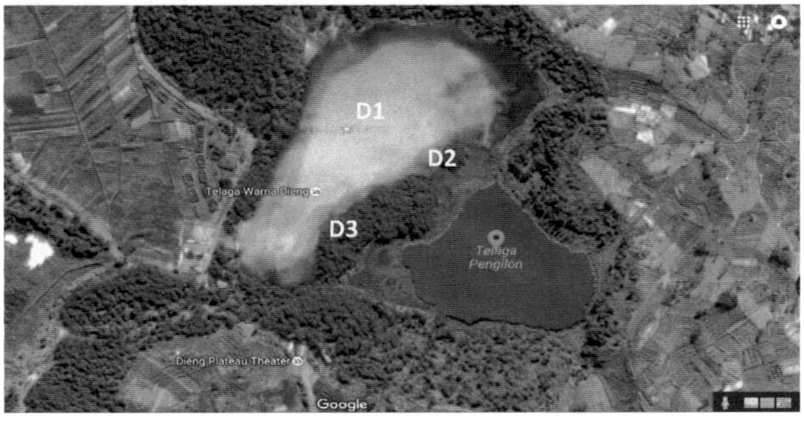

Figure 1. 4 Sites sampling in Telaga Warna Dieng.

IOP Conf. Series: Earth and Environmental Science **55** (2017) 012051 doi:10.1088/1755-1315/55/1/012051

Samples were taken at four points of the study. Measurement of water quality such as measure the temperature, pH, conductivity, turbidity and dissolved oxygen. Each parameter was measured three times as replicates. Water sampling is done at 20 cm from the surface for analysis of heavy metals Pb, Cd, Cr, Cu, silicate, chlorophyll-a, total phosphorus and total nitrogen. Coring was done by modification of sediment corer. Sediment samples were sliced every 1 cm for diatoms analysis. Separation the diatoms from the sediment carried by the addition of 10% HCl followed by 10% H_2O_2. Preparations for the diatoms uses adhesives with refractive index of 1.7, in this study used Naphrax. Diatoms identification with 1,000x magnification using mainly using Kramer-Bertalot books. Shannon Wiver Diversity, Species Richness, and Dominance Indices of diatoms were calculated.

3. Result and Discussion

A total of 59 diatoms preserved in the sediment belong to 8 groups diatoms of centric, araphid, eunotoid, monoraphid, biraphid, epithemid, nitzschioid, and surirelloid. *Eunotia, Pinnularia*, and *Melosira*were the dominant genus from Telaga Warna (Table 1).

Table 1. Diatoms preserved in the sediment of Telaga Warna

No.	Species	% Domination		
		D1	D2	D3
1	*Achnanthidium minutissimum* (Kützing)	26.49	37.60	10.05
2	*Falacea pygmaea* (Kützing)	-	0.82	0.64
3	*Amphora copulata* (Kützing)	-	1.36	5.74
4	*Amphora pedivulus* (Kützing)	-	-	3.35
5	*Amphora veneta* (Kützing)	-	-	5.58
6	*Asterionella formosa* (Hassall)	-	-	0.16
7	*Aulacoseira ambigua* (Grunow)	0.99	-	-
8	*Aulacoseira granulata* (Ehrenberg)	-	-	0.16
9	*Brachysira vitrea* (Grunow)	1.73	0.54	-
10	*Caloneis bacillum(Grunow)*	10.37	6.81	4.78
11	*Cocconeis placentula* (Ehrenberg)	1.24	1.09	-
12	*Craticula halopedia*	-	-	0.32
13	Cyclotela (Kützing)	5.45	8.72	4.78
14	*Cyclotella compta* (Kützing)	1.49	3.54	0.32
15	*Cyclotella meneghiniana* (Kützing)	-	2.72	-
16	*Diatoma hyemalis* (Roth)	-	-	1.12
17	*Diploneis finnica* (Ehrenberg)	-	-	0.64
18	*Ecynopsis microcephala*	0.99	-	-
19	*Encyonema silesiacum* (Bleisch)	-	0.82	0.64
20	*Encyonema minuta* (Hilse)	-	0.27	0.48
21	*Entomoneis alata* (Ehrenberg)	-	-	0.64
22	*Eolimnaminima* (Grunow)	0.25	0.27	-
23	*Eunotia bilunaris* (Ehrenberg)	-	-	1.44
24	*Eunotia minor* (Kützing)	-	-	2.71
25	*Eunotia monodon*(Ehrenberg)	1.98	0.27	3.51
26	*Eunotia pectinalis* (Kützing))	-	-	2.71
27	*Eunotia septentrionalis* (Strup)	-	-	0.16
28	*Eunotia sudetica* (O.F.Müller)	0.25	-	0.48
29	*Eunotia undulata* (Ralfs)	1.24	2.18	2.39
30	*Eunotia angustior* (Grunow)	-	-	1.59
31	*Eutomoneis paludosa* (W.Smith)	0.25	-	-
32	*Fallacia pygmae* (Kützing)	1.98	-	0.48
33	*Fragillaria cappucina*	3.47	0.27	3.99
34	*Gomphonema cf.affine* (Kützing)	-	-	2.07
35	*Gomphonema gracile* (Ehrenberg)	-	-	1.12
36	*Hantzschia amphioxys* (Ehrenberg)	1.73	0.27	4.47
37	*Ikeba tenuis* (Brun)	3.22	-	0.32

38	*Luticola goepeppertiana*	0.50	-	1.91
39	*Luticula mutica* (Kützing)	0.99	0.54	1.12
40	*Luticula mutica var tropica* (Kützing)	-	-	1.91
41	*Melosira varians* (O.F.Müller)	-	0.27	1.12
42	*Navicula hodgeana* (Patrick & Freese)	0.50	1.91	-
43	*Navicula lanceolata* (Ehrenberg)	0.50	-	1.12
44	*Pinnulairia gibba* (Ehrenberg)	8.42	3.54	2.07
45	*Pinnularia neomajor* (Krammer)	-	-	0.48
46	*Pinnularia stomatophora* (Grunow)	-	0.27	1.91
47	*Pinnularia sub gibba* (Ehrenberg)	-	-	1.44
48	*Pinnularia valdetolerans* (Mayama)	-	-	1.75
49	*Pinnularia viridiformis* (Krammer)	0.74	0	2.39
50	*Planothidium lanceolatum* (Brébisson)	-	1.36	1.12
51	*Rhoicosphenia curvata* (Kützing)	-	-	0.80
52	*Sellaphora bacillum* (Ehrenberg)	6.93	6.81	3.03
53	*Sellaphora pupula* (Kützing)	2.72	1.91	-
54	*Sellaphora seminulum* (Grunow)	10.15	14.71	3.99
55	*Stauroneis cohni* (Hilse)	3.47	-	4.47
56	*Stauroneis geoppertiana* (Bleish)	-	1.09	0.48
57	*Straurophora wislouchii*	0.25	-	1.28
58	*Synedra acus(Kützing)*	0.50	-	0.16
59	*Tabellaria fenestrata* (Lyngbye)	1.24	-	0.64
	Shannon Winner Index (H')	2.67	2.26	3.5
	Species richness index (R)	9.08	7.45	13.77
	Domination Index	0.11	0.18	0.03

Telaga Warna has middle diversity (Shannon Wiener Index of 2.67 – 3.5), low uniformity (Species Richness Index of 7.45- 13.77) and low dominance (Domination Index of 0.03 - 0.18, Table 1). This was meaning that no species dominance. *Achnanthidium minutissimum* (Kützing) has a highest dominance index, indicated that Telaga Warna was acid .These result in accordance with research that *A. minutissimum* tolerate in acid condition about 2.8-3.3. The abundance of species *Eunotia* and *Pinnularia* in the sediment samples indicate that the pH waters tend acidic to neutral[7]. This is supported by the results of water quality analysis of the specific lake color that is highly acidic pH (2.2 - 3.35, Table 2), so the only species that are tolerant to acid which will be able to grow in the lake of Telaga Warna.

The pH of Telaga Warna was fluctuated. In 2 October 2014, pH in Telaga Warna Dieng was in the range of 2.1 – 2.3. Based on monitoring data from Conservation Department, when dry season pH was very acid (1 – 2) due to the dry up water. Sulphate and Chloride contents in Telaga Warna moderately higher than the other telaga. The emissions show considerable fluctuations instrength. The pH of Sinila Crater Lake was 6.5 and has lower sulfate and chloride contents[3]. Telaga Pengilon is near Telaga Warna with the distance about 50 m from Telaga Warna has a higher pH (5.1- 6.6 at October 2, 2016). Dissolve oxygen at 31 August 2014 about 6.67-10.61 mg/L, at 20 February 2015 about 3.32-5.99 mg/L, and in the last monitoring at 2 October 2016 decrease about 0.8–1.7 mg/L (Table 2).

The interesting found from the research was appreence of *E. septentrionalis* a taxon that was reported by Hustedt (1937–1939) as living in acid waters (pH ca.3) in the sulphur-springs of Sumatra. The name *E.septentrionalis* var. *intermedia* was created for forms with intermediate striae counts. Negoro reported these varieties are occurring and epiphytic from many highly acidic environments in Japan[7]. From this result may the condition of Telaga Warna Dieng similar to Sulphure springs in Sumatra. The dominance of *Melosira* indicates waters rich in nitrogen. Based on the analysis of water quality, Telaga Warna in eutrophic – hipereutrophic conditions with total nitrogen concentration above 1.9 mg/L and total phosphorus concentration above 0.1 mg/L.

IOP Conf. Series: Earth and Environmental Science **55** (2017) 012051 doi:10.1088/1755-1315/55/1/012051

Based on Indonesian water quality standart regulation number 82/2001 about criteria of water quality for spesific purposes. Class I used for source of drinking water, Class II used for recreation, live stock and fish farming, Clas III used for live stock, fish farming and agriculture, and Class IV used for agriculture. From Table 2 Cadmium and Lead concentration higher than Indonesian water quality standard. For fisheries and animal husbandry, the lead content in the water should not exceed 0.1 mg/L, for agricultural alkaline or neutral soil, the lead dies not allow exceed 10 mg/L, whereas in acid soil is not more than 5 mg/L.

Table 2. Water quality of Telaga Warna Dieng

Parameter	Unit	Telaga Warna (31 Aug 2014)			Telaga Warna (20 Feb 2015)			Indonesian Water Quality Standard Government Regulation No. 82/2001	
		T1	T2	T3	T1	T2	T3	Class I-III	Class IV
pH	-	2.2	2.4	2.3	5.35	5.23	4.91		
Conductivity	mS/cm	5.08	7.9	16.02	0.83	1.13	3.92		
Turbidity	NTU		-	-	46.33	21	53.33		
Temperature	°C	20.45	26.6	21.6	15.35	19.97	19.46		
Dissolve oxygen	mg/L	6.67	10.61	9.83	3.34	5.04	5.99		
Lead (Pb)	mg/L	0.02	0.016	0.012	0.023	0.056	0.035	0.03	1
Cadmium (Cd)	mg/L	0.01	0.006	0.016	0.135	0.023	0.014	0.01	0.01
Chromium (Cr)	mg/L	0.012	0.009	0.012	0.013	0.028	0.0335	0.05	1
Copper (Cu)	mg/L	0.015	0.01	0.017	0.022	0.032	0.057	0.02	0.2
Silica (SiO_2)	mg/L	0.144	0.13	0.155	0.13	0.167	0.14	-	-
Total P	mg/L	0.24	0.161	0.182	0.284	0.185	0.267	-	-
Total N	mg/L	2089	1035	1.14	2,005	1.87	2,775	-	-
Chlorophyll a	mg/L	4,318	3.09	4.08	6.19	5,668	5,776	-	-

4. Conclusion

Based on the abundance of species *Eunotia* and *Pinnularia* in Telaga Warna Dieng indicates that water tends to be acidic. The dominance of *Melosira* indicates waters rich in nitrogen. Based on the analysis of water quality, Telaga Warna in eutrophic conditions – hipereutrophic.

Acknowledments

This research was supported by Diponegoro University Indonesia through Research Grant for International Publication (RPI), PNBP DIPA Undip Number 1052-57/UN7.5.1/PG/2016. The heartfelt thanks to Kenanga Sari, M. Hadi El Amin, Siti Mudakiroh, Alam, and Geyga for helping in the field work and lab work.

Reference

1. Sirait P, Ridwan R H, Battistelli A 2015) Reservoir Modeling for Development Capacity of Dieng Geothermal Field, Indonesia. Proceeding Fourtieth Workshop on Geothermal Reservoir Engineering..
2. Rusiah M, Satya N, and Wahyudin A 2005 Dampak aktivitas pertanian kentang terhadap kerusakan lingkungan obyek wisata dataran tinggi Dieng. Pelita 1, 1.
3. M.J. van Bergen, A. Bernard, S. Sumarti, T. Sriwana, and K. Sitorus. 2000 Crater lakes of Java: Dieng, Kelud and Ijen. Excursion Guidebook IAVCEI General Assembly, Bali.

4. Taurai B 2014 Ecological preferences of benthic diatoms in a tropical river system in São Carlos-SP, Brazil. Tropical Ecology 55, 1.
5. Smol J P 2008.Pollution Of Lakes And Rivers A Paleoenvironmental Perspective. 2nd Ed. Blackwell Publishing, USA.
6. Soeprobowati T R, Hadisusanto S, and Gell P 2012 The diatom stratigraphy of Rawapening Lake, Implying Eutrophication History. Am J Env Sci **8**, 3.
7. Denicola D 2000 A review of diatoms found in highly acidic environments. *Hydrobiologia* 433,

Improvement of Stability and Antioxidant Activities by Using Phycocyanin - Chitosan Encapsulation Technique

Meiny Suzery[1,*], Hadiyanto[2], Dian Majid[1], Deny Setyawan[1], Heri Sutanto[3]

[1]Chemistry Department, Diponegoro University, Semarang, Indonesia
[2]Chemical Engineering Department, Diponegoro University, Semarang, Indonesia
[3]Physic Department, Diponegoro University, Semarang, Indonesia
*Corresponding author: E-mail: meinysuzery10@gmail.com

Abstract. Encapsulation is a coating process to improve the stability of bioactive compounds. Phycocyanin has been encapsulated using chitosan in microcapsules form to keep its stability. This study aims to determine the optimum conditions of the encapsulation process using the extrusion method thorugh characterization of the physicochemical properties of the microcapsules, antioxidant activity test using DPPH, in vitro release performance and evaluate the storage stability against temperature. The results showed that Na-TPP provided better encapsulation performance than Na-citrate as crosslinker at 3% chitosan content. The study of antioxidant activity also showed that at 3% chitosan concentration resulted highest antioxidant activity. The morphological analysis of microcapsules showed that the beads have compact spherical shape with diameter range of 900-1000 μm. In vitro release study demonstrated a quick release in an acidic environment (SGF) during 2 hours experiments and slow release under alkaline conditions (SIF) for 8 hours experiments under constant temperature at 37°C. The encapsulation also showed that phycocyanin was more stable against temperature changes during storage.

Keywords: Encapsulation, stability, chitosan, phycocyanin

1. Introduction

Antioxidants are compounds that can counteract free radicals and an electron donating compound or a reductant. Antioxidants also are compounds that can inhibit the oxidation reaction, to scavenge free radicals and highly reactive molecules, which can inhibit cell damage [1]. Phycocyanin is a group of pigments that are bound to protein (biliprotein). Besides the potential as a natural dye, phycocyanin also known to have healing abilities, such as antioxidants [2] and anticancer [3]. But the application of phycocyanin very limited because it is vulnerable to light and temperature [4]. Phycocyanin can be damaged at temperatures above 30°C [5], and a solution of phycocyanin undergo color fading by 30% after 5 days of storage and become clear after 15 days at a temperature of 35°C [6]. Therefore, the need for technique to improve the phycocyanin stability but still maintain its antioxidant activity is required. Encapsulation is a process of coating a core material by using a specific encapsulation materials [7]. Encapsulation can be used to protect the drug from environmental influences (humidity, light, heat), as well as to control the release of antioxidants to the targeted medications [8]. In the process of encapsulation, it needs proper matrix with the core material to be encapsulated.

IOP Conf. Series: Earth and Environmental Science **55** (2017) 012052 doi:10.1088/1755-1315/55/1/012052

Chitosan has a structure similar to cellulose and capable of forming a gel that serves as the matrix in drug delivery [9]. Chitosan has many advantages such as not toxic, unstable during use, and can be used as a matrix to extract plant [10]. This study focuses on encapsulation of phycocyanin using chitosan to improve the stability of phycocyanin in order to retains its antioxidant activity.

2. Research methodology

2.1. Encapsulation by using Na-Citrate Crosslinker

Chitosan solution of 3% (w/v) was prepared by using 1% acetic acid. At the ratio of 1:1, the solution was stirred until homogeneous and dripped slowly through a 23G syringe to Na-Citrate 4% (w/v) crosslinker solution. After 45 min, the encapsulated product filtered and rinsed with distilled water. The rinse water and crosslinker solution were used to determine the encapsulation efficiency [11] and the load of encapsulation [12]. Do the procedure for other Na-citrate crosslinker concentration.

$$EE = \frac{mass\ of\ phycocyanin - mass\ of\ uncoated\ phycocyanin}{mass\ of\ phycocyanin} x100\% \quad (1)$$

$$Load = \frac{phycocyanin\ mass\ in\ the\ encapsulation}{mass\ of\ encapsulated\ product} x100\% \quad (2)$$

Determination of mass loss and phycocyanin in encapsulation (% Load) using a UV-Vis Spectrophotometer [11].

2.2. IR analysis

The interaction of physical and chemical bonds in the encapsulation process was observed using Fourier Transform Infrared Spectroscopy (FTIR) on each sample.

2.3. Determination of antioxidant activity

Each encapsulated phycocyanin was varied at concentration of 40-800 mg/mL in distilled water. Each solution was pipeted at 2 mL and added to 2 mL DPPH 50 μM. The mixture was homogenized and incubated for 30 minutes without being exposed to the light [13]. The pure phycocyanin was treated with the same procedure and used as a positive control. The percentage of inhibition was calculated according to equation 3:

$$\%\ Inhibition = \frac{Ao - As}{Ao} x100\% \quad (3)$$

where:
Ao = Absorbance of blank and As = Absorbance of samples
% Inhibition was used to determine the IC50 (anti-oxidant activity). The best antioxidant was used to determine the stability and the release test.

2.4. Phycocyanin release test

A total of 50 mg of encapsulated phycocyanin was added to 50 mL of simulated gastric fluid (SGF). The pH was set constant at 1.2 by using HCl buffer. The encapsulated phycocyanin was kept for 2 hours in acid solution before it was transferred to simulated intestinal fluid (SIF) solution. The SIF solution was prepared by using phosphate buffer at pH 7.4. The mixture was kept for 8 hours under stirring at 100 rpm at 37°C. The measurements was done by taking 4 mL solution every hour and analyzed by using UV-Vis spectrophotometer to determine the concentration of released phycocyanin.

2.5. Determination of product stability

The stability test was performed by exposing the beads to various temperature. 25 mg of sample was kept in a vial bottle at a temperature of 25, 35, 45, and 55°C with no light condition [14]. The observations were made every day during 3 days of experiment.

3. Results and Discussion

3.1. Effect of Na-Citrate and Na-TPP Crosslinkers

The encapsulation of phycocyanin was conducted by using Na-citrate and Na-TPP crosslinkers. Figure 1 shows that the beads of encapsulation by using Na-citrate crosslinker have irregular shape and softer texture. In other hands, Na-TPP crosslinker gives more uniform shape and hard texture. The encapsulation process of citric crosslinker is faster than the linking process by Na-TPP, but beads from Na-TPP crosslinker has higher mechanical properties due to stronger ionic bonds between chitosan and tripolyphosphate. In this case the charge density of tripolyphosphate is the main affect for this mechanical properties [15]. The encapsulation efficiency (EE) of both cross linkers are 40% and 60.9% for citrate and trypolyphospate, respectively (Table 1).

At variation of chitosan concentration, the encapsulation efficiency and load show a positive correlation. This result is in agreement with Yan[14] who reported that higher encapsulation efficiency was achieved at higher encapsulation concentration. Table 1 also shows that the bead sizes are influenced by chitosan concentration, and this is supported by Goncalves [16] who stated that the size of encapsulation particles was in the range of 1-1000 μm [8].

Table 1. Encapsulation efficiency and load of phyocyanin at various chitosan concentration

Encapsulation	EE (%)	LOAD (%)	Shape	Size (μm)
Chitosan 2%	51,8	16,9	Spherical, Uniform	987,8
Chitosan 2,5%	56,6	17,5	Spherical, Uniform	902,4
Chitosan 3%	60,9	22,1	Spherical, Uniform	1000

3.2. FTIR analysis

The FTIR analysis was aimed to determine the presence of functional groups in the encapsulated phycocyanin [17]. Figure 2 shows the similarities peak between phycocyanin and chitosan at wave numbers between 3200-3500 cm^{-1} which is identified as N-H bonds. Peak at a wavelength of 2850-3000 cm^{-1} indicates a C-H bond and peak at a wavelength of 1000-1150 cm^{-1} indicates the presence of C-O bond. At wavelength of 1600-1800 cm^{-1} shows the C = O bond wherein the spectra peak of phycocyanin appears at wavelength of 1550-1600 cm^{-1}. At this wavelength, there was no evidence of C = C bonds meaning that there are no presence of chitosan. A shifted peak at wavelength of 1000-1150 cm^{-1} into 1200-1250 cm^{-1} showed the presence of electrostatic interaction between chitosan and phycocyanin. FTIR analysis also showed that no new peaks appeared which indicates that no new bonds were formed during encapsulation process and only physical interactions i.e. electrostatic interactions affecting the encapsulation [18].

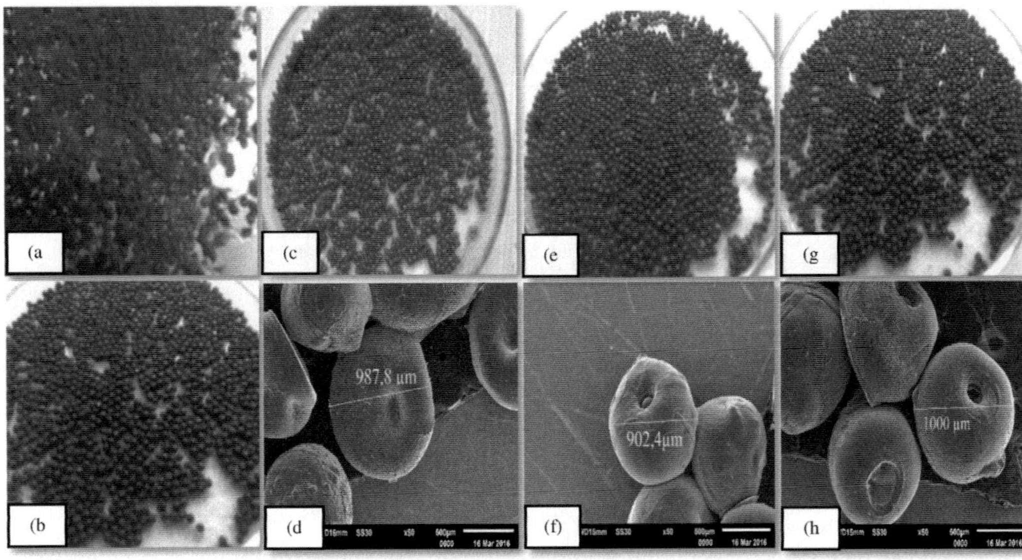

Figure 1. Results of encapsulation using: (a) Na-TPP, (b) Na-Citrate, (c, d) Encapsulation 2%, (e, f) Encapsulation 2.5%, and (g, h) Encapsulation 3%

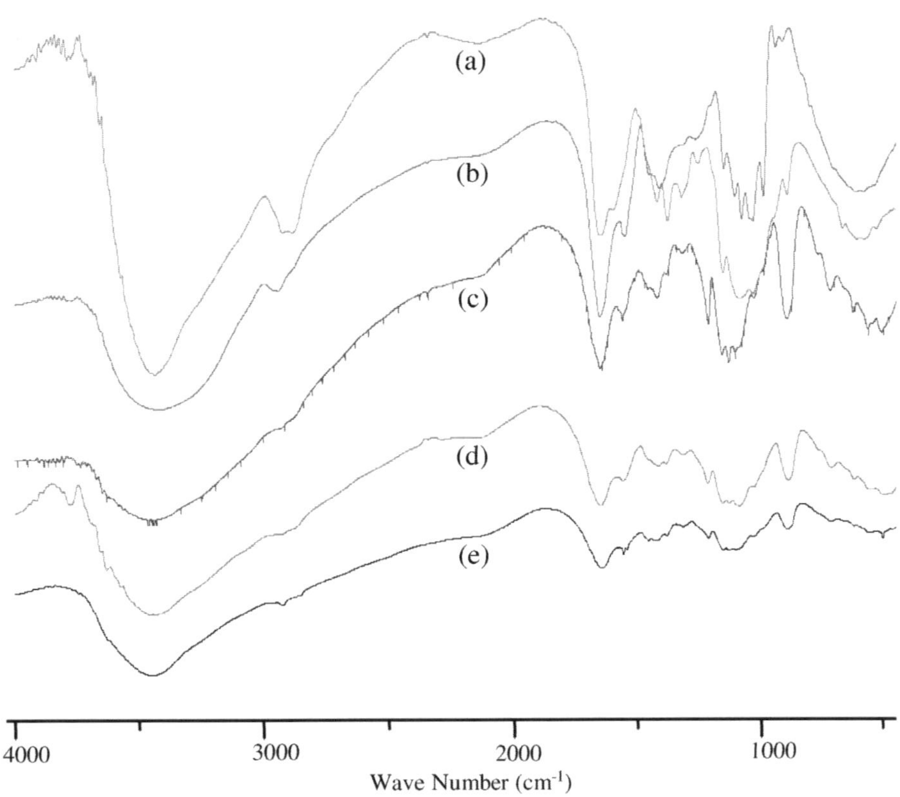

Figure 2. Results of FTIR Analysis: (a) Chitosan, (b) Phycocyanin, (c) Encapsulation of 3%, (d) Encapsulation of 2.5%, and (e) Encapsulation 2%

3.3. Antioxidant activity test

Testing antioxidant activity was determined through DPPH method and measured by using UV-Vis spectrophotometer. Measurements were carried out to determine the absorbance of DPPH remaining after the sample was added. The absorbance of DPPH will be used to determine the percentage of inhibition of free radical (% inhibition) and the result is shown in Table 2.

Table 2. IC50 of Phycocyanin

No	Sample	IC_{50} (ppm)
1	Encapsulation 2%	978.87
2	Encapsulation 2,5%	899.25
3	Encapsulation 3%	639.54
4	Phycocyanin (control)	97.44

According to the Table 2, the IC_{50} of pure phycocyanin as a control is 97,44 ppm and the highest antioxidant activity was achieved at at 3% encapsulation (IC_{50} =693,44 ppm). This experiment also shows that encapsulation will affect the stability of phycocyanin.

3.4. Phycocyanin Release Test in Encapsulation

Figure. 3 shows release of phycocyanin in the simulated stomach and intestinal. In SGF, the quick release was found between 1 to 2 hours and could achieve 53% release. The release begins with the formation of swelling in the first hour, and continued by formation of gel in 1-2 hours later which resulted in quick releaseof phycocyanin. In another hand, the testing in SIF during 3-10 hours experiments, the slow release was found and only increased 4.5%. In SIF, the hardening process occurs and the bead does not swell perfectly. This is because chitosan has less protonated amine groups, and therefore causing low-level release [20].

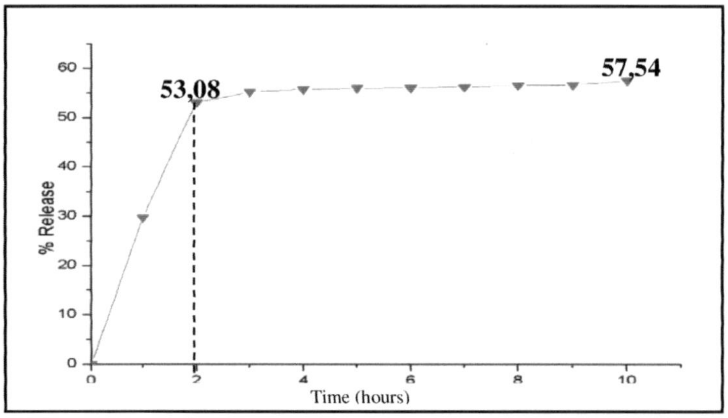

Figure 3. Phycocyanin release in SGF and SIF

3.5. Encapsulation stability against temperature exposure

Determination of stability used the beads with 3% chitosan concentration. According to Burgess and Hickey [8], Encapsulation can be used to protect the drug from environmental influences (humidity, light, heat, and oxidation). Figure 4 shows that the encapsulated phycocyanin is more stable againt temperature than phycocyanin without encapsulation treatment. In this case, the encapsulation process may formed a semipermeable membrane walls[21]. The thin membrane walls can protect the high temperatures of the environment. This is supported by the half-life constants in the degradation kinetics (Table 3).

The constant degradation has a proportional relationship to the temperature [22,23]. The value of k increases by increasing temperature, indicating that higher temperature will stimulate faster e degradation rate of phycocyanin. At temperature of 25°C, the smallest k value was observed than k at temperature of 35- 55°C.

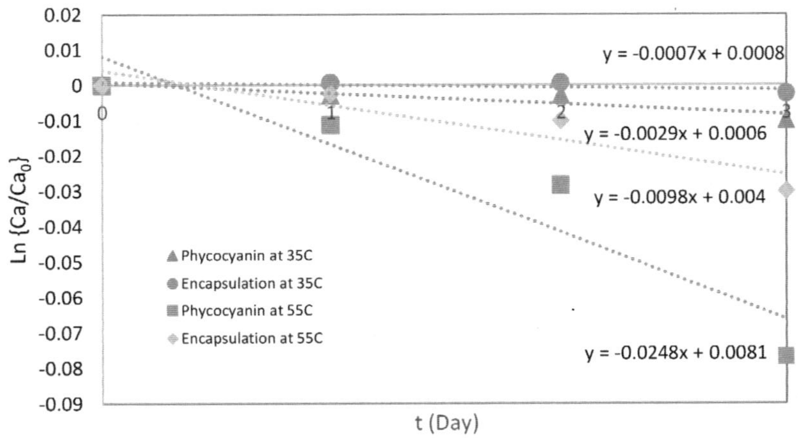

Figure 4. Stability of the encapsulated phycocyanin

Table 3. Half-life time of phycocyanin

Sample	Constant Degradation (k)	Half-life $(t_{1/2})$ (Days)
Phycocyanin at 25°C	$1{,}405 \times 10^{-3}$	493,2384
Encapasulation at 25°C	$0{,}750 \times 10^{-3}$	924,0000
Phycocyanin at 35°C	$3{,}500 \times 10^{-3}$	197,4350
Encapasulation at 35°C	$1{,}510 \times 10^{-3}$	458,9403
Phycocyanin at 45°C	$16{,}35 \times 10^{-3}$	42,38530
Encapasulation at 45°C	$7{,}700 \times 10^{-3}$	90,00000
Phycocyanin at 55°C	$32{,}850 \times 10^{-3}$	21,09580
Encapasulation at 55°C	$13{,}800 \times 10^{-3}$	50,21730

4. Conclusion

Phycocyanin encapsulation using the extrusion method obtained Na-TPP as the best crosslinker. The morphology encapsulation of Na-TPP based encapsulation was round, hard and uniform, with a value of encapsulation efficiency was 60.9778% and the maximum load was 22.19%. FTIR analysis showed that no new bond formation during encapsulation process. The highest antioxidant activity was shown at chitosan concentration of 3% with IC50 of 639 ppm. In vitro study showed that 53% of encapsulated phycocyanin could be released while a slow release has been observed at simulated intestinal pH.

Acknowledgement

We acknowledged Diponegoro University of their financial support through Riset Unggulan Universitas(RUU) 2016.

References

[1] Winarsi H 2007 Antioksidan dan radkal bebas. Yogyakarta:kanisius.

[2] Romay C, González R, Ledón N, Remirez D, Rimbau V 2003. *Current Protein and Peptide Science* **4**,207-216.

[3] Thangam R, Suresh V, Asenath Princy W, Rajkumar M, Senthilkumar N, Gunasekaran P, Rengasamy R, Anbazhagan C, Kaveri K, Kannan S 2013. *Food Chem.* **140**, 262–272.

[4] Chaiklahan R., Chirasuwan N, Bunnag B 2012. *Process Biochem.* **47**, 659–664.

[5] Sarada R, Pillai M G, Ravishankar G A 1998. *Process Biochemistry.* **34**, 795–801.

[6] Mishra S K, Shrivastav A , Mishra S 2008. *Process Biochem.* **43**, 339-345.

[7] Wu W, Roe W S, Gimino V G, Seriburi V, Martin D E and Knapp S E. 2000. Low melt encapsulation with high laurate canola oil. US. Patent 6 153 326

[8] Burgess D J, and Hickey A J 2007 Microspheres Technology and Applications. In: J. Swarbrick, and J.C. Boylan (Eds.). Encyclopedia of Pharmaceutical Technology, Ed. 3rd , New York: Informa Healthcare USA Inc., pp 2328-2338.

[9] Sutriyo S, Djajadisastra J, Novitasari A 2004 Mikroenkapsulasi propa-nolol hidroklorida dengan penyalut etil selulosa menggunakan metode penguapan pelarut. *Majalah Ilmu Kefarmasian*, **1**, 193-200.

[10] Agnihotri S A, Mallikarjuna N N, Aminabhavi T M 2004 *Journal of Controlled Release*, **100**, 5-28.

[11] Yu C Y, Yin B C, Zhang W, Si-Xue Cheng S X, Zhanga X Z, Zhu R X 2009 *Colloids and surface B:Biointerfaces.* 245-249.

[12] Ge X, Wan Z, Song N, Fan A, Wu R 2009. *J. Food Eng.* **94**, 122–128.

[13] Molyneux, P., 2004. The use of the stable free radical diphenylpicrylhydrazyl (DPPH) for estimating antioxidant activity. Technol.26(2) : 211-219.

[14] Yan M, Liu B, Jiao X, Qin S 2014.*Food and bioproducts processing* ,**9** (**2**), 89–97.

[15] Shu X Z, Zhu K J 2002. *Int J Pharm.* **233**, 1-2, 217-225.

[16] Goncalves LV, Mauro C M L, Valredo T F 2005. Effect of Crosslinking Agent in chitosan microspheres in controlled release of diclofenac sodium. *Polimeros: ciencia e tecnologia.* **15**, 6-12

[17] Zhang L L 2007. *Carbon*, **45**, 1439-1445.

[18] Suzery M, Hadiyanto H, Soetrisnanto D, Majid D, Setyawan D, and Azizah N. The improvement of phycocyanin stability extracted from Spirulina sp using extrusion encapsulation technique. *AIP Conf. Proc.* **1699**, 030011-4.

[19] Ansel H C 1989 Pengantar Bentuk Sediaan Farmasi. Edisi 4. UI Pres. Jakarta. Halaman 96, 147.

[20] Al-Assady N, Wesam K H, Jabair A H 2011. *National journal of chemistry.* Vol: 41. 113-126.

[21] Kailasapaty K 2002 Microencapsulation of probiotic bacteria: technology and potential applications. Curr. *Issues Intest. Microbiol.* **3**, 39-48.

[22] Atkins P W, 1996, Kimia Fisika, Edisi ke 4, Jilid 2, Penerjemah: Irma I. Kartohadiprodjo, Erlangga, Jakarta.

[23] Hadiyanto H., Sutanto A A and Sudarto Y. 2014. Ultrasound assisted extraction of antioxidant from Coleus tuberosus peels. Carpathian Journal of Food Science and Technology **6(1)**,58-65

Sea Water Intrusion in Kaligawe Semarang Based on Resistivity Data

Agus Setyawan[1], Najib[2], Yoga Aribowo[2], Agnis Trihadini[3], Dhana Hastuti[4], Fitra Ramdhani[4], Fajar Waskito[4], Ganap Febrika[4] and Galang Virgiawan[5]

[1] Department of Physics, Faculty of Science and Mathematics, Diponegoro University, Tembalang, Semarang, 50275, Indonesia.
[2] Department of Geology, Faculty of Engineering, Diponegoro University, Tembalang, Semarang, 50275, Indonesia
[3] Graduate Student Institute of Seismology and Volcanology, Faculty of Sciences, Kyushu University, 2-5643-29, Shin`yama, Shimabara, Nagasaki, 855-0843, Japan
[4] Undergraduate Student in Department of Physics, Faculty of Science and Mathematics, Diponegoro University, Tembalang, Semarang, 50275, Indonesia
[5] Department of geology, Faculty of Engineering, Diponegoro University, Tembalang, Semarang, 50275, Indonesia

agus.setyawan@undip.ac.id

Abstract. Semarang is a city on the north coast of the island of Java, Indonesia and it is lowland areas have experienced sea water intrusion. One of interesting area is Kaligawe which located at Eastern part of Semarang. Kaligawe has big population and industrial and it need water consumption. Excessive extraction of groundwater will be resulting height difference surface ground water to the surface of the sea water due to sea water intrusion. Electric resistivity method was used to detect for salt water intrusion. Dipole-dipole configuration was applied with 3 lines to get current, potential difference, and apparent resistivity from the field. 2D model has presented using Res2Dinv to get the true resistivity and the depth of each layer. A calibration of the model was conducted based on geological information. Result showed the subsurface area has 4 layers: sandstone, sandsilt, siltstone and clay. Moreover the sea water intrusion occurs in the Northwest, East and Southern part of the study area

1. Introduction

Kaligawe, which is located in Eastern Semarang Municipal Central Java Indonesia, is a village consisted by two main areas. The two are residencial area and industrial zone. the population of Kaligawe is 12.107 of people [1]. The high pupulation and industrial activities in that zone cause a rising in ground water consumption. Some researches show fact that withdrawing ground water excessively may induce gap between the height of surface ground water and the surface of sea water then results saline substrate of sea water, such as Chloride, infiltrates into the ground water[2].

Geographycally, Kaligawe lies on the North sea-coast of Java Island. Highly water usage in the coastal area leaves empty holes or pores in aquifer rocks, next it is potentially stimulate sea water intrusion[3][4]. Sea water infiltaration into aquifer layer in the seaboard have been main attention since iy is generally a pollutant in freshwater [5]. Thus, understanding of sea water intrusion is urgently required for water and coastal resources management [6]. The intrusive movement of sea water basically is process of salt water displacing under groundwater surface by passing aquifer layer that happens in land or in near sea region [7].

Hence the upon reasons inspired us to conduct research, which objective to determine the depth of salt water intrusion in Kaligawe based on resistivity data. Dipole-dipole configuration of the resistivity method were utilized to determine resistivity of each rock layer.

2. Geology and hydrogeology information

2.1. Geological setting

Semarang is basically underlain by two different geological units: the sedimentary rocks of the volcanic Damar formation in the South mainly consisting of tuffaceous sandstone and vulcanic breccia and alluvial sediments in the North, which basically consist of sand and clay [8]. **Figure 1** presented the stratigraphyc setting of Kaligawe is dominant of Alluvium (Qa). The Alluvium is deposit of coastal plain, river and lake. Coastal plain consists generally of clay, sand and some mixture with 50 m or mor of thickness. River and lake deposits contains pebble, cobble, clay and silt.

Damar Formation (Qtd) consists of Tuffaceous sandstone, conglomerate, volcanic breccia. Tuffaceous sandstone has color of brownished yellow with smoth to coarsed grain, the composition is mafic, feldspar, and quartz mineral with tuffa as the base mass and intermediate porocsity. Conglomerate is colored by brownished-yellow to blackish, and has component of andesite, basalt, pumice, sized of 0,5 – 5 cm, middle rounded to perfect rounded, and probably fragile. Volcanic Breccia seems to be sedimented as lahar, blackish gray color and has component of andesite and basalt, sized 1-20 cm, middle pointed-rounded and looked solid.

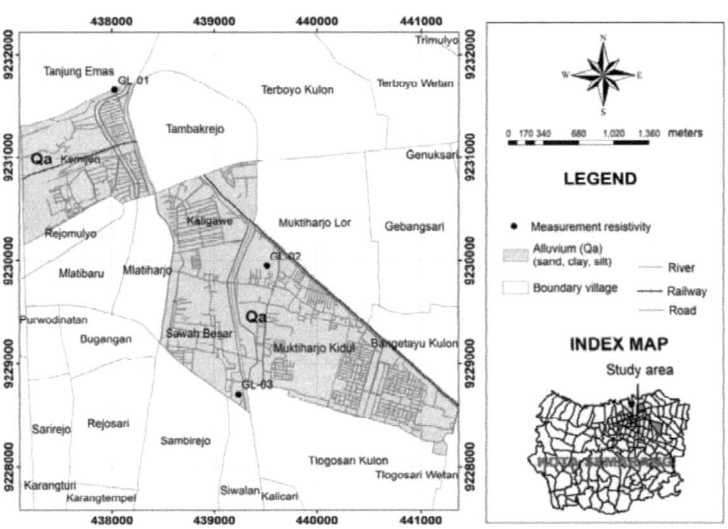

Figure 1. Geology map of research area.

2.2. Hydrologycal setting

Water potency in Semarang municipality is supplied from rivers which stream across Semarang, such as Kali Garang, Kali Pengkol, Kreo, Banjir kanal Timur, Kali Sringin, Kali Kripik, Kali

Dungadem and the others. Kali Garang is the main river that flows crossing Mount Ungaran's valleys, along the swiftly current of curved stream line [9]. Aquifer in coast of Semarang generally is alluvial sedimentation products and Kali Garang delta, that has variety depth from 60 meter to 90 meter, mainly in West Semarang District and more than 90 meter both in East Semarang District and North Semarang District [2].

3. Method

Resistivity method is measurement technique of geophysics that uses resistivity or electicity characteristic of rocks to observe subsurface of earth condition [10]. Measuerment is conducted in the land surface by injecting electricity current (in miliAmpere) along two current electrodes (C1 and C2) then gaining potential difference (in miliVolts) from two potential electrodes (P1 and P2).

The result of that quantify procedure informs about resitivity distribution in the subsurface. Resistivity value is determined from resistivity substance of each rock. Electrical conduction of rock is defined from electrolite distribution that occupies its rock pores [10]. The inner part of subsurface structure can be obtained by gradually making the space between each electode, current and potential, wider [11].

There are positve corelation between space etlectrode and deep current penetrations. The widest electrodes space could be get a dep currecnt penetration, hence the physical characteristics of deeper rock layer can be analyzed. Subsurface resistivity is linked to some geological parameters, such as mineral content, fluid, porosity and water saturation degree of rock[12]. The field data consists of electric current value (I) and electric potential or voltage (V) and leaves apparent resistivity (ρa) which is countable by this formula (1) [12].

$$\rho_a = k\frac{V}{I} \tag{1}$$

which ρ_a is apparent resistivity in Ohm, k is geometric factor relied on electrodes configuration. I is the injected electrical current while V is measured potential difference. adalah beda potensial yang terukur. Apparent resistivity has been considered as resistivity of fictional homogen medium that equivalen to the observed layered medium[12]. The configuration were applied is dipole-dipole as presented in **Figure 2.**

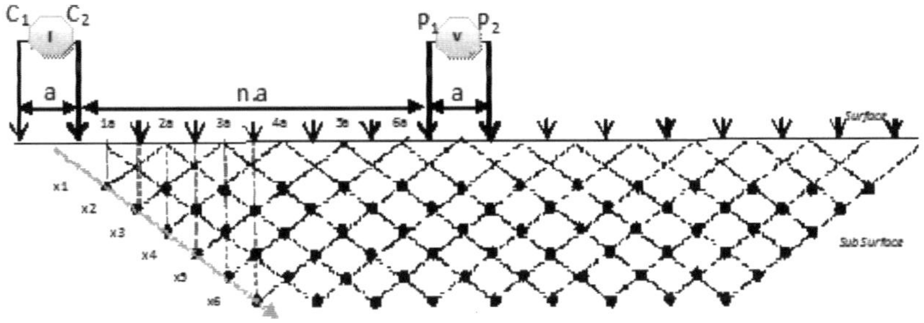

Figure 2. Dipole-dipole configuration (n =1,2,3,..) [13]

Geometric factor has given by formula 2

$$K_d=(n+1)(n+2)a \quad (2)$$

With K_d is konstanta of dipole-dipole configuration, a is the space between two same electrodes while na is the space of potential electrode and current electrode and n should be integers [13]. For getting various depth, n should be varied, the wider n the deeper depth that can be acquired. Sensitivity level of dipole-dipole configuration range is influnced by the number of a and the variety of n [14].

The geoelectric observation was done on 26 April to 09 June 2015 as three point measurement; North Western, Easten and Southern part represented of Kaligawe area as **Figure 3**

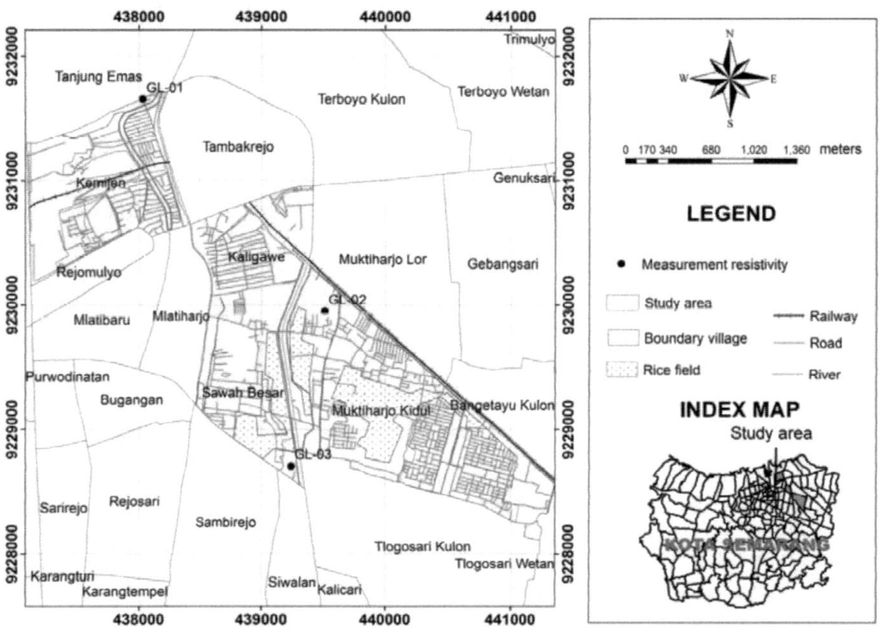

Figure 3. The geolectric point measurement; GL-01, Gl-02 and GL-03.

4. Result and discussion

4.1. GL-01 point measurement

GL-01 were located in the the North Western of research area, shows specific resistivity range of 4 rock layers as stated in **Figure 4**. The first layer has resistivity 2.07 -3.98 Ωm and at 0-35 meter of depth is considered to be sea water intrusion. We described the depth at 35-50 meter in second layer as clay stone, then in the deeper part at 50-70 meter is sandsilt while sandstone as the forth layer at 70-80 meter of depth.

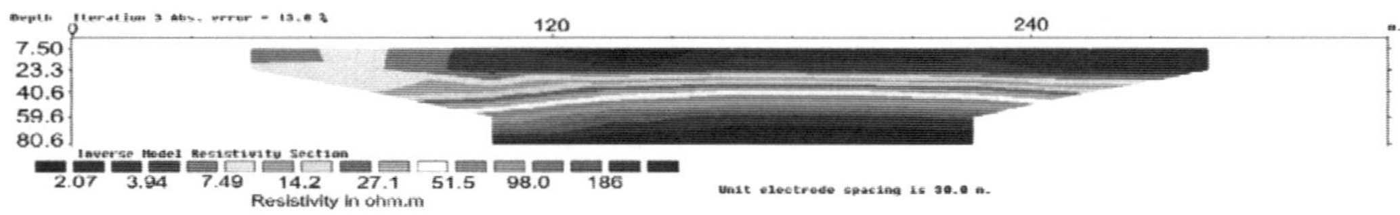

Figure 4. 2-D resistivity model of GL-01 which located in the North Western part of research area.

4.2. GL-02 point measurement

The result of GL-02 which was located in Eastern area, has specific resistivity range and similarity with the GL-01 in the number of layers as presented in **Figure 5**. In the first layer, we considered a 64 – 12.6 Ωm as saline intrusion happened in sandstone at 0 - 11.6 meter below surface. Subsequently the following layers specifications including each depth are sandsilt stone at 4-21 meter, siltstone at 22-29 meter and clay at 29-40.3 meter

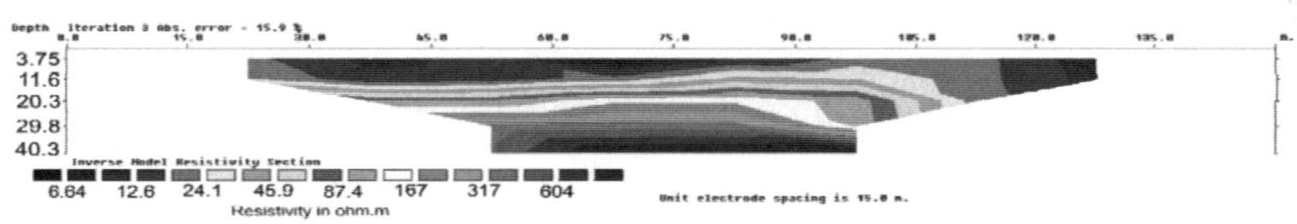

Figure 5. 2-D resistivity model of GL-02 which located in the North Western part of research area.

4.3. GL-03 point measurement

Gl-03 was located in the southern part of research area, we can see the modeling result in **Figure 5**. Succeeding of four layers, the top layer has sandstone with 4.78 – 13.2 Ωm of resistivity and depth range 0 - 11.6 meter, where the intrusion zone takes place . The three next layers including their resistiviy values are sandsilt with 22.2 – 37.2 Ωm, siltstone with 62.3 – 104 Ωm and clay as the last with 175 Ωm.

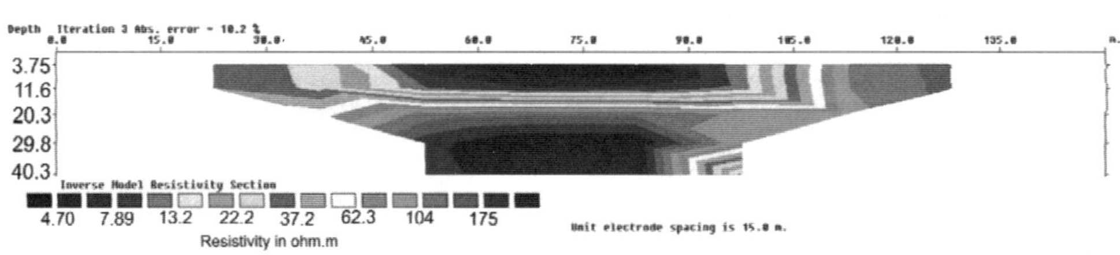

Figure 5. 2-D resistivity model of GL-02 which located in the North Western part of research area

The reason why we deduced the first layer as the place of sea water intrusion is caused by the existence of electrolyte solution comes from sea water. This condition leads to fine conductivity, thence it can transfer electricity finely. Low value in resistivity supports above statement because the contrary correlation between resistivity and conductivity. If the resistivity value of rock is lower then its conductivity will be higher resulting the ease in passing electicity. Moreover, reverse situation will be happened when the resistivity is low so the conductivity will be high.

The other side we conducted the chloride contents in several well as presented in **Table 1.** Several point has a big chloride content such as S-1, S-3 and S-7 which value of 819 -1799 mg/l. Based on the Ministry of Health regulation No. 492 Year 2010 the water can be drinked if chloride contents less than 250 mg/l. Moreover according to adhoc ommittee of salt water intrusion Jakarta, 1986 the water contain 500 – 2000 mg/l chloride has contamined .

Table 1. A table of chloride contents in research area.

Code	Longitude	Latitude	Chloride (mg/l)
S-1	439611	9230051	819.65
S-2	438899	9230209	92.15
S-3	437236	9231090	1799.35
S-4	439334	9230988	47.046
S-5	438628	9230379	153.26
S-6	439080	9228990	1697.5
S-7	439390	9229599	89.24

5. Conclusion

Based on the result analyses to investigate sea water intrusion, we conclude that the areas, where have been damaged by this, are in the North Western part, Eastern part, and Southern part of Kaligawe. This conclusion is supported by the existence of low resistivity range in the subsurface between 2.07-13.2 Ωm at 0-35 meter at deepness. Moreover in several area the water can not to drink caused the high chloride contained.

Acknowledgement

We thankfully appreciate for any supports from Department of Physics Diponegoro University's members, helped us in data measurement, processing until result discussion. Research has supported by PNBP DIPA Diponegoro University No: SP DIPA-042.01.2.400898/2026, 7 Desember 2015, Fiscal year 2016.

References

[1] Badan Pusat Statistik 2014 Profil Kependudukan Kota Semarang 2014, *Report of Statistics Indonesia*, Semarang
[2] Suhartono E, Purwanto and Suripin 2013 Kondisi Intrusi Air Laut Terhadap Air Tanah Pada Akuifer di Kota Semarang, *Proceeding of National Seminar Pengelolaan Sumber Daya Alam dan Lingkungan* Semarang
[3] Pacheco 2006 Delimitation Of Ground Failure Zones Due To Land Subsidence Using Gravity Data And Finite Element Modelling In The Queretaro Valley Mexico *Engineering Geology* Vol. 84 143-160
[4] Qahman K A and Zhou Y 2001 Monitoring of Sea water Intrusion in The Gaza Strip, Palestine, *Proceeding of First International Conference on Saltwater Intrusion and Coastal Aquifer Monitoring*, Modeling and Management Essaouira
[5] Batayneh A T 2006 Use of electrical resistivity methods for detecting subsurface fresh and saline water and delineating their interfacial configuration: a case study of the eastern Dead Sea coastal aquifers *Jordan Hydrogeology Journal* Vol 14, 1277-1283
[6] Ginzburg A dan Levanon A 1976 Determination of a saltwater interface by electric resistivity depth soundings *Hydrogeological Sciences* Vol. 21 pp 561-568
[7] Nisa K. Tony Y and Sugeng W 2012 Aplikasi Metode Geolistrik Tahanan Jenis untuk Menentukan Zona Intrusi Air Laut di Kecamatan Genuk Semarang *Journal Berkala Fisika*, Vol. 15 No.1 pp 7-14
[8] Thanden 1996 *Geological Map Of The Magelang And Semarang Sheets Jawa*, Second Edition (Bandung: Geological Research and Development Centre)
[9] Marsudi 2001 *Prediksi Laju Amblesan Tanah Di Dataran Alluvial Semarang Provinsi Jawa Tengah*, ITB, Bandung
[10] Telford W M Geldart L P and Sherif R E 1990 *Applied Geophysics* Second Edition Cambridge University Press Cambridge

[11] Broto S and Afifah R S 2008 Data Processing of geoelectric Schlumberger configuration, *Journal Teknik* Vol 29 No 2
[12] Loke M H 2000 *Electrical Imaging Surveys For Enviromental And Engineering, Studies A Practical Guide To 2-D and 3-D Surveys*, Standford University Press
[13] Reynolds J M 1997 *An Introduction to Applied and Environmental Geophysics* (England: John Wiley & Sons Ltd. England)
[14] Waluyo and Hartantyo E 2000 *Theori and Aplication of resistivity* Laboratorium of Geophysics, Gadjah Mada University Yogyakarta

IOP Conf. Series: Earth and Environmental Science **55** (2017) 012054 doi:10.1088/1755-1315/55/1/012054

In Search for Sustainable Coastal Management: A Case Study of Semarang, Indonesia

Sudharto P Hadi

Study Program of Environmental Studies, Diponegoro University, Semarang, Indonesia

sudhartophadi@yahoo.co.id and sudhartophadi@undip.ac.id

Abstract. As a coastal town, Semarang is currently facing environmental problems such as flood, tidal flood (locally called *rob*), coastal abrasion, emerging land, land subsidence and sea water intrusion. These phenomena severely affect to citizen, community and corporate, disrupting day to day activities, threatening people's health, causing economics' burden and reducing property value. Government policies in dealing with these problem are focused on its phenomena such as normalizing river for flood and building polder systems for tidal flood. Impacted people have been implementing various initiatives. People in Tanah Mas Estate set up collective efforts to reduce tidal flood by building pumping system project, while people in Kampong Tambaklorok conduct a regular mutual assistance in cleaning of waste and sedimentation, rehabilitating of local drainages and dikes, reconstructing of local streets and also maintaining of pumping system. People in Mangunharjo, the district of Tugu build a coastal belt and cultivate mangrove. Various government and local initiatives have been effective in dealing with flood and tidal flood temporarily. More comprehensive approaches and focused on the sources of problems are required to achieve sustainable coastal management.

Key words: coastal problems, government and community initiatives, sustainable coastal management

1. Introduction

Coastal area is a strategic area for various activities such as port, recreation, fishery, agriculture, industry, settlement, etc. However, coastal area is also vulnerable to change caused by activities within coastal area itself and other activities from the upstream. Local people live at coastal area also vulnerable because they rely on natural resources available at coastal area. If the area has been degraded, it will severely affect them economically and socially. As a coastal town, Semarang is encountering crucial problems such as tidal flood (locally called rob), flood and coastal abrasion. Because of various crucial problems, Semarang was chosen as one of 100 resilience city programs facilitated by Rockefeller Foundation. It is expected that the City of Semarang could formulate strategies to deal with these problems. The phenomena of tidal flood has been occured since 1980s and threaten the north coast area of Semarang from Tawang Mas, Tanah Mas, Tanjung Mas, Port area, Bandarharjo, Tambaklorok, Kemijen, Johar, and Genuk. Tidal flood is currently befalling 7 districts out of 16 districts, with the width of impacted area reach to 3,915,16 ha. Other phenomena are coastal abrasion and flood occur at Mangunharjo and Mangkang Wetan, Tugu District at the west part of coastal area of Semarang, eroding fishpond and threatening the settlements. This paper explains the driving factors of tidal flood, land

IOP Conf. Series: Earth and Environmental Science 55 (2017) 012054 doi:10.1088/1755-1315/55/1/012054

subsidence, coastal abrasion and flood, reviews the measures taken by the government and local people and explores the ideal measure for sustainable coastal management?

2. Literature Review

Tidal flood is stimulated by the effects of sea level rise and exacerbated by land subsidence. Land subsidence is caused by density of built up area and ground water exploitation. People can easily exploit groundwater and artesian well as an alternative water supply when PDAM (local water firm) do not work. Urban construction aggravates land subsidence. Sea water flows to the land subsidence causing inundated areas. The inundated area become worst during the rainy season. According to Miladan [1], during the period of 1985-2008, sea level rise in Semarang was 5,536 cm per year, while land subsidence at the port area reach 5-7,5 cm per year and has been accounted that up to 31,8 cm. Land subsidence in Tanah Mas Estate is between 4 to 9 centimeters on average per year.

The way the government responded to the coastal problems is characterized by hierarchical structure, top-down decision making, centralized authority and short-term orientation. This type of policy, according to Buchholz [2] is classified as traditional management and not eco-centric management. While Dunn [3] reiterates that what kind of decision taken by the government is linked to the value of decision-maker. In relation to the value, Fintersbusch [4] suggests that there are five principles in decision making process namely utilitarianism, libertarianism, justice rawls, democratic, and ethical pluralism. Utilitarianism is based on the principle of the greatest possible happiness for the greatest number. This type of decision is determined by expert based on quantitative technique. Libertarianism is based on the right of individualism and democratic decision making. Justice theory of Rawls relies on democratic, equity and justice. Democratic is based on supreme value of democratic and decision, not weighting. The last principle is ethical pluralism in which no value to be supreme and all value is equal. This is the ideal type of public decision making in which all stakeholders involve at the earliest stage of planning. However, utilitarianism has been widely utilized in decision making process sacrificing democracy and creating injustice #

#

3. Research Method

The type of research is analytical descriptive which is to describe the phenomena of tidal flood, land subsidence, coastal abrasion and flood, the measures taken and its effectiveness. The scope of research follows (a) the driving factors of tidal flood, land subsidence, coastal abrasion and flood. (b) the measures taken by the government and impacted people in dealing with tidal flood, land subsidence, coastal abrasion and flood (3) the ideal measures should be taken for sustainable coastal management. Two cases taken as object of studies namely tidal flood and land subsidence in Tanah Mas Estate and Kampong Tambaklorok, and coastal abrasion and flood in Kampong Mangunharjo and Mangkang Wetan. The data collection techniques consist of content analysis from mass media, analysis of secondary data, physical and social observation and informal interview with relevant interest. The data gathered is analyzed qualitatively.

4. Results and Discussions:

For local people, tidal flood disrupts day to day activities, potentially cause disease such as diarrhea and vomiting, typhoid, leptospirosis and respiratory tract infections. Miladan [1] found that when tidal flood occurs for several days, septic tanks in each house do not function and are inundated. The environmental condition are unhygienic, dirty and smells bad. This deteriorates quality of life of impacted people. In addition, the streets and walls become mossy and then ruined. Tidal flood also destroys house equipment such as cupboards, tables and vehicles. The damaged house equipment and corroded vehicles result from frequent inundations. Impacted people have to allocate budget for replacement and rehabilitation of house equipment and vehicles. Impacted people also bears burden of the rehabilitation of settlement infrastructures such as street and public facilities. Frequent inundation destroy these infrastructures. The

community collects dues for rehabilitation cost and these dues were demanded from all households in each RT (neighborhood association). Tidal flood destroys settlement lead to significant decrease of property value: land and house. Because the settlement is no longer a good place to live, the demand to live at impacted area are significantly decreased. For companies, tidal flood and land subsidence cause high economic cost. The maintenance cost increases 1,5 times than other Ports. Port company requires Rp 40 to Rp 50,- billion to elevate three docks since 2010. The old dock could not be utilized because of being inundated [5]. Coastal abrasion caused lose people's job and income. In addition, due to coastal abrasion, sea water infiltrate to people's settlement disrupting day to day activities, deteriorate community health and also reducing property value.
Coastal abrasion is also caused by reclamation done in several part of coastal area of Semarang such as in Mororejo, nearby Kampong Mangkang and Mangunharjo, Tawang Mas for PRPP (recreation and development promotion complex), Marine Recreation area and Port area. According to Anggoro et al [6] the type of reclamation done in Semarang is the one that merged with the main land. This type of reclamation is potentially causing a change of natural current pattern, deterring sediment transport, stopping existing irrigation or drainage, and in turn causing coastal abrasion, accretion and flood at the mainland. In addition, the risk of flood will occur because of the river or drainage is lengthened. Coastal abrasion has been occured at Mangunhardjo since the beginning of 1990s. Based on research done by Anggoro et al[6], coastal abrasion was stimulated by the change of current pattern caused by massive construction in the form of dock built by a factory. This activity contributes about 65% of coastal abrasion causing 65 hectare of degraded fish pond. The length of degraded coast line reach to 3,5 km, while the abrasion erodes the main land up to 1,7 km and causing the infiltration of sea water to the settlement. People in Mangunharjo and Mangkang Wetan also experience flood coming from Beringin river nearby their settlement. Every rainy season, Beringin river frequently overflow because of high volume of run off from the upstream

5. Government Policies

In dealing with the tidal flood, the government built Polder System located at Kampong Tawang, Kemijen and Bulu Drain System. The Bulu Drain system is one of the devices to control flood in Semarang coast covering three Kampongs: Panggung Lor, Panggung Kidul and Bulu Lor. Polder system consists of retention tank, pumping station and flood gate. Retention tank located in Tawang, Semarang helps reduce the volume of tidal flood and flood temporarily. While Polder in Kemijen is still under construction. The proposed location of retention tank is owned by Railway Company and up until now has not been resolved about the status of the land. Pumping station has been built and local people has a plan to voluntarily manage the Polder System by collecting dues from member of the community for operation and maintenance cost of Polder System.

In dealing with coastal abrasion in Mangunhardjo and Mangkang Wetan, the government provided a seed funding for building a coastal belt. While in responding the flood in Mangunhardjo and Mangkang Wetan, the government has planned to normalize the river by compensating people's land at the down-stream. However, up until now, negotiation regarding the land compensated has not been achieved an agreement. During the rainy season, local government is currently elevating embankment of Beringin river. From the perspective of the government, this flood is mainly caused by narrowing of Beringin river caused by sedimentation. In fact, the main contribution of flood comes from the upstream of Beringin river which is at Mijen District where land-use change has been massive since 1980s causing run off and flow to Beringin river.

6. People Initiatives

The response of impacted people to these disasters has been done individually and collective action. Miladan [1] found that in Kampong Tambaklorok, individually, each household reconstructed their houses to reduce the impact of tidal flood. People elevate their houses foundations every 5 years. They realize that this measure is only temporary but this is the only way to deal with this problem. The reconstruction of house consist of elevating the floor with piles of sand and soil or cement, terrace renovation and its roof and yard elevation. In addition, as collective action, people have been conducting a mutual assistance to reduce the impact of tidal flood by cleaning of waste and sedimentation, rehabilitating of local drainages and dikes, reconstructing of local streets and also maintaining of pump system built by government. The reconstruction work is aimed to elevate the local streets. They pile up these materials on the local streets in order to add 30 to 50 cm.

In Tanah Mas Estate, Miladan [1] further explains that individually, people have implemented mitigation efforts to reduce the impacts of tidal flood by elevating their house and streets. Each individual household constructed modest barriers around their houses, notably doors surrounding the houses. These efforts were only temporary solutions and in fact, their settlement to be inundated during the tidal flood occurrences. In 1996, people initiated to set up community collective efforts called Paguyuban Pengendali Air Pasang Panggung Lor (Association of Controlling and Handling of *Rob* (tidal flood) Risk in Panggung Lor Sub District (P5L). People consider that pumping system project was appropriated solution because the water surface of Asin River had been higher than the surface water of the settlement, so inundation discharge could be pumped. People realize that elevations of local roads and houses caused social and territorial problems

People in Mangunhardjo and Mangkang Wetan conducted a mutual assistance to build a coastal belt with the seed funding from the Government of Central Java. Up until now, they have finished building costal belt for 3,2 km out of 3,5 km of degraded coast. They also obtained funding from the Ministry of Fisheries and Marine, and other sources. To strengthen the coastal belt, local people planted mangrove parallel with coastal belt. Mangrove has a function to revitalize degraded fish pond. In fact, by having mangrove, fish pond can be planted again with shrimp and milk-fish. Coastal belt, in addition to prevent abrasion from the wave, also has a function as sediment trap to rehabilitate the degraded coast.

7. Conclusions and Recommendations

The measures taken both by the government and by local people are focused on the phenomena of tidal flood, land-subsidence, coastal abrasion and flood. These measures are successful in minimizing the impacts temporarily, however, at the long term, the measures will not able to deal with problems if the sources of problems are not dealt with. The causes of land subsidence are over burden building and ground water extraction. With regard to over burden building, it is required to remapping of the tidal flood area at north coast of Semarang and to revise spatial planning to determine the area need to be allocated for green open space. If necessary, buildings stand at the area which are not suitable with the spatial planning need to be demolished for green open space. It is also required to limit the permit issuance of ground water extraction. The permit that has been issued must be strictly controlled to comply with regulation. To prevent coastal abrasion, it is required to limit reclamation permit. The current man-made environment resulted from reclamation such as factories, recreation areas, ports need to be followed by environmental audit. This audit will show how their reclamations contribute to coastal abrasion and what measures need to be taken by them to minimize the environmental impacts. In dealing with flood in Kampong Mangunharjo and Mangkang Wetan, it is necessary to revise spatial planning at the upstream area by conducting strategic environmental assessment (SEA). SEA studies will calculate environmental carrying capacity and environmental carrying capability and recommending which area need to be used for green open space.

References

1. Miladan N 2016. *Communities' Contribution to Urban Resilience Process: a Case Study of Semarang City Toward Coastal Hydrological Risk.* Unpublished PhD Dissertation. Universite Paris-Est, French.
2. Buchholz R 1993 *Principles of Environmental Management. The Greening of Business.* New Jersey: Prentice Hall Inc.

3. Dunn W 2000 *Public Policy Analysis: An Introduction.* New Jersey: Prentice Hall
4. Fintersbush K 1989 *How Should Policy Decision be Made?* Paper presented at the Plenary Session: Socio-political Aspect of Impact Assessment of the Annual Meeting of International Association for Impact Assesment. Montreal, Canada, June.
5. Suara Merdeka Daily Newspaper 2012 *Pelabuhan Ambles 31,8 cm. (Land Subsidence at Port Area Reach to 31,8 cm)* Semarang: Suara Merdeka Daily Newspaper, February 9, 2012
6. Anggoro S 2011 *General View of Reclamation* Unpublished Paper for Reclamation Plan of Jakarta Bay, Indonesia.

# # #

Effect of economic growth and environmental quality on tourism in Southeast Asian Countries

Firmansyah

Department of Economics, Diponegoro University, Jl, Prof Sudharto SH, Tembalang, Semarang, Jawa Tengah Province, Indonesia

firmansyah@live.undip.ac.id

Abstract. The tourism is an important sector in generating income for a country, nevertheless, tourism is sensitive toward the changes in economy, as well as changes in environmental quality. By employing econometric models of error correction on annual data, this study examines the influence of environmental quality, domestic and global economic growth on foreign tourist arrivals in selected Southeast Asian countries, namely Indonesia, Malaysia, Thailand, Philippines, and Singapore. The findings of this study showed that all of countries long run model were proved statistically, indicated that world economic growth as well as environmental quality affect foreign tourism arrivals.

1. Introduction

Tourism is one of the most important activities in the world, involving millions of people and a large amount of money (foreign exchange). As a labour intensive sectors, tourism provides and expands the employment opportunities both skilled and unskilled labour, in both the developed and developing countries.

However, the tourism is vulnerable to face the change of the world economy. The contraction of the world economy due to the crisis, reduces the real per capita income of the people in every country, further causes in a tightening of public spending, including recreational expenditures. The recession such as the 2008-2009 world financial crisis which was triggered by the United States financial crisis - demonstrated by the negative growth of the world economy by 1.24 percent - causes the decrease of foreign tourists around the world by 4 percent, and the tourism expenditure lowers by 6 percent in 2009 [1].

Besides the world financial crisis, the decrease in number of tourists is also caused by some other external factors. The war in Iraq, the issue of health such as Severe Acute Respiratory Syndrome (SARS), the 2003 global economic slowdown, and the surge of world oil prices in 2005, also declines the number of foreign tourist arrivals in various countries [1]. The decrease of tourists' number also occurred in the Southeast Asia countries such as Indonesia, Malaysia, Philippines, Singapore, and Thailand. The changes in tourists number in these countries is presented in figure 1.

The influence of external factors on tourism have been widely studied, such as [1], [2], [3], [4], [5], and [6]. [1] proves that all regions of the world experienced negative growth in the number of foreign tourists, except Africa, as the impact of the 2009 world financial crisis. [2] analyses the impact of the global economic changes against the arrival of foreign tourists to Indonesia and measures the multiplier effect on other domestic industries. [3] investigates the impact of global economic crisis toward Greek tourism, [5] examines the impact of the global crisis which lowers the foreign tourists visit, and [6] discusses the impact of the crisis on the tourism decline in Hawaii.

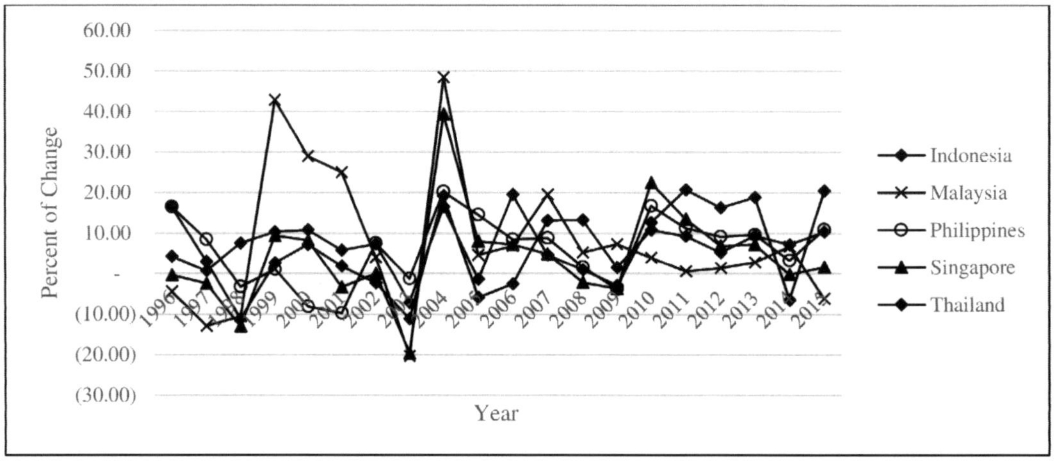

Source: [7], processed

Figure 1. The number of foreign tourists visiting the countries of Southeast Asia (1996-2014)

Besides the external factors, foreign tourist arrivals are also influenced by factors derived from the destination country itself or can be classified as internal factors. These factors include the cultural charms and beautiful nature, the cost of living, the level of security, the environmental quality, and other domestic factors. These factors may affect tourists' satisfaction and further affect tourist arrivals.

The attention to environmental factors that affect the tourism has not been widely studied. The specific destination tourist arrivals such as beach and mountains, or destination places that require the clean air and beautiful environment, so that the attention of the environmental sustainability, cleanliness, and freshness of the air are very decisive on tourism sustainability. Some studies especially on China tourism were conducted by [8], [9], and [10]. They investigates the importance of the effect of pollution on tourists' satisfaction.

According to the preconditions, this study objective is to analyse empirically the dynamic effect – long and short run – of the global economic and environmental quality changes to the tourism in five Southeast Asia countries, namely Indonesia, Malaysia, Thailand, Singapore and the Philippines. The selection of these countries based on the consideration of the geographical adjacent as placed in one region areas, the high inter-linkages in economy as the members of the Association of South East Asian Nation (ASEAN), and cultural as well as the natural contours are similar. In addition, these countries are being very intensively develop tourism in their countries.

2. Method

2.1. Data and variables

Suppose a tourist destination country in Southeast Asia is referred as country A, the definitions of the variables that are used in this study as follows:

a. The tourism demand by global to the country A is approached by the number of foreign tourist arrivals in A, the unit of measurement is the number of people per year, T.
b. The changes in the world economy that is faced by country A, is approached by the real world Gross Domestic Product, with the unit of measurements is the US dollar, WGDP.

c. The quality of environment in the country A, is approached by the level of CO2 pollutant, the unit of measurement is kilotons, Z.

The data are collected annually which are varies between countries due to limited publications. The data of Indonesian model covers 1982-2014, Malaysia (1995-2014), Thailand (1996-2014), the Philippines (1993-2014), and Singapore (1982-2014). The data collects from [11], [7], [12], [13], and other sources.

2.2. Empirical models

To achieve the research objective, this study employs a dynamic econometric model, the error correction model (ECM), which is a useful theoretical approach to estimate the effect of short run and long run of a time series with other time series variables. The model used this study is developed by Granger representation theorem, which states that if two variables X and Y are cointegrated, the relationship between the two can be expressed as ECM.

Cointegration occurs in two (or more) time series variable shows the long run relationship or balance (equilibrium) between these variables. In the short run, there may be a disequilibrium), and the "error" can be treated as an "error disequilibrium". This mechanism is called error correction. The cointegration test uses Engle-Granger (EG) and augmented Engle-Granger (AEG) test on the residual of long run equation, then the ECM equation is derived [14].

This is the long run equation in each country:

$$LT_t = \beta_0 + \beta_1 LWGDP_t + \beta_2 LZ_t + u_t \quad (1)$$

EG test is conducted to test the cointegration of the residual u_t, with using the Dicky-Fuller (DF) stationary test [14]:

$$\Delta u_t = \delta_1 u_{t-1} + v_t \quad (2)$$

Once the equation is cointegrated, it would be set up the error correction model (ECM):

$$\Delta LT_t = \alpha_0 + \alpha_1 \Delta LWGDP_t + \alpha_2 \Delta LZ_t + \alpha_6 u_{t-1} + \varepsilon_t \quad (3)$$

where L indicates the natural logarithm operator, ε_t is the error term, u_{t-1} is the error correction term of equation (1), as α_6 is a discrepancy of T – from the robust estimation – to achieve equilibrium, and subscript t indicates the time series observation.

3. Results and discussion

3.1. Estimation results of the long run model and cointegration

Table 1 displays a summary of long run model estimation. From table 1, it can be seen that all of countries show a similar pattern regarding the influence of the global economy (WGDP) and CO2 pollution (Z) to the number of tourists' arrival (T). In all countries, WGDP statistically significant affects T in a positive direction, but only in Indonesia the Z is proved affect T.

The positive magnitude of the parameter indicates that the increase in the world economy, the arrival of foreign tourists increase, vice versa. The WGDP grows by 1 percent, the number of foreign tourist arrivals in Indonesia increase by 0.712 per cent, and the increase of 1 percent of WGDP, boosts foreign tourist arrivals in Malaysia by 1.652 percent. The same explanation applies to other countries in accordance with the each county parameters. Tourists' arrival on Malaysia are the most sensitive to changes in the world economy, while Thailand's is the most inelastic. It can be said that tourism in Thailand is least affected by the world economic changes compared to other countries.

The long run results are supported by cointegration test, as shown in Table 2. From the result, it can be proved that equilibrium relationship of these variables are proved statistically, which are indicated by the stationery of the long-term residual at level 0.

Table 1. Estimation results of long-run model: Dependent variable is LT

Country	LWGDP		LZ	
	Coefficient	t statistic	Coefficient	t statistic
Indonesia	0.712	2.494*	0.489	2.178**
Malaysia	1.652	3.44*	-0.144	-0.265
Thailand	0.237	3.028*	0.573	1.686
Philippines	0.807	3.684*	0.701	0.423
Singapore	0.828	27.950*	0.038	0.816

Notes: *) significant at 1%, **) significant at 5%, ***) significant at 10% level of confident

Table 2. Estimation results of Engle-Granger Cointegration test

Country and variable	Cointegrated or not
Indonesia	Cointegrated
Malaysia	Cointegrated
Thailand	Cointegrated
Philippines	Cointegrated
Singapore	Cointegrated

3.2. Estimation results of error correction model

In general, the short run model can be proved by the ECM due to the negative significant of u_{t-1}, except for Indonesia and Thailand. Only the Malaysian model that showed the significant of individual parameters of WGDP and Z. Z is statistically significant in the Philippines in positive direction.

Table 3. Estimation results of ECM: Dependent variable is ΔLT

Country	ΔLWGDP		ΔLZ		u_{t-1}	
	Coefficient	t statistic	Coefficient	t statistic	Coefficient	t statistic
Indonesia	0.710	1.589	0.296	0.784	-0.089	-0.904
Malaysia	2.335	2.377**	-1.384	-2.713*	-0.550	-2.663*
Thailand	0.321	0.849	0.386	1.142	-0.262	-1.388
Philippines	-0.590	-0.904	1.931	3.286*	-1.022	-6.880*
Singapore	0.416	1.380	0.025	0.386	-0.524	-3.139*

Notes: *) significant at 1%, **) significant at 5%, ***) significant at 10% level of confident

From the significant value u_{t-1}, it can be demonstrated that T adjust WGDP and Z with lag. For Malaysia, it can be explained that 55 per cent discrepancy of long run and short run T can be corrected in one period. For Singapore, 52.4 percent of the discrepancy can be corrected in one period, while for the Philippines, the discrepancy were corrected during the period.

The influence of WGDP and Z to T in the short run can be explained only on ECM estimation of Malaysia; the WGDP increases by 1 percent and T increases by 2.335 percent. The increase of Z by 1 percent, lowers the T in Malaysia by 1.384 percent.

Based on the diagnostic test results on the long run and the ECM model, it can be indicated that the estimator of the long run and short run are homoscedastic and non-autocorrelation.

4. Conclusion

By employing annual data with variation of period on each countries observed, it can be empirically proven by the study that the effects of the global economic changes on tourism in Indonesia, Malaysia, Thailand, the Philippines and Singapore in the long run is real, and only on Indonesia estimation model the environmental quality affects tourism. It should be concerns by stakeholders and the policy maker of these countries, particularly Malaysia, the impact of the world economic contraction, which can suppress the number of foreign tourists visiting.

Related to factors which are influence the tourism, the policy makers has to restrain the impact of the world economic crisis by enhancing the attractiveness of the country, such as development of public facilities and other facilities related to improving the convenience of tourists, increasing the international cooperation among institutions and agents of tourism, promoting more cultural attractions and art. It is also important to be noticed by the governments, the stability of the domestic economy includes prices and costs, ensuring the availability of goods, maintaining the good level of security and the legal certainty. Especially for Indonesian government, the awareness of sustainable environmental quality including the conservation strategy should be placed as an important policy.

To produce a more comprehensive study in the future, by some limitations of method and data availability in this study, it can be done some developments. For the improvement of the data and variables, it can be extended the data coverage, and added other independent variables that are considered affecting the foreign tourists arrival, include both economic and social variables, such as exchange rates, the level of security, the cost of living and others. To investigate the tourism development in the region level, it can be developed the data panel model. For the purposes of forecasting, the future study can use the time series models that are more complicated, for example, the model which accommodate seasonal factors, volatility and so on.

References

[1] United Nation World Tourism Organization and International Labour Organization 2013 *Economic Crisis International Tourism Decline and its Impact on the Poor* (Madridd: UNWTO)
[2] Firmansyah 2013 *Seminar Nasional pariwisata hijau dan pengembangan ekonomi (Green tourism and economic development) November 2013: Pengaruh perubahan perekonomian global terhadap pariwisata Indonesia* (Mataram: Mataram University Press)
[3] Kapiki S 2012 The impact of economic crisis on tourism and hospitality: Results from a study in Greece *Central European Review of Economics and Finance* **2** 19–30
[4] Enz C A, Kosova R and Lomanno M 2011 The impact of terrorism and economic shocks on U S hotels *Cornell Hospitality Report* **11** 6–17
[5] United Nation World Tourism Organization 2009 Impact of the global economic crisis on local tourism destinations *Survey Report* http://www unwto org/destination/events/en/pdf/492_rp_e pdf
[6] Sian T L, Subramonian H , Tung L , San W H , Hui K H K and Kulampalil T T 2009 *Fundamentals of Hospitality and Tourism Management* (Malaysia: Open University)
[7] World Bank http://api worldbank org/v2/en/topic/6?downloadformat=excel
[8] Zang A, Zhong L, Xu Y, Wang H and Dang L 2015 Tourists' perception of haze pollution and the potential impacts on travel: Reshaping the features of tourism seasonality in Beijing, China *Sustainability* **7** 2397-2414
[9] Huang C 2012 The impact of local environmental quality on international tourism demand: The case of China *Master's Theses* Paper 21
[10] Basarir C and Cakir Y N 2015 Causal interactions between CO2 emissions, financial development, energy and tourism *Asian Economic and Financial Review* 5 1227-38
[11] Bloomberg 2016 Platform in Faculty of Economic Diponegoro University
[12] United Nation World Tourism Organization http://www2.unwto.org/en
[13] International Monetary Fund several years of publication *World Economic Outlook* http://www imf org/
[14] Gujarati D N and Porter D C 2009 *Basic Econometrics* 5th ed (New York: McGraw-Hill)

Isolation And Partial Characterization Of Bacteria Activity Associated With Gorgonian *Euplexaura* sp. Against Methicillin-Resistant *Staphylococcus aureus* (MRSA)

R. Kristiana[1]*, D. Ayuningrum[1], M.A. Asagabaldan[1], H. Nuryadi[1], A. Sabdono[2], O.K. Radjasa [2,3], A. Trianto[2]

[1]Coastal Resources Management, Faculty of Fisheries and Marine Sciences, Diponegoro University, 50275 Tembalang, Semarang, Indonesia
[2]Department of Marine Sciences, Faculty of Fisheries and Marine Sciences, Diponegoro University, 50275 Tembalang, Semarang, Indonesia
[3]Ministry of Research, Technology and Higher Education of the Republic of Indonesia, 10270

E-mail: rhesikristiana@gmail.com

Abstract. Methicillin-resistant *Staphylococcus aureus* (MRSA) infection has emerged in around the world and has been resistance to *ciprofloxacin, erythromycin, clindamycin.* The aims of this study were to isolate, to investigate and to characterize bacterial symbionts gorgonian having activity against MRSA. *Euplexaura* sp. was collected from Panjang Island, Jepara, Indonesia by snorkling 2-5 m in depth. Bacterias were isolated by using spesific media with dilution method. Bacterias were conducted by using the streak method. Antibacterial activity was investigated by overlay method. The potent bacteria was identified by using molecular identification (DNA extraction, electrophoresis, PCR and phylogenetic analysis using 16S rDNA genes with actinobacteria-spesific primers) and bio-chemical test (among 5 isolated bacteria from gorgonian showed activity against MRSA). The strain PG-344 was the best candidat that has an inhibition zone against MRSA. The result of sequencing bacteria is 100% closely related with *Virgibacillus salarius.* This becomes a potential new bioactive compounds to against MRSA that can be a new drug discovery.

Keywords: antibacterial, MRSA, *Euplexaura* sp., *Staphylococcus aureus*

1. Introduction

Antibiotic resistance is a problem for international hospitals when treating diseases caused by bacterial infection. Previous studies indicate that excessive use of antibiotic cause bacteria resistance [1][2]. *Staphylococcus aureus* has emerged recent decades and continues to be a threatening disease around the world. Multi-resistant cases have emerged in some parts of the world. The recent study showed that 50% of Methicillin-Resistant *Staphylococcus aureus (MRSA)* isolate have been resistant to *ciprofloxacin, gentamycin and clindamycin* [3]. MRSA medical treatments require more time and to conduct and are more expensive because of the irrational use of antibiotic.

Finding a new compounds structure to serve as an anti-MRSA agent is imperactive. Recent studies have shown that marine organisms such as sponges, bryozoans, tunicates and gorgonians produce bioactive compounds that have the potential to be used as a source of medicine. More than 1000 new bioactive compounds have been identified in the bacteria carried by these marine invertebrate. Bioactive compounds from marine invertebrate have biological properties which are essential in the development of new drugs. Bacteria associated with marine invertebrate indicated having the same compound with its host, therefore it can be an alternative in the culture and synthetic drugs in the future [4].

Gorgonian is one species of marine invertebrates that contains various critical bioactive compounds, including them as cytotoxic, anti-bacterial, anti-fungal and immunostimulan compounds [5]. Secondary metabolites that are contained in gorgonian act as a defense when disturbed by environment. Secondary metabolites play an important role in maintaining habitats in the marine environment [6]. The previous study conducted by Tanaka [7] reveals that gorgonian *Isis hippuris* has potential compounds to produce a compound steroid polioxygen.

This study aims to isolate the bacteria associated with gorgonian *Euplexaura* sp. It is estimated to produce bioactive compounds similar to the host and to identify the antibacterial association against MRSA.

2. Material and Method

2.1 Euplexaura sp. Sampling

Gorgonian *Euplexaura* sp. samples were collected from Panjang Island, Jepara Indonesia by snorklers 2-5 m depth into the ocean. The samples were found around coral reefs and were collected into a sterilized bag to be carried out a surface immediately to a handling surface [8] . The collected samples were rinsed with sterile sea water and were stored in cool box. Gorgonian identification was done by identifying the sclerit.

2.2. Isolation and microbiological analysis

One gram sample was shattered delicately. The resulting homogenate was diluted with sterile seawater at four dilutions (10^{-1}, 10^{-2}, 10^{-3}, 10^{-4}). One mililiter of each dilution was inoculated into nutrient agar (pH 7,5) medium by spread plate method. The inoculated plates were incubated for 2 days at 29 ^{0}C. The association bacteria grew selectively within the medium. Purification bacteria was executed based on different color, texture and shape of colonies. The selective colonies transferred to nutrient agar by innoculum method.

2.3. Screening bacteria for antibacterial activity against MRSA

The secondary metabolites were screened for antibacterial activity by using overlay method [9]. The activity perfomed by innoculated bacteria on nutrient agar plate against Methicillin-Resistant *Staphylococcus aureus* (MRSA). The bacteria association were incubated overnight at 28^{0}C. *S. aureus* was innoculated on nutrient broth and was grown overnight at 27^{0}C according to the shaker condition. The concentration of *S.aureus* cultur was adjusted to 0,5 McFarland standards and were poured evenly on soft agar media. Antibacterial activity was initiated by pouring *S. aureus* on nutrient agar plate which contained grown bacteria. Antibacterial activity was indicated by the emergence the inhibiton zone surounding the bacterial association after 24 hours incubation at 28^{0}C. The active isolate created a stain that identified the biological morphology. It was observed at the magnification 100 X in microscope.

2.4. PCR amplification and DNA sequencing

DNA extraction was obtained by diluting a sample with saponin, ddH_2O, chelex 20%. Supernatant which contains DNA used as a template for PCR analysis. The PCR mixture consisted of primers F (127) 1 µl, primers R (1492 R), DNA template 1 µl, PCR kit 12,5 µl, ddH_2O 9,5 µl. The total volume measured 25 µl. The PCR condition was denaturation at 45^{0}C for 1 minute, anneling at 53,9^{0}C for 1 minute, extention at 72[0]C for 30 second and post cycling at 72[0]C for 7 minute. All condition were repeated 30 times.

3. Result and discussion

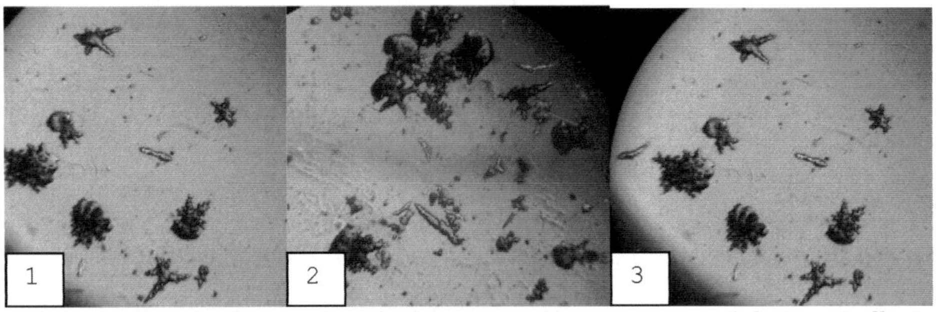

Figure 1. Sclerit of gorgonian *Euplexaura* sp. Note: 1=*strong club*; 2= *spindle*; 3= *rod*

Samples were identified by observation using a microscope at 10x magnification as presented in Figure 1. Forms sclerites observed and identified by [10]. *Euplexaura* sp. has 3 forms sclerites. They are *a strong club*, and *a spindle rod* and *a reddish color*. Each type has a gorgonian sclerites which is distinct. Basic forms sclerites include *double head*, *spindle*, *club*, *scaphoid* and *rod*. Sclerites form of a specific character possessed by each type of gorgonian can be used as a basic in determining taxonomic species. Therefore the name of the species can be recognized from the samples acquired.

Table 1. Zone of inhibition for antibacterial activity against MRSA

Bacteria	**Zone of Inhibition (diameter in mm)**										
	PG-341	PG-342	PG-343	PG-344	PG-345	PG-381	PG-382	PG-383	PG-384	PG-385	PG-386
MRSA Activiy	-	-	-	+	-	-	-	-	-	-	-

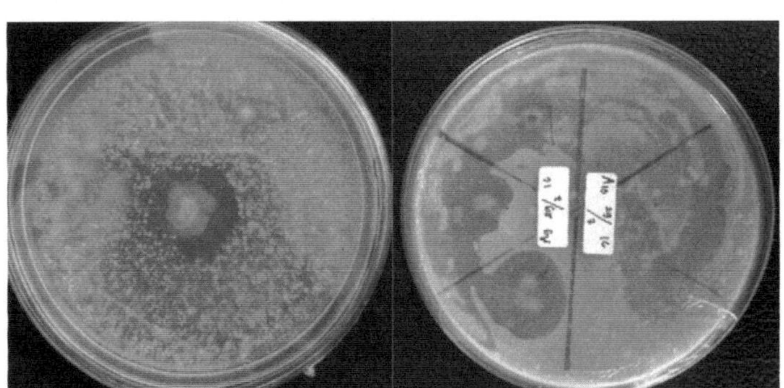

Figure 2. Antibacterial activity

The antibacterial activity of *S. aureus* was tested by overlay. The results in Table 1. showed, that one of the 11 *Euplexaura* sp. isolates possess antibacterial activity. Isolates PG-344 possess antibacterial activity as indicated by the inhibition zone around the isolates in Figure 2. Isolates PG-344 were able to inhibit the growth of *S. aureus* with an inhibition zone diameter of 21.86 mm. This is consistent with research done [11][12] that revealed that the gorgonian that contain the bioactive compound can function as an antimicrobial. Gorgonian produces a metabolism secondary to the environmental conditions in the crucial competitive interactions within a restricted environment [13].
Heindl [14] revealed that the bacteria that live in association with its host have the same activity of secondary metabolites. Isolates PG-344 have the ability to inhibit the growth of *S. aureus* activity caused by the presence of the active compounds. *Staphylococcus aureus* has experienced resistance to some antibiotics, such as *chlorampenicol*, *fosfomicin*, *tetraciclin* and *ciprofloxacine*. The clear zone that appears around the isolates indicates that there are bioactive compounds in isolates which are capable of inhibiting the growth of *S. aureus*.

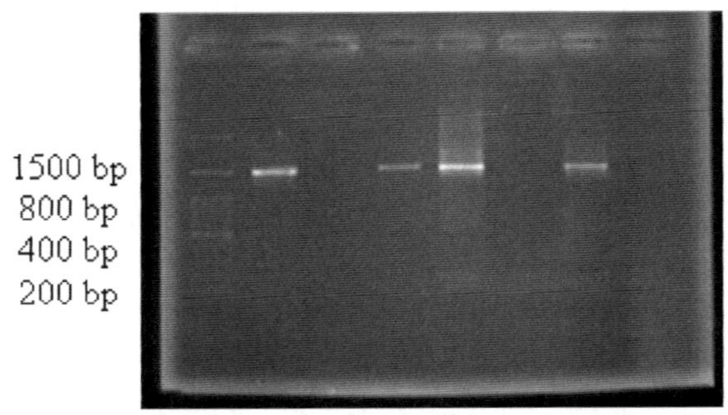

Figure 3. PCR Amplification 16S rDNA Isolate PG-344

Based on the electrophoresis results of PCR 16S rDNA, isolate PG-344 has a base length of 1500 bp. The results of DNA extraction have a band that is perfectly visible in Figure 3. Polymerase chain reaction (PCR) is a method that to detects microbial targets by amplifying the amount of target sequence inwith a relatively short time. PCR does not require a high number of DNA molecules to amplify target sequences. Only the target sequence needs to remain intact. Thus, PCR can amplify or partially damaged and denatured DNA [15].

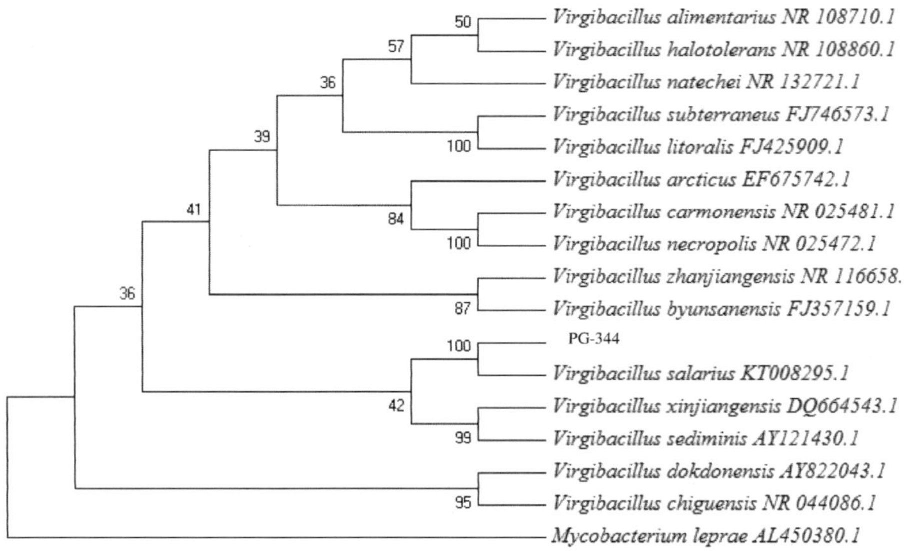

Figure 4. Filogenetic Tree

The bacteria comparator used in the phylogenetic tree are bacteria that have a closely related with the association *Euplexaura* sp. isolates. The Phylogenetic tree presented in Figure 4 was created by using the application MEGA 6.
Results of the Basic Local Alignment Search Tool (BLAST) showed that the bacterial isolates PG-34-04 have a 100% homology closely related to Virgibacillus salarius. Rosenberg [16], explains that the genus Virgibacillus is a gram-positive bacteria, with colonies that measure 0.5-5 mm. Cells occur independently, in pairs or as short chains. Endospores are round or elliptical and located in a position subterminal or terminal to sporangia. After 48 hours at 30-35 UC on a solid medium containing 10% NaCl, colonies formed that opaque and white. Additionally, these colonies were circular and covex with erose or margin that was slight stringy.

4. Conclusion

Test bacteria was selected by using a bioassay, which found one isolate, namely PG-344. PG-344 is caple of antibacterial activity that can protect against methicillin-resistant *Staphylococcus aureus* (MRSA). Selection of the test bacteria using a bioassay and gained 1 isolates, namely PG-344 which has antibacterial activity against methicillin-resistant *Staphylococcus aureus* (MRSA). Analysis of the isolate PG-344 showed that 93% of bacteria is closely related to *Virgibacillus salarius*.

Acknowledgement
Thank you to the Directorate Research and Community Services Ministry of Research Technology and Higher Education Jakarta, Indonesia for providing financial support for the completion of this research, by granting the Education of Master Degree Leasing to Doctoral Program for Excellent Graduates Scholarship.

References

[1] Arana D M, Rubio M and Alós J I 2016 Evolution of antibiotic multiresistance in Escherichia coli and Klebsiella pneumoniae isolates from urinary tract infections: A 12-year analysis. *Enferm. Infecc. Microbiol. Clin.* no. xx

[2] de Matos P D M, de Oliveira T L R, Cavalcante F S, Ferreira D C, Iorio N L P, Pereira E M, Chamon R C and Dos Santos K R N 2016 Molecular Markers of Antimicrobial Resistance in Methicillin-Resistant Staphylococcus aureus SCCmec IV Presenting Different Genetic Backgrounds. *Microb. Drug Resist.* vol. 00. no. 00, pp. 1–7

[3] Lestari E S 2009 Antimicrobial Resistance in Indonesia Prevalence, determinants and genetic basis. ISBN 978-90-8559-606-6.

[4] Mayer A M S, Abimael D R, Roberto G S B, Nobuhiro F 2011 Marine pharmacology in 2007–8: Marine compounds with antibacterial, anticoagulant, antifungal, anti-inflammatory, antimalarial, antiprotozoal, antituberculosis, and antiviral activities; affecting the immune and nervous system, and other miscellaneous mechanisms of action. Comparative Biochemistry and Physiology, Part C 153. 191–222

[5] Manzo E, Ciavatta M L, Lopez Gresa M P, Gavagnin M, Villani G, Naik C G and Cimino G 2007 New bioactive hydrogenated linderazulene-derivatives from the gorgonian Echinogorgia complexa. *Tetrahedron Lett.* vol. 48, no. 14. pp. 2569–2571

[6] Melorose J, Perroy R, and Careas S 2015 No Title No Title," *Statew. Agric. L. Use Baseline*. vol. 1

[7] Tanaka J, Trianto A, Musman M, Issa H H, Ohtani I I, Ichiba T, Higa T, Yoshida W Y, and Scheuer P J 2002 New polyoxygenated steroids exhibiting reversal of multidrug resistance from the gorgonian *Isis hippurisTetrahedron*, vol. 58, no. 32, pp. 6259–6266

[8] Sabdono A, Radjasa O K 2008 Phylogenetic Diversity of Organophophorous Pesticide-Degrading Coral Bacteria from Mid-West Coast of Indonesia. Biotechnology 7 (4): 694-701. ISSN 1682-296X. Asian Network For Scientific Information

[9] Jenkins K, Jensen P R and Fenical W 1998 Bioassays with marine microorganisms. Methods in chemical ecology. Bioassay Methods, II, pp.1–38

[10] Janes M P 2008. Laboratory methods for the identification of soft corals (Octocorallia : Alcyonacea). Advances in Coral Husbandry in Public Aquariums. Public Aquarium Husbandry Series, vol.2., 2, pp.413–426

[11] Kim S K and Van Ta Q 2012 Bioactive Sterols from Marine Resources and Their Potential Benefits for Human Health 1st ed., Elsevier Inc. Available at: http://dx.doi.org/10.1016/B978-0-12-416003-3.00017-2

[12] Habbul P, Vijayanand W, Rajesh S, Smita M, Venkatrao H K 2016 Antimicrobial metabolites from marine microorganisms.*Chinese Journal of Natural Medicines*. 14(2): 0101.0116

[13] Zhang, G 2013 Cytotoxic 9,11-secosteroids from the South China Sea gorgonian *Subergorgia suberosa*. Steroids, 78(9), pp.845–850. 10.1016

[14] Heindl H 2010 Phylogenetic diversity and antimicrobial activities of bryozoan-associated bacteria isolated from Mediterranean and Baltic Sea habitats. Systematic and Applied Microbiology, 33(2), pp.94–104. Available at: http://dx.doi.org/10.1016/j.syapm

[15] Cavanaugh C M, Mckiness Z O P, Newton I L G, and Stewart F J 2006 *The Prokaryotes*. vol. 3, no. November 1902

[16] Rosenberg E 2014 Department of Molecular Microbiology and Biotechnology. Tel Aviv University. Tel Aviv, Israel

Antioxidant and antibacterial activity of solid-liquid and enzyme-assisted extraction of phenolic compound from three species of tropical *Sargassum*

M Puspita[1,2], M Deniel[1,3], I Widowati[2], O K Radjasa[2], P Douzenel[1], G Bedoux[1] and N Bourgougnon[1]

[1] University of South Brittany. EA 3884. LBCM. IUEM. F-56000 Vannes. France.
[2] University of Diponegoro. Faculty of Fisheries and Marine Science. Marine Science Department. Semarang. 50275 Indonesia.
[3] University of West Brittany, Technopôle Brest-Iroise, Rue Dumont d'Urville, Plouzané 29280, France.

Email : maya.puspita@hotmail.co.id

Abstract. *Sargassum* has been well acknowledged for the potential natural product of its phlorotannins. Solid-liquid extraction (SLE) is the most common method used to extract them. However, this method has some drawbacks such as low yield and toxic. An alternative ecofriendly method has been proposed, i.e. enzyme-assisted extraction (EAE), proven to be more efficient. The aim is to compare the efficiency of SLE and EAE concerning their extraction yield, total phenolic content and antioxidant activity. *S. aquifolium*, *S. ilicifolium* and *S. polycystum* were extracted using water, methanol, methanol 50%, and ethanol 75% and enzymes (Viscozyme and Protamex). Total phenolic content (TPC) was analyzed by Folin-Ciocalteu and antioxidant activity via DPPH and FRAP analysis. This study implied that bioactivity of *Sargassum* extracted with enzymes is better compared to the one using organic solvents.

Keywords: Conventional extraction, alternative extraction, brown algae, bioactivities, polyphenols

1. Introduction

As an archipelagic country located in the tropical zone and the coral triangle, Indonesia has been extensively known for its remarkable diversity of marine organisms. One of them is marine seaweed. There are three groups of marine seaweed that can be found throughout Indonesian water, namely, green, red and brown seaweed. For decades, the green and red seaweed, particularly from the genera *Kappaphycus* and *Gracillaria,* have been cultivated for their carrageenan and agar [1–3].

Marine seaweed is also distinguished by its polysaccharides [4,5] but also by its biologically active compounds that have been studied and proven to have promising potential for the nutraceutical, pharmaceutical and cosmeceutical industries [6–10]. Hence, these marine plants have attracted a lot attention during the exploration of alternative natural products. The natural products derived from seaweed offer a sustainable resource with an infinite application.

Among the other three, brown seaweed from the class of Phaeophyceae is one of marine plants that has not been fully explored and exploited in Indonesia despite of their abundance and their remarkable potential [11–17]. A number of previous studies have reported that polysaccharides and bioactive compounds from the brown seaweed showed an encouraging potency as source of natural product with their antimicrobial [18], antioxidant[11,19], antifouling [20,21], antiviral [22,23], anti-proliferative [24–26], anti-inflammatory [27], tyrosinase inhibition [28] activities and many other. This potency can be further applied in foods, medicines and cosmetics in order to reduce the application of chemical products or even to fully substitute them in the future [29–33].

There are five genera of brown seaweed widely distributed in Indonesian waters which four of them are the most species-rich, namely *Dictyota, Padina, Turbinaria* and *Sargassum* with 111 species in total [34]. These four genera have not been optimally exploited; however, *Sargassum* is the most extensively studied among them concerning its alginates and bioactive compound especially the phenolic compounds also known as phlorotannins. The phlorotannin in which are mostly present in the brown seaweed [35,36] are known for its strong antioxidant activity [37–43]. Nevertheless, it also shows promising effects against cancer, allergy, diabetes, inflammation and viral and microbial infection [36]. Basically, this compound is produced as one of the components composing the cell wall of brown seaweed [35]. However, it also serves as an induced defense system as phlorotannin has plastic responses to the environmental factors for examples nutrient availability, light, ultraviolet radiation, temperature, salinity and the intensity of herbivory [44–47].

Due to the various health beneficial performed by phlorotannin, various extractants are used in order to release this compound from the algal matrix. Solid-liquid extraction (SLE) is one of the most used methods in extracting the phlorotannin of brown seaweed [48–53]. This method relies on the organic solvents during the process such as hexane, ethyl acetate, ethanol, methanol, etc. [52]. The polysaccharides' complexity of marine seaweed often becomes the hindrance during the application of SLE. Thus, it reduces the extraction efficiency causing to the low yield of extraction with only 8% to 30% of the dry material [54]. It is crucial to apply an extraction method that could enhance the target compound, improve the bioactivity, time-saving, and favorable for human and environmental. These are the necessary requirements in order to generate a 'Green concepts' as in recent trend [55].

Enzyme-assisted extraction (EAE) has been widely mentioned as a potential alternative method since it offers an eco-friendly approach, an improvement in the quantity of target compound and an enhancement in the bioactivity [56]. The advance point of EAE compared to SLE method lies on its ability to degrade or disrupt the cell walls and membranes facilitating the release of target compound [57]. The EAE method has shown to improve the extraction yields and to enhance the recovery of bioactive compound in marine seaweed [56,58–62]. Nevertheless, this method has not been extensively applied in extracting the bioactive compound from Indonesian *Sargassum.*

Consequently, the objective of this study is to evaluate the efficiency of EAE in comparison to the SLE method in extracting bioactive compound of *Sargassum* sp., particularly phenolic compounds. Later, the phenolic content and the antioxidant activity *of* Sargassum extracts from both methods will be evaluated. In additional, tyrosinase and biofilm inhibition activity of the EAE extracts will also be analyzed in order to acknowledge another possible potential from this extracts. Finally, the enzymatic extracts of *Sargassum* will be characterized using the Fourier-Transform Infra-Red (FTIR) spectroscopy. This characterization will serve as a preliminary analysis to detect the presence of the phenolic compound, phlorotannins.

2. Material and Methods

2.1 Seaweed Collection

There were three different species of *Sargassum* collected for this study. *S. polycystum* was collected from Panjang Island, Jepara in October 2013 then *S. aquifolium* and *S.ilicifollium* were collected from Telukawur, Jepara in April 2014. Both of the locations are located in the northern region of Central Java, Indonesia.

The apical and median parts of *Sargassum* thalli were cut leaving the basal part still attached to the substrate for further regeneration, rinsed with seawater and placed in closed boxes. Samples were then transferred to the Central Laboratory of Research and Services Diponegoro University (CORES-DU). They were washed with tap water to remove the remaining epiphytes and other residual sands and naturally dried away from the sunlight for seven days. After all samples had dried, they were all grinded and stored in a sealed plastic bag that was covered to avoid direct contact with the sunlight.

2.2 Solid-liquid extraction (SLE) of Sargassum

25 gr of algal dry material was diluted in 300 ml of four different solvents, i.e. water, methanol 50 % (v/v), pure methanol and ethanol 75 % (v/v). The filtered samples were then evaporated until dryness. The dried extract obtained was then added by 20 ml of H_2O in order to get the crude extract. Finally, the crude extracts were frozen and lyophilized for further analysis.

2.3 Enzyme-assisted extraction (EAE) of Sargassum

19.5 gram of dry material was diluted in 300 ml of water, then 5 % (of dry material) commercial enzymes were added. Viscozyme® – a carbohydrase, and Protamex® – a protease, were chosen. These enzymes were provided by Novozymes®, manufactured in Denmark. Extraction without enzyme served as contro, also called the aqueous extraction.

Sample solutions were incubated for 3 hours in 40 °C. After the filtration, the soluble samples were then frozen and lyophilized for further analysis. The residual materials were stored in -20 °C for TPC analysis to determine the insoluble phenolic compound that might not be digested by the enzymes.

2.4 Total Phenolic Content Analysis

Folin-Ciocalteu is a commonly used method for measuring the phenolic content. This assay is based on the reduction-oxidation (redox) reactions, which are usually considered to be relatively stoichiometric and on the redox potential of the phenolic hydroxyl group [63].

0.5 ml of samples or standard (Phloroglucinol) was added with 0.5 ml ethanol 95 %. 2.5 ml H_2O was then introduced to the samples or standard solution followed by 0.25 ml of Folin-Ciocalteu 50 %. All samples were agitated and left to stand for 5 minutes. In the end, 0.5 ml of 5 % of Na_2CO_3 (in 100 ml H_2O) was added. Then the optical density was read against the blank prepared at 725 nm. Phloroglucinol was a standard for the calibration curve prepared in different concentrations ranged from 0 – 100 µg/ml [64–66].

2.5 Antioxidant Activity

2,2-diphenyl-1-picrylhydrazyl or DPPH is a stable free radical. DPPH assay is based on the theory that a hydrogen donor is an antioxidant. It measures compounds that are radical scavengers. DPPH show a strong absorption maximum at 517 nm. This analysis is based on modified method of Yen and Chen (1995) and Chen et al. (2008).

A series of BHA, BHT and Ascorbic Acid solution – serves as standard – in different concentration is prepared (2 – 50 µg/ml). 0.25 mM of DPPH solution is introduced to 100 µL samples solution. Samples solution of *S. muticum* hydrolysates was made in different concentration by diluting the solution stock in methanol (0 – 1000 µg/mL). Before reading the optical density at 517 nm, all samples are incubated in 40 °C for 30 minutes. Then, the percentage of inhibition is calculated by following formula:

$$\boldsymbol{I}\,(\%) = \left[{}^{(\boldsymbol{A_C} - \boldsymbol{A_S})}/_{\boldsymbol{A_C}}\right] \times \mathbf{100}$$

$\boldsymbol{I}$ (%) = Inhibition (express in %)
$\boldsymbol{A_C}$ = Absorbance of control
$\boldsymbol{A_S}$ = Absorbance of samples
IC_{50} of samples is determined based on the regression obtained from dose response curve. IC is defined at the concentration sufficient to obtain 50 % of a maximum scavenging capacity.

2.6 FRAP (Ferric Reducing Antioxidant Power)

This method is based on the reduction of a ferroin analogue, the complex of tripyridyltriazine $Fe(TPTZ)^{3+}$, to the intensely bluish colour complex of $Fe(TPTZ)^{2+}$ by the presence of antioxidants in acidic medium. Then, the results are obtained from the absorbance at 593 nm and expressed as μm Fe^{2+} equivalents (Benzie & Strain 1996; 1999)

The reagent of TPTZ consisted of 300 mM of acetate buffer with pH 3.6, 10 mM TPTZ (2, 4, 6-tripyridyl-s-triazine) and 20 mM $FeCl_3$ $_6H_2O$ solution. 50 μL of *S. muticum* hydrolysates was introduced in microplate mixed with 150 μL reagent. After 15 minutes of incubation, the absorbance was read at 593 nm. The results were expressed in μM equivalent to $FeSO_{4.7}H_2O$ calculated from the calibration curve [69].

2.7 *Antibacterial Activity*

Agar diffusion method [70] was chosen as method for evaluating the antibacterial activity of *Sargassum*. One gram positive bacteria – *Bacillus subtilis* – and two gram negative bacteria – *Pseudomonas aeruginosa* and *Escherichia coli* – were chosen for this test. These bacteria were prepared in liquid nutrient broth media with bacterial density 1.5 x 10^8 cfu/ml. Bacterial suspension was poured on to solidified agar media and incubated for 1 h in 37 °C.

5 mg/ml of SL extracts and hydrolysates were prepared in sterile physiological water. 20 μL of tested samples solution – hydrolysates solution or positive control (Phosphomycin, Ampicillin and Streptomycin) – was introduced to sterile paper disc. All samples were made in triplicate. Later, the impregnated sterile paper discs were placed on the agar media in accordance with samples coding made in advance. The petri dishes were then incubated at 37 °C for 48 h. Zone of inhibition established on the agar media was then measured and expressed in cm.

2.8 *Statistical Analysis*

All results are expressed as mean ± standard deviation (SD) with n = 3. As the design of this research was a factorial design with more than one independent variable than two ways analysis of variance (ANOVA) was chosen as statistical analysis using IBM SPSS Statistics 20. Further, the statistical difference between samples were determined via Tukey Post Hoc Test with significance level at 5% ($p<0.05$).

3. Results and Discussion

3.1 *Comparison of SLE and EAE Extraction Yield, Total Phenolic Content, Antioxidant and Antibacterial activity of Sargassum*

Two-ways analysis of variance was applied to determine whether there was a significant effect or not from several independent factors towards the dependant variable. In the case of *Sargassum* yield of extraction, total phenolic content and antioxidant activity – all were the dependant variables, different species and different type of extractants served as independent factors that would influence the dependant variables.

Statistical analysis of *Sargassum* extraction yield showed that there was a significant different between the three species extracted, i.e *S. aquifolium, S. ilicifolium* and *S. polycystum*. Different type of extractants used also gave a significant effect to the extraction yield. This result implied that the extraction yield of *Sargassum* showed an inter-individually significant effect.

As seen in Table 1, the EAE yielded higher dry material compared to the SLE. The order of samples with highest yield of enzymatic extraction was *S. polycystum* > *S. aquifolium* > *S. ilicifolium*. The highest extraction yield was shown by *S. polycystum* extracted with Protamex®, 38.1 ± 6.8 % of dry material.

Table 1. Extraction yield of *Sargassum* (% of dry material) in different method of extraction (mean ± standard deviation, n=3).

Samples	Water	MeOH	MeOH 50%	EtOH 75%	Viscozyme®	Protamex®
S. aquifolium	17.3 ± 2.2*	5.7 ± 1.8*	15.8 ± 1.5*	12.3 ± 0.4*	29.5 ± 5.4*	26.2 ± 5.8*
S. ilicifolium	14.7 ± 0.8*	3.1 ± 0.7*	7.3 ± 1.0*	7.5 ± 1.3*	27.9 ± 1.4*	21.4 ± 2.0*
S. polycystum	22.8 ± 1.1*	8.3 ± 3.2*	13.4 ± 1.9*	12.5 ± 1.9*	35.7 ± 2.6*	38.1 ± 6.8*

*Significantly different at $p<0.05$.

The total phenolic content of *Sargassum* is presented in Table 2 expressed in % of dry material. Viscozyme® extracted more phenolic content in *S. aquifolium* and *S. polycstum* with 7.0 ± 1.3 and 8.1 ± 2.6 % of dry material, respectively, compared to the organic solvents. In the contrary, the highest phenolic content of *S. ilicifolium* was obtained from the methanol 7.5 ± 2.4 % of dry material. Apparently, the efficiency between Protamex and methanol 50% (v/v) in extracting the phenolic compound were the same in all *Sargassum*. As can be seen on Table 2, they only extracted between 3 up to 5 % of phenolic content.

The antiradical activity expressed in the value of IC_{50} of *Sargassum* is presented in Table 3. *S. polycystum* only showed its antiradical activity when they were extracted with ethanol 75% and Protamex® and the best activity was shown by Protamex® with IC_{50} 1.9 ± 1.1 mg/mL. The weakest antiradical activity came from the methanolic extract of *S. aquifolium* with IC_{50} 23.8 ± 9.0 mg/mL. The three *Sargassum* extracted with Protamex® gave a better antiradical activity compared to the other extractants, it can be seen by their IC_{50} that were lower as presented on Table 3. The 75% (v/v) ethanolic extracts of *S. aquifolium* and *S. ilicifolium* showed no antiradical activity. Meanwhile, the antiradical activity of all samples were still weaker compared to the standards, i.e BHA and BHT with IC_{50} $8x10^{-3}$ ± 0.5 and 12 $x10^{-3}$ ± 4.8 mg/ml, respectively. The statistical analysis of two-ways ANOVA showed that there was a significant difference between the three species of *Sargassum* and between the types of extractant that affected the antiradical activity. The statistical analysis of two-ways ANOVA showed that there was a significant difference between the three species of *Sargassum* and between the types of extractant used. Based on its phenolic content, the order of samples with highest content was *S. polycystum* > *S. aquifolium* > *S. ilicifolium.* In addition, the efficiency order of extractants in extracting the phenolic compounds of *S. polycystum* was Viscozyme® > ethanol 75% (v/v) > Protamex® > methanol > methanol 50% (v/v) > water. As for *S. aquifolium,* the efficiency order was Viscozyme® > methanol > ethanol 75% (v/v) > methanol 50% (v/v) > Protamex® > water. Then, for *S. ilicifolium,* the efficiency order was methanol > ethanol 75% (v/v) > methanol 50% (v/v) > Protamex® > Viscozyme® > water.

Table 4 presents the reducing power activity also known as FRAP of *Sargassum* expressed by a concentration in μM equivalent with Fe^{2+} (μM Eq Fe^{2+}). Vitamin C was used as standard with reducing power 58.1 ± 0.7 μM Eq Fe^{2+}. The reducing power activity that was almost near as the standard was the aqueous phase of *S. polycystum* with 43.5 ± 0.7 μM Eq Fe^{2+}. In other word, this sample had the highest reducing power activity. Meanwhile, the lowest reducing power activity was shown by *S. ilicifolium* extracted with ethanol 75%, 7.9 ± 1.2 μM Eq Fe^{2+}. In addition, it appears that Viscozyme® extracts of *S. aquifolium* and *S. polycystum* had better reducing power activity compared to methanol, methanol 50%, ethanol 75% (v/v) and even Protamex®. The statistical analysis of two-ways ANOVA showed that there was a significant difference between the three species of *Sargassum* and between the types of extractant that affected the reducing power activity.

As for the antibacterial activity of two methods of extraction, the tested bacteria were sensitive only to solid-liquid extraction (data not shown). In other words, the enzymatic extracts showed no sign of inhibition against the bacterial growth. It was assumed that the sugar contained in our enzymatic extracts might provoke the growth of bacteria instead of inhibiting them.

SLE has become the most used method during the exploration of marine bioactive compounds, in this case, phlorotannins. This method mostly relies on the polarity of the solvents used due to the chemical nature of target compound [88]. In general, the process in SLE method can be categorized in three parts; (1) changing phase of the solute, in this case the target compounds, as it dissolve in the solvent, (2) its diffusion through the solvent in the pores of the solid to the outside of the particles, and (3) the transfer of the solute from the solution in contact with the particles to the main bulk of the solution [89]. The more dispersed the solute in the solid material, the more difficult the extraction will be and the extraction rate will fall. This is because the solvent will have to penetrate further in to the layer of solid material in order to reach the solute [89]. Furthermore, the efficiency of SLE highly depends on the particles size, temperature, and agitation during the process [90]. In the extraction of phenolic compounds, the storage time of dry materials and conditions, as well as the presence of interfering substances are other important factor to consider, influencing the extraction efficiency [91]. In marine seaweed, the main drawback during the extraction of phenolic compounds occurs due to the presence of complexes polysaccharides as the main component of seaweed's cell wall. Brown seaweed's phlorotannins are known to be incorporated in its cell wall [92] covalently bonded to the polysaccharides [35] and proteins [93]. Therefore, it requires a strong condition to degrade these bonds and extract the phlorotannins. The SLE method does not meet this requirement since it does not degrade such complex bonds since it works based on the solubility of target compound in to the solvents [89]. In addition, the polarity of target compound becomes the most important factor in choosing the solvent [94]. Methanol, ethanol, acetone, ethyl acetate, hexane [50,72,75,95,96] are some examples of organic solvents mostly used for phlorotannins extraction.

Phlorotannins, class of phenolic compounds synthesized only in brown seaweed [71], is one of the most studied bioactive compounds due to their wide range of bioactivities, showing a promising potential for the nutraceutical, pharmaceutical and cosmeceutical industries. As reported by previous studies, the increasing interest towards phlorotannins occurs as they have been for their antioxidant [72–77], antimicrobial [78–80], antitumor [81,82], antidiabetic [83], antiallergic [84], anti-inflammatory [85], and antiviral effects [51,86,87].

This study concerns about the possibility in applying alternative method that is not only environmentally friendly but also more efficient in extracting the phlorotannins. For that reason, enzyme-assisted extraction (EAE) is chosen as the alternative method. Additionally, the efficacy of SLE – as the conventional method, and EAE – as the alternative one, in extracting the phlorotannin from *Sargassum* were compared. For the SLE method, four different organic solvents were chosen, i.e water, methanol, methanol 50% (v/v), and ethanol 75% (v/v). Meanwhile, for the EAE method, there were two commercial enzymes, i.e. Viscozyme® – a carbohydrase, and Protamex® – a protease.

Based on the extraction yield of these two methods, the EAE yielded the dry algal material higher compared with the SLE (Table 1). As can be seen from this table, in three hours of extraction, the SLE extraction yield ranged from 3% to 15% of dry material. In the contrary to the SLE method, the extraction yield of EAE ranged from 21% to 38%. In overall, the average increase of extraction yield when using the Viscozyme® was around 62%. As for the Protamex®, the average increase of extraction was 58%. This result had shown the efficacy of EAE method in improving the extraction yield applied under the same circumstances as the SLE method. Interestingly, the efficiency of EAE in improving the extraction yield seems to be species-dependent. As can be seen on Table 1, Protamex® yielded more dry material in *S. polycystum* with 38% than Viscozyme®. In the contrary, the extraction yield of *S. aquifolium* and *S. ilicifolium* were higher in Viscozyme®, with 29% and27%, respectively. Viscozyme® was also yielded more soluble material in *S. fulvellum* (24%), *S. horneri* (30%), *S. correanum* (30%) and *S. thunbergii* (31%) than Protamex, with 17%, 27%, 21%, and 25%, respectively [97]. In *S. muticum,* the Viscozyme managed to extract 29% of soluble material [41,98]. It appears that the cellulose and the multicarbohydrase complex action of Viscozyme® were responsible for higher solid content in lyophilized extracts [98].

IOP Conf. Series: Earth and Environmental Science **55** (2017) 012057 doi:10.1088/1755-1315/55/1/012057

Table 2. Total phenolic content (TPC) of *Sargassum* (% of dry material) (mean ± standard deviation, n=3).

Samples	Water	Methanol	Methanol 50 %	Ethanol 75%	Viscozyme®	Protamex®
S. aquifolium	1.7 ± 0.0*	6.3 ± 1.5*	3.7 ± 0.2*	5.9 ± 0.4*	7.0 ± 1.3*	3.5 ± 0.0*
S. ilicifolium	2.1 ± 0.0*	7.5 ± 2.4*	5.3 ± 0.3*	5.9 ± 0.4*	4.0 ± 0.1*	4.2 ± 0.2*
S. polycystum	1.4 ± 0.0*	4.8 ± 0.3*	4.0 ± 0.3*	5.8 ± 0.6*	8.1 ± 2.6*	5.2 ± 0.1*

*Significantly different at $p<0.05$.

Table 3. Antioxidant activity of *Sargassum* expressed in IC50 (mg/ml) (mean ± standard deviation, n=3).

Samples	DPPH (IC 50 mg/ml)					
	Water	Methanol	Methanol 50%	Ethanol 75%	Viscozyme®	Protamex®
S. aquifolium	14.6 ± 5.7	23.8 ± 9.0*	22.3 ± 8.0*	n.a	n.a	5.7 ± 7.9
S. ilicifolium	n.a	17.8 ± 9.4*	3.3 ± 1.7*	n.a	6.2 ± 3.4	2.9 ± 1.7
S. polycystum	n.a	n.a	n.a	5.2 ± 5.9	n.a	1.9 ± 1.1

*Significantly different at $p<0.05$.

Table 4. Reducing activity of *Sargassum* expressed in µM Equivalent Fe^{2+} (mean ± standard deviation, n=3).

Samples	FRAP (Concentration Equivalent Fe^{2+} (µM))					
	Water	Methanol	Methanol 50%	Ethanol 75%	Viscozyme®	Protamex®
S. aquifolium	14.9 ± 1.7*	12.6 ± 2.5*	22.4 ± 4.9*	14.4 ± 0.3*	33.2 ± 2.5*	17.0 ± 0.7*
S. ilicifolium	23.7 ± 1.6*	11.1 ± 0.8*	14.9 ± 0.8*	7.9 ± 1.24*	13.4 ± 2.2*	11.7 ± 1.2*
S. polycystum	43.5 ± 1.8*	12.8 ± 3.1*	16.9 ± 2.6*	13.8 ± 1.9*	27.7 ± 4.9*	15.9 ± 3.0*

*Significantly different at $p<0.05$.

IOP Conf. Series: Earth and Environmental Science **55** (2017) 012057 doi:10.1088/1755-1315/55/1/012057

he EAE had also recovered more of phlorotannins represented by the total phenolic content. By using Viscozyme, the total phenolic content of *Sargassum* was 28% higher than using the organic solvents. Unfortunately, the phenolic content of Protamex® extracts of *Sargassum* were not in consistent with the Viscozyme® extracts. The Protamex did not seem to recover the phenolic content of *Sargassum* as they yielded lower content compared to the organic solvents, except for water. It implied that the Viscozyme® recovered more of phenolic compounds in *Sargassum* than the Protamex®. Phlorotannins are hydrophilic cell-wall-bound compounds [99] and despite of its soluble characteristic, compounds attached to the cell wall are not easily extracted using the typical extraction methods such as the SLE [56,61]. The application of Viscozyme®, as a carbohydrase, during the extraction allowed to facilitate the recovery of phenolic compound as this enzyme degrades the cell wall of seaweed mainly composed of polysaccharides. Viscozyme® has been extensively used for the recovery of phenolic compounds of terrestrial and marine plants as reported by many studies [41,65,100–102]. However, the efficiency of extraction relies on the optimum conditions applied. Selecting the appropriate hydrolytic enzymes is important to digest specific polymer bonds present in seaweed cell wall followed by the selection of suitable process conditions for maximum recovery of active compounds [56]. Combination of incubation time-temperature, pH, size of the interested molecules and agitation are mentioned to play a critical role during the process [14].

Phlorotannins have been reported to be responsible for many bioactivities, such as antioxidant, antimicrobial, antiviral and many more, as reported by previous studies [9,29,51,96]. The key of phlorotannins important bioactivities lies to its eight interconnected ring as they contain more hydroxyl groups than other tannins – hydrolysable tannin and condensed tannins [107,108]. As a consequence, they have greater antioxidant activity [107]. Hydrogen atom transfer is one of its mechanisms against the effects of unregaluted production of free radicals [36]. As performed in this study, *Sargassum* extracts – extracted with SLE and EAE method- showed their capability in scavenging the free radicals with best activity exhibited by the Protamex® extracts in spite of Viscozyme® extracted more phenolics. It suggested the presence of other molecules like polysaccharides or proteins as shown by the FTIR since the current study used crude extracts instead of the purified. Besides free radical scavenging, our results showed that the enzymatic extracts of *Sargassum* had the ferric reducing ability – a conversion of Fe^{III} to Fe^{II}. In addition, it showed quite promising potential as skin whitening antifouling coating. Such potential was based on the tyrosinase and biofilm inhibition activities of *Sargassum* enzymatic extracts.

Antioxidant activity of *S. horneri, S. fullvelum, S. correanum,* and *S. thunbergii* have been reported by Heo et al [65]. *S. correanum* contained higher phenolics when it was extracted with Viscozyme® and higher free radical scavenging activity compared to the others. Another study also showed that Viscozyme® used to extract *S. horneri* had the ability to scavenge the free radicals until 95% of inhibition [109]. Sanchez-camargo et al. [41] stated that the total phenols and antioxidant activity of *S. muticum* extracted with Alcalase® and Viscozyme® were improved compared to control – aqueous extraction. Therefore, it is obvious that enzymes used have facilitated the release of active compounds and in the same time it enhances the bioactivities especially the antioxidant – best known activity for phlorotannins.

4. Conclusion

A strong urge to explore marine bioactive compounds has brought green chemistry concept to the surface along the process. The principle of this concept is basically to apply a method that is environmentally friendly and favorable for human interest as well. In the exploration of marine bioactive compounds that can be further applied for the nutraceutical, pharmaceutical and cosmeceutical industries, the realization of this concept is crucial. Therefore, many studies have highlighted the possibilities of enzyme-assisted extraction as an alternative method that might replace the use of organic solvent to extract the active compounds.

This study has revealed that by using enzymes, the efficiency of the extraction increased compared to the solid-liquid extraction. This method not only improves the quality and quantity of extracts but it also shows its promising potential by showing diverse bioactivities. Nevertheless, one should note that this study still works with crude extracts. Therefore, the interference of other compounds like polysaccharides and proteins should be considered. As a consequence, it is crucial to further evaluate the bioactivities of purified samples and characterize them.

5. Acknowledgment

This study was a collaboration research between University of South brittany (UBS), Vannes, France and Diponegoro University (UNDIP), Semarang, Indonesia under Double-Degree program supported by the Directorate of Higher Education (DIKTI). The author would like to sincerely thank to all members of the Laboratoire de Biotechnologie et Chimie Marines (LBCM) who had been very helpful and supporting during the research. A special gratitude was addressed to Nolwenn Terme from LBCM who helped to improve this article.

6. References

[1] Sahat H J 2013 Rumput Laut Indonesia *Minist. Trade* 1–20

[2] Nafed K 2011 Warta Ekspor: Rumput Laut dan Produk Turunannya *Minist. Trade* 1–20

[3] Neish I C 2014 Social and economic dimensions of carrageenan seaweed farming in Indonesia *Social and economic dimensions of carrageenan seaweed farming* ed D Valderrama, C Junning, N Hishamunda and N Ridler (Rome: FAO) p 204

[4] Wijesekara I, Pangestuti R and Kim S K 2011 Biological activities and potential health benefits of sulfated polysaccharides derived from marine algae *Carbohydr. Polym.* **84** 14–21

[5] Chanda S, Dave R, Kaneria M and Nagani K 2010 Seaweeds : A novel , untapped source of drugs from sea to combat Infectious diseases *Curr. Res. Technol. Educ. Top. Appl. Microbiol. Microb. Technol.* 473–80

[6] Holdt S L and Kraan S 2011 Bioactive compounds in seaweed: Functional food applications and legislation *J. Appl. Phycol.* **23** 543–97

[7] Gupta S and Abu-Ghannam N 2011 Recent developments in the application of seaweeds or seaweed extracts as a means for enhancing the safety and quality attributes of foods *Innov. Food Sci. Emerg. Technol.* **12** 600–9

[8] Michalak I and Chojnacka K 2015 Algae as production systems of bioactive compounds *Eng. Life Sci.* **15** 160–76

[9] Li Y-X, Wijesekara I, Li Y and Kim S-K 2011 Phlorotannins as bioactive agents from brown algae *Process Biochem.* **46** 2219–24

[10] Collins K, Fitzgerald G, Stanton C and Ross R 2016 Looking beyond the terrestrial: The potential of seaweed derived bioactives to treat non-communicable diseases *Mar. Drugs* **14** 60

[11] Sinurat E and Marraskuranto E 2011 Fucoidan from brown seaweed and its bioactivity *Squalen* **7** 131–7

[12] Mushollaeni W and Rusdiana E 2011 Characterization of sodium alginate from Sargassum sp., Turbinaria sp. and Padina sp. *J. Teknol dan Idustri Pangan* **XXII** 26–32

[13] Sivagnanam S, Yin S, Choi J, Park Y, Woo H and Chun B 2015 Biological Properties of Fucoxanthin in Oil Recovered from Two Brown Seaweeds Using Supercritical CO2 Extraction *Mar. Drugs* **13** 3422–42

[14] Wijesinghe W A J P and Jeon Y J 2012 Biological activities and potential industrial applications of fucose rich sulfated polysaccharides and fucoidans isolated from brown seaweeds: A review *Carbohydr. Polym.* **88** 13–20

[15] Aulanni'am, Rosdiana A and Rahmah N L 2011 Potency of Brown Seaweed (*Sargassum duplicatum* Bory) Ethanol and Ethyl Acetic Fraction to Malondialdehyde Concentration Decreasing and Histomogical Retriveal of IBD (Inflamatory Bowel Disease) Rat Small Intestinal Jejunum *J. Ilm. Kedokt.* **4** 57–64

[16] Bachtiar S Y, Tjahjaningsih W and Sianita N 2012 Pengaruh ekstrak alga coklat (*Sargassum* sp.) terhadap bakteri *Eschericia coli J. Mar. Coast. Sci.* **1** 53–60

[17] Pangestuti R and Wibowo S 2013 Prospects and health promoting effects of brown algal-derived natural pigments *Squalen* **8** 37–46

[18] Manivannan K, Karthikai devi G, Anantharaman P and Balasubramanian T 2011 Antimicrobial potential of selected brown seaweeds from Vedalai coastal waters, Gulf of Mannar *Asian Pac. J. Trop. Biomed.* **1** 114–20

[19] Je J-Y, Park P-J, Kim E-K, Park J-S, Yoon H-D, Kim K-R and Ahn C-B 2009 Antioxidant activity of enzymatic extracts from the brown seaweed *Undaria pinnatifida* by electron spin resonance spectroscopy

LWT - Food Sci. Technol. **42** 874–8

[20] Bazes A, Silkina A, Douzenel P, Faÿ F, Kervarec N, Morin D, Berge J P and Bourgougnon N 2009 Investigation of the antifouling constituents from the brown alga *Sargassum muticum* (Yendo) Fensholt *J. Appl. Phycol.* **21** 395–403

[21] Plouguerne E, Ioannou E, Georgantea P, Vagias C, Roussis V, Hellio C, Kraffe E and Stiger-Pouvreau V 2010 Anti-microfouling activity of lipidic metabolites from the invasive brown alga *Sargassum muticum* (Yendo) Fensholt *Mar. Biotechnol.* **12** 52–61

[22] Mandal P, Mateu C G, Chattopadhyay K, Pujol C A, Damonte E B and Ray B 2007 Structural Features and Antiviral Activity of Sulphated Fucans from the Brown Seaweed *Cystoseira Indica Antivir. Chem. Chemother.* **18** 153–62

[23] Thenmozhi K, Selvarani T, Prabhu B K, Clinic V and Nadu T 2013 Antiviral activity of aqueous extract from brown seaweed, *Sargassum illicifolium* (Turner) on three different types Herpes Simplex virus *Adv. Res. Pharm. Biol.* **3** 370–3

[24] Costa L S, Fidelis G P, Cordeiro S L, Oliveira R M, Sabry D A, Camara R B G, Nobre L T D B, Costa M S S P, Almeida-Lima J, Farias E H C, Leite E L and Rocha H A O 2010 Biological activities of sulfated polysaccharides from tropical seaweeds *Biomed. Pharmacother.* **64** 21–8

[25] Zorofchian Moghadamtousi S, Karimian H, Khanabdali R, Razavi M, Firoozinia M, Zandi K and Abdul Kadir H 2014 Anticancer and antitumor potential of fucoidan and fucoxanthin, two main metabolites isolated from brown algae *Sci. World J.* **2014** 10

[26] Costa L S, Fidelis G P, Telles C B S, Dantas-Santos N, Camara R B G, Cordeiro S L, Costa M S S P, Almeida-Lima J, Melo-Silveira R F, Oliveira R M, Albuquerque I R L, Andrade G P V and Rocha H A O 2011 Antioxidant and antiproliferative activities of heterofucans from the seaweed *Sargassum filipendula Mar. Drugs* **9** 952–66

[27] Ahn M J, Yoon K D, Kim C Y, Kim J H, Shin C G and Kim J 2006 Inhibitory activity on HIV-1 reverse transcriptase and integrase of a carmalol derivative from a brown Alga, *Ishige okamurae Phyther. Res.* **20** 711–3

[28] Yoon N Y, Eom T K, Kim M M and Kim S K 2009 Inhibitory effect of phlorotannins isolated from *Ecklonia cava* on mushroom tyrosinase activity and melanin formation in mouse B16F10 melanoma cells *J. Agric. Food Chem.* **57** 4124–9

[29] Balboa E M, Conde E, Moure A, Falqué E and Domínguez H 2013 In vitro antioxidant properties of crude extracts and compounds from brown algae *Food Chem.* **138** 1764–85

[30] Ale M T, Mikkelsen J D and Meyer A S 2011 Important determinants for fucoidan bioactivity: A critical review of structure-function relations and extraction methods for fucose-containing sulfated polysaccharides from brown seaweeds *Mar. Drugs* **9** 2106–30

[31] La Barre S, Potin P, Leblanc C and Delage L 2010 The halogenated metabolism of brown algae (phaeophyta), its biological importance and its environmental significance *Mar. Drugs* **8** 988–1010

[32] Maghin F, Ratti S and Corino C 2014 Biological Functions and Health Promoting Effects of Brown Seaweeds in Swine Nutrition *J. Dairy, Vet. Anim. Res. Biol.* **1** 1–3

[33] Liu L, Heinrich M, Myers S and Dworjanyn S A 2012 Towards a better understanding of medicinal uses of the brown seaweed *Sargassum* in Traditional Chinese Medicine: A phytochemical and pharmacological review *J. Ethnopharmacol.* **142** 591–619

[34] Atmadja W S and Prud'homme van Reine W F 2014 *Checklist of the Seaweed Species Biodiversity of Indonesia with their distribution and classification: green algae (Chlorophyta) and brown algae (Phaeophyceae, Ocrophyta)* (Jakarta: Coremap II-LIPI)

[35] Koivikko R 2008 *Brown algal phlorotannins: Improving and Applying Chemical Methods* (Turku: University of Turku)

[36] Kim S-K and Himaya S W A 2011 Medicinal effects of phlorotannins from marine brown algae. *Adv. Food Nutr. Res.* **64** 97–109

[37] Budhiyanti S A, Raharjo S, Marseno D W and Lelana I Y B 2012 Antioxidant activity of brown algae *Sargassum* species extract from the coastline of java island *Am. J. Agric. Biol. Sci.* **7** 337–46

[38] Prabowo A, Budhiyanti S A and Husni A 2013 *Sargassum* sp. extract as an antioxidant for fish oil emulsion system during storage at room temperature *JPB Perikan.* **8** 143–50

[39] Le Lann K, Jégou C and Stiger-Pouvreau V 2008 Effect of different conditioning treatments on total phenolic content and antioxidant activities in two Sargassacean species: Comparison of the frondose *Sargassum muticum* (Yendo) Fensholt and the cylindrical *Bifurcaria bifurcata* R. Ross *Phycol. Res.* **56** 238–45

[40] Badrianathan S, Suneeva S C, Shiju T M, Girish Kumar C P and Pragasam V 2011 Exploration of a novel hydroxyl radical scavenger from *Sargassum myriocystum J. Med. plants Res.* **5 (10)** 1997–2005

[41] Sanchez-Camargo A D P, Montero L, Stiger-Pouvreau V, Tanniou A, Cifuentes A, Herrero M and Ibanez E 2015 Considerations on the use of enzyme-assisted extraction in combination with pressurized liquids to recover bioactive compounds from algae *Food Chem.* **192** 67–74

[42] Harnita A N I, Santosa I E, Martono S, Sudarsono, Widyarini S and Harren F J M 2013 Inhibition of lipid peroxidation induced by ultraviolet radiation by crude phlorotannis isolated from brown algae *Sargassum hystrix v. buxifolium* C. Agardh *Indones. J. Chem.* **13** 14–20

[43] Nakai M, Kageyama N, Nakahara K and Miki W 2006 Phlorotannins as Radical Scavengers from the Extract of *Sargassum ringgoldianum* **8** 409–14

[44] Jormalainen V and Honkanen T 2008 Macroalgal chemical defenses and their roles in structuring temperate marine communities *Algal Chemical Defence* ed C D Amsler (Berlin, Heidelberg: Springer Berlin Heidelberg) pp 57–89

[45] Amsler C D and Fairhead V A 2005 *Incorporating Advances in Plant Pathology* vol 43(Elsevier)

[46] Jormalainen V, Honkanen T, Koivikko R and Eränen J 2003 Induction of phlorotannin production in a brown alga: defense or resource dynamics? *Oikos* **103** 640–50

[47] Hemmi A, Mäkinen A, Jormalainen V and Honkanen T 2005 Responses of growth and phlorotannins in *Fucus vesiculosus* to nutrient enrichment and herbivory *Aquat. Ecol.* **39** 201–11

[48] Nakamura T, Nagayama K, Uchida K and Tanaka R 1996 Antioxidant Activity of Phlorotannins Isolated from the Brown Alga *Eisenia bicyclis* **62** 923–6

[49] Nagayama K, Shibata T, Fujimoto K, Honjo T and Nakamura T 2003 Algicidal effect of phlorotannins from the brown alga *Ecklonia kurome* on red tide microalgae *Aquaculture* **218** 601–11

[50] Kim S M, Kang K, Jeon J-S, Jho E H, Kim C Y, Nho C W and Um B-H 2011 Isolation of phlorotannins from Eisenia bicyclis and their hepatoprotective effect against oxidative stress induced by tert-butyl hyperoxide. *Appl. Biochem. Biotechnol.* **165** 1296–307

[51] Kwon H-J J, Ryu Y B, Kim Y-M M, Song N, Kim C Y, Rho M-C C, Jeong J-H H, Cho K-O O, Lee W S and Park S-J J 2013 In vitro antiviral activity of phlorotannins isolated from *Ecklonia cava* against porcine epidemic diarrhea coronavirus infection and hemagglutination *Bioorg. Med. Chem.* **21** 4706–13

[52] Michalak I and Chojnacka K 2014 Algal extracts: Technology and advances *Eng. Life Sci.* **14** 581–91

[53] Heffernan N, Brunton N P, FitzGerald R J and Smyth T J 2015 Profiling of the molecular weight and structural isomer abundance of macroalgae-derived phlorotannins *Mar. Drugs* **13** 509–28

[54] Robic A, Rondeau-Mouro C, Sassi J-F F, Lerat Y and Lahaye M 2009 Structure and interactions of ulvan in the cell wall of the marine green algae *Ulva rotundata* (Ulvales, Chlorophyceae) *Carbohydr. Polym.* **77** 206–16

[55] Ibanez E, Herrero M, Mendiola J A and Castro-Puyana M 2012 Extraction and Characterization of Bioactive Compound with Health Benefits from Marine Sources: Macro and Micro Algae, Cyanobacteria and Invertebrates *Marine Bioactive Compounds: Sources, Characterization and Applications* vol 9781461412, ed M Hayes (Springer US) pp 55–98

[56] Jeon Y J, Wijesinghe W A J P and Kim S K 2012 Enzyme-assisted Extraction and Recovery of Bioactive Components from Seaweeds *Handbook of Marine Macroalgae: Biotechnology and Applied Phycology* ed S-K Kim (John Wiley & Sons, Ltd) pp 221–6

[57] Puri M, Sharma D and Barrow C J 2012 Enzyme-assisted extraction of bioactives from plants *Trends Biotechnol.* **30** 37–44

[58] Kim K-N, Heo S-J, Song C B, Lee J, Heo M-S, Yeo I-K, Kang K A, Hyun J W and Jeon Y-J 2006 Protective effect of Ecklonia cava enzymatic extracts on hydrogen peroxide-induced cell damage *Process Biochem.* **41** 2393–401

[59] Hahn T, Lang S, Ulber R and Muffler K 2012 Novel procedures for the extraction of fucoidan from brown algae *Process Biochem.* **47** 1691–8

[60] Hardouin K, Bedoux G, Burlot A-S, Donnay-Moreno C, Bergé J-P, Nyvall-Collén P and Bourgougnon N 2016 Enzyme-assisted extraction (EAE) for the production of antiviral and antioxidant extracts from the green seaweed Ulva armoricana (Ulvales, Ulvophyceae) *Algal Res.* **16** 233–9

[61] Wijesinghe W A J P and Jeon Y J 2012 Enzyme-assistant extraction (EAE) of bioactive components: A useful approach for recovery of industrially important metabolites from seaweeds: A review *Fitoterapia* **83** 6–12

[62] Kadam S U, Tiwari B K and O'Donnell C P 2013 Application of novel extraction technologies for bioactives from marine algae *J. Agric. Food Chem.* **61** 4667–75

[63] Hagerman A E and Butler L G 1989 Choosing appropriate methods and standards for assaying tannin. *J. Chem. Ecol.* **15** 1795–810
[64] Wang T, Jónsdóttir R, Kristinsson H G, Hreggvidsson G O, Jónsson J Ó, Thorkelsson G and Ólafsdóttir G 2010 Enzyme-enhanced extraction of antioxidant ingredients from red algae *Palmaria palmata LWT - Food Sci. Technol.* **43** 1387–93
[65] Heo S J, Park E J, Lee K W and Jeon Y J 2005 Antioxidant activities of enzymatic extracts from brown seaweeds *Bioresour. Technol.* **96** 1613–23
[66] Chandler S F and Dodds J H 1983 The effect of phosphate, nitrogen and sucrose on the production of phenolics and solasodine in callus cultures of *Solanum laciniatum Plant Cell Rep.* **2** 205–8
[67] Benzie I F and Strain J J 1996 The ferric reducing ability of plasma (FRAP) as a measure of "antioxidant power": the FRAP assay. *Anal. Biochem.* **239** 70–6
[68] Benzie I F and Strain J J 1999 Ferric reducing/antioxidant power assay: direct measure of total antioxidant activity of biological fluids and modified version for simultaneous measurement of total antioxidant power and ascorbic acid concentration. *Methods Enzymol.* **299** 15–27
[69] Shahwar D, Raza M A, Bukhari S and Bukhari G 2012 Ferric reducing antioxidant power of essential oils extracted from *Eucalyptus* and *Curcuma* species *Asian Pac. J. Trop. Biomed.* **2**
[70] Bauer A W, Kirby W M, Sherris J C and Turck M 1966 Antibiotic susceptibility testing by a standardized single disk method. *Am. J. Clin. Pathol.* **45** 493–6
[71] Stiger-Pouvreau V, Jégou C, Cérantola S, Guérard F and Le Lann K 2014 Phlorotannins in Sargassaceae Species from Brittany (France) Interesting Molecules for Ecophysiological and Valorization Purposes *Advances in Botanical Research: Sea Plants* ed N Bourgougnon (Oxford: Elsivier) pp 379–411
[72] Girija K, Hemalatha A, Saranya C, Parthiban C, Anantharaman P and Girija K, Hemalatha A, Saranya C parthiban C, Anantharaman P 2013 Extraction and Isolation of Phlorotannins From Brown Seaweed *Turbinaria Ornata* (Turner) J. Agardh and Its Antioxidant Activity *Int. J. Bioassays* **9** 1185–9
[73] Shin T, Ahn M, Hyun J W, Kim S H and Moon C 2014 Antioxidant marine algae phlorotannins and radioprotection: A review of experimental evidence *Acta Histochem.* **116** 669–74
[74] Shibata T, Ishimaru K, Kawaguchi S, Yoshikawa H and Hama Y 2008 Antioxidant activities of phlorotannins isolated from Japanese Laminariaceae *J. Appl. Phycol.* **20** 705–11
[75] Sathya R, Kanaga N, Sankar P and Jeeva S 2013 Antioxidant properties of phlorotannins from brown seaweed *Cystoseira trinodis* (Forsskal) C. Agardh *Arab. J. Chem.*
[76] Fujii Y, Tanaka R, Miyake H, Tamaru Y, Ueda M and Shibata T 2013 Evaluation for Antioxidative Properties of Phlorotannins Isolated from the Brown Alga *Eisenia bicyclis*, by the H-ORAC Method *Food Nutr. Sci.* **4** 78–82
[77] Wijesekara I, Yoon N Y and Kim S K 2010 Phlorotannins from *Ecklonia cava* (Phaeophyceae): Biological activities and potential health benefits *BioFactors* **36** 408–14
[78] Choi J S, Lee K, Lee B B, Kim Y C, Kim Y D, Hong Y K, Cho K K and Choi I S 2014 Antibacterial activity of the phlorotannins dieckol and phlorofucofuroeckol-a from *Ecklonia cava* against *Propionibacterium acnes Bot. Sci.* **92** 425–31
[79] Eom S H, Kim Y M and Kim S K 2012 Antimicrobial effect of phlorotannins from marine brown algae *Food Chem. Toxicol.* **50** 3251–5
[80] Lopes G, Sousa C, Silva L R, Pinto E, Andrade P B, Bernardo J, Mouga T and Valentão P 2012 Can phlorotannins purified extracts constitute a novel pharmacological alternative for microbial infections with associated inflammatory conditions? *PLoS One* **7**
[81] Thomas N V and Kim S K 2011 Potential pharmacological applications of polyphenolic derivatives from marine brown algae *Environ. Toxicol. Pharmacol.* **32** 325–35
[82] Yang Y I, Ahn J H, Choi Y S and Choi J H 2015 Brown algae phlorotannins enhance the tumoricidal effect of cisplatin and ameliorate cisplatin nephrotoxicity *Gynecol. Oncol.* **136** 355–64
[83] Jung H A, Yoon N Y, Woo M H and Choi J S 2008 Inhibitory activities of extracts from several kinds of seaweeds and phlorotannins from the brown alga *Ecklonia stolonifera* on glucose-mediated protein damage and rat lens aldose reductase *Fish. Sci.* **74** 1363–5
[84] Leyton A, Pezoa-Conte R, Barriga A, Buschmann A H, Mäki-Arvela P, Mikkola J P and Lienqueo M E 2016 Identification and efficient extraction method of phlorotannins from the brown seaweed *Macrocystis pyrifera* using an orthogonal experimental design *Algal Res.* **16** 201–8
[85] Jung H A, Jin S E, Ahn B R, Lee C M and Choi J S 2013 Anti-inflammatory activity of edible brown alga Eisenia bicyclis and its constituents fucosterol and phlorotannins in LPS-stimulated RAW264.7 macrophages *Food Chem. Toxicol.* **59** 199–206

[86] Kim E and Kwak J 2015 Antiviral phlorotannin from *Eisenia bicyclis* against human papilloma virus in vitro *Planta Med.* **81**PW_22
[87] Eom S, Moon S, Lee D, Kim H, Park K, Lee E, Kim T H, Chung Y, Lee M and Kim Y 2015 In vitro antiviral activity of dieckol and phlorofucofuroeckol-A isolated from edible brown alga *Eisenia bicyclis* against murine norovirus *Algae* **30** 241–6
[88] Hans-Jörg Bart 2011 Extraction of natural products from plants – An introduction. Industrial scale natural products extraction. 1–26
[89] Richardson J F, Harker J H and Backhurst J R 2002 *Coulson and Richardson's Chemical Engineering: Particle Technology and Separation Process, vol. 5.* (Oxford: Butterworth and Heinemann)
[90] Richardson J F, Harker J H and Backhurst J R 2002 *Chemical Engineering: Particle Technology and Separation Processes* vol 2(London: Butterworth-Heinemann)
[91] Naczk M and Shahidi F 2004 Extraction and analysis of phenolics in food *J. Chromatogr. A* **1054** 95–111
[92] Schoenwaelder M E A and Clayton M N 1999 The presence of phenolic compounds in isolated cell walls of brown algae *Phycologia* **38** 161–6
[93] Stern J L, Hagerman A E, Steinberg P D and Mason P K 1996 Phlorotannin-protein interactions *J. Chem. Ecol.* **22** 1877–99
[94] Azmir J, Zaidul I S M, Rahman M M, Sharif K M, Mohamed A, Sahena F, Jahurul M H A, Ghafoor K, Norulaini N A N and Omar A K M 2013 Techniques for extraction of bioactive compounds from plant materials: A review *J. Food Eng.* **117** 426–36
[95] Koivikko R, Loponen J, Pihlaja K and Jormalainen V 2007 High-performance liquid chromatographic analysis of phlorotannins from the brown alga *Fucus vesiculosus Phytochem. Anal.* **18** 326–32
[96] Le Lann K, Surget G, Couteau C, Coiffard L, Cérantola S, Gaillard F, Larnicol M, Zubia M, Guérard F, Poupart N and Stiger-Pouvreau V 2016 Sunscreen, antioxidant, and bactericide capacities of phlorotannins from the brown macroalga *Halidrys siliquosa J. Appl. Phycol.*
[97] Heo S J, Lee K W, Song C B and Jeon Y J 2003 Antioxidant Activity of Enzymatic Extracts from Brown Seaweeds *Algae* **18** 71–81
[98] Rodrigues D, Sousa S, Silva A, Amorim M, Pereira L, Rocha-Santos T A P, Gomes A M P, Duarte A C and Freitas A C 2015 Impact of enzyme- and ultrasound-assisted extraction methods on biological properties of red, brown, and green seaweeds from the Central West Coast of Portugal *J. Agric. Food Chem.* **63** 3177–88
[99] Koivikko R, Loponen J, Honkanen T and Jormalainen V 2005 Contents of soluble, cell-wall-bound and exuded phlorotannins in the brown alga Fucus vesiculosus, with implications on their ecological functions *J. Chem. Ecol.* **31** 195–212
[100] Zheng H, Hwang I-W and Chung S-K 2009 Enhancing polyphenol extraction from unripe apples by carbohydrate-hydrolyzing enzymes. *J. Zhejiang Univ. Sci. B* **10** 912–9
[101] Huynh N T, Smagghe G, Gonzales G B, Van Camp J and Raes K 2014 Enzyme-assisted extraction enhancing the phenolic release from cauliflower (*Brassica oleracea L. var. botrytis*) outer leaves *J. Agric. Food Chem.* **62** 7468–76
[102] Charoensiddhi S, Franco C, Su P and Zhang W 2014 Improved antioxidant activities of brown seaweed *Ecklonia radiata* extracts prepared by microwave-assisted enzymatic extraction *J. Appl. Phycol.* **27** 2049–58
[103] Lim S N, Cheung P C K, Ooi V E C and Ang P O 2002 Evaluation of antioxidative activity of extracts from a brown seaweed, Sargassum siliquastrum *J. Agric. Food Chem.* **50** 3862–6
[104] Ganapathi K, Subramanian V and Mathan S 2013 Bioactive potentials of brown seaweeds, *Sargassum myriocystum* J. Argadh, *S. plagiaphyllum* C. Agardh and *S. ilicifolium* (Turner) J. Argadh *Int. Res. J. Pharm. Appl. Sci.* **3** 105–11
[105] Kannan S 2014 full-textFT-IR and EDS analysis of the seaweeds Sargassum wightiiGracilaria corticata (red algae) *Int. J. Curr. Microbiol. Appl. Sci.* **3** 341–51
[106] Gomez-Ordonez E and Ruperez P 2011 FTIR-ATR spectroscopy as a tool for polysaccharide identification in edible brown and red seaweeds *Food Hydrocoll.* **25** 1514–20
[107] Wang Y, Xu Z, Bach S J and McAllister T A 2009 Sensitivity of Escherichia coli to seaweed (Ascophyllum nodosum) phlorotannins and terrestrial tannins *Asian-Australasian J. Anim. Sci.* **22** 238–45
[108] Munir N, Sharif N, Naz S and Farkhanda M 2013 Algae: A potent antioxidant source *Sky J. Microbiol. Res.* **1** 22–31
[109] Park P-J, Fereidoon S and Jeon Y-J 2004 Antioxidant activities of extracts from an edible seaweed *Sargassum horneri* using ESR spectrometry *J. Food Lipids* **11** 15–27

Low zinc serum levels and high blood lead levels among school-age children in coastal area

Adriyan Pramono[1], Binar Panunggal[1]*, M.Zen Rahfiludin[2], Fronthea Swastawati[3]

[1]Department of Nutrition/Center of Nutrition Research (CENURE), Faculty of Medicine, Diponegoro University, Jl.Dr.Sutomo No.18, Semarang 50231, Indonesia, adriyanpramono@fk.undip.ac.id
[2] Faculty of Public Health, Diponegoro University, Jl. Prof.Sudarto SH, Semarang 50277, Indonesia, rahfiludin@yahoo.com
[3] Faculty of Marine and Fisheries, Diponegoro University, Jl. Prof.Sudarto SH, Semarang 50277, Indonesia, fronthea_thp@yahoo.co.id

*Email: panunggalbinar@gmail.com

Abstract. The coverage of environmental lead toxicant was quiet wide. Lead exposure recently has been expected to be associated with zinc deficiency and blood indices disturbance. Emphasizing on children, which could absorb more than 50 % of lead that enters the body. Lead became the issue on the coastal area due to it has polluted the environment and waters as the source of fisheries products. This was a cross sectional study to determined nutritional status, blood lead levels, zinc serum levels, blood indices levels, fish intake among school children in coastal region of Semarang. This study was carried out on the school children aged between 8 and 12 years old in coastal region of Semarang. Nutritional status was figured out using anthropometry measurement. Blood lead and zinc serum levels were analyzed using the Atomic Absorbent Spectrophotometry (AAS) at a wavelength of 213.9 nm for zinc serum and 283.3 nm for blood lead. Blood indices was measured using auto blood hematology analyzer. Fish intake was assessed using 3-nonconsecutive days 24-hours food recall. The children had high lead levels (median 34.86 μg/dl, range 11.46 - 58.86 μg/dl) compared to WHO cut off. Zinc serum levels was low (median 18.10 μg/dl, range 10.25 – 41.39 μg/dl) compared to the Joint WHO/UNICEF/IAEA/IZiNCG cut off. Approximately 26.4% of children were anemic. This study concluded that all school children had high blood lead levels, low zinc serum, and presented microcytic hypochromic anemia. This phenomenon should be considered as public health concern.

Keywords: blood lead, zinc serum, coastal region, microcytic hypochromic anemia, school children

1. Introduction

Zinc (Zn) is an essential trace element plays role as a cofactor of more than 100 metaloenzym, plays an important part in cell regeneration, metabolism, growth and immune function [1]. Zinc deficiency is associated with suboptimal growth, diarrhea, and decreased of immunity [2]. *World Health*

Organization (WHO) estimates the prevalence of zinc deficiency in the world's population is ranged between 4 and 73%. Additionally, about 5 to 30% zinc deficiency has occurred in children and adolescents both in developed and developing countries [3]. Especially in the developing countries, harmful environmental exposure recently has been associated with the deficiency of Zn. Nutritional zinc deficiency can be caused from lead (Pb) exposure through dangerous cycle that increases lead absorption and increases Zn secretion consecutively [4]. Previous studies in animal experiments summarized that the conditions of marginal Zn deficiency, could increase blood lead levels in the presence of Pb exposure [5,6].

Among school children, chronic Pb exposure is correlated with nutritional deficiencies, anemia, impairments to physical growth, cognitive deficient and learning disorders [7 - 10]. Anemia is a major public health problem especially in developing countries. Additionally, WHO reported that of 25,4% anemia has been found in school children [11]. Anemia is a state of decreasing of erythrocytes and hemoglobin, whereas those condition impaired its function to carry oxygen through body tissues sufficiently [12,13]. Hemoglobin is composed of four globin chain polypeptide whereas in each chain contains a heme molecule containing Iron [14]. As well as iron, zinc also plays an important role in the formation of hemoglobin. In heme biosynthesis, δ-ALA dehydratase enzyme that catalyzes charge 2 δ-ALA into molecules that is highly dependent on zinc [15]. A study conducted in Atlanta gives us information that there is a significant correlation between serum zinc concentrations and hemoglobin levels since children who suffered iron deficiency anemia tend to be deficiency of zinc serum [16]. Normal zinc level has a consequence on erythrocytes life period, due to zinc contributes in protecting erythrocytes from oxidative stress and cell damage [17]. Increasing of blood lead levels could also interfered with erythropoiesis by inhibiting the synthesis of protoporphyrin and the absorption of iron (Fe) thus increasing the risk anemia. Furthermore, lead also affects the morphology of erythrocytes [18].

Nowadays, rapid growth of industrialization and its waste became environmental issues in coastal region of Indonesia. The industrial waste that was dumped in the coastal region induced the increasing level of lead pollutant in this area [19]. The coverage of environmental lead toxicant was quiet wide, including the exposure of fisheries products, public water source, and air pollution [20]. Lead exposure among children is an increasing problem globally, adversely affecting the quality of life among large numbers of children. The impact of lead exposure can affect all ages, whereas the children could absorbed more than 50 % of lead that enters the body, compared with adults who absorbed only 15-35 % [21]. The major vehicle of its exposure in the human body consecutively through the food consumption (65 %), the water consumption (20 %) or through the air (15 %) which were contaminated by lead [22]. In area with lead exposure, anemia and/or zinc deficiency might be occurred with or without clinical manifestation [20].

Lead exposure can not be seen as only a trivial environment and health problem. A study depicted even though given iron and zinc supplementation on lead-exposed children did not reduce blood lead levels [23]. Recently in Indonesia study about lead exposure among children mostly conducted in urban area, but not in coastal region. Especially among school children in Semarang city coastal region, the linking between blood lead levels with serum zinc level, anemia, and red blood cell indices has been not fully investigated.

2. Materials and Methods

2.1. Study population and site

A cross sectional study was conducted on an elementary school in Tambak lorok coastal region, Semarang, Central Java Indonesia. Tambak lorok is located less than 3 Kilometers from coastal java sea and near to harbor industrial region (Figure 1). About 80 subjects in this study were recruited using consecutive sampling method and met the inclusion criteria. During data collection, 8 subjects were dropped out due to did not agree to collect the blood. There were 72 children have completed data collection.

Inclusion criteria in this study: age between 8 to 12 years, and did not suffer from infectious illness (diarrhea, pneumonia, tuberculosis, and parasites) which were evidenced by looking at health records in primary health care. Demographic characteristics focused on parental educational level, parental occupation, source of water consumption, consumption of seafood products, age, and sex. Body weight was measured using calibrated digital body weight scale with accuracy 0.1 Kilogram. Body height was obtained using board stadiometer with accuracy 0.1 Centimeters. Median z score for weight for age, height for age, and BMI for age were performed to describe children nutritional status.

2.2. Biochemical collection and analysis

Biochemical data in this study consists of serum zinc level, red blood cells indices (Hb, Ht, mcv, mch, mchc), and blood lead level. About 10 mL venous blood has taken in the morning day before subjects have a breakfast using disposable plastic syringe and immediately transferred into three sterile-vacutainer. A potassium anticoagulant - EDTA (Ethylene Diamine Tetra Acetate) vacutainer was injected with 2 mL venous blood. This vacutainer was used to measure red blood cells (RBCs, in x 10^6 /mL), hemoglobin (Hb, in g/dL), hematocrit (Ht, in g/dL) mean corpuscle volume (MCV), mean corpuscle hemoglobin (MCH), and mean corpuscle hemoglobin concentration (MCHC). Blood indices were analyzed using automated Nihon Cohden Celltac E MEK-7222K hematology analyzer. Ferritin serum was analyzed using ELISA method.

The other two non-additives vacutainer were injected with 3 mL for serum zinc measurement and 5 mL for blood lead measurement. Approximately 3 mL whole blood in non-additive vacutainer was centrifuged, then the blood serum was prepared to zinc serum analysis. Substantially 5 mL whole blood was centrifuged and incorporated with APDC (Ammonium Pyrrolidine Dithiocarbamate) 2%. The incorporated solution was then re-centrifuged, allowed to stand for 5 minutes and mixed with 2 ml of butyl acetate. That mixed solution was centrifuged again until separation occurs. The reading of serum zinc level and blood lead level were operated using the Atomic Absorbent Spectrophotometry (AAS) at a wavelength of 213.9 nm for zinc serum and 283.3 nm for blood lead (Shimadzu AA6401F, Japan). Biochemical measurement was organized in the micronutrient and IDD Center laboratory, Diponegoro University, Indonesia.

Dietary assessment was conducted to determine fish intake among this population. The three-non consecutive days 24-hour food recall method was carried out to describe fish intake. In this study, local fish products have also been collected from this coastal area. Environmental assessment has been conducted on drinking water samples were collected using standard methods. Lead contents from fish products was measured using Spectrophotometry. Lead analysis of local fish products and drinking water samples were carried out in Center of Prevention on Industrial Pollution Laboratory, Ministry of Industrial, Republic Indonesia.

2.3. Statistical analysis

The SPSS for IBM version 19 and Microsoft excel software were performed for statistical analysis. All the study variables were tested for normality by the Kolmogorov-Smirnov test. The Mann Whitney test was operated to compare the blood lead levels and zinc serum levels between subjects with anemia and non anemia. The Correlation between blood lead level and serum zinc, hb, ht, RBC, mcv, mch, and ferritin were performed using Spearman rank's correlation analysis. Statistical significance was considered when p value < 0.05.

2.4. Ethics

The study protocol was approved by the Board of Medical Ethics on Faculty of Medicine Diponegoro University / Dr. Kariadi Hospital with no. 534/EC/FK-RSDK/2015. The participants, who were recruited in this study, clearly informed about the purpose of investigations and expected outcomes. Informed consent was obtained and signed by the parents of the subjects before the study began.

3. Results

3.1. Demographic characteristics of respondents

The coastal region in this study was taken place at Tambak Lorok area, the northern area of Semarang, Central Java. This area borders along with java sea. The main harbor of Central Java province has been located in this coastal area. Industrial area that was built in this region may contributed to environmental and health problems issue. Figure 1 depicts environmental situation in this coastal region.

Figure 1. Environmental situation of Semarang coastal region (authors documented).

3.2. Biochemical data analysis

Table 1 shows values of haemoglobin, hematocrit, red blood indices, zinc serum levels, blood lead levels among all children. All subjects had blood lead levels $\geq$ 10 (μg/dl), low zinc serum levels (< 65 μg/dl), and median of ferritin describes < 15 ng/ml.

Table 1. Mean, SD, median, minimum and maximum values of blood indices, zinc serum, blood lead levels, and ferritin among subjects (N=72)

Biochemical parameters	Median	Range (min – max)
Haemoglobin (g/dl)	12.85	10,00 – 15,6
Hematocrit (g/dl)	35.40	28,70 – 44,70
RBC ($\times 10^6$ mm^3)	4.82	4.21 – 6.00
MCV(μ^3)	74.45	52.10 – 81.05
MCH	27.00	18.10 – 29.70
MCHC	36.30	34.10 – 38.30
Ferritin (ng/ml)	13.56	2.28 – 44.68
Zinc serum (μg/dl)	18.10	10,25 – 41,39
Blood Lead Level (μg/dl)	34.86	11.46 – 58.86

This study was carried out to 72 children with age range from 9 to13 years with a mean value of 10.16 ± 1.02 years. Approximately 66.7% of children's age were categorized between 9 and 10 years old.

Surprisingly, about 90.3% of children were stunting, consists of mild stunting 34.7%, moderate stunting 45.8%, and severe stunting 9.7%. We can see from table 1 about 26.4% children were anemic. Father's occupation of subjects largely as fisherman (59.7%). Both father and mother of children had literate, mostly graduated from Senior High School (father 51.4%, mother 45.8%) (Table 2).

Table 2. Characteristics of school age children (N=72)

Characteristics	N (%)
Sex	
Male	39 (54.2)
Female	33 (45.8)
Age	
9 – 10	48 (66.7)
11 – 12	24 (33.3)
Height for Age Nutritional Status	
Normal	7 (9.7)
Mild Stunting	25 (34.7)
Moderate Stunting	33 (45.8)
Severe Stunting	7 (9.7)
Anemia Status	
Anemic	19 (26.4)
Non Anemic	53 (73.6)
Father literacy	
Graduated elementary school	13 (18.1)
Graduated junior high school	22 (30.6)
Graduated senior high school	37 (51.4)
Mother literacy	
Graduated elementary school	16 (22.2)
Graduated junior high school	23 (31.9)
Graduated senior high school	33 (45.8)
Father's occupation	
Salesman	5 (6.9)
Fisherman	43 (59.7)
Labor	24 (33.3)
Mother's occupation	
Household	36 (50.0)
Salesman	2 (2.8)
Labor	34 (47.2)

Comparison between mean values of different hematological parameters, blood lead levels, zinc serum levels and serum ferritin in anemic and non anemic were figured out in this study (Table 3). Regarding the hematological parameters, nearly all values were significantly lower among the anemic than the non-anemic group except for blood lead levels, which showed a highly significant elevation among the anemic group. However, according to World Health Organization (WHO) cut off for blood lead levels, all children in this study had a blood lead ≥10 µg/dl (high blood lead level group {HBLLs}). As for the RBC count, ferritin, and zinc serum, no statistically significant difference was detected between the groups.

Table 3. Comparison between mean values of different hematological parameters, zinc serum, blood lead levels, and serum level of ferritin in anemic and non-anemic groups.

Biochemical parameters	Anemic Group (N=19)	Non Anemic Group (N=53)	Test of Significance
RBC ($\times 10^6$ mm^3)	4.82 $\pm$ 0.49	4.88 $\pm$ 0.33	t = - 0.56 p = 0.58
Hb (g/dl)	11.65 $\pm$ 0.62	13.26 $\pm$ 0.77*	t = - 9.03 p = 0.000
Hct	32.66 $\pm$ 1.44	36.47 $\pm$ 1.88*	t = - 9.05 p = 0.000
MCV(μ^3)	68.43 $\pm$ 7.50	74.89 $\pm$ 3.74*	t = - 4.84 p = 0.000
MCH	24.45 $\pm$ 3.05	27.26 $\pm$ 1.65**	t = - 5.01 p = 0.000
MCHC	35.66 $\pm$ 0.92	36.39 $\pm$ 0.75*	t = - 3.09 p = 0.005
Ferritin (ng/ml)	15.59 $\pm$ 8.39	15.36 $\pm$ 8.58	t = 0.10 p = 0.92
Zinc serum (μg/dl)	20.13 $\pm$ 9.64	20.44 $\pm$ 8.31	t = - 0.12 p = 0.90
Blood lead level (μg/dl)	41.95 $\pm$ 10.61	32.03 $\pm$ 10.33**	t = 3.52 p = 0.001

**independent t-test were performed (significant if p<0.05)*
***Mann Whitney test were performed (significant if p<0.05)*

Table 4 reveals the correlation between the different red blood indices, serum zinc levels and the blood levels of lead. Blood lead levels had a significant negative correlation with haemoglobin, hematocrit, and zinc serum ($r = -0.330$, $r = -0.328$, respectively $p<0.005$ and $r = -0.265$, $p < 0.05$).

Table 4. Correlation of different hematological parameters, serum iron and ferritin levels in relation to blood lead, copper and zinc.

Hematological parameters	Blood lead levels r-value (p-value)	Zinc serum levels r-value (p-value)
RBC ($\times 10^6$ mm^3)	-0.066 (0.585)	-0.128 (0.285)
Hb (g/dl)	-0.330 (0.005)	0.119 (0.319)
Hct	-0.328 (0.005)	0.101 (0.396)
MCV(μ^3)	-0.124 (0.298)	0.130 (0.276)
MCH	-0.109 (0.363)	0.148 (0.214)
MCHC	-0.106 (0.375)	0.144 (0.227)
Ferritin (ng/ml)	0.178 (0.135)	0.075 (0.529)
Zinc serum levels	-0.265 (0.025)	-
Blood lead levels	-	-0.265 (0.025)

3.3. Environmental assessment

Environmental assessment has been conducted on drinking water samples and dietary fish intake. Local fish products were collected based on the information from dietary fish intake interview. Lead content of drinking water and local fish products can be seen in Table 5.

Table 5. Fish intake (g per day) and lead levels of drink water and local fish products

Parameters	Mean	Median	Range (min – max)
Fish intake (g/day)		40	10,0 – 128
Lead from drinking water (μg/dl)		0	0
Lead from some type of fish (mg/Kg)			
- Green mussels (Pena viridis)	1.13		
- Scallop shell (Anadara granosa)	0.76		
- Kipper fish (Scatophagidae)	0.18		
- Snapper fish (Lates calcalifer)	0.13		
- Mullets fish (Mugil cephalus)	0.18		

4. Discussion

All school children in this study (100%) had BLL ≥ 10 μg/dl and zinc serum levels < 65 μg/dl, similar to a study done by Hegazy et al [7]. who also reported about more than a half subjects (63.3%) had BLL ≥ 10 μg/dl. However the current study is similar to the results obtained for children population in India [24]. The cut off value of l0 μg/dl defined by the WHO guideline for Childhood Lead Poisoning as a limit for an elevated blood lead level primarily is based on neurological toxicity and has wide range of toxicity including neurobehavioral impairment [21]. Recently, even though blood lead levels (BLL) less than 10 μg/dl is considered safe, a study confirmed that BLL < 10 μg/dl has associated with cognitive deficits [25]. Thus our data showed all children had low zinc serum levels (< 65 μg/dl). The Joint WHO/UNICEF/IAEA/IZiNCG asserted zinc serum levels < 65 μg/dl has been recognized as serious public health problems [26].

Schwartz et al [28] reported that children living near lead smelters in the US of Idaho, had blood lead levels approximately 25 μg/dl and were correlated with anaemia. In addition, Jain et al [29]. reported that children with BLL >10 μg/dl had 1.7 times risk to be moderate anaemia, in contrast with our finding showed that 73.6% of children had normal haemoglobin. Froom et al [30]. suggested that haemoglobin level did not correlate well with BLLs. However, our finding described that Hb, Hct, MCV, MCH and MCHC values of children with anaemia lower in comparison to anemic group. Moreover high BLLs has negatively associated with lower zinc serum and some blood indices levels. Lead absorption occurs predominantly in the duodenum and jejunum. The process of lead entry into the body may carry on through food, drink or by air. Approximately 90 % of lead absorbed by the blood binds to red blood cells [27]. The children absorbed up to 50% of lead from food and/or drink through the gastrointestinal tract and will be included in the metabolic processes of the body. Those lower red blood indices is similar with hypochromic microcytic anaemia which can be caused by blood lead toxicity [7]. Lead causes anaemia by impairing heme synthesis and increasing the rate of red blood cell destruction [31]. Although a causal pathway cannot be determined, yet the study findings clearly demonstrate the differences of BLLs between anaemia status.

In the present study zinc serum level of the anemic group is insignificantly different than non-anemic group. Lead (Pb) has the same valiancy number (Pb2+) as zinc (Zn2+) and iron (Fe2+), whereas on the cellular transport by Dimetal Transporter-1 (DMT-1) competition may be occurred [32]. Lead (Pb2+) will always be dominant to carry on by DMT-1 because of its density higher than Zn2+ and Fe2+

[27]. Lead (Pb) inhibits δ-ALA dehydratase enzyme that catalyzes δ-ALA into molecules, which is highly dependent to zinc on haemoglobin synthesis [15].

In the current study high BLLs among school children may be due to air pollution, sewage contamination, and probably from food consumption which were contaminated with Lead (Pb). Based on dietary interview, children consume fish regularly (range between 10 g/day and 128 g/day). Thus based on lead measurement content on some fish products showed variety lead content (mg/Kg fish products), green mussels (Pena viridis) had the highest content of lead (1.13mg/Kg). Green mussels (Pena viridis) has been vended on street food vendor near by school building. However a causal pathway between some fish products intake and blood lead levels cannot be firmed only from this data. A large epidemiology study on general northern coastal communities should remain a concern because of the nature of accumulation.

In developing countries such as Indonesia, control of lead contamination is much slower and the negative health effects getting more sporadic. The present work revealed that lead contamination should be considered as public health concern in northern coastal population of Semarang, Central Java, Indonesia.

5. Conclusion

In the present study, all subjects has high blood lead levels (BLLs) and low zinc serum. Blood lead levels (BLLs) were negatively associated with the hematological parameters and zinc serum levels. Lead (Pb) was not presented in drinking water, but Pb was discovered in some fish products that may be regularly consume by the children. Lead exposure could be controlled and strides should be taken to reduce zinc deficiency and anaemia among children population at coastal region.

6. References

[1]. Osredkar J, Sustar N. 2011. Copper and zinc, biological role and significance of copper/zinc imbalance. J Clin Toxicol Suppl 3, 1 – 18.

[2]. Prasad AS. 2003. Zinc deficiency: Has been known of for 40 years but ignored by global health organizations. BMJ, 326(7386), 409 - 410.

[3]. Wessells KR, Brown KH. 2012. Estimating the global prevalence of zinc deficiency: results based on zinc availability in national food supplies and the prevalence of stunting. PLoS One, 7(11), e50568.

[4]. Mason LH, Harp JP, Han DY. 2014. Pb neurotoxicity: neuropsychological effects of lead toxicity. BioMed Res Int, 2014, 1 – 8.

[5]. Bushnell PJ, Levin ED. 1982. Effects of zinc deficiency on lead toxicity in rats. Neurobeh. Toxicol. Ter, 5(3), 283-288.

[6]. Ashraf MH, Fosmire GJ. 1985. Effects of marginal zinc deficiency on subclinical lead toxicity in the rat neonate. J Nutr ,115(3), 334-346.

[7]. Hegazy AA, Zaher MM, Abd el-hafez MA, Morsy AA, Saleh RA. 2010. Relation between anemia and blood levels of lead, copper, zinc and iron among children. BMC Res Notes, 3(1), 133.

[8]. Costa de Almeida GR, de Freitas Tavares CF, de Souza AM, et al. 2010. Whole blood, serum, and saliva lead concentrations in 6- to 8-year-old children. Sci. Total Environ, 408, 1551–1556.

[9]. Kordas K, Canfield RL, Lopez P, et.al. 2006. Deficits in cognitive function and achievement in Mexican first-graders with low blood lead concentrations. Environ. Res, 100(3), 371-386.

[10]. Gardner RM, Kippler M, Tofail F, et al. 2013. Environmental exposure to metals and children's growth to age 5 years: a prospective cohort study. Am. J. Epidemiol, 177(12), 1356-1367.

[11]. World Health Organization. Worldwide prevalence of anemia 1993-2005: WHO global database on anemia. 2008. Download from: http://apps.who.int/iris/bitstream/10665/43894/1/9789241596657_eng.pdf

[12]. Ferreira MU, da Silva-Nunes M, Bertolino CN, et al. 2007. Anemia and iron deficiency in school children, adolescents, and adults: a community-based study in rural Amazonia. Am J Public Health 97(2): 237-239.

[13]. World Health Organization. Haemoglobin concentrations for the diagnosis of anemia and assessment of severity. 2011. Download from: http://www.who.int/vmnis/indicators/haemoglobin.pdf
[14]. Thomas C, Lumb AB. 2012. Physiology of haemoglobin. CEACCP, 12(5), 251-256.
[15]. Scinicariello F, Murray HE, Moffett DB, et al. 2007. Lead and δ-aminolevulinic acid dehydratase polymorphism: where does it lead? A meta-analysis. Environ Health Persp, 115 (1), 35-41.
[16]. Cole CR, Grant FK, Swaby-Ellis ED, et al. 2010. Zinc and iron deficiency and their interrelations in low-income African American and Hispanic children in Atlanta. Am J Clin Nutr, 91(4), 1027 - 1034.
[17]. Ryu MS, Guthrie GJ, Maki AB, et al. 2012. Proteomic analysis shows the up regulation of erythrocyte dematin in zinc-restricted human subjects. Am J Clin Nutr, 95(5), 1096 - 1102.
[18]. Shah F, Kazi TG, Afridi HI, et al. 2010. Environmental exposure of lead and iron deficit anemia in children age ranged 1–5years: a cross sectional study. Sci Total Environ, 408(22), 5325 - 5330.
[19]. Anetor JI. 2013. Rising environmental cadmium levels in developing countries: threat to genome stability and health. Niger J Physiol Sci, 27(2), 103-115.
[20]. Kordas K, Lonnerdal B, Stoltzfus RJ. 2007. Interactions between Nutrition and Environmental Exposures: Effects on Health Outcomes in Women and Children. J Nutr, 137(12), 2794 – 7.
[21]. World Health Organization. Childhood lead poisoning. 2010. Download from: http://www.who.int/ceh/publications/leadguidance.pdf
[22]. Liu J, McCauley L, Compher C, et al. 2011. Regular breakfast and blood lead levels among preschool children. Environ Health, 10(1), 28.
[23]. Rosado JL, Lopez P, Kordas K, et al. 2006. Iron and/or zinc supplementation did not reduce blood lead concentrations in children in a randomized, placebo-controlled trial. J Nutr, 136(9), 2378-2383.
[24]. Patel AB, Williams SV, Frumkin H. 2001. Blood lead in children and its determinants in Nagpur, India. Int J Occup Environ Health, 7: 119-26.
[25]. Lanphear BP, Dietrich K, Auinger P, Cox C. 2000. Cognitive deficits associated with blood lead concentrations< 10 microg/dL in US children and adolescents. Public health reports, 115(6), 521.
[26]. De Benoist B, Darnton-Hill I, Davidsson L, Fontaine O, Hotz C. 2007. Conclusions of the joint WHO/UNICEF/IAEA/IZiNCG interagency meeting on zinc status indicators. Food and nutrition bulletin, 28(3 suppl3), S480-S484.
[27]. Kasperczyk A, Prokopowicz A, Dobrakowski M, Pawlas N, Kasperczyk S. 2012. The effect of occupational lead exposure on blood levels of zinc, iron, copper, selenium and related proteins. Biological trace element research,150(1-3), 49-55.
[28]. Schwaetz J, Landrigan PJ, Baker EL: Lead induced anemia: dose response relation and evidence for a threshold. Am J Public Health. 1990, 80: 165-8.
[29]. Jain NB, Laden F, Guller U, Shankar A, Kazani S, Garshick E. 2005. Relation between blood lead levels and childhood anemia in India. American Journal of Epidemiology, 161(10), 968-973.
[30]. Froom P, Kristal-Bonch E, Benbassat J. 1999. Lead exposure in battery-factors workers is not associated with anemia. J Occup Environ Med, 41: 120-3.
[31]. Bradman A, Eskenazi B, Sutton P, Goldman LR. 2001. Iron deficiency associated with higher blood lead level in children living in contaminated environments. Environ Health Perspect, 109: 1079-84.
[32]. Kordas K, Lonnerdal B, Stoltzfus RJ. 2007. Interactions between nutrition and environmental exposures: effects on health outcomes in women and children. J Nutr, 137, 2794.

The Need of Adequate Information to Achieve Total Compliance of Mass Drug Administration in Pekalongan

Praba Ginandjar[1*], Lintang Dian Saraswati[1], Opik Taufik[2], Nurjazuli[3] and Bagoes Widjanarko[4]

[1]Department of Epidemiology, Faculty of Public Health, Diponegoro University: Jl. Prof. Sudharto, Kampus Tembalang, Semarang 50275 – Indonesia
[2]Department of Diseases Control, City Health Office of Pekalongan: Jl. Jatayu no. 4, Pekalongan 51141 – Indonesia
[3]Master Program of Environmental Health, Faculty of Public Health, Diponegoro University: Jl. Prof. Sudharto, Kampus Tembalang, Semarang 50275 – Indonesia
[3]Department of Health Promotion, Faculty of Public Health, Diponegoro University: Jl. Prof. Sudharto, Kampus Tembalang, Semarang 50275 – Indonesia

*Corresponding author: praba@undip.ac.id

Abstract. World Health Organization (WHO) initiated The Global Program to Eliminate Lymphatic Filariasis (LF) through mass drug administration (MDA). Pekalongan started MDA in 2011. Yet the LF prevalence in 2015 remained exceed the threshold (1%). This study aimed to describe the inhibiting factors related to the compliance of MDA in community level. This was a rapid survey with cross sectional approach. A two-stages random sampling was used in this study. In the first stage, 25 clusters were randomly selected from 27 villages with proportionate to population size (PPS) methods (C-Survey). In the second stage, 10 subjects were randomly selected from each cluster. Subject consisted of 250 respondents from 25 selected clusters. Variables consisted of MDA coverage, practice of taking medication during MDA, enabling and inhibiting factors to MDA in community level. The results showed most respondents had poor knowledge on filariasis, which influence awareness of the disease. Health-illness perception, did not receive the drugs, lactation, side effect, and size of the drugs were dominant factors of non-compliance to MDA. MDA information and community empowerment were needed to improve MDA coverage. Further study to explore the appropriate model of socialization will support the success of MDA program

Keywords: filariasis, mass drug administration, compliance, side effect, health-illness perception

1. Introduction

Lymphatic filariasis (LF) is a chronic infectious disease caused by *Wuchereria Bancroft*, *Brugia malayi*, or *Brugia timori*, and transmitted by mosquito vectors [1–4]. The disease rarely causes mortality, but the effect varies from acute symptom (adenolymphangitis, recurrent fever), chronic symptom (lymphedema, hydrocele, chyluria) [2–4], to disability [2,4,5]. World Health Organization (WHO) estimated 1.4 billion people in 73 countries were at risk of LF in 2014. More than 120 million people have been infected globally with 40 million disabilities, consisted of 15 million lymphedema and 25 million hydrocele/chyluria [6]. Around 62 % cases were in Southeast Asia [7]. In Indonesia, 233 out of 404 districts (46 %) were endemic area of LF [8].

WHO launched The Global Program of Elimination of Lymphatic Filariasis (GPELF) as a Public Health Problem by the Year 2020 [9]. In accordance with The GPELF, Indonesia set Filariasis Elimination Program as national priority since 2005. The strategy for elimination is based on the WHO-recommended of applying annual single-dose mass chemotherapy, which is known as mass drug administration (MDA). MDA is addressed to the entire endemic communities [10].

Diethylcarbamazine citrate (DEC) is the drug of choice for MDA program. DEC may decrease microfilariae rate and LF transmission. However, DEC does not have macrofilaricidal effect [11]. Therefore, MDA should be continuously performed according to the duration of reproductive lifespan of the adult worm. A 5-years of MDA is believed to be sufficient to prevent filarial transmission by eliminating microfilariae in the blood of infected people [10]. Continuous MDA requires compliance of community in taking DEC appropriately. MDA is successful if the coverage reached at least 65 % annually. Maintaining the high coverage of MDA in several years is a challenge, especially in area with limited resources [12].

Pekalongan is a filariasis endemic area. The city started partial MDA in 2010. Finger blood survey in 2010 showed six villages with microfilariae rate (mf rate) more than 1 %, i.e. Bumirejo Village (5.54 %), Tegalrejo Village (2.39 %), Pabean Village (3.39 %), Bandengan Village (2.39%), and Kertoharjo Village (4,.18 %). Pekalongan started simultaneous annual MDA throughout all areas since 2011. In April-June 2015, Pekalongan carried the last round of MDA. However, the coverage of MDA has not yet been determined. This study aimed to figure out the coverage of MDA in Pekalongan, and to determine the inhibiting factors of community compliance to MDA.

2. Materials and Methods

2.1. Study design and location

This was a rapid survey with cross sectional design to describe the coverage of MDA and its inhibiting factors. The study was conducted on July-August 2015 in Pekalongan City as an endemic area of LF. Study site was divided into clusters, which was located in villages. The selected villages were Bandengan (1 cluster), Banyuurip Ageng (3 clusters), Degayu (1 cluster), Jenggot (1 cluster), Karangmalang (1 cluster), Klego (3 clusters), Kramatsari (1 cluster), Krapyak (3 clusters), Kuripan Lor (2 clusters), Panjangbaru (1 cluster), Pringrejo (2 clusters), Sampangan (1 cluster), Kebulen (1 cluster), Tirto (3 clusters), and Yosorejo (1 cluster).

2.2. Study subject

Study population was people living in areas which received MDA program. The subject was family member of selected house, who was aged $\geq$ 15 years old. We selected 10 subjects in each cluster, therefore total subject was 250 people.

2.3. Sampling technique

We used two stage random method. In the first stage we used cluster sampling with a probability proportionate to size (PPS) method according to the number of population in each village. This study had precise level ± 10 % and confidence interval 95 %. In this stage we obtained 25 clusters from 15 villages by C-Survey software. In the second stage, we selected 10 houses from each cluster using simple random sampling method. The total houses selected were 250 houses.

Village map and random table were used to determine the location of the first house. Next houses were selected based on the nearest main door from previous selected house, and willingness of its host to participate in the study.

2.4. Variables

Variables in this study consisted of the practice of taking medication during MDA, factors affecting compliance of MDA, inhibiting factors of MDA in community level, knowledge related to LF (cause, transmission, prevention, medication), information on MDA, information media that reached respondents (interpersonal, non-interpersonal).

Support of elimination officers was defined as the role of cadres or health workers in disseminating information on MDA prior to the MDA program implementation, and their role during the implementation. Inhibiting factor was defined as the obstacles faced by community and officers in the execution of MDA. MDA coverage was measured by the proportion of respondents who complied in taking medicine during MDA implementation divided by the total number of respondents received drug, and expressed in percentage.

2.5. Data analysis

Data was analyzed to describe the coverage of MDA and to identify enabling and inhibiting factors related to community acceptance toward MDA.

3. Result and Discussion

Regarding compliance with MDA, our study reported 72.8% compliance after the fifth round of MDA (Table 3). This value was higher than national rate (65.9%) in 2015, yet lower than the expected coverage rate of MDA (85.0%). Compared to other regions in Indonesia, this result was lower than Tabalong District (85.8%) and Hulu Sungai Utara (90.9%) in Kalimantan [13], Bandung District (78.0%) in West Java [14]. This finding may be the explanation why Pekalongan did not pass the pre-TAS assessment. Transmission assessment survey, abbreviated to TAS, is an activity to assess whether a series of MDA have successfully reduced the prevalence of infection to levels equal to or below the critical cut-off threshold for the various vector species and complexes, and to decide whether MDA can be stopped [15].

Many countries are now scaling down MDA activities after reaching 100% geographic coverage and instituting monitoring and evaluation processes to establish the impact of several rounds of MDA and determine that transmission had been interrupted. At the same time, countries that are yet to initiate MDA for LF elimination will adopt improved mapping and coverage assessment protocols to accelerate the efforts for achieving global elimination. This review provides an update on treatment for LF and describes the current global status of the elimination efforts, transmission control strategies, importance of determining the impact of MDA and the role of surveillance after MDA has been stopped [16].

3.1. Characteristics of subject

Table 1 shows proportion of subject decreased with age, but the proportion of non-compliance tended to increase with age, although no significant difference was found (p=0.069). Almost a quarter of subjects were female. Our result was in accordance with previous studies on MDA evaluation that most of respondent [17,18]. Men usually work at noon so more women were to be found during interview.

Tabel 1. Characteristics of subjects

Characteristics	Total (n=250)	Compliance (n=182)	Non-compliance (n=68)	p-value
Age				
- 15-30	31.2	62.8	37.2	0.069
- 31-45	37.6	79.8	20.2	
- 46-60	26.8	76.1	23.9	
- 61-75	4.4	63.6	36.4	
Sex				
- Male	26,0	83.1	17.9	0.030
- Female	74,0	69.2	30.8	
Education				
- Not attending school	14,0	65.7	34.3	0.793
- Elementary school	34,4	73.3	26.7	
- Junior high school	24,0	73.3	26.7	
- Senior high school	22,8	77.2	22.8	
- Academy/University	4,8	66.7	33.3	
Occupation				
- No occupation	2,8	28.6	71.4	0.007
- Student	2,8	42.9	57.1	
- Housewife	42,0	64.8	35.2	
- Retired	0,4	100.0	0.0	
- Government employee	1,2	100.0	0.0	
- Privat employee	5,2	93.3	7.7	
- Merchant	10,0	88.0	22.0	
- Labour	20,4	80.4	19.6	
- Entrepreneur	8,8	81.8	18.2	
- Others	6,4	75.0	25.0	
Maritas status				
- Single	15,2	73.7	26.3	0.045
- Married	80,0	74.5	25.5	
- Spouse deceased	4,8	41.7	58.3	

Proportion of non-compliance female was higher than male. There were significant difference of non-compliance to MDA between male and female in this study (p=0.030). This finding is consistent with results from Egypt [19], India [20,21], and Kenya [22], although in contrary with several previous studies [18,23]. This could be due to higher literacy level among males as compared to females, which affected exposure to the information regarding to LF. Proportion of males in our study with good knowledge of LF was higher than females. This finding is consistent with results from a study in India that found that males were better informed about the disease and its mode of transmission [24]. In this study, almost half of subjects were housewives, and one third had basic education. Subject without occupation and student are the two highest non-compliance groups. Their status is affecting lack of information regarding to LF and MDA.

3.2. Knowledge of lymphatic filariasis, the drug, and mass drug administration

Less than half of subject mentioned there was drug for LF. Yet most of them failed to name the exact type of the drug. Most subjects correctly answered the questions on MDA is designed to combat LF, subjected to all community, implemented annually, the drug should be taken during MDA, pregnant mothers and people with severe illness should not take the drug. It was interesting to see that knowledge about LF was associated with better compliance (p=0.002) compared to knowledge about MDA (p=0.151). This finding is consistent with results reported from Egypt [19]

Tabel 2. Knowledge of subjects

Knowledge of LF and DEC	Total (n=250)	Compliance (n=182)	Non-compliance (n=68)	p-value
Filariasis: definition, cause, risk factors, prevention				
- Poor	64.4	66.5	33.5	0.002
- Good	35.6	84.3	15.7	
Drug can cure LF				
- No	51.2	71.1	28.9	0..535
- Yes	48.8	74.6	25.4	
Drug to LF				
- DEC	6.0	80.0	20.0	0.727
- Other	42.8	73.8	26.2	
- Drug cannot cure	51.2	71.1	28.9	
Knowledge of MDA				
MDA is designed to cure LF				
- Yes	72.0	73.3	26.7	0.761
- No	28.0	71.4	28.6	
MDA is implemented annually				
- Yes	92.8	72.8	27.2	0.954
- No	7.2	72.2	27.8	
Drug should be taken during MDA				
- Yes	84.0	73.8	26.2	0.411
- No	26.0	67.5	32.5	
MDA is subjected to all people				
- Yes	98.4	73.2	26.8	0.302
- No	1.6	50.0	50.0	
Drug for LF patients only				
- Yes	14.4	71.0	29.0	0.125
- No	85.6	83.3	16.7	
Drug for those who aged ≥ 15				
- Yes	35.3	71.6	28.4	0.752
- No	64.7	73.3	26.5	
Pregnant mother should not take the drug				
- Yes	86.8	74.2	25.8	0.204
- No	13.2	63.6	36.4	
Children under 2 years old should not take the drug				
- Yes	40.4	81.2	18.8	0.014
- No	59.6	67.1	32.9	
People with severe illness should not take the drug				
- Yes	90.4	73.9	26.1	0.233
- No	9.6	62.5	37.5	
Level of knowledge on MDA				
- Poor	72.8	70.3	29.7	0.151
- Good	27.2	79.4	20.6	

The knowledge gap with regard to the disease may be a major factor for lower compliance. Our study showed around two third of subjects were unaware about LF, and three-quarter were unaware about MDA program. Effective community mobilization activities are essential to strengthen the people's knowledge and to change their perceptions regarding LF. Involvement of and coordination with other sectors, involvement of NGOs, local leaders, and self-help groups need to be emphasized. An effective health education campaign to make the community aware about LF and increase their participation in the MDA program is essential. The focus of the health education should be on locally-appropriate media.

Table 3. Socialization of mass drug administration received by subject

	Total (n=250)	Compliance (n=182)	Non-compliance (n=68)	p-value
Obtain socialization				
- No	28.8	18.1	81.9	0.000
- Yes	71.2	78.1	21.3	
Information detail	n=178	%		
Source of information				
- Cadre, neighbour, family	105	59,0		
- Health officer	43	24,2		
- Poster	2	1,1		
- Other	28	15,7		
Where to obtain information				
- Home	82	46,1		
- Office	1	0,6		
- Public health centre	49	27,5		
- School	2	1,1		
- Public places	4	2,2		
- Other	40	22,5		
When to obtain information				
- January	1	0,6		
- February	1	0,6		
- March	18	10,1		
- April	69	38,8		
- Do not remember	89	50,0		
Role of elimination officer	Yes		No	
	n	%	n	%
Deliver information about MDA	153	61.2	97	38.8
Deliver information that DEC must be taken	184	73.6	67	27.4
Deliver information of eligible people to take DEC	170	68.0	80	32.0
Deliver information about the benefit of MDA	181	72.4	69	27.6
Distribute DEC door to door	236	94.4	14	5.6
Entrust DEC into neighbouring	43	17.2	207	82.8
Deliver information about side effect of DEC	35	14.0	215	86.0
Directly observe when the subject take DEC	38	15.2	212	84.8

3.3. Socialization of mass drug administration and role of elimination officer

Most of the subjects claimed to have received socialization of LF and MDA (71.2%). The major source of information was disseminated by cadre, neighbour, and family (59,0%). Nearly half of subjects received the information at home (46.1%), but the majority of subjects have forgotten the time of socialization. Similar findings have been reported from other studies in India [18].

Our result reported most of elimination officers (EO) gave information about MDA to community (61.2%), that the drug must be taken (73.6%), who must take the drug (68.0%), the benefit of MDA (72.4%), and distributed the drug door to door (94.4%). However, only a few EO informed adverse reactions (14.0%). It was observed from the study that EO hardly supervised drug intake directly (15.2%) or just left the drug to the nearest neighbour (17.2%). There is an urgent need for more effective drug-delivery strategies that are adapted to local need [18]. Volunteering cadre as an unpaid EO requires an individual's commitment, and the timing of MDA needs to be compatible with their other personal and professional obligations. The seasonal availability of EO upon which NTD programs are built should be recognized as a key 'Process Indicator' of equal importance as funding, logistics, policy, guidelines, supplies and training [25].

Tabel 4. Effect of community socialization and information to knowledge of mass drug administration

Socialization and information	Knowledge		p-value
	Good	Poor	
Obtain socialization			
- Yes	42.7	57.3	0.000
- No	18.1	81.9	
Obtain information from elimination officer			
- Yes	71.1	28.9	0.535
- No	74.6	25.4	

Socialization regarding to LF in our study significantly related to knowledge (p=0.000). And although cadre is the official elimination officer closest to the community, the information did not necessarily come from cadre. Knowledge is a predisposing factor to MDA compliance as shown in our study, and previous study [19,22]. However, most subjects of this study had low levels of knowledge. In general, people with low level of knowledge are not likely to comply with the treatment, which highlighted that people with low knowledge usually also gives low priority to disease prevention.

Our result suggested socialization especially related to compliance in female (Mantel-Haenszel test, p=0.007, data not shown). Understanding the gender distribution and gender-related gaps in knowledge and experience of disease burden is important to identify the target groups for intervention and design of gender-specific education materials.

3.4. DEC consumption practice and side effect during mass drug administration

MDA program had been implemented since 2011. The program used DEC, albendazole, and paracetamol durng 2011-2014, and only DEC and albendazole since 2015. Community practice during MDA program was shown in Table 5.

Effective coverage rate is the product of coverage by the health system and compliance in the community. A sustainable high coverage of 85% or more is required for stopping transmission and elimination of disease from the community [26]. Our study showed most subjects claimed to take DEC during MDA implementation (72.8%). Nevertheless, most of them take the drug at night (87.9%) and after meal (98.4%). Consumption of the drug after meal resulted in poor compliance with drug intake [18].

Our study showed proportion of female who did not take DEC was nearly doubled that on male (Table 1). Pregnancy and breastfeeding were among the reasons for non-consumption of drug (7.4% and 22.1% respectively). It is generally expected that a higher percentage of males access MDA than females due to uncertainty related to pregnancy status [23] According to WHO, pregnant women and breast feeding mothers are not included in the MDA program [27].

Table 5. DEC consumption practice and side effect during mass drug administration

Compliance (taking DEC during MDA)	n=250	%
- No	68	27,2
- Yes	182	72,8
Reason of non-compliance	n=68	%
- Pregnant	5	7,4
- Breastfed	15	22,1
- Illness (high blood pressure, diabetes mellitus, vertigo, gout)	10	14,7
- In medication (other medicine)	2	2,9
- Do not suffer of filariasis	1	1,5
- Side effect	6	8,8
- Forget	5	7,4
- All family member non-compliance	4	4,4
- Not getting the drug	7	10,3
- Large size of the drug	1	1,5
- No reason	13	29,2
Experiencing side effect	n=182	%
- No	111	61.0
- Yes	71	39.0
Side effect	n=71	%
- Nausea	20	28.2
- Dizzy	47	66.2
- Other	4	5.6
Side effect	n=71	%
- Right after taking the drug	24	33.8
- An hour after taking the drug	26	26.6
- Two hours after taking the drug	9	12.7
- Three hours after taking the drug	9	12.7
- Four hours after taking the drug	1	1.4
- Other	2	2.8
Action when experiencing side effect	n=71	%
- Contact elimination officer	1	1.4
- Seek care in PHC	12	16.9
- Take symptomatic medicine	2	2.8
- Other	56	78.9

Overall, 10.3% subjects claimed they did not receive the drug. But even some of those who received the drug did not want consume it. It is interesting that 1.5% subjects felt unnecessary to take the drug because they were not suffer from LF, 4.4% subjects did not consume the drug because their whole family did not too, while 11.8% mentioned they just did not feel like taking the drug. A detailed probe into the matter might help us determine the other cultural and social factors related to gender which could have played a role in the compliance rate. Awareness and education certainly play a crucial role in the elimination of LF [28].

Fear of adverse reactions was also the reason our subjects did not want to consume DEC tablet (8.8%). We assessed the adverse reactions after consumption of DEC among the study population. Of 182 persons who consumed the drug, 39.0% reported adverse reactions. The most common side-effect was dizziness (66.2%), followed by nausea and/or vomiting (28.2%). These adverse reactions appeared right after and within an hour of drug intake. Theoretically, filariasis drugs were well-tolerated, and side-effects were negligible. Although the adverse reactions were insignificant, fear of adverse reactions itself a major cause of non-compliance. Therefore, it is essential to deliver appropriate health information to address people's concerns and fears about the intervention and make arrangement for the management of cases.[18]

Adverse reactions following MDA are generally mild, and the frequency decreases after the first round of MDA [29]. However, a literature review showed that fear of adverse reactions is one of the major reasons people do not comply with MDA [30]. As a result, programs should ensure that health services are prepared to handle adverse reactions as they arise, especially in the early years of MDA. A good management of adverse reactions in these first years of treatment will benefit the program in subsequent years, as this will reassure communities that the drugs are safe and that the health services can be trusted. The positive benefits of MDA need to be communicated and reiterated during the pre-MDA socialization as well as at the time of pill distribution.

4. Conclusion

Coverage of MDA in Pekalongan in 2015 was 72.8%. The major issues of MDA implementation is compliance, health education, fear of adverse reactions, community motivation and participation. People do not have enough knowledge about the benefits of MDA in LF prevention. District health office and public health center need to provide community with adequate information. Further study is needed to identify the appropriate socialization model and information distribution to the community in endemic areas of LF.

References

[1] Nutman T. 2001 Blood-borne filarial infection: Wuchereria bancrofti, Brugia malayi, Brugia timori, Loa loa, Mansonella perstans and Mansonella ozzardi. In: Gillespie SH, Pearson RD, editors. Princ. Pract. Clin. Parasitol., Chichester, UK: John Wiley & Sons, Ltd;, 433–54. doi:10.1002/0470842504.

[2] Supali T, Kurniawan A, Partono F. 2011 Wuchereria bancrofti, Brugia malayi, Brugia timori. In: Sutanto I, Ismid IS, Sjarifuddin PK, Sungkar S, editors. Parasitol. Kedokt. 4th ed., Jakarta: Balai Penerbit FKUI;, 32–40.

[3] Chandy A, Thakur AS, Singh MP, Manigauha A. 2011 A review of neglected tropical diseases: Filariasis. Asian Pac J Trop Med 4:581–6. doi:10.1016/S1995-7645(11)60150-8.

[4] Simonsen PE, Fischer PU, Hoerauf A, Weil GJ. 2014 The filariases. In: Farrar J, editor. Manson's Trop. Infect. Dis. 23rd ed., Beijing, Cina: Elsevier Saunders;, 737–765.e5. doi:10.1016/B978-0-7020-5101-2.00055-8.

[5] Zeldenryk LM, Gray M, Speare R, Gordon S, Melrose W. 2011 The emerging story of disability associated with lymphatic filariasis: a critical review. PLoS Negl Trop Dis;5:e1366. doi:10.1371/journal.pntd.0001366.

[6] World Health Organization. Lymphatic filariasis: Key facts. WHO Media Cent 2015:1. http://www.who.int/mediacentre/factsheets/fs102/en/ (accessed November 3, 2015).

[7] World Health Organization. Lymphatic filariasis: situation and trends. WHO Media Cent 2014:1. http://www.who.int/gho/neglected_diseases/lymphatic_filariasis/en/

[8] Kementerian Kesehatan RI. Laporan tahunan filariasis limfatik. Jakarta: Subdit Pengendalian Filariasis dan Kecacingan; 2013.

[9] Brady M. 2014 Seventh meeting of the Global Alliance to Eliminate Lymphatic Filariasis: reaching the vision by scaling up, scaling down, and reaching out. Parasit Vectors;7:46. doi:10.1186/1756-3305-7-46.

[10] Michael E, Malecela-Lazaro MN, Simonsen PE, Pedersen EM, Barker G, Kumar A, 2004. Mathematical modelling and the control of lymphatic filariasis. Lancet Infect Dis;4:223–34. doi:10.1016/S1473-3099(04)00973-9.

[11] Ottesen EA, Hooper PJ, Bradley M, Biswas G. 2008 The global programme to eliminate lymphatic filariasis: health impact after 8 years. PLoS Negl Trop Dis;2:e317. doi:10.1371/journal.pntd.0000317.

[12] Kolaczinski JH, Kabatereine NB, Onapa AW, Ndyomugyenyi R, Kakembo ASL, Brooker S. 2007 Neglected tropical diseases in Uganda: the prospect and challenge of integrated control. Trends Parasitol;23:485–93. doi:10.1016/j.pt.2007.08.007.

[13] Balai Besar Teknik Kesehatan Lingkungan Banjarbaru. Survei cakupan POMP filariasis di wilayah layanan BBTKL-PP Banjarbaru tahun 2015. Banjarbaru: 2015.
[14] Astuti EP, Ipa M, Wahono T, Ruliansyah A. 2015 Analisis perilaku masyarakat terhadap kepatuhan minum obat filariasis di tiga desa Kecamatan Majalaya Kabupaten Bandung. Media Heal Res Dev;24:199–208.
[15] World Health Organization. Transmission assessment surveys in the Global Programme to Eliminate Lymphatic Filariasis: WHO position statement. Wkly Epidemiol Rec 2012;48:477–92.
[16] Rebollo MP, Bockarie MJ. 2013 Toward the elimination of lymphatic filariasis by 2020: treatment update and impact assessment for the endgame. Expert Rev Anti Infect Ther;11:723–31. doi:10.1586/14787210.2013.811841.
[17] Ompusunggu S, Siswantoro H, Purnamasari T, Dewi RM. 2009 Pelaksanaan pengobatan masal filariasis d beberapadaerah dengan frekuensi pengobatan berbea. J Penyakt Menular Indones;1:10–9.
[18] Roy RN, Sarkar AP, Misra R, Chakroborty A, Mondal TK, Bag K. 2013 Coverage and awareness of and compliance with mass drug administration for elimination of lymphatic filariasis in Burdwan District, West Bengal, India. J Health Popul Nutr;31:171–7.
[19] Abd Elaziz KM, El-Setouhy M, Bradley MH, Ramzy RMR, Weil GJ. 2013 Knowledge and practice related to compliance with mass drug administration during the Egyptian national filariasis elimination program. Am J Trop Med Hyg;89:260–4. doi:10.4269/ajtmh.12-0491.
[20] Hussain MA, Sitha AK, Swain S, Kadam S, Pati S. 2014Mass drug administration for lymphatic filariasis elimination in a coastal state of India: a study on barriers to coverage and compliance. Infect Dis Poverty 3:31. doi:10.1186/2049-9957-3-31.
[21] Koradhanyamath DM, Kulkarni P, Holla R. 2012 Coverage and compliance MDA programme for lymphatic filariasis in Bidar district, Karnataka, India. Asian Pacific J Trop Dis;2:290–2. doi:10.1016/S2222-1808(12)60063-2.
[22] Njomo D, Mukoko D, Njenga S, Amuyunzu-Nyamongo M, Magambo J. 2012 Socioeconomic factors associated with compliance with mass drug administration for lymphatic filariasis elimination in Kenya: Descriptive study results. Ann Trop Med Public Heal;5:103. doi:10.4103/1755-6783.95962.
[23] Hodges MH, Sonnie M, Turay H, Conteh A, Maccarthy F, Sesay S. 2012 Maintaining effective mass drug administration for lymphatic filariasis through in-process monitoring in Sierra Leone. Parasit Vectors;5:232. doi:10.1186/1756-3305-5-232.
[24] Babu B., Hazra R., Chhotray G., Satyanarayana K. 2004 Knowledge and beliefs about elephantiasis and hydrocele of lymphatic filariasis and some socio-demographic determinants in an endemic community of Eastern India. Public Health;118:121–7. doi:10.1016/S0033-3506(03)00144-6.
[25] World Health Organization. Lymphatic filariasis: Training in monitoring and epidemiological assessment of mass drug administration for eliminating lymphatic filariasis. Geneva, Switzerland: WHO Press; 2013.
[26] Alexander NDE. 2015 Are we nearly there yet? Coverage and compliance of mass drug administration for lymphatic filariasis elimination. Trans R Soc Trop Med Hyg;109:173–4. doi:10.1093/trstmh/tru204.
[27] World Health Organization. Global programme to eliminate lymphatic filariasis; a handbook for national elimination programme. Rome, Italy: WHO Press; 2013.
[28] Cantey PT, Rout J, Rao G, Williamson J, Fox LM. 2010 Increasing Compliance with Mass Drug Administration Programs for Lymphatic Filariasis in India through Education and Lymphedema Management Programs. PLoS Negl Trop Dis;4:e728. doi:10.1371/journal.pntd.0000728.
[29] Hochberg N, Michel MC, Lammie PJ, Mathieu E, Direny AN, De Rochars MB, 2006 Symptoms reported after mass drug administration for lymphatic filariasis in Leogane, Haiti. Am J Trop Med Hyg;75:928–32.
[30] Krentel A, Fischer PU, Weil GJ. 2013 A review of factors that influence individual compliance with mass drug administration for elimination of lymphatic filariasis. PLoS Negl Trop Dis;7:e2447. doi:10.1371/journal.pntd.0002447.

Physical Characteristics of Phycocyanin from Spirulina Microcapsules using Different Coating Materials with Freeze Drying Method

E N Dewi[1], L Purnamayati[1], and R A Kurniasih[1]

[1]Faculty of Fisheries and Marine Science, Diponegoro University, Jl. Prof. Soedarto, SH, Semarang – 50275, Indonesia

Email : nurdewisatsmoko@yahoo.com

Abstract : The aim of this study was to compare the physical characteristics of phycocyanin microcapsules (F) from *Spirulina* sp. with different coating materials, such as κ-Carrageenan (C) and Na-alginate (A) in combination with maltodextrin (M) by freeze drying method. Microcapsules were prepared in three variations of coating materials i.e. maltodextrin (FM); maltodextrin and Na-alginate (FMA); and maltodextrin and carrageenan (FMC) with concentration of each materials were 10%; 9%:1.0%; and 9%:1% (w/w), respectively. The results showed that FMA with Na-alginate 1.0% produced the highest bulk density and total soluble solid, there were 0,334 g/ml and 9,067%, respectively. Color analysis by chromameter showed that FMC produced the bluest color compared to other samples. The glass transition temperature (Tg) investigated with Differential scanning calorimeter (DSC) in all of the samples.

Keywords: Carrageenan, Maltodextrin, Na-alginate, Physical Characteristics, Phycocyanin

1. Introduction

Food coloring materials both naturally or synthetic frequently used in daily life. The color is an important part which affects the product appearance, either for food, beverages, cosmetics, or drugs. Synthetic dyes often used because the price is lower than natural dyes, provides pleasing color, and more stable. However, many synthetics caused an allergic reaction when exposed to the human, even in particular cause also can be carcinogenic.

Blue synthetic dyes do not many in number, including *Brilliant Blue* and *Indigotine*. *Brilliant Blue* resistance in light and acidity, while *Indigotine* does not resist in light, heat and acidity. Both of these synthetic blue dyes do not resist to oxidation [1], so by consuming synthetic dye in high amount can harmful to health. Natural dyes can be alternative because it's safer and has beneficial effects on health.

Phycocyanin is natural blue pigment contained in *Spirulina* sp., which able to be as antioxidant [2]. Phycocyanin not stable against pH, temperature, light and humidity [3,4]. Phycocyanin has potential as a natural dye, but the effort required to maintain phycocyanin during storage.

Microencapsulation is one of an alternative to process phycocyanin into powder, which has a particle size between 1 – 1000 μm [5]. Microencapsulation process requires coating materials to forming microcapsules. The coating materials used can influence the physical properties of microcapsules. Maltodextrin is coating materials often used in microencapsulation process due to its ability to forming a film and easily soluble in water. Carrageenan able to forming a gel, coagulate and often used as stabilizing agent [6]. Alginate has biocompatibility properties and easily to forming a gel [7].

Maltodextrin will be combined using other coating materials like carrageenan and natrium alginate in order to improve its ability to protect phycocyanin released out.

Some studies using spray drying method to process microencapsulation [8,9,10,11] because of its affordable price and the availability of equipment. However, spray drying using high inlet temperature, which can reduce the phycocyanin content. One of an alternative method that can utilize is freeze drying. This process consists of freezing and sublimation, which is known as an efficient method to protect enzyme [12] and antioxidant [13]. Freeze drying method also is known to maintain the anthocyanin content rather than spray drying method [14]. The aim of this study was to compare the effect of different coating material used using freeze drying method into the physical properties of phycocyanin microcapsules.

2. Methods

2.1. Phycocyanin Microencapsulation

10% (w/v) of maltodextrin DE 10 (CV. Multi Kimia Raya, Semarang, Indonesia) mixed with phycocyanin extracted from *Spirulina* sp. powder (PT. Neoalga, Sukoharjo, Indonesia). κ-Carrageenan and Na-alginate (PT. Selalu Lancar Maju Karya, Jakarta, Indonesia) each combined with maltodextrin as coating materials then was coded as FMC (Phycocyanin microcapsules using maltodextrin and κ-Carrageenan with concentration 9%:1%) dan FMA (Phycocyanin microcapsules using maltodextrin and Na-alginate with concentration 9%:1%). The emulsion was then microencapsulated using freeze dryer (Ningbo Yinzhou Sjia Lab Equipment Co., LTD).

2.2. Yield

Yield was measured by Nunes and Mercadante [15]. The yield was calculated based on the percentage ratio of microcapsules phycocyanin mass with total solids solution of phycocyanin microparticles.

2.3. Bulk Density

Bulk density was measured by using Rao and Vidhyadhara [16] method with modification in the tube used. Phycocyanin microcapsules placed on the 10 ml tube then measured its weight. Bulk density calculated by dividing the microcapsule weight and volume and its expressed in gram per cm^3.

2.4. Color with chromameter

Phycocyanin microcapsules color was determined by chromameter CR-200 (Minolta), which measuring of L, a, and b level. The L level indicates the brightness level, a negative level indicates towards green color, b negative level indicates blue color, and b positive indicate yellow color.

2.5. Dissolved Solids

0,5 g samples and 6 mL distilled water put into 100 mL of a flask, then shaken and heated in a water bath for 5 minutes. Once cold, distilled water added up to the limit line then filtered. Next, 5 ml of substrate was taken and poured into porcelain dish which is known its weight (A) g. The filtrate evaporated in the water bath until dry. Then dried in the oven at 105° until its weight constant (B) g. The dissolved solid level calculated as follow [17]:

DS (% dB) = (B-A)/dry weight sample x 10/0,5 x 100%

2.6. Encapsulation Efficiency (EE)

The Encapsulation Efficiency was measured by Yan *et al.*, [4]. The EE was calculated based on the presentage ratio of non coated phycocyanin mass and phycocyanin mass added at the beginning of the microencapsulation process.

2.7. Differential Scanning Calorimeter (DSC)
Maltodextrin, carrageenan, and phycocyanin microcapsule were thermal analyzed using DSC. The thermogram of each sample measured at a temperature range of 30°C to 300°C with the speed of 10°C per minute [18].

2.8. Statistical Analysis
Those data were analyzed using SPSS 17 with triplication. Data were analyzed using one-way ANOVA with Tukey test.

3. Results and Discussion

Table 1. Bulk Density and DS Analysis Result

No	Sample	Yield (%)	Bulk Density (g/ml)	DS(%)	Encapsulation efficiency/EF (%)
1	FM	84,723 ± 5,738^{a}	0,193 ± 0,110^{c}	7,867 ± 0,115^{b}	30,787 ± 0,311^{a}
2	FMA	83,093 ± 2,296^{a}	0,334 ± 0,006^{a}	9,067 ± 0,306^{a}	35,917 ± 0,391^{b}
3	FMC	79,787 ± 2,678^{a}	0,306 ± 0,003^{b}	7,200 ± 0,200^{c}	39,070 ± 3,725^{c}

Note: The data was the average of triplication ± standard deviation.
Different superscript on the same column indicates significantly different at level of α 0,05

3.1. Yield
Based on phycocyanin microencapsulation yield calculation, its showed there was no significant different result between FM, FMA, and FMC treatments. Those result indicated that different coating materials used in the same concentration in freeze drying method didn't influence phycocyanin microcapsule yield (Table 1). Phycocyanin microencapsulation using freeze drying method and different coating materials could produce more than 79% of microcapsule yield. On the otherhand, the yield obtained from microencapsulation of some materials using freeze drying method was not more than 50% [19,20]. The yield finding in this research was lower than others researcher since the inlet temperature used is not stable, hence causing the product sticky on the spray dryer chamber. Some microcapsule product are also lost because it can't separate on the cyclone, so the microcapsule flies into the air and enters the filter tube. Therefore, it is leading to losses the product during the process [20,14]. In this case, it was not found on freeze drying microencapsulation because there was no drying materials flow and product that produce as spray drying methods.

3.2. Bulk Density
Table 1 showed phycocyanin microcapsule bulk density level. Among the results , FMA had high bulk density, followed by FMC and FM respectively. Alginate and Carrageenan are polysaccharides which able to form a gel, where its functional properties influenced by its unique structure [22]. These properties able to trapped phycocyanin and avoid the moisture loss during freeze drying process. The bulk density was also influenced by the water holding capacity (WHC) level from emulsion which is formed by alginate and carrageenan. Amid *et al.*, [23] explained that sodium alginate WHC level is higher than κ-carrageenan which affects the microcapsule weight particle thereby increasing the bulk density value.

3.3. Dissolved Solids (DS)
Ease of phycocyanin microcapsule water soluble indicated by dissolved solids (Table 1). The highest to the lowest dissolved solids were FMA, FM, and FMC, respectively. Alginate can dissolve in hot water [24], so the combination between maltodextrin and alginate can produce high dissolved solids. Water

solubility related to bioactive components release speed [25], so the larger dissolved solids, the faster phycocyanin released.

Bioactive release rate, also influenced by used coating materials concentration. According to Mandal *et al.* [26] the lower alginate concentration used, the faster drugs release, where 2% of alginate concentration produce the fastest drug release. Carrageenan has a water soluble properties and able to forming a gel [27] which caused FMC had the lowest dissolved solids. The low dissolved solids of FMC caused slow phycocyanin to release.

3.4. Encapsulation Efficiency (EE)

Encapsulation efficiency showed that the amount of phycocyanin that could be encapsulated, which the high of EE, the phycocyanin that could be encapsulated rise excessively (Table 1). This result indicated that in the same concentration, κ-carrageenan has higher ability to encapsulate phycocyanin using freeze drying method compared to those a-alginate and plain maltodextrin. EE influenced by the type of polymer that using as the coating materials which could affect the hydrophobic characteristic of emulsifier used. Some of the polymer types had an emulsifying ability and maintaining the viscosity of the emulsion. EE is also likely to be affected by physic-chemical factors governing the encapsulation between the biopolymers used as coating material and active compounds [28,29,30].

Phycocyanin microcapsules, there were FM, FMA, and FMC had EE value 30,787%, 35,917%, and 39,070%, respectively. This result was lower than Holkem *et al.,* [31], where the EE of alginate microcapsules with freeze drying method was 89,71%. According to Karthik and Anandharamakrishnan [32], an active compound which encapsulated using freeze drying method could produce porous microcapsule on its surface because there is ice sublimation during the drying process. Which lead the active compound oxidation. Otherwise, needs coating material which protects phycocyanin during freeze drying method.

3.5. Color

Phycocyanin microcapsule color was presented Table 2. It can be seen that FMC phycocyanin microcapsule had the bluest color compared to FM and FMA. Those finding showed that carrageenan able to protect phycocyanin during freeze drying process due to carrageenan ability to forming a gel.

Table 2. Phycocyanin Microcapsule Color Level

No	Sample	L	a	b
1	FM	$66{,}313 \pm 0{,}185^{a}$	$-4{,}397 \pm 0{,}349^{a}$	$-12{,}520 \pm 0{,}486^{b}$
2	FMA	$71{,}160 \pm 0{,}527^{b}$	$-6{,}397 \pm 0{,}067^{c}$	$-9{,}500 \pm 0{,}060^{a}$
3	FMC	$72{,}943 \pm 0{,}061^{c}$	$-5{,}133 \pm 0{,}025^{b}$	$-13{,}527 \pm 0{,}127^{c}$

Note: The data was the average of triplication ± standard deviation.
Different superscript on the same column indicates significantly different at level of $\alpha = 0{,}05$

Regarding different characteristic, some polysaccharides are able to form a gel in different type [27]. The κ-carrageenan was able to forming stable gel because of 3,6 *anhydrogalactose* and *galactose 4 sulfate* content [33]. According to Mehrad *et al.,* [25] the combination between maltodextrin and κ-carrageenan produce fish oil microcapsule with the most yellow color than the mix between maltodextrin-gelatin and maltodextrin-gelatin-κ carrageenan.

Unlike the κ-carrageenan, alginate is polysaccharides extracted from brown seaweed, which contains β-D-mannuronic acid (M) and α-L-guluronic acid (G). The comparison between M / G content of alginate will affect the gel properties result. The lower ratio of M / G of alginate will produce a weak gel. Furthermore, Fertah *et al.*, [34] reported that alginate with M / G 1,12 will bring a soft and elastic gel. In this study, FMA had the lowest blue color, so the combination between maltodextrin and alginate were less able to protect the phycocyanin because of weak gel strength.

3.6. Differential Scanning Calorimeter (DSC)

Phycocyanin microcapsule DSC Thermogram with different coating materials showed in Figure 1. FM microcapsule produces two onsets temperature of 140,93°C and 152,32°C which showed there were two components in the microcapsule of phycocyanin and maltodextrin. According to Kouassi *et al.,* [35] onset temperature showed glass transition temperature (Tg). Two onset temperature formation of DSC Thermogram showed there were two components on microcapsule. The FMA and FMC microcapsule had one onset temperature of 141,77°C and 137,04°C, respectively. One onset temperature formation revealed that phycocyanin was entirely dispersed in used coating materials. Devi and Kakati, [36] stated that there was no ascorbic acid onset temperature formation of ascorbic acid microparticles showed that ascorbic acid dispersed was into microparticles.

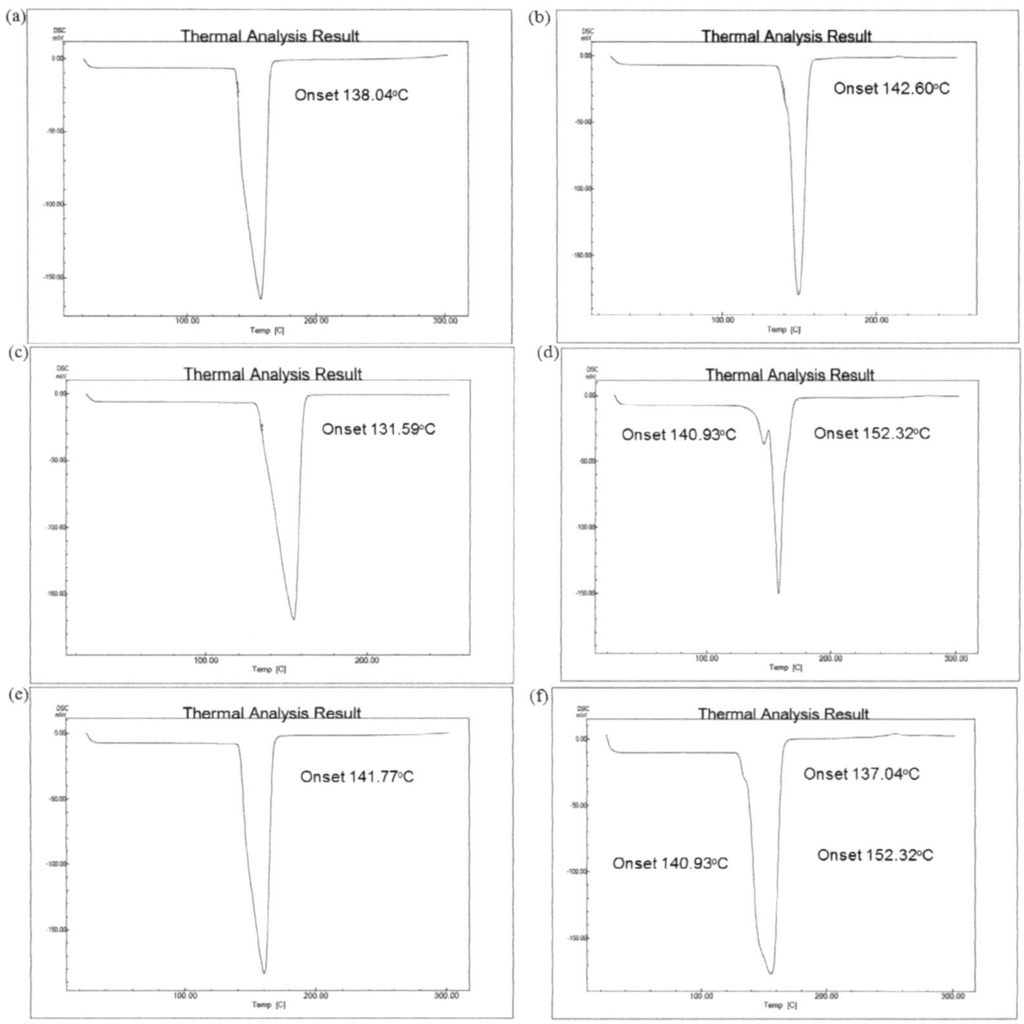

Figure 1. DSC Thermogram (a) alginate (b) κ-carrageenan (c) maltodextrin (d) FM (e) FMA (f) FMC

4. Conclusion

Phycocyanin microcapsules using maltodextrin and alginate coating materials produce the highest bulk density and dissolved solids. Phycocyanin microcapsules using maltodextrin and κ-carrageenan coating materials produce the bluest color. Based on DSC analysis, the combination between maltodextrin-alginate and maltodextrin-κ-carrageenan produce one onset temperature which showed that phycocyanin was perfectly dispersed in coating materials used. Based on that result, alginate and κ-carrageenan were potential as coating materials for phycocyanin microcapsule process.

Acknowledgement

This research was funded by Diponegoro University and Ministry of Research Technology and High Education 2016.

References

[1] Allam K V and Kumar G P 2011 Colorant-The cosmetics for the pharmaceutical dosage forms Review. *International Journal of Pharmacy and Pharmaceutical Sciences.* **3**(3) 13-21.

[2] Bertolin T E, Farias D, Guarienti C, Petry F T S, Colla L M and Costa J A V 2011 Antioxidant effect of phycocyanin on oxidative stress induced with monosodium glutamate in rats. *Brazilian Archives of Biology and Technology.* **54**(4) 733-738.

[3] Chaiklahan R, Chirasuwan N and Bunnag B 2012 Stability of phycocyanin extracted from *spirulina* sp.: influence of temperature, pH, and preservatives. *Process Biochemical.* **47** 659–664.

[4] Yan M, Liu B, Jiao X and Qin S 2014 Preparation of phycocyanin microcapsules and its properties. *Journal of Food and Bioproducts Processing.* **92** 89-97.

[5] Chanana A, Kataria M K, Sharma M and Bilandi A 2013 Microencapsulation: advancements in applications. *International Research Journal of Pharmacy.* **4**(2) 1-5.

[6] Necas J and Bartosikova L 2013 Carrageenan: a review. *Veterinarni Medicina.* **58**(4) 187-205.

[7] Lee K Y and Mooney D J 2012 Alginate: properties and biomedical applications. *Prog Polym Sci.* **37**(1) 106-126.

[8] Dewi E N, Purnamayati L and Kurniasih R A 2016 Antioxidant activities of phycocyanin microcapsules using maltodextrin and carrageenan as coating materials. *Journal of Technology.* **78**(4-2) 45-50.

[9] Ton N M N, Tran T T T and Le V V M 2016 Microencapsulation of rambutan seed oil by spray-drying using different protein preparations. *International Food Research Journal.* **23**(1) 123-128.

[10] Marcela F, Lucia C, Esther F and Elena M 2016 Microencapsulation of L-ascorbic acid by spray drying using sodium alginate as wall material. *Journal of Encapsulation and Adsorption Sciences.* **6** 1-8.

[11] Fernandez R V B, Borges S V and Botrel D A 2013 Influence of spray drying operating conditions on microencapsulated rosemary essential oil properties. *Cienc. Tecnol. Aliment., Campinas.* **33**(1) 171-178.

[12] Amid M, Manap Y and Zohdi N K 2014 Microencapsulation of purified amylase enzyme from pitaya (Hylocereus Polyrhizus) peel in arabic gum-chitosan using freeze drying. *Molecules.* **19** 3731-3743.

[13] Isailovic B, Kalusevic A, Zurzul N, Coelho M T, Dordevic V, Alves V D, Sousa I, Moldao-Martins M, Bugarski B and Nedovic V A 2012 Microencapsulation of natural antioxidants from pterospartum tridentatum in different alginate and inulin systems. *6th Central European Congress On Food.* 1075-1081.

[14] Wilkowska A, Ambroziak W, Czyzowska A and Adamiec J 2016 Effect of microencapsulation by spray-drying and freeze drying technique on the antioxidant properties of blueberry (*Vaccinium myrtillus*) juice polyphenolic compounds. *Journal of Food Nutrient and Science.* **66**(1) 11-16.

[15] Nunes I L and Mercadante A Z 2007 Encapsulation of lycopene using spray drying and molecular inclusion processes. *Brazillian Archives of Biology and Technology an International Journal.* **50**(5) 893-900.

[16] Rao T V and Vidhyadhara S 2012 Formulation and in-vitro evaluation of indomethacin microcapsules. *International Journal of Chemistry and Science.* **10**(1) 1-8.

[17] Association of Official Analytical Chemist 2005 Official Methods of Analysis. AOAC. Washington, United States.

[18] Saha A K and Ray S D 2013 Effect of cross-linked biodegradable polymers on sustained release of sodium diclofenac-loaded microspheres. *Brazilian Journal of Pharmaceutical Sciences.* **49**(4) 873-888.

[19] Krishnaiah D, Sarbatly R and Nithayanandam R 2012 Microencapsulation of morinda citrifolia L. extract by spray-drying. *Chemical Engineering Research and Design.* **90** 622–632.

[20] Tonon R V, Grosso C R F and Hubinger M D 2011 Influence of emulsion composition and inlet air temperature on the microencapsulation of flaxseed oil by spray drying. *Food Research International.* **44**(1) 282–289.

[21] Samakradhamrongthai R, Thakeow P, Kopermsub P and Utama-ang N 2015 Encapsulation of *Michelia alba* D.C. extract using spray drying and freeze drying and application on Thai dessert from rice flour. *International Journal of Food Engineering.* **1**(2) 77-85.

[22] Rhein-Knudsen N, Ale M T and Meyer A S 2015 Seaweed hydrocolloid production : an update on enzyme-assisted extraction and modification technologies. *Marine Drugs.* **13** 3340-3359.

[23] Amid B T, Mirhosseini H, Poorazarang H and Mortazavi S A 2013 Implications of partial conjugation of whey protein isolate to durian seed gum through maillard reactions: foaming properties, water holding capacity, and interfacial activity. *Molecules.* **18** 15110-15125.

[24] Moon S C and Farris R J 2009 Electrospinning of heated gelatin-sodium alginate-water solutions. *Polymer Engineering and Science.* **49**(8) 1616-1620.

[25] Mehrad B, Shabanpour B, Jafari S M and Pourashouri P 2015 Characterization of dried fish oil from menhaden encapsulated by spray drying. *Aquaculture, Aquarium, Conservation and Legislation International Journal of the Bioflux Society.* **8**(1) 57-69.

[26] Mandal S, Kumar S S, Krishnamoorthy B and Basu S K 2010 Development and evaluation of calcium alginate beads prepared by sequential and simultaneous methods. *Brazilian Journal of Pharmaceutical Sciences.* **46**(4) 785-793.

[27] Tako M 2015 The principle of polysaccharide gels. *Advances in Bioscience and Biotechnology.* **6** 22-36.

[28] Martins I M, Barreiro M F, Coelho M and Rodrigues A E 2014 Microencapsulation of essential oils with biodegradable polymeric carriers for cosmetic applications. *Chemical Engineering Journal.* **245** 191–200.

[29] Naik A, Meda V and Lele S S 2014 Freeze drying for microencapsulation of α-linolenic acid rich oil: a functional ingredient from *Lepidium sativum* seeds. *Eur. J. Lipid Sci. Technol.* **116** 837-846.

[30] Belscak-Cvitanovic A, Busic A, Barisic L, Vrsaljko D, Karlovic S, Spoljaric I, Vojvodic A, Mrsic G and Komes D 2016 Emulsion templated microencapsulation of dandelion (*Taraxacum officinale* L.) polyphenols and β-carotene by ionotropic gelation of alginate and pectin. *Food Hydrocolloids.* **57** 139-152.

[31] Holkem A T, Raddatz G C, Nunes G L, Cichoski A J, Jacob-Lopes E, Grosso C R F and de Menezes C R 2016 Development and characterization of alginate microcapsules containing Bifidobacterium BB-12 produced by emulsification/internal gelation followed by freeze drying. *LWT-Food Science and Technology.* **71** 302-308.

[32] Karthik P and Anandharamakrishnan C 2012 Microencapsulatin of docosahexaenoic acid by spray-freeze-drying method and comparison of its stability with spray-drying and freeze-drying methods. *Food Bioprocess Technology.* **6**(10) 2780-2790.

[33] Dewi E N, Darmanto Y S and Ambariyanto 2012 Characterization and quality of semi-refined carrageenan (SCR) products from different coastal waters based on fourier transform infrared technique. *Journal of Coastal Development.* **16**(1) 25-31.

[34] Fertah M, Belfkira A, Dahmane E, Taourirte M and Brouillette F 2014 Extraction and characterization of sodium alginate from moroccan laminaria digitata brown seaweed. *Arabian Journal of Chemistry.* 1-8.

[35] Kouassi G K, Teriveedhi V K, Milby C L, Ahmad T, Boley M S, Gowda N M and Terry R J 2012 Nano-microencapsulation and controlled release of linoleic acid in biopolymer matrices: effects of the physical state, water activity, and quercetin on oxidative stability. *Journal of Encapsulation and Adsorption Sciences.* **2** 1–10.

[36] Devi N and Kakati D K 2013 Smart porous microparticles based on gelatin/sodium alginate polyelectrolyte complex. *Journal of Food Engineering.* **117** 193-204.

IOP Conf. Series: Earth and Environmental Science **55** (2017) 012061 doi:10.1088/1755-1315/55/1/012061

MJO (Madden-Julian Oscillation) Analysis of the Chlorophyll-a Distribution in Western Waters Bengkulu

Y D Haryanto[1,2], N Fitrianti[1], A Hartoko[2], S Anggoro[2], M Zainuri[3]

[1] State College of Meteorology Climatology and Geophysics, Jl. Perhubungan 1 No.5 Pondok Aren, Tengerang Selatan, 15221
[2] Coastal Resource Management, Faculty of Fisheries and Marine Sciences, Diponegoro University, Jl. Imam Bardjo SH No.5, Semarang, 50241
[3] Department of Marine Science, Faculty of Fisheries and Marine Sciences, Diponegoro University, Jl. Prof.H.Soedharto, SH, Tembalang, Semarang, 50275

E-mail address: yosafatdonni@gmail.com

Abstract. The global phenomenon Madden-Julian Oscillation (MJO) is one of the dominant oscillation in the equatorial region of the Indian Ocean that oscillates between 30-60 days and experience the process of convection movement from west to east. MJO has a correlation of high intensity rainfall of the area in its path. During his journey eastward, the MJO is influenced by the position of the sun. When the sun in the equatorial MJO moves straight east. Meanwhile, when the position of the sun in the south of the equator, MJO shifted slightly to the south of the equator, known as the propagation of the south-east (south-eastern propagation). When the position of the sun is in the north of the equator, MJO shifted slightly to the north of the equator, known as the propagation of the north-east (north-east of propagation). Waters west of Bengkulu has a huge potential in the fisheries sector, which is situated overlooking the Indian Ocean. The phenomenon MJO influence on rainfall, sea surface temperature, and the concentration of chlorophyll-a. This study aims to look at the temporal distribution of sea surface temperature and chlorophyll-a and decide how MJO relationship with SST and precipitation conditions and increasing the amount of chlorophyll during the phase of the MJO in Bengkulu waters. The dataset used is data of chlorophyll-a which download in *oceancolor.gfsc.nasa.gov*, sea surface temperature data is used is a model of Kaplan Extended V2, RMM1 index data and RMM2 on www.bom.gov.au and rainfall data of Bengkulu region. The method used is descriptive statistical methods, Conditional Probability and logistics regression. From the above explanation can be said that there is a relationship between the incidence of MJO by the number of chlorophyll-a. Odds the addition of chlorophyll-a have a linear relationship with the duration of the incident MJO in Bengkulu, odds increase the amount of chlorophyll-a in Bengkulu region reaches a threshold value of 0.5, it means that the length occurrence MJO able to identify increasing the number of chlorophyll-a in Bengkulu.

1. Introduction

Bengkulu has a huge potential in the fisheries sub-sector, especially marine fisheries. The layout of the area, mostly overlooking the Indian Ocean to reach 525 km long coastline, causing Bengkulu province has a territorial sea area of 53,000 km2. In the field of marine and fisheries Bengkulu province has a potential of 145 334 tonnes.

Chlorophyll-a is an important component that boosted phytoplankton and aquatic plants, both of which are a natural food source for fish. Chlorophyll-a is a pigment active in plant cells have a role pentingterhadap the process of photosynthesis [1].

Conditions in Indonesia which is located between two continents (Asia and Australia) and two oceans (Indian and Pacific) also led to Indonesia affected by various climatic phenomena, both locally, regionally, and globally. One influential global phenomenon is the phenomenon of the Madden-Julian Oscillation (MJO). MJO first discovered by the Madden-Julian in 1971. MJO is one of the dominant oscillation in the equatorial region. This phenomenon oscillates between 30-60 days and experience the movement process of convection from west to east [2] [3].

Stated that the active phase of the MJO has a correlation of the intensity of the high rainfall of the area in its path [4]. During its journey eastward, the MJO is influenced by the position of the sun. When the sun is on the equator the MJO moves straight to the east. Meanwhile, when the position of the sun in the south of the equator, then a trip MJO slightly shifted towards the south of the equator known as the propagation of the south-east (south-eastern propagation). When the position of the sun is in the north of the equator, then a trip MJO slightly shifted to the north of the equator, known as the propagation of the north-east (north-easternpropagation).

In conducting the fisheries, the fishing area information is crucial for the efficiency and effectiveness of the arrest can be improved. Information can be obtained through the collection area of oceanographic parameters. Oceanographic parameters is one of the factors which influenced the variability in the catches, such as chlorophyll-a and sea surface temperature. Phytoplankton are located in the layer of light (photic) containing chlorophyll-a is useful for photosynthesis. Chlorophyll-a is capable of absorbing blue light and green, so the presence of phytoplankton can be detected by the ability of chlorophyll-a. Plankton, both phytoplankton and zooplankton have an important role in the marine ecosystem because plankton into food for many species of other marine animals. This oceanographic parameters can be recorded and diintrepretasi through remote sensing.

The attempt to predict the fishing area (fishing ground) can be done through physical oceanographic conditions. Almost all the fish populations that live in marine waters have a range of temperature and chlorophyll-a is optimum for life [5]. By knowing the oceanographic parameters, especially the temperature and chlorophyll-a optimum of a species of fish in a body of water, then we can surmise the existence of a group of fish and can be used for the purpose of arrest (exploitation). This study aims to look at the temporal distribution of sea surface temperature and chlorophyll-a and decide how MJO relationship with SST and precipitation conditions and increasing the amount of chlorophyll during the phase of the MJO in Bengkulu waters. The expected benefits of this research is to provide information to the relevant parties regarding the potential fishing areas in marine waters Bengkulu.

2. Material And Method

The dataset used is data of chlorophyll-a which download in *oceancolor.gfsc.nasa.gov*, sea surface temperature data is used is a model of Kaplan Extended V2, RMM1 index data and RMM2 on www.bom.gov.au and rainfall data of Bengkulu region. The method used is descriptive statistical methods, Conditional Probability and logistic regression. Logistic regression is a mathematical model approaches that can be used to describe the relationship of several variables X with a dichotomous dependent variable / politomus [6]. Binary logistic regression is a method of data analysis used to find the relationship between the response variable (y) that is binary or dichotomous with the predictor variable (x) which is polikotomus [7]. Logistic regression itself is perfect for making a probabilistic forecast, because it will never predict the probability is less than zero or greater than one [8]. Logistics equation is:

$$P = \frac{\exp(b0 + b1x)}{1 + \exp(b0 + b1x)} \tag{1}$$

Information: P = Logistic Regression probability

x = Predictor

$b0,1$ = Regression Coefficient

Logistics curve depicted in the form S sigmoid with a z value ranging from -∞ to + ∞ and the value (z) moves from 0 to 1. When the value of z approaches -∞ then the value $f\ (z)$ moves closer to a value of 0, and if the value of z approaching + ∞ then the value (z) moves closer to the value of 1. The conditional probability (Conditional Probability) is the probability of an event A with the proviso events and happening / will happen first. Notation Probabilias conditional: Pr (A | B) probability of A on condition B occurs first. Mathematical formula for conditional Probabilias are:

$$\Pr\ (AI\,B) = \frac{\Pr(A \cap B)}{\Pr(B)} \tag{2}$$

The data processing stages as follows :

1. Identification of MJO activity by collecting data using the data RMM1 MJO events and RMM2 in DJFM 2007-2012 by making spatial OLR.
2. Analysis of the current condition of SST when MJO and normal events.
3. Defining relationships SST conditions with rainfall during the MJO by way of conditional probability analysis with cumulative odds of 90%.
4. Determine the treshold values its long MJO events with the addition of chlorophyll-a at the time of the MJO with the terms, conditions, SST and precipitation predetermined cumulative odds by 90%.

3. Result And Discussion

3.1. MJO Identification

MJO position identification passing through Indonesia can be accomplished by creating composite spatial OLR of MJO phase 3, phase 4 and phase 5. The composite analysis is done by looking at the data RMM1 and RMM2 in each phase and then identify the periods of events that pass through Indonesia based on the phase 3, phase 4 and phase 5 on DJFM year period from 2007 to 2012 found that there were 223 days of activity. Figure A shows the MJO activity of phase 1 to phase 8

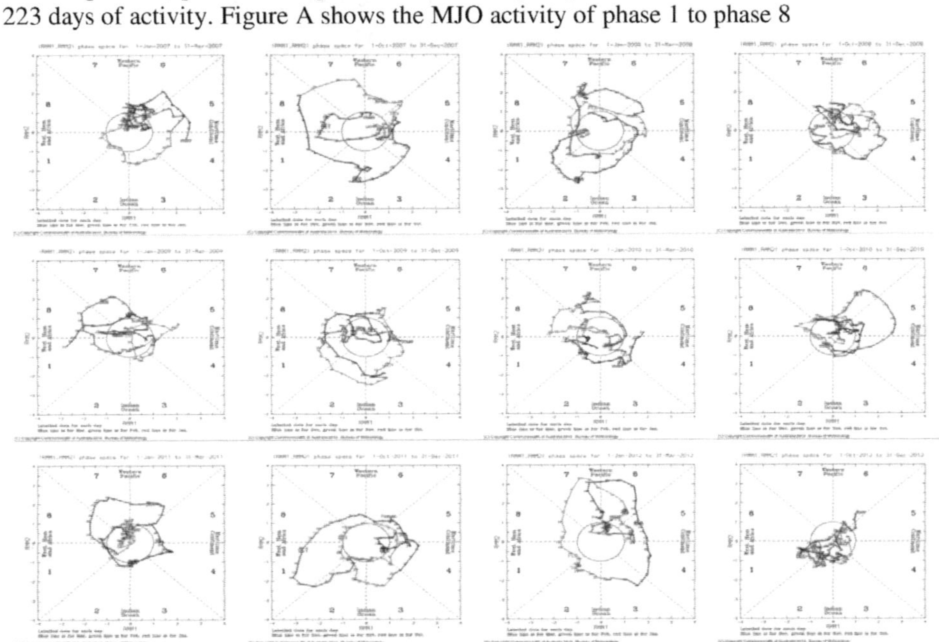

Figure 1 RMMI dan RMM2 DJFM Period 2007 – 2012

IOP Conf. Series: Earth and Environmental Science **55** (2017) 012061 doi:10.1088/1755-1315/55/1/012061

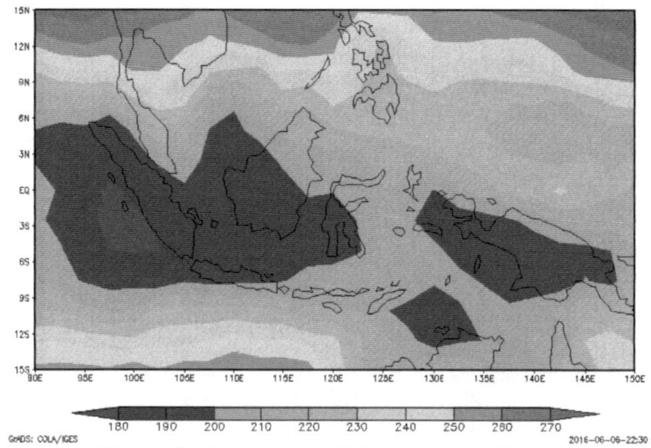

Figure 2 Composite MJO Events In Phase 3

In Figure 2 above found that in the current phase of the MJO is 3, the lowest value of OLR occurs in the southern part of Sumatra, partial ocean Java to Kalimantan. Lowest OLR value is also present in some areas of analysis, Bengkulu waters.

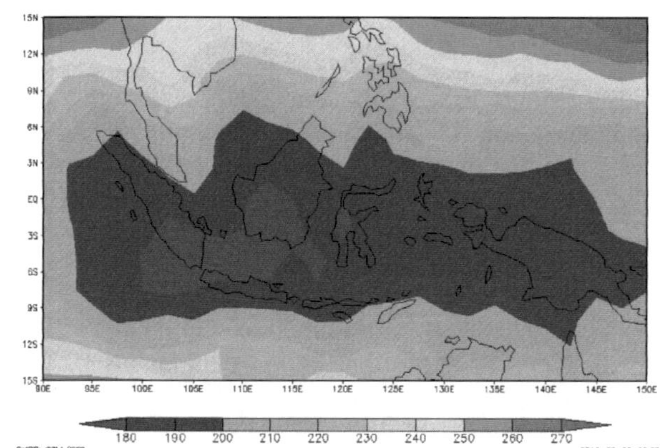

Figure 3 Composite MJO Events In Phase 4

At the time of phase 4 low OLR values move towards the east and covers an area of southern Sumatra, Java Sea and extends to South Sulawesi. While in the area of analysis is almost covered by the low OLR values.

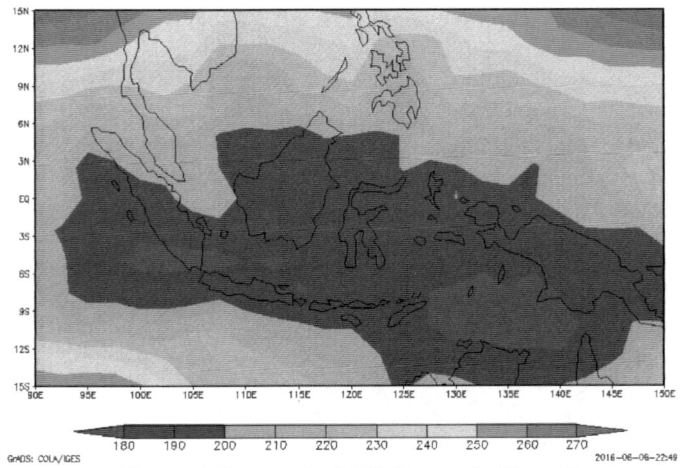

Figure 4 Composite MJO Events In Phase 5

At the time of the MJO is at phase 5, the position of the MJO are in Indonesian maritime continent. OLR and the lowest value moved eastwards and are in the northern region of Australia. OLR low value currently on the Indian Ocean region to the northeastern Australian waters. While in the area of research is still visible low OLR values, although the area coverage of the lowest OLR only cover part of research area. Based on the images A, B and C because the study area is located in Bengkulu waters, to identify the MJO influence on crops Clorofil a then danalisis MJO impact on the phase 3, phase 4 and phase 5 which in this phase is the lowest passing through the area OLR value analysis.

3.2 SST Conditions AT The MJO Events

Analysis of the condition of SST (Sea Surface Temperature) at the time of the events MJO in the period DJFM during phase 3, fase4 and phase 5 is done by classifying the data daily at the time of the MJO is then viewed condition averages are then compared to average conditions DJFM SST during the periods when the MJO event does not occur. For SST conditions in the region can be explained by the analysis of B. chart below:

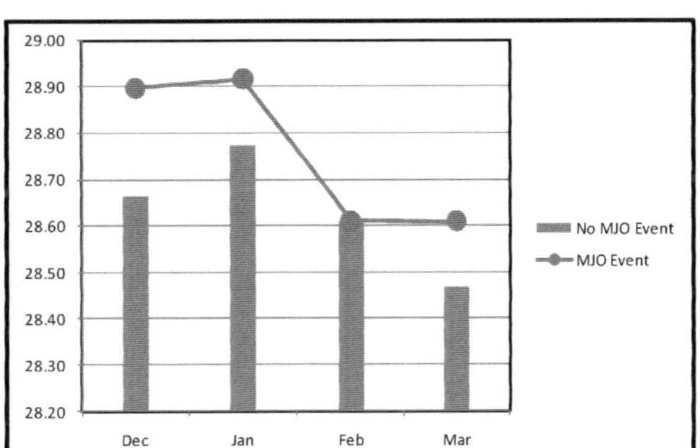

Graph 1. SST Analysis In MJO and No MJO Events

In the graph 1. above shows that when the MJO phenomenon occurred in waters Bengkulu SST conditions show improvement. In December, January and March showed an increase when compared to normal except in February where conditions at the time of occurrence of the phenomenon SST MJO and its not happening this phenomenon showed no change. It is caused by a phenomenon at least MJO in February during the 2007-2012 period.

Linkage analysis of events and conditions SST MJO is done by finding the relationship between the index and the value of SST MJO in the event MJO by using correlation method. The MJO index is calculated using data RMM1 and RMM2 through the equation $\sqrt{RMM1^2 + RMM2^2}$ where, according to [9] [10] divided into 8 categories MJO phase of strong and weak. Classification MJO index in this study were divided into three, namely, weak (0-1.5), moderate (1.5-2.5), and strong (> 2.5). Of calculation using correlation method is found that the correlation between SST and the index MJO is equal to 0.1, the value is classified as the correlation value is quite small considering the rainfall in the region is not only influenced by the SST, but if there is an increase or decrease in SST significantly will cause a change in the amount of rainfall.

3.3 Probability of Rainfall on SST

Calculations conducted by using odds Conditional Probability, namely by looking at cumulative odds that exceeds 90%, the calculation is done by dividing the SST class 5 and 4 classes of rainfall, while class divisions are as follows:

Table 1. Rainfall and SST Classsification

Class	SST	Class	Rainfall
1	0 - 25.7	1	<20
2	25.7 - 26.7	2	20 - 50
3	26.7 - 27.7	3	50 - 100
4	27.7 - 28.8	4	>100
5	>28.8		

By calculating the frequency of occurrence conditional above, where chances of rainfall on the condition that the SST condition has occurred then divided by its number of occurrences conditional lot, the importance of such probability in the following table:

Table 2. Conditional Probability Between SST and Rainfall

		SST				
		0 - 25.7	25.7 - 26.7	26.7 - 27.7	27.7 - 28.8	>28.8
Rainfall	<20	0	0	0	73%	72%
	20 - 50	0	0	0	21%	18%
	50 - 100	0	0	0	6%	7%
	>100	0	0	0	0%	3%
	Total	0	0	0	100%	100%

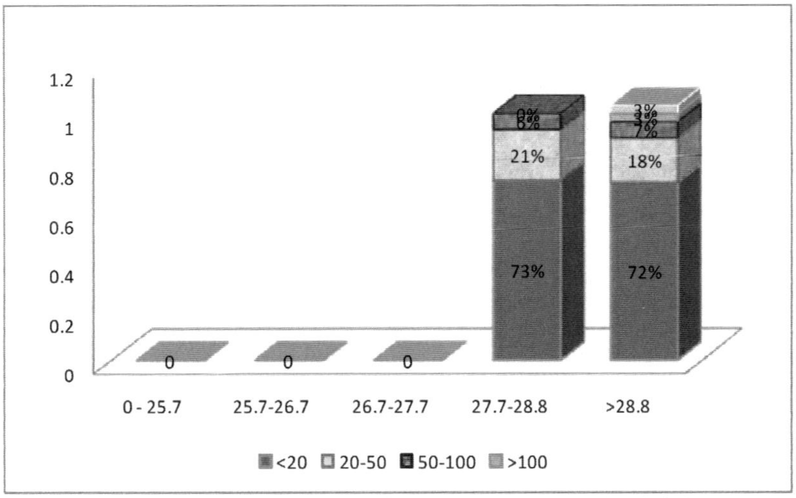

Graph 2. Conditional Probability Between SST and Rainfall

From the above table is found that the highest odds for rainfall <20 mm / day obtained on condition of SST warm and toasty, in this study were taken cumulative odds by 90% so that, got that for result rainfall ranges from 20-50 mm / day is required SST in warm conditions in terms of SST conditions> 27.7oC, given on condition of SST> 28.8 ° C was also exceeded 90% chance. Furthermore, the threshold values determined in phase 3 day duration, phase 4 and phase 5 to increase the amount of chlorophyll begins by calculating the likelihood of hot spots are generated through binary logistic regression method on any additional criteria chlorophyll, while the additional criteria ie +0.01, +0.02 , +0.05 and +0.1. The chlorophyll additional criteria adapted to the conditions warm SST (> 27.7) and rainfall of 20-50 mm / day.

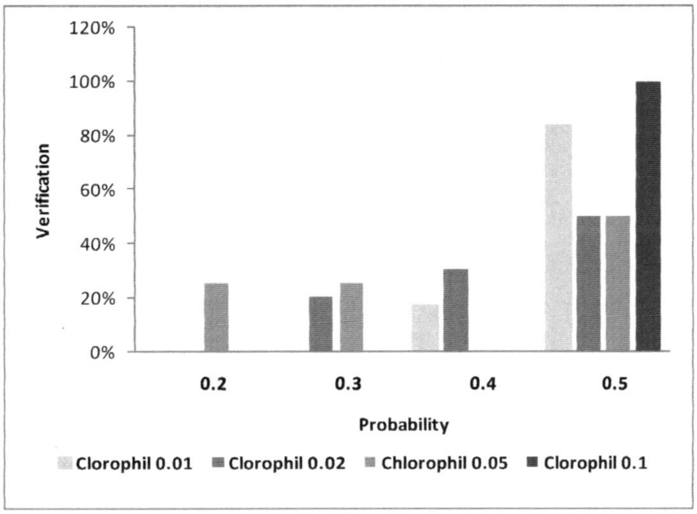

Figure 5. Frequency Distribution Probability Each of Criteria Addition Chlorophyll-a

Based on the figure 5 it can be concluded that the best verification tilapia contained in 0.5 probability model with verification value is always above 40%. Furthermore, a threshold value of 0.5 was used as odds of determining the limits increase in the number of chlorophyll which is then linked to the duration of the third phase of MJO events.

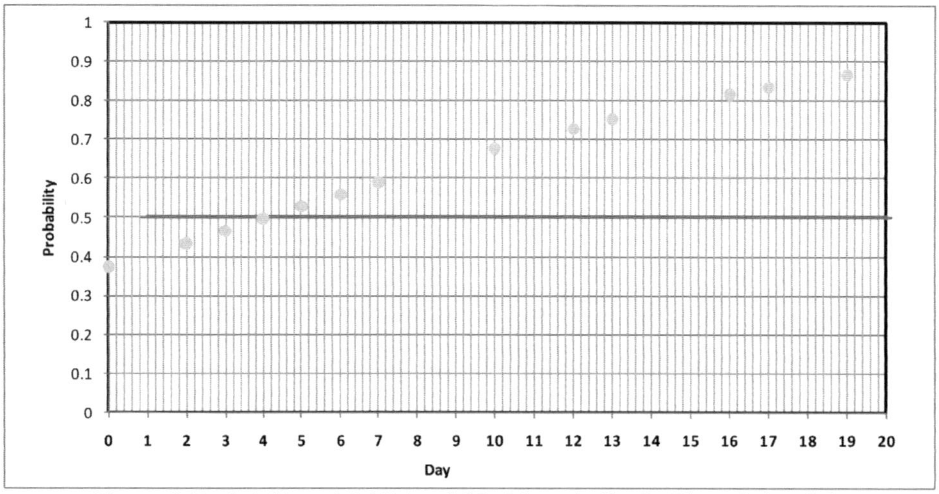

Figure 6. Probability od Addition 0,01 chlorophyll-a In Phase 3,4 and 5.

In Figure 6 describes the relationship between the odds increase in the number of chlorophyll-a with a long phase of MJO, to increase the amount of chlorophyll 0:01 on a threshold value of 0.5 occurred while longer phase are in the range of 4-7 days. This may imply that the addition amount of chlorophyll of 0.01 in Bengkulu region occurs when the incident MJO phase 3, phase 4 and phase 5 is bertutut succession occurred over four days by assuming that for 4 consecutive days average conditions SST in these waters warm and occurs rainfall 20-50 mm / day.

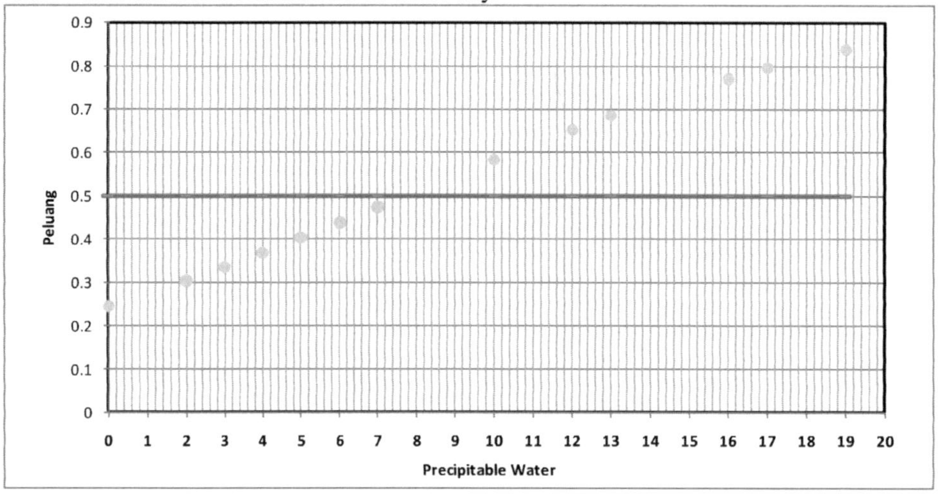

7. Probability od Addition 0,02 chlorophyll-a In Phase 3,4 and 5.

In Figure 7 describes the relationship between the odds increase in the number of chlorophyll-a with a long phase of MJO, to increase the amount of chlorophyll 0.02 on a threshold value of 0.5 occurred while longer phase is in the range 7-10 days. This may imply that the increase in the number of chlorophyll amounted to 0.02 in Bengkulu region occurs when the incident MJO phase 3, phase 4 and phase 5 is bertutut succession occurred over 7 days by assuming that for 7 consecutive days average conditions SST in these waters warm and occurs rainfall 20-50 mm / day.

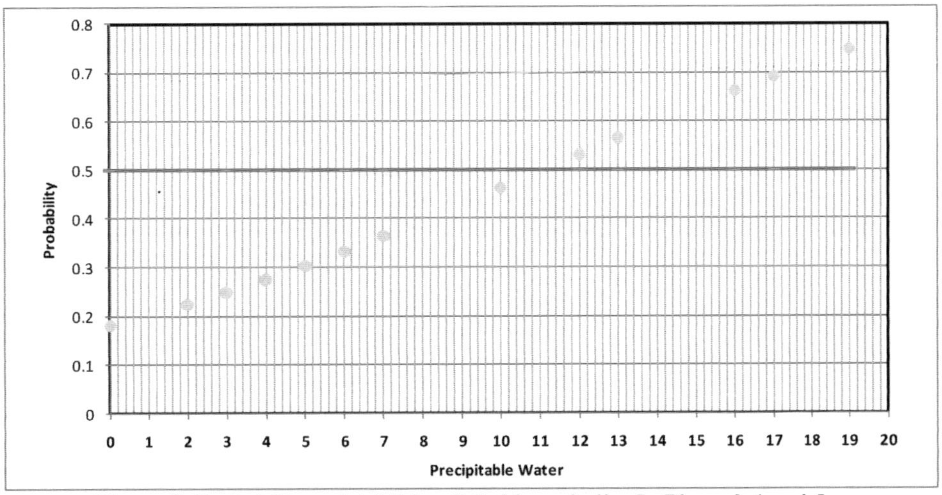

Figure 8. Probability od Addition 0,5 chlorophyll-a In Phase 3,4 and 5.

In Figure 8 describes the relationship between the odds increase in the number of chlorophyll-a with a long phase of MJO, to increase the amount of chlorophyll 0:05 on a threshold value of 0.5 occurred while longer phase is in the range 12-13 days. This may imply that the addition amount of chlorophyll of 0.05 in Bengkulu region occurs when the incident MJO phase 3, phase 4 and phase 5 is bertutut succession occurs over 12 days by assuming that for 12 consecutive days average conditions SST in these waters warm and occurs rainfall 20-50 mm / day.

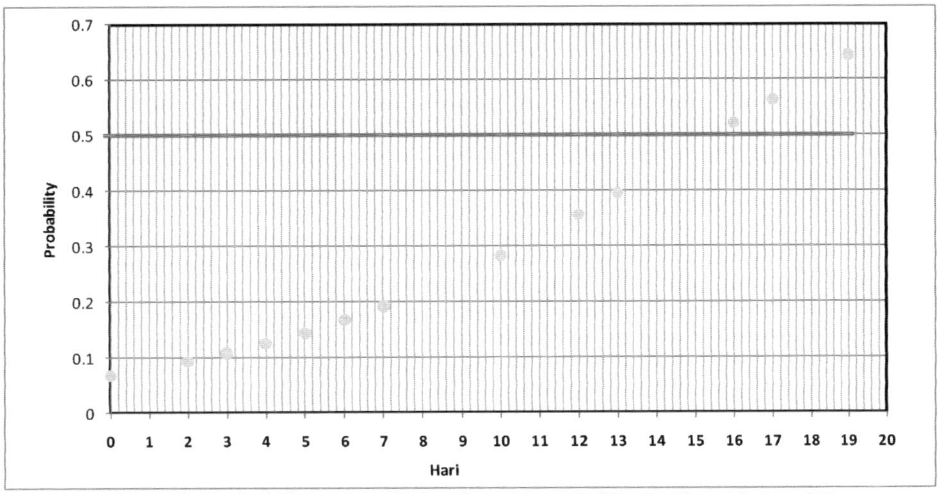

Figure 9. Probability od Addition 0,1 chlorophyll-a In Phase 3,4 and 5.

In Figure 9 describes the relationship between the odds increase in the number of chlorophyll-a with a long phase of MJO, to increase the amount of chlorophyll of 0.1 on a threshold value of 0.5 occurred while longer phase is in the range 16-18 days. This may imply that the increase in the number of chlorophyll by 0.1 in Bengkulu region occurs when the incident MJO phase 3, phase 4 and phase 5 is bertutut succession occurs over 16 days by assuming that for 7 consecutive days average conditions SST in these waters warm and occurs rainfall 20-50 mm / day. Opportunity increase in the number of chlorophyll highest reaching 70% at the time of long MJO events over 19 days.

4. Conclusion

From the above explanation can be said that there is a relationship between the incidence of MJO by the number of chlorophyll. Later in figure 4 to figure 7 has described odds the addition of chlorophyll-a have a linear relationship with the duration of the incident MJO in Bengkulu, odds increase the amount of chlorophyll in Bengkulu region reaches a threshold value of 0.5, it means that the length occurrence MJO able to identify increasing the number of chlorophyll in Bengkulu. Furthermore, the amount of chlorophyll-a at the time of the incident MJO can be used to detect areas that have the best fishing ground.

Acknowledgement

This research was supported by Indonesian Agency for Meteorology, Climatology, and Geophysics. The Authors would like to thank the institution NOAA, SST Pathfinder data, and Aqua MODIS data for providing open source data for analysis to be used in this paper. We also thank to all crews aboard Baruna Jaya I Indonesia Program Initiative On Maritime Observation and Analysis (Indonesia PRIMA/INAPRIMA). The INAPRIMA joint annual BMKG-NOAA cruise for RAMA Buoy Maintenance using an Indonesian Research Vessel providing technical training to BPPT.

References

[1] Krismono. 2010 Hubungan Antara Kualitas Air dengan Klorofil-A dan pengaruhnya terhadap populasi ikan di perairan Limboto. *LIMNOTEK (2010).* **17(2)**:170-180

[2] Madden R A and Julian P 1971 Detection of a 40±50 day oscillation in the zonal wind in the tropical Pacific *J Atmos Sci* **28** 702-708.

[3] Madden R A and Julian P 1994 Observations of the 40-50 day tropical oscillation Month Weather Rev **122** 814-837.

[4] Seto T H 2004 Pengaruh Osilasi Madden-Julian Terhadap Variabilitas Curah Hujan Tropis

[5] Laevastu and Hela I 1970 Fisheries oseanography *London: Fishing News Book Ltd.* 238p.

[6] Kleinbaum D G and Klein M 2002 Logistic Regression A Self Learning Text *Second Edition. Springer: New York.*

[7] Hosmer D W and Lemeshow S 2000 Applied Logistic Regression *2nd ed. New York: John Wiley and Sons.*

[8] Wilks D S 1995 Statistical Methods in the Atmospheric Sciences: An Introduction, Int. Geophys Ser. *Academic San Diego Calif. vol.* **59**, 467 pp.

[9] Matthews A J 2000 Propagation mechanisms for the Madden-Julian Oscillation. Quart J Roy Meteor Soc **126**, 2637-2652.

[10] Kubota H *et.al.* 2006 Observations Of The Structures Of Deep Convections And Their Environment During The Active Phase Of An Madden-Julian Ozcillation Event Over Equatorial Western Pacific *Jurnal Of The Meteorology Society Of Japan* **84 (1)** 115.

Proximate Content of "Klekap" (Microphytobenthos and Their Associated Meiofauna) from Milk-Fish Pond

Widianingsih[1], Muhammad Zainuri[1], Sutrisno Anggoro[1], Hermin Pancasakti Kusumaningrum[2]

[1] Faculty of Fisheries and Marine Sciences, Diponegoro University, Indonesia
[2]Faculty Sciences and Matematic, Diponegoro University, Indonesia
Jl. Prof. Soedharto, SH, Tembalang, Semarang. Telephone & fax : 0247474698;

widia2506@yahoo.com

Abstract. Microphytobentos and their associated mieofauna ('klekap") have important role in milkfish pond ecosystem expecially traditional fish pond. Microphytobenthos and their associated meiofauna not only consists of unicellular microalgae from class Bacillariophyceae, Cyanophyceae, Chlorophyceae but also a lot of meiofauna such as Nematoda, Annelida, Mollusc, Arthropoda, etc. Microphytobenthos and their associated meiofauna inhabit in the upper most surface sediment layer. They also have important role in the primary and secondary productivity. This research has purpose to investigate the composition of proximate in the microphytobenthos and their associated meiofauna. Microphytobenthos and their associated meiofauna were taken from three milkfish ponds in Pati area, Central of Jawa, Indonesia. Samples were taken directly by hand core from 3 stations with 3 replications. The research result showed that the value of protein content had range 6.9±0.33 - 7.73±0.37 %-dw, the value of carbohydrate content had range 10.44±1.2 - 12.59±0.15 %-dw, the value of lipid content had range 0.86±0.07 -0.96±0.07 %-dw. It concluded that the third station has higher value of protein, lipid and carbohydrate compared than the first and the second station.

Keywords: Microphytobenthos, Meiofauna, Proximate

1. Introduction

Microphytobenthos consists of unicellular eukaryotic algae and Cyanobacteria that grow within the upper several millimeters of illuminated sediment, typical appearing only as a subtle bownish or greenish shading [1] [2]. The microphytobenthos not only inhabit a thin, horizontal and time-varying region at the sediment-water interface, but they also grow in habitat ranging from wave-swept beaches to detritus-laden backwater lagoon, and includes habitat such as salt marshes, submerged aquatic vegetation beds, intertidal sand and mud flats and subtidals, illuminated sediment. They also are found at the surficial layer of the sediment that in the zone of intense microbial and geochemical activity and considerable physical reworking.

The ecology of the microphytobenthos is highly complex since light penetration into the sediment is no greater than 2 mm [3], only the diatoms in that sediment layer play a role in the fixation of inorganic carbon during photosynthesis. Furthermore, mechanism studies of microphytobenthos primary production and the coupling of primary and secondary production, are complicated by the large variability over small (mm) vertical scales and the patches over somewhat larger (cm to m) horizontal scales associated with variation in the texture and relief of the sediment surface [4].

In the benthic ecosytem, microphytobenthos has important play role as primary producer not only in the estuarine, shore water but also in the milkfish pond. In traditional of milkfish pond, microphytobenthos can grow faster. Microphytobenthos and their associated meiofauna (in Indonesian language is mentioned as "Klekap"). Milkfish can grow well and faster in pond where a lot of klekap (microphytobenthos). Milkfish likes to eat klekap (microphytobenthos and their associated meiofauna). Klekap become good natural feeding for milkfish. Klekap (microphytobenthos and their associated meiofauna) is consisted of some microalgae and benthic fauna such as Oscillatoria, Phormidium, Lyngbya, Spirulina, Navicula, Nitzschia, Amphora, Odontella, Cocconeis, Achnanthes, Pleurosigma, Protozoa (Zoothamnium, Vorticella, Epistylis, Acineta), bacteria, Rotifera, Anelida, etc. [5] [6]. As autrotrop organisms, the value of primary production by microphytobenthos in the some estuaries range from 27 – 234 g cm^{-2} / year [1] [2]. This primary and secondary production drive a rich food web, making estuaries and fish pond ecologically important areas and crucial importance for shore birds and fisheries. Besides serving as the primary food source for higher tropic level, microphytobenthos and their associated also play a very important role in sedimen dynamic. Some microphytobenthos such as`diatoms are important stabilizers of sediment through the excretion of extracellular polymeric substances [8]

As phytoplanktivorous fish such as milkfish also use resusped microphytobenthos and their associated meiofauna and thus transfer benthic production into the pelagic food web in the fish pond, estuaries, shore water and etc. In fact, a lot of klekap (microphytobenthos and their associated) in the fish pond can make milkfish grow faster and better water quality. Therefore, this research has purpose to investigate the proximate value content of microphytobenthos and their associated ("Klekap"). The information about proximate content in the klekap (microphytobenthos and their associted meiofauna) is important for necessary requirement to make artificial food for marine organisms especially sea cucumber .

2. Material and Methods

a. *Material*

Sample of microphytobenthos and their assosiated were collected by hand core (diameter 7.5 cm) from of milk-fish pond where was located at Pati, Semarang. Indonesia. Samples microphytobenthos and their associated were taken from one of milk-fish ponds in Pati, Central of Jawa, Indonesia. There were position 3 stations with 3 replications for each station.

b. *Methods*

The analyses of proximate (moisture, ash, fibers, crude fat, protein and carbohydrate) of all the samples were determined. Measuring the moisture and ash were determined using weight difference method. The nitrogen value which is the precursor for protein of substance was determined by micro Kjeldahl method[9]. Carbohydrate content was determined by Luff-Schoorf method[10].

3. Result and Discussion

According to the research result, the value of ash is higher than the other proximate compositions. The average value of ash has range 73.74±0.65 – 76.80±0.65 %-dw. The average value of fiber crude for three station is 0.7 %-dw. (Table 1.) The average value of protein content in microphytobenthos and their associated as showed in Fig.1., have range between 6.9±0.33 - 7.73±0.37 %-dw. Furthermore, the value of carbohydrate has range between 10.44±1.2 - 12.59±0.15 %-dw. According the statistical analyse, carbohydrate and protein composition for microphytobenthos and their associated meiofauna did not show any significant differences ($p > 0.05$) between station 1, 2 and 3.

Table 1. Proximate Analysis of Microphytobenthos and Their Associated from Milk-Fish Pond.

No.	Station	Protein (%-dw)	Lipid (%-dw)	Water Content (%-dw)	Ash (%-dw)	Fibre crude (%-dw)	Carbohydrate (%-dw)
1.	1-1	7.23	0.83	4.86	75.47	0.63	10.98
2.	1-2	6.9	0.94	7.34	75.05	0.7	9.07
3.	1-3	6.57	0.8	4.71	75.88	0.77	11.27
		6.9±0.33	**0.86±0.07**	**5.64±1.48**	**75.47±0.42**	**0.7±0.07**	**10.44±1.2**
4.	2-1	7.1	0.95	2.84	77.53	0.68	10.9
5.	2-2	7.27	0.86	3.27	76.55	0.75	11.3
6.	2-3	6.8	0.82	4.47	76.31	0.66	10.94
		7.06±0.24	**0.88±0.07**	**3.53±0.84**	**76.80±0.65**	**0.70±0.05**	**11.05±0.22**
7.	3-1	8.15	1.03	3.77	73.68	0.72	12.65
8.	3-2	7.6	0.94	3.71	74.41	0.63	12.71
9.	3-3	7.45	0.9	5.35	73.12	0.76	12.42
		7.73±±0.37	**0.96±0.07**	**4.28±±**	**73.74±0.65**	**0.70±0.07**	**12.59±0.15**

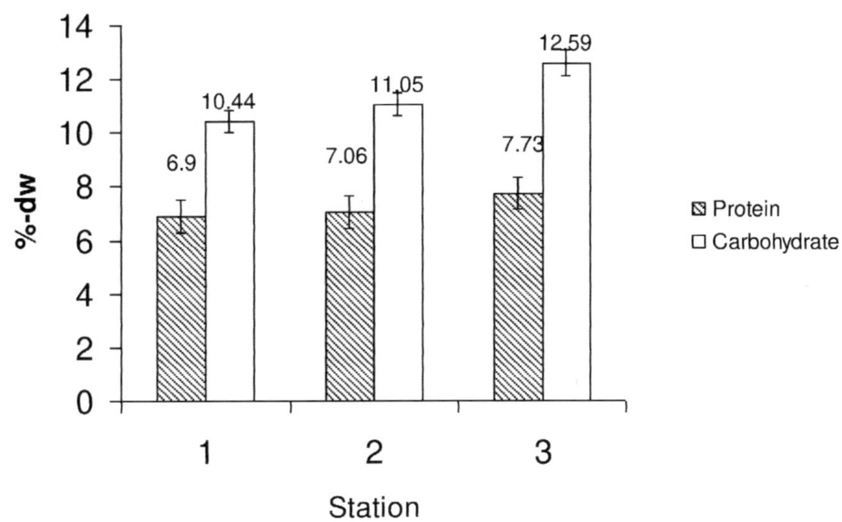

Figure 1. The value of protein and carbohydrate content in microphytobenthos and their associated (meiofauna)

The average value of lipid was exhibited in Fig 2. has range between 0.86 ± 0.07- 0.96 ± 0.07 %-dw. The smallest proximate composition was fibre crude and then followed by lipid. The highest proximate composition was reached by ash has value between 73.74±0.65 - 76.80±0.65 %-dw. According the research result, the value of phosphate for 3 stations had range 0.0125 - 0.0201 mg/L. The value of Nitrate for 3 stations had range 1.715 - 1.8235 mg/L result of measuring ammonia for 3 stations had range 0.1089 - 0.2951 mg/L (Table 2.) The value of salinity and temperature were high for 3 stations.

The value carbohydrate content has range between 10.44±1.2 - 12.59±0.15 %. This value fit with the result that the proportion of carbohydrate for varous microalgae have value 2 – 23%[12]. Comparing with the value of protein, value carbohydrate content is higher than value of protein (Table 1). High carbohydrate content in the microphytobenthos and their associated, because carbohydrate is one of a major part of the EPS (Extracellulose Polymeric Substrate) secreted by diatoms[13]. The EPS is consist of not only carbohydrate but also protein, lipid and nucleid acids[14].

Table 2. Parameters measured on 3 stations and 3 replicates at site of milk-fish pond, Pati, Central of Jawa, Indonesia.

No.	Station	Phosphate (mg/L)	Nitrate(mg/L)	Amoniak (mg/L)	Salinity (ppt)	Temperature (°C)	pH	DO (ppm)
1.	1-1	0.0227	1.8314	0.1275	63	36.8	8.5	8.1
2.	1-2	0.012	1.9686	0.0996	62	36.7	8.5	8.3
3.	1-3	0.0029	1.6706	0.0996	64	36.4	8.5	8.0
	Average	**0.0125**	**1.8235**	**0.1089**	**63**	**36.63**	**8.5**	**8.13**
4.	2-1	0.0046	1.8627	0.3063	65	38.1	8.5	6.4
5.	2-2	0.0415	1.7647	0.2696	66	38.2	8.5	6.5
6.	2-3	0.0141	1.8157	0.3093	64	38	8.5	6.6
	Average	**0.0201**	**1.8144**	**0.2951**	**65**	**38.10**	**8.5**	**6.5**
7.	3-1	0.002	1.6745	0.1246	62	39	8.5	5.3
8.	3-2	0.0083	1.698	0.1914	63	39	8.5	5.2
9.	3-3	0.0297	1.7725	0.1884	61	38.9	8.5	5.3
	Average	**0.0133**	**1.7150**	**0.1681**	**62**	**38.97**	**8.5**	**5.27**

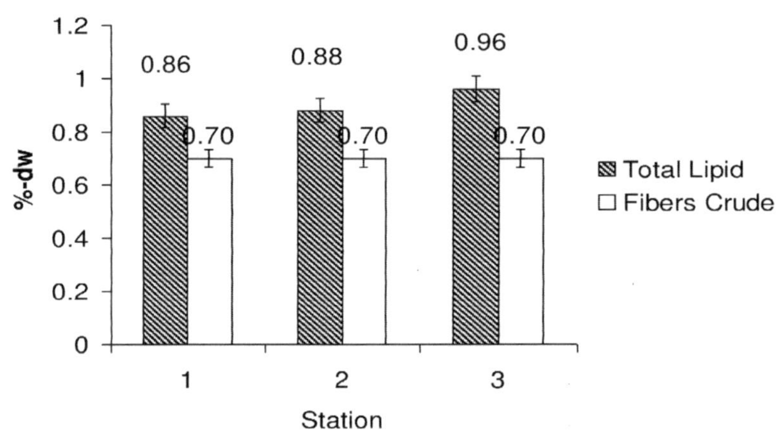

Figure 2. The value of total lipid and fibre crude in microphytobenthos and their associated (meiofauna).

The value of lipid content in the microphytobenthos and their associated was lower (between 0.86±0.07- 0.96±0.07 %-dw.). the lower of lipid content because the temperature (36.63 – 38.97 °C), in the water was enough high which did not provide the possibility of carbohydrate` to deposit or convert as lipid. High or low temperature in the water could reduce microalgae lipid production[12][13]. The highest cyanobacteria biomass when sediment temperature was low (5.3 °C)[15].

High value of ash composition in the microphytobenthos and their associated was (73.74±0.65 – 76.80±0.65 %-dw), because the ash is consisted organik matter and inorganic element. Microphytobenthos and their associated were living in the sediment and some of microphytobenthos attached in sand (episammic), living and moving in the sediment (epipelic) and epiphytic. The ecological condition was shown by the present of a lot of sand and mud in the sample of microphytobenthos. [6][14].

According the result, in the third station, the value of carbohydrate, lipid and protein content were highest than station 1 and 2. Whereas, the value of phosphate, nitrate and amoniak in the water column showed that the third station was lower than the first and the second station. It's approved that microphytobenthos used the nutrient phosphate and nitrate for growth microphytobenthos which had influent to increase the value of protein, lipid and carbohydrate content. Some previous studies found high Phosphate concentration in the Santa Cruz[15]. Phosphorus present rapid and dynamic process[15].

The high salinity value (62- 65 ppt) which found in the three station influent the abundance of benthic organism. This condition is not fullfill the requirement of microphytobenthos and their associated. Some microphytobenthos species can acclimatized according to their respective conditions[16].

4. Conclusion

In the conclussion, the average value of protein and carbohydrate in microphytobenthos and their associated have range 6.9±0.33 to 7.73±0.37 %-dw and 10.44±1.2 - 12.59±0.15 %-dw. The average value of lipid content in the microphytobenthos and their associated meiofauna was lower (between 0.86±0.07- 0.96±0.07 %-dw.). There are carbohydrate content and protein content in "klekap" (the microphytobenthos and their associated meiofauna) is good for making artificial food from "klekap" for marine organisme specially sea cucumber. According to research there are no differences lipid, carbohydrate and protein composition for microphytobenthos and their associated meiofauna did not show any significant differences ($p > 0.05$) between station 1, 2 and 3. Microphytobenthos has important role in the bentic ecosystem of milk-fish pond.

Acknowledgments

The Directorate General of Higher Education, Ministry of National Education and Culture is gratefully acknowledged for funding financial support to carry out this research work through fellowship program of BPP-DN.

References

[1] MacIntyre H L et al 1996 Microphytobenthos the ecological role of the (secret garden) of unvegetated shallow water marine habitat I Distribution abundance and primary production *Estuarine* **19** 186-201.

[2] Kromkamp JC et al 2006 Functioning of microphytobenthos in estuarine *Proceedings of the colloquium 21-23 August 2003* Amsterdam **273**

[3] Colijn F & de Jonge V N 1984 Primary production of microphytobenthos in EMS – Dollard Estuary. *Mar. Ecol. Prog. Ser.* **14** 185-196.

[4] Hardison AK et al 2013 Microphytobenthos and benthic macroalgae determine sediment organic matter composition in shallow photic sediments. *Biogeosciences* **10** 5571-5588.

[5] Gerasimyuk VP and Q A Kovtun 2014 Spesies Diversity of microphytobenthos of the Tud Sia (Egypt). **16(1)** 57-67

[6] Meleder VY et al 2007. Spatio temporal changes in microphytobenthos assembkags in a macrotidal flat (Bourgneuf. Baj. France) *Journal of phycology* **43 (6)** 1177-1190.

[7] Underwood GJC and J Kromkamp 1999 Primary production to phytoplankton and microphytobenthos in estuaries *Adv Ecol Res* **29** 93-153.

[8] Brouwer JFC et al 2006 On the function of secretion of extracellular polymeric substances by benthic diatoms and their role in intertidal mudflats: *A review of recent insight and views. Royal Netherlands of Arts and Science.*

[9] Pearson D 1979 The Chemical Analysis of Foods 7th ed. *Churchill living stone* London

[10] Brown MR et al 1997 Nutritional properties of microalgae for marine culture. *Aquaculture* **165** 281-293.

[11] Lawrence BC 2006 Upscaling primay production estimaty: Regional and Global Scale estimaty of microphy benthos production in functioning of microphytobenthos in estusries. *Royal Netherland Academy of Arts and Science*. 99-108

[12] Opute 1974 Reported that, extreme high or low temperatur could reduce microalgae lipid production
[13] Cho SH et al 2007 Optimum temperature and salinity condition for growth of green algae *Chlorella ellipsoidea* and *Nannochloropsis oculata Fish. Sci.* **73** 1050-1056.
[14] Hillebrand H & U Sommer 1997 Response of epilithic microphytobenthos of the Westren Baltic Sea to in situ experiments with nutrient enrichment *Marine Ecology Progress Series* **160** 35-46
[15] Pan JCN 2013 Intraction between Estuarine Microphtobenthos and physical Forcing. The role of Arthosperic and Sedimentary Factor. *International Jour of Geosciences.* **4** 352-361
[16] Banerjee S et al 2011 Growth and proximate composition of tropical marine *Chaeotoceros calcitrans* and *Nannocloropsis oculata* cultured outdoors and under laboratory conditions. *African Journal of Biotechnology* **10(8)** 1375-1383.

Effects of Different Heat Processing on Fucoxanthin, Antioxidant Activity and Colour of Indonesian Brown Seaweeds

Eko Susanto, A Suhaeli Fahmi, Tri Winarni Agustini, Septian Rosyadi, Ayunda Dita Wardani

Laboratory of Fish Post Harvest, Department of Fish Product Technology, Faculty of Fisheries and Marine Science, Diponegoro University
Jl. Prof. Soedarto, SH Tembalang Semarang, INDONESIA 50275

Email: ekothp@live.undip.ac.id

Abstract. Fucoxanthin (Fx) is major carotenoids in brown algae. It showed many health beneficial effects for oxidative stress. Fucoxanthin is lower stability which may cause problem in the application for functional food. The objective of this study was to evaluate the effects of various heat processing on Fx, antioxidant activity (IC_{50}), total phenolic content, and colour stability of *Sargassum ilicifolium*. The various heat processing methods showed were not significantly affected to fucoxanthin and antioxidant activities however all treatments lower affected to brown seaweeds colour. Moreover, this study showed a useful proved in the design of brown seaweeds processing which minimize Fx, antioxidant activity and colour changes.

Keywords: brown seaweed, heat processing, antioxidant activity, *Sargassum ilicifolium*, fucoxanthin

1. Introduction

Marine resources offer important bioactive molecules that have advantages on the human body. They can be applied in many fields such as the drug, cosmetic, and food industries [1]. Seaweeds have been utilized since ancient times as foods in the far East and Asia and are the raw material for industrial production of agar, carrageenan and alginates [2]. South Asian countries were the first to introduce seaweeds for their utilization for medicinal and food purposes. Brown seaweeds represent a suitable supplement and additive for food due to their high nutritional value and the health benefits they can provide [3]. Brown seaweeds contain many bioactive compunds such as polysaccarida, phlorotannin, terpenoid, and fucoxanthin [4, 5].

Brown seaweeds contain Fucoxanthin (Fx) as a major carotenoid in brown seaweeds that is responsible for its colour [4, 6]. This carotenoid has been reported showing many health benefit effects as anti-obesity, antidiabetic [7, 8], Antioxidant [9], however Fucoxantin is unstable carotenoids due to rich electron in its structure [10]. Some factors have been responsible for its instability such as pH, light exposure, oxidator and redactor, and temperature [11, 12]). Several study reported fucoxanthin are abundantly presents in sub tropic brown algae such as *Himanthalia elongata* [13], *Sargassum horneri* [14, 15], *Undaria pinnatifida* [16], *Cystoseira hakodatensis* [15], Ecklonia kurome [14]. This carotenoids also presents in tropical brown seaweeds such as. *S. crassifolium, Padina australis,*

Turbinaria ornate [14], *Sargassum cinereum, Sargassum filipendula and Sargassum echinocarpum* [17].

Processing of foods and food ingredients often exerts a major effect on their constituents, including nutraceutical products [18]. Some study reported, heat treatment in seaweeds changed their nutritional compounds and phytochemical contents [2, 19, 20]. It is known that heat treatment may also affect to fucoxanthin content, antioxidant activities and colour of brown seaweeds. For the design of an optimized processing to maximized preservation of Fx and antioxidant activity, the optimum heat treatment need to be evaluate. Therefore, the purpose of this study was to evaluate the effects of heat processing on Fx content and antioxidant activities in *S. ilicifolium* which is abundantly growth in Indonesia. The way to preserve and process of seaweeds are blanching, boiling, steaming and sterilizing. To our knowledge, there is no report on stability of Fx, phytochemical compound and their antioxidant activity after heat treatment, in Indonesian brown seaweeds.

2. Materials and Methods

2.1. Materials and sample preparation

DPPH, Folin Ciocalteu, Na_2CO_3, and gallic acids were purchased from Sigma Aldrich. Fucoxanthin standard, isolated *from Padina australis*, was purchased from MRCPP Universitas Ma Chung, Malang, Indonesia with Fx concentration >98,4%.

Fresh Sargassum ilicifolium was taken from Bandengan Beach, Jepara, Central Java, Indonesia. Immediately after cultivation, Fresh *S. ilicifolium* was transported to laboratory in low temperature condition. It was washed with fresh water to eliminate any presence of salt, and remove tiny particles also ephyphites. Afterward, *S. ilicifolium* was stored in refrigerator (-27°C) until further analysis.

2.2. Sargassum ilicifolium heat treatment processing

2.2.1. Blanching
About 150 g of *S. ilicifolium* was thawed and blanched by immersion in 40°C warm water using a water bath (Wisebath, Wenk LabTec, Germany) for 15 minutes.

2.2.2. Sterilization
About 150 g of *S. ilicifolium* were included in the Jar (250 ml) containing saturated sugar solution. About 15 minutes, seaweed was sterilized in 121°C, 1 atm air pressure.

2.2.3. Boiling
The seaweed was thawed prior to boil. Alga sample was immersed in 96°C boiled water for 15 minutes using a water bath (Wisebath, Wenk LabTec, Germany).

2.2.4. Steaming
About 150 g seaweed sample was placed on tray in a steamer and steamed over boiling water for 15 minutes.

2.3. Sample extraction
After 15 minutes of heat treatment, *S. ilicifolium* were dried with tissue and crudely cut into small pieces. Seaweed extracts were prepared with methanol solution. About 50 gr of treated seaweed

were immersed in methanol solution (1:5 w/v) for overnight. After 1 night extraction, seaweeds extract were filtered and removed organic solvent under vacuum using rotary evaporator. The extract were stored with methanol in concentration (3 mg/ml) and stored at freezer (-27°C) until further analysis.

2.4. Fucoxanthin analysis

Fx concentration was determined using a HPLC Shimadzu LC-20AT, column oven CTO-20AC system, Pump FCV-11AL, Degasser: DGU-20A, UV/Vis Detector SPD-20A,and communications bus module CBM-20A. About 20 µL methanollic seaweeds extract were used in the analysis. Before analysis, all seaweeds extract were filtrated using bulk Acrodisc PSF Syringe Filter 0,45 pM GHP membrane, 25 mm. Fucoxanthin analysis were done according to Susanto et.al [14] with slightly modification. A 5 points calibration curve was constructed with HPLC Shimadzu LC 20AT dissolved in methanol and acetonitrile (7:3 v/v) and accrued out at 28°C with octadecylsilica (ODS) column (TSK-gel ODS 80-Ts, 250 × 4,6 mm id, 5 mm particle size; Tosoh, Japan) protected with a guard column (15×3.2 mm) with the same stationary phase. The calibration formula was y=2.7977e-008X - 0.000269384 with R^2= 0.9999744.

2.5. Total Phenolic Content Analysis

Total Phenolic Content (TPC) were analyzed according to Kuda et al. [21] with slightly modification. Each 100 µL of sample solution was added to 750 µL of 10% Folin-Ciocalteu solution and incubated at 37°C. After 5 mins, 750 µL of 6% Na_2CO_3 was added and the mixture was allowed to stand at 37°C for 30 mins in the dark. The absorbance was measured at 761 nm. Total phenolics for each seaweed lipid was then calculated on the basis of a phloroglucinol standard curve and expressed as mg of galic acids equivalent (PGE)/g extract with formula Y= 0.0088x + 0.083 R^2= 0.9948.

2.6. DPPH analysis

In present study, DPPH was used to analyze antioxidant activity in treated seaweeds. The methods was carried out according to Blois [22] with slightly modification. About 0.1 ml seaweeds methanollic extracts were reacted with 3.9 ml DPPH (with variable concentration 1, 2, 3, 4, 5 ppm in ethanol), afterward homogenized using vortex for 2 minutes. After homogenized, all solutions were incubated for 30 minutes at dark room temperature. The scavenging activity of DPPH radicals was calculated as percentages by the following equation:

$$\text{DPPH radical scavenging activity (\%)} = \frac{(1 - \text{A samples})}{\text{A control}} \text{ x } 100\%$$

Where $A_{samples}$ and A_{blank} are the sample and blank absorbance in 517 nm, respectively. The IC_{50} was calculated by linier regression.

2.7. Colour intensity (L, a, b) analysis

Quantitative evaluation of color changes in methanolic extract was done using portable colorimeter (Apple, GM52223LSGJ). The extract samples were transferred into glass cell and measured with portable colorimeter (Apple, GM52223LSGJ).

The total amount of colour change to see effects of heat treatment was calculated as per equation [23].

$$\Delta E = \sqrt{(Lo - L)^2 + (ao - a)^2 + (bo - b)^2}$$

Where Lo, a_0 and b_0 are the colour parameters or fresh seaweeds. This total change takes in to account the change in each colour describes by the L^*, a^* and b^* parameters. a* positive value reflected red colour to negative value reflected green) and b* (positive value reflected yellow colour to negative value reflected blue).

2.8. FTIR analysis

Fucoxanthin structure in each treatment were analyzed with IRPRSTIGE-21 Shimadzu The analysis was based on the manual instruction of FTIR IRPRSTIGE-21.

2.9. Data analysis

The data obtained for all parameters excluded FTIR analysis were statistically treated to obtain 95 % confidence interval. The results obtained were analyzed using one way ANOVA among treatment using SPSS 16.

3. Results and discussion

3.1. Fucoxanthin Content

Fucoxanthin is xanthophyll which is abundant in brown seaweed and shows many health benefit effects. Fx were unstable due to some factors such as temperature, light, acid and pH [12, 13]). Brown seaweeds were processed in different ways before consumption. Thermal processing will affect to Fx availability in brown seaweeds. As consequence, Fx content in brown seaweeds need to evaluate after processing. Fx in *S. ilicifolium* was investigated after four different thermal processing (Fig. 1). Fx content in brown seaweeds before processing was 39.53 ppm, after processing resulted different content of Fx. The different thermal processings were not significantly different (p >0.05) to Fx content. The highest Fx content was found in blanching treatment, however sterilizing treatment showed the lowest content of Fx. These showed that Fx content in *S. ilicifolium* affected by thermal processing.

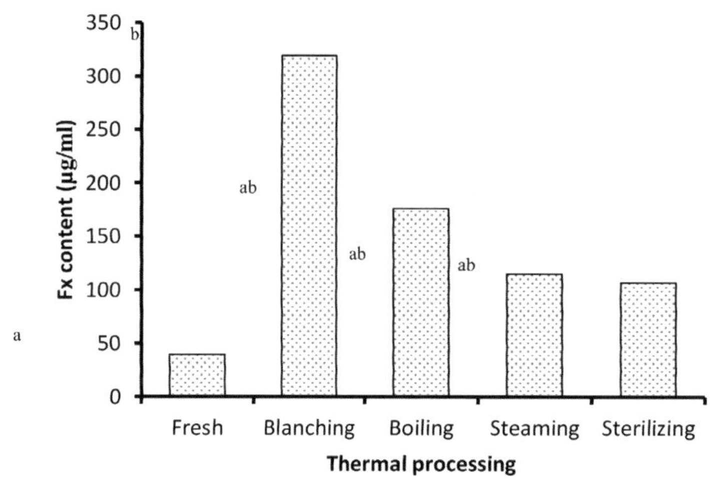

Figure 2. Fx content in fresh and thermal processing in *S. ilicifolium*. Values are express in µg/ml.

Fx content of *S. ilicifolium* were higher in thermal processing than fresh samples. The heat treatment processing might be increasing free Fx from protein during thermal processing. In oxygen-evolving photosynthetic organism, Fx is in form of Fx-chlorophyll protein (FCP) which is responsible for absorption and conversion of light energy [24]. In case of blanching treatment showed the highest Fx content among other treatment and was negatively affected by the heating,

IOP Conf. Series: Earth and Environmental Science **55** (2017) 012063 doi:10.1088/1755-1315/55/1/012063

this might be due to blanching treatment was able to inactive enzyme that degraded Fx to other compounds during preparation and extraction. This reason was supported with blanching treatment chromatogram (Fig 2.). Carrilho et al. [19], have reported that dried seaweeds followed by blanching significantly increase Fx content in *Himanthalia elongata*. In addition to the result obtained Carrilho et al. [25], which observed an increasing of Fx content *Undaria pinnatifida* and *Laminaria japonica* which were boiled for 15 minutes and 1 hour respectively. Pre-treatment of *L. japonica* (konbu) including heating, washing, cutting and freezing was able to increasing amount of Fx [26].

Fucoxanthin which is chemically structure in all-trans isomer is unstable during heat treatment. This compound might be degraded into 9'-,13- and 13'- cis Fucoxanthin which were cis isomers of Fx [27]. This isomerization is recognized Fx degradation in heat treatment. During separation of Fx from *Hijika fusimorme*, there were two other peaks along with all-trans-fucoxanthin peak. The ratio of all-trans-fucoxanthin and its two cis-isomers were 100:2:3 [28].

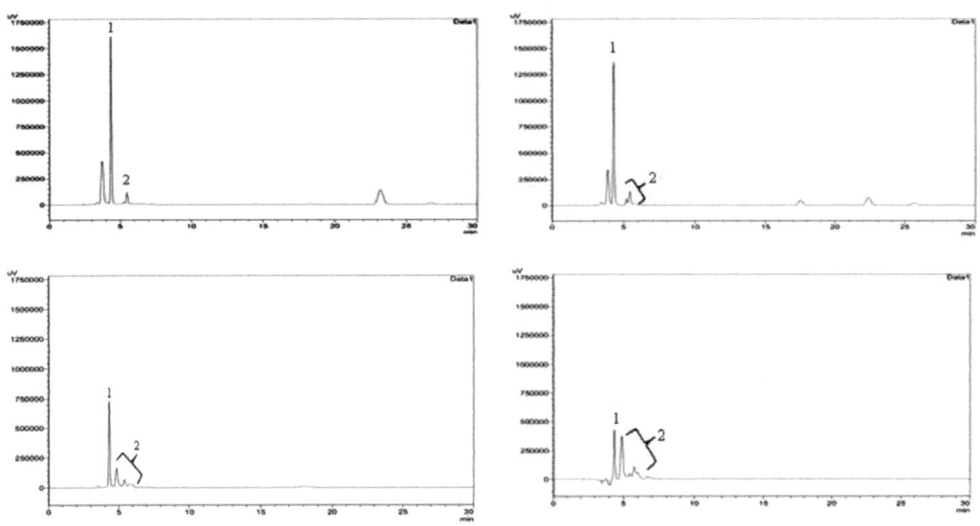

(c) (d)

Figure 2. HPLC chromatogram of blanching (a), cooking (b), steaming (c), sterilizing (d) of *S. ilicifolium*. (1) all-trans fucoxanthin (2) cis fucoxanthin.

3.2. Total Phenolic Content

The total of phenolic content (TPC) was determined using the Folin Ciocalteu Method based on Kuda et al. [21]. The total phenol content of examined *S. ilicifolium* ranged from 33.36 to 47.60 (μg GAE/mg extract) (Table 1.). Blanching treatment showed the highest content of TPC, however the lowest content found in steaming treatment.

TPC were not significantly different ($p>0.05$) in all heat treatment. This result indicated heat treatments were not significantly affected to TPC. In contrast, Dang et al., 2014 [29] who studied the effect of different drying method in *Hormosira banksii* and found that different drying method significantly affected to TPC in sample.

Table 1. TPC (μg GAE/mg extract) content in fresh and thermal processing in *S. ilicifolium*

Treatments	TPC (μg GAE/mg extract)
Fresh	47.60±15.05[a]
Blanching	42.90± 2.63[a]
Boiling	37.00± 8.46[a]
Steaming	34.08± 1.12[a]
Sterilizing	33.36± 0.11[a]

The sequence of TPC content in *S. ilicifolium* tested were blanching > boiling > steaming > sterilizing. The different total phenolic content (TPC) decreasing rate might be due to the different temperature on the treatments, TPC decreased when the heat temperature increased. Carrilho *et al.* [25], reported total phenolic content on wakame and konbu were decreasing after cooking treatment. However, TPC in *H. elongata*, *L. Saccharina*, and *L.digitata* increasing until 95°C, heating and decreased continuously until 121°C heating [29].

Brown seaweeds contain phloroglucinol phenolics, phlorotannin [31, 32, 33]. These compounds exhibit many beneficial biological activities such as antioxidant, anticancer, anti-diabetic, anti-human immunodeficiency virus, antihypertensive, matrix metalloproteinase enzyme inhibition, radioprotective, and antiallergic activities [34]. The TPC results of this study are promising on the application of food processing based algae as source of phenolic content.

3.3. DPPH radical scavenging activity of methanolic extract

DPPH reagent has been used extensively for investigating the free radical scavenging activities of compounds [35]. DPPH is primarily evaluates proton radical-scavenging ability [36]. One of parameter that has been introduced recently for the interpretation of the results from the DPPH method is efficient concentration EC_{50} or IC_{50}. EC_{50} or IC_{50} is defined as the concentration of substrate that causes 50% loss of the DPPH activity (colour) [37].

All treatments were measured for DPPH radical scavenging activity. As shown in Fig 2. All heat treatments show radical scavenging activity in *S. ilicifolium* methanolic extracts. DPPH radical scavenging activity shows various degree of antioxidant activity IC_{50} (Fig. 2.). DPPH (IC_{50}) of the methanolic extract from fresh *S. ilicifolium* was 59.14 μg/ml. Of the tested in heat treatments *S. ilicifolium* showed that, steaming treatment showed the highest antioxidant activity. In contrast, boiling and sterilizing showed the lowest antioxidant activity among treatments tested. These results indicated different heat treatment influencing their extractability. The phenolic compounds usually present in bound states of conjugates with sugar, fatty acids or protein. Thermal processing might be disrupt of cell wall and cell membrane which may produce different extractability resulting different availability of antioxidant compounds [30, 38]). In addition, heat treatment could deactivate endogenous oxidative enzymes that are responsible for the destruction of antioxidants. Therefore, its absence could increase antioxidant [30].

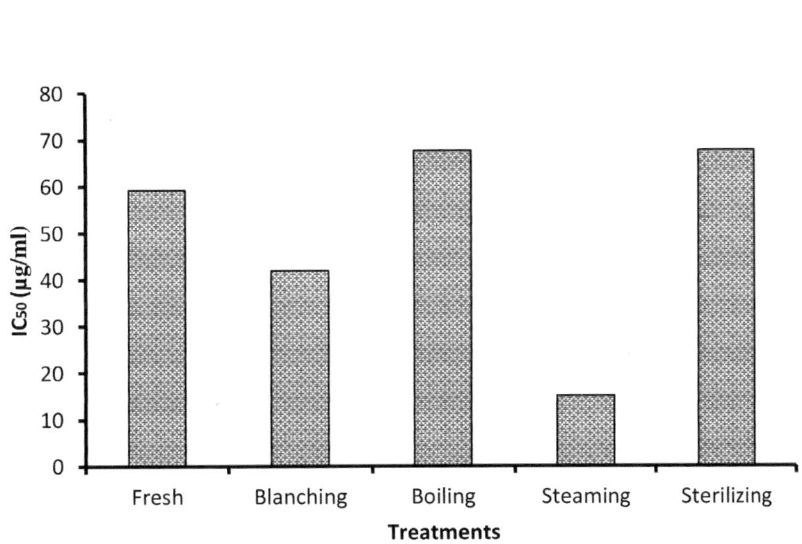

Figure 3. Antioxidant activity (IC_{50} µg/ml) in fresh and heat treatment processing of *S. ilicifolium*

The different antioxidant activity related with presence of Fx than phenolic compounds. This analysis was supported by Pearson correlation test. Fucoxanthin was found positively correlated with 0.282 besides that, TPC less correlated with 0.013. Different results found by Wang et al. [39], the extracts containing high levels of TPC were also potent DPPH radical scavenger, suggesting that algal polyphenols may be the principal constituents responsible for antioxidants. Fx was revealed as high antioxidant activity [9], however this compounds is unstable due to increasing temperature treatment (Zhao et al., 2014) [12]. The ability of phenolic compounds as antioxidant in consequence of these compounds can donate hydrogen ions or electrons and their stable radical intermediates, which prevents the oxidation of various food ingredients [40]. In addition, as antioxidant, carotenoids are able to quench singlet oxygen without changing its structure and scavenge of free radical [41]. Another antioxidant compound in brown algae is alpha tocopherol [42]. This compound is highly stable to heat [25]

3.4. Colour change

The color parameter and the change in color of methanol extract of all seaweeds treated were evaluated and calculated. The change in Hunter colour of the heating treated *S. ilicifolium* methanolic extracts are shown in Table 4. Parameter L*,a,b showed significantly different between freshand blanching treatment. Differences of L,a*,b value affected to ΔE*, the highest and the lowest ΔE* value showed in blanching and steaming treatment, respectively. This value visually meant dark and brighter color, respectively. In brown seaweeds, fucoxanthin, β carotene and chlorophyll were responsible for its colour [43].

Result showed that treatment showed low effect in the colour change among treatments, as the temperature increase the ΔE value. These results were different from Rajauria et al. [29], heat treatment strongly effect on change in color as the temperature increase. The differences on the heat treatment significantly affected ($p<0.05$) to L,a, b value on each extract.

The colour change in steaming treatment showed the lowest among other treatment. The smaller ΔE value, the closer the samples are in colour close to fresh seaweeds. In contrast, the blanching treatment revealed highest ΔE, dark green colour on this treatment due to much content of Fx.

IOP Conf. Series: Earth and Environmental Science **55** (2017) 012063 doi:10.1088/1755-1315/55/1/012063

Tabel 4. Color extract in fresh and heat treatment processing of *S. ilicifolium*

Treatment	L	a	B	ΔE
Fresh	27.700^{a}	2.208a	57.533^{a}	-
Blanching	-8.857^{b}	-31.498^{b}	24.288^{b}	77.643^{a}
Boiling	33.539^{cb}	-54.871^{bc}	64.585^{ab}	86.508^{a}
Steaming	17.84^{bd}	-42.183^{bd}	50.070^{ab}	70.906^{a}
Sterilizing	23.620^{ab}	-39.687^{be}	37.304^{ab}	76.618^{a}

ΔE in heat treated seaweeds was ranged from 70.906 to 86.508, which is indicated a huge different with fresh seaweeds. Blanching colour revealed great different (GD) to boiling and sterilizing treatment respectively; this is related with Fx content in treated seaweeds. According to Silva and Silva [24], stated that value ΔE > 12 for very great difference (VGD). Colour can be used to design adequate thermal processing conditions for maximizing final product quality. In this study, the increased ΔE value was reflected in high antioxidant activity.

3.5. FTIR

FTIR analysis was used for the characterization of the methanolic extract of fresh and treated seaweeds. The FTIR spectrum of Fx tested in seaweeds is shown in Fig 4. and it was found that different treatment have different absorption band with fresh seaweeds. From FTIR spectra we confirmed that the functional group that describe Fx is allenic bond (C=C=C) were very low in all samples. This might be due to the higher content of moisture which interfere with the infrared absorbance [44].

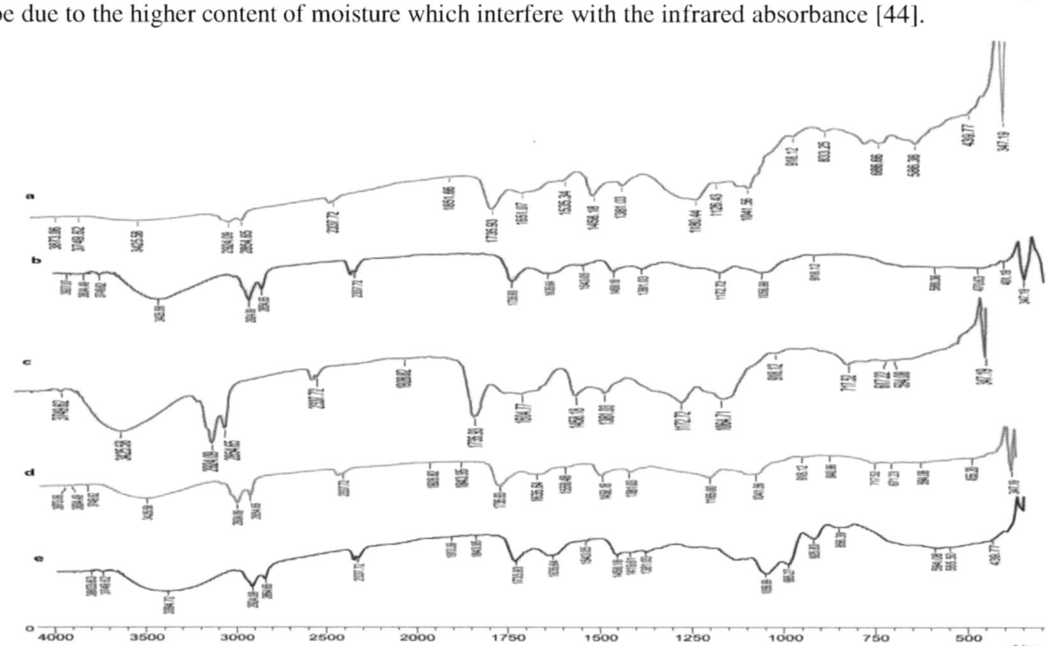

Figure 4. FTIR spectra of fresh seaweeds (a), blanching seaweeds (b), cooking seaweeds (c), steaming seaweeds (d), sterilizing seaweeds (e).

These results showed that the presence of allenic bound,which is presence in fucoxanthin is very low. This phenomenon might be due to high amount of methanol(O-H) in samples which is interfere infrared absorption [44]. The presence of alcohol (methanol) is attributed in peak 3425.28 to 3394.73. Carboxylic acids (C=O) was occurred in absorption 2924.09. Esters with -C=O bonds showed presence in absorption 1735.93. In addition, absorption of 1635.64 showed presence of alkenes (C=C) bonds. Therefore, the presence of allenic functional group in samples contain fucoxanthin as major carotenoids.

4. Conclusion

In this study, blanching treatment was efficiently inhibited degradation of fucoxanthin phenolic content compared to other treatments. The antioxidant activity showed higher content in boiling treatment. Steaming process was able to minimize colour change. This report provides useful data in the design of brown seaweeds processing which minimize Fx, antioxidant activity and colour changes.

5. Acknowledgement

This research was funded by Development and Application Research (1051-46/UN7.5.1/PG/2016, 15 April 2016), Research and Public Service Institute, Diponegoro University 2016.

6. References

[1] Hamed, I., Özogul, F., Özogul, Y., and Joe M. Regenstein. 2015. *Comprehensive Reviews in Food Scie. and Food Safety*. Vol.14, 446-465.

[2] Cox, S., Shilpi Gupta, Nissreen Abu-Ghannam. 2012. *LWT - Food Scie. Technol.* 47. 300-307

[3] Moreira, R., Francisco Chenlo, Jorge Sineiro, Santiago Arufe, and Santiago Sexto. 2016. *J Appl Phycol.* 28:2485–2494

[4] Gupta, S. and Abu-Ghanam, N. 2011. *Trends in Food Scie. Technol.* Vol.22(6):315-326.

[5] Miyashita,k., Mikami,N., and Hosokawa,M. 2013. *J. Funct. Foods.* Vol.5(4): 1507-1517.

[6] Terasaki, M., Narayan,B., kamogawa,H., Nomura,M., Stephen,V.M., Kawagoe,C., Hosokawa,M.,and Miyashita K. 2012. *J. Aquatic Food Product Technol.* Vol. 21.pp. 468-479.

[7] Maeda, H., Masashi Hosokawa, Tokutake Sashima, Katsura Funayama, and Kazuo Miyashita. (2005) *Biochemical and Biophysical Res. Com.* . 332 (2). p. 392–397.

[8] Maeda, H., Masashi Hosokawa, Tokutake Sashima, Katsura Murakami Funayama, and Kazuo Miyashita. (2009) *Mol. Med. Report.* (2). p. 897–902.

[9] Sachindra N. M. Sato,M., Maeda H., Niwano, Y., Kohno,M., Miyashita K. (2007) *J Food Sci Technol.* 55(21).p.8516–8522.

[10] Sugimura, R., Masatake Suda, Ayumi Sho, Tomoya Takahashi, Tokutake Sashima, Masayuki Abe, Masashi Hosokawa1 and Kazuo Miyashita. (2012). *Food Sci. Technol. Res.*18 (5). p.687 – 693.

[11] Hii, S-L, Pooi-Yi Choong, Kwan-Kit Woo, and Ching-Lee Wong. (2010). *Australian Journal of Basic and Applied Sciences.* 4 (10). p.4580-4584.

[12] Zhao, D., Sang-Min Kim, Cheol-Ho Pan, Donghwa Chung . 2014. *Food Chem.*, Vol. 145, 15, Pages 505-513.

[13] Rajauria,G. and Nissreen Abu-Ghannam. 2013. *International J. Analytical Chem.* Vol. 2013, Article ID 802573, pp. 1-6.

[14] Susanto, E., A Suhaeli Fahmi, M. Abe, M. Hosokawa and K. Miyashita. 2016. *Aquatic Procedia.* 7. 66 – 75.

[15] Terasaki, M., Atsushi Hirose, Bhaskar Narayan, Yuta Baba, Chikara Kawagoe, Hajime Yasui, Naotsune Saga, Masashi Hosokawa, Kazuo Miyashita. 2009. *J.Phycology.* Vol. 45 (4). 974–980.

[16] Prabhasankar, P., P. Ganesan, N. Bhaskar, A. Hirose, Nimishmol Stephen, Lalitha R. Gowda,M. Hosokawa, and K. Miyashita. 2009. *Food Chem.* 115 (2). p. 501–508.

[17] Zailanie, K. and Purnomo, H.. 2011. *J. Agri. Sci and Technol. A.* Vol. 1 (8). 1103-1105

[18] Shahidi, F.2009. *Trends in Food Scie. & Technol.* 20. 376-387.

[19] Carrilho, K.A, Maria Asunci6n Lage-Yusty & Julia L6pez-Hernández.2014. *CyTA - Journal of Food*, 12 (4). 336-339,

[20] Chan, J. C.-C. Peter C.-K. Cheung, and Put O. Ang, Jr. 1997. *Ag. J. Agric. Food Chem.*, 45, 3056−3059.

[21] Kuda, T.; Tsunekawa, M.; Goto, H.; Araki, Y. (2005). *J. Food Compos. Anal.* 18. 625–633.

[22] Blois,M.S. 1958. *Nature.* 181. 1199-1200.

[23] Silva, F.M. and Silva, C.L.M.. 1999. *Int. J. Food Scie and Technol*, 34, 87–94.

[24] Green, B.R. and Durnford, D.G. 1996. *An. Rev. Plant Bio.* Vol. 47 No. 1.pp.685-714.

[25] Amorim, K., Marıa-Asuncion Lage-Yusty and Julia Lopez-Hernandez.2012. *CyTA – Journal of Food.*Vol. 10 (4), 321–324.

[26] Kanazawa, K., Ozaki,Y., Hashimoto, T., Das, S.K., Matsushita, S., Hirang, M., Okada, T., Komoto, A., Mori, N., Nakatsuka,M. 2008. *Food Sci. Technol. Res.*, 14 (6), 573-582.

[27] Haugan, J.A. and Liasen-jansen, s. 1994. *Tetrahedron Letters.* Vol. 35 (14):2245-2248.

[28] Yan, X., Yoshihiro C., Masahiro S., and Tadahiro N.. (1999) *Bioscience, Biotechnology, and Biochemistry.* 63 (3). 605-607.

[29] Dang,T.T., Vuong, Q.V., Schreider, M.J., Bowyer, M.C., Altena, I.A.V., Scarlett, C.J. 2016. *J. Food Proc. Preserv.*

[30] Rajauria, G., Amit K. Jaiswal, Nissreen Abu-Ghannam and Shilpi Gupta. 2010. *Int. J.Food Scie and Technol.* 45, 2485–2493.

[31] de Quiro´s A. R –B. S. Frecha-Ferreiro, A. M. Vidal-Pe´rez , J. Lo´pez-Herna´ndez. 2010. *Eur Food Res Technol.* 231:495–498.

[32] Sanjeewa, K.K.A., Kim,E.A., Son, K.T., Jeon, K.Y.J. 2016. *J. Photochemistry and Photobiology B: Biology.* Vol.162:100-105.

[33] Ragan, M.A.1976. *Botannica Marina.* Vol. XIX. 145-154.

[34] Wijesekara, I., Yoon, N.Y., Kim, S.K. 2010. *Biofactor* Vol. 36 96).pp.408-414.

[35] Cox, S., Abu-Ghannam, N. and Gupta, S. 2010. *Int. Food Research Journal 17: 205-220.*

[36] Singh, N. and P.S. Rajini. 2004. *Food Chem.* Vol. 85(4). 611–616.

[37] Molyneux, P. 2004. J. Sci. Technol., 26(2) : 211-219.

[38] Gliszczyńska-Świglo, A., Ciska, E., Pawlak-Lemańska, K., Chmielewski, J., Borkowski, T., Tyrakowska, B. 2006. *Food Additves and Contaminant.* 23 (11), pp. 1088-1098.

[39] Wang, T., Rósa Jónsdóttir, and Guðrún Ólafsdóttir. 2009. *Food Chem.* Vol. 116 (1). 240–248

[40] Maillard, M.N., Soum, M.H., Boivin, P., Berset C.1996. *LWT-Food Scie. and Technol.* Vol.29(3);238-244.

[41] Miyashita, K., Sho Nishikawa, Fumiaki B., Takayuki T., Masayuki A. and Masashi H.. 2011. J Sci Food Agric 2011; 91: 1166–1174.

[42] Farvin, K.H.S.and Jacobsen, C.2013. *Food Chem.*138.1670-1681.

[43] Plaza, M., Cifuentes, A. and Iba ´n˜ez, E. 2008. *Trends in Food Sci & Technol.* 19.31-39.

[44] Mikami, K. and Hosokawa, M. 2013. *Int. J. Mol. Sci.* 14, 13763-13781.

[45] Yip, W.H., Lim S. J, Wan A W M, Mohamad Y M and Mamot S.2014. *Sains Malaysiana.* 43(9): 1345–1354

Determination and Radiocarbon Dating of Marine Mollusc Fossils in Ancient Sea Shelf of Central Java Indonesia

S Aisyah[1], D Pringgenies[2], A Hartoko[3], J T S Sumantyo[4], H Matsuzaki[5]

[1]Master of Coastal Resource Management, Diponegoro University, Semarang 50275, Indonesia
[2]Department of Marine Science, Faculty of Fisheries and Marine Science, Diponegoro University, Semarang 50275, Indonesia
[3]Marine Geomatic Center and Department of Fisheries, Faculty of Fisheries and Marine Science, Diponegoro University, Semarang 50275, Indonesia
[4]Josaphat Microwave Remote Sensing Laboratory, Center for Environmental Remote Sensing, Chiba University, 1-33 Yayoi, Inage, Chiba 263-8522, Japan
[5]Micro Analysis Laboratory, Tandem accelerator (MALT), The University Museum, The University of Tokyo, Japan

Corresponding Author: sitiaisyah859@yahoo.com

Abstract. Mollusc phylum is one of the most adaptive animal groups on Earth. They occupy and thrive in incredibly diverse habitats. Their distribution in the fossil record is equally diverse. Indonesia is one of the country with marine mollusc fossil variety, such as in the archaeological site of Sangiran, Patiayam (Ancient Muria Strait) and Grobogan – Central Java. Radiocarbon Dating is the method for dating analysis using ^{14}C. By measuring ^{14}C content, we can estimate how long ago the fossils died. Radiocarbon dating is an extremely useful technique for determining the ages of geological materials (that have some organic-derived carbon in them). Field sampling had found variety of marine mollusc fossils such us 1) Sangiran: *Dosinia* sp., *Telescopium telescopium, Anadara* sp., and *Dosinia insularum*; 2) Patiayam (Ancient Muria Strait): *Pseudodon vondenbuschianus, Elongaria orientalis, Conus (Pionoconus)* sp., *Ampullina bandongensis, Anadara pilula* and *Filopaludina javanica*; 3) Grobogan: *Antigona chemnitzii, Cultelus dilatatus, Plotia scabra* and *Tibia modesta*. Mollusc fossils sample was analysed using Accelerator Mass Spectrometer (AMS) radiocarbon system. It is a good method for dating specific samples. The result showed that Sangiran as an ancient deepsea has fossils age 38710 – 31947 years, then Patiayam (Ancient Muria Strait) with fossils age 26248 – 11994. Mollusc fossils from Grobogan was the youngest area has fossils age 25692 – 6479 years.

Keywords: Radiocarbon Dating, Mollusc Fossils, Patiayam, Grobogan, Sangiran

1. Introduction

The interrelationships and evolutionary history of molluscs have seen great advances in the last decade [1]. Indonesia is one of the country with marine mollusc fossil variety, such as in the Sangiran, Patiayam (Ancient Muria Strait) and Grobogan – Central Java. Long time periode, Indonesian archipelago had been formed through tectonic movement, volcanic eruption process, earthquake, sedimentation and erosion process simultaneously. Similar phenomenon had happened in ancient Muria Mountain with deformation phases of ocean go through with land had result sedimentation process in Demak, Jepara and Pati (the part of Central Java) to become ancient Muria Strait. Long time before Muria region had separated from Java Island. Existence of land erosion and sedimentation process from Java Island by means of river trivialisation which stream down to Muria Strait. By the time then the strait closed and become land until now.

Muria regional stratigraphy indicates that the upstream Pontang River, Muria – Kudus, approximately 50 m to the west. In this area are found rocks containing carbonate and rich in marine mollusc fossils (deposition of marine origin) [2]. Similarly, archaeological site of Sangiran and Grobogan, allegedly the area was formerly ocean which are then degraded into a land, because of the evidence of the ancient life, especially the discovery of marine fossils. Sangiran is one of the most important ancient sites in the world. Sangiran saves a lot of evidence of ancient life and formation lithology. Sangiran become data sources of archeology, geology and paleontology [3].

Understanding the ancient environment in relation to ancient climate could help us to study the evolution of fauna (marine organism) in relation to environmental changes and predict the result of changes in the distribution of fauna today [4], [5] and also it is hoped as a new marine palaentological paradigm can be used to develop the past marine ecosystem or a new site for paleo-oceanographic and marine-paleometeorological studies [6]. The area were axamined using mollusc fossils object with the aim of identifying mollusc fossils and to examine the age of mollusc fossils with radiocarbon dating method. The research was conducted in two stages; first is field study at ancient Muria Strait, ancient reef Grobogan and ancient deepsea Sangiran (Central Java) to determine the ancient coastline coordinate, sampling marine fossils and observation of sedimentary stratigraphy. Second is laboratory for marine mollusc sample analysis and radiocarbon dating.

2. Material and Methods

2.1. Sampling and Identification Mollusc Fossils
Mollusc fossils sample collected from Sangiran, Patiayam (Ancient Muria Strait), and Grobogan (Figure 1). Sampling was used purposive sampling technique based on coordination and discussion with museum staff. Mollusc fossils would have found then identified in determine genera of mollusc fossils. Identify according to marfometric identification [7].

2.1. Radiocarbon Dating Method
Radiocarbon Dating is the method for dating analysis using ^{14}C. In order to evaluate the validity of a ^{14}C age of any material, the original sources of the carbon must be known. The carbon in gastropod shell carbonate originates from as many as four different sources: atmospheric CO_2, food, water, and carbonate rocks [8]. Analysis of the age of mollusc fossils with radiocarbon dating method could estimate the ages any organic thing. ^{14}C is an unstable radioactive isotope.

Radiocarbon dating analysis using Accelerator Mass Spectrometer (AMS) system. AMS detect atoms of specific elements according to their atomic weights. They, however, do not have the sensitivity to distinguish atomic isobars (atoms of different elements that have the same atomic weight, such as in the case of carbon 14 and nitrogen 14 – the most common isotope of nitrogen). There are essentially two parts in the process of radiocarbon dating through AMS. The first part involves accelerating the ions to extraordinarily high kinetic energies, and the subsequent step involves mass analysis. There are two accelerator systems commonly used for radiocarbon dating through AMS. One is the cyclotron, and the other is a tandem electrostatic accelerator [9], [10]. This research is supported through Micro Analysis Laboratory, Tandem accelerator (MALT), The University Museum, The University of Tokyo, Japan. The protocol of the radiocarbon dating procedure from marine mollusc fossils by Matsuzaki (2015) [11].

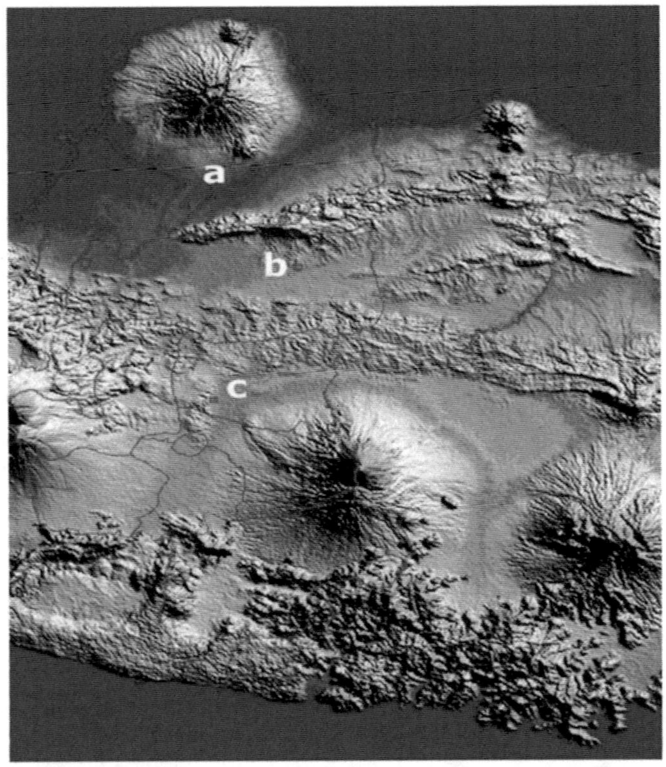

Figure 1. Sampling Site for the Collection of Mollusc Fossils from Patiayam (Ancient Muria Strait) (a), Grobogan (b) and Sangiran (c) [6]

The samples were preparation less than 10 mg and treated twice with 1.2 M HCl for 6 hr at 80°C to remove any possible carbonate contaminants. Next, the samples were treated with 0.005 M – 1.2 M Sodium hydroxide at 80°C. Then samples were washed by ultra pure water and treated again with 1.2 M HCl for 2 hr at 80 °C. To remove HCl component completely from those small samples, centrifuge and decant were repeated after adding some distilled water.

The samples were then dried in an electric oven at 90 °C. For the small sample, we added a few hundred-milligrams of CuO powder (Filamentous Copper Oxide II) as a carrier to the sample in the centrifuge tube, to collect carbonaceous materials completely. The total amount of pretreated charred samples was used to produce CO_2. The samples were placed in Vycor tubes of about 9 mm in outer diameter, with about 1 g of granular CuO. Then the tubes were connected to a vacuum line, evacuated completely, and sealed to a tube length of 300 mm. The Vycor tubes were heated to 850°C for 2 hr to completely convert carbon to CO_2. The CO_2 produced was purified cryogenically in a vacuum line and reduce to graphite on about 2 mg Fe powder in the presence of hydrogen. The graphite materials were pressed into aluminum target holders for AMS ^{14}C dating. We used the Hox-II standard as a ^{14}C-concentration reference [11]. Then, the ^{14}C ages were calibrated to a OxCal [12].

3. Result and Discussion

3.1. The Age of Marine Mollusc Fossils with Radiocarbon Dating

Analysis of marine mollusc fossils ages using Accelerator Mass Spectrometry (AMS) radiocarbon with four steps: sampel preparation, chemical pretreatment, combustion (oxidation) and sample processing with AMS. The result of marine mollusc fossils ^{14}C ages has to calibrate using OxCal Calibration (Table 1).

Table 1. ^{14}C ages and Radiocarbon Ages Calibration

No	No. Sampel	Location	^{14}C Age (BP/Before Present)	Calibrated Age (BC/ Before Christ)
1	MOL-CD1	Grobogan	23127 ± 107	25692 - 25252
2	**MOL-CD2**	**Sangiran**	**35390 ± 258**	**38710 - 37383**
3	MOL-CD3	Patiayam – Muria	23871 ± 112	26248 - 25735
4	MOL-CD4	Grobogan	7571 ± 39	6479 - 6379
5	MOL-CD5	Patiayam – Muria	12227 ± 56	12426 - 11994
6	**MOL-CD6**	**Sangiran**	**30216 ± 183**	**32672 - 31947**

***Bold: the oldest area**. Underline: the youngest area

3.2. Marfometric Identification of Mollusc Fossils

The specimens of mollusc fossils were collected from Sangiran, Patiayam (Ancient Muria Strait), and Grobogan (Figure 2, 3 and 4).

Figure 2. Collected samples from Sangiran: a) *Telescopium telescopium* Length: 111.05 mm; b) *Dosinia sp.* Length: 57.60 mm; c) *Anadara sp.* Length: 55.25 mm; d) *Dosinia insularum* Length: 42.05 mm;

The result showed that ^{14}C analysis of marine mollusc fossils collected from The Archeological Site of Sangiran, Patiayam (Ancient Muria Strait) and Grobogan. The AMS ^{14}C result has to calibarate using radiocarbon calibrated age (e.g. OxCal Calibration) [12,13,14]. The oldest age (38710 – 31947 years) was obtained for mollusc fossils sample collected from the Archeological Site of Sangiran. The result showed that marphometric identification of mollusc fossils from Sangiran are *Dosinia sp., Telescopium telescopium, Anadara sp.* and *Dosinia insularum* (Figure 2). Sangiran during the Pleistocene became one part of the Java Island have experienced several deformation due to fluctuations

in sea level surface [15]. Lithology change stems from the ocean turns into a swamp and then go trought with land by the activity of an ancient volcano with different formation lithology [3].
Based on stratighraphic analysis, Sangiran area were started with Kalibeng formation (Pliocene), Pucangan formation (early Pleistocene), Kabuh formation (middle Pleistocene) and Notopuro formation [3][6]. Marine mollusc fossil could be found near to Kalibeng formation in Puren River and Karangrejo. Field validation sampling had confirmed that the result of radiocarbon dating sample (absolute ages) and geology age are the same result. Sangiran was the oldest formation as the deepsea zone.

Figure 3. Collected samples from Patiayam – Ancient Muria Strait: a) *Pseudodon vondenbuschianus* Length: 49.80 mm; b) *Elongaria orientalis* Length: 37.95 mm; c) *Conus (Pionoconus)* sp. Length: 43,70 mm; d) *Ampullina bandongensis* Length: 22.05 mm; e) *Filopaludina javanica* Length: 27.85 mm; f) *Anadara (cunearca) pilula* Length: 28.55mm.

Fossils age from Patiayam (Ancient muria Strait) gave the result 26248 – 11994 years with mollusc fossils: *Pseudodon vondenbuschianus, Elongaria orientalis, Ampullina bandongensis, Filopaludina javanica* and *Anadara (cunearca) pilula* (Figure 3). Lithology of Ancient Muria Strait (Patiayam) divided into six formations, they are Jambe formation (Pliocene), Kancilan formation (early Pleistocene), Slumprit formation (middle Pleistocene), Kedungmojo formation, Sukobubuk formation (the end of Pleistocene) and "endapan teras" (Holosen). There are two of the period of the environment precipitation in Patiayam, they are Jambe formation in a shallow seas and four formation that the youngest area. In addition Kancilan formation was the beginning of the changes to the environment [2][16].

Figure 4. Collected samples from Grobogan: a) *Plotia scabra* Length: 14.05 mm; b) *Cultelus dilatatus* Length: 49.60 mm; c) *Antigona chemnitzii* Length: 68.05 mm; d) *Tibia modesta* Length: 19.75 mm.

While in above formation was found as the late miocene in Grobogan with limestone and identified as shallow seas. This area was the youngest area with radiocarbon dating ages 25692 – 6379 years. The zone was known in geological term as 'the north Kendeng limestone/carbonate mountain or the front-arch formation consisting with the denundation of Globigerina – limestone [6]. The field validation on the coordinate position of Grobogan had discover sampels of limestone sediment and mollusc fossils such as *Antigona chemnitzii, Cultelus dilatatus, Plotia scabra* and *Tibia modesta*.

4. Conclusion

Mollusc fossils sample was analysed using radiocarbon dating with the result that Sangiran as an ancient depsea (the oldest area) has fossils age 38710 – 31947 years, then Patiayam (Ancient Muria Strait) with fossils age 26248 – 11994. Mollusc fossils from Grobogan was the youngest area has fossils age 25692 – 6379 years.

Acknowledgments

The authors would like to thank you to Master of Coastal Resource Management, Faculty of Fisheries and Marine Science, Diponegoro University, Semarang. Thank you very much to Bapak Hariyadi – Department of Marine Science for field samples collection. Also thanks to research collaboration with Prof. Josaphat Tetuko – Chiba University, Prof. Hiroyuki Matsuzaki – Tokyo University for radiocarbon dating analysis laboratory, and all museum staff in Muria and Sangiran.

References

[1] Vinther J 2015 *Frontiers in Paleontology – The Origins of Molluscs* (Paleontology Journal) vol 58 pp 19-34

[2] Mulyaningsih S *et al* 2008 *Vulkanisme Kompleks Gunung Patiayam di Kecamatan Jekulo, Kabupaten*

Kudus, Provinsi Jawa Tengah (Jurnal Geologi Indonesia) vol 3 pp 76-86

[3] Wulandari 2012 *Topografi Situs Sangiran* (Jurnal Sangiran) vol 1 pp 14-16

[4] Louys J and Erik M 2010 *Palaeoecology of Southeast Asian megafauna-bearing sites from the Pleistocene and a review of environmental changes in the region* (Journal of Biogeography) Blackwell Publishing Ltd. Research Centre in Evolutionary Anthropology and Palaeoecology, School of Natural Sciences and Psychology, Liverpool John Moores University Liverpool L3 3AF UK. pp 18

[5] Myers C 2010 *Paleontology : The Biology of the Future* (American Paleontology. Paleontological Research Institution) vol 18 pp 27

[6] Hartoko A, Hariyadi, Petrus S, Josaphat T S 2014 *Satellite Data Spatial Based Reconstruction and Discovery of the Ancient Coastaline, Coral-fringing Reef and Mollusc Fossils at The Muria Strait – Central Java Indonesia* (The 6th Indonesia Japan Joint Scientific Symposium) pp 15

[7] Dharma B 2009 *Recent and Fossil Indonesia Shells* (Conch Book. ISBN-10: 3925919708) pp 150

[8] Pigati J S, Rech J A and Nekola J C 2006 *Radiocarbon dating of North American terrestrial gastropod shells* (National Science Foundation Sedimentary Geology and Paleobiology Competition Award #EAR 0614840) pp 1

[9] Alkass K, Hisako S, Bruce A B, Samuel B, Gunilla H, David R S, Kirsty L S and Henrik D 2013 *Analysis of Radiocarbon, Stable Isotope and DNA in Teeth to Facilitate Identification of Unknown Decedents* (www.plosone.org/journal.pone.0069597) pp 12

[10] Beta Analytic Co. Ltd. *Accelerator Mass Spectrometry Radiocarbon Dating* (www.radiocarbon.com)

[11] Matsuzaki H 2015 *Radiocarbon Dating* (Micro Analysis Laboratory, Tandem accelerator (MALT): The University Museum The University of Tokyo Japan)

[12] University of Oxford: Calibration Online *Radiocarbon Calibration* (https://c14.arch.ox.ac.uk/)

[13] Stuiver M, Pearson G W and Braziunas T 1986 *Radiocarbon Age Calibration of Marine Samples Back to 9000 Cal Yr BP* (Radiocarbon Journal) vol 28 pp 980-1021

[14] Nakamura T, Taniguchi Y, Tsuji S and Oda H 2001 *Radiocarbon Dating of Charred Residues on The Earliest Pottery in Japan* (Radiocarbon Journal) vol 43 pp 1129-1138

[15] Rosyidah M Y 2013 *Rekonstruksi Lingkungan Purba di Situs Sangiran pada Kala Plestosen Menggunakan Analisis Mesowear pada Fosil Bovidae* (Jurnal Sangiran vol 2) pp 91

[16] Zaim Y 2006 *Hominids in Indonesia: From Homo erectus (paleojavanicus) to Homo floresiensis* pp 14

Sustainable Eco Coastal Development Through Corporate Social Responsibility (CSR) Program

Arsi Rakhmanissazly[1], Yong Mursito Ardy[2], Abdullah[3]

[1] Public & Government Relations Staff, Legal & Relations Department
[2] CSR Staff, Legal & Relations Department
[3] Field Manager, PT Pertamina EP Asset 3 Tambun Field, Bekasi, West Java, Indonesia

Email: arsi.rakhmanissazly@pertamina.com, yong.ardy@pertamina.com, abdullah2@pertamina.com

Abstract. Besides technical problems the company's operational constraints that may effect high deficiency for the company is the company – community conflicts. Company – community conflict can also arise depends on the geographic conditions and characteristics of the community itself. Some studies has show that coastal community have higher level of social risk when compared to non-coastal community. Also, the coastal community ussually only rely on what sea provides as their main livelihood. Because of the level of education still contemtible the community couldn't optimized the potential of their own area. Corporate Social Responsibility (CSR) har emerged as an important approach for addressing the social and environmental impact of company activities. Through CSR program, PT Pertamina EP Asset 3 Tambun Field (PEP) try to form value integration by utilizing resources from the community and the company by making sustainable eco – coastal living in Desa Tambaksari, Karawang, one of PEP working area. Using sustainable livelihood approach begin with compiling data by doing social mapping PEP has initiate the area to becoming Fish Processing Industry Centre. By implementing PDCA in every steps of the program, PEP has multiplied some other programs such as Organic Fish Feed Processing, Seaweed Farming and Waste Bank for Green Coastal Village. These program is PEP's effort to create a sustainability environment by enhancing the community's potentials as well as resolving social problems around Tambaksari. The most important result besides getting our license to operate from the community, is the community itself can grow into an eco coastal sustainable system.

Keywords: corporate social responsibility, oil and gas industry, community program, sustainable environment, social problems.

1. Introduction

Multinational Corporations (MNCs) play an important role in global business, and the nature of the oil industry requires it to be proactive, and socially responsible alongside operating in an ethical and environmentally friendly manner [1]. Therefore, unsurprisingly when an oil and gas company operating in fields which usually located in remote area there will be a lot of expectations from community around to the company to get a direct share from company as the compensation of impact during the operational.

As for PEP, social problem that cause from company – community conflicts may generate the same broad effects as those caused by technical problems. Blockade, demonstration, damage to property, injuries or fatality, or the project delayed were common obstacle in upstream industries.

These may not only effects to operational and financial perform but also to the reputation of the company itself.

The fall of crude oil prices since mid-2014 has undermined oil & gas industry all over the world. Upstream sector suddenly sluggish due crude oil sales are no longer profitable. Inversely with the condition, the expectation from community still high about company contribution to environment and community surround.

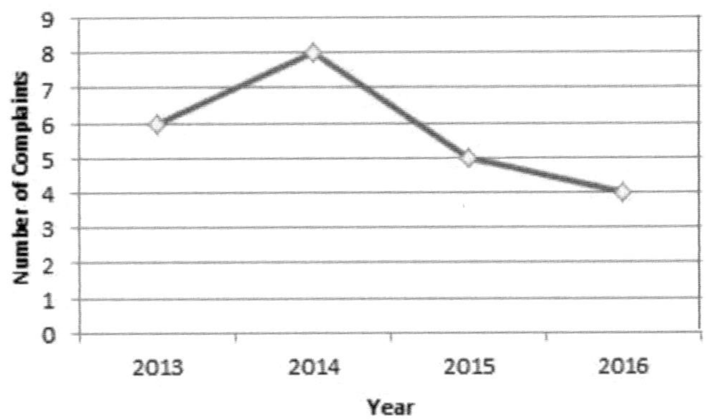

Figure 1. Trend of Complaints during Operational in Tambun Field 2013 - 2016

As we can see in Figure 1, in 2014 the level of dissatisfaction from external stakeholders refers to amount of complain was increasing while the oil crisis hit. Although, the trends show declining number of complain for the last two years when CSR program massively implemented. Corporate social responsibility (CSR) program can become a source of tremendous social progress, as the business applies its considerable resources, expertise and insights to activities that benefit society [2].

CSR also includes the responsibilities that are inherent on both side of these relationships. CSR defines society in its widest sense and on many levels, to include all stakeholders and constituent groups that maintain an ongoing interest in the organization's operation" [3]. CSR is not something that is required by the law for business to follow and implement. Companies who want to practice who want to practice business more responsibly and contribute for the better of society often choose to implement CSR strategies [4].

This paper will describe the initiatives taken by PT Pertamina EP (PEP) Asset 3 Tambun Field in coping with social issues through their strategic community development programs. Known as LARI (Laut Lestari), this strategic plan technically covers series of smaller programs to form value integration by utilizing resources based on local potential in the form of marine products and environment preservation.

2. Methodology

The term "corporate social responsibility" refers to the kinds of things companies do in their efforts to navigate these swirling currents of changing expectations, expectations that have never loomed larger in the daily lives of companies, including oil and gas companies [5]. The oil and gas sector has been among the leading industries in championing CSR. Especially in upstream which has directly and highly visible negative effects of day-to-day operations such as oil spills and the resulting protests by civil society groups and indigenous people.

For this project, PEP used social mapping and needs assessments methodology to address the problem of Desa Tambaksari's and compile the data to arrange strategies. By using these methods PEP will get more comprehensive data to be used as inputs in the formulation of policies, strategies and re-design of community development program to be better planned, focused and targeted in accordance with the principles of Community Development so that the implementation can make a better result for the society itself.

PEP use multi-information approach in social mapping which consist of:

2.1 Focus Group Discussion (FGD)

The data information obtained directly from the public in a forum, where opinions with each other reinforce each other so as to be composed of a social network mapping. This forum also provides the information about ground village map, economical society map, village potential resources map and the Venn diagram that illustrates the role and contribution of the existence of institutions, both formal and informal. The FGD followed by Tirtajaya's Sub-district Head, Tambaksari's Village, Village's Secretary, Village's Organization (BPD, RT, RW, PKK, Karang Taruna, Local NGO, KUD), opinion leader, District Representative (Bappeda, UPTD Pendidikan, UPTD Perikanan, Peternakan dan Kelautan, PERHUTANI).

2.2 In – depth Interview

After getting raw data from FGD, PEP start to collect data about cause – effect, analyze condition, influence and local tradition by doing some in-depth interview to local opinion leader in Desa Tambaksari.

2.3 Survey

PEP used this method to get the validity of the prime data, the one from FGD and in-depth interview, to the real condition.

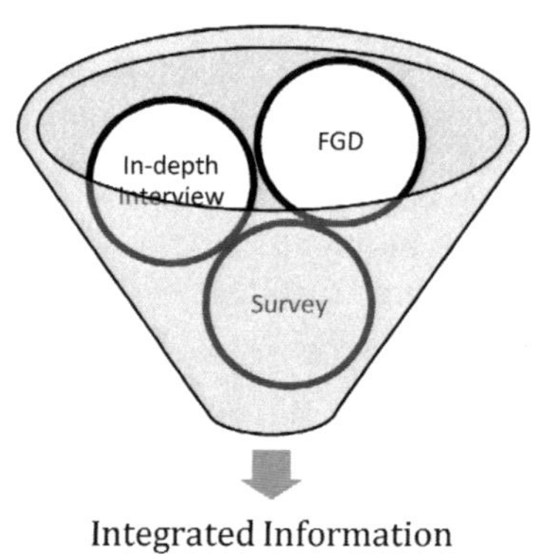

Figure 2. Social Mapping Method

Corporate Social Responsibility (CSR) is increasingly developing beyond a philanthropic activity into a cardinal business function directly linked to corporations maintaining their social license to operate. This, entails integrating social, environmental and economic concerns into a company's values and culture, and reflecting these in the company's policies, operations, strategies and decision making. Thus, CSR has got the potential to contribute positively towards the needs of disadvantaged communities in developing countries not just economically or in social and environmental terms but a combination of all these needs [6].

Practically, PEP also implemented the combinations between social condition, our mission and potential of the communities. Will be explained on the scheme below:

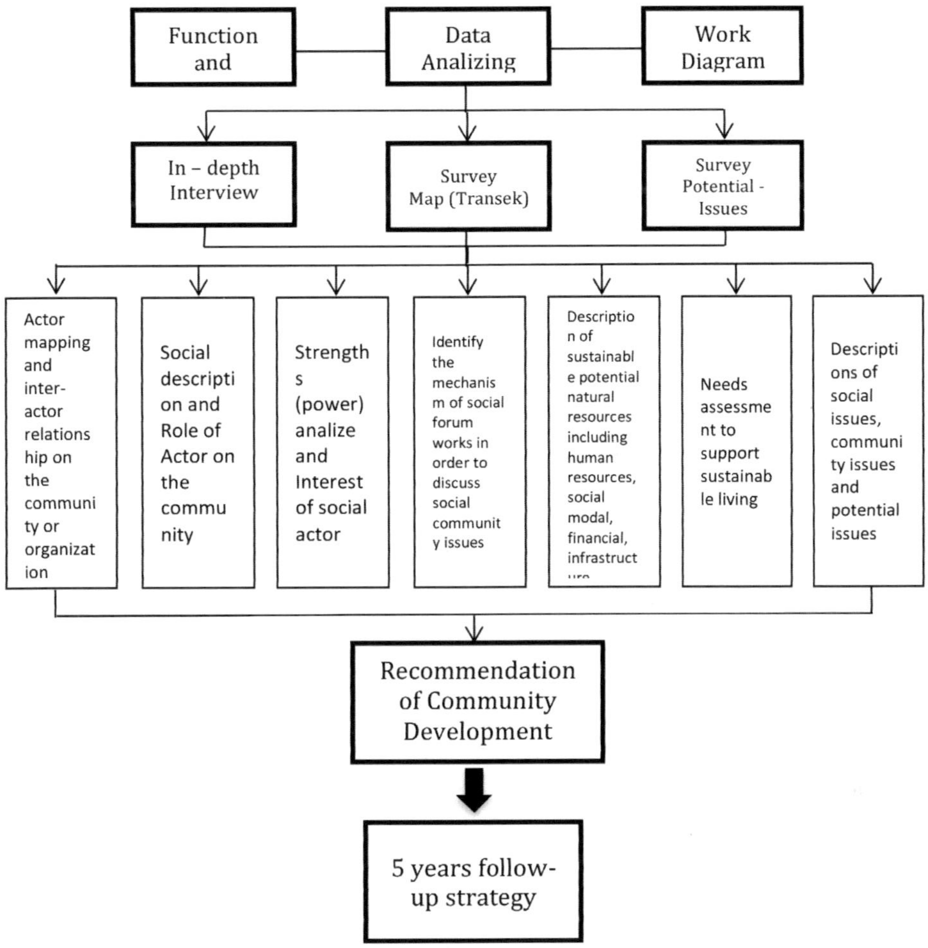

Figure 3. Social Mapping Research Design [7]

As the result of social mapping and needs assessment PEP will get the recommendation program which we use as groundwork of our CSR program in Desa Tambaksari. During 2016, PEP has developing some programs which integrate also with The Global Goals for Sustainable Development (SDGs) from ILO.

Table 1. CSR Programs on PT Pertamina EP Asset 3 Tambun Field

No	SDGs Goals	Pertamina CSR program in parthership with Karawang Government
1	No Poverty	Collaboration with West Java's P2WKSS in Rumah Tidak Layak Huni (RUTILAHU) program
		Donation for orphans and aged people
2	No Hunger	Processing food made from Milkfish and Seaweed Program, Family Food Security Program
3	Good Health	"Gerakan Makan Ikan" Program, Suplementary Feeding for Children Malnutrition

4	Quality Education	Green School SD Negeri 1 Tambaksari with Waste Bank Program
5	Gender Quality	Home industry for women in Desa Tambaksari through Processing food made from Milkfish and Seaweed Program, Waste Bank
6	Clean Water and sanitation	Zero waste processing in food processing industry
7	Clean energy	Using windmill as the substitute solar energy to fishponds water circulation
8	Economic growth	1. Processing food made from Bandeng Program
		2. Processing food made from Seaweed Program
		3. Seaweed Cultivation
		4. Mangrove Nursery
		5. Waste Bank
		6. Organic Fish Feed Manufacture Program
		All of this program is funded by Koperasi Mina Agar Makmur
9	Innovation and Infrastructure	1. Waste processing building for Waste Bank Group
		2. Milkfish's Café for Processing Food Group
		3. Development of Koperasi Mina Agar Makmur
10	Protect the planet	Mangrove nursery and trees plantation (*casuarina equisetifolia, cocos nucifera, terminalia catappa*)to prevent sea abration
11	Life below water	Mangrove plantation to preserve sea ecosystem

3. SUSTAINABLE ECO COASTAL DEVELOPMENT; KAMPUNG LAUT LESTARI (LARI)

The community empowerment and innovation program developing by PT Pertamina EP Asset 3 Tambun Field refers to the third principle of CSR (people, profit, and planet) embodied in "Laut Lestari (LARI)" strategy.

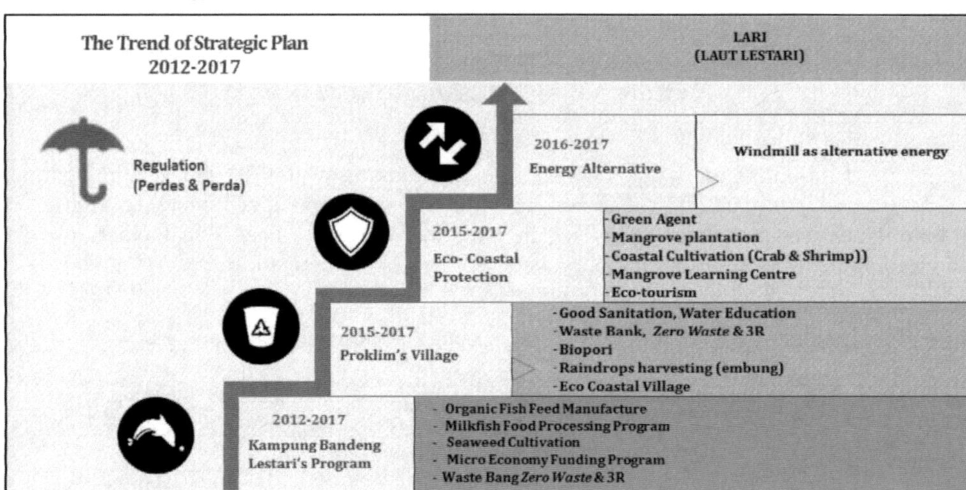

Figure 4. Grand Design LARI (Laut Lestari) Strategy [8]

By combining the three bottom lines of CSR itself, PEP is trying to create sustainable program simultaneously in Desa Tambaksari, Karawang in 2016.

3.1 Organic Fish Feed Manufacture Program

The purpose of this program is to help fishermen ponds to reduce their operational cost for Milkfish Cultivation which known as the greatest local potential of Desa Tambaksari. Lower cost production, bigger income for the community. By helping the fishermen to manufacture their own organic fish feed, PEP has applied cost- efficiency system in community activities. The fishermen also create an Eco - friendly cultivation and reduced use of chemical substances.

Using Mill's Feed Product		
Milkfish Sales	IDR	20.250.000,-
Operational Cost	IDR	13.800.000,-
Profit	**IDR**	**6.450.000,-**

Using Organic Feed Product		
Milkfish Sales	IDR	20.250.000,-
Operational Cost	IDR	8.922.500,-
Profit	**IDR**	**11.327.500,-**

Figure 5. Group's Profit Comparison [9]

3.2 Milkfish Food Processing Program

In line with SDG's goals about gender equality this program is addressed to unemployed women group in Desa Tambaksari. C73 is a social enterprise that provides alternative employment held by Mak Emat (local hero) in food processing program. By opening the chance for them to support their domestic financial, PEP also initiate the community to elevate the economical value of Milkfish itself. The innovation of this program is a Zero Waste Food Processing concept as explained on the picture below.

Existing:
1. PIRT and MUI (Halal) Ceritified
2. Packaging certified by Fisheries Department
3. 2nd Place Pekan UKM Inovatif held by Karawang Regency
4. Restaurant and Store supply

Figure 6. Zero Waste Food Processing Program [10]

3.3 Seaweed Cultivation

There are some advantages by improving seaweed cultivation. The high demand in seaweed production provides plentiful opportunities and work for the local community. The technology is simple so that production cost relatively low. Seaweed has short production cycle and also environmentally friendly. The established program has already in corporation with PT. Agarindo, one of the biggest seaweed factory in Indonesia.

IOP Conf. Series: Earth and Environmental Science 55 (2017) 012065 doi:10.1088/1755-1315/55/1/012065

3.4 Waste Bank for Green Coastal Village

This program is conducted by Green Agent which involves elementary school students in Desa Tambaksari. The Green- Agent will persuade the communities to collect, sell and recycle household waste. Our goal is that by showing communities how how employment and profit can be generated from this waste, less of it will be dismissed, thereby Desa Tambaksari will become a Green Coastal Village.

3.5 Strengthening Micro-economy Funding Program

This program was established as a source of financial services among other program. Besides the funding, Koperasi Mina Agar Makmur is expected as a place for communities to sharing knowledge and market information.

4. Result and Discussion

The oil and gas sector has been among the leading industries in championing CSR. This is at least partly due to the highly visible negative effects of day-to-day operations such as oil spills and the resulting protests by civil society groups and indigenous people [11]. The implementation of this strategy gave another opportunity for the community to develop and preserve their local potential for living. On the other hand, for PEP, the most important result is reducing the discrepancy of company – community conflicts and to achieve social stability around the operations area in Desa Tambaksari (see again Figure 1 to see the decreasing number of conflicts during program implementation).

PEP is also trying to re-connecting all the potential of social – economic network to create a sustainable eco- coastal living in Desa Tambaksari. The networking result of PEP CSR program can be explained on the picture below.

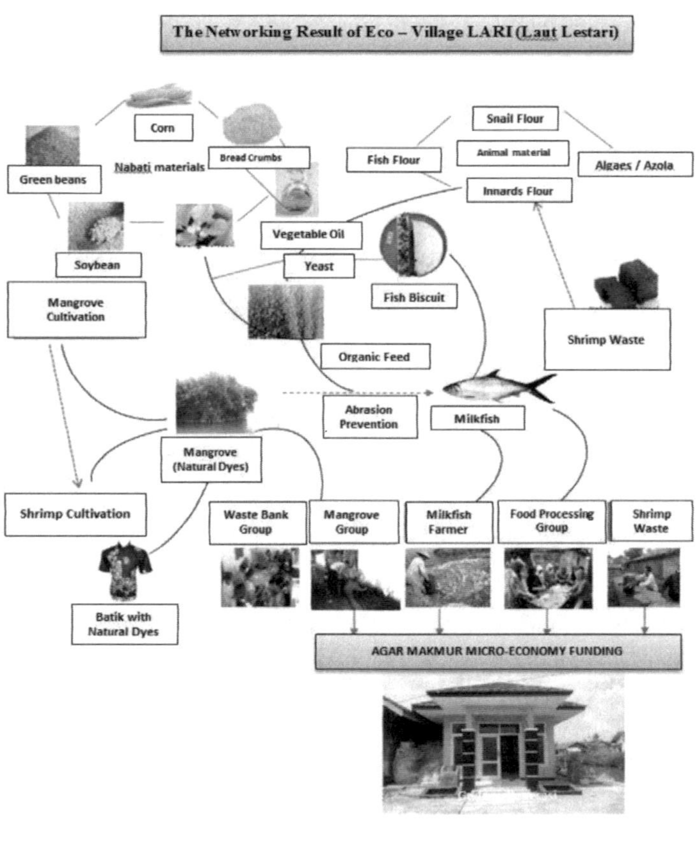

Figure 7. The Networking Result of Eco – Village LARI (Laut Lestari) [12]

5. Conclusion

In this paper, we conduct study and strategy that we implemented in our CSR program. This study uses social mapping with FGD, in-depth interview and survey as our reduction data. Then we combine the social data from social mapping with theoretical analysis to form the strategy. We try to elaborate the program we create with the program that society needs to improve their living. The synergy between company–community through CSR program somehow create social establishment and reduce negative impact from resistant group.

By providing some budget for CSR program, management realize that PEP can elude the negative impact from community while having uncertainty on social conflicts that can possibly occur at any time and also may need bigger amount of social cost (e.g. compensation etc). Often companies do decide to incorporate CSR into their business strategies because it will eventually help improve financial performance and improve the company image [13]

As for the community and natural environment, the results are:

1. CO_2 emission reduction for 3,2% from 826 tons of yam for using organic feed product
2. Reducing 15% of Methane from Green House Effect for using organic feed product
3. Increasing number of women empowerment in home industry activities
4. 88% income increase for the member of Milkfish Food Processing Program (C73)
5. Help to prevent Karawang's Coast from abrasion for Mangrove Cultivation Program
6. Increasing number of student activity as Green Agent.
7. Product innovation for local souvenirs from Seaweed Cultivation Program.

ACKNOWLEDGEMENT

This study and strategy was supported and funded by PT Pertamina EP Asset 3 Tambun Field. By implementing the program based on this study, PEP has awarded for GREEN PROPER since 2014 held by Ministry of Environment and Forestry Indonesia.

REFERENCES

[1] Olowonirejuaro R. 2010. *Implementation of 'Beyond Complience' Tools and CSR Efforts by Oil Multinationals in Brazil: Is There a Link?.*

[2] Michael E. Porter and Mark R. Kramer. December 2006. *Strategy & Society – The Link Between Competitive Advantage and Corporate Social Responsibility.*

[3] Buldybayeva G. 2014. *Both Sides of CSR Practice: A Case from O&G Industry in Kazakhstan.* Good (Positive) Side of CSR, pp: 234. Kazakhstan.

[4] Kan, Wai Ping. *Corporate Social Responsibility: A Profitable Alternative.* New York: USA.

[5] Spence, David B. 2011. *Corporate Social Responsibility in the Oil and Gas Industry: The Importance of Reputational Risk.*

[6] Olowonirejuaro, Ruqayah. 5 August, 2010. *Implementation of 'Beyond Complience' Tools and CSR Efforts by Oil Multinationals in Brazil: Is There a Link?.*

[7] PT Pertamina EP Asset 3 Tambun Field. 2016. Social Mapping Document, submitted in PROPER.

[8] PT Pertamina EP Asset 3 Tambun Field. 2016. Strategy Plan of CSR Document, submitted in PROPER.

[9] PT Pertamina EP Asset 3 Tambun Field. 2016. CSR Monitoring and Evaluation Document, submitted in PROPER.

[10] PT Pertamina EP Asset 3 Tambun Field. 2016. CSR Monitoring and Evaluation Document, submitted in PROPER.

[11] Frynas, Jedrzej George. 2009. *Corporate Social Responsibility in the Oil and Gas Sector.* Journal of World Energy Law and Business, 2 (3). Pp: 6;29. London; United Kingdom.

[12] PT Pertamina EP Asset 3 Tambun Field. 2016. CSR Monitoring and Evaluation Document, submitted in PROPER.

[13] Babiak, K., &Trendafilova, S. (2010). *CSR and Environmental Responsibility: Motives and pressures to adopt green management practices. Corporate Social Responsibility and Environment Management*, 18, 11-24. Doi: 10.1002/csr.229.

The effect of ENSO to the variability of sea surface height in western Pacific Ocean and eastern Indian Ocean and its connectivity to the Indonesia Throughflow (ITF)

H A Rejeki[1,2], Munasik[1], Kunarso[1]

[1] Faculty of Fisheries and Marine Science, Diponegoro University, Jl. Prof. Soedarto SH. UNDIP, Tembalang, Semarang 50275, Indonesia

[2]Indonesian Agency for Meteorology Climatology and Geophysics, Jl. Angkasa I No.2, Kemayoran, Jakarta 10610, Indonesia

E-mail: hasti.amrih@gmail.com

Abstract. The differences of altimetry in the Pacific Ocean and the Indian Ocean causes the Indonesia Throughflow or commonly called ITF. The altimerty will have variation when the ENSO (El Nino Southern Oscilation) occur. The altimetry data from AVISO is used to find out how much the influence of ENSO to variations of that indicator in particular the altimetry difference between the western Pacific Ocean to the eastern Indian Ocean. When El Nino occured, the altimetry in the western Pacific Ocean will be lower than the altimetry of eastern Indian Ocean while the opposite condition occurs when the La Nina happened that the differences of altimetry in western Pacific Ocean higher than the altimetry in eastern Indian Ocean. These differences will affect the transport of ITF.

1. Introduction

ENSO (El Nino Southern oscilation) is phenomenon that influence climate variability takes place in Indonesia, the equatorial Pacific and the world [1]. ENSO phenomenon consists of two events as El Nino and La Nina. El Nino occurs when an increase in sea surface temperatures in the eastern and central Pacific Ocean associated with decreased rainfall in Indonesia while when La Niña occurs indicates the opposite condition [2, 3]. The phenomenon of El Niño and La Niña can be learned from a number of indicators such as sea surface temperature anomalies, the southern oscillation index, sea surface height, and Indonesia Throughflow or commonly called the ITF [1].

El Nino and La Nina is a phenomenon in the equatorial Pacific Ocean characterized by a five consecutive 3-month running mean of sea surface temperature (SST) anomalies in the Niño 3.4 region that is above (below) the threshold of +0.5°C (-0.5°C). This standard of measure is known as the Oceanic Niño Index (https://www.ncdc.noaa.gov/teleconnections/enso/indicators/sst.php).

Indonesia Throughflow (ITF) is a system of ocean circulation in Indonesian waters where there is a current path that brings water masses from the Pacific Ocean to the Indian Ocean [4]. These currents occur due to differences in sea surface height between the Pacific Ocean and the Indian Ocean [5]. The ITF is mainly caused by differences in sea level between the Pacific Ocean and the Indian Ocean, namely the inner surface of the tropical western Pacific Ocean is higher than in the eastern Indian Ocean, resulting in a pressure gradient resulting current to flow from the Pacific to the Indian Ocean [4]. Indicators to assess the ITF is important to observe the sea surface heights in the Pacific and Indian Ocean as well as the variability of sea surface temperature in Indonesian waters [1].

ITF is an important part of the Global Conveyor Belt which connects the Pacific and Indian Oceans in the tropical zone [5]. Has many research projects on the ITF since a few hundred years ago [6]. Most publications on the ITF has investigated the variability of volume and heat flux at different time scales of decades to inter-annual and seasonal ranges as well as the phenomenon of Indian Ocean Dipole (IOD) [7] and the El Niño Southern Oscillation (ENSO) [8, 9].

Several studies have been conducted to identify the characteristics of some waters in Indonesia which is traversed by the ITF at the time of occurrence of El Niño and La Niña. The phenomenon of El Niño and La Niña effect on the ITF transport through the Strait of Lifamatola which is one of the eastern part of the ITF which has a threshold at a depth of approximately 1940 m to represent the occurrence of El Niño and La Niña [10]. The other studies in the Timor Sea about the weather phenomenon El Niño and La Niña impact on transport ITF shown variation of parameters such as sea temperature, salinity and ITF transport [4]. Sea surface height in the Pacific Ocean and the Indian Ocean is an important indicator in assessing the ITF [1]. This study will examine how different sea surface heights in the western Pacific and eastern Indian Ocean during El Niño and La Niña.

Research on the effects of El Niño and La Niña on parameters ITF has been conducted by several researchers. However, in this study focuses on sea surface height value differences between the western Pacific Ocean to the Indian Ocean at the time of the eastern part of El Niño and La Niña.

2. Data and methodology

Sea surface height data used in this study come from http://www.aviso.sea surface height.fr/en/data/products.html [11]. The data obtained are daily sea surface height data with a resolution of 0.25 ° x 0.25 ° to the length of the data is taken from the January 1993 to December 2015. The selected study area is the region of the western Pacific Ocean and the Indian Ocean to the east. The locations are at 125 ° - 145 ° E and 10 ° - 0 ° N to the Pacific Ocean to the west and 110 ° - 130 ° E and 10 ° - 20 ° S for the eastern part of the Indian Ocean region (figure 1) [12]. The location determination is based on the theory of the ITF stating that the ITF attributed to differences in sea surface height between the Pacific Ocean and the Indian Ocean [5].

To read the sea surface height data have been obtained from AVISO use ODV 4.7.8 software (Ocean Data View 4.7.8). The data is then converted into a text format that is then processed using Microsoft Excel. Processing data using Microsoft Excel is made to change the daily sea surface height data into monthly sea surface height data by way of averaging the daily data. Having obtained monthly sea surface height data, calculated the difference between sea level in the western Pacific Ocean on the sea level in the eastern part of the Indian Ocean. The calculation of the height difference is accomplished by reducing the value of sea surface height in the western Pacific Ocean with a value of sea surface height in the eastern Indian Ocean. The reason for this way is based on the theory that the ITF is caused by the difference in sea level between the Pacific Ocean and the Indian Ocean where the inner surface of the tropical Pacific Ocean West is higher than in the Indian Ocean in the east [4].

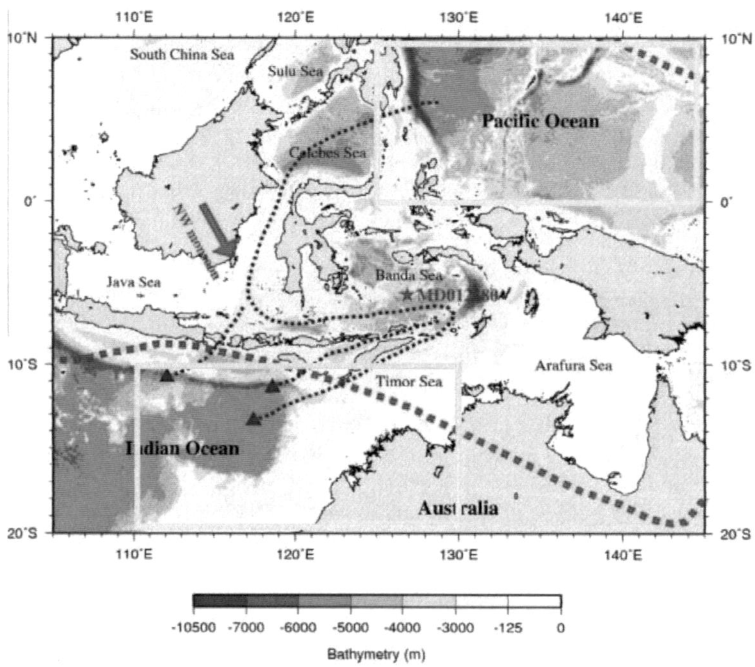

Figure 1. Research location (Source: Huang *et al.*, 2011)

To determine the condition of El Nino and La Nina, we used the sea surface temperature anomalies in the Nino 3.4 region. The anomaly data obtained from http://www.cpc.ncep.noaa.gov. If the value of Nino 3.4 SST anomalies greater than 0.5, then the condition is indicative of the occurrence of the El Nino phenomenon whereas if Nino 3.4 SST anomaly values smaller than -0.5 then indicate the occurrence of the phenomenon of La Nina. Furthermore, the period of time which indicates the occurrence of ENSO phenomenon will be compared against sea surface height conditions in both areas where he did research. Thus, it can be seen sea surface height variation among the sites to ENSO events.

3. Result

The calculation results of average difference of altimery in the western Pacific Ocean region and Indian Ocean east part shows the variation of the ENSO phenomenon. At the time of the sea surface temperature anomaly Nino 3.4 higher value of 0.5 indicating the occurrence of El Nino, sea surface height difference between the eastern Pacific Ocean and the Indian Ocean east section indicates a negative value. It shows that in the event of El Nino conditions in the Pacific Ocean sea surface height western part is lower than it would be in the eastern part of the Indian Ocean. The opposite occurs when sea surface temperature anomaly Nino 3.4 worth less than 0.5 indicating the occurrence of La Nina, sea surface height difference between the eastern Pacific Ocean and the Indian Ocean east section indicates a negative value. It shows that in the event of El Nino conditions in the Pacific Ocean sea surface height western part is lower than it would be in the eastern part of the Indian Ocean.

Table 1. The monthly difference of sea surface height to ENSO phenomena

Waktu	Jan	Feb	Mar	Apr	Mei	Jun	Jul	Agu	Sep	Okt	Nov	Des
1993	-0.06	0.00	-0.02	-0.10	-0.08	-0.06	-0.06	-0.03	-0.02	0.01	-0.01	-0.07
1994	-0.09	-0.05	0.00	-0.03	-0.05	-0.02	-0.01	0.04	0.01	0.05	0.03	-0.05
1995	-0.08	-0.08	-0.03	-0.03	-0.03	-0.01	0.03	0.06	0.06	0.05	0.05	0.01
1996	-0.03	-0.02	-0.02	0.02	-0.03	-0.04	-0.01	0.01	0.04	0.03	0.03	-0.03
1997	-0.10	-0.06	-0.09	-0.07	-0.10	-0.08	-0.07	-0.04	-0.05	-0.06	-0.07	-0.10
1998	-0.09	-0.07	-0.04	0.00	0.01	0.03	0.01	0.04	0.06	0.08	0.04	-0.01
1999	0.01	0.05	0.07	0.04	0.00	0.00	0.04	0.05	0.07	0.10	0.04	0.00
2000	0.00	0.03	0.04	0.00	-0.03	-0.01	0.02	0.03	0.05	0.03	0.01	-0.03
2001	-0.01	0.03	0.02	0.00	-0.02	-0.01	-0.02	0.01	0.04	0.05	0.01	-0.05
2002	-0.09	-0.03	-0.07	-0.06	-0.06	-0.04	-0.03	-0.04	-0.05	-0.07	-0.10	-0.14
2003	-0.12	-0.11	-0.03	-0.01	-0.01	-0.05	-0.02	0.00	0.02	0.06	-0.01	-0.04
2004	-0.04	0.00	-0.02	-0.06	-0.08	-0.06	-0.05	-0.01	0.04	0.03	0.02	-0.03
2005	-0.05	-0.05	-0.02	-0.03	-0.02	-0.01	0.01	0.01	0.04	0.03	0.01	-0.03
2006	-0.02	0.01	0.05	0.03	0.01	0.02	0.01	0.01	0.02	0.03	-0.01	-0.04
2007	-0.04	-0.05	-0.02	-0.04	-0.05	-0.01	0.03	0.05	0.08	0.11	0.09	0.06
2008	0.07	0.10	0.08	0.02	-0.03	0.00	0.05	0.09	0.08	0.05	0.03	0.05
2009	0.02	0.03	0.02	0.01	-0.03	-0.01	-0.01	0.01	0.03	-0.01	-0.05	-0.11
2010	-0.12	-0.11	-0.06	-0.01	0.01	0.04	0.04	0.04	0.06	0.02	0.03	0.00
2011	0.01	0.01	0.01	-0.04	-0.04	-0.04	-0.02	0.02	0.09	0.10	0.07	0.03
2012	0.02	0.00	0.03	-0.03	-0.02	0.02	0.04	0.08	0.07	0.06	0.00	0.00
2013	0.01	-0.01	0.02	0.05	0.02	0.00	0.00	0.02	0.05	0.04	-0.01	0.01
2014	-0.02	-0.03	-0.08	-0.09	-0.10	-0.07	-0.04	-0.02	0.00	0.01	-0.01	-0.05
2015	-0.08	-0.13	-0.12	-0.11	-0.13	-0.14	-0.11	-0.11	-0.07	-0.07	-0.09	-0.13
Rata-rata	**-0.04**	**-0.02**	**-0.01**	**-0.02**	**-0.04**	**-0.02**	**-0.01**	**0.01**	**0.03**	**0.03**	**0.00**	**-0.03**

Note: Red (El Nino) and Blue (La Nina)

Presented in a comparison chart between Nino 3.4 SST anomaly for differences sea surface height, can be seen the opposite pattern between these two parameters. When Nino 3.4 SST anomaly is positive then the difference between sea surface height is negative, whereas when the anomaly Nino 3.4 is negative then the difference between sea surface height be a positive value

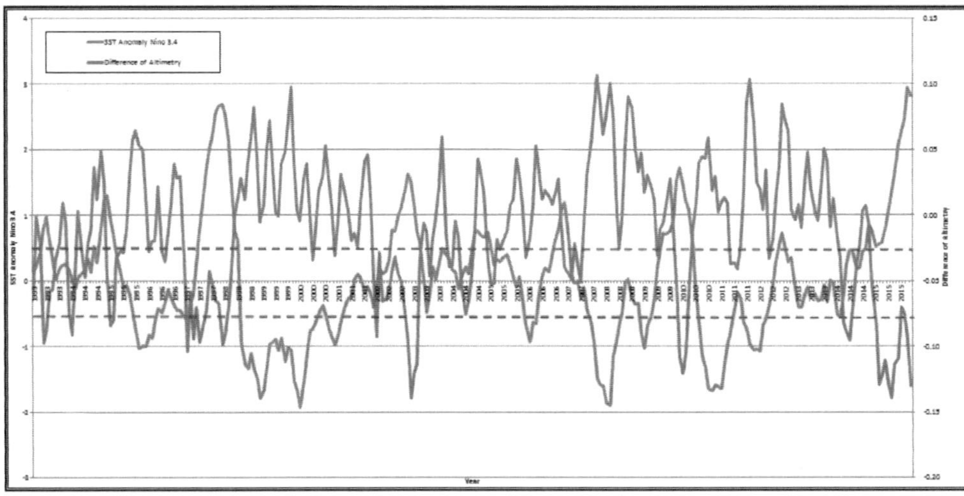

Figure 2. The comparison table between SST anomaly Nino 3.4 and difference of sea surface height (western Pacific Ocean and eastern Indian Ocean)

If it compared to the average value of the difference between sea surface height during the years 1993 - 2015 against the average difference sea surface height during El Nino and La Nina clearly indicates that at the time of El Nino in the western Pacific Ocean will be lower than the Indian Ocean to the east and conditions the opposite occurs during La Nina.

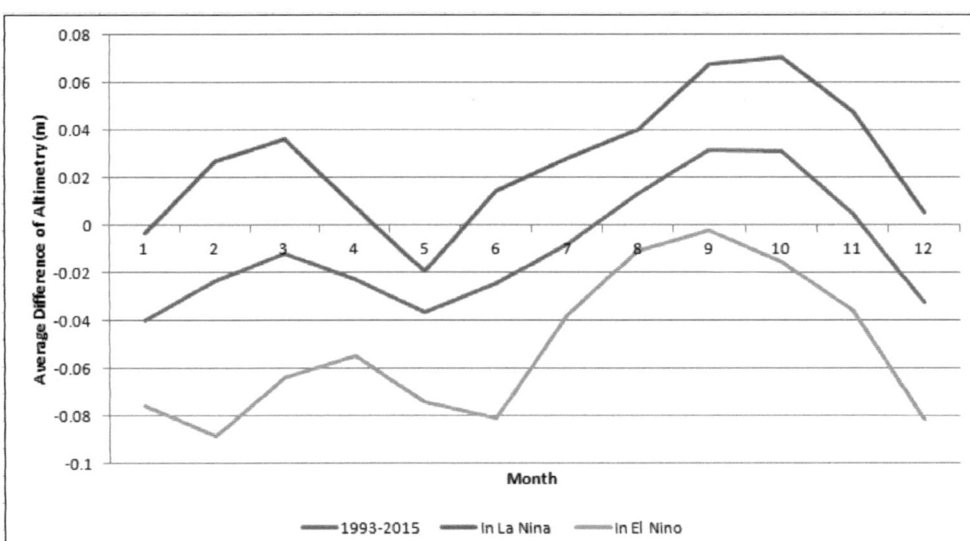

Figure 3. The comparison table between average difference of sea surface height in 1993-2015, El Nino, and La Nina

4. Discussion

From the calculation results obtained, shows that the sea level or sea surface height in the western Pacific Ocean and the Indian Ocean east section varies with ENSO events. In previous references mentioned by Hasanuddin (1998) [4] that the western Pacific ocean are higher than the Indian Ocean there are differences in the eastern part of the data processing is performed in this study. Sea surface height altitude in the western Pacific Ocean can also be lower than the height of sea surface height in the eastern Indian Ocean.

The characteristic of ITF itself throughout the current year is always moving from the Pacific Ocean to the Indian Ocean [4] if it is associated with the calculated variation sea surface height in both locations, will be related to the speed of flow and transport ITF itself. In a study conducted Safitri the ITF in 2012 in the Timor Sea, showed that when La Nina transport ITF will be higher than during El Nino. Similar to Safitri's research in 2012, previous studies during a La Nina debit ITF transport will be larger. If it is associated with the difference between sea surface height results during ENSO then there is a match against former research.

In the event of La Nina sea surface height in the western Pacific Ocean will be higher than the sea surface height in the eastern Indian Ocean. Sea surface height higher in the north will cause a larger gradient for supporting normal conditions ITF itself that moves from the Pacific Ocean to the Indian Ocean to transport ITF becoming stronger. Whereas in the event of El Nino sea surface height in the western Pacific Ocean become leih low compared with sea surface height in the eastern part of the Indian Ocean thus undermining the normal conditions of the ITF itself that causes the transport ITF becoming weaker.

5. Conclusion

ENSO effect on sea surface height variation in the western Pacific Ocean and the Indian Ocean east section which relate to the ITF transport. The size of the ITF transport associated with sea surface height conditions at both sites were impacted strengthen or weaken the normal conditions ITF. So, it makes a pressure gradient variations that cause variations in transport ITF.

References

[1] Pranowo, W. S., Kuswardhani, A. R. T. D., Kepel, T. L., Kadarwati, U. R., Makarim, S., Husrin, S. 2005. Menguak Arus Lintas Indonesia : 11-12 p. Pusat Riset Wilayah Laut & Sumberdaya Non-hayati Badan Riset Kelautan dan Perikanan Departemen Kelautan & Perikanan, Jakarta. ISBN: 979-3768-06-1.

[2] Kane, R.P. 1997. Relationship of El Niño-Southern Oscillation and Pacific Sea Surface Temperature with Rainfall in Various Region of the Globe.Mon. Wea. Rev., 125:1792-1800 p.

[3] Ropelweski, C.F dan M.S. Halpert. 1987. Global and Regional Scale Precipitation Pattern Associated with the El Niño /Southern Oscillation.Mon. Wea. Rev., 115: 1606-1626 p.

[4] Safitri, M., Cahyarini, S. Y., Putri, M. R. 2012. Variasi Arus ITF dan Parameter Oseanografi di Laut Timor Sebagai Indikasi Kejadian ENSO. Jurnal Ilmu dan Teknologi Kelautan Tropis. Departemen Ilmu dan Teknologi Kelautan FPIK IPB, Bogor.

[5] Mayer, B., Damm, P. E., Pohlmann, T., Rizal, S. 2010. What Is Driving The ITF? An Illumination of The Indonesian Throughflow with A Numerical Nested Model System. Dynamics of Atmospheres and Oceans, 50. Elsevier. doi: 10.1016/j.dynatmoce.2010.03.002.

[6] Aken, H.M., 2005. Dutch Oceanographic Research in Indonesia in Colonial Times. Oceanography 18 (4): 30–41 p.

[7] Masumoto, Y., 2002. Effects of Interannual Variability in The Eastern Indian Ocean on The Indonesian Throughflow. Journal of Oceanography, 58: 175–182 p.

[8] Gordon, A.L., Susanto, R.D., Ffield, A., 1999. Throughflow within Makassar Strait. Geophysical Research

Letters 26 (21): 3325–3328p.

[9] Potemra, J. T., Sprintall, J., Hautala, S. L., dan Pandoe, W. 2003. Observed Estimates of Convergence in The Save Sea, Indonesia. J. Geophys. Res., 108(C13): 3001 p. doi:10.1029/2002JC 0015 07.

[10] Oktaviani, A. P. P., 2008. Analisis Sinyal El Niño Southern Oscillation (ENSO) dan Hubungannya dengan Variabilitas Arus Lintas Indonesia di Selat Lifamatola. Jurnal Ilmu Kelautan. Desember 2008, 13(4): 209 – 218 p. ITB, Bandung.

[11] AVISO Archiving, Validation and Interpretation of Satellite Oceanographic. 2008. AVISO and PODAAC user Handbook IGDR and GDR Jason Product4th ed. AVISO, France, 130 p.

[12] Huang, Y. S., Lee, T. Q., Hsu, S. K. 2001. Milankovitch-scale Environmental Variation in The Banda Sea Over The Past 820 ka: Fluctuation of The Indonesian Throughflow Intensity. Journal of Asian Earth Science. Elsevier. doi: 10.1016/j.jseaes.2010.08.011.

Antioxidant activity of three microalgae *Dunaliella salina, Tetraselmis chuii* and *Isochrysis galbana* clone Tahiti

Ita Widowati[1], Muhammad Zainuri[1], Hermien Pancasakti Kusumaningrum[2], Ragil Susilowati[3], Yann Hardivillier[4], Vincent Leignel[4], Nathalie Bourgougnon[5], Jean-Luc Mouget[4]

[1]Department of Marine Science, Faculty of Fisheries and Marine Science, Diponegoro University, Jl. Prof. Soedarto SH, Tembalang, Semarang -50275 , Indonesia

[2]Department of Biology, Faculty of Sciences and Mathematic, Diponegoro University, Jl. Prof Soedarto SH, Tembalang, Semarang -50275, Indonesia

[3]Alumny Master of Coastal Ressource Management, Faculty of Fisheries and Marine Science, Diponegoro University, Jl. Prof.Soedarto SH Tembalang, Semarang-50275 , Indonesia

[4]Laboratoire du Mer Molécule et Santé, Université du Maine, Le Mans, France

[5]Laboratoire de Chimie Biologie Marine, Université de Bretagne Sud, Vannes, France

Email : ita_jusup@yahoo.co.id

Abstract. Natural alternatives antioxidant source has become a trending topic in the past decades to replace synthetic antioxidant. Microalgae have been mentioned to show interesting bioactive properties and one of them is its antioxidant activity. This study aims to evaluate the potential of three microalgae *Dunaliella salina, Tetraselmis chuii* and *Isochrysis galbanaas* new source of natural antioxidant. Proximate analysis and total phenolic content of *D. salina*, *T. chuii* and *I. galbanas* were determined. Antioxidant activity of methanolic extracts of these three species prepared in different concentration (50, 100, 250, 500, and 1000 ppm) was performed through DPPH assay. *I. galbana* clone Tahiti demonstrated a highest antioxidant potential with 61.64 of inhibition at 50 ppm followed by *D. salina* with 58.45 % of inhibition and *T. chuii* with 52.58 % of inhibition. *I. galbana* clone Tahiti was the best antioxidant with total phenol content of 17.798 mg GAE g^{-1}extract at 50 ppm; followed by *T. chuii* 16.868 mg GAE g^{-1}extract and the lowest was *D. salina* with 4.672 mg GAE g^{-1}extract. Results suggest that these microalgae posses antioxidant potential which could be considered for future applications in medicine, dietary supplements, cosmetics or food industries.

Keywords: antioxidant activity, DPPH, microalgae, total phenol.

1. Introduction

Indonesia, as an archipelagic state surrounded by ocean, presents a potential as a microalgae source because of its huge natural biodiversity. Microalgae are distributed everywhere: in salt, brackish or freshwater, in tropical to cold regions and sometimes as symbionts with other organisms, and some species are used and cultivated by farmers.

Microalgae, also known as phytoplankton, are common name of all the aquatic autotrophic organisms living in suspension in the water column. This name refers to several phyla, mostly eukaryotes but including the photosynthetic prokaryotes known as cyanobacteria. As photosynthetic organisms, this group plays an important role in the productivity of oceans and constitutes the basis of the marine food chain. Moreover, phytoplankton is responsible for half of the oxygen released in the atmosphere.

In the last decades, an increased attention has been paid to the commercial and industrial potential of microalgae. Several species are currently being studied for their ability to synthesize valuable secondary metabolites (pigments, lipids, caretonoids, etc) for biofuel production, pharmaceutical industry or aquaculture applications [1]. Other fields of investigation include nanotechnologies, environmental survey, forensic sciences and paleontology. Similarly, some micro-algae contain and/or excrete pharmacologically active compounds. For example, the dinoflagellates *Gymnodinium* sp. and *Gonyaulax* sp. produce an alkylguanidine compound that affects the central nervous system. Brominated bi-indoles of *Rivularia firma* show pharmacological activity.

Among different compounds with functional properties, antioxidants are the most widely studied since the interest in finding new, safe and powerful antioxidant from natural sources is growing nowadays. Moreover, consumers have demonstrated the important role of antioxidants in human health thus increasing the interest in such products and their demand. Most of commercially available natural antioxidants are extracted from terrestrial plants, such as cocoa, tea grape and rosemary [2]. It appears that unicellular microalgae shows to have promising potential as alternative source of antioxidants [3-5]. Microalgae have been considered as important source of bioactive compounds since they are sustainable resources easily to culture that do not require an arable land unlike the terrestrial plant. Some studies have reported the interesting and remarkable antioxidant potential of microalgae [6-8]. Additionally, microalgae species like *Dunaliella*, *Chlorella*, *Isochrysis*, *Nannochloropsis*, *Nannochloris*, *Chlamydomonas*, *Haematococcus* and *Spirulina* are available to grow in mass production [1].

In regard of these biotechnological challenges, there is a constant effort actually provided for both finding and exploiting new microalgal resources and developing their putative commercial outcomes or industrial valorizations. Thus, this research aims to characterize microalgae of interest for aquaculture for their potential biological activities, especially in antioxidant properties.

2. Material and Methods

2.1. Culture of microalgae

Three species of microalgae, *Dunaliella salina*, *Tetraselmis chuii*, and *Isochrysis galbana* clone Tahiti, were cultured in a batch culture with ratio of algae and seawater was 1:3 (volume). Walne medium was used as nutritional compound for the culture, with ratio of 1:1 (mL L^{-1}); with luminous intensity of 1500–3000 lux and temperature of 23-25^{0}C [9].

The density of the microalgae culture was measured daily with haemocytometer by using binocular microscope. The microalgae were homogenized before the measurement. The cell density was calculated by following formula:

$$\boldsymbol{N} = \frac{(N_1 + N_2)}{2} \times \frac{1}{1.02\,mm^2 \times 0.1\,mm} \times \frac{1\,mm^2}{10^{-3}ml}$$

Where N = cell density (cell.mL^{-1}), N_1 = total cell in 80 small square (replica 1), N_2 = Total cell in 80 small square (replica 2), 0.2 mm = wide of haemocytometer in 80 square, and 0.1 mm = the depth liquid on haemocytometer.

Microalgae were harvested on the stationary phase of culture. The biomass of microalgae was obtained with centrifugation at 5.000 rpm for 10 min and then left to dry for a few days to obtain the dry biomass [10].

2.2. Extraction

Dry biomass was extracted by using methanol solvent with sonication at 50 Hertz for 15 min. The solvents were evaporated by using rotary evaporation [11].

2.3. Antioxidant activity

Antioxidant activity study was divided into 2 steps: first was to identify and drying the specimen. Second step was the extraction of dry specimen and bioactive compound analysis including alkaloid, triterpenoid/steroid, saponin, phenol, flavonoid, tannin [12]. The total phenol test was performed according to [13] and antioxidant activity test with DPPH method [14].

2.4. DPPH antioxidant activity

DPPH antioxidant activity of methanol extract from *D. salina, T. chuii,* and *I. galbana* clone Tahiti were determined according to [15]. The assay was performed at 50, 100, 250, 500, and 1000 ppm. Extract solution (1 mL) (in triplicates) was mixed with 3 mL of 60 μM methanolic solution of DPPH radicals. The mixture was put in the dark for 30 min before the absorbance was taken at 517 nm. The percent inhibition was calculated using the formula:

$$\% \, \boldsymbol{of\ inhibition} = \left\{\left(\frac{\boldsymbol{A} - \boldsymbol{B}}{\boldsymbol{A}}\right) \times \mathbf{100}\right\}$$

Where, A = absorbance of the control (DPPH) and B = absorbance of test sample.

2.5. Determination of total phenol content

Total phenol content of methanol extract from *D. salina, T. chuii,* and *I. galbana* clone Tahiti were determined according to the Folin-Ciocalteu method [16]. 300 μL of extract was dispensed into test tube (in triplicates) and it was added with 1.5 mL of Folin-Ciocalteu reagent (diluted 10 times with distilled water) and a 1.2 mL of Na_2CO_3 solution (7.5 % w/v). The solution was mixed the put 30 min at room temperature before the absorbance at 765 nm. Total Plate Counted was measured in mg g^{-1} Gallic Acid Equivalent (GAE). The calibration equation for Gallic Acid was Y = 0.0645x - 0.0034.

2.6. Phytochemical content

Methanol extracts prepared from all the three different microalgae were used to screen various phytochemicals such as tannins, flavonoids, steroids, saponins, and alkaloids [17].

2.6.1. Flavonoids

The algal extract was diluted and taken some of the water, and put into a test tube. Mg was then added and three drops of HCl into a test tube. Amyl alcohol then added and beat until separated. The reaction was positive if it forms a reddish yellow color to red.

2.6.2. Quinones

Five mL of extract obtained from the experiments to extract flavonoids identification was inserted into a test tube, a few drops of a solution of NaOH 1 N then added. The formation of the red color indicates the class of quinone compound.

2.6.3. Tannins

Algal extracts added to 2-3 drops of 1% $FeCl_3$ solution. Extracts will be positive if it contains tannins greenish-black or dark blue.

2.6.4. Alkaloids

Crude extract was added with 0.5 HCL 2% and was divided into two tubes. The first tube coupled with 2-3 drops of reagent Dragendorff and the second tube was added with 2-3 drops of reagent Mayer. When an orange precipitation was formed in the first tube and yellowish precipitation was shown by second tube, it indicated the presence of alkaloid.

2.6.5. Saponins
Extract was introduced into a test tube and added with distilled water that has been warmed. It was shaken and allowed to stand for 10 min. Samples showed positive results when the foam was formed and did not disappear until 15 min after the drops of HCl.

2.6.6. Terpenoids and Steroids
The extract was added with three drops acetic anhydride and one drop of sulfuric acid (H_2SO_4). The red color indicates terpenoids, while blue color indicates steroid. Samples showed positive results when the foam was formed and did not disappear until 15 min after the addition of HCl.

3. Results and Discussion

The activity of antioxidant of methanol extracts of three microalgae *D. salina*, *T. chuii*, and *I. galbana* clone Tahiti were studied in this research at different concentrations (1000, 500, 250, 100, 50 ppm). Results showed that *I. galbana* clone Tahiti at 250 ppm showed the best result than other concentration with 67.93 % inhibition. *D. salina* gave the best result of inhibition of 62.19 % at 500 ppm. While *T. chuii* showed the best result with 71.36 % of inhibition at 1000 ppm (Table 1).

Table 1. DPPH Radical Scavenging Assay from methanolic extracts of three microalgae *D. salina*, *T. chuii*, and *I. galbana* clone Tahiti at different concentrations (1000, 500, 250, 100, 50 ppm).

Microalgae	**Concetration (ppm)**				
	1000	**500**	**250**	**100**	**50**
D. salina	53.17	62.19	58.43	59.01	58.45
T. chuii	71.36	66.18	52.38	42.19	52.58
I. galbana clone Tahiti	53.77	55.87	67.93	64.02	61.64

Note: The results in % inhibition DPPH.

Table 2. Total Phenol Content from methanolic extracts of three microalgae *D. salina*, *T. chuii*, and *I. galbana* clone Tahiti at 4 concentrations (1000, 500, 250, 100, 50 ppm).

	Concentration (ppm)				
Microalgae	**1000**	**500**	**250**	**100**	**50**
D. salina	0.539	0.839	1.223	2.336	4.672
T. chuii	0.466	1.015	2.918	8.95	16.868
I. galbana clone Tahiti	0.678	1.408	3.167	8.02	17.798

Note : The results in mg GAE g^{-1} extract

Phytochemical screening using methanol of three marine microalgae strains was evaluated. The phytochemicals present in the microalgae strains were identified as flavonoids, tannin, alkaloids, saponin and steroids; while quinon was absent (Table 3).

Table 3. Screening of phytochemicals of methanol extracts of three microalgae

Microalgae	Flavonoid	Tannin	Alkaloid	Saponin	Steroid	Quinon
D. salina	+	+	+	+	+	-
T. chuii	+	+	+	+	+	-
I. galbana clone Tahiti	+	+	+	+	+	-

The use of methanol as solvent on extract of *D. salina* and *T. chuii* has been studied before [18]. Furthermore, on the screening on antioxidant and phytochemical contents of different solvent extracts (acetone, methanol, ethanol, chloroform) of five different microalgal strains including *Tetraselmis sp.*, *Dunaliella sp*; showed that the maximum antioxidant activity were found in acetone extracts, followed by methanol extracts of *Tetraselmis sp.* and *Dunaliella sp.*[18]. The study on screening on nine Moroccan microalgae by using the ethanolic extracts showed that the antioxidant activity of *Dunalliela sp* (IC50 283 ± 0.09 μg mL^{-1})., *Tetraselmis sp* (IC_{50} 247 ± 0.01 μg mL^{-1}) were highest antioxidant potentials among nine microalgae studied. *D. salina* and *T. chuii* were two potentials microalgae on antioxidant activites and DPPH was the simple method to elucidate the activity by using different solvents [4].

The total phenolic content of methanolic extracts of *D. salina*, *T. chuii*, and *I. galbana* clone Tahiti was evaluated using the Folin–Ciocalteu method. The phenolic content varied from 4.672 to 17.798 mg GAE g^{-1}extract at concentration of 50 ppm. Among five concentrations tested (1000, 500, 250 100 and 50 ppm), 50 ppm gave the best result. The highest phenolic content was found on *Isochrysis galbana clone Tahiti* (17.798 mg GAE g^{-1}extract) followed by *Tetraselmis chuii* 16.868 mg. g^{-1} GAE and the lowest was found in *Dunaliella salina* (4.672 mg GAE g^{-1}extract) (Table 2).

The result of this study on *I. galbana* clone Tahiti was higher than those found by [4] by using the ethanolic extracts of *Isochrysis* sp (13.4 ± 0.16 mg GAE g^{-1} extract) and by [2] on extracts of *Isochrysis sp.* (4.6 mg GAE g^{-1} extract). The total phenol content of methanolic extract of *Tetraselmis sp* in this study (16.868 mg. g^{-1} GAE) was higher than showed by [2] (3.8 mg GAE g^{-1} extract); meanwhile the result of this study was lower than those showed by [4] using the ethanolic extracts (25.5±1.5 mg GAE g^{-1} extract). In addition, total phenol content of methanolic extract of *D. salina* of this study (4.672 mg GAE g^{-1} extract) was also lower than those showed by [4] by using the ethanolic extracts (14.0 ± 0.43 mg GAE g^{-1} extract). It is suggested that the different results obtained between the present study and that of by [4] was due to the type of solvent. Indeed, different solvent used will generate different content of bioactive compound. Methanol is commonly used to extract bioactive compound from any organisms as this solvent dissolves various polar compounds and certain non-group of non-polar as well. Phenolic compounds are often extracted using polar solvent yet the most suitable solvents are aqueous mixtures containing ethanol, methanol, acetone and ethyl acetate [19][20]. In general, methanol is favourable for lower molecular weight phenolic compounds [21].

5. Conclusion

The methanolic extracts of three-tested microalgae *I. galbana clone Tahiti*, *T. chuii* and *D. salina* showed high antioxidant activities. The higher antioxidants potentials were obtained in *Dunalliela sp.*, *Tetraselmis chuii*. These microalgae posses antioxidant potential, which could be considered for future applications in medicine, dietary supplements, cosmetics or food industries. In the future however it is crucial to perform characterization of phenolic compounds from *D. salina, T. chuii, and I. galbana* clone Tahiti.

Acknowledgement

The authors wish to thank the Ministry of Research Technology and Higher Education, for funding this research in accordance with the Implementation Agreement Assignment Research Collaboration and International Publication, No. DIPA: -023.04.1.673453/2015, Date of 14 November 2014, DIPA Revision 01, Date of 29 February 2015.

References

[1] Talero E, García-Mauriño S, Ávila-Román J, Rodríguez-Luna A, Alcaide A, Motilva V. 2015. Bioactive Compounds Isolated from Microalgae in Chronic Inflammation and Cancer. Mar Drugs 13(10):6152–6209.

[2] Goiris K, Muylaert K, Fraeye I, Foubert I, De Brabanter J, De Cooman L. Antioxidant Potential of

Microalgae in Relation to Their Phenolic and Carotenoid Content. J Appl Phycol [Internet]. 2012 Dec 22 [cited 2017 Jan 26];24(6):1477–1486. Available from: http://link.springer.com/10.1007/s10811-012-9804-6

[3] Shanab SMM, Mostafa SSM, Shalaby EA, Mahmoud GI. Aqueous Extracts of Microalgae Exhibit Antioxidant and Anticancer Activities. Asian Pac J Trop Biomed [Internet]. 2012 Aug [cited 2017 Jan 26];2(8):608–615. Available from: http://www.ncbi.nlm.nih.gov/pubmed/23569980

[4] Maadane A, Merghoub N, Ainane T, El Arroussi H, Benhima R, Amzazi S, et al. Antioxidant activity of some Moroccan marine microalgae: Pufa profiles, carotenoids and phenolic content. J Biotechnol [Internet]. 2015 Dec 10 [cited 2017 Jan 26];215:13–9. Available from: http://www.ncbi.nlm.nih.gov/pubmed/26113214

[5] Lee S-H, Lee J-B, Lee K-W, Jeon Y-J. Antioxidant Properties of Tidal Pool Microalgae, *Halochlorococcum porphyrae* and *Oltamannsiellopsis unicellularis* from Jeju Island, Korea. ALGAE [Internet]. 2010 Mar 1 [cited 2017 Jan 26];25(1):45–56. Available from: http://koreascience.or.kr/journal/view.jsp?kj=JORHBK&py=2010&vnc=v25n1&sp=45

[6] Elsayed H, Ali A, Mahmoud S, Shanab M, Aly M, Abo-State M, et al. Screening of Microalgae for Antioxidant Activities, Carotenoids and Phenolic Contents.

[7] Guedes AC, Gião MS, Seabra R, Ferreira ACS, Tamagnini P, Moradas-Ferreira P, et al. Evaluation of the Antioxidant Activity of Cell Extracts from Microalgae. 2013. Mar Drugs. 11(4):1256–1270.

[8] Ahmed F, Fanning K, Netzel M, Turner W, Li Y, Schenk PM. Profiling Of Carotenoids and Antioxidant Capacity of Microalgae from Subtropical Coastal and Brackish Waters. Food Chem [Internet]. 2014 Dec 15 [cited 2017 Jan 26];165:300–6. Available from: http://www.ncbi.nlm.nih.gov/pubmed/25038679

[9] Andersen R. 2005. Algal Culturing Techniques, 1st Edition. Academic Press. 596 pp.

[10] Mishra, S. K, A Shrivastav, R Maurya, K Shailesh, Patidar S, Haldar S. Mishra. 2012. Effect of Light Quality on the c-phycoerythrin Production in Marine Cyanobacteria *pseudanabaena sp.* Isolated from Gujarat Coast, India. Protein Expression and Purification (80): 234-238. doi: 10.1016/j.pep.2011.08.011.

[11] Trianto, A, Hermawan, I, Suzuka, T, Tanaka, J. 2011. Two New Cytotoxic Candidaspongiolides from an Indonesian Sponge. Int'l Scholarly Res Network. doi: 10.5402/2011/852619.

[12] Harborne, J.B. 1973. Phytochemical Methods. London. Chapman and Hall, Ltd. pp. 49-188.

[13] Yangthong M, Hutadilok-Towatana N, Phromkunthong W. 2009. Antioxidant Activities of Four Edible Seaweeds from The Southern Coast of Thailand. Plant Foods Hum Nutr. 2009 Sep;64(3):218-23. doi: 10.1007/s11130-009-0127-y.

[14] Aranda RS, Lopez LAP, Arroyo JL, Garza BAA, Torres NW. 2009. Antimicrobial and Antioxidant Activities of Plants from Northeast of Mexico. *Evidence-Based Complementary and Alternative Medicine* 20(11): 1-6.

[15] Saranya, C, Hemalatha, A, Parthiban, C, Anantharaman, P. 2014. Evaluation of Antioxidant Properties, Total Phenolic and Carotenoid Content of Chaetoceros calcitrans, Chlorella salina and Isochrysis galbana. Int. J. Curr. Microbiol. App. Sci. 3 (8): 365-377.

[16] Irondi AE, Oboh G, Akintunde JK. 2012. Comparative And Synergistic Antioxidant Properties Of Carica Papaya And Azadarichta Indica Leaves. Int J Phar Sci Res. Vol. 3 Issue 12.

[17] Sanjeet K, Kabi M, Kumari M. 2010. Study on phytochemicals analysis from leaves of Bixaorellana. Emerging Science 2:5.

[18] Rajendran N, K Selvan B, S Piriya P, V Logeswari, Kathiresan E. Tamilselvi A, J.Vennison S 2014.Phytochemicals, Antimicrobial and Antioxidant Screening from Five Different Marine Microalgae. J Chem and Pharmaceut Scie. : 78 - 85.

[19] Do QD, Angkawijaya AE, Tran-Nguyen PL, Huynh LH, Soetaredjo FE, Ismadji S, et al. 2014. Effect of Extraction Solvent on Total Phenol Content, Total Flavonoid Content, and Antioxidant Activity of *Limnophila aromatica*. J Food Drug Anal 22(3):296–302.

[20] Hadiyanto H. , Sutanto A A, Suharto Y. 2014 Ultrasound assisted extraction of antioxidant from Coleus tuberosus peels. Carpathian Journal of Food Science and Technology.6(1),58-65

[21] Dai J, Mumper RJ. 2010. Plant Phenolics: Extraction, Analysis and Their Antioxidant and Anticancer Properties. Molecules 15(10):7313–7352.

Study on the effect of different concentration of *Spirulina platensis* paste added into dried noodle to its quality characteristics

T W Agustini[1], W F Ma'ruf[1], Widayat[2], B A Wibowo[1] and Hadiyanto[2]

[1]Fisheries Department, Faculty of Fisheries and Marine Science, Diponegoro University, Indonesia

[2]Chemical Engineering Department, Faculty of Engineering, Diponegoro University, Indonesia

E-mail: tagustini@yahoo.com

Abstract. High nutritional content of microalgae *S. platensis* is very good for human health especially for its protein and β-carotene. These compounds can be used for improving quality of food from nutritional point of view. This study was designed to observe the effect of different concentration of *S. platensis* paste to dried noodle quality based on physical, chemical and sensory perspectives. Material used in this study was *S. platensis* powder obtained from CV Neoalgae, Sukoharjo. The best treatment observed from preliminary study was then used for further study and compared to control treatment (without added *S. platensis* paste). The dried noodle resulted was then analyzed for elasticity, chemical (β-carotene, water, protein, ash, fat and carbohydrate) and sensory (hedonic). Parametric data obtained were analyzed using t-Test while non-parametric data was analyzed using *Kruskal-Wallis* followed by *Multiple Comparison* test. The study showed that dried noodle added with 9% S. *platensis* paste give significant effect ($P<0.05$) on elasticity; β-carotene, water, protein, ash, fat and carbohydrate and also sensory (hedonic). Dried noodle without added *S. platensis* paste has lower in β-carotene and protein compared to that of added *S. platensis* paste with the value of 0.06 mg/100 gr and 17.51 mg/100gr (β-carotene) and 8.88% and 38.6% (protein), respectively. Those, chemical composition of dried noodle added with *S. platensis* comply with Indonesian National Standard in term of protein, water, and ash content with sensory of $7.26 \leq \mu \leq 7.63$. Addition of *S. platensis* can significantly increase the nutritional of dried noodle especially protein (by 4 times) and β-carotene.

Key words: dried noodle, *S. platensis*, β-carotene, protein

1. Introduction

Malnutrition experienced by Indonesian people reached 50% or more than 100 million people. Nutrition problems are often missed and take a long time to overcome, which can bring a big problem. Generally there is a big macro nutrition problems could be identified, include Malnutrition disorder, Animea, and lack of vitamin A (Vitamin A deficiency).

Spirulina is a multicellular and one type of blue-green algae that belong to class of *Cyanobacterium* and *Spirulina* sp. as the species. Spirulina contains a lot of proteins, vitamins, essential amino acids and essential fats. Spirulina contains 60-70% protein, a lot of vitamin B12 dan pro-vitamin A (β-carotene) and mineral, especially iron (Fe) [1].

Some people often consume noodles as a substitute of rice because it is simple, easy to be processed and fast to be served. Nowadays noodle products are lack of nutrition. The proportion of wheat, which contains a lot of carbohydrate, are more dominant than the proportion of other nutrition, therefore we need to improve the noodle's nutrition, especially the proteins and minerals [2]. Many studies have been conducted on quality and processing improvement of dried noodle including efficiency on energy consumption for drying process [3], effect of additive and steaming process [4], the use of buckwheat flour as antioxidant [5] and wheat bran on instant and dried noodles [6], application on functional properties of chitosan-xylose that produce Maillard reaction product to improve quality of semi-dried noodle [7], . However, there is no study yet with concern on its quality improvement especially for protein and β carotene of dried noodle.

The addition of *Spirulina sp.* in producing dried noodle can increase its nutritional quality, because *Spirulina* contain high protein and β-carotene. Carotenoids which have β-carotene as the main compounds are zeaxanthin, echinenone, myxoxanthophyl, dan β-cryptoxanthin that present in *Spirulina*. β-carotene as an antioxidant, convert pro-vitamin A into vitamin A in human body. Nowadays *Spirulina* is available in high doses supplement. People believe that *Spirulina* can decrease the risk of cancer because of its antioxidant [8].

This study was aimed to know the effect of *S.platensis* paste addition to dried noodle's nutrition and to compare the quality of dried noodle with *S.platensis* paste addition and dried noodle without *S.platensis* paste addition, in physical, chemistry and hedonic perspectives.

2. Materials and Methods

2.1. Experimental Design

Experimental design used was experimental laboratories. According to Nazir (2005), experimental laboratories is designed to know whether there is a relation or not between each treatment and how much the relation influence to the treatment [9].

This study used paired T-test because comparing the 2 groups, dried noodle control, without the addition of *S.platensis* paste and dried noodle with the addition of 9% *S.platensis* paste. Paired T-test is a methods where the data used is paired [10].

The treatments were repeated three times for each parameter. The parameter observed on this research include β-carotene content test, proximate analysis (moisture, ash, protein, fat, and carbohydrate), and elasticity test as well as sensory analysis on dried noodle without addition of *S.platensis* paste and with addition of 9% *S.platensis* paste.

2.2. Materials

Material used for producing dried noodle consist of *S.platensis* paste made from *S.platensis* powder produced by CV Neoalgae Technology, Sukoharjo and hard wheat flour produced by PT Indofood Sukser Makmur. The other materials such as salt, CMC (carboxymethyl celullosa), basil leaf, egg and water were also used. The materials used for analysis of each parameters observed are presented in Table 1.

Table 1. Materials used for parameters analysis of dried noodle

No.	Parameters	Materials
1.	β-carotene	- Dried noodle, DMSO, Heptane, saturated KOH and methanol
2.	Moisture content	- Dried noodle
3.	Ash content	- Dried noodle
4.	Protein content	- Dried noodle - H_2SO_4, Selenium, NaOH 40%, H_3BO_3 2%, aquadest, MR and BCG, and HCl 0.1 N
5.	Fat content	- Dried noodle, aquadest, and Ethyl ether
6.	Carbohydrate content	- Dried noodle
7.	Elasticity	- Dried noodle
8.	Hedonic	- Dried noodle

2.3. Methods

2.3.1. Production of S.platensis Paste

Dried *S.platensis* was mashed in the mortar and filtered with mesh size of 80 mm to get the same size of dried *S.platensis*. Fresh basil was selected for fresh green and perfect leaves. Fresh basil was washed to clean the leaves and extraction was done by grounding the basil leaves using mortar without the addition of water.

The *S.platensis* paste was made by adding the *S.platensis* powder as much as the basil extracts concentrate. The ratio between *S. platensis* powder and basil extract is 1:3, for example 7% concentration was made by mixing 16.24 g of *S.platensis* powder and 48.78 g of basil extract. Analog for 9% and 11% concentration.

2.3.2. The Procedure of dried noodles manufactured

Manufacture of dried noodle added with *S.platensis* paste was conducted as follow: Preparation of *S.platensis* powder, mix with basil leaf to form paste. The basil leaf was used in order to eliminate fishy flavor of *S.platensis*. The paste was mixed and kneaded with wheat flour, salt, egg and water. The dough was mixed until homogenous, shiny appearance, soft, smooth, and not sticky. The formulation of the dried noodles was presented in Table 2. It was followed by forming and cutting of homogenous dough to produce noodle. The noodle was steamed at 90°C for 5 minutes and then dried at 50°C for 90 minutes. Pressing and rolling were done by put the dough to the roller press. The aim of this process was to smoothing the gluten fiber and to dividing the dough into pieces. The distance between the rollers can be set to get the desire thickness. The following steps were forming and steaming. Thin sheet noodles were put to the hammer grill that can change the noodles sheet into the wavy noodles shape, then the wavy noodles shape were steamed in the steamer in the boiling water (100°C) for 5 minutes. Drying was conducted after the steamed noodles were drained then it was dried in the vacuum drying for 90 minutes at 50°C then cooled it down.

Table 2. The Formulation of Dried Noodles

Ingredients	Composition (gr)			
	Control	7%	9%	11%
S. Platensis powder	-	16.24	20.88	25.52
Basil extract	-	48.72	62.64	76.56
Wheat flour	200	200	200	200
Egg	30	30	30	30
Salt	1	1	1	1
Baking powder	1	1	1	1

Note: *S.platensis* powder percentage was calculated from total weight of the control

3. Results and Discussion

3.1. Preliminary Study

The processing of dried noodles consists of some ingredients: wheat flour, salt, baking powder and eggs with the addition of 0%, 7%, 9% and 11% of *S.platensis* paste. The aim of adding the different concentration of *S.platensis* in dried noodles is to observe the highest concentration that acceptable by panelist based on hedonic score covering appearance, odor, flavor, color and texture.

3.2. Main Study

The main study was purposed to compare the quality and β-carotene content between dried noodle without addition of 9% *S.platensis* paste, and dried noodle with addition of 9% *S.platensis* paste. The addition of 9% *S.platensis* paste (w/v) was obtained from the preliminary study.

3.3.Parameters of Product Quality

3.3.1.Hedonic Test

The quality test of dried noodles was referring to SNI-01-2987-1992, this method was used to measure the panelists' acceptance of the products by using the score sheet. The panelists' acceptance can be different based on the quality range. The test can be changed to numbers and do the statistics analysis to make the conclusion.

3.3.2.β-carotene Test [11]

Extraction

The sample was weighed about 30 g and put to the test tube and add 3 g of some piece of glass then 2.5 mL DMSO in each test tube. Vortex each tube for 30 seconds and then place the tube into 50°C water bath for 30 mins, in every 10 mins remove the tubes from the water bath and vortex for 30 seconds until end. Add 5 mL of ethanol to each tube, close the tube and vortex it for 30 seconds. Centrifuge the tubes at 4200 rpm for 3 mins, with dropper, take the supernatant from each tube and put it into the funnel. Add 4 mL of methanol, closed the tube, vortex it for 15-30 seconds and centrifuge for 3 mins. Collect all the supernatant into the funnel. Add the methanol to cover some piece of glass. Close the tube and vortex it for 30 seconds. Remove the cover and add 4 mL of methanol. Close the tubes and vortex for 30 seconds, do until the methanol is evident. The remaining color in the methanol is β-carotene. After all supernatant are collected add the methanol until 25 mL and centrifuge when it is required.

β-carotene Analysis

Take 8 mL of 25 mL methanol extract in the funnel with dropper with dropper and put it into the centrifuge tube of 115 mL. Add 5 mL of heptane and 1.5 mL of saturated KOH in methanol. Close the tube without mixing. Place the tubes in a dark place to saponification for 15 mins, vortex the tube lightly. After 30 mins, vortex tube for 15 seconds and centrifuge for 3 mins at 4200 rpm. Take heptane layer with a dropper and put it into 10 mL cylinder. Add 1 mL of fresh heptane into the tube and wash the interphase. Add 3 mL more of fresh heptane into the tube. Close the tube and shake lightly 8 times to make sure the heptane was added to the remaining β-carotene. Let heptane separated from methanol in 2 mins then take heptane with dropper to 10 mL cylinder. Then pour about 5 mL heptane extract into a clean tube then add some water up to maximum level and vortex for 5 seconds. Centrifuge the tubes for 3 mins at 4200 rpm. With the spectrophotometer, read the absorbance at 436 nm of the extract of the blank (heptane).

$$\beta\text{-carotene \%} = \frac{\text{Abs } 436 \times 25 \text{ mL} \times 1.25 \times 100 \times 0.84}{196 \times (\text{weight (mg)} \times \text{dry weight})}$$

Ash Content Test [12]

Put 4 gram wet sample to porcelain crucible and put it to the oven at 60-105°C for 8 hours. Dried sample was burned using a hot plate until smokeless time for ± 20 minutes. The sample was ash in a furnace at 600°C for 3 hours and weight it.

$$\text{\% Ash Content} = \frac{\text{ash weight (g)} \times 100\%}{\text{Sample weight}}$$

Protein Content Test [12]

There are 3 steps in protein analysis, destruction, distillation, and titration. Protein was determined using micro Kjeldahl methods. Put 0.25 gram sample to 100 ml Kjeldahl tube, add 0.25 gram selenium and 3 ml concentrated H_2SO_4. Destruction the sample at 410°C for about an hour then cool it down. Add 50 ml aquadest to Kjeldahl tube and 20 ml NaOH 40 %, then distillation the sample at 100°C. Put the distillated sample in a 125 ml flask that contain 10 ml boric acid (H_3BO_3) 2% and 2 drops of methyl red indicator. If the volume of distillate reached 40 ml and bluish green, stop the distillation process, then distillate was titrated with 0.1N HCl until it changes color into pink.

$$\text{\%N} = \frac{(\text{ml HCl sample} - \text{ml HCl blank}) \times \text{N HCl} \times 14.007 \times 100}{\text{mg sample}}$$

Protein content (%) = Conversion Factor x %N

Fat Content Test [12]

Fat content used 0,5 ml sample, covered with filter paper and placed on the extraction tool. Pour hexane solvent into the flask accordance with the size of the soxhlet reflux for at least 16 hours until the solvent drops back into the tube. The solvent in the flask was distilled and collected. Dried the tube that contain fat extraction in an oven at 105°C for 5 hours. Cool it down in a desiccator for about 20-30 mins.

$$\text{\% Fat Content} = \frac{\text{Fat weight (g)} \times 100\%}{\text{Sample weight}}$$

Carbohydrate Test [12]
Carbohydrate analysis was calculated by different method of reduction of 100% with known moisture content, ash content, protein content and fat content. Carbohydrate affect other nutrients compounds.

% Carbohydrate = 100% - (Moisture content + Ash content + Protein content + Fat content)

Elasticity Test [12]
Elasticity is the rate of material change from its original shape after the external force is applied. Cut the sample with a diameter of 3 cm and a thickness of 3 cm. Put the sample under the probe and press the sample twice. The pace of decline is 3 mm per minute compared to 1: 1 at a rate of graph paper. Elasticity was calculated by divided the second graph high pressure (H2) with the top height chart after first pressure (H1).

RESULTS AND DISCUSSION

Preliminary Study

The first step of this study was making dried noodle with the adding of 0%; 7%; 9%; and 11% *S.platensis* paste. The aim of adding the different concentrate of *S.platensis* in dried noodles is to know the highest concentrate that acceptable by panelist using organoleptic score sheet. The result of hedonic test of *S. platensis* paste added into dried noodle was presented in Table 3.

Dried noodle with addition of 9% *S. platensis* paste performed highest hedonic value of $7.26 \leq \mu \leq 7.63$. Hedonic test is conducted to evaluate consumer's acceptance comprehensively on the product attributes, and usually consumer tend to select food that has attractive performance as a whole product [13]. Thus based on preliminary study, among the concentration treatments applied, it was selected that 9% addition of *S.platensis* paste will be applied on main study.

Table 3. Hedonic Test of Dried Noodle with Different concentration of *S.platensis* paste

Concentration of *S. platensis* paste	Hedonic Scale
0%	$6.21 \leq \mu \leq 7.24$
7%	$7{,}08 \leq \mu \leq 7{,}31$
9%	$7{,}26 \leq \mu \leq 7{,}63$
11%	$6{,}23 \leq \mu \leq 6{,}50$

Main study

The chemical parameters of dried noodle with the addition of 9% *S.platensis* pasta compared to that of 0% S.platensis was presented in Table 4.

Table 4. The chemical parameters of dried noodle with the addition of different concentration of *S.platensis* paste

Parameters	*Concentration of S.platensis* Paste	
	0 %	9 %
β-carotene (mg/100g)	0.06 ± 0.007	1.51±0.305
Moisture (%)	4.25 ± 0.12	5.61±0.28
Ash (%)	1.37 ± 0.115	2.46±0.155
Protein (%)	8.88±0.39	28.603±0.33
Fat (%)	1.76±0.11	2.18±0.04
Carbohydrate (%)	83.72±0.38	61.12±0.55

β-carotene Content

Based on paired T-test, the result of β-carotene content showed significantly different result. Based on this result, the addition of *S.platensis* pasta showed significantly different result to β-carotene content in dried noodle. *S.platensis* contained rich in β-carotene with the amount of 58 mg/100 g [14]. Basil leaf has been proven not contained β-carotene as it has only 0.0006 mg/100 g of β-carotene. Moreover, highest carotenoid containing in *S.platensis* is β-carotene with the value of 26.74% [15]. β-carotene containing in dried noodle added with *S platensis* paste exceed the limit of β-carotene suggested by WHO. High consumption of natural β-carotene can be converted into free Vitamin A (retinol). Daily β-carotene consumption for supply nutrition ideally is 4 mg/100 g [16]. Decrease in β-carotene on dried noodle was due to heating process applied during processing as well as concentration used of *S. platensis* paste. High temperature process can reduce β-carotene and also possibility for oxygen exposure of the product which results in enzymic oxidation of β-carotene by lipoxygenase enzyme that produce all trans- β-carotene [17]. Moreover, β-carotene is unstable to light and heat at temperature more than 60°C and convert the form into cis- β-carotene that is unstable due to isometric change [18].

Moisture Content

The paired T-test showed that P (T<=t) one-tail (0,001) < (p=0,005), the addition of *S.platensis* pasta showed significantly different result to moisture content in dried noodle. Water content is an important parameter to determine the characteristic of product, because water content influence appearance, texture, flavor of food [19]. Dried noodle added with 9% *S.platensis* has higher water content comparing to that of 0% concentration of *S.platensis* paste. This phenomenon was due to addition of *S.platensis* paste which has basil leaf extract containing water. *S.platensis* powder has low water content of <10%, however it has hygroscopic and non-polar properties that can react with water so that addition of 9% *S.platensis* paste has increased water content of dried noodle.

Ash Content

The result of ash content in dried noodle without addition of *S.platensis* paste is 1.37% and dried noodle with addition of 9 % *S.platensis* paste is 2.46% and this profound that there is an increase in ash content. Based on this result, the addition of *S.platensis* paste showed significant different to ash content of dried noodle. Ash content is an important parameter to food nutrition. Ash is an inorganic substances resulted from the combustion of an organic material. The high result of ash content was affected by high mineral content of *S.platensis* or the materials have been contaminated. Ash content can be used as the contamination indicator such as sand and stone [12]. The maximum ash content of dried noodle is 3% [20].

Protein Content
Protein is an important component that affecting the quality of noodles both nutrition and elasticity. The paired T-test showed that P (T<=t) one-tail (0,000) < (p=0.005). The addition of *S.platensis* paste showed significant different to protein content of dried noodle. Dried *S.platensis* contains 55-75% of protein [14]. Protein is important parameters for enzyme formation, antibody and hormones. Protein can be used as energy. Protein complex is proteins that contain all the essential amino acid that suitable for the growth purposes and needs [21].

Fat Content
The paired T-test showed that P (T<=t) one-tail (0,000) < (p=0,005), the addition of *S.platensis* pasta showed significantly different result to fat content in dried noodle. The fat function in human body is as a source of energy, a source of essential fatty acids, fat soluble vitamin transportation, to save the use of proteins in the body, to give a sense of satiety and delicacy and to protect organs [21]. Moreover, the fat function in food is to give savory flavor and as an energy savory in human body.

Carbohydrate Content
Based on this result, the addition of *S.platensis* pasta showed not significantly different result to carbohydrate content in dried noodle. The low moisture content cause the high carbohydrate content. Carbohydrate content was determined by difference of total percentage (100%) reduced by moisture content, ash content, fat content and protein content [22].The physical and sensory result of dried noodle with the addition of 9% *S.platensis* pasta presented in Table 4.

Table 4. The physical result of dried noodle with the addition of different concentration of *S.Platensis* paste

Parameters	Cocentration of *S.platensis* Paste	
	0 %	9 %
Elasticity (gf)	2157.91±25.64	2430.187±16.89
Hedonic	$6.89 \leq \mu \leq 7.27$	$7.10 \leq \mu \leq 7.49$

Elasticity
Based on this result, the addition of *S.platensis* pasta showed significantly different result to elasticity in dried noodle. The elasticity of dried noodle was affected by the moisture content, so the dried noodle becomes more lenient. *S.platensis* is one of microalgae that are hygroscopic, therefore it is important to know the best drying methods to obtain constant dry weight [22]. The elasticity of wet noodle is depending on its protein [24]. Moreover not only the adding of *S.platensis* paste influence dried noodle elasticity, but also the flour that used in the making proses of dried noodle. The functions of wheat flour are to form the noodle structure, protein and carbohydrate source. Gluten that contained in wheat flour can influence the noodle elasticity and structure.

Hedonic Score
Hedonic is a subjective test intended to determine the level of preference of the panelists to a product that is tested. Each panelist rates the dried noodle parameters such as appearance, odor, flavor, and elasticity with score sheet. The panelist numbers are 30 people. Hedonic test is conducted to evaluate

consumer's acceptance comprehensively on the product attributes and usually consumer tend to select food that has attractive performance as a whole product [15]. The hedonic result showed that the panelists prefer dried noodle with the *S.platensis* paste addition.

Conclusion

Based on the study, it can be concluded that addition of *S.platensis* paste into dried noodle can improve its nutritional value, especially on protein and β-carotene. Dried noodle added with *S. platensis* comply with Indonesian National Standard in term of protein, water, and ash content with acceptable sensory score of $7.26 \leq \mu \leq 7.63$. Protein content of dried noodle added with *S. platensis* paste was 4 times higher than without *S.platensis*

References

[1] Belay A 2002 *J. Amer. Nutr. Association* 5 (2) : 27-48.
[2] Agustin I *et al* 2002 *Med. Pet.* 26 (2) : 52-59.
[3] Wang Z *et al* 2017 *Applied Thermal Engineering* 110: 941-948
[4] Gatode A A and Sahoo, A K 2015 *Food Sci.and Tech.* 52 (12) : 8395-8402
[5] Choy A L *et al* 2013 *J.of Cereal Sci.* 57: 281-287.
[6] Song X *et al* 2013 *J.of Cereal Sci.* 58:400-407
[7] Zhu K *et al* 2013 *Carbohydrate Polymers* 92: 1972-1977
[8] Ramadhani S 2010 Gramedia Pustaka Utama. Jakarta.
[9] Nazir M 2005 Research Method. Ghalia Indonesia, Bogor
[10] Kurniawan T 2008 Application of gelatin made of red snapper bone for making jelly candy (Thesis) (in Indonesian)
[11] Kristianingrum S 2010 Study on different method of Beta carotene analysis of food. (in Indonesian) Proceeding of National Seminar on Research, Education and Applied Science. Yogyakarta
[12] Sudarmadji S, Haryono B, Suhardi 1997 Analysis Procedure for Food and Agricultural Material. Liberty, Yogyakarta (in Indonesian)
[13] Soekarto 1985 Sensory assessment for Food industry and Agriculture Products. Jakarta : Bharata Karya Aksara (in Indonesian)
[14] Agustini T W *et al* 2015. *Procedia Environmental Sciences* 23 : 282-289.
[15] Sethu K M P 1996 Studies On Important Phytochemicals and Genetic Transformation of The Cyanobacterium *Spirulina*. [Thesis]. Plant Cell Biotechnology Department. Central Food Technological Research Institute Mysore. India
[16] World Health Organization (WHO) 2012 Technical Note: Supplementary Foods for the Management of Moderate Acute Malnutrition in Infants and Children 6-59 months of age. Geneva, World Health Organization
[17] Aisiyah I N 2012 Beta carotene, Protein, Iron and hedonic test of cracker substituted with *Ipomoea batatas l.*and anchovy (*Stolephorus sp.*) (Thesis) (in Indonesian)
[18] Hock Eng, C *et al* 2005 *American Journal of Environmental Sciences,* 1 (4) : 264-269.
[19] Afrianto E and E Liviawaty 1989 Fish Processing and Preservation (in Indonesian). Kanisius. Yogyakarta.
[20] National Standardization Agency 1992 SNI 01-2987-1992. Wet Noodle. (in Indonesian)
[21] Almatsier 2004 Basic Principle of Food Nutrition. PT Gramedia Pustaka Utama. Jakarta. (in Indonesian)
[22] Widyaningtyas M and Wahono H S 2014 *J. of Food and Agroindustry* (in Indonesian)
[23] Prabakusuma A S and Tri SU 2009 *J. Sains MIPA* 15 (3) : 167 – 176 (in Indonesian)
[24] Pomeranz Y and Shellenberger J A 1971 Bread Science and Technology. The AVI Publishing Company, Westport, Connecticut.

Study on Anticancer Activity of Extracts of Sponges Collected from Biak Water, Indonesia

A.Trianto[1,2], A. Ridhlo[2], D.W. Triningsih[2], and J. Tanaka[3]
[1] Natural Product Laboratory, Centre of Research and Services (Laboratorium Terpadu) Diponegoro University,
[2] Department Marine Science, Faculty of Fisheries and Marine Science, Diponegoro University, Semarang, Indonesia
[3] Department Chemistry, Biology, and Marine Science, University of The Ryukyus, 1-Senbaru, Nishihara, Okinwa, Japan.

E-mail:
agustrianto.undip@gmail.com

Abstract. Indonesia is center of biodiversity where marine sponges are abundant. a source of bioactive compounds with various pharmaceutical properties such as anticancer, antifungal, antibacterial, antioxidants, anti-inflammatory, and anti-malarial. In a continuation of a search for biologically active molecules from marine organisms we investigated the potency of marine sponges as anticancer. A total of 106 sponge specimens were collected between 3-40 m depths by SCUBA diving in Biak waters during August 2005. The specimens were extracted with methanol to provided crude extracts. The methanolic extracts were tested against NBT-T2 cell line. The assay result showed that 8.5 %, 29.2 % and 46.2 % of the extract have activity against the cell line at 0.1, 1.0 and 10.0 μg/mL. While, a 16.0 % of the extract did not showed activity against the cell line.

Keywords : sponge, anti-cancer, NBT-T2, screening, extract

1. Introduction

Sponges are the most primitive of multicellular invertebrate that distribute widespread in the marine ecosystem. Sponges are sessile animal with commonly have soft body that easy to be predated by various animal such as such as fishes, turtles, sea urchin, nudibranches, and flat worms [1][2]. Sponges also compete for space either with other sessile organisms or fouling organisms. For their survival sponges produce various classes of secondary metabolites such as alkaloid, steroid, glycosides, macrolides, and terpenoid [3].

Many researches showed that the secondary metabolites from sponges have pharmacological properties as anticancer, antibacterial, antifungal, antioxidant and other activities [4][5][6][7].
Cancer cells develop the resistance against the drug basically in two method i.e. by develop the membrane cell and development pumping system or known as mdr type 1 and mdr type 2. On the mdr type 1 the cell membrane will selectively pass the substrate that enter to cell. Once the cell recognize the anticancer drug the membrane will refuse it [8].

Marine organisms have proven to be an exceptionally valuable resource for bioactive natural products discovery efforts, particularly in the area of potential anticancer compounds. Metabolites isolated from marine samples often possess unique structural features and incorporate new or unusual assemblages of functional groups. They have demonstrated a broad range of potent biological activities, and there are currently at least 11 marine-derived compounds under clinical development in the U.S. and Europe [9]. In addition, there are currently two approved marine-derived drugs, the cone snail peptide Prialt for chronic nerve pain and the ascidian alkaloid Yondelis for soft tissue sarcoma treatment [10].

Indonesia is located in tropical area, blessed with a huge marine biodiversity. A number of bioactive compounds derived from Indonesian marine organisms were reported i.e. manadoperoxides originally isolated from a sponge *Plakortis* cfr. *simplex* posessed antimalarial activity [11]. A review on marine natural products in the period of 2001-2010 revealed that Indonesia is the second country that contribute on the finding of the new compounds after Japan [12]. This report indicate that Indonesia provide a good source for the drug development.

In our continuing search on bioactive compounds we collected sponges sample from Biak water. In this paper we described the bioactivity of the methanolic extracts from the marine sponges against NBT-T2 cells.

2. Materials and methods

2.1 Animal Material

The sponges were collected at 10-35 m depths with SCUBA in Biak, Papua-Indonesia on August 2008. The specimens were kept frozen until extraction [13].

2.2 Extraction and Isolation

After cutting into small pieces, the sponges specimens were soaked with methanol (laboratory grade) for 24 h. The extraction process was repeated for three times. Combined methanol extracts were concentrated by a rotary evaporator under vacuum to gave crude materials [13]. .

2.3 Bioassay for Screening process

NBT-T2, a cell line derived from chemically induced rat bladder carcinoma cells [13] was purchased from RIKEN and used for cytotoxicity screening and MTT assay. The sponge extracts were tested at concentrations of 0.1, 1 and 10 μg/mL in triplicates. The NBT-T2 cells were cultured in DMEM (Dulbecco's Modified Eagle Medium) in the presence of antibiotic-antimycotic (Sigma Aldrich, US), fetal bovine serum (Biowest, Germany), MEM non essential amino acid (Gibco, US) on 24 or 48 well plate (Falcon, US). After adding an aliquot of extracts, cells were incubated for 24 h under 5% CO_2 at 36°C. Then, the cells were observed under a microscope to evaluate cytotoxicity of the extracts [14].

3. Result and Discussion

Marine sponges produce secondary metabolite that can be utilized as prolific source of anticancer agents [4]. A cytotoxic screening of 106 marine sponges methanolic extracts were conducted by using MTT assay method against NBT-T2 cells. The result is displayed on Figure 1, about 84% extract was cytotoxic at a series of concentration levels, the rest 16% extract did not show citotoxicity. Among the active extracts, 46% showed moderate activity at 10.0 μg/mL, 29% showed strong activity at 1.0 μg/mL and 9% showed potent activity at 0.1 μg/mL on the cytotoxixity testing. These results suggested that marine sponges offer a promising anticancer candidates.

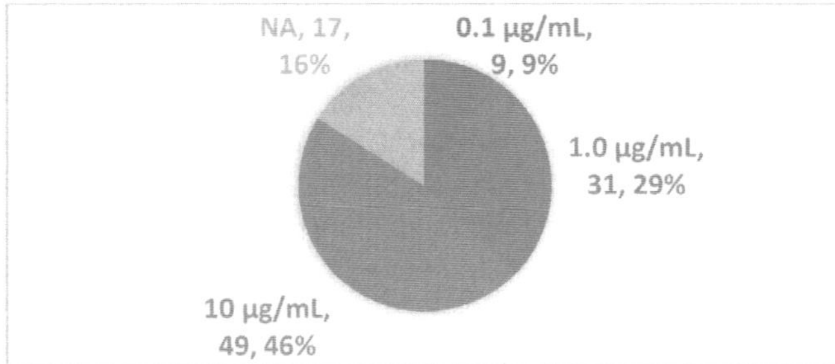

Figure 1. The percentage of the toxicity test result of the Biak collection sponges extracts against the NBT-T2 cells

Aaptos sp. *Strongylophora* sp.

Lyosina sp. *Phylospongia* sp.

Agelas sp. *Dysidea* sp.

Figure 2. Few examples of the sponges collected from Biak water.

Biak Water is located in Coral Triangle, an area housing of an abundant marine organisms. In this study, eighteen sponge genera were revealed from about 22 identified sponge specimens. Figure 2 showed some identified sponges that collected from Biak water during this study. Dysidea is one of the genus sponge that can be found in this area. Many studies reported that the genus of Dysidea is rich of bioactive compounds.

In our previous paper, we reported two new and four known bioactive compounds isolated from the sponge Dysidea sp originally from Biak Water (Figure 3.). Debromoantazirine showed moderate cytotoxicity against NBT-T2 cells with IC_{50} 4.7 µg/mL, while, the dyzidazirine was reported cytotoxic to be L1210 cells and inhibited *Pseudomonas aeruginosa*, *Candida albicans*, and *Saccaromyces cereviceae* [16] Furthermore, frondosin, neoavarol and antazirine were reported to be cytotoxic to PKC [17], inhibitors of IL-8 Rα and IL-8 Rβ receptors [18], and piscicidal [19] respectively.

Among the genera, *Theonella* extract showed significant cytotoxic activity which active in the lowest concentration 0.1 µg/mL. Marine organisms are recognized as a rich source of compounds that can inhibit HIV infection or modulate specific molecular targets associated with viral entry or replication.1,2 Within this group of antiviral marine metabolites is a family of cyclic sponge peptides with potent anti-HIV properties, Papuamides A-D were obtained from Papua New Guinea *Theonella mirabilis* and *T. swinhoei* samples [20]. Theopederins A-E were antitumor originally isolated from *Theonella* sp. against P388 murine leukemia cells [21]. Moreover, member of genus *Theonella* are recognized to have sterols that exibited anticancer activities [22].

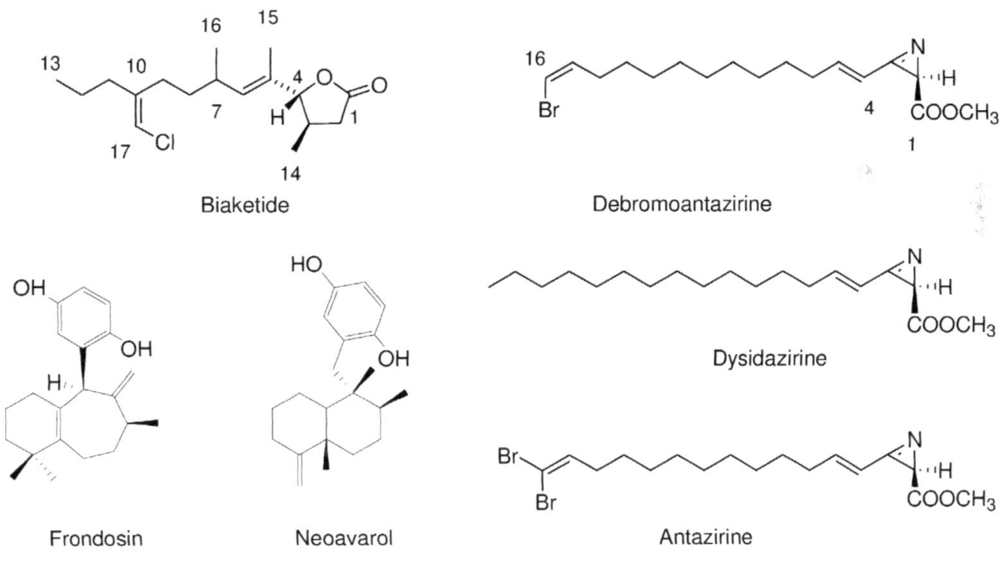

Figure 3. Some bioactive compounds isolated from the sponge *Dysidea* sp. collected from Biak, Papua [15]

4. Conclusion

In conclusion, the preliminary screening on cytotoxic assay of some marine sponges suggested that marine ecosystem is still providing a desirable drug mine. The sponge *Dysidea* sp is one of the promising candidate as drug source.

5. Acknowledgement

This research was partially funded by PharmaMar, S.A., Spain. We thank Mr. Eric Farwas of Biak DIVING for his assistance during collection.

References

[1] Braekman, J.C. and Dazole, D., 1986. Chemical defence in sponges. *Pure & Appl Chem.*, 58 (3): 357—364.
[2] Proksch P. 1994. Defensive roles for secondary metabolites from marine sponges and sponge-feeding nudibranchs. *Toxicon*. 32(6):639-655.
[3] Faulkner, D.J. 2000. Highlights of Marine Natural Products Chemistry. *Natural Product Report* 17: 1-6.
[4] Russo, P.; Nastrucci, C.; Cesario, A. 2011. From the sea to anticancer therapy. *Curr. Med. Chem. 18*: 3551-3562.
[5] Hughes, C.C.; Fenical, W. 2010. Antibacterials from the Sea. *Chemistry* 16 12512–12525.
[6] Kobayashi, M.; Kondo, K.; Kitagawa, I. 1993. Antifungal Peroxyketal Acids from an Okinawan Marine Sponge of *Plakortis* sp. *Chem.Phar. Bull.* 41 1324-1326.
[7] Liu, Y., H. Ji, J. Dong, S. Zhang, K. J. Lee and S. Matthew. 2008. Antioxidant Alkaloid from the South China Sea Marine Sponge *Iotrochota* sp. *Z. Naturforsch.* 63c: 636-638.
[8] Ullah, M.F. 2008. Cancer Multidrug Resistance (MDR): A Major Impediment to Effective Chemotherapy. *Asian Pacific J. Cancer Prev.* 9: 1-6.
[9] Butler, M. F. 2005. Natural products to drugs: natural product derived compounds in clinical trials. *Nat. Prod. Rep.* 22: 162–195.
[10] Mayer AMS, Glaser KB, Cuevas C, *et al.* 2010. The odyssey of marine pharmaceuticals: a current pipeline perspective. *Trends Pharmacol Sci.* 31:255–265.
[11] Fattorusso, C., Persico,M.,Calcinai, B. *et al.* 2010. Manadoperoxides A-D from the Indonesian Sponge Plakortis cfr. simplex. Further Insights on the Structure-Activity Relationships of Simple 1,2-Dioxane Antimalarials. *J. Nat. Prod.* 73: 1138-1145
[12] Mehbub M.H., Lei, J., Franco, C., and Zhang, W. 2014. Marine Sponge Derived Natural Products between 2001 and 2010: Trends and Opportunities for Discovery of Bioactives. *Mar Drugs*. 12(8): 4539–4577.
[13] Trianto, A., Ambariyanto and R. Murwani, 2004, Skrining Bahan Anti Kanker pada berbagai jenis Sponge dan Gorgonian terhadap L1210 cell line, Majalah Ilmiah Imu Kelautan. Vol. 9 (3) : 124-128.
[14] Herman, C, J., Vegt, P. D., Debruyne, F. M., Vooijs, G. P., Ramaekers, F. C. 1985. Squamous and transitional elements in rat bladder carcinomas induced by N-butyl-N-4-hydroxybutyl-nitrosamine (BBN). A study of cytokeratin expression. *Am. J. Pathol.* 120: 419-426.
[15] Trianto, A., de Voodg, N. J., and Tanaka, J. 2014. Two new compounds from an Indonesian sponge *Dysidea* sp. *Journal of Asian Natural Products Research*. 16(2): 163–168.
[16] Salomon C. E., Williams D.H., Faulkner, D.J., 1995. New azacyclopropene derivatives from Dysidea fragilis collected in Pohnpei. *J Nat Prod.* 58(9):1463-1466.
[18] Freyer, A.J.; Patil, A.D.; Killmer, L.; Offen, P.; Carte, B.; Jurewicz, A.J.; Johnson, R.K. 1997. Frondosins, five new sesquiterpene hydroquinone derivatives with novel skeletons from the sponge *Dysidea frondosa*: Inhibitors of interleukin-8 receptors. *Tetrahedron*. 53: 5047–5060.
[19] Iguchi, K., Sahashi, A., Kohno, J., Yamada, Y., 1990. New sesquiterpenoid hydroquinone and quinones from the Okinawan marine sponge (*Dysidea* sp.). *Chem. Pharm. Bull.* 38: 1121-1123.
[20] Ford, P. W., Gustafson, K. R., McKee, T. C. *et al.*1999. Papuamides A–D, HIV-Inhibitory and Cytotoxic Depsipeptides from the Sponges Theonella mirabilis and Theonella swinhoei Collected in Papua New Guinea. *J. Am. Chem. Soc.*, 121 (25): 5899–5909.
[21] Fusetani, N., S. Matsunaga, and T. Sugawara. 1992. Theopederins A-E, potent antitumor metabolites from a marine sponge, *Theonella* sp. *The Journal of Organic Chemistry* 57(14).
[22] Guo, J., C.Y. Chiang, M.C. Lu, W.B. Chang and J.H. Su. 4-Methylenesterols from a Sponge *Theonella swinhoei*, *Mar. Drugs* 2012, *10*(7), 1536-1544.

The Application of Ozon and Chitosan as Microbial Inhibitor Prawn Larvae Rearing

Delianis Pringgenies, Muhammad Nur, Rosti Angelia

Department of Marine Science, Diponegoro University, Jl. Prof. Soedarto SH, Tembalang, Semarang 50275, Indonesia

*Email: pringgenies@yahoo.com

Abstract. The application of ozon and chitosan solution was found to be effective means in hampering microbial growth since the introduction of ozon can inhibit microbial growth in aquatic environment and the fact that chitosan can act as a coagulant, effectively increasing water quality as prawn breeding medium and increasing the survival rate of prawn larvae. This research aims to measure the efficacy of ozon and chitosan usage in prawn larvae rearing, and to measure the survial rate of prawn larvae during the rearing stage. The study was carried out using experimental method in laboratory, with factorial research design using 3 treatment combinations and 1 control groups. The chitosan dose administered in this research was 25 ml in 25 L of sea water, euating to 10 ppm. The dissolved ozon in this research was measured at the concentration of 8.245 – 13.748 ppm. Weekly measurement of water quality in terms of temperature, pH, salinity, dissolved oxygen and dissolved ozon were carried out throughout the course of the research. Total microbial population in the water was measured by means of Total Plate Count (TPC) methiod in the Institute for Natural Medicines of Diponegoro University. Statistics figure from "ANOVA" test suggested that the application of ozon and the introduction of chitosan in prawn larvae rearing media gave impact to the microbial population in the media. Results of BNT test showed that there was a significant difference between the measurement results of the 3 treatment combination groups and that of the control group. The highest prawn larvae survival rate was found in the media with combined ozon and chitosan treatment, which was recorded at 100%. The second highest survival rate was recorded in the treatement combination group of ozon and chitosan with 80%, and the lowest survival rate was attributed to the control group with 20% prawn larvae survival rate. It is concluded that treatment combination of 10 ppm chitosan and 8.245 – 13.748 ppm of ozon showed significantly positive results in its application in giant tiger prawn farming media.

Keywords: Ozon, Chitosan, Microbe, Prawn Larvae.

1. Introduction

Bacteria can hamper aqua farming process, as exemplified in prawn aqua farms. Prawn, as an export commodity, can contribute to national income from the non-energy sector. In addition to being sought after by many for its taste, prawn has also been proven to have high protein content, making it one of the major export commodities in fisheries. Recently, however, prawn aqua farm suffers due to widespread pathogenic bacteria infestations. To suppress the growth of harmful microbial, formalin, antibiotics, ozone as well as chitosan are commonly used. The use of formalin and antibiotics is highly discouraged considering the detrimental impact caused to environment and natural aquatic ecosystems. Therefore, ozone and chitosan provide viable alternatives to the former antimicrobial agents since the latter are organic and environmentally safe.

Based on the study by Summerfelt *et al.*[1], ozone is formed from nonthermal plasma, raised by feeding air or oxygen gas through coronal discharge as well as dielectric barrier discharge or silent plasma. Ozone is beneficial in that it can increase water quality, suppress the growth of pathogenic microbes, save for fish, has high reaction time, and its reaction results merely in oxygen which is safe for farm ecosystem. The use of ozone also do not produce malodorous or foul residue compared to the use of chlorine [2]. However, utilization of ozone still has drawbacks which are attributed to the inability of the treatment process in transferring diluted minerals and salt, high cost from electricity consumption of the device, and the dangers from excessive concentration of bromate from the treatment process used ozone in water sterilization, fruits and fresh product [3]. In the experiment, tomatoes washed with water could only be kept for a week, whereas those which were sprayed with ozone could be stored for 27 days and was proven to be free of harmful bacteria.

In addition to ozone, chitosan has been proven to suppress the growth of harmful bacteria and has seen commercial application in commercial biomedicines. Chitosan is fiber made of polysaccharides derived from marine animals such as oysters, shrimps, crabs and etc. The benefits of chitosan are attributed to its organic nature, efficiency in usage (concentrated extract) making it economical in its use, its efficient application in detoxification, its use in bacterial inactivation, its non-toxic nature and its biodegradability [4].

Based on the information on ozone and chitosan above, that both agents has been proven to inhibit the growth of harmful bacteria, this research aims to measure the efficiency of the application of ozone and chitosan in prawn larvae rearing treatment, as well as to measure the survival rate of prawn larvae during the experiment.

2. Research Method

2.1 Research Materials

Giant tiger prawn *Penaeus monodon*, numbering forty in total, in post-larval stage (PL 30) was selected as the test species. The containers used in the experiment were 40-Liter glass aquariums, with 25 Liters of water as rearing media in each aquarium. Four aquariums were used in this research, with each aquarium subjected to one type of treatment. Ten prawn larvae were placed in aquarium.

Ozone acted as antibacterial agent in this study. During the course of the experiment, ozone was discharged into the rearing media by means of circulation. Ozonizers were turned on for 10 minutes to discharge the ozone, with 40 minutes interval for each discharge. One of the main ozone discharging devices was the ozone reactor. The main characteristics of the ozone reactor is spiral configuration of its electrodes. Each spiral electrode is made of copper wire and aluminum foil cylindrical electrode. Dielectric materials used in this research were 500 mm *Pyrex* tubes. A schematic of the ozone reactor used in this study is displayed in Figure 1 and the configuration of the devices can be seen in Figure 2.

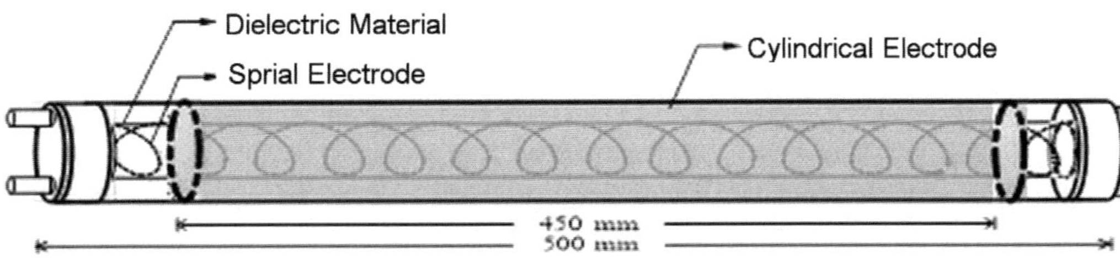

Figure 1. Schematics of an ozone reactor.

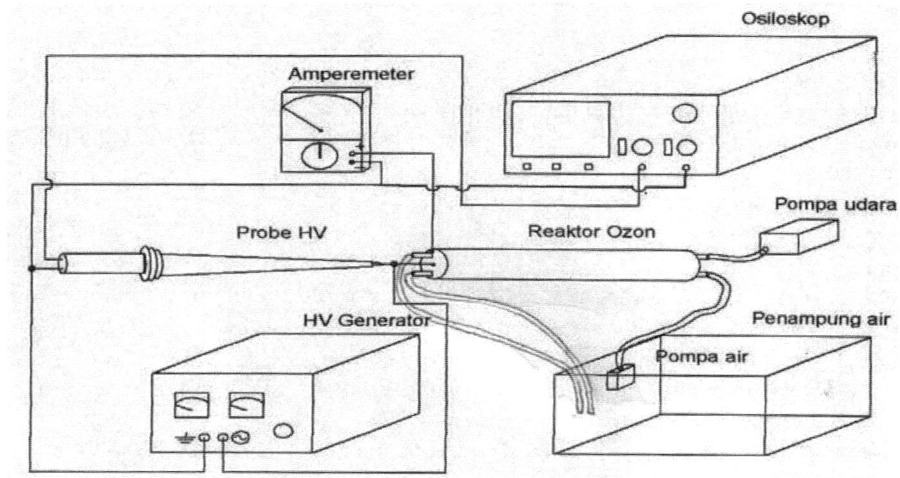

Figure 2. Schematics of the research equipment.

In order to form nonthermal plasma, the spiral electrodes were connected to a high-voltage AC power source, whereas the cylindrical electrodes were connected to ground. Water was fed from the intake of the reactor simultaneously as the reactor form ozone which resulted in diluted ozone. The diluted ozone was then dispersed through the output port.

The chitosan used in this study was made of chitin taken from crabs of *Portunidae* family. The chitosan materials were procured from Laboratory of Microbiology, Institute to Natural Medicines, Diponegoro University [5], with 25 ml of chitosan diluted in 25 L of sea water. The mixture resulted in a diluted chitosan in sea water with 10 ppm concentration. Chitosan in this research was given as bacterial inactivation agent in the rearing media. The administration of chitosan was performed every two days, after replacement of rearing media.

The samples were fed with *Artemia* brine shrimp. The food was given in the ration of 20 : 1, or 20 brine shrimps were provided for each sample for every feeding [6]. Feeding was performed three times in one day, at 06:00 hrs (after media replacement) , at 12:00 hrs and at 18:00 hrs [7].

The experiment was divided into two phases, namely the preparation and the execution phases. The preparation process of the experiment is explained in the preparation phase sub-heading, and the series of experiments conducted are explained in the execution phase sub-heading.

2.1.1 Preparation Phase

The preparation phase began with procuring the rearing container; 4 aquariums with 45 cm of length and 30 cm of width. The placement layout of the giant tiger prawn (*P. monodon*) larvae was determined randomly by drawing lots. Sea water as rearing medium was treated with 5 gr of chlorine to kill harmful microbes and parasites. One to two days after the chlorine treatment, the medium was further treated with 10 gr of thiosulfate to remove chlorine odor and improve the usability of the media. Nutrient Agar was provided as bacterial culture media.

2.1.2 Execution stage

Each rearing container was filled with 25 liter of sea water, after which 10 giant tiger prawn larvae were placed. Chitosan was diluted every two days during the course of the experiment. Ozonizer was regularly operated for 10 minutes with 40 minutes interval between ozonization of media. Measurements of the water quality were taken every week for 6 times (45 days). Temperature, acidity and salinity of water were measured using mercury thermometer, pH paper, and refractometer respectively. Dissolved oxygen concentration in the rearing media with chitosan and ozone was measured weekly, using Winkler titration method. Before being tested in the UV-Vis spectrophotometer, rearing media sample was added with binding solution KH_2PO_4, Na_2PO_4 and KI 0.06 M. Absorbance value of the rearing media was then calculated. Growth of colony bacteria was measured weekly by dilution method factoring 10^{-1} to 10^{-4}. Rearing media was replaced by water siphoning, as much as one third of the initial volume. *Artemia* brine shrimps as food were given three times a day, everyday during the course of the experiment, with regular feeding times in the morning, at noon, and in the evening. At the end of the experiment, the number of living larvae was recorded to determine survival rate. The rearing media was also examined to determine microbial capable of surviving the treatment used.

2.2 Identification of Bacteria

Biochemical tests performed in this study referred to the methods in *Cowan and Steel's manual for the identification of medical bacteria* [8]. Based on the observation made on bacterial shape during the incubation period, 20 bacterial isolates were identified in control media and 8 bacterial isolates were found in the treated media. However, in this study only the most dominant species will be identified.

2.3 Data Collection

After 45 days of experiment, the data collected were:

- Survival rate data from observation of each groups [9]
- Water quality and dissolved ozone which was measured throughout the course of the experiment, data on bacterial colony growth and identified bacterial species.

2.4 Data Analysis

The primary aim of this research was to prove whether the treatment can yield significant or insignificant results. The results of this research was tesetd using analysis of variance (ANOVA) method, after they went through normality test by Kolmogorov-Smirnov method.

1.5. Analysis of Variance (ANOVA)

The hypothesis formed were H_0: The four different treatments (ozone, chitosan, ozone and chitosan, and control) do not give significant difference in resulting bacterial population. H_1: The four different treatments (ozone, chitosan, ozone and chitosan, and control) (ozone, chitosan, ozone and chitosan, and control) give significant difference in resulting bacterial population.

If H_0 was rejected and H_1 was accepted, then the next step was to conduct multiple comparative test (further testing) to determine which treatment attributed to the rejection of H_0. One of the suitable test method to determine the validity of the result is the LSD (Least Significant Difference) test.

3. Results

3.1 Water Quality Parameters

Dissolved oxygen measurement results (in ppm) from each treatment, which were regularly taken from week one, showed the following results: 6.12; 6.51; 5.89; 4.8; 5.12; 5.78 for control; 6.14; 6.65; 6.61; 5.37; 6.39; 6.64 for chitosan treatment; 7.68; 7.52; 7.24; 7.13; 7.89; 7.76 for ozone treatment; and 7.48; 7.29; 7.5; 7.07; 7.64; 7.66 for chitosan & ozone treatment. Figure 4a below provides the temporal graphs for the results

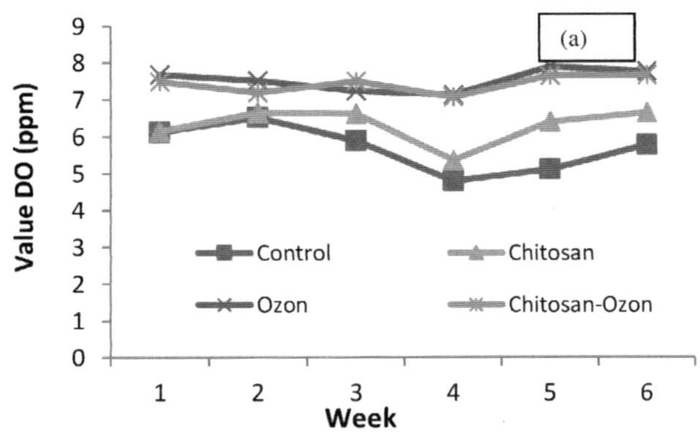

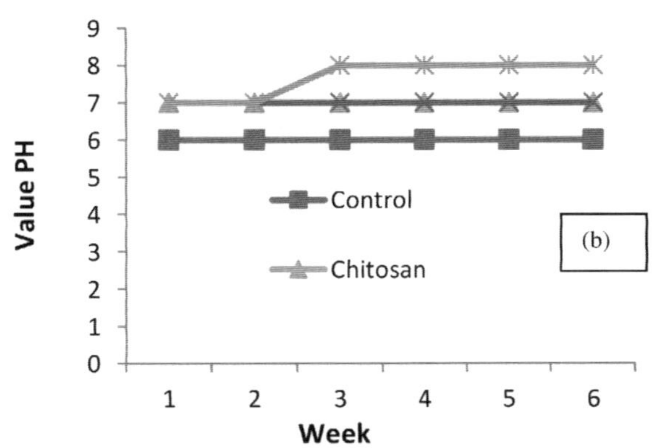

Figure 3. Graphic charts of dissolved oxygen measured during the experiment (a) and Graphic chart of water acidity measured during the experiment (b)

Measurement results of water acidity for each treatment during the course of the experiment showed the following figures: 6 for control media, 7 for chitosan media, 7 for ozone media, and 7 for chitosan & ozone media in the first two weeks, in which between the third and the sixth week rose to 8, as seen in Figure 3b.

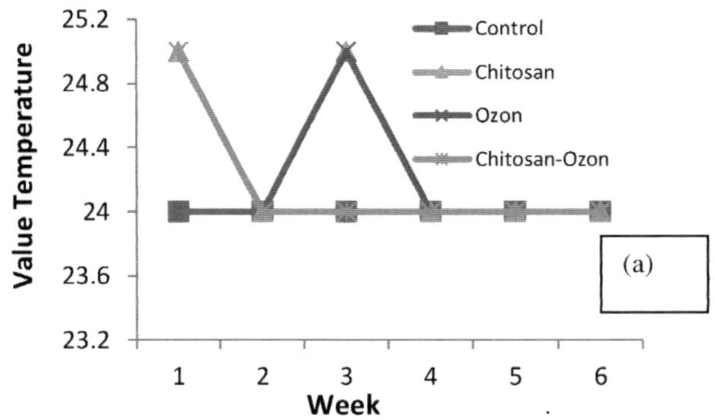

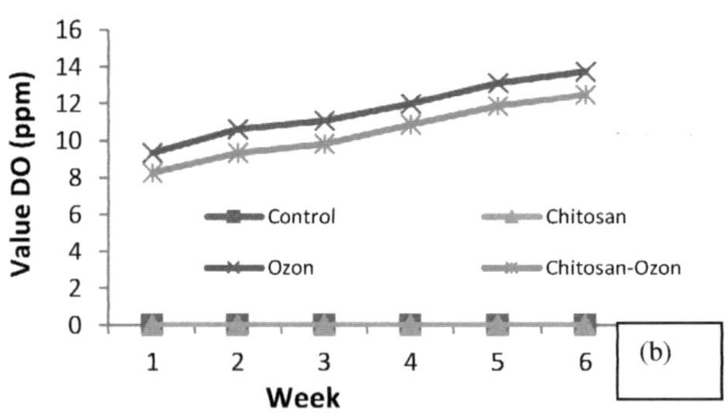

Figure 4. Graphic charts of water temperature measured during the experiment (a), and Graphic charts of dissolved ozone measured during the experiment (b

The water temperature measurements for control and all treatment groups showed steady results between 24 – 25 ^{0}C, as seen in Figure 4a. Measurements for dissolved ozone could only be retreived from media with ozone treatments, making this measurement not applicable for chitosan and control media. The dissolved ozone (in ppm) through the course of the experiment were measured at 9.316; 10.588; 11.065; 11.998; 13.127; 13.748 for ozone treatment and at 8.245; 9.312; 9.798; 10.842; 11.857; 12.478 for chitosan & ozone treatment. Figure 4a above details the graphic charts of these results.

3.2 Growth of Bacterial Colony in Each Media
The observation on the growth of bacterial colony for control and all treatment media showed that the highest Colony-forming Unit (CFU) found in control media was 1.5 x 10^6, in chitosan & ozone media was 1 x 10^5, in chitosan media was 2.8 x 10^4, and in ozone media was 1 x 10^4, as described in Figure 6.

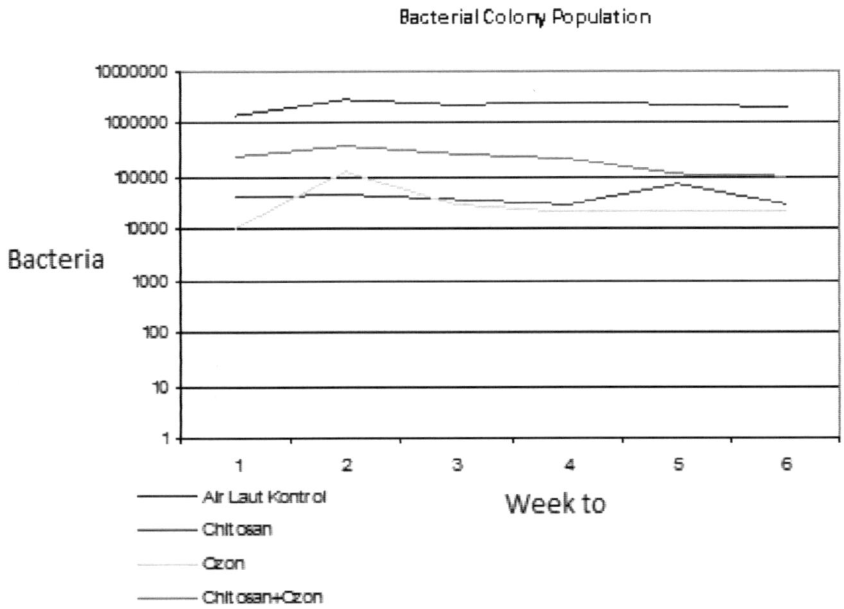

Figure 6. Bacterial seeding in the first week of observation

3.3 Statistical Calculation of Bacterial Colony

To prove that the data distribution was normal, a Kolmogorov-Smirnov test was carried out on the data. The test criteria were reject H_0 if sig- or P-value at $< \alpha = 5\%$. The test result proved that H_0 was accepted which means that the data residue showed normal distribution, since data output displayed Kolmogorov-Smirnov value of Sig-P from the transformation logarithm = 0.150. The result of Kolmogorov-Smirnov test can be seen in Figure 7

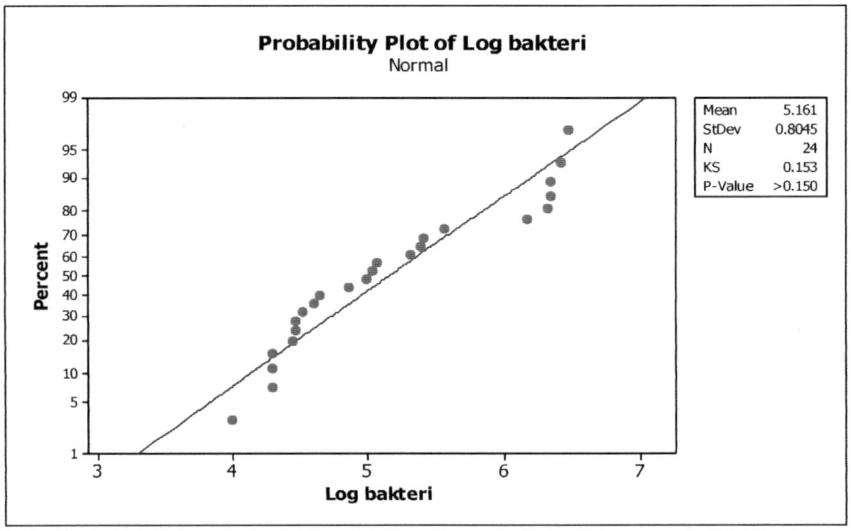

Figure 6. Normality test by Kolmogorov-Smirnov method

3.4 ANOVA test and LSD (Least Significant Difference) test

Statistical test using ANOVA showed significant difference between each treatment group. With significance rate $\alpha = 0.05$, F-Value = 106.29 with Pr>F = 0.0001 was obtained.

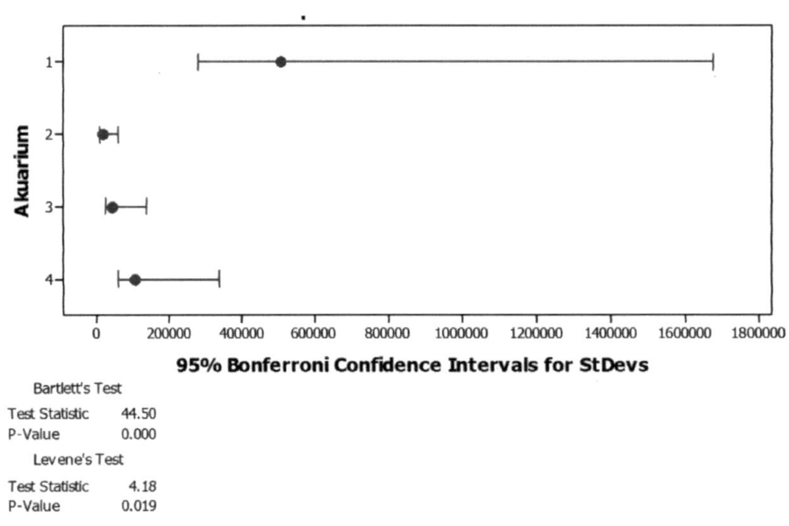

Figure 7. Hypothesis testing on variance similarity by LSD method

IOP Conf. Series: Earth and Environmental Science **55** (2017) 012070 doi:10.1088/1755-1315/55/1/012070

The results yielded a P-Value for Levene's test = $0.019 < \alpha = 0.05$, thereby H_0 was rejected. The difference in median of all the treatment groups were shown to be less than the BNT value of 5%. Hypothesis test results on similarity between variance and BNT is presented in Figure 8.

3.5 Survival Rate of Giant Tiger Prawn (P. monodon)

The final survival rate was obtained at day 45 of the rearing stage. Ten prawn larvae were introduced into each container at the beginning of the experiment. The highest survival rate was observed in two treatment media, chitosan treatment and ozone treatment, with both media yielding 100% larvae survival rate. The ozone and chitosan combination treatment was capable of resulting 80% of survival rate, whereas the control media larvae population had the lowest survival rate of all with 20% rate (Table 1)

Table 1. Survival rate of tiger prawn (*P. monodon*) larvae

Treatment	**No (prawn)**	**Nt (prawn)**	**SR (%)**
Ozone Medium	10	10	100%
Chitosan Medium	10	10	100%
Chitosan and Ozone Medium	10	8	80%
Control Medium	10	2	20%

Table 2. Biochemical test of bacterial species in control and ozone media

Biochemical Test	**Bacterial Isolate Code**	
	E. 09.166	E. 09.167
Gram	-	+
Shape	Rod	Rod
Acid Fast	v	-
Spore	-	+
Spore position	v	VX
Cell length > 3μm	v	-
Motility	+	+
Aerobic	+	+
Anaerobic	-	+F
Catalase	+	+
Oxidase	+	+
Glucose Acid	+	+
Carbohydrate (OF)	O	O
Biochemical Test	**Bacterial Isolate Code**	**Biochemical Test**
Nitrate reduced	+	+
Gas from glucose	-	-
Indol	-	-
ONPG	v	-
VP	+	+
Hydrolysis of: - starch - urea - casein	 + v -	 + v +

Moller's decarboxylase: - Arginine - Lysine - Ornithine	 v v v	 - - -
Gelatinase	-	+
Growth on 50^{0}C	v	-
Growth on 42^{0}C	-	V
Acid from ASS medium - L-arabinose - Salicin - Sucrose - Xylose - Celibiose - Galactose - Rafinose	 - - - + - + -	 v - V - + + -
Pigment production	-	-
Utilization of Citrat	+	+
Flagellum	Polar	V
Growth with 10% NaCl	v	-

Legend: VX = central oval, v= n/a, + ^{F} = Facultative anaerobic, O = Oxidative

3.6 Identification of Bacteria

The identification of bacteria in the control media and ozone media was carried out up to the species level. The biochemical test method used in the identification referred to Cowan and Steel's Manual for The Identification of Medical Bacteria. Biochemical test results are displayed in Table 2.

The biochemical test result showed that one of the two most dominant bacteria was identified as *Pseudomonas stuszeri* with code E.09.166. This species is gram negative, rod-shaped, with no spore, has polar flagella, motile, aerobic, capable of producing catalase and oxidase enzymes, capable of using glucose, capable of oxidizing carbohydrate, incapable of forming pigment, capable of reducing nitrates, incapable of producing gas from glucose, incapable of forming indole, capable of hydrolyzing starch, incapable of hydrolyzing casein, incapable of producing L-arabinose acid, incapable of synthesizing acid from salicin, incapable of synthesizing acid from sucrose, capable of synthesizing acid from xylose, incapable of synthesizing acid from cellobiose, capable of synthesizing acid from galactose, incapable of synthesizing acid from rafinose, and do not survive at 42 ^{0}C temperature.

Isolate code E.09.167 was identified as gram positive, rod-shaped, has spore with central oval position, motile, aerobic and facultative anaerobic in nature, is capable of producing catalase and oxidase enzymes, capable of utilizing glucose, capable of oxidizing carbohydrate, capable of reducing nitrate, incapable of synthesizing gas from glucose, incapable of forming indole, capable of hydrolyzing starch, capable of hydrolyzing casein, incapable of synthesizing acid from salicine, incapable of synthesizing acid from xylose, capable of synthesizing acid from cellobiose, capable of synthesizing acid from galactose, incapable of synthesizing acid from rafinose, incapable of decarboxylation of arginine, incapable of decarboxylation of lysine, incapable of decarboxylation of ornithine and cannot thrive in 50 ^{0}C of temperature and in 10% concentration of NaCl. This species was later identified as *Bacilllus thuringiensis*.

4. Discussion

The water quality parameters being measured covered dissolved oxygen, pH, temperature, salinity and dissolved ozone. Results of the measurements showed that the rearing media were capable of supporting the ecosystem of the test subjects. Generally, water quality is a determining factor in survival and immunity formation respond rate. Despite the fact that only temperature directly influence the prawn population, DO, pH, and salinity also play important roles in immunity system formation of the prawns. According to Amri (2003) [10], the water quality early in the experiment was still suitable for prawn rearing, because salinity levels were at 19 – 25 $^{0}/_{00}$ pH, temperatures were in 25 – 32 ^{0}C, and DO rates were at 4 – 8 ppm. All these readings met the requirement for the suitable rearing media.

The dissolved oxygen (DO) parameter was found to be a limiting factor in the survival rate of the prawn larvae. The minimum DO rate for larvae survival is 3 ppm [11]. Dissolved oxygen from each treatment media during the course of the study was measured at between 4.8 – 7.89 ppm. Ideal DO rate spans around 4 – 7 ppm. However, minimum DO during night time is suggested to be more than 3 ppm [6].

The highest DO rate was found in ozone-treated media, measured at 7.13 – 7.89 ppm. The capability of ozone to enhance the availability of oxygen in water was attributed to the high DO in the media. This was made possible due to the oxygen released by ozone reactor whenever silent plasma was formed in dielectric barrier-equipped plasma discharge, which was dissolved in the water. When free electron has sufficient energy, ozone molecule is believed to undergo dissociation into oxygen molecule and atom [12]. The formed oxygen atom collides with ozone to form two other oxygen molecules. In addition, ozone is a highly reactive and unstable gas, with fast decay rate before reforming into oxygen [13].

The dissolved oxygen levels of the chitosan and ozone media placed second after that of the ozone-treated media, measured at 7.07 – 7.66 ppm. Chitosan is a natural coagulant, which precipitate all particles both essentials and pollutants dissolved in the water. This nature of chitosan was believed to be the cause of increasing thickness of the water, forming sediments that slows oxygen reaction in the media. Therefore, it was concluded that the combined usage of chitosan and ozone is not the best treatment to support optimum dissolved oxygen. The majority of chitosan usage is aimed at water treatment to separate radioactive heavy metal. Chitosan is one of the most efficient chemical agents to bind dissolved substances such as dissolved solids, dissolved colloids, and suspended solids in a body of water by separating and forming sediments or floating crusts.

Chitosan was shown to lower dissolved oxygen in the research, and to that extent the use of chitosan was limited to antimicrobial agent for media treatment. Furthermore, that the dissolved oxygen figures in chitosan media was measured at 5.37 – 6.65 ppm supported the statement that chitosan is capable of dissolving organic contaminants [5].

The control medium had the lowest dissolved oxygen rate of all the media in the research, measured at 4.8 – 6.51 ppm. The use of only two aerators was believed to be the main cause of this condition. These aerators might not supply sufficient oxygen in the water, which in turn reduced the availability of dissolved oxygen in the water. The lack of oxygen in the water was also exacerbated by oxygen consumption by the prawn larvae. Romimohtarto [14] stated that oxygen is created from photosynthesis process of plants and animals and that its availability is vital for respiration. Low larvae survival rate in the control medium may also be attributed to the cannibalistic nature of prawn larvae. It is known that larvae may cannibalize their own during the molting state, where larvae are in their weakest physical condition.

Judging from the DO rate of the control medium, it was concluded that control medium was not the most efficient medium to be used in larvae rearing. Romimohtarto [14] stated that low DO rate can reduce oxygen absorption efficiency of marine animals, reducing their survival rate in such environment.

Acidity (pH) of the rearing media of this research was still within the acceptable limits, measured at approximately 6 – 8. According to Amri (2003) [10], water acidity for prawn rearing ecosystem should ideally be kept between 6 – 9. This is also supported by Anggoro [15] who stated that the ideal pH value for post larvae stage rearing media is between 6.5 – 8.5.

Changes in pH may also affect the survival rate of prawn larvae. During the course of the experiment, the lowest pH was measured at 6, in the control media. Low acidity was identified with very low survival rate, at 20%, of the medium, followed by the 2nd lowest survival rate measured from that of the chitosan medium. Royce [16] stated that for medium with pH 6.4, although the prawn larvae managed to survive, the growth of the larvae was slowed down to 60%. Furthermore, Romimohtarto [14] revealed that changes in acidity may be directly and indirectly detrimental to marine ecosystem. Change in water acidity directly impact marine animals in that fish, spawn, eggs and other kinds of animals may die from the change. Another direct impact is that the change in acidity may reduce primary productivity. Even a slight change from the natural acidity may hint disruption in the biology support system.

The temperature parameter for all media in this research showed almost similar readings at 24 – 25 ^{0}C. The data obtained from all media showed no significant difference in temperature between each medium. According to Nurdjana, M. L., Martosudarmo, B dan Anindyastuti [17], the optimal water temperature for prawn rearing is identified at 25 – 32 ^{0}C. However, temperature is also believed to have a significant relationship with dissolved ozone.

Lenntech [18] explained that water temperature is a determining factor in the levels of dissolved oxygen in water. Generally, rate of oxygen dissolvance will slow down in 40 ^{0}C of water temperature. Starrs. D *et al.* [19] stated that within certain limits, the rate at which prawn grows is linear with the increase of temperature, albeit conversely, the survival rate of prawn is inversely proportional o the rise of temperature.

The concentration of dissolved ozone is highly determined by the amount discharge, reaction speed and contact time of ozone in the water, and water quality. One of the parameters of water quality which determines dissolved ozone is temperature. The lower the water temperature, the more ozone concentration there is in the water and *vice versa*. Other variable which may affect dissolved ozone in water is acidity (pH must be close to 7) [20]

Salinity of all the media during the treatment were measured between 30 to 31 $^0/_{00}$. Accoring to Amri [10], the most ideal water salinity to support prawn growth is around 19 – 35 $^0/_{00}$. On the other hand, Tricahyo [19] explained that the most ideal salinity for prawn ecosystem is at 15 – 30 $^0/_{00}$. Sudden change in salinity may result in high prawn mortality rate. Prawn is generally more resistant to low salinity during its larval stage compared to its adult stage [21].

Medium treated with combination of chitosan and ozone showed significant increase in dissolved ozone from week to week during the experiment. The most effective dissolved ozone rate in water obtained was between 9.316 – 13.748. The highest dissolved ozone concentration was found in ozone-treated medium, a condition believed to be the result of reaction between formed oxygen atoms with other oxygen atoms which in turn formed more ozone. Kogelschatz *et all* [22] stated that ozone is created by feeding oxygen gas through electrical discharge. Oxygen ions types are O^+, O_2^+, O^-, O^{2-} and O_3^-, which all of their combination results in ozone. Nitrogen molecule can also react into nitrogen dioxide which is unstable in nature and can react with oxygen to produce ozone. Combined chitosan and ozone medium had dissolved ozone concentration of 8.245 – 12.478. This media showed the lower dissolved ozone figure out of the two ozone-treated media, with larvae survival rate observed at 80%. Low dissolved ozone in combined chitosan and ozone treatment was believed to be caused by high water turbidity, owing the the nature of chitosan as coagulant, which prevented the collision of oxygen molecules with electron, other atoms or molecules. Chitosan is one of the substances capable of binding other fine materials such as dissolved solids, colloidal solids, and suspended solids in a body of water which it then separates by forming sediment or crusts. Furthermore, Lentech [18] stated that measurement of ozone can only be taken in contact with water. This is made possible by the fact that contact with water made it safe for surface contact for both human and prawn. Survival rate in the ozone-treated water was also found to be the highest of all the media, at 100%. The most immeadiate hint for ozone safety levels is whenever water becomes malodorous, then the ozone levels in the water has become unsafe [23].

Control medium showed the highest microbial population among the media in this research. Weekly interval of microbial population count of the control media was measured at 1.5 x 10^6 – 3.0 x 10^6. This was believed to be caused by aeration treatment of the control medium, by simply discharging oxygen into the water. The bacterial colony count of ozone media was significantly different to that of the control media. This result is in line with Lentech [18], which mentioned germicidal, bactericidal and fungicidal properties of oxygen is far weaker than that of ozone. The media with combined chitosan and ozone treatment had the second highest bacteria colony count, whereas the weekly interval growth of the colony was measured at 1 x 10^5– 3.7 x 10^5. The result shown by chitosan media was not as good as that shown by ozone media. This is believed to be caused by concetration of dissolved chitosan in chitosan treated media, at 10 ppm (25 ml in 25 L of sea water), might not be comparable to the dissolved ozone concentration in the ozone treated media, at 9.316 – 13.748 ppm, 10 ml of chitosan solution in 1 L of water resulted in 10 ppm of dissolved chitosan.

The relatively low growth of bacterial colony found in chitosan media might be attributed to the nature of chitosan as antimicrobial agents, which is capable of attacking the membrane of bacteria. Roberts [24] explained that chitosan is cationic polyelectrolyte (positively charged) and readily attract opposite negative ions. In addition, the positively charged polycation of chitosan is proven to be capable in hampering bacterial growth [5]. These properties play an important part in keeping the bacterial growth to a very low level and was believed to be the main factor in the lowest bacterial growth rate compared with control and chitosan and ozone combination treatment media, being measured at 2.8 x 10^5 – 7.4 x 10^5 in weekly interval. Ozone treatment showed the most efficient result. From week to week, colony in the media showed interval of 1 x 10^4 – 3 x 10^4. Ozone treatment and the high concentration of dissolved oxygen was believed to keep bacterial growth at minimum. In aquarium, these properties are also utilized to kill various parasites, such as bacteria, virus, and spores as well as to neutralize other pollutants. Furthermore, the efficacy of ozone in clearing away odor and turbidity is one of the reasons why ozone is preferable as sea water aquarium treatment [18].

Through oxidation, ozone is capable of killing various microorganisms such as *Eschercia coli, Salmonella enteriditis, Hepatitis A* and many other pathogenic bacteria. Direct oxidation is highly damaging to outer membrane of cells (cell lysis), which result in death of the cell [25]. A Team of Unibraw Medical School [26] wrote that in the event of cell death, important metabolites escape. In addition, ozone technology is time and cost efficient as well as practical in that it does not require much space for its installation. The bacterial growth in ozone medium was observed at 1 x 10^4 on the initial week, due to lag phase of the bacterial growth. On the second week, bacterial growth in the medium was measured at 1.2 x 10^4, which was believed to be the initial phase of growth. Measurements on the third week showed 3 x 10^4, a relatively rapid growth which was believed to be the exponential phase. Between the fourth and the final week of the experiment, bacterial growth in the medium showed a constant value of 2 x 3 x 10^4, which was believed to be the static phase. These results are in line with Lud Waluyo [27] who proposed four phases in bacterial growth. Soon after being dispersed in a medium, bacteria begin to adapt with substrate and environment. No reduplication takes place in this phase since several enzymes may not be synthesized yet. This phase is also called lag phase. After the adaptation has been completed, cells begin to multiply in low rate. This phase is called initial growth phase. After the previous phase, cells begin to multiply rapidly at an exponential rate, often following logarithmic patterns. It is believed that this phase is highly affected by the properties of the environment in which the bacteria grows, such as acidity, nutrient contents and temperature. This phase is also known as logarithmic growth phase (exponential phase or rapid growth phase).

Between the fourth and the final week of this research, there was little or no bacterial growth observed in the ozone medium. This phenomenon is believed to be caused by the change in the environment which may be disadvantageous to bacterial growth and development. Bacterial cell population was observed to be constant, which might also be attributed by the balance in the amount of new and dying cells. This phase is referred as the static growth phase. When observed by visual plot residual (respond of the bacterial population), linear reading was obtained. It was assumed visually that at this phase the growth had met normality pattern (normal distribution). Mathematical calculation of P Value (significant) for Kolmogorov-Smirnov method yielded > α, in which $\alpha = 5\%$. It was calculated that P Value = 0.150 making H_0 accepted and reinforced the assumptions for residual respond (bacterial population normality). The result was then further analyzed by ANOVA method.

By using ANOVA checking model, it was concluded that there were differences among the 4 treatment media, namely ozone, chitosan, chitosan and ozone, and control, in bacteria population (respond/dependent variable). This was shown by the results, where the value of $Pr > F = 0.0001$. The value $0.0001 < \alpha = 0.05$. This meant that H_0 was false and that there were significant influence among 4 treatment media toward the bacteria population (respond/dependent variable). $Pr > F$ is less than α, then H_0 is rejected.

The value obtained on P-Value Levene's Test = $0.019 < \alpha = 0.05$, which meant that H_0 was rejected. This result validated the suspected difference in influence between the three treatment media and the control medium. According to Srigandono [28], P-Value Levene's Test $< \alpha$ meant that there is a significant difference among 3 or more treatment groups. This also meant that there were differences in the variance of all four rearing media. In conclusion, the results showed that there was significant difference among the three treatment media and control medium, as well as between all the treatment media and the control medium.

When hypothesis zero (H_0) was rejected through Analysis of Variance procedure, then it can be concluded that there was significant difference in at least one of the observed treatment. Srigandono [27] stated that to identify the most significant difference, further test on mean value of the treatments must be carried out. Survival rate of the tiger prawn (*P. monodon*) showed different value for each medium. The highest survival rate was observed in both ozone treatment and chitosan treatment media with a perfect 100% figure. The second highest survival rate was found in the combined ozone and chitosan treatment medium with 80% and the lowest was identified in the control medium with 20%. However, the survival rate for each medium was deemed to be good. Biochemical test of 2 different test media, ozone treatment and control, showed different results. The most dominant surviving bacteria in control medium was identified as *Pseudomonas stutzeri* and the dominant bacteria capable of survival in ozone-treated ecosystem was identified as *Bacillus thuringiensis*. Each species bacteria had different survival and adaptation capacity. Bacteria can quickly develop new defense mechanism against external threats that may harm them. Bacteria incapable of neutralizing harmful substances in an ecosystem cannot survive in the environment. In the control ecosystem, the most dominant surviving bacteria was *Pseudomonas stutzeri* which was identified to be gram negative, has polar flagella, and has no spore. The survival of this species was believed to be made possible by the abundance of oxygen in the water of the control medium. According to Rheinmer [29], *Pseudomonas stutzeri* is a facultative anaerobic species which thrive in oxygen-rich environment.

On the other hand, the most dominant surviving bacteria on the ozone treatment medium was *Bacillus thuringiensis*, a species known to be both aerobic and facultative anaerobic. The survival of this species bacteria was made possible by the fact that the species is known to be resistant to low-oxygen environment, extreme shift in acidity, salinity or dissolved oxygen, such as during the ozone discharge into the medium, adapting by forming spores. Microbiology Team of Unibraw [26] reported that the formation of spores in this species takes place whenever the available nutrition can no longer sustain bacterial growth. According to Rheinmer [29], *Bacillus thuringiensis* is a chemoorganotrophic species with fermentation metabolism or respiration that is sensitive to heat, acidity and salinity. The species is proven to be pathogenic only to a handful number of vertebrates and invertebrates. Bacteria of the *Bacillus* species can be terminated on 0.2 ppm concentration of ozone with 30 seconds contact time, whereas *Pseudomonas* is easily affected or penetrated by ozone [30]. This fact was believed to be the main key in keeping the population of *Pseudomonas* low in ozone-treated medium, owing to the fact the species easily undergoes lysis by the introduction of ozone. However, this bacteria can still survive ozone treatment due to the fact that it has spores.

4. Conclusion

This research found that the most effective medium to be used in prawn larvae rearing is ozone-treated medium. In order of magnitude, the highest to lowest number of bacterial colony found was in control medium (1.5 x 10^6 CFU), in chitosan and ozone combined treatment (1 x 10^5 CFU), in chitosan treatment (2.8 x 10^4 CFU), and in ozone treatment (1 x 10^4 CFU). The use of ozone and chitosan proved to be significant in suppressing the bacterial population in the prawn larvae rearing media. Significant difference was found between the three treatment media and the control medium. The highest survival rate was found in ozone treatment and chitosan treatment, both measured at 100%, which was followed by combined chitosan and ozone treatment with 80% and control medium with 20%.

References

[1] Summerfelt, S.T. (2002). Ozonization and UV Irradiation an Introduction and Examples of Current Application, Aquaculture Engineering The Conservation Fund's Freshwater Institute, Shepherdstown USA.
[2] Surawiria, Unus. (2003). Mikrobiologi Air. Alumni. Bandung. Cetakan ke -3.
[3] Gonçalves. (2009). Ozone – an emerging technology for the seafood industry. Braz. arch. biol.technol. vol.52 no.6 Curitiba (http://dx.doi.org/10.1590/S1516-89132009000600025, December 6, 2011).
[4] Kaho, Angela Riwu. (2007). Chitosan Magic of Nature (http://keberhasilantotal.blogsome.com/2007/11/18/chitosan-magic-of-nature/trackback/, April 16, 2008).
[5] Pringgenies.d., A. Hartoko. (2008). Development and Application of Chitosan as Anti-bacterial Agent on Fabric. USAID-SENADA, June 16, 2008. Laporan Penelitian.
[6] Haliman, Rubiyanto Widodo dan Dian Adijaya. (2005). Udang Penaeus. Penebar Swadaya. Jakarta.
[7] Murtidjo, B. (1992). Tambak Air Payau Budidaya Udang dan Bandeng. Kanisius. Yogyakarta. 138 hlm.
[8] Cowan, S.T and Steel, K.J. (1993). Manual for the Identification of Medical Bacteria. Edition 3. Cambrige University Press. New York. Cambridge. Hal 42-45.
[9] Efendie, M.I. (2002). Biologi Perikanan. Edisi Revisi. Yayasan Pustaka Nusatama. Jakarta. 163 hlm.
[10] Amri. (2003). Budi Daya Udang Windu. PT Agro Media Pustaka. Jakarta.
[11] Soetomo, H.A. (1990). Morfologi Udang. Edisi Revisi. Bina Adiaksara. Jakarta.
[12] Blumler, Jay G., & Michael Gurevitch. (1995). The Crisis of Public Communications. New York: Routledge.
[13] Farajzadeh D, Qorbanpoor A, Rafati H, Isfeedvajani MS, J Res Med Sci. (2013). Reduction of date microbial load with ozone, 18(4):330-4.
[14] Romimohtarto K. (1985). Kualitas Air dalam Budidaya Laut. Badan Penyimpanan Dokumen FAO Laporan Kerja Budidaya Laut, Bandar Lampung, 28 Oktober – 1 November 1985.
[15] Anggoro, Sutrisno. (1992). Efek Osmotik Berbagai Tingkst Salinitas Media Terhadap Daya Tetas Telur dan Vitalitas Larva Udang Windu (*Pennaeus monodon fab*). Program Pascasarjana. Institut Pertanian Bogor. Bogor.
[16] Royce, W.F. (1982). Introduction to the Fishery Sciences. Academic Press, Inc.
[17] Nurdjana. M.L, B. Martosudarmo, dan Anindiastuti. (1980). Pedoman Pembenihan Udang Penaeid. Direktorat Jendral Perikanan Depertemen Pertanian Yakarta.
[18] Lenntech. (2008). Ozone Decomposition (http://www.lentech.com/ozone/ozone-transfer-mechanism.htm, May 1, 2009).
[19] Starrs. D., Jacqueline T. Davis, Jodie Schlaefer, Brendan C. Ebner, Stephen M. Eggins, and Christopher J. Fulton. (2013). Maternally transmitted isotopes and their effects on larval fish:a validation of dual isotopic

marks within a meta-analysis context. Can. J. Fish. Aquat. Sci (http://www.water-pollution.org.uk/ozonewastewatertreatment.htm, November 1, 2014).

[20] Yanong, Roy. (2005). Fish Health Management Considerations in Recirculating Aquaculture Sytems-Part2: Pathogens.pdf (www.ecosensors.com, July 7, 2007).

[21] Darmono. (1991). Budidaya Udang Penaeus. Kanisius. Yogyakarta.

[22] Kogelschatz. U., B. Eliassin and M. Hirth. (1988). Ozone Generation From Oxigen and Air: Discharge Physics And Reaction Mechanisms. Ozone Science & Enginering. Vol. 10. Pp. 367-378. USA.

[23] Saha. U., Leticia Sonon, Mark Rissel and David Kissel. (2014). Water Quality and Common Treatments for Private Drinking Water Systems. Water Quality and Common Treatments for Private Drinking Water Systems. Agricultural and Environmental Services Laboratories 1 Department of Biological and Agricultural Engineering. UGA Extension Bulletin 939. P. 30.

[24] Roberts GAF (1992) Chitin Chemistry. Houndmills UK: Macmillan Press.

[25] Wijaya, Marvin Chandra. (2005). Prototipe Pengukur Konsentrasi Ozon. Fakultas MIPA. Universitas Indonesia (http://educate-yourself.org/ozone/ozoneart2.shtml, May 1, 2009).

[26] Team of Unibraw Medical School. (2003). Bakteriologi Medik. Bayumedia Publishing. Malang. Jawa Timur.

[27] Waluyo Lud. (2007). Mikrobiologi Umum Edisi Revisi. Malang: UMM Press. h. 319 dan 330.

[28] Srigandono, B. (1987). Rancangan Percobaan (Eksperimen Design). Fakultas Peternakan Universitas Diponegoro. Semarang.

[29] Rheinmer, G. 1992. Aquatic Microbiology 4th Edition. John Wiley and Sons. New York. 363 hlm.

[30] Lagzi, I., and Mészáros, R.. (2013). Atmospheric Chemistry. Eötvös Loránd University Facultity of Science Institute of Geography and Earth Science.

Production of Chitosan from *Amusium sp* Scallop Shell Waste

Nur Rokhati[1,2], Titik Istirokhatun [2,3], Dwi Titik Apriyanti[1,2], Heru Susanto[1,2]

[1]Chemical Engineering Department, Engineering Faculty, Diponegoro University
[2]Membrane Research Centre (Mer-C), Center of Research and Services-Diponegoro University
[3]Environmental Engineering Department, Engineering Faculty, Diponegoro University
Jl. Prof Soedharto, Tembalang, Semarang, Indonesia 50275
*Corresponding author, telephone: +62 24 76480678, fax +62 24 76918157,

nur_r81@undip.ac.id

Abstract. Chitosan is one of the natural polysaccharides, which is produced from chitin by deacetylation process. In this study, chitosan was produced from *Amusium sp* scallop shell waste. First, chitin was isolated by extraction via deproteinization using alkaline solution followed by demineralization using acid solution. Thereafter, chitosan was resulted from deacetylation of chitin using a high concentration of alkaline solution. The chemical structure of chitin and chitosan products was characterized using fourier transform infrared spectroscopy (FTIR).

Keywords: chitin, chitosan , *Amusium sp* scallop shell waste

1. Introduction

Indonesia is an archipelago with abundant marine products. Along with the increasing use of marine resources, solid wastes as by products cause new problems. Therefore, it should be properly handled to avoide environmental damage. One of the fishery product found in Indonesia is *Amusium sp* scallop. It was reported by Susilowati et al. [1] that *Amusium sp* scallop constists of meat (19-28% wt.), shell (53-65% wt.) and liquids (9-25% wt.).

Chitosan has wide range of industrial applications, e.g. in cosmetics, agriculture, food, pharmacy, biomedical, paper industries and as absorbent materials for wastewater treatment [2-4]. Chitosan is generally prevalent in marine invertebrates, terrestrial, and fungi of the genus *Mucor, Phycomyces, and Saccharomyces*. The existence of chitosan in nature is generally bound with protein, minerals and various pigments. It is soluble in aqueous acidic medium due to the presence of amino groups [5].

Many publications have been reported about chitosan production, but most of them used shrimp shell as raw material. The objective of the present work is to isolate the useful polymers chitin from the waste shell of *amusium sp* scallop and further used for chitosan production. Considering the content of scallop shell, it is posible to be used for chitin prodution and further processed into chitosan.

2. Materials and Methods

2.1. Materials

The scallop shell waste was obtained from Tambak Lorok Fish Market, Semarang, Indonesia. NaOH, CH_3COOH, and HCl were purchased from Merck (Hohenbrunn, Germany). Commercial chitosan as reference was purchased from Biotech Surendo, Cirebon, Indonesia. Distilled water for all experiments was produced from a home-made pure water unit.

2.2. Methods

The scallop shell waste was initially washed and dried until a constant weight has been reached. It was then grounded followed by sieving using a 250 μm sieve resulting in scallops shell powder (SSP). Thereafter, deproteinization was performed by dissolving SSP in NaOH solution (ratio: 1 g to 4 ml) with different NaOH concentrations (1-5%), time of deproteination (20-140 minutes) and temperature (30-90°C). The solution was then filtered to obtain the remaining solid, which was further processed by washing using distilled water until a neutral pH was reached. The solid was dried in oven at 40°C for 24 hours.

Demineralization was performed after deproteination has completed. Deproteinated-SSP was dissolved in acid solution (ratio: 1 g to 5 ml) with different HCl concentrations (0.5-4M), time demineralization (15-120 minutes) and temperature (30-90°C). The solution was then filtered to obtain the remaining solid, which was chitin. Chitin was further processed by washing using distilled water until a neutral pH was reached. The chitin was dried in oven at 40°C for 24 hours. In order to obtain chitosan, deacetylation followed the method proposed by Kurita [6] was performed. Chitin was dissolved in 50% NaOH solution (ratio: 1 g to 20 ml) at 90°C. Chitosan was then washed with distilled water until a neutral pH was obtained. Finally, it was dried until reached a constant weight.

2.3 Analysis

Protein content was determined by proximate analysis using Kjedahl method [7]. Ash content of dried chitin was performed by burning in a muffle furnace at 700 °C for 6 hours and weighing after it has been cooled to room temperature and placed in desiccator. The degree of deacetylation (DD %) of chitosan was determined using a FTIR. It was calculated from the intensity of the absorption band at 1655 cm-1 (amide I band) with the band at 3450 cm^{-1} (OH band) as an internal standard. The DD % was calculated using Baxter's equation [8], DD %= 100 – [(A1655/A3450) x 115].

3. Results and discussions

3.1 Chitin Isolation from Scallop Shell Waste

The results from deproteination experiments are presented in Figures 1-3. Figure 1 shows that the increase in NaOH concentration decreased protein content of SSP indicating deproteination increased. It can be explained that the higher NaOH concentration the higher ability to break protein-SSP powder bond. Nevertheless, beyond 4% the increase in NaOH concentration did not increase deproteination anymore. Figure 2 shows the effect of temperature on protein content of SSP. It is seen that protein content of SSP decreased significantly when the temperature was increased from 30-40°C. A plateau condition of protein content was observed for the temperature increase higher than 40°C. The increase in time of deproteination decreased the protein content indicating more protein has been released (Figure 3). However increasing deproteination time higher than 60 minutes did not decrease protein significantly. These 60 minutes seems to be the optimum deproteination time.

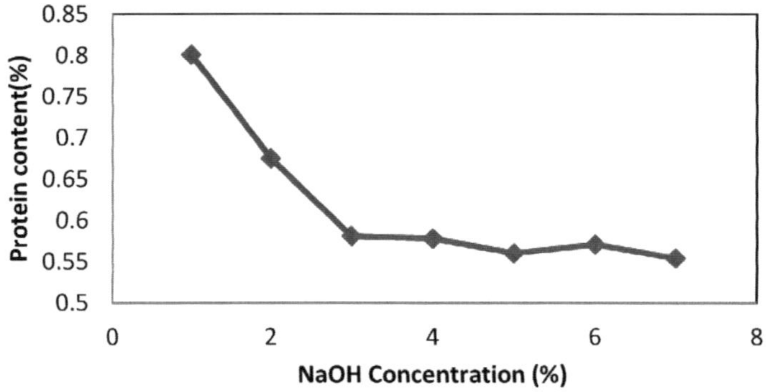

Figure 1. The effect of NaOH concentration on protein content of SSP (t=120 min, T=70°C)

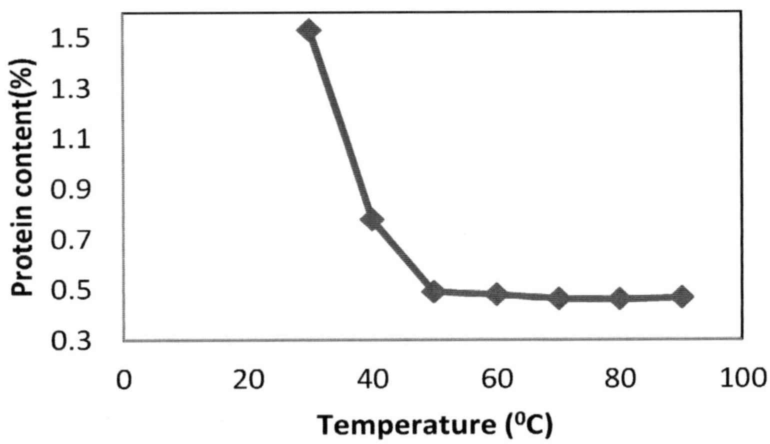

Figure 2. The effect of temperature on protein content of SSP (t=120 min, c=4%)

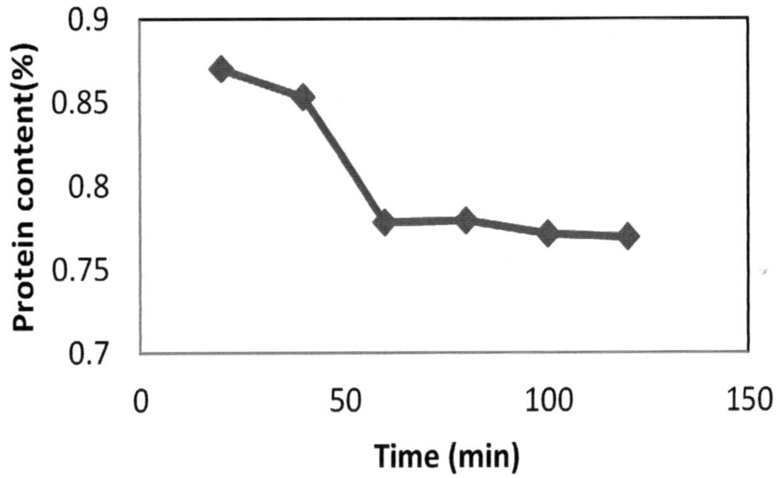

Figure 3. The effect of deproteination time on protein content of SSP (c=4%, T=60°C)

3.2 Demineralization

Deproteinated-SSP was further demineralized using HCl solution. The results are presented Figures 4-6. As HCl concentration was increased the ash content of SSP decreased indicating more minerals react with HCl forming chloride salt, which was dissolved in water (Figure 4). The increase in temperature decreased ash content indicating more minerals have been released (Figure 5). The ash content decreased with increasing demineralization time. However, beyond 60°C the decrease in ash content could not be seen anymore (Figure 6).

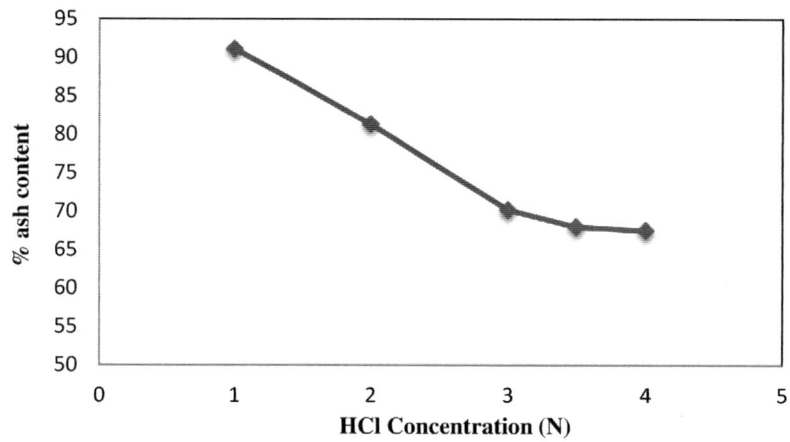

Figure 4. The effect of HCl concentration on ash content (t=120 min, T=70°C)

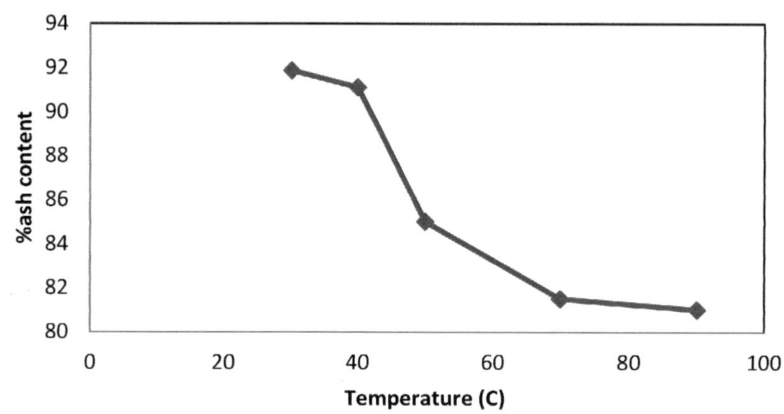

Figure 5. The effect of temperature on ash content (c= 3.5 M, t=120 min)

3.3 Chitin Deacetylation

Deacetylation of chitin resulted in chitosan with yield was 9.7% and degree of deacetylation was 71.8%. Further characterization using FTIR is shown by Figure 7. In general, chitin and chitosan showed similar IR spectra. Specific peak, which characterize both of them is around 3400-3450 cm^{-1} attributed to –NH2 and –OH groups. This peak also indicates intermolecular hydrogen bonding [9]. The two bands at around 1680 and 1625 cm^{-1} indicate amide I and C=O groups were observed in chitin spectra. In addition peak at 1380 indicates amide III (C-N stretch). After deacetylation of chitin, peak at 1680 disappeared and new peak at 1565 was observed. Further, peak shifting from 1380 to 1430 was also observed. This suggests that deacetylation process was confirmed and chitosan was really produced. Further characteristics of chitosan are peak at 1030-1050 cm^{-1} indicating CO stretching, peak at ~1430 cm^{-1} indicating CH_2 bonding, peak at ~1560 cm^{-1} indicating amide II band, peak at 1620 cm-1 indicating amide I band, peak at 2860 indicating CH stretching were observed.

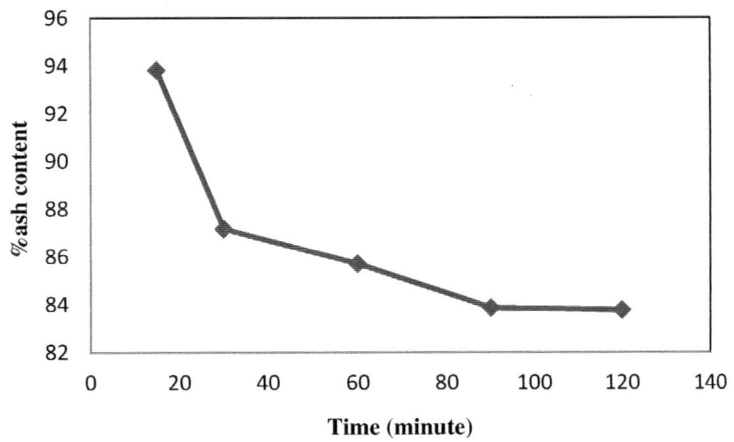

Figure 6. The effect of demineralization time on ash content (c= 3.5 M, T=70°C)

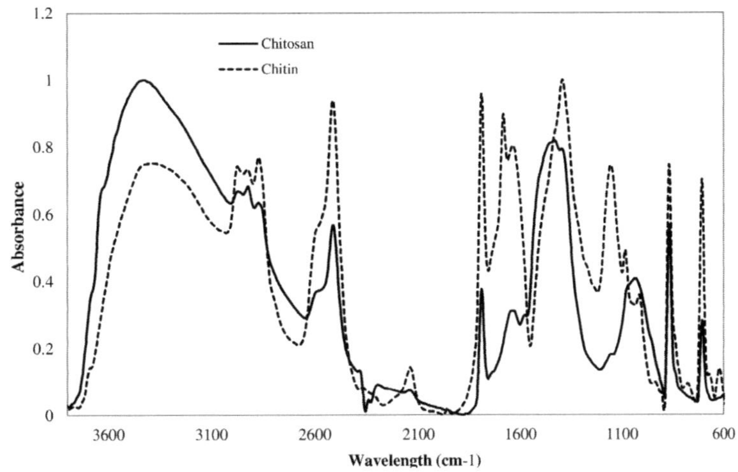

Figure 7. FTIR spectra of chitin and chitosan from scallop shell waste

4. Conclusion

Synthesis of chitosan from *Amusium* sp scallop shell waste was performed. Deproteination and demineralization of SSP were influenced by the concentration of alkaline or acid solution, temperature, and deproteination/demineralization time. The yield obtained for the synthesis of chitosan was 9.7% with DD was 71.8%. The chemical characteristic was similar with the commercial chitosan.

Acknowledgments

The authors wish to acknowledge Diponegoro University for the financial support under the Project number SP DIPA-042.01.2.400898/2016.

References

[1] Susilowati, I., Widowati, I., Agustini, T.W. and Raharjo, A.B., 2008. *Report of Research Institutes of Diponegoro University*, Semarang, Indonesia.

[2] Bautista-Baños, S., Hernández-Lauzardo, A. N., Velázquez-del Valle, M. G., Hernández-López, M., Ait Barka, E., Bosquez-Molina, E., et al., 2006. *Crop. Protection*, 25, 108–118.

[3] Rashidova, S. Sh., Milusheva, R.Y., Voropaeva, N. L., Pulatova, S. R., Nikonovich, G.V., and Ruban, I. N., 2004. *Chromatographia*, 59, 783-786.
[4] Sashiwa, H. and Aiba, S., 2004. *Prog. Polym. Sci.*, 29, 887-908.
[5] Al-Sagheer,F.A., Al-Sughayer, M.A., Muslim, S. and Elsabee, M.Z., 2008. *Carbohydrate Polym.*, 77, 410-419.
[6] Kurita, K., 2001. *Progr. Polym. Sci.*, 26, 1921-1971.
[7] Shahidi, F. and Synowiecki, J., 1991. *J. Agricul. Food Chem.*, 39, 1527-1532
[8] Zhanga Y., Zhanga, X., Dinga, R., Zhanga, J. and Liub, J., 2011. *Carbohydrate Poly.*, 83, 813-817.
[9] Dahmane, E.M., Taourirte, M., Eladlani, N., Rhazi, M., 2014. *International J. Polym. Anal. Charact.*, 19, 342-351.

Achieving Research University: Indonesian Case

Yos Johan Utama[1], Ambariyanto[2]

[1]Faculty of Law, Diponegoro University, Jl. Prof. H. Soedarto, SH, Tembalang Semarang. 50275
[2]Faculty of Fisheries and Marine Science, Diponegoro University, Jl. Prof. H. Soedarto, SH, Tembalang, Semarang. 50275

yosjohanutama@yahoo.com; ambariyanto.undip@gmail.com

Abstract. Today many universities have the vision to become a research university, including in Indonesia. It is based on the desire to play a role in advancing science for the benefit of humanity as well as to enhance the university reputation at the international level. However, in the case of Indonesia, it can only be done by several universities, given the large number of universities with very different capabilities. In addition, another problem is human resources, infrastructure, and research funding. Various targets indicator used to determine its success include the number of publications, patents and industrial products. There is an urgent need to improve all factors that can accelerate the increase in research in Indonesia universities, and has been started by the policy of the current government.

Keywords: research university, Indonesia, government policy

1. Introduction

Rapid development of higher education is currently happening around the world [1][2]. Changes in the form of a university education in the past become a research university is not inevitable. The public demands of university is not only producing graduates but also expected to provide new alternative findings of the research. The fast development of lifestyle demands a new technology capable of meeting those needs [3]. University and industry partnership is very important and play important role in this matter [4][5].

These conditions make the universities transform itself into a research university that is able to produce the technology needed by the community. Furthermore, changes continued after a research university and then apply knowledge and become an entrepreneur university [6]. This development is increasingly affecting many universities in the world, including in Indonesia.

This brief paper discusses the state of research universities in Indonesia. Given that there are still many obstacles faced in order to become a research university, it will also discuss what policies are needed to support the development of research universities in Indonesia.

2. Development of university in Indonesia

Indonesia has several types of higher education institutions, those are university, institute, polytechnic, college, academy, and community college. University is higher education institutions that produce graduates in various fields up to the doctoral level. Institute is also institution of higher education which only on specific field up to doctoral level, while college also has specialized field, but only up to bachelor level. Polytechnic is a higher educational institution that produces diplomas in various fields up to Diploma level 4, while the academy only to the Diploma level 3 on specific field. For a community college that has recently been developed in Indonesia, only up to Diploma level 1 or 2.

All universities in Indonesia initially were intended to produce graduates with degree or diploma. Its main target is to produce people who are experts in a particular field as much as possible with a good quality according to the needs of society and the market. Based on the development of higher education in the world, several major universities and institute in Indonesia began to transform themselves into research universities[7]. However, considering the real situation of universities in Indonesia, only a few are able to run the process as well as a research university.

The number of universities in Indonesia is currently more than 4,000 institutions. This number is very large, but has not been able to accommodate all existing high school graduates annually. Most universities in Indonesia are private, only about 300 are state university. The ability and capacity of these universities is very different both in financial, management, infrastructure, the number of students and professors, as well as other things are concerned.

3. Constraints research university

To become a research university, it takes the ability and adequate institutional capacity. Research universities require various aspects that are not easily met by universities in Indonesia. Being a successful research institutions not only need financial capabilities, but also support human resources and infrastructure facilities [8]. On top of that, there is a need to have good management in higher education management for its sustainability [9], with university strong and visionary leadership[10].

Initially, Indonesian government giving financial support to public universities only, because the existing rules do not allow the state to help private institutions. Therefore, private universities should seek their own funding from both student fees and from other businesses. Currently, the government has begun providing assistance to private universities, with methods permitted by the rules and regulations in Indonesia. For example, the government has been providing a wide range of research funding schemes may be accessed by anyone, from both public and private universities. However, given that government funds are limited, the support given to college do not meet all of the needs. Therefore, to become a good research university, generally financial constraints being faced by these universities.

In terms of infrastructure, the general conditions of all the universities are very different. Even among public universities was also not the same. Several universities in the top 10 in Indonesia, is generally adequate and able they are expected to develop into a research university. However, laboratory and other supporting facilities at the university also need to be improved, as well as other universities. On the other hand, some major private university established by large companies, generally have adequate infrastructure. Some universities also have a laboratory facility, but still needs improvement.

Among all universities in Indonesia today the availability of human resources with high research ability has not been evenly distributed. The number of researchers who holds a doctor or professor is still small. On the other hand, research universities require human resources with high research capabilities. They must also be able to follow the development of science through articles published in good scientific journals in order to develop cutting-edge research topic

4. Necessary policies

To support the increase of nation competitiveness, the universities have to work hard in order to carry out cutting edge research to deliver the products required by the community and the market. Research and education is the main foundation of the progress of a nation[11]. It is necessary for the central and local government policy, in addition to college policy to encourage the performance of research at each university.

In regard to financial policy, consistency is required of the central government in allocating research funds to universities. Allocation of funds should be given to universities on a competitive basis to produce good research results based on the criteria of a good proposal. The central government should have specific targets of those funded universities. Currently the central government, through the Ministry of Research Technology and Higher Education has set a target of 7 + 1 field of research, namely the areas of food, renewable energy, health, transportation, information technology and communication, defence and security, materials, and maritime. It takes courage for the central government to focus on the few universities that has been able to become a research university grow faster with the allocation of special funds with specific targets.

Government policies are needed to be able to encourage some universities selected through the allocation of special funds for the development of the infrastructure necessary for the development of research. In general, some universities have had the basic infrastructure that has can be used for research, but for certain things required special infrastructure.

In order to improve the ability of the existing human resources, it is also required special government policies to encourage this. Among others is by providing scholarships to students who have the ability to develop research and education levels. Related to this, the government has provided many scholarships, for example through the program LPDP. To enhance the experience of Indonesian academics to conduct research, government policy is needed to provide an adequate and flexible funding to develop research collaboration with other researchers from developed countries. Accelerate the recruitment of human resources who have the capacity and capability of high research also must be implemented immediately.

5. Conclusion

Improving the ability of universities in Indonesia in the research development, the central government policy is needed in particular to improve the competitiveness of the nation. The policy is primarily in financial assistance for the implementation of research, development of infrastructure and improvement of human resource capability.

References

[1] Lozano, R., Ceulemans, K., Alonso-Almeida, M., Huisingh, D., Lozano, F.J., Waas, T., Lambrechts, W., Lukman, R. and Hugé, J., 2015. A review of commitment and implementation of sustainable development in higher education: results from a worldwide survey. Journal of Cleaner Production, 108, pp.1-18.

[2] Nicholls, G., 2014. Professional development in higher education: New dimensions and directions. Routledge.

[3] Axsen, J., TyreeHageman, J. and Lentz, A., 2012. Lifestyle practices and pro-environmental technology. Ecological Economics, 82, pp.64-74.

[4] Gagnon, R. J., & Morgan, S. D. 2014. Remanufacturing scheduling systems: an exploratory analysis comparing academic research and industry practice. International Journal of Rapid Manufacturing, 4(2-4), 179-198.

[5] McDonald, C. C., &Rivara, F. 2015. Capitalising on research-industry partnerships to advance injury prevention.Injury Prevention, 21(3), 214-214.

[6] Walker, K., 2012. The technopreneurship process: Academic entrepreneur university spin-offs. RIThink, 2, pp.11-22.

[7] Angwarmase, A.Y., 2014. Implementasi Cyber Public Relations Dalam Meningkatkan Reputasi Universitas Gadjah Mada Yogyakarta Sebagai Universitas Riset Berkelas Dunia (Doctoral Dissertation, Universitas Atma Jaya Yogyakarta.).

[8] McCammon, M.G., Conrad, C.M., Klug, Z.T., Myers, C.D., Watkins, M.L., Wiley, J.W. and Bower, C.L., 2013. From an Infrastructure to a Service-based Business Model: 5 Years of Mobile Clinical Research at the University of Michigan. Clinical and translational science, 6(1), pp.3-6.

[9] Figueiró, P. S., &Raufflet, E. 2015. Sustainability in Higher Education: A systematic review with focus on management education. Journal of Cleaner Production. 106, 22–33.

[10] Crevani, L., Ekman, M., Lindgren, M., &Packendorff, J. 2015. Leadership cultures and discursive hybridisation: On the cultural production of leadership in higher education reforms. International Journal of Public Leadership, 11(3/4).

[11] Matthews, D. 2012. A Stronger Nation through Higher Education: How and Why Americans Must Achieve a Big Goal for College Attainment. A Special Report from Lumina Foundation. Lumina Foundation for Education

Institute of Physics
Dirac House, Temple Back
Bristol BS1 6BE UK

ISSN: 1755-1307
ISBN 978-1-5108-3711-9

CURRAN ASSOCIATES INC.
proceedings.com